Transport in Nanostructures

Second Edition

Providing a much-needed update on the latest experimental research, this new edition has been thoroughly revised and develops a detailed theoretical framework for understanding the behavior of mesoscopic devices.

The second edition now contains greater coverage of the quantum Hall effect, in particular, the fractional quantum Hall effect; one-dimensional structures, following the growth of research in self-assembled nanowires and nanotubes; nanoscale electronic devices, due to the evolution of device scaling to nanometer dimensions in the semiconductor industry; and quantum dots.

The authors combine reviews of the relevant experimental literature with theoretical understanding and interpretation of phenomena at the nanoscale. This second edition will be of great interest to graduate students taking courses in mesoscopic physics or nanoelectronics, and researchers working on semiconductor nanostructures.

DAVID K. FERRY is Regents' Professor in the Department of Electrical Engineering at Arizona State University. His areas of research include nanoelectronic devices, quantum transport, and nonequilibrium transport. He is a Fellow of the American Physical Society, the Institute of Electrical and Electronics Engineers, and of the Institute of Physics (UK).

STEPHEN M. GOODNICK is Director of the Arizona Institute for Nanoelectronics and Professor of Electrical Engineering at Arizona State University, where he researches transport in semiconductor devices, computational electronics, quantum and nanostructured devices, and device technology.

JONATHAN BIRD is a Professor in the Department of Electrical Engineering at the University of Buffalo and a Visiting Professor in the Graduate School of Advanced Integration Science, Chiba University. His research interests lie in the area of nanoelectronics. He is the co-author of more than two hundred peer-reviewed publications, and of undergraduate and graduate textbooks.

Transport in Nanostructures

Second Edition

David K. Ferry
Arizona State University

Stephen M. Goodnick
Arizona State University

Jonathan Bird
University at Buffalo

CAMBRIDGE
UNIVERSITY PRESS

University Printing House, Cambridge CB2 8BS, United Kingdom

One Liberty Plaza, 20th Floor, New York, NY 10006, USA

477 Williamstown Road, Port Melbourne, VIC 3207, Australia

314-321, 3rd Floor, Plot 3, Splendor Forum, Jasola District Centre, New Delhi - 110025, India

103 Penang Road, #05-06/07, Visioncrest Commercial, Singapore 238467

Cambridge University Press is part of the University of Cambridge.

It furthers the University's mission by disseminating knowledge in the pursuit of education, learning and research at the highest international levels of excellence.

www.cambridge.org
Information on this title: www.cambridge.org/9780521877480

© D. Ferry, S. Goodnick, and J. Bird 2009

This publication is in copyright. Subject to statutory exception and to the provisions of relevant collective licensing agreements, no reproduction of any part may take place without the written permission of Cambridge University Press.

First published 2009
Reprinted 2017

A catalogue record for this publication is available from the British Library

Library of Congress Cataloging in Publication data
Ferry, David K.
  Transport in nanostructures / David Ferry, Stephen M. Goodnick. – 2nd ed. / Jonathan Bird.
    p. cm.
  Includes index.
  ISBN 978-0-521-87748-0 (hardback)
  1. Nanostructures. 2. Mesoscopic phenomena (Physics) 3. Solid state electronics.
I. Goodnick, Stephen M. (Stephen Marshall), 1955- II. Bird, Jonathan P. III. Title.
  QC176.8.N35F47 2009
  530.4´1–dc22

2009015597

ISBN 978-0-521-87748-0 Hardback

Cambridge University Press has no responsibility for the persistence or accuracy of URLs for external or third-party internet websites referred to in this publication, and does not guarantee that any content on such websites is, or will remain, accurate or appropriate.

Contents

Preface	*page* vii
Acknowledgements	x

1 Introduction	1
1.1 Nanostructures: the impact	2
1.2 Mesoscopic observables in nanostructures	9
1.3 Space and time scales	17
1.4 Nanostructures and nanodevices	19
1.5 An introduction to the subsequent chapters	23
1.6 What is omitted	25

2 Quantum confined systems	28
2.1 Nanostructure materials	29
2.2 Quantization in heterojunction systems	35
2.3 Lateral confinement: quantum wires and quantum dots	52
2.4 Electronic states in quantum wires and dots	58
2.5 Magnetic field effects in quantum confined systems	66
2.6 Screening and collective excitations in low-dimensional systems	76
2.7 Homogeneous transport in low-dimensional systems	83

3 Transmission in nanostructures	116
3.1 Tunneling in planar barrier structures	117
3.2 Current in resonant tunneling diodes	123
3.3 Landauer formula	136
3.4 The multi-channel case	140
3.5 Transport in quantum waveguide structures	155

4 The quantum Hall effects	193
4.1 The integer quantum Hall effect in two-dimensional electron systems	194
4.2 Edge-state propagation in nanostructures	210
4.3 The fractional quantum Hall effect	220
4.4 The many-body picture	235

5 Ballistic transport in quantum wires — 248
5.1 Conductance quantization in quantum point contacts — 249
5.2 Non-integer conductance quantization in quantum point contacts — 271
5.3 Some ballistic device concepts — 290

6 Quantum dots — 299
6.1 Fundamentals of single-electron tunneling — 300
6.2 Single-electron tunneling in semiconductor quantum dots — 338
6.3 Coupled quantum dots as artificial molecules — 361
6.4 Quantum interference due to spatial wave function coherence in quantum dots — 388

7 Weakly disordered systems — 413
7.1 Disordered semiconductors — 414
7.2 Conductivity — 427
7.3 Weak localization — 439
7.4 Universal conductance fluctuations — 459
7.5 Green's functions in disordered materials — 467

8 Temperature decay of fluctuations — 491
8.1 Temperature decay of coherence — 493
8.2 The role of temperature on the fluctuations — 503
8.3 Electron–electron interaction effects — 511
8.4 Conductivity — 554

9 Nonequilibrium transport and nanodevices — 563
9.1 Nonequilibrium transport in mesoscopic structures — 566
9.2 Semiconductor nanodevices in the real world — 593
9.3 Quantum simulations via the scattering matrix — 610
9.4 Real-time Green's functions — 620

Index — 653

Preface

The original edition of this book grew out of our somewhat disorganized attempts to teach the physics and electronics of mesoscopic devices over the past decade. Fortunately, these evolved into a more consistent approach, and the book tried to balance experiments and theory in the current, at that time, understanding of mesoscopic physics. Whenever possible, we attempted to first introduce the important experimental results in this field followed by the relevant theoretical approaches. The focus of the book was on electronic transport in nanostructure systems, and therefore by necessity we omitted many important aspects of nanostructures such as their optical properties, or details of nanostructure fabrication. Due to length considerations, many germane topics related to transport itself did not receive full coverage, or were referred to only by reference. Also, due to the enormity of the literature related to this field, we did not include an exhaustive bibliography of nanostructure transport. Rather, we tried to refer the interested reader to comprehensive review articles and book chapters when possible.

The decision to do a second edition of this book was reached only after long and hard consideration and discussion among the authors. While the first edition was very successful, the world has changed significantly since its publication. The second edition would have to be revised extensively and considerable new material added. A decision to go ahead was made only after welcoming Jon Bird to the author's team. Once this was done, we then carefully discussed the revison and its required redistribution of material among several new chapters. Even so, the inclusion of this considerable material has meant that a lot of material has been left out of this second edition, in order to bring it down to a tractable size. This even included a considerable amount of material that was in the first edition, but no longer appears. We hope that the reader is not put off by this; as the first edition was so successful, we anticipate many of the readers will already have that tome. But, it was essential to include more up-to-date material and topics while maintaining a rational size for the book. Thus, the decision was in principle already made for us.

Currently, we still are teaching a two semester graduate sequence (at ASU) on the material contained in the book. In the first course, which is suitable for first-year graduate students, the experiments and simpler theory, such as that for tunneling, edge states, quantum Hall effect, quantum dots, and the Landauer–Büttiker method, are introduced. This covers parts of each of the chapters, but does not delve into the topic of Green's functions. Rather, the much more difficult treatment of Green's functions is left to the second course, which is intended for more serious-minded doctoral students. Even here, the developments of the zero-temperature Green's functions now in Chapter 3, followed by the Matsubara Green's functions in Chapter 8, and the nonequilibrium (real-time) Green's functions in Chapter 9, are all coupled closely to the experiments in mesoscopic devices.

In spite of the desire to consistently increase the level of difficulty and understanding as one moves through the book, there remain some anomalies. We have chosen, for example, to put the treatment of the recursive Green's functions in the chapter with waveguide transport and the recursive solutions to the Lippmann–Schwinger equation, since these two treatments of quantum transport are closely coupled. Nevertheless, the reader would be well served to go through the introduction of the Green's functions prior to undertaking an in-depth study of the recursive Green's function. This, of course, signals that topics have been grouped together in the chapters in a manner that relies on their connection to one another in physics, rather than in a manner that would be optimally chosen for a textbook. Nevertheless, we are convinced that one can use this book in graduate coursework, as is clear from our own courses.

Chapter 1 is, of course, an introduction to the material in the entire book, but new material on nanodevices has been added, as progress in silicon technology has brought the normal metal-oxide-semiconductor field-effect transistor (MOSFET) into the mesoscopic world. Additionally, a new introduction to nanowires and carbon nanotubes is given. Chapter 2 remains a discussion of quantum confined systems, but now focuses more on the one-dimensional structures rather than the two-dimensional ones. New material here includes a discussion of numerical solutions to the Schrödinger equation and Poisson's equation as well as non-self-consistent Born approximations to scattering in quasi-one-dimensional systems. Chapter 3 remains focused upon very low temperature transport, but now includes the introduction to the different approaches to (equilibrium) quantum transport. Chapter 4 is a completely new chapter focused upon the quantum Hall effect and the fractional quantum Hall effect.

Chapter 5 is also a new chapter which focuses upon quantum wires, and includes some of the modern investigations into various effects in these wires. Chapter 6 focuses upon quantum dots with new material on the role of spin. The focus upon single electron tunneling remains, but considerable new material on coupled dots has been added. Then, Chapter 7 discusses weakly disordered systems. New material on (strong) localization is presented so that weak localization can be placed in its context in relation to this material.

Chapter 8 is mostly carried over from the first edition and discusses the role of temperature. Here, the Matsubara Green's functions are introduced in addition to the semiclassical approach. Finally, Chapter 9 discusses nonequilibrium transport and nanodevices. Here, considerable new material on semiconductor nanodevices has been added. Device simulation via the scattering matrix implementation based upon the Lippmann–Schwinger equation appears as well as the treatment with nonequilibrium Green's functions.

Acknowledgements

The authors are indebted to a great number of people, who have read (and suggested changes) in all or parts of the manuscript during its preparation, and particularly since the first edition appeared. In addition, there are the students who suffered through the courses mentioned above without an adequate textbook. In particular, we wish to thank (in alphabetical order) Richard Akis, Joy Barker, Paolo Bordone, Adam Burke, Kevin Connolly, Neil Deutscher, Manfred Dür, Matthew Gilbert, Allen Gunther, Irena Knezevic, Anu Krishnaswamy, David Pivin, Chetan Prasad, Jo Rack, Steve Ramey, Lucian Shifren, Dragica Vasileska, and Christoph Wasshuber. In particular, a considerable amount of the material on quantum transport has come from their experiments and theory, as well as from the Purdue group.

The authors are also indebted to several groups and institutes who supported the original writing of this manuscript. These include Tom Zipperian at Sandia Laboratories, Peter Vogl at the Technical University of Munich, and Chihiro Hamaguchi at Osaka University. Finally, the authors would like to thank Larry Cooper, formerly at the Office of Naval Research, for his support in the publication of the original version of this book.

1
Introduction

Nanostructures are generally regarded as ideal systems for the study of electronic transport. What does this simple statement mean?

First, consider transport in large, macroscopic systems. In bulk materials and devices, transport has been well described via the Boltzmann transport equation or similar kinetic equation approaches. The validity of this approach is based on the following set of assumptions: (i) scattering processes are local and occur at a single point in space; (ii) the scattering is instantaneous (local) in time; (iii) the scattering is very weak and the fields are low, such that these two quantities form separate perturbations on the equilibrium system; (iv) the time scale is such that only events that are slow compared to the mean free time between collisions are of interest. In short, one is dealing with structures in which the potentials vary slowly on both the spatial scale of the electron thermal wavelength (to be defined below) and the temporal scale of the scattering processes.

Since the late 1960s and early 1970s, researchers have observed quantum effects due to confinement of carriers at surfaces and interfaces, for example along the Si/SiO_2 interface, or in heterostructure systems formed between lattice-matched semiconductors. In such systems, it is still possible to separate the motion of carriers parallel to the surface or interface, from the quantized motion perpendicular, and describe motion semiclassically in the unconstrained directions. Since the 1980s, however, it has been possible to pattern structures (and devices) in which characteristic dimensions are actually smaller than the appropriate mean free paths of interest. In GaAs/AlGaAs semiconductor heterostructures, it is possible at low temperature to reach mobilities in excess of 10^7 cm^2/Vs, which leads to a (mobility) mean free path on the order of 100 μm and an inelastic (or phase-breaking) mean free path even longer. (By "phase-breaking" we mean decay of the energy or phase of the "wave function" representing the carrier.) This means that transport in a regime in which the Boltzmann equation is clearly invalid becomes easily accessible. Each of the assumptions detailed above provides a factor that is neglected in the usual Boltzmann transport picture. Structures (and devices) can readily be built with dimensions that are much smaller than these dimensions, so new physical processes become important in the overall transport. These devices have come to be called nanostructures, nanodevices, or mesoscopic devices, where the latter term is used to indicate structures that are large compared to the microscopic (atomic) scale but small compared to the macroscopic scale upon which normal Boltzmann transport theory has come to be applied.

A simple consideration illustrates some of the problems. If the basic semiconductor material is doped to 10^{18} cm^{-3}, then the mean distance between impurity atoms is 10 nm, so that any discrete device size, say 0.1 μm, spans a countably small number of impurity atoms. That is, a cubic volume of 0.1 μm on a side contains only 1000 atoms. These atoms are not uniformly distributed in the material; instead they are randomly distributed with large fluctuations in the actual concentration on this size scale. The variance in the actual number N in any volume (that is, the difference from one such volume to another) is roughly $\sqrt{N}$, which in this example is about 32 atoms (or 3.3% of the doping). Since these atoms often comprise the main scattering centers at low temperatures, the material is better described as a highly conducting but disordered material, since the material is certainly not uniform on the spatial scale of interest here. As the current lines distort to avoid locally high densities of impurities, the current density becomes non-uniform spatially within the material; this can be expected to lead to new effects. Since the dimensions can be smaller than characteristic scattering lengths, transport can be ballistic and highly sensitive to boundary conditions (contacts, surfaces, and interfaces). To complicate the problem, many new effects that can be observed depend upon the complicated many-body system itself, and simple one-electron theory no longer describes these new effects. Finally, the size can be small compared to the phase-breaking length, which nominally describes the distance over which the electron wave's phase is destroyed by some process. In this case, the phase of the particle becomes important, and many phase-interference effects begin to appear in the characteristic conductance of the material.

Our purpose in this book is twofold. First, we will attempt to review the observed experimental effects that are seen in nanoscale and mesoscopic devices. Second, we want to develop the theoretical understanding necessary to describe these experimentally observed phenomena. In the remainder of this chapter, the goal is simply to give an introduction into the type of effects that are seen and to discuss why these effects will be important to future technology, as well as for their interesting physics.

1.1 Nanostructures: the impact

1.1.1 Progressing technology

Since the introduction of the integrated circuit in the late 1950s, the number of individual transistors that can be placed upon a single integrated circuit chip has approximately quadrupled every three years. The fact that more functionality can be put on a chip when there are more transistors, coupled to the fact that the basic cost of the chip (in terms of \$/cm^2) has changed very little from one generation to the next (until recently), leads to the conclusion that greater integration leads to a reduction in the basic cost per function for high-level computation as more functions are placed on the chip. It is this simple functionality argument that has driven device feature reduction according to a complicated scaling relationship [1]. In 1980, Hewlett-Packard produced a single-chip microprocessor containing approximately 0.5 M devices in its 1 cm^2 area [2]. This chip was produced with transistors having a nominal 1.25 µm gate length and was considered a remarkable step forward. In contrast, by 2007, the functionality of the dynamic random access memory (DRAM) is on the order of 2 Gbit, a number which is expected to double by 2010 [3]. The printed gate length of production microprocessor transistors in this same year was 48 nm and the physical gate length closer to 25 nm. Research devices have been demonstrated down to 10 nm gate length or less. Clearly, current integrated circuit manufacturing is truly a nanoscale technology.

For a 25 nm gate length Si device, the number of atoms spanning the channel is on the order of a 100 or less. Hence, one can reasonably ask just how far the size of an individual electron device can be reduced, and if we understand the physical principles that will govern the behavior of devices as we approach this limiting size. In 1972, Hoeneisen and Mead [4] discussed the minimal size expected for a simple MOS gate (as well as for bipolar devices). Effects such as oxide breakdown, source–drain punch-through, impact ionization in the channel, and so on were major candidates for processes to limit downscaling. Years later, Mead [5] reconsidered this limit in terms of the newer technologies that have appeared since the earlier work, concluding that one could easily downsize the transistor to a gate length of 30 nm if macroscopic transport theory continued to hold. The current ITRS roadmap (2007) now predicts scaling solutions down to 10 nm gate length before a serious "brick wall" is encountered, and 15 nm gate lengths are scheduled for production by 2010. Laboratory MOSFET devices with gate lengths down to 15 nm have been reported by Intel [6] and AMD [7] which exhibit excellent I–V characteristics, and 6 nm gate length p-channel transistors have been reported by IBM [8].

Given this rapid scaling of device technology towards 1 nm feature sizes, it becomes obvious that we must now ask whether our physical understanding of devices and their operation can be extrapolated down to very small space and time scales without upsetting the basic macroscopic transport physics – or do the underlying quantum electronic principles prevent a down-scaling of the essential semiclassical concepts upon which this macroscopic understanding is based? Preliminary considerations of this question were presented more than two decades ago [9]. If transport is ballistic, meaning carriers suffer few or no scattering as they traverse the channel, quantum effects are expected to play a major role. Ballistic (and therefore coherent and unscattered) transport was already observed in the base region of a GaAs/AlGaAs hot electron transistor [10]. From this, it is estimated that the inelastic mean free path for electrons in GaAs may be as much as 0.12 μm at room temperature. Simulation results in Si indicate that at room temperature, the ballistic mean free path may be much smaller, only a few nanometers [11], which may partly explain the success in scaling of Si MOSFETs discussed above. The inelastic mean free path is on the order of (and usually equal to) the energy relaxation length $l_e = v\tau_e$, where τ_e is the energy relaxation time and v is a characteristic velocity (which is often the Fermi velocity in a degenerate system).[1] Since the phase will likely remain coherent over these distances, it is quite natural to expect phase interference effects to appear in the transport, and to expect most of the assumptions inherent in the Boltzmann picture to be violated. A small device will then reflect the intimate details of the impurity distribution in the particular device, and macroscopic variations can then be expected from one device to another. These effects are, of course, well known in the world of mesoscopic devices. Thus, the study of mesoscopic devices, even at quite low temperatures, provides significant insight into effects that may well be expected to occur in future devices.

Consider, as an example, a simple MOSFET with a gate length of 50 nm and a gate width of 100 nm. If the number of carriers in the channel is 2×10^{12} cm^{-2}, there are only about 100 electrons on average in the open channel. If there is a fluctuation of a single impurity, the change in the conductance will not be 1%, but will be governed by the manner in which the phase interference of the carriers is affected by this fluctuation. This effect is traditionally taken to be of order e^2/h, which leads to a fluctuation in conductance of about 40 μS. If our device were to exhibit conductance of 1 S/mm (of gate width), the absolute conductance would only be 100 μS, so that the fluctuation is on the order of 40% of the actual conductance. This is a very significant fluctuation, arising from the lack of ensemble averaging in the limited number of carriers in the device. In fact, this may well be a limiting mechanism for the down-scaling of individual transistor sizes, when trying to realize circuit architectures involving 100s of millions of transistors that have to perform within a relatively narrow range of tolerance, necessitating entirely new types of fault tolerant designs to accommodate such fluctuations.

---

[1] There is some ambiguity here because the energy relaxation time is usually defined as the effective inverse decay rate for the mean electron energy or temperature. The definition here talks about a mean free path for energy relaxation, which is not quite the same thing. This is complicated by the fact that, in mesoscopic systems, one really talks about a phase-breaking time, which is meant to refer to the average time for relaxation of the coherent single-particle phase of a charge carrier. Again, this is a slightly different definition. This ambiguity exists throughout the literature, and although we will probably succumb to it in later chapters, the reader should recognize these subtle differences.

1.1.2 Some physical considerations

In macroscopic conductors, the resistance that is found to exist between two contacts is related to the bulk conductivity and to the dimensions of the conductor. In short, this relationship is expressed by

$$R = \frac{L}{\sigma A}, \tag{1.1}$$

where σ is the conductivity and L and A are the length and cross-sectional area of the conductor, respectively. If the conductor is a two-dimensional conductor, such as a thin sheet of metal, then the conductivity is the conductance per square, and the cross-sectional area is just the width W. This changes the basic formula (1.1) only slightly, but the argument can be extended to any number of dimensions. Thus, for a d-dimensional conductor, the cross-sectional area has the dimension $A = L^{d-1}$, where here L must be interpreted as a "characteristic length." Then, we may rewrite (1.1) as

$$R = \frac{L^{2-d}}{\sigma_d}. \tag{1.2}$$

Here, σ_d is the d-dimensional conductivity. Whereas one normally thinks of the conductivity, in simple terms, as $\sigma = ne\mu$, the d-dimensional term depends upon the d-dimensional density that is used in this definition. Thus, in three dimensions, σ_3 is defined from the density per unit volume, while in two dimensions σ_2 is defined as the conductivity per unit square and the density is the sheet density of carriers. The conductivity (in any dimension) is not expected to vary much with the characteristic dimension, so we may take the logarithm of the last equation. Then, taking the derivative with respect to $\ln(L)$ leads to

$$\frac{\partial \ln(R)}{\partial \ln(L)} = 2 - d. \tag{1.3}$$

This result is expected for macroscopic conducting systems, where resistance is related to the conductivity through Eq. (1.2). We may think of this limit as the *bulk* limit, in which any characteristic length is large compared to any characteristic transport length.

In mesoscopic conductors, the above is not necessarily the case, since we must begin to consider the effects of ballistic transport through the conductor. (For ballistic transport we generally adopt the view that the carrier moves through the structure with very little or no scattering, so that it follows normal phase space trajectories.) However, let us first consider a simpler situation. We have assumed that the conductivity is independent of the length, or that σ_d is a constant. However, if there is surface scattering, which can dominate the mean free path, then one could expect that the latter is $l \sim L$. Since $l = v_F \tau$, where v_F is the Fermi velocity in a degenerate semiconductor and τ is the mean free time, this leads to

$$\sigma_d = \frac{n_d e^2 \tau}{m} = \frac{n_d e^2 L}{m v_F}. \tag{1.4}$$

Hence the dependence of the mean free time on the dimensions of the conductor changes the basic behavior of the macroscopic result (1.3). This is the simplest of the modifications. For more intense disorder or more intense scattering, the carriers are localized because the size of the conductor creates localized states whose energy difference is greater than the thermal excitation, and the conductance will be quite low. In fact, we may actually have the resistance only on the order of [12]

$$R = e^{\alpha L} - 1, \tag{1.5}$$

where α is a small quantity. (The exponential factor arises from the presumption of tunneling between neighboring sites: the factor of -1 is required for the proper limit as $\alpha L \to 0$.) We think of the form of Eq. (1.5) as arising from the localized carriers tunneling from one site to another (hence the exponential dependence on the length), with the unity factor added to allow the proper limit for small L. Then, the above scaling relationship (1.3) becomes modified to

$$\frac{\partial \ln(R)}{\partial \ln(L)} \approx \alpha L. \tag{1.6}$$

In this situation, unless the conductance is sufficiently high, the transport is localized and the carriers move by hopping. The necessary value has been termed the minimum metallic conductivity [12], but its value is not given by the present arguments. Here we just want to point out the difference in the scaling relationships between systems that are highly conducting (and bulk-like) and those that are largely localized due to the high disorder.

In a strongly disordered system, the wave functions decay exponentially away from the specific site at which the carrier is present. This means that there is no long-range wavelike behavior in the carrier's character. On the other hand, by bulk-like extended states we mean that the carrier is wavelike in nature and has a well-defined wave vector **k** and momentum $\hbar \mathbf{k}$. Most mesoscopic systems have sufficient scattering that the carriers do not have fully wavelike behavior, but they are sufficiently ordered so that the carriers are not exponentially localized. Thus, when we talk about diffusive transport, we generally mean almost-wavelike states with very high scattering rates. Such states are neither free electron-like nor fully localized. We have to adopt concepts from both areas of research.

The rationale for such a view lies in the expectations of quantization in such small, mesoscopic conductors. We assume that the semiconductor sample is such that the electrons move in a potential that is uniform on a macroscopic scale but that varies on the mesoscopic scale, such that the states are disordered on the microscopic scale. Nevertheless, it is assumed that the entire conduction band is not localized, but that it retains a region in the center of this energy band that has extended states and a non-zero conductivity as the temperature is reduced to zero. For this material, the density of electronic states per unit energy per unit volume is given simply by the familiar dn/dE. Since the conductor has a finite volume, the electronic states are discrete levels determined by the size of this volume. These individual energy levels are sensitive to the boundary conditions applied to the ends of the sample (and to the "sides") and can be shifted by small amounts on the order of $\hbar/\tau$, where τ is the time required for an electron to diffuse to the end of the sample. In essence, one is defining here a broadening of the levels that is due to the finite lifetime of the electrons in the sample, a lifetime determined not by scattering but by the carriers' exit from the sample. This, in turn, defines a maximum coherence length in terms of the sample length. This coherence length is defined here as the distance over which the electrons lose their phase memory, which we will take to be the sample length. The time required to diffuse to the end of the conductor (or from one end to the other) is L^2/D, where D is the diffusion constant for the electron (or hole, as the case may be) [13]. The conductivity of the material is related to the diffusion constant (we assume for the moment that $T = 0$) as

$$\sigma(E) = \frac{n_d e^2 \tau}{m} = e^2 D \frac{dn}{dE}, \tag{1.7}$$

where we have used the fact that $n = (2/d)(dn/dE)E$, where d is the dimensionality, and $D = v_F^2 \tau/d$. If L is now introduced as the effective length, and t is the time for diffusion, both from D, one finds that

$$\frac{\hbar}{t} = \frac{\hbar}{e^2} \frac{\sigma}{L^2} \frac{dE}{dn}. \tag{1.8}$$

The quantity on the left side of Eq. (1.8) can be defined as the average broadening of the energy levels ΔE_a, and the dimensionless ratio of this width to the average spacing of the energy levels may be defined as

$$\frac{\Delta E_a}{dE/dn} = \frac{\hbar}{e^2} \sigma L^{-2}. \tag{1.9}$$

Finally, we change to the total number of carriers $N = nL^d$, so that

$$\frac{\Delta E_a}{dE/dN} = \frac{\hbar}{e^2} \sigma L^{d-2}. \tag{1.10}$$

This last equation is often seen with an additional factor of 2 to account for the double degeneracy of each level arising from the spin of the electron.

Another method of looking at Eq. (1.8) is to notice that a conductor connected to two metallic reservoirs will carry a current defined by the difference in the Fermi levels between the two ends, which may be taken to be eV. Now there are $eV(dn/dE)$ states contributing to the current, and each of these states carries a current e/t. Thus, the total conductance is $(e^2/t)(dn/dE)$ which leads directly to Eq. (1.8).

The quantity on the left side of Eq. (1.9) is of interest in setting the minimum metallic conductivity. In a disordered material, the ratio of the overlap energy between different sites and the disorder-induced broadening of the energy levels is important. The former quantity is related to the width of the energy bands. If this ratio is small, it is hard to match the width of the energy level on one site with that on a neighboring site, so that the allowed energies do not overlap and there is no appreciable conductivity through the sample. On the other hand, if the ratio is large, the energy levels easily overlap and we have bands of allowed energy, so that there are extended wave functions and a large conductance through the sample. The ratio (1.9) just expresses this quantity. The factor $e^2/\hbar$ is related to the fundamental unit of conductance and is just 2.43×10^{-4} siemens (the inverse is just 4.12 kΩ).

It is now possible to define a dimensionless conductance, called the Thouless number by Anderson and coworkers [14] in terms of the conductance as

$$g(L) = \frac{2\hbar}{e^2} G(L), \tag{1.11}$$

where $G(L) = \sigma L^{d-2}$ is the actual conductance in the highly conducting system. These latter authors have given a scaling theory based upon renormalization group theory, which gives us the dependence on the scale length L and the dimensionality of the system. The details of such a theory are beyond the present work. However, we can obtain the limiting form of their results from the above arguments. The important factor is a critical exponent for the reduced conductance $g(L)$ which may be defined by

$$\beta_d \equiv \lim_{g \to \infty} \frac{d[\ln g(L)]}{d \ln(L)} \to d - 2, \tag{1.12}$$

which is just Eq. (1.3) rewritten in terms of the conductance rather than the resistance. By the same token, one can rework Eq. (1.6) for the low-conducting state to give

$$\beta_d \equiv \lim_{g \to \infty} \frac{d[\ln g(L)]}{d \ln(L)} \to -\alpha L. \tag{1.13}$$

What the full scaling theory provides is the connection between these two limits when the conductance is neither large nor small.

For three dimensions, the critical exponent changes from negative to positive as one moves from low conductivity to high conductivity, so that the concept of a *mobility edge* in disordered (and amorphous) conductors is really interpreted as the point where $\beta_3 = 0$. This can be expected to occur about where the reduced conductance is unity, or for a value of the total conductance of $e^2/\pi\hbar$ (the factor of $\pi/2$ arises from a more exact treatment). In two dimensions, there is no critical value of the exponent, as it is by and large always negative, approaching 0 asymptotically. Instead of a sharp mobility edge, there is a universal crossover from logarithmic localization at large conductance to exponential localization at small conductance. This same crossover appears in one dimension as well, except the logarithmic localization is much stronger. Hence, it may be expected that all states will be localized for $d < 2$ if there is any disorder at all in the conductor. This is the source of the size dependence that is observed in mesoscopic structures. In fact, we note that in the case of surface scattering (1.4), the additional factor of L in the conductivity leads immediately to variation in the conductance as L^{d-1}, which gives the value arising from (1.13) immediately for two dimensions.

1.2 Mesoscopic observables in nanostructures

1.2.1 Ballistic and quasi-ballistic transport

In the simplest case of transport in small structures, one may assume that the particles move through the active region without scattering. This transport is termed *ballistic transport*. In this approach, we assume that the "device" region is characterized by the transmission and reflection of the incoming waves at both sides; essentially, this is a scattering matrix approach [15]. Particles flow through the active region without scattering (except for a possible reflection from a barrier) and move elastically. Thus, if the reservoirs are described by the Fermi levels E_{Fl} and E_{Fr}, which are separated by a small applied bias energy $e\delta V$, only those electrons flowing from filled states on the left (which is assumed to lie at higher energy) to the empty states on the right contribute to the current (see Fig. 1.1). This is easily seen for $T = 0$, where the states are completely filled up to the Fermi level and completely empty for energies above the Fermi level. Since the carriers move elastically, electrons leaving the left reservoir at the Fermi energy arrive at the right reservoir with an excess energy $e\delta V$, which must be dissipated in the latter reservoir (contact). If there is no barrier to this flow, the current is determined solely by the number of electrons which can leave the source contact, the left reservoir (actually the number which leave per unit time). When a barrier is present, a portion of the leaving electrons are reflected back into the source contact, and a fraction is transmitted through the barrier, based upon tunneling. We take the view of describing the properties of the barrier region in terms of the incoming currents from the contacts.

Fig. 1.1 The currents flowing into, and out of a scattering barrier.

The process is described by the incoming particle currents j_i and j'_i, from the left and the right sides, respectively. The Fermi level on the left side of the scatterer is raised slightly by an applied bias of $e\delta V$. This results in electrons appearing in the i channel, but those returning in the reflected r channel are also absorbed in the left reservoir described by its Fermi level. On the left side, there is an extra density of electrons in the levels near the Fermi surface (over the equilibrium values), and this density is given by

$$\delta n = \frac{dn}{dE} e \delta V. \tag{1.14}$$

At the same time, the extra density of electrons on the left side will be given by the sum of the magnitudes of the particle currents on the left minus those on the right, with each divided by the respective velocities,

$$\delta n = \frac{j_i + j_r}{v_l} - \frac{j'_i + j_0}{v_r} = \hbar R \frac{j_i - j'_i}{dE/dk_x} \tag{1.15}$$

where j_0 is the current transmitted from the left to the right, and we have assumed the velocity is the same on both sides of the barrier. Here, we have used the facts that $j_i - j_0 = (1 - T)j_i = Rj_i$ and $j'_i - j'_r = (1 - T)j'_i = Rj'_i$. The total current into the system is just $I = e(j_i - j_r) = e(j_0 - j'_i) = eT(j_i - j'_i)$, so that

$$G = \frac{I}{\delta V} = \frac{e^2}{\hbar} \frac{T}{R} \frac{dn}{dE} \frac{dE}{dk_x}. \tag{1.16}$$

Now, in one dimension we can write the density of states as

$$dn = N(k)\left(\frac{dk}{dE}\right) dE = \frac{1}{\pi}\left(\frac{dk}{dE}\right) dE, \tag{1.17}$$

where a factor of 2 has been added for spin degeneracy. This may be combined with Eq. (1.16) to give

$$G = \frac{e^2}{\pi \hbar} \frac{T}{R} = \frac{2e^2}{h} \frac{T}{R}, \tag{1.18}$$

which is known as the *Landauer formula*. Actually, this latter form is a special one for four-terminal measurements, where the potentials are measured adjacent to the scattering site rather than in the reservoirs. In the case where the reservoirs are the source for the potential measurements, contact potentials would be possible, and the factor of R^{-1} is deleted. It should be noted that the transmission coefficient T is that for all possible energy states that can tunnel through the structure from one side to the other. When there are many such channels possible, the single factor of T is usually replaced with a summation over the individual T_i, as we shall see.

The Landauer formula is a generalization of a formula that is normally obtained in three dimensions for tunneling structures. Generally, one can describe the tunneling probability (the transmission coefficient) $T(E)$ for an arbitrary barrier structure. Then the current through this structure may be written semiclassically as

$$I = 2eA \int \frac{d^3 \mathbf{k}}{(2\pi)^3} v_z(k_z) T(E_z)[f_l(E) - f_r(E + \delta V)], \tag{1.19}$$

where the two distribution functions represent those on the left and right sides of the barrier. It will be shown in a later chapter that this can be rewritten as

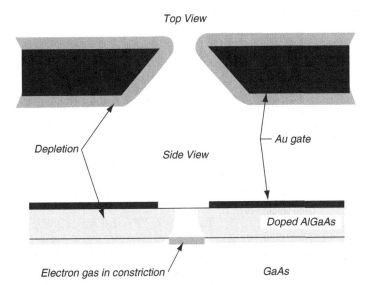

Fig. 1.2 Structure of the metal gates used to define a constriction to observe quantized conductance.

$$I = \frac{2e^2}{h} \bar{T}(E_F) N_\perp, \qquad (1.20)$$

where we have introduced an average tunneling coefficient and where $N_\perp$ is the number of transverse modes, or

$$N_\perp = A \int \frac{d^2 \mathbf{k}}{(2\pi)^2} = \frac{mA}{2\pi \hbar^2} \int dE_\perp. \qquad (1.21)$$

This means that the simple Landauer formula is a result of a more complex three-dimensional theory when the latter is reduced to small structures with lateral quantization.

The significance of the Landauer formula, for a system such as a mesoscopic system of small size where phase coherence can be maintained over the tunneling distance, is that the conductance is a constant value that includes a fundamental value multiplied by the number of modes that are transmitting. This has been detected in simple but elegant sets of experiments by van Wees *et al.* [16] and by Wharam *et al.* [17]. In those of the former group, a very-high-mobility two-dimensional electron gas, formed at the interface between GaAs and doped AlGaAs, is used. A pair of gates is used to create a constriction between two parts of the electron gas, and the conductance between these two parts is measured. The structure of the gates is illustrated in Fig. 1.2, and the measurements are illustrated in Fig. 1.3. The gates form a short one-dimensional channel with several allowed modes of propagation between the two parts of the broad-area electron gas. Here, the transmission coefficient is either 0 or 1, depending upon whether the channel subband energy lies above or below the Fermi energy. The gates may be assumed to introduce a harmonic potential, so that the subbands

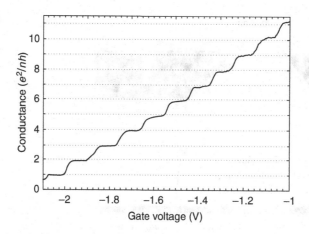

Fig. 1.3 The conductance obtained from a typical GaAs/AlGaAs structure such as that of Figure 1.2. (Reprinted with permission from van Wees *et al.* [16] Copyright 1988 The American Physical Society.)

are equally spaced. As the bias on the gates is varied, the number of channels below the Fermi level is changed as the saddle potential between the gates is raised or lowered. For a sufficiently large negative gate bias, all channels are pinched off and the conductance drops to zero. These results are a very dramatic verification of the Landauer formula and ballistic transport in mesoscopic systems discussed in detail in Chapter 3.

1.2.2 Phase interference

The relevant quantity for discussion of quantum interference effects is the phase of the carrier as it moves through the semiconductor. Interference between differing waves can occur over distances on the order of the coherence length of the carrier wave, and the latter distance is generally different from the inelastic mean free path for quasi-ballistic carriers (those with weak scattering). The latter is related to the energy relaxation length $l_e = v\tau_e$, where τ_e is the energy relaxation time and v is a characteristic velocity (typically the Fermi velocity in degenerate material and the thermal velocity in nondegenerate material). The inelastic mean free path can be quite long, on the order of several tens of microns for electrons at low temperatures in the inversion channel of a high-electron-mobility transistor in GaAs/AlGaAs. On the other hand, the coherence length is usually defined for weakly disordered systems by the diffusion constant (as we discuss below). Here, we are interested in the quasi-ballistic regime (nearly free, unscattered carriers) and so will use the inelastic mean free path as the critical length.

Consider two waves (or one single wave which is split into two parts which propagate over different paths) given by the general form $\psi_i = A_i e^{j\varphi_i}$. Then, when the two waves are combined, the probability amplitude varies as

$$P = |\psi_1 + \psi_2| = |A_1|^2 + |A_2|^2 + 4|A_1^* A_2|\cos(\varphi_1 - \varphi_2). \tag{1.22}$$

1.2 Mesoscopic observables in nanostructures

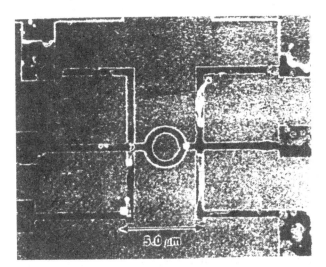

Fig. 1.4 Micrograph of the etched ring structure. (After Mankiewich *et al.* [22].)

The probability can therefore range from the sum of the two amplitudes to the differences of the two amplitudes, depending on how the phases of the two waves are related. In most cases, it is not important to retain any information about the phase in device problems because the coherence length is much smaller than any device length scale and because *ensemble averaging* averages over the phase interference factor so that it smooths completely away in macroscopic effects. This ensemble averaging requires that a large number of such small phase coherent regions are combined stochastically. In small structures this does not occur, and many observed quantum interference effects are direct results of the lack of ensemble averaging [18].

A particularly remarkable illustration of the importance of the quantum phase is the magnetic Aharonov–Bohm effect [19], as may be seen in quasi-two-dimensional semiconductor systems. The basic structure of the experiment is illustrated in Fig. 1.4. A quasi-one-dimensional conducting channel is fabricated on the surface of a semiconductor. This channel is usually produced in a high-electron-mobility heterostructure in which the channel is defined by reactive-ion etching [20,21,22], or by electrostatic confinement [23]. In either case, it is preferable to have the waveguide sufficiently small so that only one or a few electron modes are possible. The incident electrons, from the left of the ring in Fig. 1.4, have their waves split at the entrance to the ring. The waves propagate around the two halves of the ring to recombine (and interfere) at the exit port. The overall transmission through the structure, from the left electrodes to the right electrodes, depends upon the relative size of the ring circumference in comparison to the electron wavelength. If the size of the ring is small compared to the inelastic mean free path, the transmission depends on the phase of the two fractional paths. In the Aharonov–Bohm effect, a magnetic field is passed through the annulus of the ring, and this magnetic field will modulate the phase interference at the exit port.

The vector potential for a magnetic field passing through the annulus of the ring is azimuthal, so that electrons passing through either side of the ring will travel either parallel or antiparallel to the vector potential, and this difference produces the phase modulation. The vector potential will be considered to be directed counterclockwise around the ring. (We adopt cylindrical coordinates, with the magnetic field directed in the z-direction and the vector potential in the θ-direction.) The phase of the electron in the presence of the vector potential is given by the Peierl's substitution, in which the normal momentum vector $\mathbf{k}$ is replaced by $(\mathbf{p} + e\mathbf{A})/\hbar$,

$$\varphi = \varphi_0 + \frac{1}{\hbar}(\mathbf{p} + e\mathbf{A}) \cdot \mathbf{r}, \tag{1.23}$$

so that the exit phases for the upper and lower arms of the ring can be expressed as

$$\varphi_{up} = \varphi_0 + \int_{\pi}^{0} \left(\mathbf{k} + \frac{e}{\hbar}\mathbf{A}\right) \cdot \mathbf{a}_\vartheta r d\vartheta$$

$$\varphi_{lo} = \varphi_0 - \int_{-\pi}^{0} \left(\mathbf{k} - \frac{e}{\hbar}\mathbf{A}\right) \cdot \mathbf{a}_\vartheta r d\vartheta \tag{1.24}$$

and the phase difference is just

$$\delta\varphi = \frac{e}{\hbar} \int_0^{2\pi} \mathbf{A} \cdot \mathbf{a}_\vartheta r d\vartheta = \frac{e}{\hbar} \int_{ring} \mathbf{B} \cdot \mathbf{n} dS = 2\pi \frac{\Phi}{\Phi_0}, \tag{1.25}$$

where $\Phi_0 = h/e$ is the quantum unit of flux and Φ is the magnetic flux coupled through the ring. The phase interference term in Eq. (1.22) goes through a complete oscillation each time the magnetic field is increased by one flux quantum unit. This produces a modulation in the conductance (resistance) that is periodic in the magnetic field, with a period h/eS, where S is the area of the ring. This periodic oscillation is the Aharonov–Bohm effect, and in Fig. 1.5 results are shown for such a semiconductor structure. While these oscillations are obvious in such a constructed ring, mesoscopic devices are described by a great many accidental rings that come and go as the electrochemical potential is varied.

1.2.3 Carrier heating in nanostructures

One of the classic problems of semiconductors and semiconductor devices is carrier heating in the high electric fields present in the device. In general, low

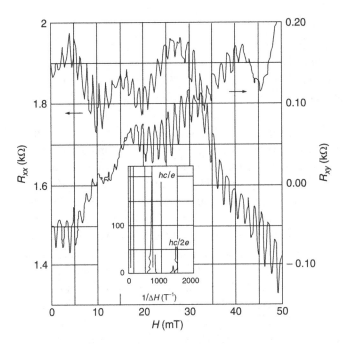

Fig. 1.5 Magnetoresistance and Hall resistance at 0.3 K for the ring of Fig. 1.4. The inset is the Fourier transform of the data showing the main peak at h/e. (After Mankiewich et al. [22].)

electric field transport is normally treated in the relaxation time approximation through a rigid shift of the distribution function in momentum space so that the carriers acquire a net velocity while maintaining a distribution function characterized by thermal equilibrium and the normal Fermi energy [24]. With the problem of carrier heating, however, this is no longer the case. The input of energy from the applied fields and potentials leads to an increase of the kinetic motion of the particles as well as a directed drift velocity. The distribution function is no longer the equilibrium one, and usually the major problem is finding this new distribution function. For this purpose one must rely upon a transport equation.

The Boltzmann transport equation (BTE) has been a cornerstone of semi-classical transport theory for many years. However, as we pointed out above, there are a great many assumptions built into this equation, and some of these will no longer be applicable in nanostructures. One of the strongest limitations on the BTE arises from the restrictions that the field and the scattering events are treated as noninteracting perturbations, or that each scattering event is treated as a separate entity which does not interact with other scattering events or processes. This result cannot be expected to hold either in high electric fields or in constrained geometries arising from nanostructures. This means that a new kinetic equation must he obtained to replace the BTE of semiclassical transport theory.

Virtually all quantum kinetic theories are based upon reductions of the Liouville–von Neumann equation for the density matrix ρ, with appropriate boundary conditions,

$$i\hbar \frac{\partial \rho}{\partial t} = [H_0, \rho] + [V, \rho] + [F, \rho], \tag{1.26}$$

where H_0 is the Hamiltonian for the carriers plus phonons plus impurities, V is the electron-scatterer interaction, and F is the Hamiltonian that couples the electrons (and scatterers in some cases) to external fields. In nanostructures, the latter can arise not only from applied fields but also from boundary and confining potentials such as surface roughness scattering and quantum confinement. The density matrix equation expressed here is one form of a quantum kinetic equation, but others exist. For example, the density matrix (and the Wigner distribution obtained from it) builds in correlation between different points, but it is a single-time function. In many cases, the correlation between different times is very important, and other approaches based upon nonequilibrium real-time Green's functions are required. One resultant problem in the case of quantum kinetic theories is that there are a variety of equations that result, as well as a diverse set of definitions of the so-called distribution function describing the one-electron probability in momentum and real space. It is not at all clear that quantum kinetic theories reduce to the semiclassical BTE, and there has been considerable work to try to establish this fact [25]. In most cases, however, additional assumptions, beyond those inherent in the BTE, must be made to connect a quantum kinetic equation to the BTE, even in the limiting case of low fields and macroscopic dimensions. Once spatial and temporal correlations are introduced into the quantum kinetic theory, these correlations must be addressed in the limiting process required to achieve any connection with the BTE, which is an equation for a local quantity. These correlations introduce retardation and memory effects into the equivalent one-electron quantum kinetic equation.

If the applied fields are time dependent, further retardation and memory effects become important. These are related first to the inertial response of the electron plus scatterer system, and second to the role the oscillating potentials can play in the extraction of energy in quanta, either from the field by the system or by the field from the system. The temporal nonlocality is important as well if the collision process cannot be completed within a single cycle of the oscillating potential. Although this normally is not a large problem, it can be important in systems with long phase memory, and is enhanced in those cases where the applied field can actually weaken the collisions (such as with impurity scattering).

Inertial effects are also anticipated as a consequence of the failure of the lattice to keep pace with the driven electron subsystem. Hence, spatial and temporal nonlocality can arise from the effect of a nonequilibrium phonon distribution. The extreme nonlinearity of the coupled electron and phonon systems at high driving fields makes the existence of a shock structure in the coupled distributions a distinct possibility. The self-consistent screening of the carriers can also lead to an indirect field-dependent effect, in which the high fields lead to a nonlinear distribution function that results in de-screening of the electron–phonon interaction. This variety of nonlinear effects just illustrates the difficulty of treating nonequilibrium transport in small structures.

Consequently, one of the central features of the nanostructure is that its transport cannot be treated in isolation. The quantum kinetic equations are strongly coupled to the boundaries, and the overall environment, through the boundary conditions on the equation. The basic Liouville equation and its causal boundary conditions are basically modified due to the strong nonlinear interaction between them. Moreover, the nonlinear system in the presence of the fields is strongly impacted by the boundaries, so that the resulting distribution function will largely be determined by these boundaries as well as the internal driving fields. While the incoherent coupling of the device to its environment and boundaries has traditionally been treated through weak scattering processes such as surface-roughness scattering, the effects of the boundaries on nanostructures can be much stronger and far more complicated to treat theoretically.

1.3 Space and time scales

In the previous sections, we have discussed briefly the major new effects that cause the transport in nanostructures to differ significantly from that of macroscopic semiconductor devices. Effects such as phase interference, with the consequent universal conductance fluctuations and weak localization, and carrier heating all depend upon the relative sizes of important correlation lengths and the device dimensions. The key lengths of the above sections have been taken to be the inelastic and elastic mean free paths. It is worthwhile at this point to make a few estimates of some of the key parameters that will he utilized in the remainder of the book, and to assure ourselves that no fundamental limits of understanding are being violated. Some of these key parameters are listed in Table 1.1. In the case for which these numbers are tabulated, it is assumed that we have a quasi-two-dimensional electron gas localized at an interface. For the GaAs case, this assumption is predicated upon a modulation-doped AlGaAs/GaAs heterostructure with the dopant atoms in the AlGaAs and the free electrons forming an inversion layer on the GaAs side of the interface. For the Si case, it is assumed that electrons are introduced at an interface either between Si and an oxide or between Si and, for example, a strained SiGe layer utilizing modulation doping. However, we take the lower mobility of the former interface for the example. In each case, a modest inversion density (equal for the two examples) is chosen, and the characteristic lengths are worked out from these assumptions.

The density is the sheet density of carriers in the quasi-two-dimensional electron gas at the interface discussed above. The actual density can usually be varied by an order of magnitude on either side of this value: for example,

Table 1.1 *Some important parameters at low temperature.*

Parameter	GaAs	Si	Units
Density	4.0	4.0	10^{11} cm^{-2}
Mobility	10^5	10^4	cm^2/Vs
Scattering time	3.8	1.1	10^{-12} s
Fermi wave vector	1.6	1.6	10^6 cm^{-1}
Fermi velocity	2.76	0.97	10^7 cm/s
Elastic mfp	1.05	0.107	10^{-4} cm
Inelastic mfp	5.0	0.5	10^{-4} cm
Phase-breaking time	1.8	0.57	10^{-11} s
Diffusion constant	1.45	0.52	10^3 cm^2/s
Phase coherence length	1.62	0.54	10^{-4} cm

10^{11}–4×10^{12} cm^{-2} are possible in Si and somewhat lower at the upper end in GaAs. Again the mobility is a typical value, but in high-mobility structures, higher than 10^7 cm^2/Vs has been achieved in GaAs, and 2×10^5 cm^2/Vs has been achieved in Si (modulation doped with SiGe at the interface). Finding the scattering time from the mobility is straightforward, and masses of 0.067 m_0 and 0.19 m_0 are used for GaAs and Si, respectively. The Fermi wavevector is determined by the density through $k_F = (2\pi n_s)^{1/2}$. The Fermi velocity is then $v_F = \hbar k_F/m$. The elastic mean free path is then defined by the scattering time and the Fermi velocity (it is assumed that the quasi-two-dimensional electron gas is degenerate at low temperature) as $l_e = v_F \tau_{sc}$. The inelastic mean free path is estimated for the two materials based upon experiments that will be discussed in later chapters, but it also should be recognized that there will be a range of values for this parameter. The inelastic mean free time, or phase-breaking time, is found through the relationship $l_{in} = v_F \tau_\varphi$, and the inelastic, or phase-breaking, time τ_φ, is found from the previous estimate of the inelastic mean free path. The diffusion constant is just $D = v_F^2 \tau_{sc}/d$ (in d dimensions). Finally, the phase coherence length is the normally defined Thouless length $l_\varphi = (D\tau_\varphi)^{1/2}$. It should he noted that the *inelastic mean free path* and the *phase coherence length* are two different quantities, with the latter being about a factor of three smaller than the former for the parameters used in the table, *even though these two quantities are often assumed to be interchangable and indistinguishable.* If this were the case, then the inelastic mean free time would he clearly defined to be one-half the scattering time from the definitions introduced here. Usually, in fact, the inelastic mean free time is larger than the scattering time, so that the inelastic mean free path is larger than the phase coherence length. Both of these quantities are important in phase interference, and we will try not to confuse the issue by using the two names inappropriately. But they do differ.

The inelastic mean free path describes the distance an electron travels *ballistically* in the phase-breaking (or energy relaxation) time τ_φ. Thus, this quantity is useful in describing those processes, such as tunneling, in which the dominant transport process is one in which the carriers move ballistically with little scattering through the active region of interest. On the other hand, the coherence length l_φ, is defined with the *diffusion* constant D. This means that the carriers are in a region of extensive scattering, so that their transport is describable as a diffusion process in which the coherence length is the equivalent diffusion length defined with the phase-breaking time. In some small nanostructures, the transport is neither ballistic nor diffusive, but is somewhere between these two limits. For these structures, the effective phase relaxation length is neither the inelastic mean free path nor the coherence length. These structures are more difficult to understand, and they are usually still quite sensitive to the boundary conditions. Care must be taken to be sure that the actual length (and the descriptive terminology) is appropriate to the situation under study.

We will see in later chapters that there are two other lengths that are important. These are the *thermal length* $l_T = (\hbar D/k_B T)^{1/2}$ and the *magnetic length* $l_m = (\hbar/eB)^{1/2}$. The latter is important as it relates to the filling factor of the Landau levels and to the cyclotron radius at the Fermi energy $r_c = \hbar k_F/eB = k_F l_m^2$.

1.4 Nanostructures and nanodevices

As the previous sections have illustrated, transport in nanoscale systems is defined by the characteristic length scales associated with the motion of the carriers. A *nanodevice* is a functional structure with nanoscale dimensions which performs some useful operation, for example a nanoscale transistor. One can generically consider the prototypical nanodevice illustrated in Fig. 1.6. The "device" can be considered an active region coupled to two contacts, left and right, which serve as a source and sink (drain) for electrons. Here the contacts are drawn as metallic-like reservoirs, characterized by chemical potentials μ_S and μ_D, and are separated by an external bias, $qV_A = \mu_S - \mu_D$. The current flowing through the device is then a property of the chemical potential difference and the transmission properties of the active region itself. A separate gate electrode serves to change the transmission properties of the active region, and hence controls the current. This gate coupling may be through electric fields (field effect), another junction (bipolar operation), or even chemical control. This separation of a nanodevice into ideal injecting and extracting contacts, an active region which limits the transport of charge, and a gate(s) which regulate current flow, is a common way of visualizing the transport properties of nanoscale devices. However, it clearly has limitations, the contacts themselves form/become a part of the active system, and are driven out of equilibrium due

Fig. 1.6 Schematic of a generic nanoelectronic device consisting of sources and sinks for charge carriers (the source and drain contacts), and a "gate" which controls the transfer characteristics of the active region.

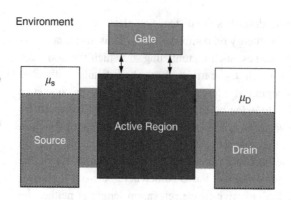

to current flow, as well as coupling strongly to the active region through the long-range Coulomb interaction of charge carriers.

The nature of transport in a nanodevice such as that illustrated in Fig. 1.6, depends on the characteristic length scales of the active region of the device, L, as discussed in the previous section and in later chapters. If scattering events are frequent as carriers traverse the active region of the device, carrier transport is *diffusive* in nature, and is reasonably approximated by the semiclassical BTE. Energy dissipation occurs throughout the device, and the contacts are simply injectors and extractors of carriers near equilibrium. In contrast, if little or no scattering occurs from source to drain, transport is said to be *ballistic*, and the wave nature of charge carriers becomes important in terms of quantum mechanical reflection and interference *from the structure itself*, and the overall description of transport is in terms of quantum mechanical fluxes and transmission as discussed earlier. Energy is no longer dissipated in the active region of the device, rather, it is dissipated in the contacts themselves.

To fabricate nanodevices beyond current limits dictated by scaling, CMOS technology is rapidly moving towards quasi-3D structures such as dual-gate, tri-gate, and FinFET structures [26], in which the active channel is increasingly a nanowire or nanotube rather than bulk region. Such 3D gate structures are needed to maintain charge control in the channel, as channel lengths scale towards nanometer dimensions. This trend is illustrated in Fig. 1.7. The traditional materials used in the semiconductor industry (Si substrates, SiO_2 insulators, metal or poly-Si gates) are rapidly being replaced by higher performance alternatives. Due to the rapid increase of tunneling current as the gate oxide is reduced by scaling below 1.5 nm, alternative high dielectric constant insulators are presently being introduced that provide the same capacitance with a much thicker insulator to mitigate gate leakage. The heavily doped Si substrate is increasingly being replaced by SiGe and Si on insulator technology as well, and currently III–V materials are being considered as alternatives for the channel to increase mobility and enhance drive current [27]. We will return to this subject in Chapter 9.

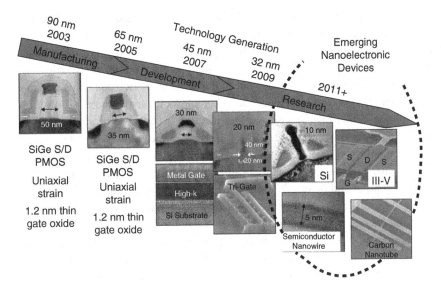

Fig. 1.7 Scaling of successive generations of MOSFETs into the nanoscale regime and emerging nanoelectronic devices (from R. Chau, presented at INFOS 2005, with permission).

Beyond field effect transistors, there have been numerous studies over the past two decades of alternatives to classical CMOS at the nanoscale. As dimensions become shorter than the phase-coherence length of electrons, the quantum-mechanical wave nature of electrons becomes increasingly apparent, leading to phenomena such as interference, tunneling, and quantization of energy and momentum as discussed earlier. Indeed, for a one-dimensional wire, the system may be considered a waveguide with "modes," each with a conductance less than or equal to a fundamental constant $2e^2/h$, as discussed earlier in this chapter. While various early schemes were proposed for quantum interference devices based on analogies to passive microwave structures (see for example [28,29,30]), most suffer from difficulty in control of the desired waveguide behavior in the presence of unintentional disorder. This disorder can arise from the discrete impurity effects discussed earlier, as well as the necessity for process control at true nanometer scale dimensions. More recently, promising results have been obtained on ballistic Y-branch structures [31], where nonlinear switching behavior has been demonstrated even at room temperature [32], which we discuss in later chapters.

As mentioned earlier, the role of discrete impurities as an undesirable element in the performance of nanoscale FETs due to device-to-device fluctuations. However, the discrete nature of charge in individual electrons, and control of charge motion of single electrons has in fact been the basis of a great deal of research in single electron devices and circuits (see for example [33]), discussed extensively in Chapter 6. The understanding of single electron behavior is most easily provided in terms of the capacitance, C, of a small tunnel junction, and the corresponding change in electrostatic energy, $E = e^2/2C$, when an electron tunnels from one side to the other. When physical dimensions are sufficiently small, the corresponding capacitance (which is a geometrical quantity in general) is correspondingly small, so that the change in energy is greater than the thermal energy, resulting in the possibility of a "Coulomb blockade," or suppression of tunnel conductance due to the necessity to overcome this electrostatic energy. This Coulomb blockade effect allows the experimental control of electrons to tunnel one by one across a junction in response to a control gate bias. Single electron transistors [34], turnstiles [35,36], and pumps [37] have been demonstrated, even at room temperature [38]. As in the case of quantum interference devices, the present-day difficulties arise from fluctuations due to random charges and other inhomogeneities, as well as the difficulty in realizing lithographically defined structures with sufficiently small dimensions to have charging energies approaching kT and above.

Throughout this chapter, we have emphasized the issue of device to device fluctuations due to structural inhomogeneities, either inherent or produced through the device fabrication process. There has been rapid progress over the last decade in realizing functional nanoscale electronic devices based on self-assembled structures such as semiconductor nanowires (NWs) [39] and carbon nanotubes (CNTs) [40], where reduced dimensionality is achieved through bottom-up growth rather than top-down manufacture. The limits of transistor scaling shown in Fig. 1.7 illustrates both types of materials. Semiconductor nanowires have been studied over the past decade in terms of their transport properties, and for nanodevice applications such as resonant tunneling diodes [41], single electron transistors [42,43], and field-effect structures [39]. Recently, there has been a dramatic increase in interest in NWs due to the demonstration of directed self-assembly of NWs via in-situ epitaxial growth [44,45]. Such semiconductor NWs can be elemental (Si, Ge) or III–V semiconductors, where it has been demonstrated that such wires may be controllably doped during growth [46], and abrupt compositional changes forming high-quality 1D heterojunctions can be achieved [47,48]. Nanowire FETs, bipolar devices and complementary inverters have been synthesized using such techniques [49,50]. The ability to controllably fabricate heterostructure nanowires has led to demonstration of nanoelectronic devices such as resonant tunneling diodes [51] and single electron transistors [52]. The scalability of arrays of such nanowires to circuits and architectures has also begun to be addressed [53], although the primary difficulty at present is in the ability to grow and orient NWs with desired location and direction.

Single-walled (SW) carbon nanotubes (CNTs), are a tubular form of carbon with diameters as small as 1 nm and lengths of a few nm to microns. CNTs have received considerable attention due to the ability to synthesize NTs with metallic, semiconducting, and insulating behavior, depending on the diameter, and particularly on the chirality (i.e. how the graphite sheets forming the structure of the CNT wrap around and join themselves) [54]. Due to their remarkable electronic and mechanical properties, CNTs are currently of interest for a number of applications including interconnects, CNT-based molecular electronics, AFM-based imaging, nanomanipulation, nanotube sensors for force, pressure, and chemical nature, nanotube biosensors, molecular motors, nanoelectromechanical systems (NEMS), hydrogen and lithium storage, and field emitters for instrumentation including flat panel displays. In terms of transport, measurements have demonstrated very high mobilities and nearly ballistic transport [55,56]. Complementary n and p-channel transistors have been fabricated from CNTs, and basic logic functions demonstrated [57]. Again, as with semiconductor nanowires, the primary difficulty faced today in a manufacturable device technology is the directed growth of CNTs with the desired chirality, and positioning on a semiconductor surface, suitable for large-scale production.

1.5 An introduction to the subsequent chapters

This book is of two general themes. The first is a systematic review of the experimental status of mesoscopic systems and devices, in particular the experiments that have been carried out to understand the relevant physics and the manner in which this physics affects the transport. We apologize up front that this review is not an exhaustive review. The second theme is the general development of quantum transport theory based primarily upon the Green's function (and quantum field theory) solutions to the Schrödinger equation in confined systems. To be sure, there are a great many review articles in the literature that cover mesoscopic systems and nanostructures [58,59,60,61]. Most of these review the experimental results for particular measurements or systems, but we do not feel that there is a systematic treatment of this at present. And, for sure, there is no systematic development of the theory to cover the range of measurements in mesoscopic systems, particularly in a manner suitable to use as a textbook. On the other hand, there are several good books on the use of quantum field theory and Green's functions in treating condensed matter systems [62,63,64,65]. It is not our aim to supplant these, but we will instead rely upon them as needed background on the general techniques. Here we want to develop a systematic treatment of the theory underlying mesoscopic processes in semiconductor nanostructures, which provide the basis for most of the interesting experiments.

In Chapter 2 we begin by reviewing important aspects of quantization and modes of nanostructure systems including heterojunctions, nanowires and quantum dots, as well as self-assembled structures such as CNTs and self-assembled nanowires. We then examine the effect of a magnetic field. Finally, we briefly review homogeneous transport theory as applied to low-dimensional structures.

The general topics of Chapter 3 are tunneling and the extension of the Landauer formula to realistic, many-terminal mesoscopic devices. First, the general idea of resonant tunneling through a bound state in planar barrier structures, and the experiments which probe this effect is introduced. The extension of these experiments to many barriers and multiple wells will also be discussed. Then, the theoretical basis of wave transmission and reflection will be introduced so that the basis of the equations above can be established. We first discuss the experiments that describe the transition from a short gated region (the quantum point contact of Fig. 1.2) to a long waveguide in which the carriers move diffusively. We discuss the confined solutions to Schrödinger's equation through the mode picture of the quantum waveguide in analogy to a microwave waveguide. In general, the experiments tell us that the concept of coherent individual modes is rapidly eliminated in long guides, and the role of disorder in this process is discussed next. The role of contacts in modifying the basic forms of the Landauer equation leads us naturally into multi-terminal descriptions and the Büttiker scattering matrix approach with a magnetic field. A discussion of zero-temperature Green's functions for waveguides, and the development of Dyson's equation for their solution is given as well.

Chapter 4 then picks up from this introduction to describe the quantum Hall effect and the gated quantum Hall effect in local and nonlocal geometries. An introduction to many-body effects and the fractional quantum Hall effect is then presented.

Chapter 5 follows from the previous two chapters, in which the important experimental results associated with coherent transport in nanowires of various geometries, as well as more detail of the quantized conductance in quantum point contacts, are discussed.

The general structure of a quantum dot is developed in Chapter 6 through the use of the two-dimensional harmonic oscillator. This introduces angular momentum states and the role of the magnetic field in producing a complex spectrum. Experiments on magneto-tunneling have probed the delicate details of these spectra, and these are discussed and reviewed. We discuss the limit of almost nontransparent barriers for which single-electron tunneling becomes the norm. This single-electron tunneling discussion begins with the experiments on metallic islands. We then consider the double junction, then turn to semiconductor dots in which quantized levels become important. The Russian orthodox theory is then discussed. The role of spins in quantum dots is considered as well as applications to information processing using quantum dots as qubits for quantum computing. We also discuss fluctuations observed in open dots, and the transition from quantum to classical behavior in such systems.

Chapter 7 begins our discussion of higher-order transport properties through weak localization and the universal conductance fluctuations (UCF) and their experimental scaling with size and temperature. This leads us to develop the higher-order perturbation theory necessary to incorporate the long-range correlations responsible for these effects. Here we will introduce the impurity ladder diagrams and the complex multi-particle interactions that lead to diffusons and cooperons.

In Chapter 8, the temperature Green's functions are introduced to investigate the temperature dependence of the UCF, weak localization, and other quantum interference. It is necessary here to give a deeper treatment of the electron–electron interaction to handle the cooperons and "screening." The latter is then treated through the random-phase approximation, which also leads us to discuss the lifetimes of single-particle states.

Finally, in Chapter 9, we return to the more complex picture of the nanodevice, which leads us into the real-time Green's functions. The more extensive family of these functions is necessitated by the fact that the system is now out of equilibrium so that the "distribution function" is no longer known in the system under bias. Unfortunately, considerably less is known about the application of these functions to meaningful discussions of the various effects in mesoscopic devices. However, there are experiments that illustrate the presence of carrier heating and the need for the more extensive description. Along the way, this approach allows us to connect to simulations that have been carried out in reduced descriptions with the density matrix and the Wigner distribution function (reduced in the sense that the latter are single-time functions which ignore temporal correlations). We end with an attempt to loop back to semiclassical transport concepts.

1.6 What is omitted

In the previous section, a short introduction to the material that is included in this book was presented. The reader will note that the list is quite selective, in spite of covering perhaps too much material. However, this book cannot be a self-contained book on either the theory or the experiments in nanostructures. The former would be much too long by the time we introduce the new concepts that have appeared since [60–63] (and comparable other books). Equally, it is not our desire to exhaustively review the experimental literature, since there are many reviews which address this task. Rather, we have tried to select a systematic set of experiments that leads to an understanding of what must be described by theory, and then to develop the required theory. Consequently, much is omitted, among which is a comprehensive review of the literature for either theory or experiment.

References

[1] G. Baccarani, M.R. Wordeman, and R.H. Dennard, *IEEE Trans. Electron Dev.* **31**, 452 (1984).
[2] J.M. Mikkelson, L.A. Hall, L.A. Malhotra, S.D. Seccombe, and M.S. Wilson, *IEEE J. Sol. State Circuits* **16**, 542 (1981).

[3] *International Technology Roadmap*, www.itrs.net.
[4] B. Hoeneisen and C.A. Mead, *Sol.-State Electron.* **15**, 819, 891 (1972).
[5] C.A. Mead, *J. VLSI Signal Processing* **8**, 9 (1995).
[6] R. Chau, D. Doyle, M. Doczy, *et al.*, in the Proceedings of the 61st Device Research Conference, pp. 123–26, Salt Lake City, 23–25 June 2003.
[7] B. Doris *et al.*, Technical Digest of the IEEE International Device Meeting, 2001, 937.
[8] B. Yu *et al.*, Technical Digest of the IEEE International Device Meeting, 2002, 267.
[9] J.R. Barker and D.K. Ferry, *Sol.-State Electron.* **23**, 519, 531 (1980).
[10] These are reviewed by M. Heiblum, in *High Speed Electronics*, eds. B. Kallback and H. Beneking (Berlin, Springer-Verlag, 1986), and by J.R. Hayes, A.J.F. Levi, A.C. Gossard, and J.H. English, in *High Speed Electronics*, eds. B. Kallback and H. Beneking (Berlin, Springer-Verlag, 1986).
[11] M.J. Gilbert, R. Akis, and D.K. Ferry, *J. Appl. Phys.* **98**, 094303 (2005).
[12] N.F. Mott, *Philos. Mag.* **22**, 7 (1970).
[13] D.J. Thouless, *Phys. Rept.* **13C**, 93 (1974); J.T. Edwards and D.J. Thouless, *J. Phys. C* **5**, 807 (1972); D.C. Licciardello and D.J. Thouless, *J. Phys. C* **8**, 4157 (1975); D. J. Thouless, *Phys. Rev. Lett.* **39**, 1167 (1977).
[14] E. Abrahams, P.W. Anderson, D.C. Licciardello, and T.V. Ramakrishnan, *Phys. Rev. Lett.* **42**, 673 (1979).
[15] R. Landauer, *IBM J. Res. Develop.* **1**, 223 (1957); *Philos. Mag.* **21**, 863 (1970).
[16] B.J. van Wees, H. van Houten, C.W.J. Beenakker, *et al.*, *Phys. Rev. Lett.* **60**, 848 (1988).
[17] D.A. Wharam, T.J. Thornton, R. Newbury, *et al.*, *J. Phys.* **C 21**, L209 (1988).
[18] Y. Imry, in *Directions in Condensed Matter Physics*, eds. G. Grinstein and E. Mazenko (Singapore, World Scientific Press, 1986), pp. 103–163.
[19] Y. Aharonov and D. Bohm, *Phys. Rev.* **115**, 485 (1959).
[20] R.A. Webb, S. Washburn, C.P. Umbach, and R.B. Laibowitz, *Phys. Rev. Lett.* **54**, 2596 (1985).
[21] K. Ishibashi, Y. Takagaki, K. Gamo, *et al.*, *Sol. State Commun.* **64**, 573 (1987).
[22] P.M. Mankiewich, R.E. Behringer, R.E. Howard, *et al.*, *J. Vac. Sci. Technol.* **B6**, 131 (1988).
[23] C.J.B. Ford, T.J. Thornton, R. Newbury, *et al.*, *J. Phys.* **C 21**, L325 (1988).
[24] D.K. Ferry, *Semiconductors* (New York, Macmillan, 1991).
[25] D.K. Ferry and H. Grubin, in *Solid-State Physics*, eds. H. Ehrenreich and W. Turnbull (New York, Academic Press, in press).
[26] M. Leong, H.-S. Wong, E. Nowak, J. Kedzierski, and E. Jones, *ISQED2002*, **492** (2002).
[27] R. Chau, *et al.*, *IEEE Trans. Nanotechnol.* **4**, 153 (2005).
[28] F. Sols, M. Macucci, U. Ravaioli, and K. Hess, *J. Appl. Phys.* **66**, 3892 (1989).
[29] S. Datta, *Superlatt. Microstruct.* **6**, 83 (1989).
[30] A. Weisshaar, J. Lary, S.M. Goodnick, and V.K. Tripathi, *IEEE Electron Device Lett.* **12**, 2 (1991).
[31] L. Worschech, B. Weidner, S. Reitzenstein, and A. Forchel, *Appl. Phys. Lett.* **78**, 3325 (2001).
[32] K. Hieke and M. Ulfward, *Phys. Rev. B* **62**, 16727 (2000).
[33] K.K. Likharev, *Proc. IEEE* **87**, 606 (1999).
[34] K. Likharev, *IBM J. Res. Develop.* **32**, 144 (1988).
[35] L.J. Geerligs, V.F. Anderegg, P.A.M. Holweg, *et al.*, *Phys. Rev. Lett.* **54**, 2691 (1990).
[36] L.P. Kouwenhouven, A.T. Johnson, N.C. van der Vaart, C.J.P.M. Harmans, and C.T. Foxon, *Phys. Rev. Lett.* **67**, 1626 (1991).

References

[37] H. Pothier, P. Lafarge, C. Urbina, D. Esteve, and M.H. Devoret, *Europhysics Lett.* **17**, 249 (1992).

[38] D.H. Kim, S.-K. Sung, K.R. Kim, et al., *IEEE Trans. ED* **49**, 627 (2002).

[39] Y. Cui and C.M. Lieber, *Science* **291**, 851 (2001).

[40] R. Martel, V. Derycke, C. Lavoie, et al., *Phys. Rev. Lett.* **87**, 256805 (2001).

[41] M.A. Reed, J.N. Randall, R.J. Aggarwal, et al., *Phys. Rev. Lett.* **60**, 535 (1988).

[42] L. Zhuang, L. Guo, and S.Y. Chou, *Appl. Phys. Lett.* **72**, 1205 (1998).

[43] D.H. Kim, S.-K. Sung, K.R. Kim, et al., *IEEE Trans. ED* **49**, 627 (2002).

[44] K. Hiruma, M. Yazawa, T. Katsuyama, et al., *J. Appl. Phys.* **77**, 447 (1995).

[45] H. Dai, E.W. Wong, Y.Z. Lu, S. Fan, and C.M. Lieber, *Nature* **375**, 769 (1995).

[46] Y. Cui, X. Duan, J. Hu, and C.M. Lieber, *J. Phys. Chem.* **B104**, 5213 (2000).

[47] M.T. Björk, B.J. Ohlsoon, T. Sass, et al., *Appl. Phys. Lett.* **80**, 1058 (2002).

[48] M.T. Björk, B.J. Ohlsoon, T. Sass, et al., *Nano Lett.* **2**, 87 (2002).

[49] Y. Cui, Z. Zhong, D. Wang, W.U. Wang, and C.M. Lieber, *Nano Lett.* **3**, 149 (2003).

[50] X. Duan, C. Niu, V. Sahl, et al., *Nature* **425**, 274 (2003).

[51] M.T. Björk, B.J. Ohlsoon, C. Thelander, et al., *Appl. Phys. Lett.* **81**, 4458 (2002).

[52] C. Thelander, T. Martensson, M.T. Björk, et al., *Appl. Phys. Lett.* **83**, 2052 (2003).

[53] Z. Zhong, D. Wang, Y. Cui, M.W. Bockrath, and C.M. Lieber, *Science* **302**, 1377 (2003).

[54] M.S. Dresselhaus, G. Dresselhaus, and P.C. Eklund, *Science of Fullerenes and Carbon Nanotubes* (New York, Academic Press, 1996).

[55] T. Dürkop, S.A. Getty, E. Cobas, and M.S. Fuhrer, *Nano Lett.* **4**, 35 (2004).

[56] A. Javey, J. Guo, Q. Wang, M. Lundstrom, and H. Dai, *Nature* **424**, 654 (2003).

[57] See for example P.L. McEuen, M.S. Fuhrer, and H. Park, *IEEE Trans. on Nanotechnology* **1**, 78 (2002).

[58] M.A. Reed, ed., *Nanostructured Systems* (New York, Academic Press, 1992).

[59] H.A. Cerdeira, F. Guinea Lopez, and U. Weiss, eds., *Quantum Fluctuations in Mesoscopic and Macroscopic Systems* (Singapore, World Scientific Press, 1990).

[60] S. Namba, C. Harnaguchi, and T. Ando, eds., *Science and Technology of Mesoscopic Structures* (Tokyo, Springer-Verlag, 1992).

[61] B.L. Altshuler, P.A. Lee, and R.A. Webb, eds., *Mesoscopic Phenomena in Solids* (Amsterdam, North-Holland, 1991).

[62] A.L. Fetter and J.D. Walecka, *Quantum Theory of Many-Particle Systems* (New York, McGraw-Hill, 1971).

[63] L.P. Kadanoff and G. Baym, *Quantum Statistical Mechanics* (New York, Benjamin, 1962).

[64] G.D. Mahan, *Many-Particle Physics* (New York, Plenum Press, 1981).

[65] A.A. Abrikosov, L.P. Gorkov, and I.Ye. Dzyaloshinskii, *Quantum Field Theoretical Methods in Statistical Physics*, 2nd edn. (Oxford, Pergamon Press, 1965).

2
Quantum confined systems

As discussed in the previous chapter, there are two issues that distinguish transport in nanostructure systems from that in bulk systems. One is the granular or discrete nature of electronic charge, which evidences itself in single-electron charging phenomena (see Chapter 6). The second involves the preservation of phase coherence of the electron wave over short dimensions. Artificially confined structures are now routinely realized through advanced epitaxial growth and lithography techniques in which the relevant dimensions are smaller than the phase coherence length of charge carriers. We can distinguish two principal effects on the electronic motion depending on whether the carrier energy is less than or greater than the confining potential energy due to the artificial structure. In the former case, the electrons are generally described as bound in the direction normal to the confining potentials, which gives rise to quantization of the particle momentum and energy as discussed in Section 2.2. For such states, the envelope function of the carriers (within the effective mass approximation) is localized within the space defined by the classical turning points, and then decays away. Such decaying states are referred to as *evanescent states* and play a role in tunneling as discussed in Chapter 3. The time-dependent solution of the Schrodinger equation corresponds to oscillatory motion within the domain of the confining potential.

The second type of motion we will be concerned with is that associated with propagating states of the system. Here the carrier energy is such that it lies above that of the confining potentials, or that the potentials are limited sufficiently in extent so that quantum mechanical tunneling through such barriers can occur. In this regime, where potential variations occur on length scales smaller than the phase coherence length, transport is more readily defined in terms of reflection and transmission of matter waves. This leads to definition of conduction in terms of the Landauer formula discussed in Chapter 3. In the present chapter, we will concern ourselves with quantum confined structures and the so-called reduced-dimensionality systems associated with such confinement of the particle motion. The topic of electronic states in quantum confined systems, and the host of fundamental studies and device applications related to this subject, is quite extensive. Since the purpose of this book is to elucidate transport in nanostructures, we will focus only on those aspects of quantum confined systems that relate to electronic transport, and thus for example ignore the well-studied optical properties of these systems except as they pertain to the main theme. There are a number of excellent reviews on the topic such as that of Ando, Fowler, and Stern [1] on the pioneering research in this field during the 1970s related to quantization at the Si/SiO_2 interface in Si MOS (metal oxide semiconductor) devices. More recent books devoted to this topic in semiconductor heterojunction systems are now available [2,3].

2.1 Nanostructure materials

In the context of semiconductor materials and fabrication based on planar integrated circuit technology, quantum confinement has traditionally been realized in roughly two different ways: (i) through the growth of inhomogeneous layer structures resulting in quantization perpendicular to the substrate surface, and (ii) through lateral patterning using ultrafine lithography techniques. Historically, the development of quantum confined systems was realized in heterolayer structures grown on semiconducting substrates. Here, the discontinuities in the conduction and valence band edges between different materials behave as potential discontinuities within the framework of the effective mass approximation discussed in more detail later. The first demonstration of quantization of semiconductor states due to artificial confinement was in the inversion layers of Si MOS structures [4]. In this system, quantization of the carrier motion is due to the confining potential of the Si/SiO_2 interface barrier and the potential well in the other direction due to band bending. Part of the success of measurements made in the Si/SiO_2 system are due to the relatively ideal interface formed between Si and its oxide. Later, with the development of precision epitaxial growth techniques such as molecular beam epitaxy (MBE) [5] and metal organic chemical vapor deposition (MOCVD), high-quality lattice-matched heterojunction systems could be realized. These systems exhibit quantum confinement effects far superior to those in the Si MOS system due to several factors, including the low surface state density at the interface of lattice-matched materials such as GaAs and $Al_xGa_{1-x}As$, and the lower conduction band mass of III–V compound materials in general (compared to Si, which increases the energy spacing of the quantized levels). Figure 2.1 shows a transmission-electron-microscope (TEM) image of the heterointerface of a GaAs/AlGaAs heterostructure, grown by MBE. Since the lattice constants of GaAs and AlAs are very close, there is very little interface strain between the two lattice-matched binary systems. Roughness on the order of one or two monolayers is readily apparent in this image.

A further innovation which is an essential component of a large fraction of the semiconductor nanostructures studied today is the technique of modulation doping (MD) [6], in which the dopants that provide free carriers in the

Fig. 2.1 Image of a GaAs/AlGaAs superlattice, illustrating the near atomic-level quality of the material. Picture provided courtesy of A. Trampert and O. Brandt, Paul-Drude Institute, Berlin.

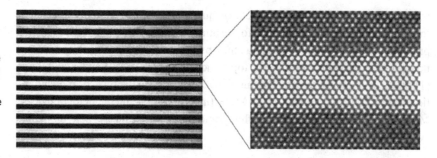

heterostructure are spatially separated from these carriers themselves. This scheme usually involves the growth of an undoped active layer with a smaller bandgap such as GaAs, onto which a doped, wider-bandgap, material such as $Al_xGa_{1-x}As$ is grown with an undoped spacer layer. In high-electron-mobility transistors (HEMTs) based on the principle of MD, a layer of unintentionally doped GaAs (typically with carrier density between 10^{14}–10^{15}/cm^3) is grown on a semi-insulating GaAs substrate. Then, an undoped spacer layer of $Al_xGa_{1-x}As$ is grown, followed by a thicker heavily doped layer of the same material. A cap layer of doped or undoped GaAs is usually grown on top to facilitate Ohmic-contact formation and to reduce oxidation of the AlGaAs layer. Since the free carriers forming a channel at the GaAs/AlGaAs interface are spatially remote from their parent donors, ionized impurity scattering is greatly reduced. Mobilities in excess of 1×10^7 cm^2/Vs are obtainable in such structures, as discussed in Section 2.7.3 [7].

2.1.1 Recent developments – self-assembled nanostructures

In the time since the writing of the first edition of this book, there has been an explosion in the development of nanostructures that are realized by exploiting natural processes of self-assembly. The study of this area actually has a long history, and was driven for many years by the study of self-assembled semiconductor quantum dots, realized by suitable modifications to MBE growth. The leading example of this approach is provided by the self-assembly of InAs, or InGaAs, quantum dots that form on a GaAs substrate via the Stransky–Krastinov growth process [8]. In this mode of growth, a thin layer of InAs is grown on top of a GaAs substrate, but, if the layer is sufficiently thin, the strain will cause the InAs to agglomerate into small three-dimensional quantum dots. These dots have been of interest, since they are in an optically active media, and therefore are of interest for quantum-dot lasers [9]. From the perspective of electron transport, however, the main topic of interest here, the difficulty in making controlled electrical contacts to these structures has tended to limit studies of their electrical characteristics. Having said that, various authors have succeeded in implementing vertical tunnel devices, in which the application of a potential difference between top and substrate electrodes allows for control of the charge number in the quantum dot [10,11,12,13,14,15]. Such structures have recently become of particular interest for the controlled injection of *excitons*, and for the generation of single, and entangled, photons for application in quantum computing [16,17].

Overtaking interest in the properties of self-assembled quantum dots have been a series of remarkable developments over the past decade in the synthesis of various nanotubes and nanowires. Carbon nanotubes (CNTs) [18] are probably the most well known of these, being structures that are essentially comprised of rolled graphene sheets. The nanotubes have typical diameters in the range of a few, to a few tens of, nanometers, and may exist as either single- or multi-walled structures, dependent upon the number of graphene layers that are rolled to form the nanotube structure. The remarkable feature of CNTs is that they may be either metallic or semiconducting, depending on the manner in which they are rolled relative to the two-dimensional lattice structure of graphene. This has led to interest in their use as highly conducting nanoscale interconnects (in the metallic phase) for ultra-high density integrated circuits [19], or as the conducting channel of FET structures (in the semiconducting phase) [20]. Electron transport through these nanotubes can be ballistic, even at room temperature [21], and the strong one-dimensional character of these structures has resulted in the observation of a variety of novel transport phenomena, including evidence for Luttinger-liquid formation [22], enhanced electron correlations [23], Aharonov–Bohm oscillations [24], and Fabry–Perot resonances [25] due to coherent quantum interference. Figure 2.2 shows a scanning-electron-microscope image of a single-walled CNT and its contacts as well as the measured magneto-conductance oscillations associated with the AB effect [26].

One of the problematic issues with CNTs is the development of systematic methods for their synthesis. The most commonly used methods are *arc-discharge* [27], in which carbon is evaporated in a helium atmosphere by producing a large electrical discharge between two carbon electrodes, and *laser ablation* [28], in which intense laser pulses are used to ablate a carbon source containing a small proportion of metal catalyst. In either method, nanotubes are produced along with other debris and considerable effort is required to refine individual nanotubes for device processing. While there have been many efforts to explore the directed growth of nanotubes used in patterned catalyst sites [29], or electric-field-induced alignment [30], there has also been a huge parallel effort to develop other nanowire systems using more controlled epitaxial processes. Indeed, this field has now grown to become so large that it is impossible to provide any comprehensive review here. We therefore only briefly focus on a few examples, including a discussion of the electronic states of CNTs in Section 2.4.2, reflecting our personal bias from the field of conventional semiconductor nanoelectronics, and refer the reader instead to Refs. [31,32] for more thorough overviews of this field.

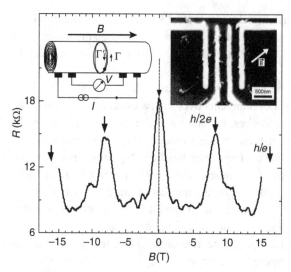

Fig. 2.2 The upper image is a scanning-electron-microscope image of a single-walled CNT that is electrically contacted by four different electrodes. At the bottom the magneto-resistance of a multi-walled CNT is shown. The magnetic field is oriented along the axis of the CNT, as indicated in the inset, and the magneto-resistance shows oscillations associated with the AB effect ([26], with permission).

One of the major broad techniques used for the growth of semiconducting nanowires is *vapor-phase synthesis*, in which nanowires are grown by starting from appropriate gaseous components. The so-called *vapor-liquid-solid* (VLS) mechanism uses metallic nanoparticles as seed sites to stimulate the self-assembled growth of nanowires. The desired semiconductor system is introduced in terms of its gaseous components and the entire assembly is heated to a temperature beyond the eutectic temperature of the metal/semiconductor system. Under these conditions, the metal forms a liquid droplet, with a typical size of a few nanometers. Once this droplet becomes supersaturated with semiconductor, it essentially nucleates the growth of the nanowire from the base of the droplet. Figure 2.3 shows examples of Si nanowires grown by the VLS method using gold nanoparticles as the seeding droplets. The high crystalline integrity of this nanowire can be clearly seen in this image, which also makes clear how the diameter of the nanowire is connected to the size of the catalyst droplet [33]. The wire shown here was grown by using chemical-vapor deposition (CVD) to generate the semiconductor precursors, a popular approach to VLS. Other methods may also be used, however, including laser ablation and MBE. The VLS method has emerged as an extremely popular method for the fabrication of a variety of nanowires. As described more extensively in Ref. [34], it has also been used to realize various III–V (GaN, GaAs, GaP, InP, InAs) and II–IV (ZnS, ZnSe, CdS, CdSe) semiconductor nanowires, as well as several different wide-bandgap oxides (ZnO, MgO, SiO_2, CdO).

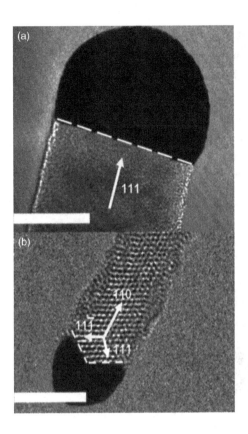

Fig. 2.3 Shown in (a) is an image highlighting the interface between a catalyst droplet and a Si nanowire with a <111> growth axis. The scale bar indicates 20 nm. In (b) a similar image is shown but the growth axis of the nanowires is <110> and the scale bar denotes 5 nm ([33], with permission).

From the perspective of transport studies, Lieber and co-workers have applied the VLS method to realize semiconducting Si nanowires, metallic nickel silicide nanowires, and Si/NiSi nanowire heterostructures [34]. The silicide nanowires showed clear metallic behavior, with linear I–V characteristics and room-temperature resistivities of order 10 μΩcm, indicating their high electronic quality. The silicide nanowires were formed by coating silicon nanowires realized by VLS with a thin layer of nickel, and then reacting at 550 °C to form the silicide. By selective masking of individual Si nanowires prior to nickel evaporation, they were able to therefore realize Si/NiSi nanowire heterostructures with atomically sharp heterointerfaces. The authors also applied this technique to realize NiSi/Si/NiSe FETs and were able to explore their resulting electrical characteristics.

Samuelson and co-workers have also had enormous success in developing nanoscale electronic devices that utilize VLS-formed, III–V semiconductor, nanowires as their active elements [35], as shown in Fig. 2.4. They have demonstrated that heterostructure nanowires of InAs and InP, as well as GaAs and InAs, can be realized that have very sharp heterointerfaces [36]. They have subsequently used this technique to implement a variety of nanoscale devices,

Fig. 2.4 An InAs nanowhisker, containing several InP heterostructures. Shown in (a) is a high-resolution transmission-electron-microscope image of a whisker with a diameter of 40 nm. The power spectrum of this image is shown in (b). In (c), the bright regions indicate InP embedded in the nanowire ([36], with permission).

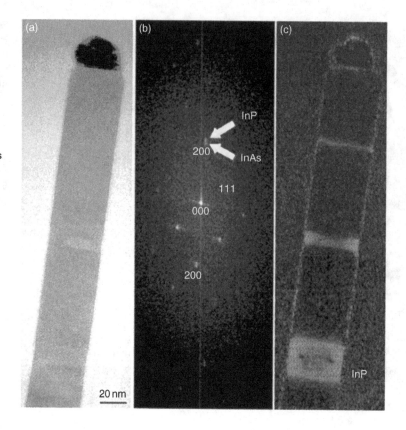

such as resonant-tunneling diodes [37], single- [38], and multiply-coupled [39,40] quantum dots. The strong lateral confinement generated in these structures, combined with their high crystalline quality, endows them with robust quantum-transport characteristics. Quantum dots realized using these structures show very clear single-electron tunneling signatures, with evidence that the g-factor of the electrons can be tuned over a very wide range [41]. The ability to arbitrarily introduce serial heterointerfaces into such nanowires should offer huge potential in the future for the further development of novel nanodevices.

Finally, at the time of preparing this edition of our book, it had become evident that a further class of self-assembled nanostructures was preparing to play a major role in nanoelectronics research in the next decade or more. Section 2.4.2 briefly discusses the electronic structure of graphene. The discovery that single-atomic-layer sheets of graphene may be successfully isolated, and that the electrical properties of their two-dimensional electrons or holes studied, is causing a renewed interest in two-dimensional electron gas (2DEG) quantum systems, with the potential of quantum effects at room temperature [42,43,44,45,46].

2.2 Quantization in heterojunction systems

For simplicity, we will restrict our discussion of transport in nanostructures primarily to the conduction band states of the GaAs/AlGaAs system, in which the vast majority of experiments have been performed. Here we consider the form of the electronic states in a layered heterojunction system introduced in the previous section. The electronic states in GaAs/AlGaAs systems are most easily described within the effective mass approximation (EMA). The criterion for the validity of this approach in heterojunctions has been discussed in detail [3]. As long as we associate ourselves with states of the same symmetry across the heterojunction, such as Γ valley electrons in both the GaAs and $Al_xGa_{1-x}As$ (x less than 0.44 for direct gap behavior), the motion of particles relative to the band edge is found by solving the envelope-function equation

$$\left[\frac{\hbar^2}{2}\frac{\partial}{\partial z}\frac{1}{m(z)}\frac{\partial}{\partial z} + \frac{\hbar^2}{2m_\parallel}\nabla_\mathbf{r}^2 + V_{eff}(z)\right]\psi(\mathbf{r},z) = E\psi(\mathbf{r},z), \qquad (2.1)$$

where $m(z)$ is the effective mass perpendicular to the heterointerface, $m_\parallel$ is the mass parallel to the interface, where $\mathbf{r}$ is the position vector parallel to the interface, and z is the direction perpendicular to the interface. The validity of the EMA requires that the envelope function $\psi(\mathbf{r}, z)$ be slowly varying over dimensions comparable to the unit cell of the crystal. The form for the kinetic energy operator for the motion perpendicular to the interface has been chosen to satisfy the requirement that the Hamiltonian be Hermitian. In Eq. (2.1), $V_{eff}(z)$ is the effective potential energy normal to the interface,

$$V_{eff}(z) = E_c(z) + V_D(z) + V_{ee}(z), \qquad (2.2)$$

where $E_c(z)$ is the heterojunction conduction band discontinuities, $V_D(z)$ is the electrostatic potential due to ionized donors and accepters, and $V_{ee}(z)$ is the one-electron self-consistent Hartree and exchange-correlation potentials due to free carriers within the local density approximation. Since for the heterojunction system the potential variation is only in the z-direction, the solution is separable as

$$\psi(\mathbf{r},z) = \frac{1}{A^{1/2}}e^{i\mathbf{k}\cdot\mathbf{r}}\varphi_n(z), \qquad (2.3)$$

corresponding to free-electron motion in the plane parallel to the interface, where $\mathbf{k}$ is the wavevector in the plane parallel to the interface; A, the normalization factor, is the lateral area of the system; and n labels the eigenstates in the normal direction. The one-dimensional eigenfunctions $\varphi_n(z)$ satisfy

$$\left[\frac{\hbar^2}{2}\frac{\partial}{\partial z}\frac{1}{m(z)}\frac{\partial}{\partial z} + V_{eff}(z)\right]\varphi_n(z) = E_n\varphi_n(z). \qquad (2.4)$$

The total energy relative to the band minima (maxima) is thus:

$$E_{n,k} = \frac{\hbar^2 k^2}{2m_\parallel} + E_n, \qquad (2.5)$$

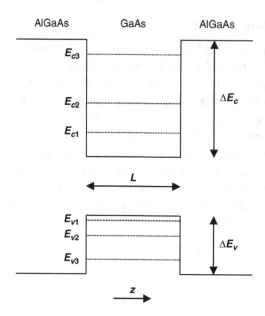

Fig. 2.5 Schematic of a Type I AlGaAs/GaAs/AlGaAs quantum well.

assuming parabolic bands for simplicity. Continuity requires that the envelope function $\varphi_n(z)$ be continuous. For abrupt variations in the material, conservation of probability current requires that $m(z)^{-1}\partial\varphi_n(z)/\partial z$ be continuous, in analogy to the continuity of the electric field across an abrupt interface between two materials of differing dielectric constants.

2.2.1 Quantum wells and quasi-two-dimensional systems

The solution of Eq. (2.4) for $\varphi_n(z)$ allows both bound-state and propagating solutions depending on the energy and the detailed form of the potential. As the simplest example of bound-state behavior, consider the *quantum well* formed by a thin GaAs layer of thickness L sandwiched between two larger bandgap $Al_xGa_{1-x}As$ layers, as shown in Fig. 2.5. In this figure, the bandgap difference is distributed between the valence and conduction bands in such a way that both electrons and holes are confined to the smaller bandgap GaAs layer. Such a heterojunction is referred to as a Type I system, to which we will primarily confine ourselves through the rest of the book.

The 300-K bandgap (in eV) of $Al_xGa_{1-x}As$ as a function of the Al mole fraction is given empirically [47] as $E_g^{\Gamma}(x) = 1.424 + 1.245x$ for $x < 0.45$. The actual fraction of the bandgap difference that appears across the conduction band, ΔE_c, compared to that in the valence band, ΔE_v, has long been studied (see for example [48]). Measurements based on photoluminesence studies of quantum well systems find that roughly 65% of the bandgap difference appears as the conduction band discontinuity [49]. Thus we see that for $x = 0.3$, for example, the bandgap discontinuity would be 243 meV according to the empirical relation above. If we assume a low free carrier density (less than 10^{10} cm^{-2}) and low doping, then the terms $V_D(z)$ and $V_{ee}(z)$ in Eq. (2.2) are negligible, and the solution to Eq. (2.4) is that of a simple finite square well, a classic problem in elementary quantum mechanics.

As shown in Fig. 2.5, discrete levels form in the GaAs conduction and valence bands due to the confinement associated with the bandgap discontinuities. In the valence band, bound states associated with both light holes and heavy holes are possible. The quantum confinement further mixes the light and heavy hole bulk states so that the bound states in the valence band have the character of both types.

If we assume that the potential barriers are large compared to the bound state energies E_n, one can consider the even simpler solution corresponding to infinite potential barriers. Choosing the left interface as the origin, the boundary conditions in this case are that the envelope function vanishes at the points $z = 0, L$. The normalized solutions satisfying these conditions are simple sinusoids

$$\varphi_n(z) = \sqrt{\frac{2}{L}} \sin\frac{n\pi z}{L} \quad n = 1, 2, 3, \ldots \qquad (2.6)$$

The corresponding eigenenergies are given by

$$E_n = \frac{n^2\pi^2\hbar^2}{2m_z L^2}. \qquad (2.7)$$

The term E_n varies from several hundred meV for $L < 50$ nm to several meV as L approaches 100 nm. The motion in the plane parallel to the interface is free, whereas that normal to the epitaxial growth direction is confined. Such a system is referred to as a quasi-two-dimensional electron gas (2DEG) structure and plays a central role in the many nanostructure devices. For each solution n there exists a continuum of two-dimensional states called *subbands*, although later, in connection with quantum waveguide structures discussed in the next chapter, these solutions are sometimes referred to as *modes* in analogy to the standing wave solutions one would encounter for electromagnetic waves propagating between metallic plates.

To make a connection to the introductory remarks of the previous chapter, we note that the solutions of Eq. (2.6) represent states that give rise to a coherent superposition of single-particle wave functions. Other possible solutions destructively interfere and thus decay with time. Obviously, the preservation of phase coherence is necessary to establish the quantization condition given by Eq. (2.7). As the length L becomes longer than the phase coherence length introduced in the first chapter, the particle loses its phase "memory" of the boundaries and thus behaves in a bulk-like fashion rather than as a quantized system.

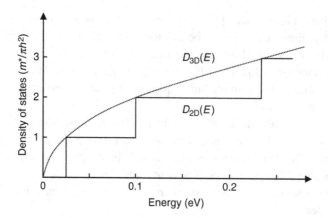

Fig. 2.6 Density of states versus energy for a 15-nm GaAs infinite quantum well (solid curve). The equivalent 3D density of states in the well are included here for comparison.

It is instructive to consider the density of states (per unit area) of the 2DEG system above. The density of states may be calculated directly from

$$D(E) = \sum_{n_s, n_v, n, \mathbf{k}} \delta(E - E_{n,\mathbf{k}}), \qquad (2.8)$$

where n_s and n_v are the spin and valley degeneracy (for a multi-valley minimum), respectively. Assuming a simple spherical conduction band minimum such that $E_{n,\mathbf{k}}$ is given by Eq. (2.5), the sum over the quasi-continuous wavevectors $\mathbf{k}$ may be converted to an integration, and the corresponding angular integration performed as

$$\sum_{n_s, n_v, n, \mathbf{k}} \to 2n_v \sum_n \frac{1}{2\pi} \int_0^\infty dk\, k\, \delta\left[E - E_n - \frac{\hbar^2 k^2}{2m_\parallel}\right], \qquad (2.9)$$

assuming two-fold spin degeneracy per state. The integration of k is easily performed to obtain

$$D(E) = \sum_n \frac{n_v m_\parallel}{\pi \hbar^2} \vartheta(E - E_n) = \sum_n D_0 \vartheta(E - E_n), \qquad (2.10)$$

where ϑ is the unit step function. The density of states is thus a staircase in which the density of states for individual subbands is constant in energy characteristic of a two-dimensional system. Figure 2.6 shows the density of states for a 15-nm AlGaAs/GaAs quantum well versus energy. As shown by comparison to the parabolic 3D density of states in the well, the 2D density of states is a piecewise-continuous approximation to the 3D case.

The 2D density of electrons in the well in equilibrium may be calculated by integrating the density of states (2.10) with the Fermi function for electrons to give

$$n_{2D} = \int_0^\infty dE\, D(E) f_0(E) = \sum_n N_n = k_B T D_0 \sum_n \ln(1 + e^{(E_F - E_n)/k_B T}), \qquad (2.11)$$

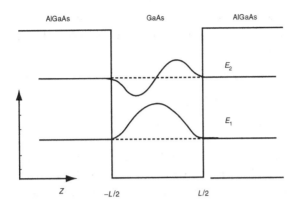

Fig. 2.7 Conduction band profile of a finite square well system showing the ground and the first excited subband.

where N_n is the 2D density of carriers in the nth subband, E_F is the Fermi energy of the system, and T is the temperature of the electron gas. As the well width decreases, the energy spacing of the levels increases, and the system behaves increasingly like a pure two-dimensional system. The *extreme quantum limit* corresponds to the case in which only the ground ($n = 1$) subband is occupied by electrons. This limit is realized at low temperatures where the electron density is sufficiently small such that E_F is below the second subband. Then Eq. (2.11) reduces to

$$n_{2D} = D_0 E_F, \tag{2.12}$$

and the system appears metallic with a well-defined Fermi surface (in this case, perimeter), with the Fermi wavevector given by

$$k_F = \sqrt{2\pi n_{2D}}. \tag{2.13}$$

For the problem of a finite-barrier-height quantum well, consider the conduction band profile shown in Fig. 2.7 with $V_0 = \Delta E_c$. Generally, when considering such piecewise-continuous one-dimensional potentials, one would write the solution in each region in terms of the two independent solutions of the wave equation for a constant potential. States with energies below the potential barrier V_0 are denoted *bound states,* and those above are known as *propagating* or *continuum states.* In characterizing the latter types of solutions, we would use the whole set of independent states from each region, using the matching conditions discussed earlier, and characterize the solutions in terms of their respective transmission and reflection characteristics (discussed in the next chapter). Of course, the bound states are important in connection with our discussion of quantum confined systems. First, we can simplify the solutions by recognizing the inversion symmetry around the center of the well and choosing the origin of our system there, as shown in Fig. 2.7. The solutions are thus symmetric or antisymmetric with respect to the origin, which may be chosen as sine or cosine functions depending on the symmetry. The solutions in the classically forbidden regions are real exponential solutions (evanescent states), which must remain bounded away from the well. Therefore, only states that decay in space away from the well are allowed, and the solution in the three regions can be written as

$$\varphi_n(z) = \begin{cases} Be^{\kappa_z z}, & z < -L/2 \\ A\cos(k_z z)\{\sin(k_z z)\}, & -L/2 < z < L/2, \\ Be^{-\kappa_z z}, & z > L/2 \end{cases} \quad (2.14)$$

where $k_z = \sqrt{2m_I E/\hbar^2}$ and $\kappa_z = \sqrt{2m_{II}(V_0 - E)/\hbar^2}$, where m_I and m_{II} are the effective masses in the well and barrier materials respectively. Since we have used symmetry to equate the coefficient on the left and right evanescent states, we need only apply the matching conditions for $\varphi_n(z)$ and $m(z)^{-1}\partial\varphi_n(z)/\partial z$ at either boundary. The energy eigenvalues are then given for the even cosine solutions by the transcendental equations

$$\frac{k_z}{m_{II}} \tan\frac{k_z L}{2} = \frac{\kappa_z}{m_I}, \quad (2.15)$$

and similarly for the odd sine solutions as

$$\frac{k_z}{m_{II}} \cot\frac{k_z L}{2} = -\frac{\kappa_z}{m_I}. \quad (2.16)$$

The corresponding energies must be solved graphically or numerically for all possible values lying in the well. (At least one bound state exists for any thickness and barrier height.) As is expected, the difference between the finite well solutions and the infinite well case are smallest for the ground ($n = 1$) solutions and become increasingly worse as one goes up in energy.

2.2.2 Coupled wells and superlattices

When we went from the case of the infinite well to a finite barrier well, one of the main differences is the existence in the latter case of decaying states in the barrier regions. If we have several quantum wells that are separated well beyond the characteristic decay length of a particular state, then they are expected to behave as isolated single wells. Such a system of uncoupled quantum wells is often referred to as *multiple quantum wells* (MQWs); the system would be characterized by N degenerate states for each level, where N is the number of wells.

If the barrier thickness between two wells is decreased, the overlap of the envelope function from each well into the other increases. If the overlap is small, one may treat this interaction using degenerate perturbation theory (see for example [50]). As is well known from this formalism, the degeneracy of the states of the two identical wells is split by the presence of the other, with the degree of splitting depending on the matrix element connecting the two wells

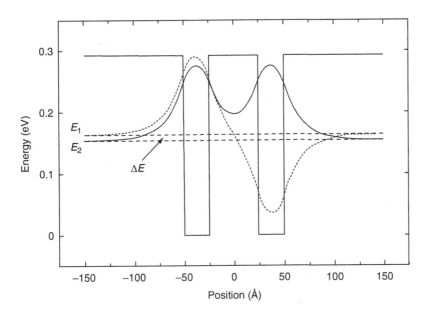

Fig. 2.8 Envelope functions of the lowest two conduction band states for two symmetric coupled AlGaAs/GaAs quantum wells.

which increases as the well separation is decreased. Figure 2.8 shows an exact numerical calculation of Eq. (2.4) for two 2.5-nm GaAs quantum wells separated by a 10-nm $Al_{0.35}Ga_{0.65}As$ barrier for the two lowest states. As was discussed, the lowest two states are split by an amount ΔE. The two solutions for this symmetric well correspond to envelope functions that are symmetric and antisymmetric with respect to the two wells as shown. Figure 2.9 shows the calculated energy eigenvalues for two 10-nm quantum wells separated by various barrier widths. As expected, the splitting of the finite well states increases as the barrier separating the wells decreases. Significant splitting of the ground subband level does not occur until the barrier thickness is less than 4 nm, whereas for higher levels the splitting is more pronounced. This fact reflects the larger penetration [smaller κ_z in Eq. (2.14)] of the upper subband envelope functions into the AlGaAs due to the smaller effective barrier. The splitting of the levels of the symmetric coupled well may also occur in asymmetric wells when, for example, an external electric field is applied. Figure 2.10 shows a wide and narrow quantum well under the influence of a uniform electric field. Experimentally, this situation has been realized by imbedding the wells in the intrinsic region of a reverse-biased *p-i-n* diode [51]. As shown in Fig. 2.10, the two wells have been biased in such a way that the $n = 1$ state of the narrow well is in resonance with the $n = 2$ state of the wide well. The two levels split by an amount dependent on the overlap of the envelope functions, and again they have a symmetric/antisymmetric character in the two wells. Note that the $n = 1$ solution of the wide well is localized entirely in that well with little communication with the narrow well. The coupled-well structure illustrates nicely the

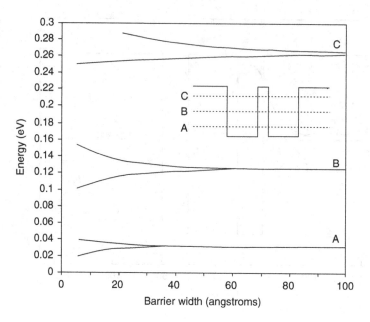

Fig. 2.9 Subband minima for two coupled 100-nm finite square wells, separated by an AlGaAs barrier of variable width (after J. Lary, Ph.D. dissertation).

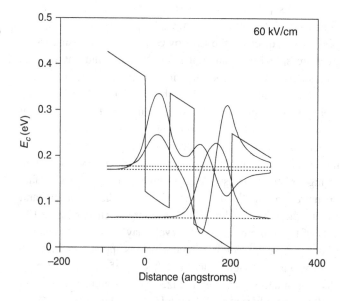

Fig. 2.10 Asymmetric coupled wells biased into resonance.

concept of resonant tunneling, which we will come back to in the next chapter. For the asymmetric coupled-well system, a particle existing in the $n = 1$ state of the wide well has very little probability of existing at some later time in the narrow well, as can be seen from the decay of the envelope function there. However, when two states coincide, a resonance condition exists in which there is a greatly enhanced probability of a particle tunneling from one well to another.

Such a situation is referred to as *resonant tunneling,* to be discussed in detail in Chapter 3. Experimentally the resonant transfer of carriers from one well to another in coupled-well systems has been observed both in optical studies [52] and from transport measurements [53].

The time-dependent solutions of such resonant coupled-well systems have an interesting behavior. If one were able to prepare the system initially with an electron on one side or the other (i.e., by a mixed superposition of the two solutions), then the time-dependent solution of the Schrödinger equation predicts that the wavepacket will oscillate back and forth between the two wells with a characteristic frequency $\omega = \Delta E/\hbar$ until the phase coherence of the packet is destroyed through inelastic scattering, as discussed previously. Evidence for such coherent oscillations have in fact been reported in ultrafast pump and probe optical experiments in which the time resolution is sufficiently short to resolve the wavepacket motion [52]. Such charge oscillations have also been shown to generate terahertz frequency radiation characteristic of the oscillation frequencies predicted by the simple splitting of the levels, which may have practical applications as well [54]. This concept of resonant behavior also extends to coupled quantum dots as discussed later in Chapter 6.

We can extend the coupled-well case to consider many identical wells. Figure 2.11 illustrates the behavior of the level splitting versus barrier width for five wells. Now the five-fold degeneracy of each level is split by the perturbation of the neighboring wells, giving rise to a band of discrete levels, with one state in each band arising from each well. Of course, this picture looks very similar to the broadening of the atomic levels in a solid to form energy bands within the tight binding or LCAO (linear combination of atomic orbitals) model discussed in solid state physics textbooks (see for example [55,56]). If we increase the number of wells indefinitely, we form a semiconductor *superlattice*. One can impose periodic boundary conditions and express the solutions of Eq. (2.4) for the one-dimensional periodic potential in the form of Bloch functions. Again, if the doping and free carrier contributions to the potential are negligible, the solution for a Type I superlattice such as shown in Fig. 2.11 (extended infinitely) may be obtained from the textbook Kronig–Penney model (see for example [55]), generalized to the case of different effective masses in the barriers and wells. Thus the envelope function solutions to Eq. (2.4) can be expressed as

$$\varphi_n(z) = e^{ik_z z} u_{n,k_z}(z), \quad u(z+L) = u(z), \qquad (2.17)$$

where L is the period of the superlattice, and k_z is the wavevector along the growth direction of the multilayer structure. Figure 2.12 shows a calculation based on the Kronig–Penney model for the allowed energies as a function of barrier and well width for a fixed barrier height of $V_0 = 0.4$ eV [57]. The discrete bands of states shown in Fig. 2.11 now become a continuum of states (*minibands*),

Fig. 2.11 Same calculation as Figure 2.9, but for five coupled wells (after J. Lary, Ph.D. dissertation).

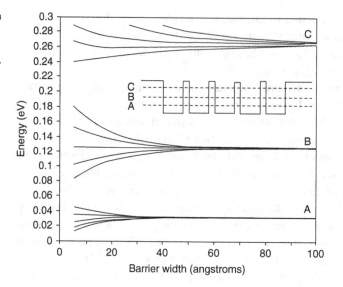

Fig. 2.12 Allowed energy bands versus well and barrier width for a superlattice with barrier height of $V_0 = 0.4$ eV.

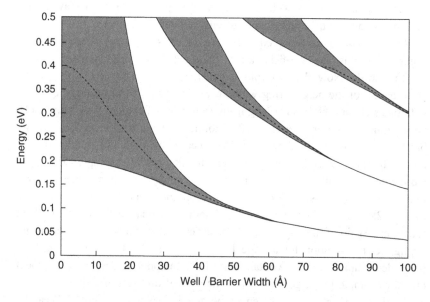

with *minigaps* separating them. Such minibands exist both above and below the energy barrier V_0. As the barrier width increases, the minibands shrink in width back to the discrete states of the individual wells, becoming a system of multiple quantum wells rather than a superlattice.

2.2.3 Doped heterojunction systems and self-consistent solutions

In the preceding discussion, we have ignored the effects of ionized donors and acceptors, as well as the Coulomb interaction of free electrons in the solution of the bound states in heterojunction systems. Undoped quantum well systems in which such effects are negligible are often used in optical studies and in optoelectronic devices. However, in electronic devices, free carriers are necessary for charge transport to occur. In particular, the modulation-doped structures discussed in Section 2.1 find widespread use in semiconductor mesoscopic devices due to the high mobilities, and hence long elastic mean free paths, one of the critical length scales discussed in the introduction. Such high-electron-mobility heterojunction material very often forms the basis for fabricating the quantum waveguide and quantum dot structures discussed in the subsequent chapters, in which a host of rich physical phenomena are observed.

In order to calculate the electronic states of doped heterolayer systems, one must include the additional terms in Eq. (2.2) neglected in the previous two sections, namely $V_D(z)$ and $V_{ee}(z)$. The many-body contribution to the one-electron potential energy, $V_{ee}(z)$, may be approximated as

$$V_{ee}(z) = V_h(z) + V_{xc}(z), \qquad (2.18)$$

where $V_{xc}(z)$ is the exchange and correlation contributions to the one-electron potential energy, while $V_h(z)$ represents the *Hartree* contribution, the electrostatic potential energy arising from the average charge density of all the other electrons in the system. Since $V_h(z)$ is electrostatic in origin, we may combine it with the energy $V_D(z)$ obtained from the solution to Poisson's equation

$$\nabla^2 \varphi_e = \frac{q(\rho_I(z) - qn(z))}{\varepsilon_s}, \quad \varphi_e = V_h + V_D, \qquad (2.19)$$

where $\rho_I(z)$ is the charge density due to ionized donors and acceptors, $\varepsilon_s = \kappa_s \varepsilon_0$ is the permittivity of the semiconductor, q is the magnitude of the electric charge, and $n(z)$ is the 3D particle density of the 2DEG. This density is given by:

$$n(z) = \sum_i N_i |\varphi_i(z)|^2, \qquad (2.20)$$

where N_i is the 2D density of carriers in each subband given by Eq. (2.11). In writing Eq. (2.19), we have neglected the variation of the dielectric constant across the interface of a given heterojunction. For an abrupt interface between two dielectric materials, an additional image potential should be added to Eq. (2.18) to satisfy the boundary conditions of the normal electric field at a dielectric discontinuity. However, this leads to a nonintegrable singularity in the energy when the envelope function penetration into the barrier cannot be neglected. This singularity can be avoided by including a finite grading at the interface, which must exist at the atomic scale, as discussed by Stern [58,59]. Fortunately, in the case of the AlGaAs/GaAs system, the difference in dielectric constants is relatively small, and thus the image potential is usually ignored.

The exchange-correlation contribution to the one-electron energy is usually included through the density-functional method using the Kohn–Sham equation with the local density-functional approximation [60,61]. Various parameterizations for V_{xc} within the local density-functional approximation give similar results. Das Sarma and Vinter [61] have used [62]:

$$V_{xc}(z) = -(1 + 0.7734x \ln(1 + x^{-1}))(2/\pi\alpha r_s)Ry^*$$

$$\alpha = (4/9\pi)^{1/3}, \ x \equiv x(z) = r_s/21, \ r_s \equiv r_s(z) = \frac{1}{a^*}\left[\frac{4}{3}\pi n(z)\right]^{-1/3}, \quad (2.21)$$

where $a^* = \kappa_s \hbar^2/m^* q^2$ is the effective Bohr radius, $n(z)$ is given by Eq. (2.20), and the energy is measured in units of effective Rydbergs (approximately 5 meV for GaAs). The above equation assumes a uniform dielectric constant across the interface, which as mentioned earlier is not an unreasonable approximation for the GaAs/AlGaAs system.

The inclusion of many-body effects due to V_h and V_{xc} couples the nonlinear envelope function equation (2.4) to Poisson's equation through $n(z)$. The simultaneous solutions of these two equations are referred to as *self-consistent* solutions, solved numerically in an iterative fashion [63]. The usual procedure is to begin the calculation with a trial solution for the 1D Schrödinger equation, $\varphi_n^0(z)$, such as the solution in the absence of the many-body potential or a solution derived from some model potential. The initial particle density $n^0(z)$ is calculated from Eq. (2.20), and Poisson's equation is solved numerically using for example finite differences. Once the electrostatic potential is calculated together with the exchange-correlation potential (2.21) using $n^0(z)$, the Schrödinger equation is solved numerically using, for example, the Numerov method [64] or finite elements techniques [65] to obtain the new solutions, $\varphi_n^1(z)$ and E_n^1. These solutions are used as an approximation for the second iteration, and the procedure continues until the nth iteration converges to within some acceptable numerical error.

We return now to the problem of the modulation-doped single-heterojunction system. Figure 2.13(a) shows the energy band diagram of a modulation-doped heterojunction between two semi-infinite regions of GaAs and AlGaAs. A thin spacer layer of thickness d_s separates a region of uniform n-type doping of concentration N_d (Si for example) in the AlGaAs from the 2DEG at the interface. The term ΔE_c characterizes the effective potential due to the band offset of the heterojunction, which as discussed earlier is a function of the mole fraction of Al in the AlGaAs. The GaAs is assumed to be lightly p-doped with an acceptor concentration of N_a. The charge density of either side of the junction decays away into the bulk regions where charge neutrality exists. Electrons from ionized donors in the AlGaAs transfer to the lower-energy GaAs layer. Due to the confinement associated with the band-bending in the GaAs and the band offset, ΔE_c, the electrons are localized at the interface and form a quantum confined system.

2.2 Quantization in heterojunction systems

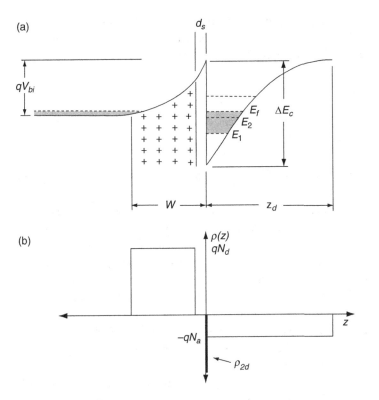

Fig. 2.13 (a) Conduction band profile through a modulation-doped heterojunction system. (b) Charge density versus distance due to ionized donors and acceptors.

To calculate the equilibrium charge density in the 2DEG, we can make the approximation that the free carrier concentration is zero inside of the depletion region of width W on the left, and of width z_d on the right (depletion approximation), except in the region close to the interface. Then the charge density due to ionized donors in the AlGaAs, and to ionized acceptors in the GaAs, is as shown schematically in Fig. 2.13(b). The depletion width in the GaAs may be given approximately by the inversion condition for the formation of the 2DEG at the interface:

$$z_d = \left[\frac{2\varepsilon_s \psi_{s,inv}}{qN_a}\right]^{1/2}, \quad \psi_{s,inv} \approx 2(E_i - E_{F_p})/q, \tag{2.22}$$

where $\psi_{s,inv}$ is the total electrostatic band-bending when electrons start to populate the inversion layer 2DEG, and $(E_i - E_{F_p})$ is the difference between the intrinsic Fermi energy and the Fermi level in the GaAs bulk, calculated as discussed in the usual treatises on semiconductor physics. The depletion width W on the n-side and the sheet density of carriers in the 2DEG, n_{2D}, are coupled together through Gauss' law:

$$(W - d_s)N_d = n_{2D}(E_F) + n_{depl}, \tag{2.23}$$

where $n_{depl} = N_d z_d$ assuming overall charge neutrality. The total potential variation in the AlGaAs, V_{bi}, is also related to the depletion width W and is found by integrating the charge distribution of Fig. 2.13(b) from $-\infty$ to 0 twice to obtain:

$$V_{bi} = \frac{qN_d(W^2 - d_s^2)}{2\varepsilon_s}. \qquad (2.24)$$

From the band diagram, Fig. 2.13, we can deduce the following relation between V_{bi} and the Fermi energy:

$$q\psi_n + \frac{q^2 N_d(W^2 - d_s^2)}{2\varepsilon_s} = \Delta E_c - E_F, \qquad (2.25)$$

where $q\psi_n = E_c - E_{Fn}$ is the difference between the conduction band edge and the Fermi energy in the bulk AlGaAs far from the interface. Further, n_{2D} is an explicit function of the Fermi energy through (2.11) once the subband energies E_n are given. Thus Eqs. (2.23) and (2.25) may be solved simultaneously to obtain W and E_F.

In order to obtain the correct subband energies and charge transfer in the well, the iterative self-consistent procedure described earlier is followed. However, now the total charge in the 2DEG, n_{2D}, must be updated according to the solution for E_F described above after every iteration step in which E_F is recalculated. Such an analysis can be further extended to the case in which the AlGaAs layer is finite in extent, terminated with a metal gate as in an actual HEMT. (For more details of charge control in HEMTs, the reader is referred to [66], Chapter 13.) Figure 2.14 shows the result of such a calculation for the probability density of the first three confined states superimposed on the conduction band profile. Here only the Hartree contribution has been included, and V_{xc} has been neglected. The parameters used in the calculation are shown in the figure key. There are several interesting features to notice about the solutions. The first three solutions are localized primarily in the GaAs with an exponential tail in the AlGaAs due to the finite barrier. However, if a lower conduction band offset had been chosen (for example $\Delta E_c = 0.125$ eV), the $n = 3$ state would become a bound state located in the AlGaAs barrier rather than the GaAs. Such solutions localized in the barrier are associated with an important effect in the transport properties of actual quantum confined systems, that of real space transfer [67]. If the carriers in the 2DEG of the channel are given sufficient energy, for example, via a d.c. electric field applied in the plane of the heterojunction, electrons may populate higher levels and effectively transfer into the AlGaAs layer. Charge transfer into the AlGaAs also gives rise to a problematic effect in HEMTs sometimes referred to as the parasitic MESFET effect [66]. As the gate voltage is made more positive, the Fermi energy in the 2DEG moves up in energy. Eventually, with sufficient forward bias, strong charge accumulation occurs in the AlGaAs. Since the transport properties in the heavily doped AlGaAs layer are far inferior to that of the 2DEG in the undoped GaAs, a noticeable saturation of channel current with gate bias occurs. A further point to note in the

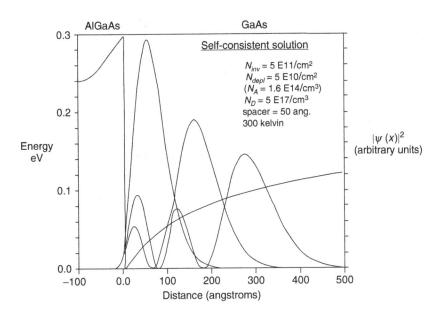

Fig. 2.14 Probability density versus distance for the first three levels using self-consistent solutions.

solutions shown in Figure 2.14 is that the number of lobes in the probability density corresponds to the subband index n, as was the case of the square quantum well. However, in the present case, the effective width of the nth level:

$$\langle z_n \rangle = \int_{-\infty}^{\infty} z \varphi_n^2(z) dz \qquad (2.26)$$

increases with increasing subband index.

Before leaving the topic of the modulation-doped, single-heterojunction system, we should mention one model solution that is often employed, the so-called triangular-well approximation. One can observe in Figures 2.13 and 2.14 that the potential variation on the GaAs side in the vicinity of the interface is roughly linear. Thus we can assume that the potential may be written:

$$V(z > 0) = qF_{eff}z, \quad F_{eff} = \frac{q}{\varepsilon_s}\left[n_{depl} + \frac{1}{2}n_{2D}\right], \qquad (2.27)$$

where F_{eff} is an effective surface field and the factor of $1/2$ accounts for the average field in the channel due to the presence of the free carrier charge, n_{2D}. If the barrier height ΔE_c is sufficiently large, we can neglect the penetration of the envelope function in that region and assume that $\varphi_n(z) = 0$ for $z < 0$. Exact solutions exist for the linear or triangular potential in the form of Airy functions:

$$\varphi_n(z) = C_n Ai\left[\left(\frac{2m_z}{\hbar^2 q^2 F_{eff}^2}\right)^{1/3}(qF_{eff}z - E_n)\right], \qquad (2.28)$$

where C_n is the normalization constant determined from the recursion relations for the Airy function, $Ai(z)$. The energy eigenvalues E_n are determined from the

Table 2.1 *Comparison of the self-consistent (SC) and the triangular potential (TP) solutions for an AlGaAs/GaAs single heterojunction. Shown are the calculated energy eigenvalues and percent occupancy of each level. The calculation is at 300 K for a 2D density of 5×10^{11} cm^{-2} and a depletion density of $n_{depl} = 5 \times 10^{10}$ cm^{-2}. The donor density in the AlGaAs is 5×10^{17} cm^{-3}, the spacer layer 5 nm, and the conduction band offset $\Delta E_c = 0.3$ eV.*

Level	Energy (SC) (meV)	Occupancy (SC) (%)	Energy (TP) (meV)	Occupancy (TP) (%)
1	55.3	60.2	50.0	70.4
2	87.9	19.8	87.3	20.0
3	107.0	9.9	118.0	6.4
4	120.0	6.0	145.0	2.3
5	130.0	4.1	170.0	0.9
Fermi	38.1		37.9	

Dirichlet condition at $z = 0$, which to lowest order may be expressed analytically from the asymptotic expansion of the zeros of the Airy function (see [68]):

$$E_n = \left[\frac{\hbar^2}{2m_z}\right]^{1/3} \left[\frac{3\pi q F_{\textit{eff}}}{2}\left(n - \frac{1}{4}\right)\right]^{2/3}, \quad n = 1, 2, 3, \ldots \quad (2.29)$$

Table 2.1 shows a comparison of the first five energy eigenvalues, subband occupancy, and Fermi energy for the triangular-well approximation and the self-consistent calculation in the Hartree approximation for the parameters indicated. As expected, as the energy increases, the discrepancy between the two solutions increases due to the softening of the actual self-consistent potential with increasing energy.

2.2.4 Other heterojunction systems

In our discussion of heterostructure systems thus far we have focused exclusively on the GaAs/AlGaAs system, which represents the most widely studied, and well understood, of heterostructures. Heterostructures formed from different combinations of other semiconductors are also of increasing technological interest, however, due to the unique applications that their specific combinations can enable. In this section, we very briefly review some of these heterostructure systems and their applications.

There are numerous heterostructures based on combinations of the groups III and V semiconductors, just one example of which is provided by the well-known GaAs/AlGaAs system. Several factors determine the choice of a particular pair of materials for use in a heterostructure, namely: the extent to which the materials are lattice matched, which determines the strain and disorder that will result when the materials are grown on top of each other; and the size of the conduction- and valence-band discontinuities between the semiconductors, which will determine the strength of the resulting electron and hole confinement in the quantum well. Since semiconductor bandgaps may be varied by alloying, as discussed already for the $Al_xGa_{1-x}As$ system, there can be enormous flexibility in the choice of material systems to realize heterostructures with specific applications.

Heterostructures based on the narrow-bandgap semiconductor InAs use wider-gap materials, such as AlSb and GaSb, and are of interest [69,70,71, 72,73] due to the high mobility that arises, predominantly due to the small effective mass of InAs. (Due to non-parabolicity of the InAs conduction band, the InAs effective mass varies with the electron density, but is typically always smaller than that of GaAs.) It is well established that, in nominally undoped InAs/AlSb structures, a 2DEG is formed in the InAs well with a sheet density on the order of 10^{12} cm^{-2}. The proposed sources of these carriers include donor-like defects in the AlSb barriers, donors at the InAs/AlSb interfaces, and surface donors in the GaSb cap layer. Doping of these materials is also possible, however, and can increase the sheet density by a further order of magnitude.

InAs quantum wells have emerged as an important potential system for the development of *spintronic* [74] devices, which seek to utilize the electron spin as the basis of their operation. InAs is of interest here since the high electron densities, and associated Fermi velocities in excess of 10^6 m/s, allow for spin-dependent transport via the *Rashba effect* [75]. This effect arises when electrons move in an asymmetric potential generated near the interface of a heterojunction. The net electric field that exists perpendicular to the interface corresponds to an effective magnetic field in the rest frame of the moving electron. This magnetic field lies in the plane of the two-dimensional electron gas, and lifts the spin degeneracy of the carriers. It also gives rise to *precession* of the spin vector, the frequency of which may be modulated by using a gate voltage to tune the confinement potential at the heterointerface.

In addition to narrow-gap semiconductors, there has also recently been much interest in the properties of nitride-based heterostructures. These utilize the wide bandgap of materials such as GaN, AlN, and InN, whose bandgaps range from 2–6 eV, spanning the visible and ultraviolet ranges. As direct bandgap materials, these systems therefore allow the realization of LEDs and laser diodes capable of providing full-color output [76].

While silicon is the dominant material in modern microelectronics, it suffers from low mobility compared to III–V semiconductors that hinders its application to high-frequency electronics. SiGe alloys offer a means to circumvent this problem, and to realize high-speed (>100 GHz) electronics compatible with standard silicon processes. Si/SiGe heterostructures can also be implemented and in these systems the 4% lattice mismatch of Si and Ge gives rise to significant strain effects that may be utilized to improve device performance [77].

2.3 Lateral confinement: quantum wires and quantum dots

2.3.1 Nanolithography

In Section 2.2 we were concerned with both the quantization of energy states and the corresponding reduced dimensionality of the free carrier system due to quantum confinement in layered heterojunction systems. In such layered systems the motion of carriers in the plane perpendicular to the material growth direction is unconstrained. The next advance historically began with early attempts to confine the lateral motion in the plane of the 2DEG using advanced lithographic techniques beginning in the late 1970s and early 1980s. *Lithography* refers to the transference of a desired pattern onto a substrate. The primary technique to accomplish this in modern IC technology involves a resist, which when selectively exposed to a particular radiation source, changes its chemical bonding in such a way that the irradiated portion dissolves when immersed in a developer, while the unexposed resist remains (positive resist), or vice versa (negative resist). To realize quantum confinement in the lateral directions, the dimensions must be such that the quantized energies are resolvable from thermal broadening and broadening due to the existing disorder in the system. As a rule of thumb, we want lithography tools capable of transferring patterns with critical dimensions less than 0.1 μm, although this is not absolutely necessary (discussed below). As one might expect, diffraction effects impose limitations on the wavelength of the radiation source employed to expose the resist, although clever schemes such as the use of phase-shift masks and immersion lithography have extended conventional optical lithography down to dimensions less than 100 nm line width. Beyond this, however, extensive research and development is directed towards shorter-wavelength electromagnetic sources such as deep-ultraviolet and X-rays. The current challenges are not related to the sources themselves, but rather to the associated optics and mask materials that allow delineation of small features.

The most popular technique to date for realizing nanometer structures in the research laboratory setting has been the use of electron-beam (e-beam) lithography. The characteristic wavelength for electrons at the energies used (typically 30–50 keV) is on the order of interatomic distances. The actual limitation on the resolution is related to scattering due to the inelastic loss processes as the high-energy electrons lose energy to or are reflected by the substrate. Such factors limit the ultimate resolution to around 10 nm using current technology. Much of the current popularity of e-beam lithography can be ascribed to the fact that it can be performed with relatively simple modification to a conventional scanning electron microscope (SEM) through computer

2.3 Lateral confinement: quantum wires and quantum dots

control of the position of the beam. Thus the fabrication of individual device structures for transport experiments may be accomplished with relatively little capital investment. E-beam lithography is typically performed in a direct write fashion in which the electron beam is rastered over a resist material such as PMMA (poly-methyl-methacrylate), which is spin-cast as a thin coating over the substrate. The developed PMMA can be used as an etch mask for wet chemical or dry etching, or metal can be evaporated and a lift-off process employed to transfer metal lines to the semiconductor substrate. Interestingly enough, the idea of direct write e-beam lithography predated the use of optical lithography [78] at the time in which SEM technology was evolving. However, the limitations of the vacuum technology and material quality precluded effective use of this technology until the 1980s. A related technique is focused ion-beam (FIB) technology, in which charged ions replace electrons as the high-energy incident particles. FIB technology has the additional advantage of allowing direct in-situ etching with 10^{-2} μm resolution, as well as the possibility of implanting donor and acceptor species in the semiconductor, allowing for the definition of nanometer-scale doped regions.

The ultimate limit of nanolithography is the control of the placement of individual atoms on the surface of a material. Many groups have been researching the techniques of scanning tunneling lithography and atomic force lithography, both of which in fact do allow such atomic control. Both techniques have evolved from the remarkable development in the 1980s of the scanning tunneling microscope (STM) for which researchers at IBM Zurich were awarded the Nobel Prize in physics [79]. The basic principle of the STM is that of the exponential relation between tunneling current and the thickness of the barrier separating the electrodes, as discussed earlier in connection with the double-well problem, and as will be discussed in more detail in Chapter 3. In the case of the STM, one electrode is the semiconductor substrate while the other is a fine metallic tip brought in close proximity to the semiconductor surface. As the tip is moved across the surface, the tunnel current provides a sensitive measure of the proximity of the tip to the charge density on the substrate. This technique has allowed atomic level mapping of the atomic structure on material surfaces, most notably in the beautiful work by the Ruger group at IBM on quantum corrals [80].

Recently, there have been several studies that have succeeded in achieving a unique marriage of conventional semiconductor processing, with scanning-probe-based nanofabrication, to realize highly flexible nanostructures. The first of these techniques is remarkably simple and uses the process of *oxidation lithography* [81,82,83,84,85,86,87]. In this approach, the nanofabrication steps are actually performed under ambient conditions, rather than in the ultra-high-vacuum environment typical of electron-beam lithography, and one seeks to locally oxidize the surface of some substrate of interest. This is achieved by applying a sufficient voltage between the tip of an atomic force microscope

(AFM) and the substrate, to break down surface-adsorbed water and generate free radicals that locally oxidize the surface (the substrate forms the anode in this process, which is consequently often referred to as *anodic* oxidation lithography). Scanning the AFM tip across the surface in a predetermined manner therefore allows an oxidized line with a specific form to be transferred to the surface. When the oxidized line is formed on the GaAs cap layer of a GaAs/AlGaAs heterojunction, and the 2DEG is sufficiently close to (less than 40 nm from) the surface, the oxidation can cause a complete depletion of the 2DEG immediately underneath it. Simply by drawing the AFM tip across the top surface, it is therefore possible to electrically isolate regions of 2DEG from each other. This approach provides for huge flexibility, since nanostructures of arbitrary design may be realized simply by varying the scanning path of the AFM tip. The Ensslin group, in particular, has had huge success with this method, utilizing it to realize quantum point contacts [88], single [89] and coupled [90] quantum dots, Aharonov–Bohm rings [91], and other more complicated structures [92]. Rokhinson has also successfully applied the technique to realized quantum point contacts for studies of the so-called 0.7 feature (see Chapter 5) [93,94]. In Fig. 2.15, we show an example where AFM anodic oxidation has been used to realize a quantum point contact, a short bottleneck for the flow of current that couples two regions of 2DEG. Simply by drawing two oxidized lines at right angles to each other, separated by a gap of order a hundred nanometers in length, the authors have been able to realize source and drain connections to the point contact. The horizontal line in the figure completely isolates the regions of 2DEG on either side of it, so that the lower area can be used as an in-plane gate. By biasing this gate at a negative potential with respect to the other areas of 2DEG, the effective conducting width of the point contact channel can be decreased, causing the distinctive step-like variation of the conductance shown in the figure (the origins of which will be discussed in detail in Chapter 5).

While the surface oxidation generated during anodic lithography corresponds to a *permanent* structural change, in the process of *erasable electrostatic lithography* (EEL) nanostructures are realized through scanning-probe surface modification, but these changes are reversible [95]. The principle of this approach involves using a scanning probe to induce local charging of an insulating substrate under low-temperature/ultra-high vacuum conditions. By charging the surface as the scanning probe is rastered over it, charged lines with specific geometries can again be achieved. Under conditions where a 2DEG is sufficiently close to the charged surface, its carriers can be depleted, allowing for the isolation of specific regions as in the case of AFM oxidation lithography. The great advantage of EEL, however, is that, as its name suggests, the surface charging induced by the scanning probe can be eliminated, either by reversing the polarity of the scanning probe or by illuminating the 2DEG with uniform light from an LED. Crook and his colleagues have used this approach successfully,

2.3 Lateral confinement: quantum wires and quantum dots

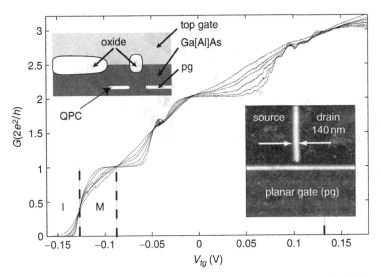

Fig. 2.15 The lower inset shows a quantum point contact formed by AFM oxidation. The point contact is formed by the two oxide lines (which appear bright). The upper inset shows a schematic cross section of the device (which also features a deposited top gate that allows for further control of the conductance). The main figure shows the conductance as a function of the top gate voltage at temperatures between 1.7 K and 20 K ([88], with permission).

to study a variety of quantum-transport effects in quantum point contacts [95,96] and quantum dots [97]. In Fig. 2.16, we illustrate the use of EEL to realize a quantum dot. The device is formed by using EEL to form the two pairs of lines that connect to preformed metallic electrodes [97]. The lower part of the figure shows conductance fluctuations in a magnetic field that are well known to be a universal characteristic of quantum dots, although these are normally realized by different techniques (as we shall discuss in Chapter 6).

2.3.2 Quantum wire and quantum dot structures

With the use of nanoscale lithography techniques, an infinite variety of structures can be envisioned. Assuming we start with a 2DEG heterostructure, confinement of the 2DEG in another direction, say the y-direction, forms what is called a *quantum wire*, which corresponds to a quasi-one-dimensional system. Alternatively, as we have seen already earlier in this chapter, such quantum wires may be realized as self-assembled structures from carbon nanotubes or epitaxially formed nanowires. Here translational invariance exists along one axis, while the electronic states are well confined in the other two. Free electron motion is still possible along the axis of the wire, and a continuum of one-dimensional states exists. Additional confinement in the remaining direction completely confines the motion of charge carriers, giving rise to a discrete

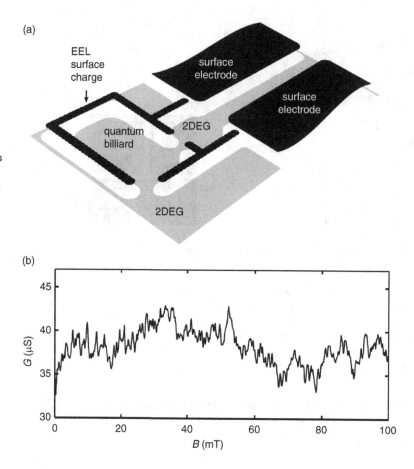

Fig. 2.16 (a) The schematic shows how a quantum dot can be formed by EEL by using erasable electrostatic lithography to charge spots on the device surface, thereby forming a micron-scale quantum dot. (b) The figure shows magneto-conductance fluctuations measured in an EEL-defined quantum dot at 20 mK ([97], with permission).

spectrum of bound states much like an isolated atom or molecule. Such systems are called *quantum dots*, or sometimes *artificial atoms*, and will be discussed in detail in Chapter 6.

In traditional semiconductor processing, quantum wires are typically realized by a combination of high-resolution lithography and physical etching, as shown in Fig. 2.17. In Fig. 2.17(a), the layers are etched below the 2DEG as shown. The dominant problem with this method (as well as all lithographic processes) for forming a quantum wire is that the quality of the sidewall/air interface is vastly inferior to that existing between the AlGaAs and GaAs layers. At best, roughness fluctuations on the order of 2–5 nm exist using present-day wet or dry etching techniques (compared with the 0.2–0.5 nm fluctuations of the heterojunction interface). In addition, the free surface of GaAs is invariably pinned upon exposure to ambient conditions due to the large density of surface states that form through defect formation [98]. Thus, instead of sharp confining potentials, the actual potential is that of lateral depletion layers, which looks more like a harmonic rather than a "hard wall" potential. This lateral depletion,

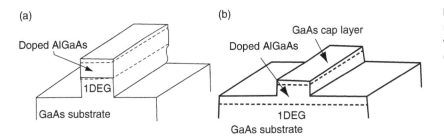

Fig. 2.17 Two different realizations of quantum wire structures by wet or dry etching.

and the "soft" potential associated with it, limits the energy splitting between levels in the transverse direction. If the wire is made too thin, the space charge regions from either side overlap and deplete the free carriers from the wire, making them useless for transport measurements. Typically the lateral dimensions of such wires must be greater than 0.1–0.2 μm. Figure 2.17(b) shows a modification of the etched structure in which the AlGaAs layer, or even just the heavily doped GaAs cap layer, is partially etched. If a metal electrode is deposited covering the entire structure, the partially etched region is depleted first as the electrode bias is made increasingly negative, leaving behind a 1DEG in the wire region. Since the etching does not extend all the way to the active region of the wire, the detrimental effects associated with the poor sidewalls are less severe.

A popular technique in transport studies for realizing 1D channels is the so-called *split-gate technique,* shown in Fig. 2.18 [99]. Here metal Schottky gate electrodes are deposited on the surface of a 2DEG heterostructure between Ohmic source and drain contacts, as normally required in fabricating HEMTs. However, the gate is split, leaving a narrow channel between the two electrodes. In contrast to etched wire structures, the metal electrodes (usually defined with e-beam lithography) are negatively biased to turn off or deplete the 2DEG under the gates, leaving a narrow region of undepleted channel between the two electrodes. The 2DEG is also depleted laterally from the gate electrodes as the gate bias is made more negative, in effect squeezing the 1D channel until eventually all the carriers are emptied from the region. In the intermediate regime, carriers in the channel feel a confining potential due to electrostatic potential of the reverse-biased Schottky contacts. A similar concept was in fact utilized some time earlier in studies of quantized electrons in Si/SiO$_2$ inversion layers using narrow channel MOSFETs [100]. *pn* junctions were implanted adjacent to the channel of the MOSFET and reverse biased to laterally confine electrons in the channel in the direction perpendicular to current flow between the usual source and drain contacts of the device. The limitation of the split-gate structure is the soft potential due to the lateral depletion, which limits the energy separation due to lateral confinement and limits the distance in which the electrodes may be spaced.

Fig. 2.18 Sketch of a split-gate Schottky contacts above a 2DEG structure. Negative potential applied to the gates depletes the 2DEG below the gates and laterally into the ungated regions.

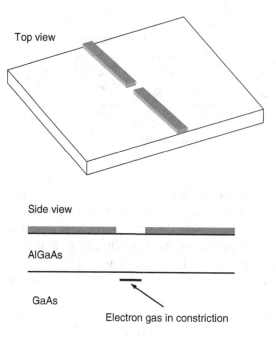

The fabrication techniques discussed above are similarly used to define more general geometries as will be discussed in later chapters. Quantum dots are realized by extending the lateral confinement of the wire structures to the remaining degree of freedom of the carriers. Figure 2.19 shows a micrograph of quantum dot structures realized by etching narrow trenches into heterojunction material. Figure 2.20 shows an array of coupled quantum dots defined using electrostatic confinement as discussed in the split-gate quantum wires.

2.4 Electronic states in quantum wires and dots

2.4.1 Semiconductor nanowires and quantum dots

In an ideal quantum wire, the system is assumed sufficiently long along the axis of the wire that translational invariance holds. Within the EMA for a single band system, we can again write the envelope function as the product of one-dimensional free carrier solutions and confined solutions as

$$\psi(\mathbf{r},z) = \frac{1}{L}\varphi_{n,m}(\mathbf{r})e^{ik_x x}, \qquad (2.30)$$

where $\mathbf{r}$ is the position vector in the y–z plane parallel to the wire cross section, and L is the normalization length of the wire. The function $\varphi_{n,m}(\mathbf{r})$ now satisfies the two-dimensional Schrödinger equation:

$$\left[\frac{\hbar^2}{2m^*}\nabla_\mathbf{r}^2 + V_{\mathit{eff}}(\mathbf{r})\right]\varphi_{n,m}(\mathbf{r}) = E_{n,m}\varphi_{n,m}(\mathbf{r}), \qquad (2.31)$$

2.4 Electronic states in quantum wires and dots

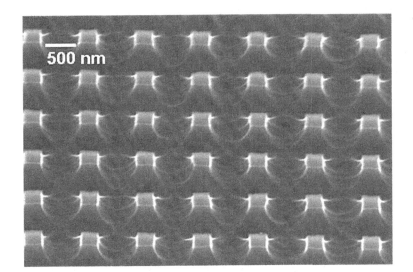

Fig. 2.19 Scanning-electron-microscope image showing a large-area array of quantum dots fabricated by etching trenches into a GaAs/AlGaAs substrate with a high mobility 2DEG. *Image courtesy of Dr. N. Kabir.*

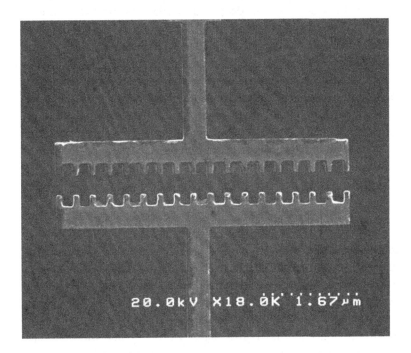

Fig. 2.20 A scanning-electron-microscope image showing the use of the split-gate method to realize a linear chain of 17 coupled quantum dots. *Image courtesy of Dr. K. Ishibashi, RIKEN.*

where again $V_{eff}(\mathbf{r})$ contains the potential of the band discontinuities, the electrostatic potential of the ionized donors and acceptors, the many-body contribution due to free carriers, and the image potential (which is particularly important when considering freestanding wires, or wires with etched sidewalls). An isotropic effective mass, m^*, has also been assumed for simplicity, corresponding to a spherically symmetric, parabolic energy dispersion relation for the band minima.

For simplicity, we can consider special cases as before. First consider a wire with a rectangular cross section and with infinite potential outside of the region defining the wire (hard walls). For the rectangular symmetry of this case, the solution of Eq. (2.31) is separable into y- and z-dependent solutions as

$$\varphi_{n,m}(y,z) = \left[\frac{4}{L_y L_z}\right]^{1/2} \sin\frac{n\pi y}{L_y} \sin\frac{m\pi z}{L_z}, \quad n, m = 1, 2, 3, \ldots, \quad (2.32)$$

where L_y and L_z are the dimensions in the y- and z-directions, and the origin of the coordinate system has been placed at one corner of the wire. The corresponding eigenenergies are given by

$$E_{k_x,n,m} = E_{k_x} + E_{n,m} = \frac{\hbar^2 k_x^2}{2m^*} + \frac{n^2 \pi^2 \hbar^2}{2m^* L_y^2} + \frac{m^2 \pi^2 \hbar^2}{2m^* L_z^2}. \quad (2.33)$$

In analogy to the 2DEG discussion earlier, we now have a continuum of one-dimensional states associated with each pair of integers n and m, called subbands, modes, or channels, depending on the context. In particular, the solutions of Eq. (2.33) are identical to the allowed solutions of electromagnetic waves in a metallic rectangular waveguide for the transverse magnetic and electric solutions. In Chapter 3, we will connect different regions of such 1D waveguides together and look at the general propagation characteristics through the composite structure. Such systems are aptly named *quantum waveguides*.

The density of states for the quasi-1D system described above are again derivable using Eq. (2.8). Assuming the case of bands above, the density of states per unit length of the wire system is given by

$$D(E) = n_v \frac{2m^*}{\pi \hbar^2} \left[\frac{\hbar^2}{2m^*}\right]^{1/2} \sum_{n,m} (E - E_{n,m})^{-1/2} \vartheta(E - E_{n,m}), \quad (2.34)$$

where the two-fold degeneracy for spin has been included. Figure 2.21 shows the calculated density of states for a GaAs/AlGaAs rectangular quantum wire, illustrating the diverging density of states as the energy approaches that of one of the subband minima. Such divergences may lead to interesting effects in the transport properties and collective excitations of such systems.

Numerical self-consistent solutions have been performed for electrons in Si quantum wires by Laux and Stern [101] and in GaAs quantum wires by Laux *et al.* [102] and Kojima *et al.* [103]. Such calculations typically use a similar iterative procedure to treat the many-body contributions within various approximation schemes. The calculated potential for a GaAs/AlGaAs wire formed in a split-gate structure is shown in Fig. 2.22. It is interesting to note that due to the depletion region confinement, the potential appears more parabolic than hard walled, particularly for increasing negative gate bias. Typical subband spacings are on the order of 5 to 10 meV.

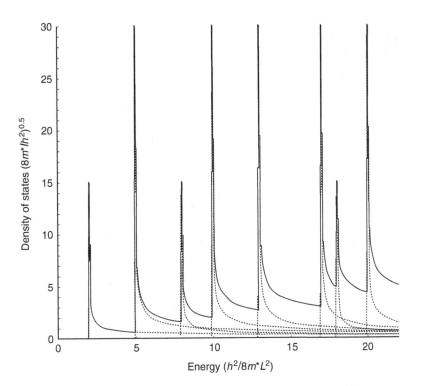

Fig. 2.21 Calculated density of states per unit length of a 15 nm-by-15 nm GaAs quantum wire. The solid curve is the total density of states while the broken lines show the contributions from the various subbands. The discontinuities in the density of states have been artificially truncated to allow the reader to identify those subbands with single and double degeneracy.

A crude description of the states of a quantum dot can of course be obtained by extending the hard wall confinement in all three directions, such that the energy is given by

$$E_{l,n,m} = \frac{l^2\pi^2\hbar^2}{2m^*L_x^2} + \frac{n^2\pi^2\hbar^2}{2m^*L_y^2} + \frac{m^2\pi^2\hbar^2}{2m^*L_z^2}, \quad (2.35)$$

where l, n, and m are positive, non-zero integers. The spectrum of energies and thus the density of states is completely discrete, with degeneracies due to spin and multiple valleys. In Section 6.6 we consider in more detail the states in a parabolic confining potential. However, as will be discussed in much more detail in Chapter 6, the states in real quantum dot systems are complicated by the granular nature of electronic charge. The change of one electron in a quantum dot system has a large influence on the electrostatic potential and many-body ground state of the system. Such effects are of great interest in themselves in the so-called *Coulomb blockade* effect, which is the basis of a number of interesting experiments and proposals for ultra-small electronic devices, and thus we will defer full discussion of such systems until later.

2.4.2 Electronic states in graphene and carbon nanotubes

The electronic states in carbon nanotubes (CNTs) cannot be simply described within the effective mass approximation as described above. Rather, the electronic

Fig. 2.22 Numerical self-consistent potential for a split-gate GaAs/AlGaAs structure (After Laux et al., *Surf. Sci.* **196**, 101 (1988), with permission).

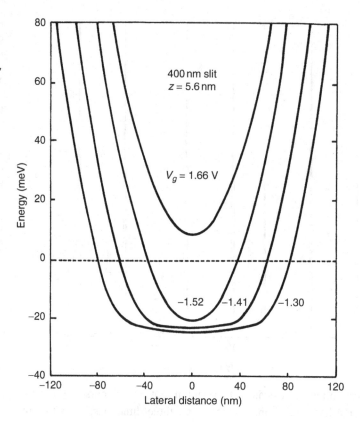

states of CNTs are most simply thought of in terms of the states associated with graphene (the extreme single monolayer limit of graphite), when rolled in a tube, and periodic boundary conditions imposed [18,104]. Hence, in order to discuss the electronic states of CNTs, we first start with a brief discussion of the electronic structure of graphene, using a simple tight binding model.

As mentioned earlier, at the time of the writing of this book, graphene in and of itself has become a system of great scientific interest because of its interesting transport properties associated with its massless, Dirac-like particle behavior at the Fermi energy associated with the linear dispersion there. Single monolayer graphene sheets have been prepared by exfoliation of graphite [42] and vacuum graphitiziation of SiC [46]. Magneto-transport studies on back gated structures show a number of interesting properties, such as a cyclotron mass for massless carriers that varies in a relativistic way as $E = m_c c_{eff}^2$ (where c_{eff} is the effective speed of light, approximately 10^6 m/s), and that the carrier density may be electrostatically contolled, and changed from n-type to p-type as the gate voltage is swept from positive to negative [44]. The ability to tune the carrier type from n-type to p-type, has led to fabrication of planar bipolar junction devices [105,106]. Further, graphene may be patterned at the nanoscale to realize nano-ribbons [46], which show quantum confinement with very long mean free paths.

2.4 Electronic states in quantum wires and dots

There is considerable interest presently for fabrication of a variety of electronic devices formed through gated control of the carrier density, the unusual transport properties, and control of the electronic states through lateral patterning.

In terms of the bonding model of C, the underlying structure of graphene arises from the hybridization of the s and p_x, p_y atomic states into sp^2 molecular orbitals in the plane of the graphene forming directed σ bonds between nearest neighbors, with three nearest neighbors in the plane, resulting in the hexagonal lattice structure illustrated in Fig. 2.23(a). The remaining p_z orbitals are relatively weakly bound, contributing one electron per atom, and their overlap to form π-bonds within the graphene lattice is primarily responsible for the higher lying conduction and valence bands associated with transport and the position of the Fermi energy.

As shown in Fig. 2.23(a), the primitive lattice vectors of this hexagonal lattice are associated with a 2-atom C basis at each point (i.e. two equivalent sublattices), and any equivalent point within the graphene lattice is obtained by translation in terms of integer multiples of these vectors, given by (relative to the x and y axes shown).

$$\mathbf{a}_1 = a_0 \left(\frac{\sqrt{3}}{2}; \frac{1}{2} \right)$$
$$\mathbf{a}_2 = a_0 \left(\frac{\sqrt{3}}{2}; -\frac{1}{2} \right), \tag{2.36}$$

where a_0 is the lattice constant, $a_0 = \sqrt{3} a_{cc} \approx 0.246$ nm, where a_{cc} is the C–C bond length of 0.142 nm. The corresponding reciprocal lattice is also hexagonal, as shown in Fig. 2.23(b), with reciprocal lattice vectors given by the condition $\mathbf{a}_i \cdot \mathbf{b}_j = 2\pi \delta_{ij}$:

$$\mathbf{b}_1 = \frac{2\pi}{a_0} \left(\frac{1}{\sqrt{3}}; 1 \right)$$
$$\mathbf{b}_2 = \frac{2\pi}{a_0} \left(\frac{1}{\sqrt{3}}; -1 \right). \tag{2.37}$$

High symmetry points correspond to the $\mathbf{\Gamma}$, $\mathbf{M}$, and $\mathbf{K}$ points, the latter of which are the corners of the hexagonal first Brillouin zone in k-space, as shown in Fig. 2.23(b). These also correspond to the so-called *Dirac* points, where the conduction and valence bands touch for the infinite 2D lattice.

Within the tight binding model, the electronic states of the σ bonds are tightly bound, and do not interact strongly with the π bonds. Hence, it is sufficient to consider just the p_z orbitals within this model [107]. Since the graphene lattice has two electrons per unit cell, one writes the linear combination of atomic orbitals arising from each sublattice separately, leading to a 2×2 diagonalization for the energy eigenvalue, E (see for example [108])

$$\begin{vmatrix} H_{AA} - E & H_{AB} \\ H_{BA} & H_{BB} - E \end{vmatrix} = 0, \tag{2.38}$$

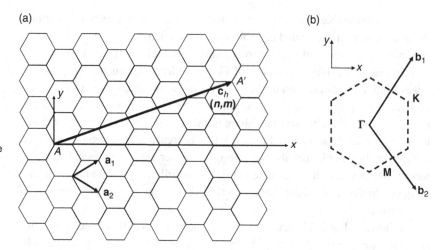

Fig. 2.23 (a) The hexagonal lattice of two-dimensional graphene, the primitive lattice vectors $\mathbf{a}_1$ and $\mathbf{a}_2$, and the chiral vector $\mathbf{c}_h$ characterizing the corresponding carbon nanotube structure. (b) The first Brillouin zone of graphite showing the reciprocal lattice vectors and high symmetry points.

where A and B refer to the different sublattices, and H refers to the matrix elements taken between the respective sublattices. Restricting the interaction to only nearest neighbor atoms, and choosing the reference of energy to be that of the site energy itself, H_{AA} (H_{BB}), the diagonal terms are zero. The corresponding energy has two solutions, the lower corresponding to the filled valence band, and the upper the unfilled conduction band

$$E^{\pm}(\mathbf{k}) = \pm t\sqrt{3 + 2\cos(\mathbf{k} \cdot \mathbf{a}_1) + 2\cos(\mathbf{k} \cdot \mathbf{a}_2) + 2\cos[\mathbf{k} \cdot (\mathbf{a}_2 - \mathbf{a}_1)]}, \quad (2.39)$$

where t is the overlap integral, or hopping matrix element between nearest neighbor p_z orbitals, equal to approximately 3.0 eV. In Fig. 2.23(b), the high symmetry $\mathbf{K}$ points occur at

$$\mathbf{K} = \left(j \pm \frac{1}{3}\right)\mathbf{b}_1 + \left(k \mp \frac{1}{3}\right)\mathbf{b}_2, \quad (2.40)$$

where j and k are integers. Substitution of Eq. (2.40) into Eq. (2.39) results in E going to zero, corresponding to the Dirac point in which the two solutions are the same. Hence, there is no gap, and the Fermi energy lies at the intersection of the upper and lower bands for charge neutrality. Expanding Eq. (2.39) around one of the equivalent Dirac points, the dispersion relation is linear given by (see for example [109])

$$E^{\pm}(\mathbf{k}) \approx \pm \frac{\sqrt{3}}{2} a_0 t |\mathbf{k} - \mathbf{K}|. \quad (2.41)$$

Hence, the constant energy surfaces are conical in shape, and the effective mass is zero due to the linear dispersion. The effective speed of light, c_{eff}, corresponds to the gradient of Eq. (2.41)

$$v = \frac{1}{\hbar}\nabla_{\mathbf{k}}E^{\pm}(\mathbf{k}) \approx \pm\frac{\sqrt{3}}{2}a_0 t = 0.97 \times 10^6 \text{ m/s} \qquad (2.42)$$

similar to that reported experimentally [44].

As mentioned earlier, a carbon nanotube may be considered a rolled graphene sheet, with the chirality defined by the direction in which the sheet is rolled. In Fig. 2.23(a), the vector $\mathbf{c}_h = n\mathbf{a}_1 + m\mathbf{a}_2$, or (n,m), joins two crystallographically equivalent sites, A and A', where n and m are integers. The CNT can then be viewed as joining all equivalent sets of points, A and A', to form the nanotube. The length of $\mathbf{c}_h$ then corresponds to the circumference of the nanotube, so that the diameter of the tube is given as

$$d = |\mathbf{c}_h|/\pi = a_0\sqrt{n^2 + nm + m^2}/\pi. \qquad (2.43)$$

One special case corresponds to the so called *zigzag* nanotube, when $m = 0$, $(n,0)$. Another is the *armchair* CNT corresponding to $n = m$.

The joining of the sheet into a cylinder imposes periodic boundary conditions on the wave functions, which means that the projection of the wavevector along the chiral vector $\mathbf{c}_h$ is quantized according to

$$\mathbf{k} \cdot \mathbf{c}_h = 2\pi q, \qquad (2.44)$$

where q is an integer. Electron motion perpendicular to the chiral vector is unconstrained, leading to a quasi-1D system while the transverse modes are defined by Eq. (2.44).

To see the effect on the energy dispersion relation, we can look at the E–k relation close to one of the Dirac points of the graphene lattice about $\mathbf{K}$ [109]

$$E^{\pm}(\mathbf{k}) \approx \pm\frac{\sqrt{3}}{2}a_0 t|\mathbf{k} - \mathbf{K}| = \pm\frac{\sqrt{3}}{2}a_0 t\sqrt{k_{c,q}'^2 + k_t'^2}, \qquad (2.45)$$

where k_t' is the component of momentum relative to $\mathbf{K}$ transverse to the chiral vector, and $k_{c,q}'$ is the component along $\mathbf{c}_h$ relative to $\mathbf{K}$. Using the definition of the $\mathbf{K}$ points given by Eq. (2.40), one can show that

$$k_{c,q}' = \frac{(\mathbf{k} - \mathbf{K}) \cdot \mathbf{c}_h}{|\mathbf{c}_h|} = \frac{1}{3d}[3q - (n - m)], \qquad (2.46)$$

where d is the diameter of the tube defined in Eq. (2.43). The minimum energy in Eq. (2.45) defines the bandgap for the nanotube. One can see that if $n - m$ is a multiple of 3 (since q is an integer, negative or positive, including zero), that the minimum energy is zero, and the dispersion relation corresponds to a metallic nanotube with a linear 1D dispersion around $\mathbf{K}$ equivalent to Eq. (2.41). This corresponds to the armchair nanotube since $n = m$. The calculated bandstructure using the full π-band model for an armchair nanotube (7,7) is shown in Fig. 2.24(a) [110], where as predicted, there is no gap and the lowest subband is metallic with a linear dispersion. For a zigzag lattice for which n is not a multiple of 3, the

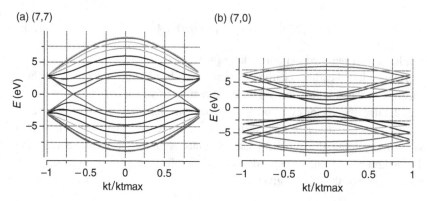

Fig. 2.24 Calculated subband structure for the valence and conduction bands of carbon nanotubes for: (a) a (7,7) armchair CNT and (b) a (7,0) zigzag CNT (simulations were provided by the Network for Computational Nanotechnology (NCN) at nanoHUB.org using CNTbands 2.0 [110]).

nanotube is semiconducting with a non-zero bandgap, as shown in Fig. 2.24(b), using the same π-band model for the case of (7,0).

The minimum value of the $k'_{c,q}$ corresponds to $k'_{c,q} = 2/3d$ from Eq. (2.46), which when substituted into Eq. (2.45), and setting k'_t to zero, gives the gap at the **K** point as

$$E_g = \frac{2\sqrt{3}a_0 t}{d} \approx \frac{0.8\ eV}{d}, \qquad (2.47)$$

with d given in nanometers. Hence the bandgap for a semiconducting nanotube scales inversely proportional to the tube diameter. The corresponding density of states corresponds to that of a quasi-1D system, as given by Eq. (2.34), diverging as $E^{-1/2}$, but with different prefactor and subband energy spacing.

2.5 Magnetic field effects in quantum confined systems

In nanostructure systems, a static magnetic field may have a profound effect on their electronic and transport properties. External application of high magnetic fields to nanostructures is an invaluable degree of freedom available to the experimentalist in probing the system. In and of itself, magnetic fields give rise to new fundamental behavior not observed in bulk-like systems, such as the quantum Hall effect (QHE, discussed in Chapter 4). The fundamental quantity characterizing a magnetic field is the magnetic flux density, **B**, which in mks units is measured in Teslas. In the study of semiconductor nanostructures, low fields usually correspond to fields less than 1 T, which is the regime in which low-field magneto-transport experiments such as Hall effect measurements are usually performed. Magnetic intensities of 10–20 T are obtainable in the usual university and commercial research environments, using superconducting alloy coils with high critical fields immersed in a liquid He dewar. Higher magnetic fields are obtainable only at a few large-scale facilities scattered around the world.

2.5.1 Magnetic field in a 2DEG

When we consider nanostructure systems such as quantum wells, wires, and dots, the effect of a magnetic field may be roughly separated into two cases. In the first, the magnetic field is parallel to one of the directions of free-electron propagation; in the other, it is perpendicular to the free-electron motion of the system. Qualitatively, when free particles are subject to a magnetic field, they experience a Lorentz force

$$\mathbf{F} = q\mathbf{v} \times \mathbf{B}, \tag{2.48}$$

where $\mathbf{v}$ is the velocity of the carrier and q is the charge. Since the force is always perpendicular to the direction of travel of the particle, its motion in the absence of other forces is circular, with angular frequency given by the *cyclotron frequency*, which for electrons with charge e may be written as

$$\omega_c = \frac{eB}{m_c}, \tag{2.49}$$

where B is the magnitude of the magnetic flux density and m_c is the cyclotron mass, which is an average over the mass as the electron performs circular orbits in k-space over a constant energy surface (it does not gain or lose energy due to the angular acceleration of the magnetic field). For isotropic systems like those we have considered so far, the cyclotron mass becomes simply the electron effective mass, m^*. Quantum mechanically, the circular orbits associated with the Lorentz force must be quantized in analogy to the orbital quantization occurring about a central potential, for example, about the nucleus of an atom. Since the particles execute time-harmonic motion similar to the motion in a harmonic oscillator potential, the energy associated with the motion in the plane perpendicular to the magnetic field is expected to be quantized. If we now consider the magnetic field applied perpendicular to the plane of a 2DEG, then the entire free-electron-like motion in the plane parallel to the interface is quantized, and the energy spectrum becomes completely discrete.

In Hamiltonian mechanics, the equations of motion are formulated in terms of potentials rather than in terms of the electric and magnetic fields directly. The potential associated with a magnetic field is a vector rather than a scalar quantity, defined from the curl equation

$$\nabla \times \mathbf{A} = \mathbf{B}, \tag{2.50}$$

where $\mathbf{A}$ is the *vector potential*. The choice of $\mathbf{A}$ for a given $\mathbf{B}$ is not unique, for Eq. (2.50) allows for the *gauge transformation* $\mathbf{A} \to \mathbf{A} + \nabla V(\mathbf{R})$ (where $V(\mathbf{R})$

is a scalar field) since $\nabla \times \nabla V \equiv 0$. The physical properties of the system, of course, cannot be affected by this choice of gauge. The Hamiltonian in the presence of a vector potential is obtained by making the Peierl's substitution $\mathbf{p} \rightarrow \mathbf{p} + q\mathbf{A}$, where $\mathbf{p}$ is the momentum operator. Consider now a quasi-two-dimensional system with a static magnetic field in the z-direction, $\mathbf{B} = (0, 0, B)$. One choice of a vector potential satisfying Eq. (2.50) is the so-called *Landau gauge*, $\mathbf{A} = (0, Bx, 0)$. The single-band effective mass Hamiltonian of Eq. (2.1) may be generalized to include this vector potential as

$$\left[-\frac{\hbar^2}{2m_z} \frac{\partial^2}{\partial z^2} + \frac{1}{2m^*} \left(\frac{\hbar}{i} \nabla_\mathbf{r} + q\mathbf{A} \right)^2 + V_{eff}(z) \right] \psi(\mathbf{r},z) = E\psi(\mathbf{r},z), \quad (2.51)$$

where m^* is the effective mass in the plane of the 2DEG. Again, the solution is separable as $\psi(\mathbf{r}, z) = \varphi(z)\chi(x, y)$, where $\varphi(z)$ is the solution of Eq. (2.4), and where $\chi(x, y)$ satisfies

$$\left[-\frac{\hbar^2}{2m^*} \frac{\partial^2}{\partial x^2} + \frac{m^*\omega_c^2}{2}(x-x_0)^2 \right] \chi(x,y) = E_n \chi(x,y), \quad x_0 = \frac{1}{eB} \frac{\hbar}{i} \frac{\partial}{\partial y}, \quad (2.52)$$

where E_n is the energy associated with the transverse motion. If we assume a solution for $\chi(x, y)$ of the form

$$\chi(x,y) = \chi(x)e^{ik_y y}, \quad (2.53)$$

then in Eq. (2.52) x_0 may be written $x_0 = \hbar k_y/eB$, which is referred to as the *center coordinate*. With this substitution, Eq. (2.52) becomes simply the harmonic oscillator equation for $\chi(x)$, with eigenfunctions given by

$$\chi_n(x) = (2^n n! \sqrt{\pi} l_m)^{-1/2} \exp\left[-\frac{(x-x_0)^2}{2l_m^2} \right] H_n\left[\frac{x-x_0}{l_m} \right], \quad n = 0, 1, 2, \ldots, \quad (2.54)$$

where $H_n(z)$ is the Hermite polynomial of order n, and $l_m = (\hbar/eB)^{1/2}$ is the *magnetic length*, which is basically the cyclotron radius of the ground state. The expectation value of the position operator for the eigenfunctions of Eq. (2.54) may be shown to be $<\chi_n|x|\chi_n> = x_0$, and therefore the center coordinate represents the average position of the magnetic state. The corresponding energy eigenvalues are independent of the center coordinate (and thus the transverse momentum, k_y) and are given by the well-known expression for a harmonic oscillator

$$E_n = \left[n + \frac{1}{2} \right] \hbar\omega_c, \quad n = 0, 1, 2, \ldots \quad (2.55)$$

The cyclotron radius corresponds to the classical orbit associated with the energy of Eq. (2.55). Since the carrier velocity in a circular orbit of radius r_n

2.5 Magnetic field effects in quantum confined systems

is given as $v = r_n\omega_c$, we can equate the kinetic energy $\frac{1}{2}m^*v^2$ with Eq. (2.55) to give

$$r_n = \left[\frac{2\hbar}{eB}\left(n+\frac{1}{2}\right)\right]^{1/2}. \tag{2.56}$$

For $n = 0$, we have $r_0 = l_m$, the magnetic length as asserted above.

As in the 3D case of magnetic quantization, each integer n corresponds to a different *Landau* level associated with a magnetic subband. In contrast to the bulk situation where a remaining degree of freedom exists due to motion parallel to the magnetic field, the total energy spectrum of the quasi-2D system is discrete, given by

$$E = E_i + \left[n+\frac{1}{2}\right]\hbar\omega_c, \tag{2.57}$$

where E_i are the energy eigenvalues associated with the perpendicular confining potential. Even though the Hamiltonian contains the y-momentum, k_y, the total energy given by Eq. (2.57) is independent of k, and therefore of the group velocity

$$v = \frac{1}{\hbar}\frac{\partial E}{\partial k_y}. \tag{2.58}$$

In the above discussion, we have neglected the effect of the magnetic field on the spin variables of the electrons. In fact, an extra term should be added to the Hamiltonian of the envelope of Eq. (2.51)

$$H_s = g^*\mu_B\boldsymbol{\sigma}\cdot\mathbf{B}, \tag{2.59}$$

where $\boldsymbol{\sigma}$ is the electron spin operator (spinor), $\mu_B = (e\hbar/2m_0c)$ is the Bohr magneton, and g^* is the effective Landé g factor (which is equal to 2 in vacuum but differs in semiconductors). Electrons with spin up are raised in energy by $g^*\mu_B B/2$ and those with spin down are lowered by the same amount. The result is a splitting of the spin degeneracy, and thus there is a further splitting of the levels given by Eq. (2.57) by an amount $\pm g^*\mu_B B/2$.

The density of states associated with a quasi-2D system in the presence of a perpendicular magnetic field is singular as shown in Fig. 2.25, where the density of states in the presence of a magnetic field is shown in comparison to that of an ideal, single subband 2D system. The lowest Landau level is shifted up $\hbar\omega_c/2$ with respect to the subband edge of the 2D system. The singular density of states corresponds to a zero-dimensional system, much like the states of the quantum dot systems (that will be discussed in Chapter 6). However, in contrast to the states of an isolated atomic-like quantum dot, each Landau level is highly degenerate, corresponding to all the 2D states with a range $\pm\hbar\omega_c/2$ that collapse into the level. The density of 2D states for a single subband system per unit area and energy was given by Eq. (2.10) as $D_0 = n_s n_v m^*/2\pi\hbar^2$.

Fig. 2.25 Density of states versus energy of a two-dimensional electron gas in the presence of a perpendicular magnetic field. In (a) the solid curve is the density of states in an ideal 2DEG with no impurities or other inhomogeneities. The dotted curve is the density of states for an ideal 2D system corresponding to $B = 0$. (b) shows how the delta-function density of states of an ideal 2DEG is modified due to scattering from impurities and other inhomogeneities, causing a broadening of the Landau levels.

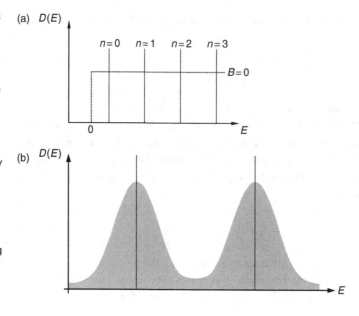

Thus, the total number of states per unit area in each Landau level is given by this expression multiplied by $\hbar\omega_c$ yielding $D_0 = n_s n_v eB/2\pi\hbar$. The degeneracy of each Landau level increases linearly as the magnetic field increases. As the density of carriers in the system is increased (for example, through gate bias or optical excitation), the Fermi level is pinned in the highest occupied magnetic subband until it is completely filled, then jumps discontinuously to the next Landau level. The highest occupied magnetic subband (at $T = 0$ K) corresponds to the total density divided by the density per Landau level as

$$n_{\max} = \text{Int}\left[\frac{2\pi\hbar n_{2D}}{n_s n_v eB} + 1\right], \quad (2.60)$$

where Int() signifies taking the integer part. Correspondingly, if the carrier density is fixed but the magnetic field increases, the Fermi level moves with the highest occupied magnetic subband. However, as the magnetic field increases, the density of states in lower-lying subbands increases, and at a certain critical field, the highest occupied Landau level becomes depopulated and the Fermi level jumps discontinuously to the next lower level. Qualitatively, this discontinuity in the Fermi energy with either density or magnetic field shows up in the magneto-resistance measured externally and results in oscillatory behavior known as *Shubnikov–de-Haas oscillations* (discussed in more detail in Chapter 4). Measurement of the period of the conductance or resistance as a function of magnetic field allows direct determination of the density of the occupied 2D subbands of the system, which is a powerful experimental technique used frequently to characterize quasi-2D systems.

If impurities or other nonidealities are present in the system, the delta-function density of states of the ideal system is broadened as shown in the lower part of Fig. 2.25. Using the so-called self-consistent Born approximation in perturbation theory, Ando [1] has shown that the density of states in the presence of short-range potential impurities is

$$D(E) = \frac{1}{2\pi l_m^2} \sum_n \left[1 - \left(\frac{E - E_n}{\Gamma_n}\right)^2\right]^{1/2}, \quad (2.61)$$

where Γ_n is a broadening factor associated with the short-range impurities. If the magnetic length is such that $d < l_m/(2n+1)^{1/2}$, where d is the range of the potential, then the broadening factor is simply given by

$$\Gamma^2 = \frac{2}{\pi} \hbar \omega_c \frac{\hbar}{\tau_f}, \quad (2.62)$$

where τ_f is the scattering time for the same impurities at the Fermi energy for $B = 0$. If the scattering rate $(1/\tau_f)$ is sufficiently large, the Landau levels merge into one another, and the oscillatory behavior observed in Shubnikov–de-Haas measurements is damped. Qualitatively, such oscillations will be observed only if distinct Landau levels exist, which implies that the broadening should be less than the Landau level spacing.

2.5.2 Magnetic field and one-dimensional waveguides: edge states

We now consider a more complicated problem, that of potential barriers such that the 2DEG is confined in one direction. As we saw in Section 2.4, such confinement leads to the formation of a quasi-one-dimensional system or quantum wire. For simplicity, assume for example a quantum wire formed by etching sidewalls in a 2DEG structure. Assume again that the direction perpendicular to the heterointerface is the z-direction, and that a magnetic field *is* applied in this direction. The confining potential due to the sidewall is assumed to be in the x-direction, and the axis of the wire is in the y-direction. The envelope function equation is again separable, and the z-dependent equation may be solved independently. The remaining equation in the xy-plane has the form as before

$$\left[\frac{1}{2m^*}\left(\frac{\hbar}{i}\nabla_r + q\mathbf{A}\right)^2 + V(x)\right]\chi(x,y) = E_n\chi(x,y), \quad (2.63)$$

where $V(x)$ is the confining potential due to lateral confinement and spin splitting is ignored for simplicity. In the Landau gauge, where $A = (0, Bx, 0)$, Eq. (2.63) may be written:

$$\left[-\frac{\hbar^2}{2m^*}\frac{\partial^2}{\partial x^2} + \frac{m^*\omega_c^2}{2}\left(x - \frac{1}{eB}\frac{\hbar}{i}\frac{\partial}{\partial y}\right)^2 + V(x)\right]\chi(x,y) = E_n\chi(x,y). \quad (2.64)$$

If we again write the solution as in Eq. (2.53), then Eq. (2.64) becomes:

$$\left[-\frac{\hbar^2}{2m^*}\frac{\partial^2}{\partial x^2} + \frac{m^*\omega_c^2}{2}(x-x_0)^2 + V(x)\right]\chi(x) = E_n\chi(x). \quad (2.65)$$

A somewhat illustrative solution may be obtained by assuming that the lateral confining potential is parabolic of the form $V(x) = \frac{1}{2}m^*\omega_0^2$. For the etched quantum wire structure used as the model here, the assumption of a parabolic potential is not inappropriate, since the effect of surface states on either sidewall gives rise to such a quadratic potential due to band-bending. Expanding the quadratic term in the Hamiltonian operator appearing in Eq. (2.65) and combining with $V(x)$, we obtain

$$H = \frac{p_x^2}{2m^*} + \frac{m^*(\omega_c^2+\omega_0^2)}{2}x^2 - m^*\omega_c^2 xx_0 + \frac{m^*\omega_c^2}{2}x_0^2. \quad (2.66)$$

By defining a new center coordinate, $x'_0 = x_0\omega_c^2/\omega^2$ we can complete the square involving x and thus rewrite the Hamiltonian as

$$H = \frac{p_x^2}{2m^*} + \frac{m^*\omega^2}{2}(x-x'_0)^2 + \frac{\hbar^2 k_y^2}{2M}, \quad (2.67)$$

where

$$\frac{\hbar^2 k_y^2}{2M} = \frac{m^*\omega_c^2\omega_0^2}{2\omega^2}x_0^2 \quad \text{and} \quad M = m^*\frac{\omega^2}{\omega_0^2} = m^*\frac{\omega_0^2+\omega_c^2}{\omega_0^2}. \quad (2.68)$$

Comparing this to the case of an infinite two-dimensional gas, there is an additional term on the right-hand side of Eq. (2.67) which looks like the free-electron dispersion of a one-dimensional particle with an effective mass M given by Eq. (2.68). This term may be combined with the energy E_n on the right side of Eq. (2.65) to obtain a harmonic oscillator-like equation in the form of Eq. (2.57). The energy becomes

$$E = E_{n,i}(k_y) = E_i + \left[n+\frac{1}{2}\right]\hbar\omega_c + \frac{\hbar^2 k_y^2}{2M}, \quad (2.69)$$

where E_n is again the quantized energy due to the perpendicular confinement. The Landau levels are no longer degenerate, being spread in energy by the momentum in the y-direction. Likewise, the density of states is no longer discrete as in the purely two-dimensional case, but rather corresponds to a quasi-one-dimensional system as shown in Fig. 2.21. Also, the group velocity is no longer zero in the y-direction; instead, $v = \hbar k_y/M$. For $\omega_c \gg \omega_0$, the mass M goes to zero, giving the infinite two-dimensional gas case. For the other extreme $\omega_c \ll \omega_0$, M goes to the effective mass, m^*, which is just the limit of a quantum wire with no magnetic field present, given by Eq. (2.33). It is interesting to note in this model that wave functions associated with the harmonic-oscillator-type solutions in the transverse direction are localized on one side or the other of the wire, depending on the center coordinate, x'_0, which in turn depends on k_y, the momentum along the axis of the wire. For positive k_y, the envelope function is shifted to the right side of the wire; for negative k_y, the envelope function is shifted to the left side. Therefore, probability flux in one direction is localized on one side of the wire, while states with flux propagating in the opposite direction are localized on the opposite side.

The physical picture associated with the states in a quantum wire in a magnetic field is perhaps clearer if we consider a much less simple model, that of hard wall confinement (i.e., semi-infinite potential barriers) in the transverse x-direction. For this example, assume that the potential $V(x) = 0$ for $0 < x < W$ and is infinite otherwise outside in Eq. (2.65). The exact solution cannot be expressed analytically, and therefore the dispersion relation for energy and momentum has to be determined numerically. To calculate the energy numerically, it is convenient to expand the solution in the basis set of the infinite potential well as

$$\chi(x) = \sum_n a_n \sin\frac{n\pi x}{W}, \qquad (2.70)$$

which satisfies the boundary conditions at 0 and W. Substituting into Eq. (2.65), we obtain

$$\sum_n a_n \left[\frac{\hbar^2}{2m^*}\left(\frac{n\pi}{W}\right)^2 - E\right] \sin\frac{n\pi x}{W} + \frac{m^*\omega_c^2}{2}(x - x_0)^2 a_n \sin\frac{n\pi x}{W} = 0. \qquad (2.71)$$

Multiplying this equation by $(2/W)\sin(m\pi x/W)$ and integrating over the width of the wire (using the orthogonality of the infinite well eigenfunctions), we obtain a coupled set of linear equations in the expansion coefficient a_m as

$$\frac{\hbar^2}{2m^*}\left(\frac{m\pi}{W}\right)^2 a_m + \sum_n F_{nm} a_n = E a_m, \qquad (2.72)$$

where

$$F_{nm} = \frac{m^*\omega_c^2}{W}\int_0^W dx (x-x_0)^2 \sin\frac{m\pi x}{W}\sin\frac{n\pi x}{W}. \qquad (2.73)$$

Equation (2.73) can be rewritten as

$$F_{nm} = \frac{m^*\omega_c^2 W^2}{2\pi^3}\int_0^\pi dy(y-y_0)^2 (\cos((n+m)y)),$$

with $y_0 = \pi x_0/W$, which is easily evaluated as

$$F_{nm} = \frac{m^*\omega_c^2 W^2}{\pi^2}\left[\left(1 - \frac{x_0}{W}\right)\left(\frac{\cos(\pi(n-m))}{(n-m)^2} - \frac{\cos(\pi(n+m))}{(n+m)^2}\right) + \frac{x_0}{W}\frac{4nm}{(n^2-m^2)^2}\right], \qquad (2.74)$$

for $n \neq m$, and

$$F_{nm} = \frac{m^*\omega_c^2 W^2}{\pi^2}\left[\frac{\pi^2((1-x_0/W)^3 + (x_0/W)^3)}{6} - \frac{1}{4m^2}\right], \qquad (2.75)$$

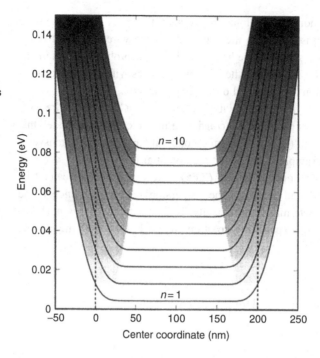

Fig. 2.26 Calculated energy versus center coordinate x_0 for a 200-nm-wide wire and a magnetic field intensity of 5 T. The shaded regions correspond to skipping orbits associated with edge-state behavior.

for $n = m$. The eigenvalues of the determinant formed from the matrix equation (2.72) correspond to the allowed energies for each value of x_0, which are proportional to k_y. A plot of the dispersion relation obtained from this calculation is shown in Fig. 2.26, where we plot E versus x_0. Although x_0 represents the momentum in reciprocal space, it is instructive to plot the energy versus the center coordinate itself, since this coordinate is a measure of the relative position of the envelope function associated with a state of momentum k_y. As can be seen from Fig. 2.26, the dispersion relation is almost flat in the middle of the waveguide and then shows strong dispersion close to either side. A simple physical interpretation of this behavior is possible from the classical motion of the particle in the waveguide, as shown in Fig. 2.27 (see for example [111]). The relevant parameters in interpreting this behavior are the cyclotron radius for a particular level (Eq. (2.56)), and the center coordinate, giving the center position of the circular cyclotron orbit of radius r_n. Three possible types of orbits in the confined structure are possible, based on the cyclotron radius and the center coordinate: pure cyclotron orbits, skipping orbits, and traversing orbits. In the case of pure cyclotron orbits, the distance from the center coordinate to one wall or the other is less than r_n, so that no scattering of the particle off the wall occurs, and pure cyclotron-type motion is possible, as shown by the circular orbits in Fig. 2.27. In this case, we expect the states to resemble those of simple magnetic quantization in a 2DEG, which is dispersionless. The flat regions of the dispersion curves shown in Fig. 2.26

2.5 Magnetic field effects in quantum confined systems

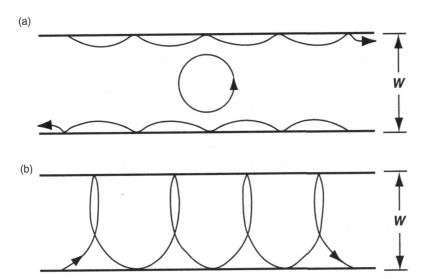

Fig. 2.27 Classical motion of a particle in a magnetic field for different energies and center coordinates.

correspond to these orbits. If the distance between the center coordinate to either wall is less than r_n, then boundary scattering occurs, giving rise to the skipping orbits shown in Fig. 2.27. If boundary scattering is assumed to be specular (rather than diffuse), the reflected momentum is such that the electron has a net momentum in the plus or minus y-direction, depending whether x_0 (and thus k_y) is positive or negative. The states associated with these skipping orbits are referred to as *edge states,* and they correspond to the shaded region of Fig. 2.26 (the extent of which is determined by the classical overlap argument of the cyclotron radius about the center coordinate relative to the boundary). Because the edge states have a net velocity in the $\pm y$-direction, the dispersion relation is no longer flat, as observed in the shaded regions of Fig. 2.26. The slope of the dispersion relation is related to the group velocity, and as we can see from the figure, edge states on the right side have positive velocity while those on the left have negative velocity. Finally, for large energy, the cyclotron orbit is sufficiently large that the electrons perform traversing orbits in which the electron interacts with both boundaries. As one may qualitatively observe, specular scattering from either boundary results in a net momentum along the axis of the waveguide, which corresponds to a transition from dispersionless behavior to more 1D waveguide behavior for the higher magnetic subbands.

This concept of edge states that propagate in opposite directions on opposite sides of a 1D waveguide is sometimes compactly shown by the representation given in Fig. 2.28. The flow lines represent the direction and relative center coordinate of the edge state with respect to the waveguide boundaries. We will come back to this description later in Chapters 3 and 4 when we discuss mode matching and the general behavior of edge states in inhomogeneous structures.

Fig. 2.28 Schematic representation of edge states in a quantum waveguide.

2.6 Screening and collective excitations in low-dimensional systems

So far we have discussed the electronic properties of quantum confined systems in terms of a single-particle or one-electron picture. The other electrons in the system were either ignored or treated as an average static potential, as in the case of self-consistent solutions discussed in Section 2.2.3. However, in transport and other nonequilibrium phenomena, one has to consider the dynamic response of the system of electrons to the presence of an external perturbation. The most familiar example is the dielectric response of a material to an external, time-varying electric field. Due to the presence of the electric field in the solid, charges rearrange themselves in response to the external force, resulting in an additional induced charge in the solid. This induced charge gives rise to a polarization field in addition to the external field, and the net effect is described in terms of the frequency- (and wavevector-) dependent dielectric function of the material.

In a bulk semiconductor, we generally can separate various frequency-dependent contributions to the dielectric response due to the lattice (in polar semiconductors), the valence electrons (which determines the electronic polarizability of the material and thus the "high-frequency" dielectric constant), and finally the free mobile carriers in the conduction band and unoccupied states in the valence band (holes). In a bulk system, the dielectric response of the system may be represented in the Fourier domain with respect to time and space by the *dielectric function*, $\varepsilon(\mathbf{Q}, \omega)$. For an external potential, $V^{ext}(\mathbf{R}, t)$, the effective, or total, potential in the system is the sum of the external potential and the potential due to the induced charge by the external potential, which in the Fourier domain is given by

$$V^{tot}(\mathbf{Q}, \omega) = \frac{V^{ext}(\mathbf{Q}, \omega)}{\varepsilon(\mathbf{Q}, \omega)}. \tag{2.76}$$

In a bulk system, the total dielectric function may be expressed as (see for example [112])

$$\varepsilon(q, \omega) = \varepsilon(\infty) + \varepsilon_l(\omega) + \frac{e^2}{\varepsilon_0 q^2} \sum_{\mathbf{k}} \frac{f(\mathbf{k}) - f(\mathbf{k}+\mathbf{Q})}{E(\mathbf{k}+\mathbf{Q}) - E(\mathbf{k}) - \hbar\omega - i\alpha\hbar} \tag{2.77}$$

where ε_0 is the permittivity of free space, $f(\mathbf{k})$ is the one-particle distribution function for a particle of wavevector $\mathbf{k}$, $E(\mathbf{k})$ is the energy of state $\mathbf{k}$, and α is a small convergence parameter. The first term on the right is the relative permittivity due to core and valence electrons, the second term is the contribution due to the lattice in polar materials, and the third term is the so-called *Lindhard dielectric function* for the free carriers (electrons and holes in general). The latter contribution due to free carriers is responsible for effects such as free carrier absorption of photons, screening of external potentials such as those responsible for scattering in the system, and collective excitations of the free-electron gas, which in the parlance of many-body theory are referred to as plasmons.

In this section, we look at the dielectric response of free carriers residing in quantum confined systems such as quantum wells and quantum wires. In particular, we are interested in how the reduced dimensionality of the system influences the collective behavior of the electron gas as compared to a bulk system, and also the consequences this has for transport in such systems. The main difference of the quantum confined systems discussed in this chapter compared to bulk systems is the quantization of the perpendicular motion resulting in a series of quantum subbands, each with free-electron behavior in the parallel direction. While the dimensionality of the quantum confined system is reduced due to the confining potential, perturbations of the system generally are three-dimensional. An example is the 3D potential due to an ionized impurity located in the vicinity of a quantum well or wire. The quantum confined system itself responds (polarizes) to the presence of this potential, both in the free direction of the system (intrasubband polarization) and in the confined directions (intersubband polarization). The consequence of this is that instead of being able to define a single dielectric function of the system, as in the bulk case, the dielectric response in a quantum confined system must be defined in terms of a dielectric matrix [111] rather than a simple (although already complicated) scalar function. Here we discuss the form of the dielectric matrix for quasi-2D and quasi-1D systems derived from time-dependent perturbation theory similar to that used to derived the Lindhard dielectric function, Eq. (2.77), for bulk 3D systems (see for example [113]).

2.6.1 Dielectric function in quasi-2D systems

We begin with the case of a quasi-2D system as discussed earlier. We assume that the ideal system with no external perturbation is described by an unperturbed Hamiltonian whose properly normalized eigenstates are given by

$$\psi_{\mathbf{k},n}(\mathbf{r},z) = \varphi_n(z)e^{i\mathbf{k}\cdot\mathbf{r}} \Rightarrow |\mathbf{k},n\rangle, \qquad (2.78)$$

as given in Eq. (2.3), with the normalization factor set equal to unity for convenience. To the ideal system we add an external, generally time-varying potential, $V^{ext}(\mathbf{r},z)e^{-i\omega t}e^{\alpha t}$ where α is a small parameter corresponding to the adiabatic "turning on" of the perturbation. Because of the multi-subband nature of dielectric response, there are contributions from each subband separately (intrasubband) and between subbands (intersubband), which results in a generalization of Eq. (2.76) to replace the dielectric function with a dielectric matrix

$$V_{mm'}^{ext}(\mathbf{q}) = \sum_{n,n'} \left[\delta_{nm}\delta_{n'm'} - \frac{e^2}{2\varepsilon_s q} F_{nn'}^{mm'}(\mathbf{q}) L_{nn'}(\mathbf{q},\omega) \right] V_{nn'}^{tot}(\mathbf{q})$$

$$= \sum_{n,n'} \varepsilon_{nn'mm'}(\mathbf{q},\omega) V_{nn'}^{tot}(\mathbf{q}),$$

$$V_{nn'}^{tot}(\mathbf{q}) = \sum_{n,n'} \varepsilon_{nn'mm'}^{-1}(\mathbf{q},\omega) V_{nn'}^{ext}(\mathbf{q}), \quad (2.79)$$

where $\varepsilon_{nn'mm'}(\mathbf{q},\omega)$ defines the four-dimensional dielectric matrix, with $\mathbf{q}$ the 2-dimensional wavevector. The potential $V_{nn'}^{ext}(\mathbf{q})$ represents the matrix element of the external potential (similarly for the total potential $V_{nn'}^{tot}(\mathbf{q})$)

$$V_{nn'}^{ext}(\mathbf{q}) = \langle \mathbf{k}+\mathbf{q}, n' | V^{ext}(\mathbf{r},z) | \mathbf{k}, n \rangle = \int d\mathbf{r} \int_{-\infty}^{\infty} dz\, \varphi_{n'}^*(z)\varphi_n(z) V^{ext}(\mathbf{r},z) e^{-i\mathbf{q}\cdot\mathbf{r}}. \quad (2.80)$$

$L_{nn'}(\mathbf{q},\omega)$ is related to the state occupancy and has the form of the Lindhard dielectric function, Eq. (2.77)

$$L_{nn'}(\mathbf{q},\omega) = \frac{f_n(\mathbf{k}) - f_n(\mathbf{k}+\mathbf{q})}{E_{n'}(\mathbf{k}+\mathbf{q}) - E_n(\mathbf{k}) - \hbar\omega - i\alpha\hbar}, \quad (2.81)$$

where $f_n(\mathbf{k})$ is the distribution function normalized such that

$$2\sum_{\mathbf{k},n} f_n(\mathbf{k}) = n_{2d}, \quad (2.82)$$

where the factor of 2 is for spin degeneracy and the valley degeneracy is unity. The form factor in (2.79) is given by

$$F_{nn'}^{mm'}(\mathbf{q}) = \int_{-\infty}^{\infty} dz \int_{-\infty}^{\infty} dz'\, \varphi_{m'}^*(z)\varphi_m(z)\varphi_n^*(z')\varphi_{n'}(z') e^{-q(z-z')}. \quad (2.83)$$

Since the envelope functions are orthonormal functions with respect to the subband indices,

$$\lim_{q \to 0} F_{nn'}^{mm'}(\mathbf{q}) = \delta_{nm}\delta_{n'm'}. \quad (2.84)$$

The long-wavelength limit, $\mathbf{q} \to 0$, is reached when the wavelength $(2\pi/q)$ is large compared to the spatial extent of the subband envelope functions. Thus, in the limit of a purely 2D system of zero width, Eq. (2.84) holds for all $\mathbf{q}$. As a consequence of (2.84), intersubband transitions are completely unscreened in the long-wavelength limit

$$V_{mm'}^{tot}(\mathbf{q}) = V_{mm'}^{ext}(\mathbf{q}), \quad (2.85)$$

whereas, for intrasubband transitions Eq. (2.79) reduces to

$$V_{mm'}^{ext}(\mathbf{q}) = V_{mm'}^{tot}(\mathbf{q}) + \sum_n \frac{e^2}{2\varepsilon_s q} L_{nn}(\mathbf{q},\omega) V_{mm'}^{tot}(\mathbf{q})$$

$$= V_{mm'}^{tot}(\mathbf{q}) + \sum_n \chi_{nn}(\mathbf{q},\omega) V_{mm'}^{tot}(\mathbf{q}), \quad (2.86)$$

2.6 Screening and collective excitations

where $\chi_{nn}(\mathbf{q}, \omega)$ represents the intrasubband polarizability of subband n. The matrix equation (2.86) may be cast in the form of a matrix equation

$$\mathbf{V}^{ext} = \bar{\varepsilon} \cdot \mathbf{V}^{tot}; \quad \bar{\varepsilon} = \bar{\mathbf{I}} + \bar{\mathbf{u}} \cdot \bar{\mathbf{v}}, \tag{2.87}$$

where the dielectric matrix may be factored into the identity matrix plus the product of an identity column vector and a polarizability vector

$$\bar{\mathbf{v}} = [\chi_{11} \quad \chi_{22} \quad \chi_{33} \cdots], \tag{2.88}$$

A matrix in this symmetric form may be inverted analytically [114], and thus (2.86) is inverted to give

$$V^{tot}_{mm}(q) = \frac{V^{ext}_{mm}(\mathbf{q}) + \sum_{n \neq m} \chi_{nn}(\mathbf{q}, \omega)\left(V^{ext}_{mm}(\mathbf{q}) - V^{ext}_{nn}(\mathbf{q})\right)}{\varepsilon_D(\mathbf{q}, \omega)}, \tag{2.89}$$

where the denominator is the determinant of $\bar{\varepsilon}$, which is the scalar dielectric function associated with the independent polarizabilities in each subband,

$$\varepsilon_D(\mathbf{q}, \omega) = 1 + \sum_n \chi_{nn}(\mathbf{q}, \omega). \tag{2.90}$$

The second term in (2.89) is expected to be small, and thus under the conditions of neglecting the intersubband polarizabilities and the correction terms due to the intrasubband polarizabilities of the other subbands, the multisubband dielectric matrix reduces to a scalar relation as in the bulk case

$$V^{tot}_{mm}(\mathbf{q}) = \frac{V^{ext}_{mm}(\mathbf{q})}{\varepsilon_D(\mathbf{q}, \omega)}. \tag{2.91}$$

Close to equilibrium, the intraband polarizability is a function of the distribution function $f_n(E)$, which is a function only of the energy. In this case, Maldague [115] has shown that the polarizability at finite temperature may be calculated as an integral over the zero temperature function as

$$\chi_{nn}(\mathbf{q}, \omega; T, E_f) = \int_0^\infty dE'_f \frac{\chi_{nn}\left(\mathbf{q}, \omega; 0, E'_f\right)}{4k_B T \cosh^2\left(\frac{E_f - E'_f}{2k_B T}\right)}, \tag{2.92}$$

where E_f is the Fermi energy, which appears as a parameter in the distribution function.

2.6.1.1 Static screening

If the potential is static (i.e. $\omega = 0$), such as the potential due to an ionized impurity atom in the system, some further simplification is possible. Taking the long-wavelength limit again, the polarization function $L_{nn}(\mathbf{q}, 0)$, defined in (2.81), reduces to

$$\lim_{q\to 0} L_{nn}(\mathbf{q},0) = \lim_{q\to 0} 2 \sum_{\mathbf{k}} \frac{f_n(\mathbf{k}) - f_n(\mathbf{k}+\mathbf{q})}{E_n(\mathbf{k}+\mathbf{q}) - E_n(\mathbf{k})}$$

(2.93)

$$= \frac{-2}{4\pi^2} \int_0^{2\pi} d\vartheta \int_0^{\infty} dk\, k \frac{\partial f_n(\mathbf{k})}{\partial E_n(\mathbf{k})}.$$

If we assume a simple parabolic dispersion relation for each subband, then the integral over k may be converted to one over energy to yield

$$\lim_{q\to 0} L_{nn}(\mathbf{q},0) = \frac{m_n^*}{\pi\hbar^2} f_n(0) = D_n f_n(0), \qquad (2.94)$$

where D_n is the 2D density of states from Eq. (2.10) (generalizing to different masses in different subbands), and $f_n(0)$ is the occupancy of the bottom of subband n. From Eq. (2.94), the scalar dielectric function Eq. (2.90) may be written

$$\lim_{q\to 0} \varepsilon_D(\mathbf{q},0) = 1 + \frac{e^2}{2\varepsilon_s q} \sum_n D_n f_n(0) = 1 + \frac{q_s}{q}, \qquad (2.95)$$

where q_s is the 2D Thomas–Fermi screening constant. At low temperature in the extreme quantum limit, only the $n=1$ state is occupied, and the occupancy $f_n(0) = 1$, so that $q_s = e^2 D_1/(2\varepsilon_s)$. For a one-subband system at high temperature, Eq. (2.11) may be inverted to solve for the Fermi energy. This is then substituted in the high-temperature distribution function, $f_n(0) = e^{(E_f - E_0)/k_B T}$ and assuming $n_{2D}/(k_B T D_1) \ll 1$, the screening constant is $q_s = e^2 n_{2D}/(2\varepsilon_s k_B T)$. Typically the screening length, q_s^{-1}, representing the fall off of the potential with distance, is shorter than the corresponding quantity in a bulk 3D system. In 3D, however, the real-space decay of the screened Coulomb potential found by inverse transforming the 3D equivalent of Eq. (2.91) is exponential, whereas the corresponding 2D decay goes as r^{-3}, which is weaker. Thus one cannot say quantitatively whether screening is stronger or weaker in low-dimensional systems.

2.6.1.2 Collective excitations

In a bulk system, the dielectric function may vanish for certain $\mathbf{q}$ and ω. This implies that a potential of wavevector $\mathbf{q}$ and frequency ω gives rise to an infinite response in the system. Thus for certain frequencies, potential oscillations are sustained in the absence of an external potential. Such self-oscillations are referred to as plasma oscillations and correspond to the collective modes of vibration of the entire electron gas about its positive background. In a bulk system, the zeros of the Lindhard dielectric function, Eq. (2.77), give the dispersion relation for the long-wavelength plasma oscillations as

$$\omega_p = \left(\frac{n_{3D} e^2}{\varepsilon_s m^*} \right)^{1/2}. \qquad (2.96)$$

For long-wavelength oscillations, the plasma frequency is independent of wavevector.

2.6 Screening and collective excitations

For a multi-band system, the plasma frequency is complicated by the matrix nature of the dielectric function as given, for example, by Eq. (2.79). The zeros of the dielectric matrix correspond to the zeros of the determinant of the matrix. Solving the determinant equation is of course quite complicated, and may sustain both intraband plasma oscillations as well as intersubband oscillations. However, if we again limit ourselves to $\mathbf{q} \to 0$, the dielectric matrix is two dimensional, and the determinant may be written from Eq. (2.80)

$$\det|\varepsilon| = 1 + \chi_{11}(\mathbf{q},\omega) + \chi_{22}(\mathbf{q},\omega) + \chi_{33}(\mathbf{q},\omega) + \cdots$$

and thus the plasma frequencies are given by the zeros of the above relation. Let us now consider the long-wavelength plasma frequency ($\mathbf{q} \to 0$) by looking at

$$\begin{aligned}0 &= 1 + \sum_i \chi_{ii}(\mathbf{q},\omega_p) = 1 + \frac{e^2}{2\varepsilon_s q} \sum_i L_{ii}(\mathbf{q},\omega_p) \\ &= 1 + 2\frac{e^2}{2\varepsilon_s q} \sum_{\mathbf{k},i} \frac{f_i(\mathbf{k}) - f_i(\mathbf{k}+\mathbf{q})}{(E_i(\mathbf{k}+\mathbf{q}) - E_i(\mathbf{k}) - \hbar\omega_p - i\alpha\hbar)},\end{aligned} \quad (2.97)$$

where ω_p is the plasma frequency. Unfolding the summation and combining terms, Eq. (2.97) becomes

$$\begin{aligned}0 &= 1 + 2\frac{e^2}{2\varepsilon_s q} \sum_{\mathbf{k},i} \left\{ \frac{f_i(\mathbf{k})}{E_i(\mathbf{k}+\mathbf{q}) - E_n(\mathbf{k}) - \hbar\omega_p - i\alpha\hbar} \right. \\ &\quad \left. - \frac{f_i(\mathbf{k})}{E_n(\mathbf{k}) - E_n(\mathbf{k}-\mathbf{q}) - \hbar\omega_p - i\alpha\hbar} \right\} \\ &= 1 + \frac{e^2}{\varepsilon_s q} \sum_{\mathbf{k},i} \frac{f_i(\mathbf{k})[2E_i(\mathbf{k}) - E_i(\mathbf{k}+\mathbf{q}) - E_i(\mathbf{k}-\mathbf{q})]}{(E_i(\mathbf{k}+\mathbf{q}) - E_i(\mathbf{k}) - \hbar\omega_p - i\alpha\hbar)(E_n(\mathbf{k}) - E_n(\mathbf{k}-\mathbf{q}) - \hbar\omega_p - i\alpha\hbar)}\end{aligned} \quad (2.98)$$

In the limit that $\mathbf{q}$ becomes small, the denominator goes to $(\hbar\omega)^2$ and the term in brackets in the numerator goes to $-\hbar^2 q^2/m_i^*$. The sum over $\mathbf{k}$ over the subband distribution function gives n_{2D}^i, the 2D sheet density in subband i. Thus the long-wavelength plasma dispersion relation becomes

$$0 = 1 + \frac{e^2 q}{\varepsilon_s \omega_p} \sum_i \frac{n_{2D}^i}{m_i^*}. \quad (2.99)$$

If the subband masses are assumed to be the same, the 2D plasma frequency becomes

$$\omega_p^{2d} = \left(\frac{n_{2d} e^2 q}{\varepsilon_s m^*}\right)^{1/2}, \quad (2.100)$$

where n_{2D} is the total 2D density. In contrast to the dispersionless 3D relationship, ω_p goes as $q^{1/2}$ and thus the plasma frequency goes to zero as q approaches zero. As q becomes larger, the q dependence of the form factor (2.83) can no longer be neglected, and a more complicated dependence on q arises.

As in the bulk case, the plasma-like oscillations of the 2D electron gas may be described using collective coordinates for a boson-like system of quasi-particles called *plasmons* as described by the dispersion relation above (which is analogous to the collective representation of the vibrational modes of the lattice in terms of quasi-particles called phonons). The plasmons may interact with the free electrons in the system or with incident electromagnetic energy, emitting and absorbing energy in quanta corresponding to Eq. (2.100). In transport, plasmons may evidence themselves as an additional energy loss mechanism for electrons; this effect should be considered along with the other scattering mechanisms in the system. Evidence for 2D plasmons was observed early on in the Si/SiO$_2$ system in far-infrared transmission experiments [116,117] and in inversion layers excited by an electric current [118].

2.6.2 Dielectric function for quasi-1D systems

The derivation of the matrix dielectric function for a quasi-1D system follows the basic one described above for quasi-2D systems. The differences are that the wavevector for propagation in the parallel direction is now restricted to one dimension, and that the envelope function describing the confined states is a function of two perpendicular directions, as given by Eq. (2.30). To simplify the index notation, we express the transverse modes of the quantum wire in terms of a single subband index it and write the envelope function (setting L to unity for convenience) as

$$\psi(\mathbf{r},x) = \varphi_n(\mathbf{r})e^{ik_x \cdot x}, \tag{2.101}$$

where x is the direction parallel to the axis of the quantum wire, and $\mathbf{r}$ is the y, z confinement direction perpendicular to the axis of the wire. Similar to the 2D case discussed earlier, the dielectric matrix may be written

$$V^{ext}_{nn'}(q_x) = \sum_{n,n'} \left[\delta_{nm}\delta_{n'm'} - \frac{e^2}{2\pi^2 \varepsilon_s} F^{mm'}_{nn'}(q_x) L_{nn'}(q_x,\omega) \right] V^{tot}_{nn'}(q_x),$$
$$= \sum_{n,n'} \varepsilon_{nn'mm'}(q_x,\omega) V^{tot}_{nn'}(q_x), \tag{2.102}$$

where

$$L_{nn'}(q_x,\omega) = \frac{f_n(k_x) - f_n(k_x + q_x)}{E_{n'}(k_x + q_x) - E_n(k_x) - \hbar\omega - i\alpha\hbar}, \tag{2.103}$$

and the form factor in the quasi-1D case is given by

$$F^{mm'}_{nn'}(\mathbf{q}) = \int d\mathbf{r} \int d\mathbf{r}' \varphi_n^*(\mathbf{r})\varphi_{n'}(\mathbf{r})\varphi_{m'}^*(\mathbf{r})\varphi_m(\mathbf{r}) K_0(|q_x||\mathbf{r}-\mathbf{r}'|). \tag{2.104}$$

The form factor (2.104) is complicated by the Bessel function which arises from the 1D Fourier transform of the Coulomb potential, and which diverges logarithmically as $K_0(x) \sim -\ln(x)$ as $x \to 0$. However, as is discussed for the 2D case, it is expected that the intersubband contributions to the polarizability should be small. For the intrasubband case, Hu and Das Sarma [119] have shown that if one assumes that the width of the wire is infinitesimally narrow in one transverse direction (for example, in the growth direction), and assumes a square quantum well of width a bounded by infinite potential barriers in the other transverse direction, the intrasubband form factor for the ground subband reduces to

$$F^{11}_{11}(q_x) = K_0(|q_x|a) + 1.9726917\ldots \quad \text{for} \quad |q_x|a \to 0. \tag{2.105}$$

In the long-wavelength, static limit, $L_{nn}(q_x, 0)$ in the 1D case may be written

$$\lim_{q_x \to 0} L_{nn'}(q_x, 0) = \frac{-2}{2\pi} \int_{-\infty}^{\infty} d(k_x) \frac{\partial f_n(k_x)}{\partial E_n(k_x)}. \tag{2.106}$$

If we consider the quantum limit at low temperature, the derivative of the Fermi–Dirac function becomes a delta function, $-\delta(E - E_f)$, and therefore (2.106) is simply the definition of the 1D density of states at the Fermi energy given by Eq. (2.34),

$$L_{nn'}(q_x, 0) = D_{1D}(E_f) = \frac{2m^*}{\pi^2 \hbar} E_f^{-1/2}, \tag{2.107}$$

where the Fermi energy is measured with respect to the lowest subband minima, and a single-valley system is again assumed ($\eta_v = 1$). For long-wavelength, static screening in the quantum limit, the dielectric function in 1D becomes

$$\varepsilon_{1D}(q_x) = 1 - \frac{e^2}{2\pi^2 \varepsilon_s} \ln(|q_x|a) D_{1d}(E_f), \tag{2.108}$$

where a may be taken as the average width of the 1D gas.

The long-wavelength limit of the plasmon dispersion relation is again derived in an analogous fashion to the 2D case given by Eq. (2.100). Using the same unfolding as was used in Eq. (2.98), the zeros of the 1D dielectric function for an infinite square well become

$$\lim_{q_x \to 0} \varepsilon_{1D}(q_x, \omega_p) = 0 = 1 + \frac{e^2 n_{1d} q_x^2}{\pi^2 \varepsilon_s m^* \omega_p^2} \ln(|q_x|a), \tag{2.109}$$

and thus the plasmon dispersion goes as

$$\omega_p^{1D} = \left(\frac{e^2 n_{1D}}{\pi^2 \varepsilon_s m^*}\right)^{1/2} q_x \sqrt{-\ln(|q_x|a)} \quad \cdots |q_x|a \ll 1. \tag{2.110}$$

2.7 Homogeneous transport in low-dimensional systems

In the quasi-two-dimensional system discussed earlier, one degree of freedom of the electrons is restricted by the confinement potential associated with an interface(s) leading to quantization of the momentum in that direction, while motion in the remaining two directions is unconstrained. Likewise, in a quasi-one-dimensional system such as the quantum wire discussed above, the motion along the axis of the wire is free electron-like. If we restrict ourselves to carrier transport parallel to confining potentials, then we can discuss the homogeneous transport properties of such reduced-dimensionality systems in the same context as homogeneous transport in bulk systems, that is, in terms of macroscopic phenomenological parameters such as mobility, conductivity, thermopower, and so on, which may be measured in an appropriately designed experiment. Transport in this context is quite different from the case of transport perpendicular to the confining barriers discussed in Chapter 3 in which quantum mechanical reflection and transmission from the confining barriers themselves play a central role.

The same framework of theoretical techniques in nonequilibrium statistical mechanics used to explain the bulk transport properties of materials may he applied to the homogeneous properties of the quasi-2D and quasi-1D systems. Indeed, an enormous volume of experimental and theoretical literature has been published on the parallel transport properties of quasi-2D systems, beginning in the late 1960s with studies of the Si/SiO$_2$ system and continuing with the heterojunction semiconductor systems starting in the early 1980s until today. Perhaps the most striking manifestation of reduced dimensionality is in the transport properties of the 2DEG subject to a perpendicular magnetic field. Measurement of the Hall coefficient versus magnetic field or Fermi energy of a 2DEG under strong magnetic fields at low temperature reveals that the transverse resistivity, ρ_{xy}, is quantized extremely accurately in integer multiples of $h/e^2(1/n)$, where n is an integer. The discovery of the integer *quantum Hall effect* (QHE) [120] led to the Nobel prize in physics for Klaus von Klitzing in 1985. While elegant arguments based on gauge invariance have related this phenomenon to the bulk properties of the 2DEG [121], a simple alternative model associated with lateral confinement of the 2DEG and the formation of so-called edge states has been given by Büttiker [122], which we will revisit in Chapter 4. At even lower temperatures and in very-high-mobility samples, new plateaus appear at fractional values of h/e^2, in which the fraction is a ratio of two integers [123]. This *fractional quantum Hall effect* (FQHE) is theorized to arise from the condensation of the interacting electron system into a new many-body ground state characteristic of an incompressible fluid [124]. A more detailed discussion of the FQHE will be given in Chapter 4.

The transport properties of quantum wire systems has become a very active area of investigation over the past decade, due both in part to improvements in top down nanoscale fabrication techniques that allow nanowire-like transistors to be fabricated, as well as the enormous growth of research on self-assembled 1D conductors such as carbon nanotubes (CNTs) and semiconductor nanowires (SNWs) grown by vapor-solid-liquid epitaxy, as discussed earlier in Section 2.1.1. Very high mobilities have been reported in such systems, particularly in CNTs, due to the reduction of process-induced defects in such self-assembled structures.

The transport properties of 1D systems are of particular interest, as many interesting predictions of unusual many-body behavior have been suggested, such as the formation of a Luttinger liquid [125] or the presence of a charge-density-wave ground state due to the lattice Peierls distortion [126]. However, evidence for such behavior in present quantum wire systems is lacking due to the energy broadening effects of disorder in present technology today, which leads to normal Fermi liquid-like behavior [119]. Thus, in the following section we will discuss parallel transport in quantum wires on the same footing as parallel transport in quantum wells in the context of semiclassical transport.

2.7.1 Semiclassical transport

In discussing semiclassical transport in low-dimensional systems, we first begin by assuming that the description of the system in terms of the solutions of the effective mass equation is still a "good" solution to the problem in the presence of disorder. Use of the effective mass approximation is probably valid for semiconductor nanowires down to 5 nm in diameter, below which atomistic electronic structure models need to be employed. In low-dimensional systems, we consider the perfect or unperturbed system to be defined by the static lattice, with perfectly smooth boundaries defining the system, free of impurities or other random inhomogeneities. A "good" solution basically implies that the broadening of the energy levels due to disorder (i.e. the real part of the self-energy) is small, so that crystal momentum conservation is approximately preserved. In this weak coupling limit, we can then construct a kinetic equation in which the particle density function (or distribution function) evolves in time under the streaming motion of external forces and spatial gradients, and the randomizing influence of nearly point-like (in space-time) scattering events. Let $f_i(\mathbf{r},\mathbf{k})$ represent the one-particle distribution function in a $2n$-dimensional phase space, where n is the dimensionality of the system and i labels the subband index. Here $\mathbf{r}$ and $\mathbf{k}$ refer to the position and wavevector in the propagating direction(s) of the system (i.e., parallel to the quantum wire axis or interface in a quantum well). By describing the system in terms of $f_i(\mathbf{r},\mathbf{k})$, which gives the probability of finding a particle in an infinitesimal volume $d\mathbf{r}d\mathbf{k}$ around $\mathbf{r}$ and $\mathbf{k}$, we implicitly neglect the fact that $\mathbf{r}$ and $\mathbf{k}$ are quantum-mechanically noncommuting variables. This assumption does not pose a serious problem in considering homogeneous transport in a semi-infinite medium, as discussed in the present section, further assuming that the phase information carried by the carrier is lost between subsequent collisions. As we will see in Chapters 5 and 6, conditions exist where the phase is preserved between elastic scattering events giving rise to experimentally observable phenomena such as negative magnetoresistance.

Under the influence of a driving force such as an electric field applied in the plane parallel to the interface for 2D systems (or parallel to the longitudinal axis in the 1D case), the in-plane crystal momentum of carriers in the unperturbed system evolves according to the acceleration theorem

$$\frac{d(\hbar\mathbf{k})}{dt} = \mathbf{F} = e(\mathbf{E} + \mathbf{v} \times \mathbf{B}), \qquad (2.111)$$

where $\mathbf{F}$ is the force acting on the particle, $\mathbf{E}$ is the electric field, $\mathbf{B}$ is the magnetic flux density, and $\mathbf{v}$ is the particle velocity given by the group velocity

$$\mathbf{v} = \frac{1}{\hbar}\nabla_k E(\mathbf{k}), \qquad (2.112)$$

where E is the energy associated with state $\mathbf{k}$. The *Boltzmann transport equation* (BTE) may be derived by writing the continuity equation in the $2n$-dimensional phase space in terms of the particle flux through a small hypervolume of this space centered around $\mathbf{r}$ and $\mathbf{k}$

$$\frac{\partial f_i}{\partial t} = -\frac{1}{\hbar}\nabla_k E(\mathbf{k}) \cdot \nabla_r f_i - \frac{1}{\hbar}\mathbf{F} \cdot \nabla_k f_i + \left.\frac{\partial f_i}{\partial t}\right|_{collisions}, \qquad (2.113)$$

where the last term represents the rate of change of the distribution function due to scattering. Assume instantaneous (again in both space and time) phase-randomizing collisions, in the classical sense as well as the quantum mechanical, meaning that the particle loses any correlation it has with other particles, allowing one to decouple higher-order two-, three-, etc., particle distribution functions from (2.113). Under this assumption, the last term may be written as a detailed balance of in-scattering and out-scattering events as the so-called *collision integral*

$$\left.\frac{\partial f_i}{\partial t}\right|_{collisions} = \sum_{j,k'} S_{j,i}(\mathbf{k}',\mathbf{k})\left[f_j(\mathbf{k}')(1-f_i(\mathbf{k}))\right] - S_{i,j}(\mathbf{k},\mathbf{k}')\left[f_i(\mathbf{k})(1-f_j(\mathbf{k}'))\right], \qquad (2.114)$$

where $S_{j,i}(\mathbf{k}',\mathbf{k})$ represents the scattering rate from a state in subband j of wavevector $\mathbf{k}'$ to a state in subband i with wavevector $\mathbf{k}$. The case of $i=j$ refers to intrasubband scattering whereas $i \neq j$ is denoted intersubband scattering. For a quasi-1DEG, $\mathbf{k}$ is no longer a vector quantity, but rather a scalar. The only possible states after scattering are either in the forward or the backward direction, thus the sum over $\mathbf{k}'$ reduces to just two terms.

The assumption of instantaneous collisions in space and time allows us to write the detailed balance above as affecting only the $\mathbf{k}$-dependent part of f_i in the infinitesimal volume in phase space that we consider in the general continuity equation (2.113). However, when we look at the problem of transport through generalized structures with dimensions that vary on the order of the de Broglie wavelength of carriers (i.e. nanostructures), the nonlocality of the particle invalidates this simplifying approximation. The collision integral in general couples the distribution function from subband j into the BTE for subband i through the *intersubband* scattering terms involving $S_{i,j}$, $i \neq j$, thus leading to the necessity of solving a set of coupled partial differential equations for $f_i(\mathbf{r},\mathbf{k})$. In reality, one has the same complication in a bulk 3D system if one considers transport including multiple bands (e.g., light-hole/heavy-hole transport), or multiple valleys such as the problem of intervalley transfer in high-field transport in the conduction bands. However, in the case of quasi-2D or quasi-1D systems, the spacing of the subbands is usually tenths of an electron volt, and thus multiband transport is not negligible except in the extreme quantum limit in which only one subband is occupied.

To determine phenomenological transport parameters such as the carrier mobility and diffusion coefficient in low-dimensional systems, the relaxation time approximation may be employed as in bulk systems (for a derivation of the relaxation time approximation in bulk systems, see for example [112]). First consider the system to be stationary, so that the time derivative in Eq. (2.113) is zero, and bring the streaming terms involving gradients in phase space to the left side of the BTE. If the system is not driven far from equilibrium, as in the case of measuring the low-field mobility, the collision integral (2.114) may be written in terms of a relaxation time for the ith subband as

$$\left. \frac{\partial f_i}{\partial t} \right|_{collisions} = -\frac{f_i(\mathbf{k}) - f^0(E)}{\tau_i}, \qquad (2.115)$$

where $f^0(E)$ is the equilibrium Fermi–Dirac function, and τ_i is the relaxation time. The energy E is understood to be the total energy for a particle in a given subband, $E = E_i + E_\mathbf{k}$ where $\mathbf{k}$ is the crystal momentum in subband i. Again for the case of a 1DEG, $\mathbf{k}$ is no longer a vector quantity but rather a scalar along the propagation direction of the system. In the linear response regime, only the lowest-order solution is kept on the LHS of the BTE

$$LHS = \frac{\mathbf{F}}{\hbar} \cdot \nabla_\mathbf{k} f^0(E) = \frac{\mathbf{F}}{\hbar} \cdot \nabla_\mathbf{k} E_\mathbf{k} \frac{\partial f^0}{\partial E} = \mathbf{F} \cdot \mathbf{v_k} \frac{\partial f^0}{\partial E}. \qquad (2.116)$$

The nonequilibrium distribution function may thus be written

$$f_i(\mathbf{k}) = f^0(E) - \tau_i \mathbf{F} \cdot \mathbf{v}_k \frac{\partial f^0}{\partial E} = f^0(E) + f_i^1(\mathbf{k}), \qquad (2.117)$$

where $f_i^1(\mathbf{k})$ represents the perturbation to the equilibrium distribution due to the applied field. To evaluate τ_i, we further restrict ourselves to elastic scattering, so that $S_{i,j}(\mathbf{k}, \mathbf{k}') = S_{j,i}(\mathbf{k}', \mathbf{k})$. Substituting (2.117) into (2.114) and observing that in equilibrium the collision integral must vanish, (2.114) becomes

$$\left. \frac{\partial f_i}{\partial t} \right|_{collisions} = \sum_{j,\mathbf{k}'} S_{i,j}(\mathbf{k}, \mathbf{k}') \left(f_i^1(\mathbf{k}) - f_j^1(\mathbf{k}') \right). \qquad (2.118)$$

Using (2.117), (2.118) may be written

$$\left.\frac{\partial f_i}{\partial t}\right|_{\text{collisions}} = -f_i^1(\mathbf{k}) \sum_{j,\mathbf{k}'} S_{i,j}(\mathbf{k},\mathbf{k}') \left(1 - \frac{\tau_j(E)}{\tau_i(E)} \frac{\mathbf{v_k} \cdot \mathbf{v_{k'}}}{v^2}\right). \tag{2.119}$$

If we assume for simplicity that the states are derived from a spherically symmetric constant energy minimum, then the velocities may be written as $\mathbf{v_k} = \hbar \mathbf{k}/m^*$. Then comparing (2.119) to (2.117) and (2.115), the relaxation time τ_i is given by

$$\frac{1}{\tau_i(E)} = \sum_{j,\mathbf{k}'} S_{i,j}(\mathbf{k},\mathbf{k}') \left(1 - \frac{\tau_j(E)}{\tau_i(E)} \cos\vartheta\right), \tag{2.120}$$

where ϑ is the angle between $\mathbf{k}$ and $\mathbf{k}'$. Again, for a quasi-1D system, the only possible choice of scattering is either forwards or backwards, so that $\cos\vartheta$ is either 1 or -1. As can be seen from Eq. (2.120), the solution for τ_i is coupled to the relaxation times of all the other participating subbands, and thus (2.120) represents a set of simultaneous equations which must be solved for each energy, E. In the case that the intersubband scattering rate is negligible (i.e., $S_{i,j} \cong 0$ for $i \neq j$), or when only one subband is occupied (the extreme quantum limit), a simple momentum relaxation time in closed form may be defined

$$\frac{1}{\tau_i(E)} = \sum_{\mathbf{k}'} S_{i,i}(\mathbf{k},\mathbf{k}')(1 - \cos\vartheta). \tag{2.121}$$

To calculate the mobility, we assume that the system is driven by a small external electric field. Thus for electrons, $\mathbf{F} = -q\mathbf{E}$, where $\mathbf{E}$ is the electric field. The current carried by the ith subband in steady state is determined only by the antisymmetric part of the distribution function, $f_i^1(\mathbf{k})$, given by (2.117). For electrons, the current density is an average over the velocity and particle density,

$$\mathbf{J}_i = -q \int d^d\mathbf{k} N(\mathbf{k}) \mathbf{v_k} f_i^1(\mathbf{k}) \tag{2.122}$$

where d is the dimensionality of the system and $N(\mathbf{k})$ is the density of states per unit volume in k-space,

$$N(\mathbf{k}) = \left(\frac{1}{2\pi}\right)^d. \tag{2.123}$$

The integration over $\mathbf{k}$ in Eq. (2.122) may be converted to an integration over energy as

$$\mathbf{J}_i = -q^2 \int_0^\infty dE D_i(E) \tau_i(E) \mathbf{v_k}(\mathbf{E} \cdot \mathbf{v_k}) \left(\frac{\partial f^0}{\partial E}\right), \tag{2.124}$$

where $D_i(E)$ is the density of states per unit energy of subband i, and the energy $E = E_\mathbf{k}$ is the kinetic energy of the particle in subband i relative to the subband minima. Factoring out the electric field and the density per subband, n_i, Eq. (2.124) becomes

$$\mathbf{J}_i = -q^2 n_i \mathbf{E} \frac{\int_0^\infty dE D_i(E) \tau_i(E) v_E^2 (\partial f^0/\partial E)}{\int_0^\infty dE D_i(E) f^0(E)}, \qquad (2.125)$$

where v_E is the magnitude of the velocity in the direction of the field. By arguing that the drift velocity is much smaller than the thermal velocity for the linear response regime considered here, we can replace component v_E^2 with the total velocity as $v_E^2 = v^2/d = 2E/(m^*d)$, where again d is the dimensionality of the system. Further, since $D_i(E) \propto E^{d/2-1}$, we may write

$$\mathbf{J}_i = -\frac{q^2 n_i \mathbf{E}}{m^*} \frac{\int_0^\infty dE E^{d/2} \tau_i(E) (\partial f^0/\partial E)}{\int_0^\infty dE E^{d/2} (\partial f^0/\partial E)} = qn_i \left(\frac{q\langle \tau_i \rangle}{m^*}\right) \mathbf{E} = qn_i \mu_i \mathbf{E} \qquad (2.126)$$

where integration by parts was used to write the denominator of (2.125) on the same footing as the numerator (which also contributes a factor of $2/d$), and is the subband mobility related to the averaged relaxation time, given by

$$\langle \tau_i \rangle = \frac{\int_0^\infty dE E^{d/2} \tau_i(E) (\partial f^0/\partial E)}{\int_0^\infty dE E^{d/2} (\partial f^0/\partial E)}. \qquad (2.127)$$

The total current is given by the sum over the current of the individual subbands

$$\mathbf{J}_i = q \sum_i n_i \mu_i \mathbf{E} = qn_T \bar{\mu} \mathbf{E}, \qquad (2.128)$$

where n_T is the total density, and the average mobility $\bar{\mu}$ is defined by

$$\bar{\mu} = \left(\sum_i n_i \mu_i\right) \bigg/ n_T. \qquad (2.129)$$

2.7.2 Scattering mechanisms in low-dimensional systems

In the weak coupling limit discussed above, in which the time between collisions is relatively long, the transition rate or scattering rate may be calculated from Fermi's golden rule. The rate is derived from first order using time-dependent perturbation theory (see for example [50]) as

$$S_{i,j}(\mathbf{k},\mathbf{k'}) = \frac{2\pi}{\hbar} |\langle \mathbf{k'},j|V_s(\mathbf{r})|\mathbf{k},i\rangle|^2 \delta(E_{\mathbf{k'}} + E_j - E_{\mathbf{k}} - E_i \mp \hbar\omega_\mathbf{q}), \qquad (2.130)$$

where $V_s(\mathbf{r})$ is the potential associated with a particular scattering mechanism, and $\omega_\mathbf{q}$ is the frequency associated with a harmonic time-dependent perturbation such as the normal modes of the crystal lattice (*phonons*) or a time-varying electromagnetic field, with the upper sign referring to absorption (and the lower sign, emission) of a quasi-particle excitation of the field. As is well known from the standard derivation of Fermi's rule, the delta function in (2.130) is only approximately true as the time between collisions approaches infinity, with an uncertainty in the final energy after the collision that decreases as $\hbar/\tau_c$, where τ_c is the time after the collision.

For many scattering mechanisms, the scattering rate in a low-dimensional system differs from that in 3D only in the different initial and final states, which become increasingly restricted as the dimensionality is reduced. In particular, in a quasi-1D system there are only two possible types of scattering, forward and backward with respect to the wire axis. The density of states tends to be reflected in the scattering rates. As the dimensionality is reduced, the density of states at the subband edge increases. In 1D, the scattering rates can be divergent at low energies because of the singularity in the 1D density of states of the ideal system at the subband minimum, as seen in Section 2.4. New mechanisms that are not present per se in bulk systems, such as surface or boundary roughness scattering, interface states, interface phonons, remote impurities, etc., also must be included. In addition, the nature of the allowed phonon modes themselves are modified by the inhomogeneous structure associated with low-dimensional systems. Folding of the acoustic branches is possible, giving rise to non-zero frequencies for long-wavelength excitations. The optical branches may be confined as well, giving rise to interface modes and "guided" modes, in analogy to electromagnetic waveguide modes. In the following sections, we briefly summarize the important scattering mechanisms in quasi-2D and quasi-1D systems, and later connect this to the phenomenological transport parameters of the homogeneous systems.

We first consider two important scattering mechanisms that limit the mobility in nanostructure systems, particularly at low temperature: Coulomb scattering due to ionized impurities, and surface roughness scattering associated with the unavoidable atomic level fluctuations of the interface between two dissimilar materials, or the process-induced roughness associated with nanoscale lithography or self-assembled growth. Both are treated as *elastic* scattering processes, which means that the energy before and after scattering is the same, only the momentum is changed. This leads to the simplification in Fermi's rule, Eq. (2.130), that $\hbar\omega_\mathbf{q} = 0$, i.e. there is no energy exchange with the scatterer itself.

2.7.2.1 Coulomb scattering

Coulomb scattering is generally associated with ionized impurities in bulk systems, and it is dominant at low temperatures. Additional Coulomb scatterers exist due to the presence of the interface in the form of surface states and fixed charges at or near the interface. In self-assembled nanowires these occur due to the additional bonds on the surface that are dangling, or terminated with hydrogen, an oxide or some other dissimilar material from the semiconductor, giving rise to chargeable states. In CNTs, charge centers may be introduced through phys-absorption of for example oxygen, or chemical functionalization of other organic molecules to the

2.7 Homogeneous transport in low-dimensional systems

tube, giving rise to a change in electronic structure, often in the form of charge centers and associated Coulomb scattering, and significant reduction in conductivity due to the small diameters (a desirable property for sensor applications). In modulation-doped heterojunction systems, the distribution of impurities is inhomogeneous, with high doping occurring in a region spatially separated from the 2DEG. In general, then, we have to consider a variety of distributions of impurities when discussing Coulomb scattering in low-dimensional systems.

The general scattering rate in a 2DEG due to point charges is discussed in detail in [1] with regards to transport in the Si/SiO$_2$ system, which is applicable to the heterojunction case as well. The unscreened potential energy due to a charge located at $\mathbf{r}$ and z_i is given as

$$V_i^0(\mathbf{r},z) = \frac{Z_i e^2}{4\pi\varepsilon_s\left\{(\mathbf{r}-\mathbf{r}_i)^2 + (z-z_i)^2\right\}^{1/2}} + V_{image}, \tag{2.131}$$

where ε_s is the permittivity of the semiconductor and Z_i is the charge state of the impurity. As discussed earlier, the image potential is necessary to satisfy the usual electrostatic boundary conditions at the interfaces between layers of different permittivity. Such conditions may be satisfied by adding additional image charges outside the domain of solution which, combined with the potential of the source charge, satisfy the required boundary conditions. If the system consists of a single interface at $z=0$ between two materials of differing permittivity, and if the impurity lies in the region 1 corresponding to $z_i < 0$ and the scattering particle is in region 2 corresponding to $z > 0$, V_{image} may be accounted for by replacing ε_s with the average dielectric constant, $\bar{\varepsilon} = (\varepsilon_1 + \varepsilon_2)/2$. For an impurity charge in the same layer as the scattering electron (for a single interface) in region 2, the image potential corresponds to a charge located at $-z_i$ as

$$V_{image}(\mathbf{r},z) = \frac{(\varepsilon_1 - \varepsilon_2)Z_i e^2}{4\pi\varepsilon_s(\varepsilon_1+\varepsilon_2)\left\{(\mathbf{r}-\mathbf{r}_i)^2 + (z+z_i)^2\right\}^{1/2}}. \tag{2.132}$$

For the GaAs/AlGaAs system, the difference in permittivity is small, and we can neglect V_{image} in Eq. (2.131). For simplicity we will assume such a system in the following.

Assuming normalized envelope function type solutions for the confined states in the 2DEG system as given in Eq. (2.78), the matrix element for scattering introduces the 2D Fourier transform of the Coulomb potential

$$\frac{1}{4\pi\varepsilon_s\left\{(\mathbf{r}-\mathbf{r}_i)^2 + (z+z_i)^2\right\}^{1/2}} = \int d\mathbf{q}\,\frac{e^{-q(z-z_i)}e^{i\mathbf{q}\cdot(\mathbf{r}-\mathbf{r}_i)}}{2q\varepsilon_s}. \tag{2.133}$$

Therefore, the matrix element for scattering in Eq. (2.130), may be written (neglecting V_{image}) as

$$\langle \mathbf{k}',n|V_i^0(\mathbf{r},z)|\mathbf{k},m\rangle = \frac{Z_i e^2}{2\varepsilon_s q} e^{i\mathbf{q}\cdot\mathbf{r}_i}\left(e^{qz_i}\int_{-\infty}^{\infty}dze^{-qz}\rho_{mn}(z)\right) = V_{i,nm}^0(q,z_i)e^{i\mathbf{q}\cdot\mathbf{r}_i},$$
$$\rho_{mn}(z) = \varphi_n(z)\varphi_m(z),$$
(2.134)

where $\varphi_n(z)$ is the envelope function solution from (2.78), m and n refer to the initial and final subbands, respectively, and $\mathbf{q} = \mathbf{k} - \mathbf{k}'$ is the scattered wavevector with $q = |\mathbf{k} - \mathbf{k}'|$. The matrix element above relates only to the bare potential of the impurity, and does not account for the additional potential due to the polarization of the 2DEG, which is included through the wavevector-dependent dielectric function. As discussed in Section 2.6, the screened potential is a complicated matrix function due to the multisubband nature of the low-dimensional system. Under the simplifying assumptions discussed in Section 2.6.1, the screened potential for intrasubband scattering ($n = m$) may be obtained by dividing (2.134) by the scalar static dielectric function given by (2.83). For intersubband scattering, the screening function is unity to first order.

Equation (2.134) represents the scattering matrix element due to a single impurity. The potential due to all the impurities in the system may be written as

$$V_{nm}^{ii}(q) = \sum_i e^{i\mathbf{q}\cdot\mathbf{r}_j}\frac{V_{i,nm}^0(q,z_i)}{\varepsilon_D(q)}.$$
(2.135)

To calculate the scattering rate from Fermi's rule (2.130), the square of the matrix element is needed. As a first approximation, we assume that in any given plane parallel to the interface the position of the impurities is completely uncorrelated. The cross terms arising from squaring (2.135) then cancel on the average (similar to the random phase approximation), and using (2.130) and (2.135), we can write the elastic scattering rate for impurity scattering as

$$S_{nm}^{ii}(\mathbf{k},\mathbf{k}') = \frac{2\pi}{\hbar}|V_{nm}^{ii}(q)|^2\delta(E_{\mathbf{k}'} - E_{\mathbf{k}} + E_n - E_m)$$
$$= \int_{-\infty}^{\infty}dz_i N_i(z_i)\left|\frac{V_{i,nm}^0(q,z_i)}{\varepsilon_D(q)}\right|^2\delta(E_{\mathbf{k}'} - E_{\mathbf{k}} + E_n - E_m),$$
(2.136)

where $N_i(z_i)$ is the density of impurities per unit volume as a function of z_i. If intersubband scattering is weak, the momentum relaxation time for each subband is given by Eq. (2.121). Writing the matrix element explicitly using (2.134), we get

$$\frac{1}{\tau_n^{ii}(E)} = \frac{\pi Z_i^2 e^4}{2\hbar\varepsilon_s^2}\sum_{\mathbf{k}'}\int_{-\infty}^{\infty}dz_i N_i(z_i)\frac{e^{2qz_i}}{q^2\varepsilon_D^2(q)}F_{nn}^2(q)(1-\cos\vartheta)\delta(E_{\mathbf{k}'} - E_{\mathbf{k}}),$$
(2.137)

where ϑ is the angle between $\mathbf{k}$ and $\mathbf{k}'$, and

$$F_{nn}(q) = \int_{-\infty}^{\infty}dze^{-qz}\rho_{nn}(z),$$
(2.138)

with $\rho_{nn}(z)$ defined as in (2.134). The sum over $\mathbf{k}'$ may be converted to an integral; assuming parabolic bands,

$$\sum_{\mathbf{k}'} \to \frac{1}{2\pi^2} \int_0^{2\pi} d\vartheta \int_0^\infty dk' k' = \frac{m^*}{2\pi^2 \hbar^2} \int_0^{2\pi} d\vartheta \int_0^\infty dE. \qquad (2.139)$$

The delta function over energy can be used to reduce the latter integral, giving $|\mathbf{k}| = |\mathbf{k}'|$, and therefore Eq. (2.137) may be written

$$\frac{1}{\tau_n^{ii}(E)} = \frac{\pi Z_i^2 e^4 m^*}{8\pi^2 \hbar \varepsilon_s^2} \int_0^{2\pi} d\vartheta \int_{-\infty}^\infty dz_i N_i(z_i) \frac{e^{2qz_i}}{q^2 \varepsilon_D^2(q)} F_{nn}^2(q)(1 - \cos\vartheta), \qquad (2.140)$$

where

$$q = |\mathbf{k} - \mathbf{k}'| = \sqrt{k^2 + k'^2 - 2kk' \cos\vartheta} = \sqrt{2k^2(1 - \cos\vartheta)} = 2k \sin\frac{\vartheta}{2}. \qquad (2.141)$$

The momentum relaxation time (2.140) is written assuming one type of impurity. For multiple types of impurity distributions, for example from interface, remote impurities, and bulk impurities, the momentum relaxation time may be written as

$$\frac{1}{\tau_n^{ii}(E)} = \frac{1}{\tau_{1,n}^{ii}(E)} + \frac{1}{\tau_{2,n}^{ii}(E)} + \cdots, \qquad (2.142)$$

where (2.140) is used to calculate the relaxation time for each individual impurity distribution.

Further simplification of (2.140) is possible if we consider impurity scattering due to a 2D sheet of impurities, $N_i(z_i) = N_{ss}\delta(z_i)$ at the interface $z_i = 0$, which may for example exist due to interface states at the hetero- or oxide–semiconductor interface. If the 2DEG is assumed to be extremely narrow, and we neglect screening altogether ($\varepsilon_D(q) = 1$), then (2.140) reduces to

$$\frac{1}{\tau^{ii}(E)} = \frac{Z_i e^4 N_{ss}}{8\hbar \varepsilon_s^2 E}. \qquad (2.143)$$

The mobility limited by interface impurity scattering under the above approximations is calculated using the average (2.127). At low temperatures, $E \to E_f$; at high temperatures, the average (2.127) results in the substitution $E \to k_B T$ which gives the mobility

$$\mu_{ii} = \frac{8\hbar \varepsilon_s^2 k_b T}{Z_i^2 e^3 m^* N_{ss}}. \qquad (2.144)$$

This equation shows that the impurity limited mobility decreases as $1/N_{ss}$ and increases with increasing temperature. In reality, the inclusion of screening through the temperature-dependent dielectric function (2.92) gives a more complicated temperature dependence, since screening tends to decrease with increasing temperature, resulting in an increased scattering rate from consideration of screening alone.

If we wish to consider scattering in a quantum wire due to an impurity, the Coulomb matrix element for scattering is calculated between the quasi-one-dimensional electronic states labeled $\varphi_n(\mathbf{r})e^{ik_x x}$, where x is the longitudinal direction of the wire. For an impurity located at point x_i, y_i, z_i, the unscreened matrix element for scattering from an electron initially with wavevector k_x in subband m is written

$$H(q_x) = \left|\langle k'_x, n | V_i^0(\mathbf{r}, x) | k_x, m\rangle\right|^2$$

$$= \frac{e^2}{4\pi\varepsilon_s} \int dz \int dy \varphi_n^*(y,z)\varphi_m(y,z) \frac{1}{L} \int dx \frac{e^{-i(k_x - k'_x)x}}{\sqrt{(x-x_i)^2 + (y-y_i)^2 + (z-z_i)^2}} \quad (2.145)$$

$$= \frac{e^2 e^{-iq_x x_i}}{2\pi\varepsilon_s L} \int dz \int dy \varphi_n^*(y,z)\varphi_m(y,z) K_0\left(|q_x|\sqrt{(y-y_i)^2 + (z-z_i)^2}\right),$$

where $q_x = k_x - k'_x$, L is the normalization length of the 1D wire, and K_0 is the zeroth-order modified Bessel function. The total scattering rate is calculated from a sum over all final states

$$S_m^{ii}(k_x) = \frac{2\pi}{\hbar} \sum_n \sum_{k_x} |H(q_x)|^2 \delta\left(\frac{\hbar^2 k'^2_x}{2m_n} + E_n - \frac{\hbar^2 k_x^2}{2m_m} - E_m\right), \quad (2.146)$$

where $m_{m,n}$ refer to masses in the individual subbands m and n, and $E_{m,n}$ are the respective 1D subband minima. In 1D, the delta function only allows two possible solutions for the final wavevector and subband

$$k'_x = \pm\sqrt{k_x^2 + \frac{2m_m}{\hbar^2}(E_m - E_n)} \quad (2.147)$$

Assuming we have n_i impurities per unit length uniformly distributed along the wire, the total scattering rate may therefore be written [127]

$$S_m^{ii}(k_x) = \frac{e^4 n_i}{4\pi^2 \varepsilon_s^2 \hbar^3} \sum_n \left[\frac{m_n}{|k'_x|}\left(|H(k_x - k'_x)|^2 + |H(k_x + k'_x)|^2\right)\right]. \quad (2.148)$$

In the case that the initial and final subbands are the same, only one scattering process is possible that changes the momentum of the particle, that is complete backscattering from k_x to $-k_x$ (the second term above). If we want to include screening, then we need to replace $H(q_x) \to H(q_x)/\varepsilon(q_x)$ where ε is the one-dimensional dielectric function defined in Section 2.6.2.

2.7.2.2 Surface roughness

Surface roughness is a term generically applied to the random fluctuations of the boundaries that nominally form the confining potential to low-dimensional systems. On a microscopic level, roughness appears as atomic layer steps in the interface between two differing materials, even those which are lattice matched. As evidenced by the high-resolution transmission electron microscopy

2.7 Homogeneous transport in low-dimensional systems

(HRTEM) image shown in Fig. 2.1, the extent of the roughness fluctuations is only one or two monolayers at the growth interface between two materials. In contrast, the roughness evident in lateral patterning using present-day lithographic techniques is at least an order of magnitude larger due to the "Neolithic" nature of the pattern transfer process which usually involves etching of some sort. Even in self-assembled semiconductor nanowires, roughness may be considerably more pronounced than in an epitaxially grown interface, as shown in Fig. 2.3. As a consequence, while roughness scattering may be relatively weak in heterojunction 2DEG systems, in 1D wires it is still a dominant effect.

To date, a truly microscopic model of interface roughness scattering in semiconductor nanostructures has not been attempted. Instead, most treatments of roughness scattering assume that the fluctuations of the interface from its ideal flat boundary are described by a two-dimensional roughness function, $\Delta(\mathbf{r})$, where as before, $\mathbf{r}$ is the two-dimensional position vector in the plane of the interface, illustrated in Fig. 2.29. The potential associated with the roughness $\Delta(\mathbf{r})$ can be viewed as a combination of a boundary perturbation which causes the envelope functions to be displaced from their unperturbed values, and electrostatic contributions due to the imposed fluctuation of the electric fields and charge density at the rough interface. The combined effect on a two-dimensional system has been considered in detail elsewhere [128].

For simplicity, assume that in a two-dimensional system the potential may be expanded as

$$V^{sr}_{nm}(\mathbf{r},z) = V_{eff}(\mathbf{r},z+\Delta(\mathbf{r})) - V_{eff}(z) \approx \Delta(\mathbf{r})\frac{\partial V_{eff}(z)}{\partial z}, \qquad (2.149)$$

where V_{eff} is the one-dimensional potential including many-body and image potential corrections. The scattering matrix element of the roughness potential (2.137) is

$$V^{sr}_{nm}(\mathbf{k}-\mathbf{k}') = \int d\mathbf{r} e^{-i(\mathbf{k}-\mathbf{k}')\mathbf{r}}\Delta(\mathbf{r})\int_{-\infty}^{\infty}dz\varphi^*_m(z)\frac{\partial V_{eff}(z)}{\partial z}\varphi_n(z) = e\Delta(\mathbf{q})F^{nm}_{avg}, \qquad (2.150)$$

where F^{nm}_{avg} is the average surface field defined from the z integration above, and the two-dimensional Fourier transform of the roughness function has been introduced

$$\Delta(\mathbf{q}) = \int d\mathbf{r}\Delta(r)e^{i\mathbf{q}\cdot\mathbf{r}}, \qquad (2.151)$$

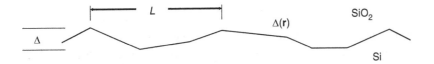

Fig. 2.29 Schematic representation of the surface roughness function, including the rms height and autocovariance length L.

where $\mathbf{q} = \mathbf{k} - \mathbf{k}'$. The intraband momentum relaxation time or scattering rate is again calculated from Fermi's rule,

$$\frac{1}{\tau_{sr}^{nn}(E)} = \frac{2\pi}{\hbar} \left|F_{avg}^{nn}\right|^2 \sum_{\mathbf{k}'} S(\mathbf{q})(1 - \cos\vartheta) \left|\frac{\Gamma(q)}{\varepsilon_D(q)}\right|^2 \delta(E_{\mathbf{k}'} - E_{\mathbf{k}}), \qquad (2.152)$$

where the scalar static screening function has been introduced, and where $\Gamma(\mathbf{q})$ represents corrections for image potential and electric field modification at the deformed interface [128] which are on the order of unity. The function $S(\mathbf{q}) = |\Delta(\mathbf{q})|^2$ is the power spectrum of the roughness fluctuations, which depends only on the magnitude of the fluctuations, and thus the phase information can be neglected. As before, the sum over final states can be converted to an integral, and the delta function can be used to reduce the integration of the magnitude of $\mathbf{k}'$, again assuming parabolic bands, to give

$$\frac{1}{\tau_{sr}^{nn}(E)} = \frac{e^2 \left|F_{avg}^{nn}\right|^2 m^*}{2\pi\hbar^3} \int_0^{2\pi} d\vartheta S(\mathbf{q})(1 - \cos\vartheta) \left|\frac{\Gamma(q)}{\varepsilon_D(q)}\right|^2, \quad q = 2k\sin(\vartheta/2). \qquad (2.153)$$

The statistical properties of the roughness function $\Delta(\mathbf{r})$ are contained in the autocovarance function

$$C(\mathbf{r}) = \langle \Delta(\mathbf{r}')\Delta(\mathbf{r}' - \mathbf{r}) \rangle, \qquad (2.154)$$

where the brackets denote the ensemble average of the random variable $\Delta(\mathbf{r})$. The autocovariance function measures the probability that given a certain value of the roughness function at $\mathbf{r}'$, the function has the same value at $\mathbf{r}' - \mathbf{r}$. It is apparent that this probability should decay in some fashion as the distance $\mathbf{r}' - \mathbf{r}$ increases due to the random nature of $\Delta(\mathbf{r})$. In early works, it was usual to choose a Gaussian decay for the autocovariance function (2.154) as

$$C(\mathbf{r}) \approx \Delta^2 e^{-r^2/L^2}, \qquad (2.155)$$

where Δ is the rms value of the roughness fluctuations and L is the autocovariance length which roughly (no pun intended) may be interpreted as the mean distance between "bumps" along the surface as shown in Fig. 2.29. The main justification for the use of a Gaussian autocovariance function is the rather simple mathematical connection to the power spectrum, $S(\mathbf{q})$. By the Wiener–Kitchine theorem, the Fourier transform of the autocovariance function of a random variable is the power spectrum, $S(\mathbf{q})$. The Gaussian function has the desirable property that the Fourier transform of a Gaussian is again a Gaussian, which in 2D gives

$$S(\mathbf{q}) = \pi \Delta^2 L^2 e^{-q^2 L^2/4} \qquad (2.156)$$

which may be substituted into (2.153) to facilitate evaluation of the scattering rate. However, as was shown by Goodnick et al. [129], the actual autocovariance function is well fit by an exponential rather than Gaussian autocovariance function. Figure 2.30 shows the estimated power spectrum associated with the

2.7 Homogeneous transport in low-dimensional systems

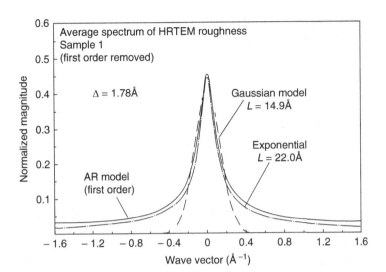

Fig. 2.30 Power spectrum corresponding to the 1D roughness function of the Si/SiO$_2$ interface from HRTEM measurements (after Goodnick *et al.* [129]).

HRTEM images of the Si/SiO$_2$ interface. The power spectrum has been fit assuming a Gaussian autocovariance function and an exponential autocovariance function, where the latter is clearly a much better representation of the power spectrum. The roughness measured in HRTEM is always a 1D roughness rather than the true 2D roughness function, and thus consideration must be given to the actual form of the 2D roughness estimated from measurement of the 1D function. For an exponential model, the autocovariance function is given by

$$C(\mathbf{r}) \approx \Delta^2 e^{-\sqrt{2}\, r/L}, \tag{2.157}$$

where the parameters Δ and L have the same meaning as the Gaussian model. The fact that an exponential model provides a better fit to the autocovariance function is perhaps no surprise as an exponential autocovariance is characteristic of a Markov process, which in the analysis of stochastic processes is the lowest-order model for a random process. The 1D transform corresponding to the fit in Fig. 2.30 is a Lorentzian function, whereas if an isotropic roughness in 2D is assumed, the power spectrum in 2D is given by the 2D transform of (2.157) as

$$S(\mathbf{q}) = \frac{\pi \Delta^2 L^2}{\left(1 + (q^2 L^2/2)\right)^{3/2}}. \tag{2.158}$$

We will revisit the topic of surface roughness scattering in Chapter 9 in connection with nonequilibrium transport in inversion layers, in which the influence of the different roughness power spectrum models is compared in a more formal manner.

Surface roughness scattering in 1D wires has also been treated in a similar fashion in terms of a parameterized roughness power spectrum associated with the different boundaries [130]. In that work a Gaussian autocovariance was employed. As mentioned earlier, the roughness in lateral boundaries defined by present-day lithographic techniques typically results in rms heights Δ which are an order of magnitude larger than those found at the oxide–semiconductor or heterointerface. The effect on the calculated mobility in 1D can therefore be severe and limit this quantity even at room temperature.

2.7.2.3 Lattice scattering

Phonons themselves are associated with the propagating coupled vibrational modes of the individual atoms of the crystal lattice, which when Fourier transformed to so-called "normal coordinates," form an independent collection of harmonic oscillators, one for each mode $\mathbf{q}$ and branch μ. For the semiconductors of interest in this book, the two-atom basis of the diamond or zincblende crystal lattice results in three acoustic branches and three optical branches. The three branches roughly correspond to one longitudinal and two transverse modes relative to the direction of propagation of the lattice wave. Acoustic phonons are associated with the low-lying vibrational modes of the lattice whose energy $\hbar\omega_\mathbf{q}$ goes to zero as the wavevector $\mathbf{q}$ goes to zero. The dispersion relation is linear close to $\mathbf{q}=0$, with group velocity v_s, which is the sound velocity in the crystal. On the other hand, the optical modes go to a non-zero value, ω_{lo} or ω_{to}, as $\mathbf{q}=0$ associated with the longitudinal optical or transverse optical modes. The optical modes are almost dispersionless which means zero group velocity. Thus it is convenient to consider optical phonons as interacting with electrons and holes at one frequency, ω_{lo} or ω_{to}, exchanging fixed quanta of energy $\hbar\omega_0$.

As mentioned above, the modes of vibrations in normal coordinates correspond to independent harmonic oscillators, each with energy

$$E_{\mathbf{q},\mu} = \hbar\omega_{\mathbf{q},\mu}\left(n_{\mathbf{q},\mu} + \frac{1}{2}\right), \quad (2.159)$$
$$n_{\mathbf{q},\mu} = 0, 1, 2, \ldots$$

The excitation state index, $n_{\mathbf{q},\mu}$, is given the physical picture of representing the number of quasi-particles called phonons in each mode $\mathbf{q}$ and μ. Thus, the total energy of a particular mode is the zero-point energy plus $\hbar\omega_{\mathbf{q},\mu} n_{\mathbf{q},\mu}$, the number of phonons multiplied by the energy per phonon. The excitation state index, $n_{\mathbf{q},\mu}$, can be greater than 1, as the phonons are a system of bosons, obeying Bose–Einstein statistics in equilibrium

$$n_{\mathbf{q},\mu} = \frac{1}{e^{\hbar\omega_{q,\mu}/k_B T} - 1}. \quad (2.160)$$

When we discuss phonons in nanostructures, we have to pay attention to the fact that the modes themselves are modified by the geometry. In particular, lattice vibrational modes may be localized in the various layers existing in the system, as well as forming surface or interface modes at the boundaries between different regions. In general, microscopic calculations are necessary to truly account for the behavior of lattice vibrations in general nanoscale structures, although such methods are usually expensive computationally. Some features are obvious, however, from experimental studies as well as simple analytical models. One effect is folding or confinement of the acoustic modes due to breaking of the translational invariance in a nanostructure. Due to this folding, modes with frequency not equal to zero at $\mathbf{q} = 0$ are possible. Evidence for zone-folded acoustic phonons is readily apparent from inelastic light scattering experiments [131]. For optical phonons in polar materials, it is possible to calculate the possible modes in a quantum confined structure from electrostatic considerations using the dielectric continuum model (see [132] and references therein) by solving Laplace's equation in different regions and applying the appropriate electrostatic boundary conditions at the interfaces. This model has been found to compare well to first principles calculation of the lattice displacements in quantum well structures [133]. The main picture that emerges from this model, at least for a simple quantum well system, is that two types of modes appear. One type of mode corresponds to guided wave modes existing in the layer of quantum confined structures, sometimes referred to as slab modes. A second type of solution is associated with surface or interface modes, which decay spatially away from an interface. The interaction of electrons with all these possible modes is rather detailed, and significant effects due to the confinement of the phonon modes do not occur until very small dimensions are reached. Therefore, for simplicity, we assume in the following that electrons interact with bulk-like modes unless otherwise stated.

2.7.2.4 Acoustic phonons

The interaction of acoustic modes with electrons and holes in semiconductors is usually treated using the deformation potential Ansatz. Here, the local deformation of the crystal lattice due to an acoustic vibration is associated with a shift of the band edges, which is the acoustic deformation potential. For a spherical constant energy surface, the square of the matrix element for scattering in a quasi-2DEG system may be written [134,135]

$$|\langle \mathbf{k}', n | V^{ac} | \mathbf{k}, m \rangle|^2 = \sum_{q_z} \frac{\hbar D_{ac}^2 q^2}{2\omega_q \rho V} \left(n_q + \frac{1}{2} \mp \frac{1}{2} \right) \delta_{q_t, \mathbf{k}-\mathbf{k}'} |G_{nm}(q_z)|^2, \quad (2.161)$$

where D_{ac} is the acoustic deformation potential, the upper sign is for absorption and the lower sign emission of a phonon, and $q^2 = q_t^2 + q_z^2$ where z is the direction perpendicular to the interface. The form factor is given by

$$G_{nm}(q_z) = \int_{-\infty}^{\infty} dz \varphi_n^*(z) \varphi_m(z) e^{\mp i q_z z} = \int_{-\infty}^{\infty} dz \rho_{nm}(z) e^{\mp i q_z z}. \quad (2.162)$$

For longitudinal acoustic modes in an isotropic system, $\omega_\mathbf{q} = u_l q$ where u_l is the longitudinal sound velocity. If we assume high temperatures, the equipartition limit of (2.160) may be used:

$$n_q \approx n_q + 1 \approx \frac{k_B T}{\hbar \omega_q}. \tag{2.163}$$

With these simplifications, the matrix element squared may be written as

$$|\langle \mathbf{k}', n | V^{ac} | \mathbf{k}, m \rangle|^2 = \sum_{q_z} \frac{D_{ac}^2 k_B T}{u_l^2 \rho V} \int_{-\infty}^{\infty} dz \rho_{nm}(z) \int_{-\infty}^{\infty} dz' \rho_{nm}(z') e^{\mp i q_z (z-z')} dz. \tag{2.164}$$

The sum over q_z may be converted to an integral

$$\sum_{q_z} \to \frac{L}{2\pi} \int_{-\infty}^{\infty} dq_z e^{\mp i q_z (z-z')} = L\delta(z-z'), \tag{2.165}$$

where L is the length of the system in the z-direction, which reduces (2.164) to

$$|\langle \mathbf{k}', n | V^{ac} | \mathbf{k}, m \rangle|^2 = \frac{D_{ac}^2 k_B T I_{ac}^{nm}}{u_l^2 \rho A}. \tag{2.166}$$

The overlap factor is given by

$$I_{ac}^{nm} = \int_{-\infty}^{\infty} dz \rho_{nm}^2(z), \tag{2.167}$$

which for an infinite square well of width W is easily integrated to give

$$\begin{aligned} I_{ac}^{nm} &= \frac{3}{2W}, \quad n = m \\ &= \frac{1}{W}, \quad n \neq m. \end{aligned} \tag{2.168}$$

Thus, the form factor for intrasubband scattering for the acoustic phonons is only 1.5 times that of the intersubband rate. To complete the derivation of the intrasubband relaxation time, use Fermi's rule and sum over final states as before:

$$\frac{1}{\tau_{ac}^{nm}} = \frac{2\pi}{\hbar} \frac{D_{ac}^2 k_B T I_{ac}^{nm}}{u_l^2 \rho A} \sum_{\mathbf{k}'} \delta(E_{\mathbf{k}'} - E_{\mathbf{k}})(1 - \cos \vartheta). \tag{2.169}$$

In the energy-conserving delta function, the energy of the acoustic phonon has been assumed small compared to the initial and final energies since the energy of the phonon goes to zero for long-wavelength phonons. In this approximation then, acoustic phonon scattering is quasi-elastic. The scattering rate is independent of angle; hence the factor $1 - \cos \vartheta$ may be removed, and therefore the momentum relaxation time is equal to the total inverse scattering rate. The sum over the delta function just gives the density of states of the final subband times the area of the 2D system, and thus (2.169) becomes

$$\frac{1}{\tau_{ac}^{mm}} = \frac{D_{ac}^2 k_B T m^* I_{ac}^{mm}}{u_l^2 \rho \hbar^3}. \tag{2.170}$$

Since (2.170) is independent of energy, the average relaxation time is the same as the above expression. The scattering rate increases with increasing

2.7 Homogeneous transport in low-dimensional systems

lattice temperature due to the phonon occupancy factors which gave rise to the factor $k_B T$ above. The overlap integral decreases as the width of the system increases, decreasing the scattering rate. Of course, as the width of the well increases, the number of available subbands within a given energy increases, increasing the total scattering rate due to intra- and intersubband scattering, which approaches the bulk acoustic phonon scattering rate as the width becomes infinite.

Another form of interaction with acoustic phonons in polar materials is through the piezoelectric interaction. This mechanism is often found to be important at intermediate temperatures in bulk materials before the onset of strong scattering due to polar optical phonons at higher temperatures as discussed in the next section. Piezoelectric scattering in 2DEG structures has been considered for example by Price [136].

As before with impurity scattering, for 1D phonon scattering within Fermi's golden rule, the only modification is to account for the transverse modes of the quantum wire, with the states again assumed of the form $\varphi_n(y,z)e^{ik_x x}$. The 1D matrix element for acoustic mode scattering may be written [137]

$$|\langle k'_x, n|V^{ac}|k, m\rangle|^2 = \sum_{\mathbf{q}_t} \frac{\hbar D_{ac}^2 q^2}{2\omega_q \rho L}\left(n_q + \frac{1}{2} \mp \frac{1}{2}\right)\delta_{\mathbf{q}_t, k_x - k'_x}|I_{nm}(q_y, q_z)|^2, \qquad (2.171)$$

where the variables have the same meaning as in Eq. (2.161). The overlap integral is now given by

$$I_{nm}(q_y, q_z) = \iint dy\,dz\, \varphi_n(y,z)\varphi_m(y,z)e^{i(q_y y + q_z z)}. \qquad (2.172)$$

Assuming the equipartition limit again, and assuming a linear phonon dispersion for acoustic phonons near the zone center, Eq. (2.171) reduces to

$$|\langle k'_x, n|V^{ac}|k, m\rangle|^2 = \sum_{\mathbf{q}_t} \frac{D_{ac}^2 k_B T}{u_l^2 \rho L}\delta_{\mathbf{q}_t, k_x - k'_x}|I_{nm}(q_y, q_z)|^2. \qquad (2.173)$$

Substituting the 1D acoustic scattering matrix into Fermi's golden rule in the elastic scattering limit results in

$$\frac{1}{\tau_{ac}^{nm}} = \Gamma_{nm}(k_x) = \frac{2\pi}{\hbar}\frac{D_{ac}^2 k_B T}{u_l^2 \rho L}\sum_{\mathbf{q}_t, k'_x}|I_{nm}(q_y, q_z)|^2 \delta(E_{k'_x} + E_n - E_{k_x} - E_m), \qquad (2.174)$$

where $\Gamma_{nm}(k_x)$ is the scattering rate. Converting the sums to integrals, the total scattering rate becomes

$$\Gamma_{nm}(k_x) = \frac{D_{ac}^2 k_B T \sqrt{m^*}}{\sqrt{2}u_l^2 \rho}D_{nm}\frac{1}{\sqrt{E_{k_x} - \Delta_{nm}}}\Theta(E_{k_x} - \Delta_{nm}), \qquad (2.175)$$

where $\Delta_{nm} = E_n - E_m$ is the difference in 1D subband energies, and the confinement factor, D_{nm}, is given by

$$D_{nm} = \frac{1}{4\pi^2} \iint dq_y dq_x |I_{nm}(q_y, q_z)|^2 = \iint dy dz |\varphi_n(y,z)|^2 |\varphi_m(y,z)|^2. \qquad (2.176)$$

There are two main observations about the form of the 1D scattering rate above. The first is that the scattering rate goes as $E_{k_x}^{-1/2}$, which diverges as the energy goes to zero, which mirrors the 1D density of states. The scattering rate reflects the final density of states, and hence the dimensionality of the system, as expected. Second, the scattering rate depends on the confinement factor given by Eq. (2.176); the stronger the confinement, the higher the scattering rate. This tends to increase the scattering as the dimensions of the nanowire dimensions are reduced.

Figure 2.31 shows the calculated acoustic phonon scattering rate versus energy for electrons in the lowest subband due to both bulk phonons (dashed curve) from Eq. (2.175), and for confined phonons (solid curve) for an 8 nm × 8 nm rectangular Si nanowire [137]. The bulk phonon rate shows a series of spikes due to the 1D density of states and its effect on scattering, as each successive 1D subband is crossed in energy. Confinement of the acoustic modes results in a zone folding of the acoustic phonon dispersion, and therefore a series of guided modes with non-zero energy at the zone center. The electrons interact with long-wavelength phonons with a spectrum of confined energies, giving rise to the series of peaks shown, centered around the bulk peak at the subband edge. One set corresponds to absorption, the other to emission, due to the finite energy of the confined phonon modes at the zone center.

2.7.2.5 Polar optical phonons

For III–V compound materials such as GaAs and AlGaAs, the dominant energy relaxation mechanism for electrons in the central valley below the Γ–L threshold is that of polar-optical-phonon (POP) emission. This mechanism is usually treated within the continuum model described by the Fröhlich interaction relating the polarization field associated with the charge separation between cation and anion to the scattering potential due to the optical-mode displacement. For bulk-like polar optical phonons, the scattering rate for an electron initially in subband n to a final subband m may be written [138, 139, 140]

$$S_{nm}^{pop}(k) = \frac{eE_0}{2\hbar} \left[\left(n_{\omega_{lo}} + \frac{1}{2} \mp \frac{1}{2} \right) \int_0^\infty d\vartheta \frac{H_{nm}(q_\pm)}{q_\pm} \right], \qquad (2.177)$$

where $n_{\omega_{lo}}$ is the longitudinal optical phonon occupancy, $q_\pm$ is the scattered wavevector in the plane parallel to the interface, and the upper and lower signs correspond to phonon absorption and emission, respectively. The effective field eE_0, is given by

$$eE_0 = \frac{m^* e^2 \hbar \omega_{lo}}{4\pi \hbar^2} \left(\frac{1}{\varepsilon_\infty} - \frac{1}{\varepsilon_0} \right), \qquad (2.178)$$

where ε_0 and ε_∞ are the low- and high-frequency permittivities of the material. The function $H_{nm}(q_\pm)$ is given by

2.7 Homogeneous transport in low-dimensional systems

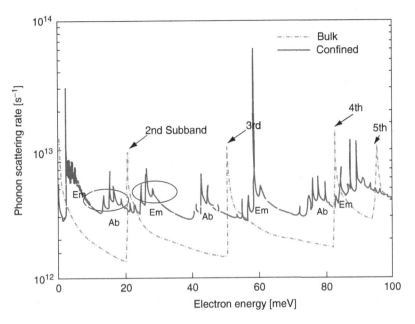

Fig. 2.31 Electron-acoustic phonon scattering rate for the lowest electron subband of an 8 × 8 nm Si nanowire, calculated assuming bulk (dashed) and confined (solid) phonons. (After Ramayya et al. [137].)

$$H_{nm}(q\pm) = \int_{-\infty}^{\infty} dz \rho_{nm}(z) \int_{-\infty}^{\infty} dz' \rho_{nm}^*(z') e^{-q_\pm |z-z'|}, \qquad (2.179)$$

where the scattered wavevector is fixed by energy conservation to be

$$q_\pm = |\mathbf{k} - \mathbf{k}'| = \left[2k^2 \pm \frac{2\omega_{nm}^* m^*}{\hbar} - 2k \left(k^2 \pm \frac{2\omega_{nm}^* m^*}{\hbar} \right)^{1/2} \cos\vartheta \right]^{1/2}, \qquad (2.180)$$

$$\hbar\omega_{nm}^* = \hbar\omega_{lo} \pm (E_n - E_m),$$

where $\hbar\omega_{nm}^*$ represents an effective phonon energy, which may be zero or negative. For intersubband scattering, the rate given by (2.177) is a maximum when $q_\pm$ in the denominator is a minimum, which occurs when $\hbar\omega_{lo} = E_n - E_m$, resulting in a resonance which may be observed experimentally.

Figure 2.32 shows the calculated LO phonon scattering rate versus energy in a 15-nm GaAs quantum well. In comparison to the 3D rate, the 2D rate shows a much sharper emission threshold, and sharp discontinuities corresponding to the onset of emission and absorption for higher subbands. The sharpness of the onsets is related to the discontinuous density of states of a 2D versus 3D system. For quantum wires, these discontinuities are even more pronounced due to the divergent density of states.

Figure 2.33 shows the calculated intersubband scattering rates from the second subband to the first subband due to LO phonons, acoustic phonons, and impurity scattering in a 27-nm wide GaAs well. For low temperatures, the acoustic rate cannot be calculated using the equipartition approximation. The rate shown in Fig. 2.33 is calculated numerically directly from the matrix element (2.169) without assuming elastic scattering. As can be seen, the LO phonon emission rate

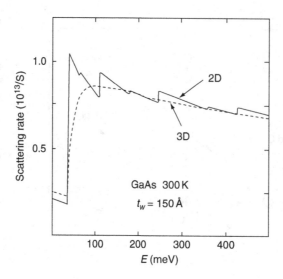

Fig. 2.32 Total scattering rate due to LO phonon scattering at 300 K for a 15-nm GaAs quantum well for electrons in the first subband (solid line). The dashed line corresponds to the 3D scattering rate. (After Goodnick and Lugli, *Phys. Rev. B* **37**, 2578 (1988).)

is much larger than that due to acoustic phonons. However, since the subband spacing in this case (14 meV) is less than the optical phonon energy, electrons below this threshold cannot emit to the ground subband, thus leading to a bottleneck in the intersubband transition rate governed by the long-time acoustic phonon rather than LO phonon rate [141]. With impurities present, however, another channel is present for electrons to scatter from the higher to lower subband.

Various authors have considered the corresponding 1D polar optical phonon scattering rates, for example Jovanovic and Leburton [142]. The rates are essentially calculated from the Fröhlich interaction discussed above, with the matrix elements taken between the initial and final quasi-1D envelope functions. The phonons in these studies are taken as bulk-like while the electronic states are confined. The calculated rates for acoustic, piezoelectric, and polar optical phonons as a function of energy are shown in Fig. 2.34. As expected, once the threshold for optical phonon emission is exceeded, the polar optical rate is largest. The 1D density of states leads to the peaked behavior in the scattering rate when the energy corresponds to one of the subband minima of the multi-subband structure. These rates were used in a Monte Carlo simulation of transport in quasi-1D systems.

2.7.2.6 Optical deformation potential and intervalley scattering

In Si, the nonpolar optical deformation potential is the predominant interaction of carriers with optical phonons. This may occur through intravalley scattering, or via intervalley scattering between equivalent minima. For the case of GaAs, the intravalley scattering contribution is forbidden by symmetry for zero-order scattering. However, as carriers are heated, or in cases of large carrier

2.7 Homogeneous transport in low-dimensional systems

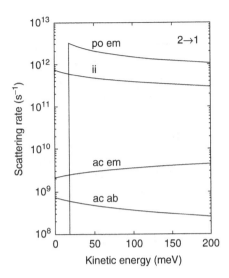

Fig. 2.33 The total scattering rates for electrons from the second to the first subband due to polar optical and acoustic phonon modes (both emission and absorption), and ionized impurities ($n_i = 1.5 \times 10^{11}$ cm^{-1}) for a 27-nm-wide GaAs well at 10 K.

confinement so that the satellite valleys are accessible, the intervalley scattering mechanism is important.

For 2D electrons, if we assume that the states are quantized for all valleys, we may write the scattering rate for nonpolar optical deformation scattering as [140]

$$S_{nm}^0 = \sum_{vf} m_f^* \frac{E_0^2 \left(n_{\omega_0} + \frac{1}{2} \mp \frac{1}{2}\right) I_{nm}^0}{2\rho \omega_0 \hbar^2}, \qquad (2.181)$$

where the sum is over the final valleys if intervalley scattering is considered, m_f^* is the mass of electrons in the final valley, and ω_0 is the relevant phonon frequency. For simple nonpolar optical scattering, ω_0 is either the longitudinal optical or transverse optical frequency. For intervalley scattering, this frequency corresponds to the wavevector of the phonon that couples one valley to another, and may correspond to several different values. In (2.181) E_0 is the optical deformation potential in the case of intervalley scattering, whereas it is the intervalley deformation potential for intervalley scattering. The overlap integral in (2.175) is given by

$$I_{nm}^0 = \int_{-\infty}^{\infty} dz |\varphi_{v_i,n}(z)|^2 |\varphi_{v_f,m}(z)|^2, \qquad (2.182)$$

where v_i and v_f refer to the initial and final valley envelope functions.

For intervalley scattering in 1D, the matrix element for optical intervalley rate may be written in a similar fashion to the acoustic phonon rate

$$|\langle k_x', n, v'|V^{iv}|k, m, v\rangle|^2 = \frac{\hbar E_0^2}{2\omega_0 \rho V}\left(n_{\omega_0} + \frac{1}{2} \mp \frac{1}{2}\right) \delta_{\mathbf{q}_t, k_x - k_x'} |I_{nm}^{v'v}(q_y, q_z)|^2, \qquad (2.183)$$

where v and v' label the initial and final valleys, and $I_{nm}^{v'v}(q_y, q_z)$ has the same definition as Eq. (2.172), except generalized to different initial and final valleys

Fig. 2.34 Scattering rates for polar optical, acoustic, and unscreened piezoelectric phonon scattering at 30 K relative to the subband minimum. (After Jovanovic and Leburton, from *Monte Carlo Device Simulation: Full Band and Beyond*, ed. K. Hess (Boston, Kluwer Academic Publishers, 1991), by permission.)

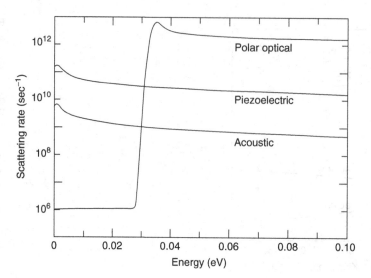

$$I_{nm}^{v'v}(q_y, q_z) = \iint dy dz \varphi_n^{v'}(y,z) \varphi_m^v(y,z) e^{i(q_y y + q_z z)}. \quad (2.184)$$

Following the same procedure as for the acoustic rate, and accounting for the inelastic nature of optical phonon scattering, the optical phonon scattering rate may be written

$$\Gamma_{nm}^{v'v}(k_x) = \frac{E_0^2 \sqrt{m^*}}{2\sqrt{2}\hbar\rho\omega_0} D_{nm}^{v'v}\left(n_{\omega_0} + \frac{1}{2} \mp \frac{1}{2}\right) \frac{1}{\sqrt{E_{k_x} - \Delta_{nm}^{v'v} \pm \hbar\omega_0}} \Theta\left(E_{k_x} - \Delta_{nm}^{v'v}\right), \quad (2.185)$$

where $\Delta_{nm}^{v'v} = E_n - E_m + E_{v'} - E_v$ is the difference in 1D subband energies and intervalley separation, and the intervalley confinement factor, $D_{nm}^{v'v}$, is given by

$$D_{nm}^{v'v} = \frac{1}{4\pi^2} \iint dq_y dq_x |I_{nm}^{v'v}(q_y,q_z)|^2 = \iint dy dz |\varphi_n^{v'}(y,z)|^2 |\varphi_m^v(y,z)|^2. \quad (2.186)$$

2.7.3 Experimental mobility in 2D and 1D systems

2.7.3.1 Mobility in 2DEG heterostructures

As mentioned in Section 2.1, the inclusion of a spacer layer separates the 2DEG from the heavily doped AlGaAs region. This gives rise to an exponential attenuation of the scattering rate for a given q, as seen in (2.140), since z_i is negative for an impurity located in the AlGaAs layer, with the interface located at $z = 0$. Figure 2.35 shows the measured mobility versus temperature for a number of different modulation-doped samples for different spacer layer thicknesses. As expected, the low-temperature mobility increases with increasing spacer layer thickness compared to the bulk case, which is dominated by ionized impurity scattering.

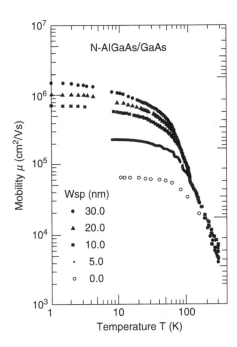

Fig. 2.35 Hall mobility versus temperature of *n*-modulation doped AlGaAs/GaAs heterostructures of varying spacer layer thickness (after Solomon *et al.*, *IEEE Elec. Dev. Lett.* EDL-5, 379 (1984), by permission).

Another limiting scattering mechanism at low temperature is the background impurity concentration. Unintentional impurities are incorporated into the epitaxial layers during growth due to the residual background concentration of impurities such as carbon and oxygen in the growth chamber itself. As the quality of epitaxial growth has improved over the years, so has the low-temperature mobility in modulation-doped structures. Figure 2.36 illustrates the mobility versus temperature reported by various groups which is quite similar in trend to the data shown in Fig. 2.35. However, here the increase in low-temperature mobility is not due to increasing spacer layer thickness, but rather to decreasing background impurity concentration as the quality of epitaxial growth systems improves [7]. For the best data shown in Fig. 2.33, electron mobilities in excess of 1×10^7 cm^2/Vs are evident at low temperature, which correspond with an estimated background ionized impurity concentration in the channel of 2×10^{11}/cm^3.

For the data shown in Figs. 2.35 and 2.36, the 2DEG mobility in heterolayers shows a strong temperature dependence at high temperature that eventually plateaus with very little temperature dependence at low temperature. Figure 2.37 shows a comparison of the experimental mobility as a function of temperature for one sample reported by Lin *et al.* [143]. The dashed lines in this figure compare the calculated contributions due to the remote impurities with the various lattice scattering mechanisms discussed in Section 2.7.2 treated within the relaxation time approximation. Although it is not proper to treat inelastic mechanisms such as LO phonon scattering within this approximation, the results are still

Fig. 2.36 Hall mobility versus temperature of modulation-doped AlGaAs/GaAs structures by various groups, showing the continued improvement in epitaxial material quality with time (after Pfeiffer *et al.*, *Appl. Phys. Lett.* **55**, 1888 (1989), by permission).

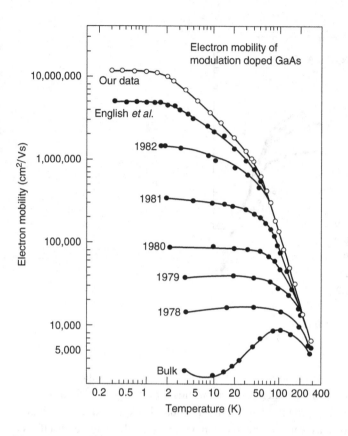

illustrative of the dominant contributions within different temperature ranges. As the temperature approaches room temperature, the contribution due to polar optical (LO) phonon scattering is dominant, as in bulk materials. Close to the knee of the mobility curve around 90 K, the mobility is more dominated by the combined contributions of acoustic and piezoelectric scattering. Finally, at low temperature, the mobility is entirely dominated by impurities, here assumed to be all in the form of remote impurities. Comparing Fig. 2.37 to the more recent data shown in Fig. 2.36, it is clear that the contributions due to acoustic and piezoelectric scattering have been overestimated in order to fit the experimental data (since the mobilities limited by these mechanisms are predicted to be less than 1×10^7). Part of the problem is choosing the exact deformation potential to use in describing the acoustic interaction.

2.7.3.2 Low field transport in quasi-1D systems

Interest in the transport properties of quantum wires was spurred by predictions of enhanced mobility due to the reduction in scattering associated with the 1D density of states, and reduced possible final states after scattering [144]. However, the actual mobilities measured in nanowires fabricated from top-down patterned 2DEG structures are typically less than that of the underlying 2DEG

2.7 Homogeneous transport in low-dimensional systems

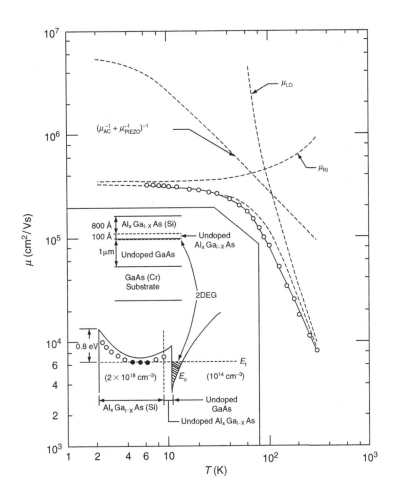

Fig. 2.37 Temperature dependence of the 2DEG mobility in a sample with $n_{2D} = 5.35 \times 10^{11}$ cm^{-1}. The open circles are experimental data, and the dashed lines are the various theoretical contributions to the mobility [after Lin et al., *Appl. Phys. Lett.* **45**, 695 (1984), by permission].

mobility itself, as shown in several studies [145],[146],[147],[148],[149]. As mentioned earlier, the primary reduction in mobility is the process-induced damage associated with the lateral patterning process, resulting in induced side-wall roughness and charged surface states, contributing both to increased roughness and impurity scattering. Improved processing, particularly in terms of nanometer scale Si CMOS processing, has resulted in steady improvement of performance. As mentioned in Chapter 1, non-planar CMOS technologies such as FinFETs [150] and tri-gate transistors [151] are essentially Si nanowire technologies as the channel is typically quite narrow.

The fabrication of as-grown or self-assembled nanowires helps mitigate the process-induced degradation of transport properties. Cleave edge epitaxial growth for example has been used to grow nanowires with very long mean free paths at low temperature [152]. More recently, there has been considerable study on the transport properties of self-assembled semiconductor nanowires grown by VLS and similar techniques as discussed in Section 2.1.1. Figure 2.3 shows and electron micrograph of a Si nanowire, where the wire cross section shows a high degree of crystallinity, with an amorphous surface layer evident. A certain degree of surface roughness is present as well, although some of this is associated with the imaging process itself.

Nanowire field effect transistors (FETs) were synthesized using the above growth techniques by Cui and Lieber [153] by dispersing the nanowires on a substrate and patterning contacts to two ends of the nanowire. Part (a) of Fig. 2.38 shows a schematic and electron micrograph of the nanowire transistor structure. Transport in such semiconductor nanowires is quasi-1D, depending on the diameter of the wire. For Si wires greater than 20 nm, it is expected that transport will be more bulk-like due to many occupied subbands, whereas smaller diameters should exhibit significant quantization of motion. Due to the small diameter in self-assembled structures, transport is very sensitive to the structure of the surface, the degree of roughness, the presence of traps or interface charges due to dangling bonds, etc. Contacts and contact resistance is another issue which has to be separated from transport in the nanowire itself. Cui and Lieber studied a number of different passivation techniques to improve the mobility; Fig. 2.38 (b) shows the measured mobility before (left set of data), and after modification with 4-nitrophenyl octadecanoate. Clearly an enormous improvement in mobility is observed, presumably due to passivation of dangling bonds by the annealing process, suggestive of the strong role played by the surface in transport.

The electronic and transport properties are strongly dependent on the chirality and the diameter of the CNT. As discussed in Section 2.4.2, for a given chiral vector $\mathbf{c}_h$, if $n - m$ is a multiple of 3, then the nanotube is metallic, which includes the armchair ($n = m$) structure shown in Fig. 2.27(a). The conductivity of metallic CNTs is quite high due to the high mobility and 1D carrier density. Other chiral structures not satisfying this equality behave as semiconductors. In terms of transport, measurements have demonstrated very high mobilities and nearly ballistic transport [108,154,155]. In this context, a diffusive picture of transport in CNTs is not appropriate, rather a treatment in terms of quantum fluxes as discussed in Chapter 3, and transport may be dominated by the contacts. However, dopants and defects can lead to scattering. CNTs are inherently p-type, but by annealing in vacuum or doping with electropositive element (e.g. K), they can be doped n-type. Electron–electron (e–e) interactions can contribute to scattering as well. While normally e–e scattering conserves the net momentum of the two particles, and hence does not relax the net momentum, Umklapp processes are possible within the reduced zone of the CNT bandstructure, which do lead to a net backscattering [156]. There are various other mechanisms that limit transport, particularly at high fields when electrons are accelerated above the threshold for various types of phonon scattering. In particular, because of the unique hollow structures of CNTs, there are torsional modes of vibration, similar to molecular chains, such as twistons, which are

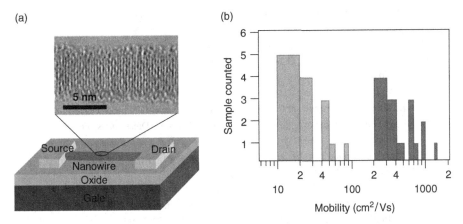

Fig. 2.38 Si nanowire field effect transistor structure. The left panel shows a schematic and electron micrograph of the transistor structure. The right side shows the measured mobility before (data, left side), and after (data, right side) surface modification (after Cui and Lieber [153], with permission).

essentially long-wavelength acoustic phonon modes [157]. Optical modes associated with the in-plane modes of graphene, and zone boundary phonons coupling different Fermi wavevectors are also believed to be important [158]. Studies of high field transport in metallic SW CNTs with low contact resistance compared with theory indicate that the saturation of current at high bias is associated with optical and zone boundary phonon emission [158].

References

[1] T. Ando, A. Fowler, and F. Stern, *Rev. Mod. Phys.* **54**, 437 (1982).
[2] C. Weisbuch and B. Vinter, *Quantum Semiconductor Structures* (Boston, Academic Press, 1991).
[3] G. Bastard, J. A. Brum, and R. Fetreira, *Sol. State Phys.* **44**, 229 (1991).
[4] A. B. Fowler, F. F. Fang, W. E. Howard, and P. J. Stiles, *Phys. Rev. Lett.* **16**, 901 (1966).
[5] A. Y. Cho and J. R. Arthur, *Prog. Solid State Chem.* **10**, 157 (1975).
[6] R. Dingle, H. L. Störmer, A. C. Gossard, and W. Wiegmann, *Appl. Phys. Lett.* **33**, 665 (1978).
[7] L. Pfeiffer, K. W. West, H. L. Störmer, and K. W. Baldwin, *Appl. Phys. Lett.* **55**, 1888 (1989).
[8] See, for example: D. Bimberg, M. Grundmann, and N. N. Ledentsov, *Quantum Dot Heterostructures* (Chichester, Wiley, 1999), and references therein.
[9] See, for example: M. Grundmann and D. Bimberg, *Jpn. J. Appl. Phys.* **36**, 4181 (1997), and references therein.
[10] H. Drexler, D. Leonard, W. Hansen, J. P. Kotthaus, and P. M. Petroff, *Phys. Rev. Lett.* **73**, 2252 (1994).
[11] R. J. Warburton, C. S. Dürr, K. Karrai, *et al.*, *Phys. Rev. Lett.* **79**, 5282 (1997).

[12] M. C. Bödefeld, R. J. Warburton, K. Karrai, et al., *Appl. Phys. Lett.* **74**, 1839 (1999).
[13] S. K. Jung, C. K. Hyon, J. H. Park, et al., *Appl. Phys. Lett.* **75**, 1167 (1999).
[14] J. Seufert, M. Rambach, G. Bacher, et al., *Appl. Phys. Lett.* **82**, 3946 (2003).
[15] C. H. Li, G. Kioseoglou, O. M. J. van 't Erve, et al., *Appl. Phys. Lett.* **86**, 132503 (2005).
[16] Z. Yuan, B. E. Kardynal, R. M. Stevenson, et al., *Science* **295**, 102 (2002).
[17] K. Karrai, R. J. Warburton, C. Schulhauser, et al., *Nature* **427**, 135 (2004).
[18] For a comprehensive overview, see: M. S. Dresselhaus, G. Dresselhaus, and Ph. Avouris, eds., *Carbon Nanotubes: Synthesis, Structure, Properties, and Applications*, Springer Topics in Applied Physics, vol. 80, Berlin/Heidelberg (2001).
[19] J. Robertson, *Mat. Today* **10**, 36 (2007).
[20] S. J. Tans, A. R. M. Verschueren, and C. Dekker, *Nature* **393**, 49 (1998).
[21] S. Frank, P. Poncharal, Z. L. Wang, and W. A. de Heer, *Science* **280**, 1744 (1998).
[22] M. Bockrath, D. H. Cobden, J. Liu, et al., *Nature* **397**, 598 (1999).
[23] S. J. Tans, M. H. Devoret, R. J. A. Groeneveld, and C. Dekker, *Nature* **394**, 761 (1998).
[24] A. Bachtold, C. Strunk, J.-P. Salvetat, et al., *Nature* **397**, 673 (1999).
[25] W. Liang, M. Bockrath, D. Bozovic, et al., *Nature* **411**, 665 (2001).
[26] C. Schönenberger, A. Bachtold, C. Strunk, J.-P. Salvetat, and L. Forro, *Appl. Phys. A* **69**, 283 (1999).
[27] C. Journet, W. K. Maser, P. Bernier, et al., *Nature* **388**, 756 (1997).
[28] L.-C. Qin and S. Ijima, *Chem. Phys. Lett.* **269**, 65 (1997).
[29] A. Cassel, N. Franklin, T. Tombler, et al., *J. Amer. Chem. Soc.* **121**, 7975 (1999).
[30] Y. Zhang, A. Chang, J. Cao, et al., *Appl. Phys. Lett.* **79**, 3155 (2001).
[31] M. Law, J. Goldberger, and P. D. Yang, *Ann. Rev. Mat. Sci.* **34**, 83 (2004).
[32] H. J. Fan, P. Werner, and M. Zacharias, *Small* **2**, 700 (2006).
[33] W. Lu and C. M. Lieber, *J. Phys. D: Appl. Phys.* **39**, R387 (2006).
[34] Y. Wu, J. Xiang, C. Yang, W. Lu, and C. M. Lieber, *Nature* **430**, 61 (2004).
[35] L. Samuelson, *Mat. Today* **6**, 22 (2003).
[36] M. T. Björk, B. J. Ohlsson, T. Sass, et al., *Nano Lett.* **2**, 87 (2002).
[37] M. T. Björk, B. J. Ohlsson, C. Thelander, et al., *Appl. Phys. Lett.* **81**, 4458 (2002).
[38] C. Thelander, T. Mårtensson, M. T. Björk, et al., *Appl. Phys. Lett.* **83**, 2052 (2003).
[39] A. Fuhrer, C. Fasth, and L. Samuelson, *Appl. Phys. Lett.* **91**, 052109 (2007).
[40] A. Fuhrer, L. E. Froberg, J. N. Pedersen, et al., *Nano Lett.* **7**, 243 (2007).
[41] M. T. Björk, A. Fuhrer, A. E. Hansen, et al., *Phys. Rev. B* **72**, 201307 (2005).
[42] K. S. Novoselov, A. K. Geim, S. V. Morozov, et al., *Science* **306**, 666 (2004).
[43] K. S. Novoselov, D. Jiang, F. Schedin, et al., *Proc. Nat. Acad. Sci.* **102**, 10451 (2005).
[44] K. S. Novoselov, A. K. Geim, S. V. Morozov, et al., *Nature* **438**, 197 (2005).
[45] Y. Zhang, Z. Jiang, J. P. Small, et al., *Phys. Rev. Lett.* **96**, 136806 (2006).
[46] C. Berger, Z. Song, X. Li, et al., *Science* **312**, 1191 (2006).
[47] S. Adachi, *J. Appl. Phys.* **58**, 62 (1985).
[48] F. Capasso and G. Margaritondo, *Heterojunction Band Discontinuities: Physics and Device Applications* (Amsterdam, Elsevier Science Publishing Company, 1987).
[49] D. J. Wolford, T. F. Kuech, J. A. Bradley, et al., *J. Vac. Sci. Technol. B* **4**, 1043 (1986).
[50] L. I. Schiff, *Quantum Mechanics* (New York, McGraw-Hill, 1955).
[51] D. Y. Medi, J. Shah, T. C. Damen, et al., *Phys. Rev. B* **40**, 3028 (1989).
[52] K. Leo, J. Shah, E. O. Gabel, et al., *Phys. Rev. Lett.* **66**, 201 (1991).

References

[53] J. P. Eisenstein, L. N. Pfeiffer, and K. W. West, *Phys. Rev. Lett.* **26**, 3804 (1992).
[54] N. M. Froberg, B. B. Hu, X.-C. Zhang, and D. H. Auston, *IEEE J. Quan. Elec.* **28**, 2291 (1992).
[55] N. W. Ashcroft and N. D. Mermin, *Solid State Physics* (Holt, Rinehart and Winston, New York, 1976).
[56] D. K. Ferry and J. P. Bird, *Electronic Materials and Devices* (San Diego, Academic Press, 2001).
[57] L. Esaki, in *Recent Topics in Semiconductor Physics*, eds. H. Kamimura and Y. Toyozawa (Singapore, World Scientific Press, 1983).
[58] F. Stern, *Phys. Rev. B* **17**, 5009 (1978).
[59] F. Stern and S. Das Sarma, *Phys. Rev. B* **30**, 840 (1984).
[60] T. Ando, *Phys. Rev. B* **13**, 3468 (1976).
[61] S. Das Sarma and B. Vinter, *Phys. Rev. B* **23**, 6832 (1981).
[62] L. Hedin and B. I. Lundqvist, *J. Phys. C* **4**, 2064 (1971).
[63] F. Stern, *J. Comput. Phys.* **6**, 56 (1970).
[64] P. C. Chow, *Am. J. Phys.* **40**, 730 (1972).
[65] S. Bhobe, W. Porod, S. Bandyopadhyay, and D. J. Kirkner, *Surf. Interface Anal.* **14**, 590 (1990).
[66] M. Shur, *GaAs Devices and Circuits* (New York, Plenum Press, 1987).
[67] K. Hess, *Physica B* **117B-118B**, 723 (1983).
[68] M. Abramowitz and I. A. Stegun, *Handbook of Mathematical Functions*, National Bureau of Standards Applied Mathematics Series, No. 55 (Washington, U.S. Government Printing Office, 1964).
[69] G. Tuttle, H. Kroemer, and J. H. English, *J. Appl. Phys.* **65**, 5239 (1989).
[70] I. Lo, W. C. Mitchell, M. O. Manasreh, C. E. Stutz, and K. R. Evans, *Appl. Phys. Lett.* **60**, 751 (1992).
[71] C. Nguyen, B. Brar, H. Kroemer, and J. H. English, *Appl. Phys. Lett.* **60**, 1854 (1992).
[72] Ch. Gauer, J. Scriba, A. Wixforth, *et al.*, *Semicond. Sci. Technol.* **8**, S137 (1993).
[73] F.-C. Wang, W. E. Zhang, C. H. Yang, M. J. Yang, and B. R. Bennet, *Appl. Phys. Lett.* **69**, 1417 (1996).
[74] I. Zutic, J. Fabian, and S. Das Sarma, *Rev. Mod. Phys.* **76**, 323 (2004).
[75] Yu. A. Bychkov and E. I. Rashba, *J. Phys. C: Sol. St. Phys.* **17**, 6039 (1984).
[76] J. W. Orton and C. T. Foxon, *Rep. Prog. Phys.* **61**, 1 (1998).
[77] D. J. Paul, *Semicond. Sci. Technol.* **19**, R75 (2004).
[78] R. F. W. Pease, *J. Vac. Sci. Technol. B* **10**, 278 (1992).
[79] G. Binnig, H. Rohrer, Ch. Gerber, and E. Weibel, *Phys. Rev. Lett.* **49**, 57 (1982); C. Binnig and H. Rohrer, *Rev. Mod. Phys.* **59**, 615 (1987); J. A. Stroscio and W. J. Kaiser, eds., *Scanning Tunneling Microscopy* in Methods of Experimental Physics, vol. 27 (Boston, Academic Press, 1993).
[80] www.almaden.ibm.com/vis/stm/gallery.html.
[81] R. S. Becker, J. A. Golavchenko, and B. S. Swartzentruber, *Nature* **325**, 419 (1987).
[82] J. A. Dagata, J. Schneir, H. H. Harary, and C. J. Evans, *Appl. Phys. Lett.* **56**, 2001 (1990).
[83] H. Sugimura, T. Uchida, N. Kitamura, and H. Masuhara, *Appl. Phys. Lett.* **63**, 1288 (1993).
[84] E. S. Snow, D. Park, and P. M. Campbell, *Appl. Phys. Lett.* **69**, 269 (1996).
[85] E. S. Snow and P. M. Campbell, *Appl. Phys. Lett.* **64**, 1933 (1994).
[86] K. Matsumoto, M. Ishii, K. Segana, *et al.*, *Appl. Phys. Lett.* **68**, 34 (1996).
[87] R. Held, T. Vancura, T. Heinzel, *et al.*, *Appl. Phys. Lett.* **73**, 262 (1998).
[88] V. Senz, T. Heinzel, T. Ihn, *et al.*, *J. Phys.: Condens. Matt.* **13**, 3831 (2001).

[89] R. Schleser, E. Ruh, T. Ihn, et al., *Appl. Phys. Lett.* **85**, 2005 (2004).
[90] M. Sigrist, T. Ihn, K. Ensslin, et al., *Phys. Rev. Lett.* **96**, 036804 (2006).
[91] A. Fuhrer, T. Ihn, K. Ensslin, W. Wegscheider, and M. Bichler, *Phys. Rev. Lett.* **93**, 176803 (2004).
[92] B. Grbic, R. Leturcq, K. Ensslin, D. Reuter, and A. D. Wieck, *Appl. Phys. Lett.* **87**, 232108 (2005).
[93] L. P. Rokhinson, V. Larkina, Y. B. Lyanda-Geller, L. N. Pfeiffer, and K. W. West, *Phys. Rev. Lett.* **93**, 146601 (2004).
[94] L. P. Rokhinson, L. N. Pfeiffer, and K. W. West, *Phys. Rev. Lett.* **96**, 156602 (2006).
[95] R. Crook, A. C. Graham, C. G. Smith, et al., *Nature* **424**, 751 (2003).
[96] R. Crook, J. Prance, K. J. Thomas, et al., *Science* **312**, 1359 (2006).
[97] R. Crook, C. G. Smith, A. C. Graham, et al., *Phys. Rev. Lett.* **91**, 246803 (2003).
[98] W. E. Spicer, Z. L. Liliental-Weber, E. Weber, *J. Vac. Sci. Technol. B* **6**, 1245 (1988).
[99] T. J. Thornton, M. Pepper, H. Ahmed, D. Andrews, and G. J. Davies, *Phys. Rev. Lett.* **56**, 1198 (1986).
[100] A. B. Fowler, A. Hartstein, and R. A. Webb, *Phys. Rev. Lett.* **48**, 196 (1982).
[101] S. E. Laux and F. Stern, *Appl. Phys. Lett.* **49**, 91 (1986).
[102] S. E. Laux, D. I. Frank and F. Stern, *Surf. Sci.* **196**, 101 (1988).
[103] K. Kojima, K. Mitsunaga, and K. Kyuma, *Appl. Phys. Lett.* **55**, 862 (1989).
[104] R. Saito, M. Fujita, G. Dresselhaus, and M. S. Dresselhaus, *Appl. Phys. Lett.* **60**, 2201 (1992).
[105] J. R. Williams, L. DiCarlo, and C. M. Marcus, *Science* **317**, 638 (2007).
[106] A. Abanin and L. S. Levitov, *Science* **317**, 641 (2007).
[107] P. R. Wallace, *Phys. Rev.* **71**, 622 (1947).
[108] J.-C. Charlier, X. Blasé, and S. Roche, *Rev. Mod. Phys.* **79**, 677 (2007).
[109] Jing Guo Ph.D., Purdue University, August, 2004. Carbon Nanotube Electronics: Modeling, Physics, and Applications, www.nanohub.org/resources/1928/.
[110] Simulation services for results using CNTbands 2.1 were provided by the Network for Computational Nanotechnology (NCN) at nanoHUB.org; Y. Yoon, K. D. Kienle, J. K. Fodor, G. Liang, A. Matsudaira, G. Klimeck, and J. Guo (2006), "CNTbands," doi: 10254/nanohub-r1838.3.; p_z-orbital model: L. Yang, M. P. Anantram, and J. P. Lu, *Phys. Rev. B* **60**, 13874 (1999).
[111] C. W. J. Beenakker, H. van Houten, and B. J. van Wees, *Superlatt. Microstruct.* **5**, 127 (1989); C. W. J. Beenakker and H. van Houten, "Quantum transport in semiconductor nanostructures," in *Solid State Physics* **44**, H. Ehrenreich and D. Turnbull, eds. (Boston, Academic Press, 1991), pp. 1–228.
[112] D. K. Ferry, *Semiconductors* (Macmillan, New York, 1991).
[113] E. D. Siggia and P. C. Kwok, *Phys. Rev. B* **B2B**, 1024 (1970).
[114] CRC Handbook of Mathematical Sciences, 6th edn. (West Palm Beach, CRC Press, 1987).
[115] P. F. Maldague, *Surf. Sci.* **78**, 296 (1978).
[116] S. J. Allen, Jr., D. C. Tsui, and R. A. Logan, *Phys. Rev. Lett.* **38**, 980 (1977).
[117] T. N. Theis, J. P. Kotthaus, and J. P. Stiles, *Solid State Comm.* **26**, 603 (1978).
[118] D. C. Tsui, E. Gornik, and R. A. Logan, *Solid State Comm.* **35**, 875 (1980).
[119] B. Yu-Kuang Hu and S. Das Sarma, *Phys. Rev. B* **48**, 5469 (1993).
[120] K. von Klitzing, G. Dorda, and M. Pepper, *Phys. Rev. Lett.* **45**, 494 (1980); K. von Klitzing and G. Ebert, *Springer Ser. Solid State Sci.* **59**, 242 (1984).

[121] R. B. Laughlin, *Phys. Rev. B* **25**, 5632 (1981); R. B. Laughlin, *Springer Ser. Solid State Sci.* **59**, 272 288 (1984).
[122] M. Büttiker, *Phys. Rev. B* **38**, 9375 (1988).
[123] D. C. Tsui, H. L. Störmer, and A. C. Gossard, *Phys. Rev. Lett.* **48**, 1559 (1982).
[124] R. B. Laughlin, *Phys. Rev. Lett.* **50**, 1395 (1983).
[125] J. M. Luttinger, *Phys. Rev.* **121**, 924 (1961).
[126] R. E. Peierls, *Quantum Theory of Solids* (Oxford, Clarendon, 1955).
[127] L. Rota, F. Rossi, S. M. Goodnick, *et al.*, *Phys. Rev. B* **47**, 1632 (1993).
[128] T. Ando, *J. Phys. Soc. Jpn* **43**, 1616 (1977).
[129] S. M. Goodnick, D. K. Ferry, C. W. Wilmsen, *et al.*, *Phys. Rev. B* **32**, 8171 (1985).
[130] H. Akera and T. Ando, *Phys. Rev. B* **41**, 11967 (1990); *Phys. Rev. B* **43**, 11676 (1991).
[131] M. Cardona, *Superlatt. Microstruct.* **7**, 183 (1990).
[132] K. W. Kim and M. A. Stroscio, *J. Appl. Phys.* **68**, 6289 (1990).
[133] H. Rücker, E. Molinari, and P. Lugli, *Phys. Rev. B* **45**, 6747 (1992).
[134] S. Kawajii, *J. Phys. Soc. Jpn* **27**, 906 (1969).
[135] H. Ezawa, T. Kuroda, and K. Nakamura, *Surf. Sci.* **27**, 218 (1971).
[136] P. J. Price, *Surf. Sci.* **143**, 145 (1984).
[137] E. B. Ramayya, D. Vasileska, S. M. Goodnick, and I. Knezevic, *J. Appl. Phys.* **7**, 319–23 (2008).
[138] P. J. Price, *Ann. Phys.* **133**, 217 (1981).
[139] F. A. Riddoch and B. K. Ridley, *J. Phys. C* **16**, 6971 (1983); *Physica* **134B**, 342 (1985).
[140] S. M. Goodnick and P. Lugli, *Phys. Rev. B* **37**, 2578 (1988).
[141] M. Duer, S. M. Goodnick and P. Lugli, *Phys. Rev. B* **54**, 17794 (1996).
[142] D. Jovanovic and J. P. Leburton, "Monte Carlo simulation of quasi-one-dimensional systems," in *Monte Carlo Device Simulation: Full Band and Beyond*, ed. K. Hess (Boston, Kluwer Academic Publishers, 1991).
[143] B. J. F. Lin, D. C. Tsui, M. A. Paalanen, and A. C. Gossard, *Appl. Phys. Lett.* **45**, 695 (1984).
[144] H. Sakaki, *Jpn J. Appl. Phys.* **19**, L735 (1980).
[145] T. J. Thornton, M. L. Roukes, A. Scherer, and B. P. Van de Gaag, *Phys. Rev. Lett.* **63**, 2128 (1989).
[146] K. Ismail, D. A. Antoniadis, and H. I. Smith, *Appl. Phys. Lett.* **54**, 1130 (1989).
[147] N. Sawaki, R. Sugimoto, and T. Hori, *Semicond. Sci. Technol.* **9**, 946 (1994).
[148] C. Wirner, H. Momose, and C. Hamaguchi, *Physica B* **227**, 34 (1996).
[149] A. A. Talin, F. Léonard, B. S. Swartzentruber, X. Wang, and S. D. Hersee, *Phys. Rev. Lett.* **101**, 076802 (2008).
[150] Y.-K. Choi, T.-J. King, and C. Hu, *IEEE Elec. Dev. Lett.* **23**, 25 (2002).
[151] B. S. Doyle, S. Datta, M. Doczy, *et al.*, *IEEE Elec. Dev. Lett.* **24**, 263 (2003).
[152] R. de Picciotto, L. N. Pfeiffer, K. W. Baldwin, and K. W. West, *Phys Rev. Lett.* **92**, 036805 (2004).
[153] Y. Cui and C. M. Lieber, *Science* **291**, 851 (2001).
[154] T. Dürkop, S. A. Getty, E. Cobas, and M. S. Fuhrer, *Nano Lett.* **4**, 35 (2004).
[155] A. Javey, J. Guo, Q. Wang, M. Lundstrom, and H. Dai, *Nature* **424**, 654 (2003).
[156] L. Balents and M. P. A. Fisher, *Phys. Rev. B* **55**, 11973 (1997).
[157] C. L. Kane and E. J. Mele, *Phys. Rev. Lett.* **78**, 1932 (1997).
[158] Z. Yao, C. L. Kane, and C. Dekker, *Phys. Rev. Lett.* **84**, 2941 (2000).

3
Transmission in nanostructures

In Chapter 2, we introduced the idea of low-dimensional systems arising from quantum confinement. Such confinement may be due to a heterojunction, an oxide–semiconductor interface, or simply a semiconductor–air interface (for example, in an etched quantum wire structure). When we look at transport *parallel* to such barriers, such as along the channel of an HEMT or MOSFET, or along the axis of a quantum wire, to a large extent we can employ the usual kinetic equation formalisms for transport and ignore the phase information of the particles. Quantum effects enter only through the description of the basis states arising from the confinement, and the quantum mechanical transition rates between these states are due to the scattering potential. This is not to say that quantum interference effects do not play a role in parallel transport. As we will see in the later chapters, several effects manifest themselves in parallel transport studies such as *weak localization* and *universal conductance fluctuations*, which at their origin have effects due to the coherent interaction of electrons.

In contrast to transport parallel to barriers, when particles traverse regions in which the medium is changing on length scales comparable to the phase coherence length of the particles, quantum interference is expected to be important. By "quantum interference" we mean the superposition of incident and reflected waves, which, in analogy to the electromagnetic case, leads to constructive and destructive interference. Such a coherent superposition of states is of course what leads to the quantization of momentum and energy in the formation of low-dimensional systems discussed in the previous chapter. However, we now want to look at the general case of transport in *open systems* when carriers are incident on such barriers, and how the reflected and transmitted amplitudes of the particle waves through such structures determine transport. Prior to the early 1980s, most artificially engineered structures for studying such quantum transport were in the form of planar barrier structures, in which, for example, one thin layer of material is grown on top of another, such as the simple AlGaAs/GaAs heterostructure discussed in the previous chapter. The change in potential is therefore in only one dimension (the vertical growth direction), with the other two degrees of freedom unconstrained. We will begin our discussion of transmission in nanostructures by first discussing transport in these structures, with particular emphasis on the so-called *resonant tunneling* diode discussed in Sections 3.1–3.2. From the mid 1980s up to the present, researchers became increasingly engaged in the fabrication of fully three-dimensional quantum interference structures through the rapidly evolving technologies related to lateral patterning and ultra-small device fabrication discussed in Chapter 2. The generalization of quantum transmission concepts to these more complicated 3D structures will be pursued in the remaining sections of this chapter.

3.1 Tunneling in planar barrier structures

Perhaps the most well studied transport phenomenon associated with quantum transmission is that of tunneling. In general, the term "tunneling" refers to particle transport through a classically forbidden region, where we mean a region in which the total energy of a classical point particle is less than its potential energy. This is illustrated in Fig. 3.1, where a particle of energy E is incident on an arbitrary-shaped potential energy barrier of height $V_0 > E$. In classical mechanics a particle is completely reflected at the so-called *classical turning points* labeled A and B in Fig. 3.1, that is, points where the total energy equals the potential energy. Quantum mechanically, the underlying equation of motion is the Schrödinger equation, in which the role of the potential is analogous in electromagnetics to that of a spatially varying permittivity. In electromagnetics, the solution of the wave equation must satisfy certain boundary conditions at the abrupt interface between two dielectrics of different permittivity, which leads to a certain portion of an incident wave being transmitted and a certain portion reflected. Likewise, in quantum mechanics the wave function and its normal derivative must be continuous across a boundary of two regions of different potential energy, which similarly leads to reflection and transmission of probability waves at the boundary. As shown in Fig. 3.1, the wave function associated with a particle incident from the left on the potential barrier has non-zero solutions inside the barrier and on the right side. Because the square of the wave function represents the probability density for finding a particle in a given region of space, it follows that quantum mechanically a particle incident on a potential bather has a finite probability of *tunneling* through the barrier and appearing on the other side.

Historically, the phenomenon of tunneling was recognized soon after the foundations of quantum theory had been established in connection with field ionization of atoms and nuclear decay of alpha particles. Shortly thereafter, tunneling in solids was studied by Fowler and Nordheim [1] in the field emission of electrons from metals, that is, the electric field-aided thermionic emission of electrons from metal into vacuum. Later, interest developed in tunneling through thin insulating layers (such as thermally grown oxides) between metals (MIM), and between semiconductors and metals (MIS). After the development of the band theory of solids, Zener [2] proposed the concept of *interband* tunneling, in

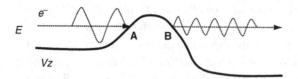

Fig. 3.1 Quantum mechanical tunneling through a potential barrier. The points A and B correspond to the classical turning points.

which electrons tunnel from one band to another through the forbidden energy gap of the solid. The time period of the late 1940s and early 1950s saw tremendous breakthroughs in the development of semiconductor device technology, and conditions favorable to the experimental observation of Zener tunneling in p–n junction diodes were realized. In the late 1950s, Esaki [3] proposed the so-called Esaki diode, in which negative differential resistance (NDR) is observed in the I–V characteristics of heavily doped p–n diodes due to interband Zener tunneling between the valence and conduction bands. The Esaki diode continues to be important technologically and finds many applications in microwave technology. In the 1960s, a flourish of activity developed related to measurements of tunneling between superconductors and normal metals [4] and between superconductors themselves separated by thin insulating layers [5], which reveal striking evidence of the superconducting density of states and the associated superconducting gap. Theoretically, this led to development of perturbative many-body theories of tunneling embodied by the *tunneling* or *transfer Hamiltonian method* [6]. Recent theories associated with *single electron charging* in quantum dots are descendants of the transfer Hamiltonian method and will be revisited in a later chapter in connection with the single electron phenomenon. The transfer Hamiltonian model was also utilized quite extensively in the study of independent particle tunneling [7,8,9], which became the basis for interpreting a host of experimental tunneling studies in normal metals and semiconductors. A thorough review of the status of experimental and theoretical tunneling related research prior to the 1970s is given by Duke [10].

During the 1970s, advances in epitaxial growth techniques such as MBE increasingly allowed the growth of well-controlled heterostructure layers with atomic precision and low background impurity densities. In their pioneering work in this field, Tsu and Esaki [11,12] at IBM predicted that when bias is applied across the structure, the current-voltage (I–V) characteristics of GaAs/Al$_x$Ga$_{1-x}$As double and multiple barrier structures should show NDR similar to that in Esaki diodes. However, NDR in this case occurs due to resonant tunneling through the barriers within the same band. Resonant tunneling refers to tunneling in which the electron transmission coefficient through a structure is sharply peaked about certain energies, analogous to the sharp transmission peaks as a function of wavelength evident through optical filters, such as a Fabry–Perot étalon consisting of two parallel dielectric interfaces. A schematic representation of the resonant tunneling process and the associated NDR in the I–V characteristics of such a *resonant tunneling diode* (RTD) is shown for the structure reported by Sollner *et al.* in Fig. 3.2 [13]. This particular structure consists of two $Al_xGa_{1-x}As$ bathers ($x \sim 0.25$–0.30) separated by a thin GaAs quantum well. For the Al mole fraction of this structure, the estimated barrier height due to the conduction band offset, ΔE_c, is approximately 0.23 eV. The thickness of the barriers (here 5 nm) is sufficiently thin that tunneling through the barriers is significant. The energy E_1 corresponds to the lowest resonant energy, which as discussed above is the energy where the transmission coefficient is very peaked, and in fact may approach unity in some cases. This energy may qualitatively be thought of as the bound state associated with the quantum well formed between the two confining barriers. However, since the electron may tunnel out of this bound state in either direction, there is a finite lifetime τ associated with this state, and the width of the resonance in energy (i.e., the energy range in which the transmission coefficient is sizable) is inversely proportional to this lifetime, approximately as $\hbar/\tau$. Depending on the well width and barrier heights, there may exist several such *quasi-bound states* in the system. The double barrier structure is surrounded by heavily-doped GaAs layers, as shown in Fig. 3.2, which provide low-resistance emitter and collector contacts to the tunneling region, forming the RTD structure. With a positive bias applied to the right contact relative to the left, the Fermi energy on the left is pulled through the resonant level E_1. As the Fermi energy passes through the resonant state, a large current flows due to the increased transmission from left to right. At the same time, the back flow of carriers from right to left is suppressed as electrons at the Fermi energy on the right see only a large potential barrier, as shown in the figure. Further bias pulls the bottom of the conduction band on the left side through the resonant energy, which cuts off the supply of electrons available at the resonant energy for tunneling. The result is a marked decrease of the current with increasing voltage, giving rise to a region of NDR as shown schematically by the I–V characteristics in Fig. 3.2.

The first experimental evidence for resonant tunneling in double barrier structures was reported by the IBM group in MBE grown structures [12] where weak NDR was observed in the I–V characteristics. The difficulty at the time in realizing pronounced NDR as observed in interband Esaki diodes was due to the difficulty in achieving low background impurity concentrations in epitaxial layers at the time. The effect of such inhomogeneities is to broaden the expected resonances, in effect washing them out. It was not until the 1980s that epitaxial material quality had improved sufficiently to observe marked resonant tunneling behavior in III–V epitaxial layer structures. Sollner *et al.* at MIT reported pronounced NDR in the low temperature I–V characteristics of the double barrier structure [13] shown in Fig. 3.2. Figure 3.3 shows the I–V characteristics for both forward and reverse bias at three different temperatures. The I–V curves are approximately symmetric about the origin although it is clear that the peak current is slightly lower in the negative bias direction. Such asymmetry in RTDs is typical in many structures and may arise from several sources such as

Fig. 3.2 Energy band diagram and I–V characteristics of a GaAs/AlGaAs resonant tunneling diode. The upper part shows the electron energy as a function of position in the quantum well structure. The parameters are $N_{D1} = N_{D3} = 10^{18}$ cm^{-3}, $N_{D2} = 10^{17}$ cm^{-3}, and $W_1 = W_2 = W_3 = 5$ nm. The aluminum mole fraction in the barriers is $x \sim 25\%$–30% (after Sollner *et al.*, *Appl. Phys. Lett.* **43**, 588 (1983), by permission).

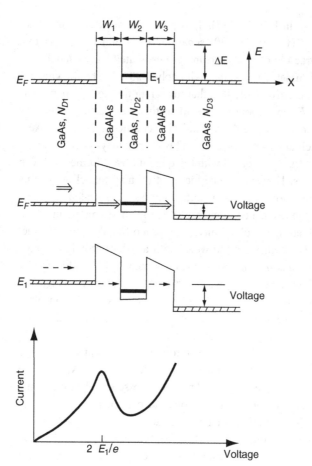

out-diffusion of impurities during MBE growth, or due to differences in interface roughness between AlGaAs grown on GaAs versus GaAs grown on AlGaAs, where the latter is believed to be rougher. For many applications, NDR devices should have a large peak current and a small valley current, where the latter is the minimum current following the peak current as the magnitude of the voltage increases. Therefore an important figure of merit for an NDR device such as the RTD is the *peak to valley ratio* (PVR). For the data in Fig. 3.3, the PVR is 6:1 at low temperature (< 50 K), while at increasingly higher temperatures the PVR decreases until the NDR effect completely vanishes. The reduction of the PVR as temperature increases is simply related to the increase in off-resonant current over the barriers due to thermionic emission, as well as the spreading of the distribution function around the resonant energy, which decreases the peak current as discussed in more detail below. The valley current is also increased at higher temperatures due to inelastic phonon-assisted tunneling that further degrades the PVR of the diode.

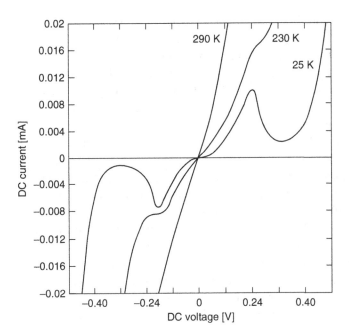

Fig. 3.3 Current–voltage characteristics of the RTD structure shown in Fig. 3.2 at three different temperatures (after Sollner *et al.*, *Appl. Phys. Lett.* **43**, 588 (1983), by permission).

A considerable volume of research into RTDs has evolved since the MIT group's successful demonstration of NDR effects. Interest stems not only from the fundamental physics aspects of this deceptively simple structure, but also from the potential practical applications in high-speed microwave systems and novel digital logic circuits. One advantage of the RTD for electronic applications is that the fundamental time associated with the intrinsic tunneling process itself may be quite short, often taken as the lifetime of the quasi-bound state (i.e., the inverse of the resonance width). In their original work, Sollner *et al.* demonstrated NDR up to frequencies of 2.5 THz, which qualitatively implies charge transport on the order of $\tau = 6 \times 10^{-14}$ s. In reality, there are several time constants that come into play in the frequency response of an RTD, including the transit time across the nontunneling regions of the device and the RC time constant associated with the capacitance of the structure. With proper design, the various time constants can be minimized, and high-frequency performance may be obtained in analog applications such as oscillators operating up to 420 GHz in GaAs/AlAs double-barrier structures [14].

Improvements in material growth, as well as an understanding of the physics and device design of RTDs, has led to ever improving performance. Room-temperature NDR was soon obtained in similar structures to Fig. 3.2 [15]. There have been systematic studies of the dependence of the *I–V* characteristics upon structural parameters of $Al_xGa_{1-x}As$/GaAs RTDs, such as the GaAs well width [16], the $Al_xGa_{1-x}As$ barrier height [17], and thickness [18,19]. To minimize impurity scattering due to dopant interdiffusion into the barrier regions, thin spacer layers of undoped or reduced-doping material are usually included on either side of the tunneling barriers (see for example [16]). Studies of the dependence of RTD performance on the thickness of such spacers were reported by Yoo *et al.* [20]. For good devices with thin AlAs barriers, PVRs close to 4:1 and peak current densities in excess of 1×10^5 A/cm^2 [21] may be obtained at 300 K, although not in the same structures, since there is usually a trade-off between these two parameters in terms of device design.

Alternative material systems to the AlGaAs/GaAs system have yielded greatly improved performance benchmarks. In$_{0.53}$Ga$_{0.47}$As lattice matched to InP has been employed as the narrow gap material in double-barrier structures. PVRs of 14:1 at room temperature using an InGaAs well have been reported [22]. Even better PVRs of 30:1 and 63:1 at 300 K and 77 K were reported using a strained InAs well [23] between AlAs barriers. More recently, attention has focused on the use of Type II staggered bandgap systems such as the InAs/AlSb system for resonant tunneling applications [24]. In contrast to a Type I heterojunction system (e.g. GaAs/AlGaAs), the valence band edge of the wider bandgap AlSb lies higher in energy than that of InAs, so that electrons are confined in the InAs layer, but holes are not (see Fig. 3.4). Söderström *et al.* reported room-temperature PVRs of 3.2 and peak current densities of 3.7×10^5 A/m^2 in double barrier structures [25] consisting of InAs wells and cladding layers and AlSb bafflers (grown on GaAs substrates). High-frequency oscillations were measured up to 712 GHz [26], the highest frequency reported to that date of any solid state oscillator. Part of the improvement in performance in this material system is associated with the favorable transport and Ohmic contact properties of InAs compared to GaAs, reducing parasitic contributions to the total delay time across the device.

One final and interesting innovation in resonant tunneling diodes has been the successful development of resonant interband tunneling diodes (RITs) in which both electrons and holes participate in transport [27]. Figure 3.4 shows the band diagram (for the Γ point) of an RIT showing the staggered alignment of the Type-II system. Of particular importance is the fact that between InAs and GaSb, the band offset is so large that the top of the valence band of GaSb is coincident with the bottom of the conduction band of InAs, here separated by thin AlSb barriers. When bias is applied, the Fermi energy in the InAs emitter aligns with the quasi-bound state associated with the quantized hole state in the GaSb well, and electrons may tunnel through the valence band of the GaSb well into the InAs collector on the opposite side. As further bias is applied, this resonant energy drops below the CB edge of the emitter, and current flow is quickly suppressed. In contrast to the conventional intra-band tunneling diodes discussed so far, in which the barrier is lowered with increasing bias, the barrier to electrons tunneling from the InAs actually becomes larger as the electrons have to tunnel not only through the AlSb barriers but also through the bandgap of the GaSb. The advantage is that the valley current is greatly suppressed

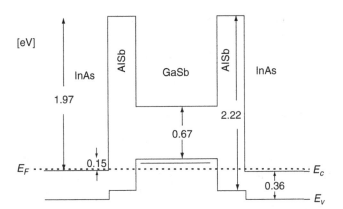

Fig. 3.4 Band-edge diagram (Γ point) for a resonant interband diode (RIT) at room temperature. The valence-band edge of the GaSb and the quantized hole state of the well are both above the InAs conduction band edge at zero bias (after Söderström *et al.*, *Appl. Phys. Lett.* **55**, 1094 (1989), by permission).

compared to a conventional RTD. In fact, the device operation may be thought of qualitatively as a combination of an Esaki diode (where tunneling is from valence to conduction band across the bandgap) and an RTD. This fact is reflected in the large PVRs measured in these and similar structures (see e.g. [27,28]), which are on the order of 20:1 at room temperature and 150:1 at low temperature.

In recent years, there has been an interest in spin transport, or the so-called area of spintronics [29]. In this regard, it has become of interest to use the asymmetric properties of the quantum well in the RTD to give some spin polarization, either by the inversion asymmetry of the crystal [30] or the asymmetry of the structure itself under bias [31]. Both of these effects arise from the spin-orbit coupling in the energy bands. In principle, there should be a spin splitting naturally in the RTD under bias, as has been proposed [32,33]. While some spin polarization can be obtained in the normal RTD, especially if the quantum well is composed of a material with a high value for g, such as InAs, a much larger effect can be obtained in the RIT, where tunneling is through e.g. the valence band of GaSb. The spin splitting is naturally larger in the valence band, and this can lead to a relatively efficient spin polarization [34,35]. Indeed, the RIT structure is sufficiently good in this regard that it can be effectively utilized as a spin filter [36,37].

3.2 Current in resonant tunneling diodes

In real RTD structures, the actual potential contains contributions from the ionized dopants, free carriers, and the applied potential itself, in addition to the band offsets of the heterojunctions. We may cascade a large number of piecewise-continuous potentials together to approximate the actual potential in a real RTD structure to an arbitrary accuracy as the basis for a numerical solution of the transmission and reflection coefficients. Given that we can, in general, calculate the transmission and reflection properties through such a structure, we now want to turn to the problem of using these results to calculate the current flow through a three-dimensional device such as an RTD.

3.2.1 Coherent tunneling

When we look at the tunneling process, we generally consider a single one-dimensional traveling wave that propagates through the barrier region and out the other side. As such, we view tunneling as an elastic process involving no loss of energy of the particle. In a real structure such as an RTD, the ideal problem consists of a plane wave incident on a barrier potential that is semi-infinite in extent in the two transverse directions and that varies only in the third direction. For almost all practical planar barrier devices, this variation in potential is in the growth direction due to the bandgap discontinuities of the heterojunction interfaces, and the space charge due to doping and the applied bias. The plane wave has some component of its wavevector (and hence momentum) in the transverse direction parallel to the barrier. Along with our assumption that tunneling is an energy conserving process, we will further assume that the transverse momentum is conserved, that is, that it remains the same before and after tunneling. This latter assumption is violated in real structures if random inhomogeneities exist in the lateral direction, such as interface roughness and ionized impurities. The main effect is to broaden the effective transmission resonance, reducing the PVR in measured structures compared to the ideal model.

To connect the quantum mechanical fluxes to charge current, we need to introduce the statistical mechanical distribution function to tell us the occupancy of current-carrying states incident and transmitted on the barriers. Exactly what distribution function to use is perhaps one of the central issues of describing non-stationary transport in a phase-coherent system such as the nanostructures discussed in this book. The starting model we will use assumes that we have contacts or reservoirs on the left and right sides of a barrier structure that are essentially in equilibrium and are described by a single-particle distribution function such as the Fermi–Dirac distribution characterized by a Fermi energy. However, since current is flowing, the distribution function cannot truly be characterized by the equilibrium distribution function. We will consider the consequences of this fact later.

The general problem is shown in Fig. 3.5 for a generic tunneling barrier. The applied bias separates the Fermi energies on the left and right by an amount eV. The Hamiltonian on either side of the barrier is assumed separable into perpendicular (z-direction) and transverse components. If we choose the zero-reference of the potential energy in the system to be the conduction band minimum on the left, $E_{c,l} = 0$, the energy of a particle before and after tunneling may be written as

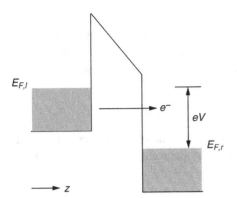

Fig. 3.5 Band diagram for a tunnel barrier under bias, illustrating charge flow.

$$E = E_z + E_t = \frac{\hbar^2 k_{z,l}^2}{2m^*} + \frac{\hbar^2 k_{t,l}^2}{2m^*} \qquad (3.1)$$

on the left side, and

$$E = E_z + E_t = \frac{\hbar^2 k_{z,r}^2}{2m^*} + \frac{\hbar^2 k_{t,r}^2}{2m^*} + E_{c,r} \qquad (3.2)$$

on the right side, where $E_{c,r}$ is the conduction band minimum on the right side and k_z and k_t are the longitudinal and transverse components of the wavevector relative to the barrier. A single, parabolic, isotropic conduction band minimum has been assumed for simplicity. Since the transverse momentum is assumed to be conserved during the tunneling process, $k_{t,l} = k_{t,r}$, and the transverse energy $E_{t,l} = E_{t,r}$ is the same on both sides for the tunneling electron. Therefore, the z-component of the energy is

$$E_z = \frac{\hbar^2 k_{z,l}^2}{2m^*} = \frac{\hbar^2 k_{z,r}^2}{2m^*} + E_{c,r} \qquad (3.3)$$

on the left and right sides of the barrier.

As a further approximation, necessary to introduce irreversibility into the formalism, we assume that the contacts are perfectly absorbing. This means that when a particle injected from one side reaches the contact region of the other side, its phase coherence and excess energy are lost through inelastic collisions with the Fermi sea of electrons in the contact. Thus we assume that an electron injected from one contact at a certain energy E has a certain probability of being transmitted through the boundary determined by $T(E)$, exits the barrier with the same energy and transverse momentum, and finally is absorbed in the opposite contact, where it loses the energy and memory of its previous state. Current flow in this picture is essentially the net difference between the number of particles per unit time transmitted to the right and collected versus those transmitted to the left. This view of tunneling is referred to as *coherent* since the particles maintain their phase coherence across the whole structure before losing energy in the contacts.

To proceed with this picture, consider the current density perpendicular to the barrier in the z-direction for a given energy E with corresponding z-component E_z. The incident current density on the barrier from the *left* due to particles in an infinitesimal volume of momentum space dk, around k, may be written

$$j_l = -e\rho(\mathbf{k}_l)f_l(\mathbf{k}_l)v_z(\mathbf{k}_l)d\mathbf{k}_l, \quad \rho(\mathbf{k}_l) = \frac{2}{(2\pi)^3}, \tag{3.4}$$

where f_l is the distribution function on the left side of the barrier, $\rho(\mathbf{k})$ is the density of states in **k**-space, and the velocity perpendicular to the barrier from the left is

$$v_z(\mathbf{k}_l) = \frac{1}{\hbar}\frac{\partial E(\mathbf{k}_l)}{\partial k_{z,l}} = \frac{\hbar k_{z,l}}{m^*} \tag{3.5}$$

using the parabolic relation (3.3). Here, we neglect the possibility that the energy states on the left or right side of the barriers may be quantized due, for example, to band bending, and we therefore treat the states as three-dimensional. The transmitted current density from the left to right is simply Eq. (3.4) weighted by the transmission coefficient

$$j_l = -\frac{2e\hbar}{(2\pi)^3 m^*} T(k_{z,l}) f_l(\mathbf{k}_t, k_{z,l}) k_{z,l} dk_{z,l} d\mathbf{k}_t, \tag{3.6}$$

where $T(k_{z,l})$ is the transmission coefficient which, for the ideal case, is only a function of the longitudinal momentum and energy. Similarly, the transmitted current from right to left may be written for the same energy E and E_z

$$j_r = -\frac{2e\hbar}{(2\pi)^3 m^*} T(\mathbf{k}_{z,r}) f_r(\mathbf{k}_t, k_{z,r}) k_{z,r} dk_{z,r} d\mathbf{k}_t. \tag{3.7}$$

At a given longitundinal energy E_z, the transmission coefficient is symmetric so that $T(E_{z,l}) = T(E_{z,r})$. Further, $k_{z,l}dk_{z,l} = k_{z,l}dk_{z,l} = m^*dE_z/\hbar^2$ if we differentiate both sides of (3.3). Therefore, the net current in the direction of the voltage drop is the difference between the left and right currents integrated over all **k**, or

$$J_T = \frac{2e}{(2\pi)^3\hbar}\int_0^\infty dE_z \int_0^\infty dk_t k_t \int_0^{2\pi} d\vartheta T(E_z)[f_l(E_z, k_t) - f_r(E_z, k_t)], \tag{3.8}$$

where the integration over E_z is from zero to infinity because tunneling from right to left below $E_z = 0$ is forbidden.

At this point, no further reduction to Eq. (3.8) can be made unless we make assumptions concerning the nature of the distribution functions on the left and right sides. The lowest-order approximation is to assume that these distributions are given by the equilibrium Fermi–Dirac functions determined by the bulk Fermi levels on the respective sides of the barrier,

$$f_{l,r}(E_z, E_t) = \frac{1}{1 + \exp\left(\frac{E_z + E_t - E_{F,l,r}}{k_B T}\right)}, \qquad (3.9)$$

where T is the lattice temperature and $E_{F,l,r}$ is the Fermi energy on the left and right side, respectively. The difference between the two is just the applied bias, $E_{F,l} = E_{F,r} + eV$. Since the Fermi function is isotropic, the angular integration gives 2π. Likewise, the integration over the transverse wavevector may be converted to an integral over energy. Assuming parabolic bands, Eq. (3.8) becomes

$$J_T = \frac{4\pi e m^*}{(2\pi)^3 \hbar^3} \int_0^\infty dE_z T(E_z) \int_0^\infty dE_t [f_l(E_z, E_t) - f_r(E_z, T_t)]. \qquad (3.10)$$

For the Fermi function (3.9), the integration over energy is easily evaluated to give

$$J_T = \frac{e m^* k_B T}{2\pi^2 \hbar^3} \int_0^\infty dE_z T(E_z) \ln\left[\frac{1 + e^{(E_{F,l} - E_z)/k_B T}}{1 + e^{(E_{F,l} - eV - E_z)/k_B T}}\right], \qquad (3.11)$$

sometimes referred to as the *Tsu–Esaki formula*, where the particular form was popularized [11] in connection to resonant tunneling diodes. (Similar equations appear much earlier in single-particle tunneling; see for example [10].) The logarithmic term is sometimes called the *supply function* [38], since it more or less determines the relative weight of available carriers at a given perpendicular energy. At low temperature, the supply function becomes step-like, and Eq. (3.11) becomes simply

$$J_T = \frac{e m^*}{2\pi^2 \hbar^3} \left[\int_0^\infty dE_z T(E_z)(E_F - E_z) \right.$$
$$\left. - \int_0^\infty dE_z T(E_z)(E_F - E_z - eV) \right]. \qquad (3.12)$$

If we now consider current through a resonant structure such as an RTD the current density is dominated by the resonant portion of the transmission coefficient. For example, if the transmission coefficient is assumed to be very peaked around $E_z = E_n$ using a Lorentzian form, we can approximate it as a *delta* function so that at low temperature, Eq. (3.12) may be integrated to give

$$J_T = \frac{e m^* T_{res} \Gamma_n}{4\pi \hbar^3} (E_F - E_n), \; 0 < E_n < E_F, \qquad (3.13)$$

where the asymptotic approximation for the delta function has been employed:

$$\delta(E_z - E_n) = \frac{1}{\pi} \lim_{\Gamma_n \to 0} \frac{\Gamma_n/2}{(\Gamma_n/2)^2 + (E_z - E_n)^2}. \qquad (3.14)$$

The voltage dependence enters through E_n, the bound state energy. If we assume that the voltage drop is equally divided between the two barriers, the well is lowered in potential energy by an amount $eV/2$ with respect to the emitter (the left side), and is higher in energy than the collector by the same amount. The term E_n may therefore be replaced in (3.13) by $E_{n,0} - eV/2$, where $E_{n,0}$ is the quasi-bound state energy relative to the well bottom. This gives a sudden turn-on of current when $eV = 2(E_{n,0} - E_F)$, and cuts off when $eV = 2E_{n,0}$ giving rise to NDR. The peak occurs when $E_n = 0$, giving a peak current density

$$J_T = \frac{em^* T_{res} \Gamma_n E_F}{4\pi\hbar^3}. \qquad (3.15)$$

As can be seen, the peak current depends physically on the Fermi energy in the emitter, and hence the doping there, as well as on the product of the peak transmission probability and resonance width. Since both the resonance width and the resonant transmission amplitude increase as the barrier thickness decreases, thin barriers are essential for high peak current densities.

Valley current represents contributions due to scattering, both elastic and inelastic. Both types of scattering allow a relaxation of the parallel-momentum conservation rule and thus increase the amount of current which may flow off-resonance. We may nominally associate elastic scattering in RTDs with interface roughness at the heterojunction interfaces, unintentional doping in the tunneling region, and alloy disorder when barrier materials like $Al_xGa_{1-x}As$ are employed. The first two depend strongly on the quality of epitaxial material growth, which helps explain the dramatic improvements in PVRs in the GaAs/AlGaAs system over time. Calculations that incorporate roughness effects have been reported by [39] and [40], for example. Another factor is the limitation of the single-band model used here. As bias is increased, carriers are injected at higher and higher energies into the AlGaAs barriers and GaAs well. Ultimately, high conduction band minima are accessible to these carriers, particularly the X-valley minimum in AlGaAs, which provide additional channels for current to flow [41] and necessitates a multiband calculation of the tunnel current [42].

Inelastic scattering via phonons and collective excitations not only break the assumption of transverse momentum conservation but also lead to loss of phase coherence, as mentioned in the introductory chapter. As a consequence, if such interactions are strong, the tunneling process is no longer characterized as coherent, since the phase relationship leading to the buildup of the resonant amplitude is only partially preserved. We will defer a rigorous discussion of this phenomenon until a later chapter, when inelastic effects are included formally into the transport model. Here we basically consider inelastic scattering from a phenomenological point of view in regards to planar barrier tunneling, and we discuss some of the consequences of this interaction on the tunneling process in RTDs.

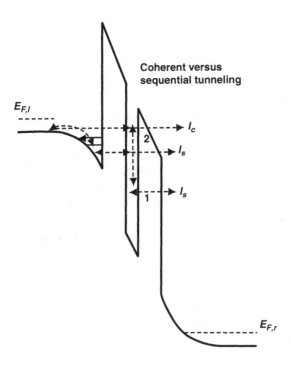

Fig. 3.6 Illustration of incoherent (sequential) tunneling in an RTD structure. I_c corresponds to the coherent component of the current, while I_s represents incoherent components.

3.2.2 Incoherent or sequential tunneling

Inelastic or phonon-assisted tunneling has long been recognized as an important process in tunneling, as discussed in the review of pre-1970 tunneling by Duke [10]. Luryi [43] pointed out in connection with double barrier RTD structures that NDR could be explained with a *sequential* tunneling process in which phase coherence is lost. In this model, shown schematically in Fig. 3.6, carriers tunnel through the first barrier into the quasi-2DEG residing in the well and subsequently lose their phase memory through a scattering process there. The phase-randomized carriers tunnel out of the second barrier through a second uncorrelated tunneling process. Luryi argued that in order to explain NDR in an RTD structure, it is sufficient to consider the tunneling of a 3D electron from the emitter into a 2DEG under the assumption of transverse momentum conservation. Assuming that the excess kinetic energy in the quantum confined well is associated only with the transverse motion, we may equate the total energy of a tunneling electron from the emitter into the well as

$$E = E_c + \frac{\hbar^2 k_z^2}{2m^*} + \frac{\hbar^2 k_{t,e}^2}{2m^*} = E_n + \frac{\hbar^2 k_{t,w}^2}{2m^*}, \quad (3.16)$$

where $k_{t,e}$ and $k_{t,w}$ denote the transverse wavevector in the emitter and well, respectively. If tunneling itself is energy-conserving, and the transverse moment is also conserved, the condition for tunneling to occur becomes

$$E_c + \frac{\hbar^2 k_z^2}{2m^*} = E_n. \tag{3.17}$$

Since the longitudinal wavevector k_z is real, no electrons can tunnel and conserve transverse momentum when $E_c > E_n$, that is, when the conduction band edge in the emitter rises above the bound state energy in the well. At low temperature, no electrons tunnel until the Fermi energy crosses the bound state energy. Then the supply of electrons available for tunneling rises as $E_F - E_n$ until the criterion (3.17) is violated and the current drops abruptly, giving rise to NDR. The qualitative shape of the I–V relationship is the same as that predicted in the coherent model of the previous section, although at first glance the current predicted by the two should be quite different. However, as was subsequently shown [44,45,46], the peak current density of the two models is in fact the same and cannot be used to distinguish between coherent and incoherent tunneling.

To illustrate this, we use the simple model given by Stone and Lee [47] to include inelastic scattering in the resonant tunneling amplitude. There it is assumed within the context of the Breit–Wigner formula that part of the incident flux is absorbed by inelastic processes that can be described by adding an additional imaginary part [48], $i\Gamma_i/2$, to the complex energy eigenvalue. This imaginary part may be qualitatively associated with the scattering time, $\tau_s = \hbar/\Gamma_i$, due to inelastic processes. The transmission coefficient in the general asymmetric barrier case then becomes

$$T_c(E) = T_{res} \frac{\Gamma_n^2/4}{\Gamma_n^2/4 + (E - E_n)^2} = T_{res} \left(\frac{\Gamma_n}{\Gamma_T}\right)^2 \frac{\Gamma_T^2/4}{\Gamma_T^2/4 + (E - E_n)^2}, \tag{3.18}$$

where $\Gamma_T = \Gamma_n + \Gamma_i$. Here we use T_c to denote the coherent transmission since the damping term Γ_i represents just the loss of flux from that transmitted coherently. Because of this flux loss, probability current is no longer conserved, i.e., $T_c + R_c < 1$. However, since charge is not created or destroyed, this lost flux must appear somewhere else as transmitted flux through another channel in order to conserve the total flux. This extra flux may be thought as that lost due to scattering in the well to lower energy (as envisioned through the sequential model) and subsequently transmitted in a separate process at a different energy. This inelastic fraction is therefore that lost from the coherent part

$$T_i + R_i = 1 - T_c - R_c = T_{res} \frac{\Gamma_i \Gamma_n/2}{\Gamma_T^2/4 + (E - E_n)^2}, \tag{3.19}$$

where T_i and R_i are the inelastic forward and backward scattering probabilities, and the expression for the coherent reflection coefficient has been similarly expanded about the resonance energy. If these are assumed to be the same, the total transmission coefficient including both elastic and inelastic scattering

$$T(E) = T_i + T_c = T_{res}\left(\frac{\Gamma_n}{\Gamma_T}\right)\frac{\Gamma_T^2/4}{\Gamma_T^2/4 + (E - E_n)^2}. \qquad (3.20)$$

Here, the maximum transmission is reduced due to inelastic scattering by the ratio $\Gamma_n/(\Gamma_n + \Gamma_i)$ and the resonance is broadened compared to the purely coherent case. A similar form was obtained by Jonson and Grincwajg [46] using ray-tracing arguments and by Büttiker [49,50] using the multi-channel Landauer–Büttiker formula, which will be discussed in Section 3.4 below.

If we go back to our derivation of the peak current density leading to Eq. (3.15), and if the same assumptions are made at low temperature and for peaked transmission coefficients, it is apparent that in the more general case

$$J_P \sim T_{res}\left(\frac{\Gamma_n}{\Gamma_T}\right)\Gamma_T = T_{res}\Gamma_n, \qquad (3.21)$$

which is independent of the inelastic scattering time. Therefore, as long as the resonance is not broadened too much by inelastic scattering, the peak currents are expected to be the same in the coherent and incoherent cases.

As mentioned earlier, the characteristic time for tunneling associated with the coherent width Γ_n may be quite long compared with that due to inelastic scattering. Thus, pure coherent transport in RTDs probably does not occur except in very thin barrier structures such as those characteristic of state-of-the art devices. The role of such inelastic scattering on the frequency response of RTDs is still unclear. For more discussion in this regard, the reader is referred to the review by Liu and Sollner [51].

3.2.3 Space charge effects and self-consistent solutions

The ideal model we developed so far neglected the electrostatic potential arising from ionized donors and acceptors as well as the free carrier charge distribution. In reality, space charge layers form in the device, leading to additional potential drops inside the device other than that across the barriers themselves. Further, free carriers may be stored in the well, particularly on resonance, which modifies the potential distribution across the RTD structure. Evidence for charge storage is observed in magneto-tunneling experiments in which a magnetic field is applied perpendicular to the plane of the tunnel barrier [52], and Shubnikov–de Haas (SdH)-like oscillations in the conductance are observed related to magnetic quantization of the quasi-2D states in the well [53]. Space charge buildup has also been probed optically [54] where luminescent transitions between the quasi-bound electron and hole states in the well of the RTD are observable. The storage of free carriers in the well as the device is biased through resonance has been suggested as a source of *intrinsic bistability*, giving rise to the experimentally observed hysteresis in the I–V characteristics of RTDs in the NDR regime [55]. Practically, however, there is great difficulty in separating bistability due to physical charge storage in the quantum well from external measurement circuit effects when probing the unstable NDR regime of an RTD [56,57] unless asymmetric RTD structures are prepared which accentuate charge storage effects [58].

In Chapter 2, we discussed the necessity for self-consistent solutions in which the coupled Schrödinger–Poisson equations are solved with appropriate approximations for the many-body potential for modulation-doped structures. For an RTD structure, similar self-consistent solution methods must be employed to account for the space charge layer that forms in the diode under bias, and to account for the pile-up of free charge in the well and in the accumulation layer adjacent to the double barrier structure on the emitter side. However, in contrast to the quantum well case, an RTD represents an *open* structure in which the wave function does not vanish on the boundaries (at least when we conceptually separate the microscopic system from the macroscopic external environment). In the open system, particles move in and out of the boundary of the microscopic system into the macroscopic external circuit, which we have modeled as ideal absorbing and emitting contacts. Therefore, care must be taken in specifying the boundary conditions in such a system and the statistical weighting associated with normalization of the single-particle wave functions in order to conserve charge, and to give the appropriate charge density (when summed over all states) in the contact regions. In this sense, for an open system, one cannot formulate the problem purely in a reversible quantum mechanical framework. Rather, nonequilibrium statistical mechanics also plays a central role in the system description.

In order to move beyond the simple rectangular barrier approximations used above, we can still assume that flat band conditions are reached at some point far away from the tunneling region on the right and left sides. These somewhat arbitrary points serve as the boundaries of the system across which the interior solution is matched to asymptotic plane-wave solutions in the contact regions. In the interior, we have not only the potential due to the band discontinuities, $E_c(z)$, but also the potential due to ionized donors and free carriers,

$$\frac{d}{dz}\left(\varepsilon(z)\frac{d\varphi}{dz}\right) = -e\left[N_D^+(z) - N_A^-(z) - n(z)\right], \qquad (3.22)$$

where $\varphi(z)$ is the electrostatic potential, $N_D^+(z)$ and $N_A^-(z)$ are the ionized donor and acceptor concentrations respectively, and $n(z)$ is the free carrier concentration. In the n-type asymptotic contact regions, charge neutrality is assumed so that $n = N_D - N_A$ independent of position (assuming complete ionization).

A semiclassical approximation for the free carrier density is to assume the Thomas–Fermi screening model in which the density is determined by the position of the Fermi energy relative to the band edge

$$n(z) = N_c F_{1/2}(E_F - e\varphi - E_c), \qquad (3.23)$$

where $F_{1/2}$ is the Fermi–Dirac integral, and N_c is the effective density of states in the conduction band. When substituted into Eq. (3.22), the resulting nonlinear Poisson equation must be solved numerically. The Thomas–Fermi model is strictly valid only in equilibrium, but it is typically extended to nonequilibrium cases by defining quasi-Fermi energies governing the local density. In an RTD structure, this is usually accomplished by partitioning the device into regions governed either by the Fermi energy of the left contact or that of the right contact. Of course, this is arbitrary to a large extent, and so accordingly are the solutions of the problem one obtains.

Within the Hartree approximation (see Chapter 2), we need to combine the one-electron envelope functions with the appropriate distribution function. Due to the flux of particles from the left, the density on the left of the barrier may be written

$$n_l(z) = \frac{2}{(2\pi)^3} \int d\mathbf{k}_t \int_0^\infty d\mathbf{k}_z |\psi_l(k_z)|^2 f_l(\mathbf{k}_t, k_z), \tag{3.24}$$

where the prefactor is the density of states in k-space with a factor 2 for spin degeneracy, and where $f_l(\mathbf{k}_t, k_z)$ is the one-particle distribution function on the left side. The integration over k_z is only over positive values because the envelope function, $\psi_l(k_z)$, includes both the incident and reflected amplitudes. In various treatments of self-consistent solutions in RTD structures [59,60,61,62], the distribution function is assumed to be a Fermi–Dirac function in the contact region characterized by an effective temperature, T. Then the easily performed integration over the transverse motion gives

$$n_l(z) = \frac{m^* k_B T}{2\pi^2 \hbar^2} \int_0^\infty d\mathbf{k}_z |\psi_l(k_z)|^2 \ln\left[1 + e^{(E_{F,l} - E_z)/k_B T}\right], \tag{3.25}$$

where $E_{F,l}$ is the Fermi energy on the left. A similar equation may be written for the density due to carriers flowing from the right, with the appropriate change of $l \to r$ above. The total density is the sum of the two.

However, the assumption of equilibrium distribution functions in the contacts under nonequilibrium conditions when current is flowing leads to an unphysical pile-up or depletion of charge in the contact region unless an artificial readjustment of charge is made [63]. This fact is apparent if one looks at the asymptotic density on the left side; for example,

$$n_l = \int_0^\infty dk_z \{[2 - T(E_z)] f_l(E_z) + T(E_z) f_r(E_z)\}, \tag{3.26}$$

where f_r is the averaged distribution function in the right contact, with $E_{F,r} = E_{F,l} - eV$, with V the applied voltage. The first term in the integral is the incident and reflected density due to carriers injected from the left contact, $1 + R = 2 - T$

(invoking current conservation). The second term is the transmitted density from the right (which properly should be written over negative wavevectors, but, assuming f_r is symmetric, can be combined above). Here the incident waves from left and right have been normalized to unity. In equilibrium ($V = 0$), the terms involving $T(E_z)$ cancel, and the bulk carrier density is recovered. Under forward bias, the right term vanishes, and the density reduces below its equilibrium value, causing a depletion in the contact. Correspondingly, there is a pile-up of charge in the right contact above the equilibrium value under forward bias given by

$$n_r = \int_0^\infty dk_z \{[2 - T(E_z)] f_r(E_z) + T(E_z) f_l(E_z)\}. \tag{3.27}$$

This evident rearrangement of free charge due to the barrier region, and the inconsistency of the simple use of the equilibrium contact distribution function, is intimately related to the derivation of the *Landauer formula* (see Section 3.3 below). In the present context of self-consistent solutions in RTDs, Pötz [63] used a drifted Fermi–Dirac distribution for the contacts in order to conserve charge, which when integrated over the transverse momentum gives:

$$f_i(E_z) = \frac{m^* k_B T}{2\pi^2 \hbar^2} \ln\left[1 + \exp\left(\frac{E_{F,i} - \hbar^2 (k_{z,i} - k_{0,l})^2 / 2m^*}{k_B T}\right)\right], \tag{3.28}$$

where $i = l, r$ refers to the left or right side, T is the electron temperature in the corresponding region, and $\hbar k_{0,i}/m^*$ is the drift velocity in region i. The drift wavevectors and/or electron temperatures in the respective contact regions are used as parameters to establish charge neutrality in the contacts.

In order to calculate the current self-consistently, we still need the transmission coefficient through the tunneling region. For quasi-one-dimensional problems like planar barrier tunneling, a transfer matrix technique is frequently employed [64,65]. Here, a one-dimensional grid is defined, and the transfer matrix from one grid point to the next calculated. The simplest approximation is just to assume constant potential in each grid, so that the transfer matrix is just that due to a finite step at every grid point. A better approximation is to assume that the potential varies linearly between grid points, in which the solution in each segment is given by Airy functions [59], and to define the transfer matrices accordingly. The total potential used to solve the transfer matrix problem is the sum of the electrostatic potential and that due to the band offsets in the simple one-band model. Since the electrostatic potential also depends on the solution of the Schrödinger equation through the density (3.26), the total solution must be found iteratively [61] in a similar fashion to that discussed in Chapter 2.

Figure 3.7 shows an example of the calculated band profile of an RTD structure at several bias voltages. The RTD structure has 5-nm spacer layers adjacent to the barrier on either side which give rise to band bending even for

3.2 Current in resonant tunneling diodes

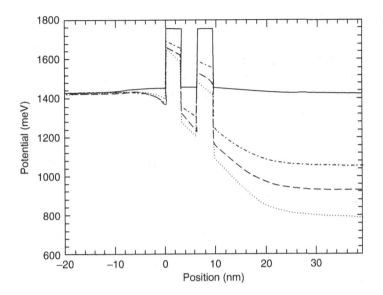

Fig. 3.7 Effective potential for a double barrier structure with 5-nm undoped spacer layers and $N_D = 2 \times 10^{18}$ cm^{-3} for various applied biases. Solid line, 0 V; dot-dashed line, 0.375 V (close to resonance); dashed line, 0.500 V (above resonance); and the dotted line, 0.625 V (after W. Pötz, *J. Appl. Phys.* **66**, 2458 (1989), by permission).

zero applied bias. For increasing bias, it is apparent that a substantial fraction of the bias drops across the region outside of the double barrier structure in contrast with the simple model introduced in the previous sections. One consequence is that the voltage corresponding to the peak current is pushed to higher voltages than that predicted by the ideal model, which in some sense resembles a series resistance effect. Another interesting feature is the presence of a potential notch on the emitter side of the double barrier structure. This notch resembles the quantum confining potential on a single-heterojunction system discussed in Chapter 2 and may trap charge there in quasi-bound states (quasi-bound as they still have a finite lifetime) forming an accumulation layer adjacent to the tunnel barrier. In fact, the existence of such a layer is evident in magneto-tunneling experiments of the type discussed earlier in this section [66] where SdH-like oscillations are observed associated with quantum confined electrons there. The contribution to the current due to tunneling out of these states is quantitatively different than that due to electrons in the continuum above [67] and not accounted for in the heretofore described flux formalism. In fact, in our ballistic paradigm developed so far, there is no possibility of charge becoming trapped there because this requires some sort of inelastic process to relax carriers into the accumulation layer. This paradox really points to the limitations of a description in which dissipation occurs only in the contacts. In reality, the distribution function for electrons cannot be described by a local semiclassical function across the device. This has led to consideration of quantum distribution functions such as the Wigner distribution [68] for RTD modeling [69,70]. We will come back to issues regarding dissipative nonequilibrium transport in more detail in a later chapter.

3.3 Landauer formula

In the earlier sections of this chapter, we introduced quantum transport through a discussion of tunneling in planar barrier structures. We now extend these concepts to more complicated geometrical structures in which phase coherence is preserved and the description in terms of quantum flux is important. In this sense, we want to talk about transmission instead of merely tunneling. Laterally confined systems may have quantized conductance in phase coherent regions, and our description of transmission, rather than tunneling, will emphasize that the approach introduced here goes far beyond mere tunneling structures. Hence, our discussion can include quantum point contacts as well as more complicated quantum waveguide structures in two and three dimensions. In fact, once we have generalized our view to that of transmission, the treatment can follow regions which are not phase coherent, but have considerable scattering processes present. Indeed, there is no limitation to the transmission view to phase coherent structures, since we will talk about the transmission being determined between a mode in the input region to a mode in the output region, without any regard to how the coupling from one such mode to another actually was realized. To generalize our formalism for treating current flow through such systems, we will use an early formalism due to Landauer [71,72] which has now become standard in the parlance of nanostructure transport.

To begin, we note that in our previous derivation of the tunneling current through planar barrier structures, inconsistencies arise associated with current being injected from ideal equilibrium contacts, and with the space charge that forms as a result of current flow through the structure. In semiconductors, charge fluctuations near the barriers may substantially affect the potential near the barriers. Also, with current flow, the distribution function itself may be significantly perturbed from the equilibrium. The spatial inhomogeneity resulting from current flow around obstacles was recognized early on by Landauer [71]. He formulated the problem in a somewhat different way by considering a one-dimensional system in which a constant current was forced to flow through a structure containing scatterers, and asking the question of just what the resulting potential distribution will be, due to the spatially inhomogeneous distribution of scatters. In this context, our planar barrier of the previous section may be considered as a general scattering center. The result in one dimension is the so-called *Landauer formula,* which is derived below for the *single-channel case* and then will be extended, in the next section, to the general *multi-channel* case [73].

To begin, consider the general barrier problem for a 1D conductor shown in Fig. 3.8. As we discussed in Chapter 2, an ideal 1D conductor is realized by a quantum wire in the extreme quantum limit where only one subband (channel) is occupied. Ideal (i.e., without scattering) conducting leads connect the scattering region to reservoirs on the left and right characterized by quasi-Fermi energies

3.3 Landauer formula

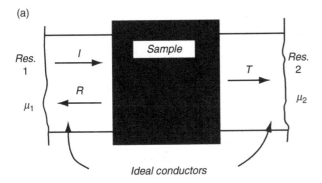

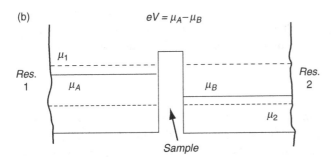

Fig. 3.8 Schematic illustration of the conductance of a one-dimensional sample: (a) the ideal decomposition of the structure into the sample where scattering occurs, and ideal 1D leads connecting the sample to reservoirs on either side; (b) the redistribution of charge due to the reservoirs resulting in new Fermi energies μ_A and μ_B on the left and right sides, respectively.

μ_1 and μ_2, respectively, corresponding to the electron densities there. As we discussed in the previous sections, these reservoirs or contacts randomize the phase of the injected and absorbed electrons through inelastic processes such that there is no phase relation between particles. For such an ideal 1D system, the current injected from the left and right may be written as an integral over the flux as we did in the previous sections:

$$I = \frac{2e}{2\pi} \left[\int_0^\infty dk v(k) f_1(k) T(E) - \int_0^\infty dk' v(k') f_2(k') T(E') \right], \quad (3.29)$$

where the constant is the 1D density of states in k-space, $v(k)$ is the velocity, $T(E)$ is the transmission coefficient, and f_1 and f_2 are the reservoir distribution functions characterized by their respective Fermi energies introduced above. The integrations are only over positive k and k' relative to the direction of the injected charge. If we now assume low temperatures, electrons are injected up to an energy μ_1, into the left lead and injected up to μ_2 into the right one. Converting to integrals over energy, the current becomes

$$I = \frac{e}{\pi}\left[\int_0^{\mu_1} dE\left(\frac{dk}{dE}\right)v(k)f_1(k)T(E)\right.$$
$$\left. - \int_0^{\mu_2} dE\left(\frac{dk'}{dE}\right)v(k')f_2(k')T(E)\right] \quad (3.30)$$
$$= \frac{e}{\pi\hbar}\int_{\mu_1}^{\mu_2} dE\, T(E).$$

Note that in 1D, the energy-dependent velocity and density of states factor arising from the change of variables cancel to a simple constant, an important result in our discussion of quantized conductance in point contacts. If we further assume that the applied voltage is small (i.e., in the linear response regime) so that the energy dependence of $T(E)$ is negligible, the current becomes simply

$$I = \frac{2e}{h}T(\mu_1 - \mu_2). \quad (3.31)$$

As a result of transmission and reflection around the barrier with current flowing, there is a reduction in carrier density on the left side of the barrier and pile-up of charge on the right side. Assuming we are still in the linear response regime, we can approximate this charge rearrangement by an average density in the ideal leads on either side of the scattering structure characterized by different Fermi energies μ_A and μ_B as shown in Fig. 3.8. The actual voltage drop V across the scattering structure is then given by

$$eV = \mu_A - \mu_B, \quad (3.32)$$

which is less than the voltage between the reservoirs, $\mu_1 - \mu_2$, the difference representing a *contact* potential drop. We need to find the potential difference (3.32) in terms of the current flowing through the structure given by (3.31). To do this, we can write the 1D density in the ideal lead on the left side as

$$n_a = \frac{1}{\pi}\int_{-\infty}^{\infty} dk f_a(E) = \frac{1}{\pi}\int_{-\infty}^{\infty} dk\{(2-T)f_1(E) + Tf_2(E)\}, \quad (3.33)$$

where $f_a(E)$ represents the near-equilibrium distribution function in the left lead, characterized by a Fermi energy μ_A. The integral over $\pm k$ considers carriers traveling in both directions. On the right side, we have written the average density in terms of the injected carriers from the left and right reservoirs into the left lead, where again current conservation has been used to write $1 + R = 2 - T$. It is implicitly assumed that the Friedel oscillations of the charge density around the barrier damp out sufficiently rapidly to use the asymptotic form of the charge density there. Similarly, the density in the right lead is given by

3.3 Landauer formula

$$n_b = \frac{1}{\pi} \int_{-\infty}^{\infty} dk f_b(E) = \frac{1}{\pi} \int_{-\infty}^{\infty} dk \{(2-T)f_2(E) + Tf_1(E)\}. \quad (3.34)$$

Now, taking the low-temperature limit, we can subtract (3.34) from (3.33) to give

$$2\int_{\mu_B}^{\mu_A} dE \left(\frac{dk}{dE}\right) = \int_{\mu_2}^{\mu_1} dE \left(\frac{dk}{dE}\right)(2-T) - \int_{\mu_2}^{\mu_1} dE \left(\frac{dk}{dE}\right)T. \quad (3.35)$$

If we now assume that the difference between the Fermi energies is sufficiently small that we may neglect the energy dependence of T and the inverse velocity, dk/dE, then (3.35) may be simply integrated to give

$$\mu_A - \mu_B = (1-T)(\mu_1 - \mu_2). \quad (3.36)$$

It is obvious, however, that the distribution of carriers in the leads is not a Fermi–Dirac, particularly when the bias is non-zero. Carriers do not populate all the states between μ_2 and μ_1, for example, on the right side. If we use the Fermi energy μ_B only as a counting scheme for the density of carriers on the right side, it can be defined such that the number of occupied states above this level is equal to the number of unoccupied states (holes) below [73]. The number of states occupied above μ_B on the right side due to injection from the left is $TD(E)$ $(\mu_1 - \mu_B)$, where $D(E)$ is the density of states in energy corresponding to positive k only (1/2 the total density of states). Likewise, the number of unoccupied states below μ_B is given by $2D(E)(\mu_B - \mu_2) - TD(E)(\mu_B - \mu_2)$, the states below μ_2 being completely filled. Equating the two gives

$$T(\mu_1 - \mu_B) = (2-T)(\mu_B - \mu_2). \quad (3.37)$$

Similarly, on the left side, μ_A is defined from

$$(1+R)(\mu_1 - \mu_A) = [2 - (1+R)](\mu_A - \mu_2), \quad (3.38)$$

where the left side is the number of occupied states above μ_A, and the right side is the number of unoccupied states below. Combining the two equations gives (3.36) above.

Substituting (3.36) into (3.31) gives

$$I = \frac{2e}{h} \frac{T}{1-T}(\mu_A - \mu_B), \quad (3.39)$$

or, in terms of the conductance using (3.32),

$$G = \frac{I}{V} = \frac{2e^2}{h}\left(\frac{T}{1-T}\right) = \frac{2e^2}{h}\frac{T}{R}, \quad (3.40)$$

which is finally the single-channel *Landauer formula* [71], [72]. We see that the conductance is given very simply by the product of the ratio of the transmission and reflection at the Fermi-energy and the fundamental conductance, $2e^2/h = 7.748 \times 10^{-5}$ mhos. As discussed in the introduction, these conductance quanta (corresponding to a resistance of 12,907 Ω) play a fundamental role in the physics of mesoscopic systems, being the conductance associated with a single one-dimensional channel. One characteristic of mesoscopic systems is that their conductance properties may be governed by transport through a few such channels, giving rise to characteristic changes of precisely $2e^2/h$.

The factor $1 - T$ appearing in the denominator has resulted in some confusion in the literature, which has been discussed in detail in subsequent publications [74,75,76,77,78]. Basically the problem is where you actually measure the voltage. If one applies current through a pair of contacts and then measures the voltage difference in the ideal leads noninvasively through a separate pair of contacts (four-terminal measurement), then Eq. (3.40) would provide the appropriate conductance. The catch is how one would devise leads to actually measure this voltage without influencing the conductance of the structure. In a two-terminal measurement, the voltage and current are measured through the same set of leads. In this case, the voltage measured is

$$eV = \mu_1 - \mu_2, \qquad (3.41)$$

which from (3.31) clearly gives

$$G = \frac{I}{V} = \frac{2e^2}{h} T, \qquad (3.42)$$

without the factor $1 - T$. The reason is that now in addition to the potential drop across the scattering structure, we are measuring a contact potential drop across the ideal leads due to the self-consistent charge build-up characterized by a contact resistance

$$R_c = \frac{h}{2e^2}. \qquad (3.43)$$

Of course, if T is very small, then the series resistance of the barrier dominates, and the two-terminal and four-terminal measurement give identical results.

3.4 The multi-channel case

We want to generalize now to the case in which multiple independent conducting channels are present [73]. This case corresponds, for example, to the low-temperature situation in which N 1D subbands (modes) are populated at the Fermi energy, all of which may contribute to the current. Let us be a little more specific. Here, we assume that there are several connection leads to the mesoscopic system. Each of these leads may have several modes populated at the Fermi energy. Alternately, one can view the situation at finite temperature as a multi-channel situation in which each channel represents a particular narrow range of energy. On the left side, the stationary solutions in the ideal lead may be written as, for example,

$$\psi_i^l(\mathbf{r}_t, z) = \sum_{j=1}^{N} \left(A_j e^{ik_{j,z}z} + B_j e^{-ik_{j,z}z}\right) \varphi_j(\mathbf{r}_t), \quad (3.44)$$

where $\mathbf{r}_t$ represents the position vector in the transverse direction, j labels the transverse solutions, and $k_{j,z}$ is the wavevector corresponding to this mode. Here, the subscript i labels a particular lead. A similar such solution may be written on the right side of the scatterer, as

$$\psi_n^r(\mathbf{r}_t, z) = \sum_{j=1}^{N'} \left(C_j e^{ik_{j,z}z} + D_j e^{-ik_{j,z}z}\right) \varphi_j(\mathbf{r}_t). \quad (3.45)$$

A scattering matrix connects the coefficients on the left, $2N$ in this case, with $2N'$ on the right side. The elements of the scattering matrix describe the connection of a mode in the input lead with a mode in the output mode, and these two modes do not necessarily need to have the same transverse modal behavior.

3.4.1 An introduction to the multi-channel case

For simplicity, consider the case in which the input and output leads are equivalent, in that the number of channels on either side is the same value, N. An incoming wave in mode i on the left has a certain probability, $T_{ji} = |t_{ji}|^2$, to be transmitted into mode j on the right side, and probability $R_{ji} = |r_{ji}|^2$ of being reflected back. The transmission and reflection amplitudes, t_{ji} and r_{ji} are the components of the $2N \times 2N$ scattering matrix

$$\mathbf{S} = \begin{bmatrix} r & t' \\ t & r' \end{bmatrix}, \quad (3.46)$$

where r, t, r', and t' all represent $N \times N$ arrays rather than individual elements. The scattering matrix in the case of equivalent input and output leads is again both unitary and symmetric.

Carriers are fed equally into all the modes in each lead from the reservoirs, and fill these modes up to the Fermi energy μ_1 on the left and μ_2 on the right. For a particular mode j on the left side, the current injected into channel i on the right side between μ_2 and μ_1 is again independent of the velocity and density of states as in (3.31). Since each channel is assumed to be fed equally, the total current due to charge injected into mode i on the right side between μ_2 and μ_1 is now given by

$$I_i = \frac{2e}{h} \sum_{j=1}^{N} T_{ij}(\mu_1 - \mu_2) = \frac{2e}{h} T_i(\mu_1 - \mu_2), \quad (3.47)$$

where T_i is the shorthand notation for the sum over j on the left. Since all channels are independent, the total current is given by

$$I_{tot} = \sum_{i=1}^{N} I_i = \frac{2e}{h} \sum_{i=1}^{N} T_i(\mu_1 - \mu_2) = \frac{2e(\mu_1 - \mu_2)}{h} Tr(tt^+), \quad (3.48)$$

where on the right side the sum over V is written in terms of the transmission submatrices

$$\sum_{i=1}^{N} T_i = \sum_{i,j=1}^{N} T_{ij} = \sum_{i,j=1}^{N} |t_{ij}|^2 = \sum_{i,j=1}^{N} t_{ij} t_{ji}^* = Tr(tt^+). \quad (3.49)$$

Similarly, the current into channel i on the left side may also be expressed in terms of the reflection coefficients, R_{ij}, starting with

$$I_i = \frac{2e(\mu_1 - \mu_2)}{h}\left[1 - \sum_{j=1}^{N} R_{ij}\right] = \frac{2e(\mu_1 - \mu_2)}{h}[1 - R_i]. \quad (3.50)$$

Therefore, the total current becomes

$$I_{tot} = \sum_{i=1}^{N} I_i = \frac{2e(\mu_1 - \mu_2)}{h} \sum_{i=1}^{N} [1 - R_i]. \quad (3.51)$$

Comparing to Eq. (3.48), this gives the current continuity relation between all channels as

$$\sum_{i=1}^{N} T_i = \sum_{i=1}^{N} [1 - R_i]. \quad (3.52)$$

Again, as in the single-channel case, the ideal leads are labeled by different Fermi energies μ_A and μ_B to account for the self-consistent pileup of charge on either side of the scatterer. To calculate these, the number counting argument of the last section is extended to look at the total number of occupied states on the right side in all channels as

$$n^r_{occ} = \sum_{i=1}^{N} T_i D_i(E) (\mu_1 - \mu_B), \quad (3.53)$$

where $D_i(E)$ is the density of states with positive velocity for channel i. The total number of unoccupied states is just the total density of states (both velocities) minus the injected density from the left

$$n^r_{unocc} = \sum_{i=1}^{N} D_i(E)(\mu_B - \mu_2) - \sum_{i=1}^{N} T_i D_i(E)(\mu_B - \mu_2). \quad (3.54)$$

Equating the two again gives μ_B in terms of the reservoir Fermi energies as

$$\sum_{i=1}^{N} T_i D_i(E)(\mu_1 - \mu_B) = \sum_{i=1}^{N} (2 - T_i) D_i(E)(\mu_B - \mu_2). \quad (3.55)$$

Rearranging to solve for μ_B,

3.4 The multi-channel case

$$\mu_B = \mu_2 + \frac{\sum_{i=1}^{N} T_i D_i(E)(\mu_1 - \mu_2)}{2\sum_{i=1}^{N} D_i(E)} = \mu_2 + \frac{\sum_{i=1}^{N} T_i v_i^{-1}(\mu_1 - \mu_2)}{2\sum_{i=1}^{N} v_i^{-1}}, \quad (3.56)$$

where the density of states in 1D has been written $D_i = 1/\pi\hbar v_i$. By similar generalization of Eq. (3.38), the Fermi energy on the left is determined from

$$\sum_{i=1}^{N}(1+R_i)D_i(E)(\mu_1 - \mu_A) = \sum_{i=1}^{N}(1-R_i)D_i(E)(\mu_A - \mu_2), \quad (3.57)$$

which upon rearranging gives μ_A as

$$\mu_A = \frac{\mu_1 + \mu_2}{2} + \frac{\sum_{i=1}^{N} R_i v_i^{-1}(\mu_1 - \mu_2)}{2\sum_{i=1}^{N} v_i^{-1}}. \quad (3.58)$$

The voltage drop across the scatterer is therefore given as the difference of these two potential energies, and

$$eV = \mu_A - \mu_B = \left(\frac{\mu_1 - \mu_2}{2}\right)\frac{\sum_{i=1}^{N}(1+R_i-T_i)v_i^{-1}}{\sum_{i=1}^{N} v_i^{-1}}. \quad (3.59)$$

Writing the conductance of the scatterer as I/V, using (3.48) and (3.59), we finally obtain

$$G = \frac{I_{tot}}{V} = \frac{2e^2}{h}\sum_{i=1}^{N} T_i \frac{2\sum_{i=1}^{N} v_i^{-1}}{\sum_{i=1}^{N}(1+R_i-T_i)v_i^{-1}}. \quad (3.60)$$

As before, if we performed a two-terminal measurement such that we measure the potential drop across both the scatterer and the leads, then the conductance reduces simply to

$$G_{2term} = \frac{2e^2}{h}\sum_{i=1}^{N} T_i = \frac{2e^2}{h} Tr(tt^+). \quad (3.61)$$

If we had considered an unequal number of channels on either side of the scatterer (e.g., if the width of the ideal 1D conductor was different on the two sides), Eq. (3.60) may be generalized to

$$G = \frac{2e^2}{h}\sum_{i=1}^{N'} T_i \frac{2}{1 + \frac{1}{g_l}\sum_{i=1}^{N} R_i v_{l,i}^{-1} - \frac{1}{g_r}\sum_{i=1}^{N'} T_i v_{r,i}^{-1}}, \quad (3.62)$$

where $v_{l,i}$ and $v_{r,i}$ are the velocities on the left and right side, respectively, and

$$g_l = \sum_{i=1}^{N} v_{l,i}^{-1}, \; g_r = \sum_{i=1}^{N'} v_{r,i}^{-1}. \qquad (3.63)$$

Equation (3.62) obviously reduces to the symmetric result (3.60) when the velocities are the same on both sides. If the reflection coefficients are nearly unity and the transmission coefficients small, (3.62) reduces to the two-terminal result (3.61).

3.4.2 The generalized multi-channel case

We now consider the various conductances for the *multi-probe* case from the point of view of the single- and multi-channel formulas of the previous sections, primarily as derived by Büttiker [79,80]. As a generalization of the two-contact case, current probes are considered phase-randomizing agents that are connected through ideal leads to reservoirs (characterized by a chemical potential μ_i, for the ith probe) which emit and absorb electrons incoherently. In general, the "probes" need not be physical objects. They can be any phase-randomizing entity, such as inelastic scatterers distributed throughout the sample [80]. Here, we will assume that voltage probes are phase randomizing, although there is some disagreement over the validity of this assumption of phase randomization.

To simplify the initial discussion, first assume that the leads contain only a single channel with two states at the Fermi energy corresponding to positive and negative velocity. The various leads are labeled $i = 1, 2, \ldots$, each with corresponding chemical potential μ_i. A scattering matrix may be defined which connects the states in lead i with those in lead j. Thus, $T_{ij} = |t_{ij}|^2$ is the transmission coefficient into lead i of a particle incident on the sample in lead j, while $R_{ii} = |t_{ii}|^2$ is the probability of a carrier incident on the sample from lead i to be reflected back into that lead. In the general case, there may be a magnetic flux Φ present which penetrates the sample. The elements of the scattering matrix must satisfy the following reciprocity relations due to time reversal symmetry:

$$t_{ij}(\Phi) = t_{ji}(-\Phi). \qquad (3.64)$$

As a result, the reflection and transmission coefficients satisfy the symmetry relations

$$R_{ij}(\Phi) = R_{ji}(-\Phi), \; T_{ij}(\Phi) = T_{ji}(-\Phi). \qquad (3.65)$$

In order to simplify the discussion, an additional chemical potential is introduced, μ_0, which is less than or equal to the value of all the other chemical potentials so that all states below μ_0 can be considered filled (at $T_L = 0$ K). With reference then to μ_0, the total current injected from lead i is given by Eq. (3.31)

3.4 The multi-channel case

as $(2e/h)\Delta\mu_i$, where $\Delta\mu_i = \mu_i - \mu_0$ (again assuming the difference in chemical potentials is sufficiently small that the energy dependencies of the transmission and reflection coefficients may be neglected). As we have mentioned, a fraction R_{ii} of the current is reflected back into lead i. Similarly, according to (3.31), lead j injects carriers into lead i as $(2e/h)T_{ij}\Delta\mu_j$. The net current flowing in lead i is therefore the net difference between the current injected from lead i to that injected back into lead i due to reflection and transmission from all the other leads:

$$I_i = \frac{2e}{h}\left[(1 - R_{ii})\Delta\mu_i - \sum_{j \neq i} T_{ij}\Delta\mu_j\right]. \tag{3.66}$$

Conservation of particle flux requires that

$$1 - R_{ii} = T_{ii} = \sum_{i \neq j} T_{ji} = \sum_{j \neq i} T_{ij}. \tag{3.67}$$

Therefore, the reference potential μ_0 drops out to yield

$$I_i = \frac{2e}{h}\left[(1 - R_{ii})\mu_i - \sum_{j \neq i} T_{ij}\mu_j\right]. \tag{3.68}$$

The generalization of (3.68) to the multi-channel case is obtained by simply considering that in each lead there are N_i channels at the Fermi energy μ_i. A generalized scattering matrix may be defined which connects the different leads and the different channels in each lead to one another. The elements of this scattering matrix are labeled $t_{ij,mn}$, where i and j label the leads and m and n label the channels. The probability for a carrier incident in lead j in channel n to be scattered into channel m of lead i is given by $T_{ij,mn} = |t_{ij,mn}|^2$. Likewise, the probability of being reflected within lead i from channel n *into* channel m is given by $R_{ii,mn} = |t_{ii,mn}|^2$. The reciprocity of the generalized scattering matrix with respect to magnetic flux now becomes $t_{ij,mn}(\Phi) = t_{ji,nm}(-\Phi)$. The reservoirs are assumed to feed all the channels equally up to the respective Fermi energy of a given lead. If we define the reduced transmission and reflection coefficients as

$$T_{ij} = \sum_{mn} T_{ij,mn}, \; R_{ij} = \sum_{mn} R_{ij,mn}, \tag{3.69}$$

then the total current in lead i becomes

$$I_i = \frac{2e}{h}\left[(N_i - R_{ii})\mu_i - \sum_{j \neq i} T_{ij}\mu_j\right], \tag{3.70}$$

which differs from (3.68) by the factor N_i, the number of channels in that lead.

Note that we may associate a voltage with each Fermi energy $eV_i = \mu_i$. Thus we can rewrite (3.70) in matrix form as

$$\mathbf{I} = \mathbf{GV}, \tag{3.71}$$

where **I** and **V** are column vectors for the probe currents and voltages respectively, and **G** is an $N \times N$ conductance tensor

$$\mathbf{G} = \begin{bmatrix} N_1 - R_{11} & -T_{12} & \cdots & -T_{1N} \\ -T_{21} & N_2 - R_{22} & \cdots & -T_{2N} \\ \cdots & \cdots & \cdots & \cdots \\ -T_{N1} & -T_{N2} & \cdots & N_N - R_{NN} \end{bmatrix}, \qquad (3.72)$$

where, as before, N is the number of probes. Due to flux conservation, both the rows and the columns add to zero. We note that the components of the conductance tensor contain the same symmetry with respect to the magnetic field as the reflection and transmission coefficients:

$$G_{ij}(\Phi) = G_{ji}(-\Phi). \qquad (3.73)$$

Thus the conductance matrix is equal to its transpose under reversal of the magnetic flux. Of course, such relations are expected in linear response theory for a system governed by a local conductivity tensor satisfying the Onsager–Casimir symmetry relation. Equation (3.73), derived for the nonlocal conductance in a mesoscopic system, shows that such relationships extend to systems governed by nonlocal transport due to the underlying symmetries of the scattering matrix.

The resistance matrix may be defined from the inverse of the conductance matrix

$$\mathbf{RI} = \mathbf{V}, \quad \mathbf{R} = \mathbf{G}^{-1}. \qquad (3.74)$$

Under magnetic flux reversal, the new resistance matrix is the inverse of the transpose of **G**. Assuming the conductance matrix is nonsingular, the inverse of the transpose of a matrix is equal to the transpose of the inverse [81]. Thus the components of the resistance matrix also satisfy

$$R_{ij}(\Phi) = R_{ji}(-\Phi). \qquad (3.75)$$

The above result may be used to prove an important reciprocity relationship concerning a four-terminal transport measurement. Consider the resistance $R_{mn,kl}$, defined as the ratio of the voltage measured between leads k and l, when current $I = I_{mn}$ is driven into contact m and taken out from contact n:

$$R_{mn,kl} = \frac{V_k - V_l}{I}. \qquad (3.76)$$

In terms of the resistance matrix, the voltage V_k is associated with the elements of the kth row, and V_l with the lth row of the matrix. The only non-zero current elements correspond to $I_m = I$ and $I_n = -I$, which couple to columns m and n of the resistance matrix, so that (3.76) may be written

$$\begin{aligned} R_{mn,kl}(\Phi) &= \frac{(R_{km}I - R_{kn}I) - (R_{lm}I - R_{ln}I)}{I} \\ &= R_{km} + R_{ln} - R_{kn} - R_{lm} \end{aligned} \qquad (3.77)$$

in the presence of the magnetic flux Φ. If the current leads and voltage leads are exchanged (i.e., current forced between k and l while the voltage is measured between m and n), then the new resistance is defined letting $m \to k$, $n \to l$, $k \to m$, and $l \to n$ as

$$R_{kl,mn}(\Phi) = R_{mk} + R_{nl} - R_{ml} - R_{nk}. \qquad (3.78)$$

If the magnetic flux is simultaneously reversed, (3.75) shows that we may write

$$R_{kl,mn}(-\Phi) = R_{km} + R_{ln} - R_{kn} - R_{lm} = R_{mn,kl}(\Phi), \qquad (3.79)$$

which, in the absence of magnetic flux, reduces to the simpler *reciprocity theorem*

$$R_{kl,mn} = R_{mn,kl}. \qquad (3.80)$$

This well-known experimental result basically states that in the absence of a magnetic field, the resistance measured by passing current through one pair of contacts and measuring the voltage between the other two is identical to that measured if the voltage and current contacts are swapped. In the presence of a magnetic field, the more general result (3.79) requires that this be done while simultaneously reversing the flux through the sample.

3.4.3 Some specific examples

3.4.3.1 Two-terminal conductance

An expression for the conductance measured in a two-probe conductor was given by Eq. (3.61), where the current and voltage are simultaneously measured through the same pair of contacts. Equation (3.70) also yields the same expression if we restrict the sum over the probe indices i and j to 2. In the two-terminal case, set $I = I_1 = -I_2$ as the current flowing into terminal 1. By conservation of particle flux, $N_1 = R_{11} + T_{12}$ in terminal 1, and $N_2 = R_{22} + T_{21}$ in terminal 2. Using these relations in (3.70) above gives

$$I = \frac{2e}{h} T_{12}(\mu_1 - \mu_2) = -\frac{2e}{h} T_{21}(\mu_2 - \mu_1), \qquad (3.81)$$

which implies that the transmission coefficient in the two-probe case, $T = T_{12} = T_{21}$, is symmetric with respect to the magnetic flux, $T(\Phi) = T(-\Phi)$. The measured voltage in the two-terminal case is just the difference in Fermi energies of the two contacts, so that the conductance is identical to (3.61)

$$G = \frac{Ie}{\mu_1 - \mu_2} = \frac{2e^2}{h} T, \qquad (3.82)$$

which is symmetric with respect to magnetic field reversal.

3.4.3.2 Three-terminal conductance

Consider now the three-probe situation shown in Fig. 3.9. In this example, we wish to consider the various resistances in the case that current flows between

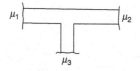

Fig. 3.9 Generic representation of a three-terminal structure.

contacts 1 and 2, and for which no net current flows through contact 3 (i.e., $I_3 = 0$). Thus contact 3 represents an ideal voltage probe that draws no current.

First consider the resistance $R_{12,13}$, which signifies the ratio of the voltage measured between contacts 1 and 3 for a certain current I flowing from contact 1 into contact 2. Using (3.70), the condition of zero current flowing in contact 3 gives

$$0 = \frac{2e}{h}[(N_3 - R_{33})\mu_3 - T_{31}\mu_1 - T_{32}\mu_2]. \tag{3.83}$$

Since $N_3 - R_{33} = T_{31} + T_{32}$ by current conservation

$$\mu_3 = \frac{T_{31}\mu_1 + T_{32}\mu_2}{T_{31} + T_{32}}. \tag{3.84}$$

The current flowing into contact 1 is again $I = I_1 = -I_2$. Starting with the equation for I_2, we may write

$$-I = \frac{2e}{h}[(N_2 - R_{22})\mu_2 - T_{21}\mu_1 - T_{23}\mu_3]. \tag{3.85}$$

Writing $N_2 - R_{22} = T_{21} + T_{23}$ and using a rearranged version of (3.84) to eliminate μ_2, this equation becomes

$$-I = \frac{2e}{h}\left[\frac{(T_{21} + T_{23})(T_{31} + T_{32}) - T_{23}T_{32}}{T_{32}}\mu_3 - \frac{T_{31}(T_{21} + T_{23}) + T_{21}T_{32}}{T_{32}}\mu_1\right]. \tag{3.86}$$

A little algebra shows that the two numerator terms are identical, and we may write these as

$$T_{31}T_{21} + T_{31}T_{23} + T_{21}T_{32} = D, \tag{3.87}$$

and the resistance $R_{12,13}$ is given by

$$R_{12,13} = \frac{(\mu_1 - \mu_3)}{eI} = \left(\frac{h}{2e^2}\right)\frac{T_{32}}{D}. \tag{3.88}$$

By an analogous procedure, we may calculate the resistance $R_{12,32}$ (starting from the equation for $I_1 = I$), which is the ratio of the voltage between probes 2 and 3, for a current I between 1 and 2:

$$R_{12,32} = \frac{(\mu_3 - \mu_2)}{eI} = \left(\frac{h}{2e^2}\right)\frac{T_{31}}{D}. \tag{3.89}$$

The two-terminal resistance may be calculated by combining Eqs. (3.88) and (3.89) to yield

$$R_{12,12} = \frac{(\mu_1 - \mu_2)}{eI} = \frac{(\mu_1 - \mu_3)}{eI} + \frac{(\mu_3 - \mu_2)}{eI}$$
$$= \left(\frac{h}{2e^2}\right)\frac{T_{32} + T_{31}}{D}. \tag{3.90}$$

Now, as D is the sum of products of pairs of transmission coefficients, it will be invariant under reversal of any magnetic flux. Since $T_{31} + T_{32} = 1 - R_{33}$

3.4 The multi-channel case

is invariant with respect to flux reversal, the two-terminal resistance (or conductance) is invariant with respect to magnetic flux reversal even in the presence of a third non-current-carrying terminal. This symmetry may be further extended to many probes as shown by Büttiker [80]. The two-terminal conductance given by the inverse of (3.90), even in the presence of the third non-current-carrying probe, is

$$G_{12,12} = \frac{2e^2}{h} \frac{D}{T_{32} + T_{31}}. \tag{3.91}$$

In the limit that the transmission coefficients into and out of probe 3 from the other two probes are much smaller than the transmission coefficient from 1 to 2 (i.e., $T_{23}, T_{31} \ll T_{21}$), Eq. (3.91) reduces to the two-probe formula (3.42) with $T = T_{21}$. The fact that, in general, the two-terminal conductance, with a third non-current-carrying probe, is different from the simple two-terminal conductance Eq. (3.91) demonstrates that for mesoscopic systems it is difficult to separate the effect of the measurement probes from the measurement itself. Such uncertainty in separating a measured quantity from the measurement apparatus itself is of course fundamentally related to the question of measurement in quantum mechanics and phase coherence in ballistic structures.

An interesting interpretation of the effect of the passive third probe in a two-terminal measurement has been suggested by Büttiker [50], [80], where it was proposed that the additional third terminal could represent the effect of phase-randomizing inelastic scattering. Electrons that are transmitted into probe 3 lose their phase coherence by thermalizing in the reservoir. Since no net current was assumed to flow into the probe, the same flux of carriers are reemitted into the sample, but without any phase relation to the particles incident into the probe. Identifying the elastic contribution between probes 1 and 2 as T_{21}, then, using (3.91), we can write

$$T = T_{coh} + T_{sc} = T_{21} + \frac{T_{23}T_{31}}{T_{31} + T_{23}}, \tag{3.92}$$

where the last term on the right side is the inelastic transmission coefficient (if probe 3 is in fact a randomizing contact). In the limit that the transmission coefficients into and out of probe 3 become small, T becomes $T_{coh} = T_{21}$. In the other extreme limit of no coherent transmission (i.e. $T_{21} \sim 0$), the resistance becomes

$$R_{12,12} = \left(\frac{h}{2e^2}\right)\left(\frac{1}{T_{31}} + \frac{1}{T_{32}}\right), \tag{3.93}$$

which is nothing more than the addition of two series resistance with no coherence between them. Büttiker used this idea to develop a simple model for the effect of inelastic scattering in double barrier planar structures.

3.4.3.3 Four-terminal conductance

As a final example, we wish to make a connection back to the earlier discussion concerning the interpretation of the Landauer formula in terms of a two-terminal versus a four-terminal measurement. As discussed earlier, if a two-terminal measurement is performed on a sample using the same pair of contacts to measure both the current and voltage, then we expect that the conductance is proportional simply to the transmission coefficient as described above and not to the ratio of transmission to reflection coefficients which is more appropriate for the four-terminal conductance. The latter form arises if one considers the pile-up of charge on either side of the barrier, yet measures the potential drop only across the active part of the sample. Thus, the conductance in the two-terminal measurement is always finite, even when transmission through the sample is unity (implying no resistance due to the contact resistance associated with connecting from a reservoir to the leads). As suggested by Engquist and Anderson [74], if noninvasive voltage probes rather than the current contacts are introduced to measure the potential distribution at the sample, one should in principle measure the Landauer conductance through a four-terminal measurement, taking the resistance as the ratio of the voltage difference of the two voltage probes over the injected current through the other two contacts. We now consider such a measurement within the context of the multi-probe formula developed above and follow the development of Büttiker in arriving at the Landauer conductance to lowest order in the coupling of the probes to the system.

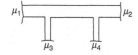

Fig. 3.10 Schematic of a four-probe experiment in which current is forced between contacts 1 and 2 while weakly coupled probes (3 and 4) measure the potential distribution along the sample.

Consider the four-probe configuration shown in Fig. 3.10. A current I is passed between contacts 1 and 2 such that $I_1 = -I_2 = I$. The two probes 3 and 4 are weakly coupled to the sample. The weak coupling could be due, for example, to tunnel barriers isolating the probes from the sample. The voltage probes are assumed to draw no net current, such that $I_3 = I_4 = 0$. The condition of zero net current in probes 3 and 4 gives

$$I_3 = 0 = \frac{2e}{h}[(1 - R_{33})\mu_3 - T_{31}\mu_1 - T_{32}\mu_2 - T_{34}\mu_4]$$
$$I_4 = 0 = \frac{2e}{h}[(1 - R_{44})\mu_4 - T_{41}\mu_1 - T_{42}\mu_2 - T_{43}\mu_3]. \quad (3.94)$$

Rewriting the above two equations we get

$$(1 - R_{33})\mu_3 - T_{34}\mu_4 = T_{31}\mu_1 + T_{32}\mu_2$$
$$(1 - R_{44})\mu_4 - T_{43}\mu_3 = T_{41}\mu_1 + T_{42}\mu_2. \quad (3.95)$$

Here we consider a single channel in each lead for simplicity, but the procedure is easily extendible to the multi-channel, multi-probe case.

Now, following Büttiker, we assume that the coupling into the voltage probes is characterized by a small parameter, ε, which attenuates the flux into probes 3 and 4 due to tunnel barriers across the contacts for example. The transmission coefficients may thus be expanded in powers ε as

$$T_{12} = T_{12}^{(0)} + \varepsilon T_{12}^{(1)} + \cdots$$
$$T_{13} = \varepsilon T_{13}^{(1)} + \cdots \quad (3.96)$$
$$T_{34} = \varepsilon^2 T_{34}^{(2)} \cdots,$$

where we interpret $T_{12}^{(0)}$ as the transmission coefficient from 1 to 2 in the absence of the other two probes, and the higher-order terms are small corrections arising from the presence of these probes. As such, $T_{12}^{(0)}$ is symmetric with respect to flux reversal as we saw above. The coefficient T_{12} is zero order in ε as flux may go from probe 1 to 2 without crossing a tunnel barrier, while T_{13} is first order in ε as flux traverses the tunnel barrier once, while T_{34} is quadratic in ε since the flux must traverse two such tunnel barriers.

We now proceed to calculate the two-terminal resistance, $R_{12,12}$ (where the first two indices represent the current leads, the second two the voltage leads) measured in the presence of the two voltage leads 3 and 4. To do this, we must therefore eliminate the chemical potentials μ_3 and μ_4 by expressing them in terms of the chemical potentials μ_1 and μ_2. Eliminating μ_4 from the first of equations (3.95) gives

$$[(1 - R_{33})(1 - R_{44}) - T_{34}T_{43}]\mu_3 = [T_{31}(1 - R_{44}) + T_{34}T_{41}]\mu_1 \quad (3.97)$$
$$+ [T_{32}(1 - R_{44}) + T_{34}T_{42}]\mu_2.$$

Using flux conservation, the left side of this equation may be written

$$[(1 - R_{33})(1 - R_{44}) - T_{34}T_{43}] = [(T_{31} + T_{32})(T_{41} + T_{42})] \quad (3.98)$$
$$+ [T_{34}(T_{41} + T_{42}) + T_{43}(T_{31} + T_{32})].$$

The first term in square brackets on the right-hand side is proportional to ε^2 while the last term is proportional to ε^3, and we will ignore this last term, keeping only the leading order in the small parameter. Using this expansion, we may rewrite the equation for μ_3 as

$$\mu_3 \approx \frac{T_{31}(1 - R_{44}) + T_{34}T_{41}}{(T_{31} + T_{32})(T_{41} + T_{42})}\mu_1 + \frac{T_{32}(1 - R_{44}) + T_{34}T_{42}}{(T_{31} + T_{32})(T_{41} + T_{42})}\mu_2. \quad (3.99)$$

If we had started with the above equations, but eliminated μ_3 instead and solved for μ_4, an identical second-order analysis yields

$$\mu_4 \approx \frac{T_{41}(1 - R_{33}) + T_{31}T_{43}}{(T_{31} + T_{32})(T_{41} + T_{42})}\mu_1 + \frac{T_{42}(1 - R_{33}) + T_{32}T_{43}}{(T_{31} + T_{32})(T_{41} + T_{42})}\mu_2. \quad (3.100)$$

We may now write the current through the remaining two contacts to get our working equation. We will use the current through lead 1 for this, and

$$I_1 = I = \frac{2e}{h}[(1 - R_{11})\mu_1 - T_{12}\mu_2 - T_{13}\mu_3 - T_{14}\mu_4]. \quad (3.101)$$

We now substitute Equations (3.99) and (3.100) to yield

$$I = \frac{2e}{h}\left\{\left[(1-R_{11}) - T_{13}\frac{T_{31}(1-R_{44}) + T_{34}T_{41}}{(T_{31}+T_{32})(T_{41}+T_{42})} - T_{14}\frac{T_{41}(1-R_{33}) + T_{31}T_{43}}{(T_{31}+T_{32})(T_{41}+T_{42})}\right]\mu_1 \\ - \left[T_{12} + T_{13}\frac{T_{32}(1-R_{44}) + T_{34}T_{42}}{(T_{31}+T_{32})(T_{41}+T_{42})} + T_{14}\frac{T_{42}(1-R_{33}) + T_{32}T_{43}}{(T_{31}+T_{32})(T_{41}+T_{42})}\right]\mu_2\right\}. \quad (3.102)$$

We see that, while the denominator of the 2nd and 3rd terms in the square brackets is of order ε^2, the numerator of both terms is at least of order ε^3. The leading terms, $T_{12} \sim (1-R_{11}) \sim T_{12}^{(0)}$, are zero order in ε and thus dominate over the other terms, so that

$$I \approx \frac{2e}{h}T_{12}^{(0)}(\mu_1 - \mu_2), \quad (3.103)$$

and thus the two-terminal resistance is given by

$$R_{12,12} = \frac{\mu_1 - \mu_2}{eI} = \frac{h}{2e^2}\frac{1}{T_{12}^{(0)}}, \quad (3.104)$$

which is symmetric with respect to flux reversal or reversal of the measurement leads (i.e., $R_{21,21}$). Thus we have recovered to lowest order in the coupling constant the two-terminal Landauer conductance given by (3.42).

Now we want to calculate the four-terminal resistance, $R_{12,34}$, in which current is passed between terminals 1 and 2, while the voltage drop along the sample is measured independently in the weakly coupled probes 3 and 4. It is now necessary to eliminate the chemical potentials μ_1 and μ_2 from (3.94) above, following an analogous procedure to that used to eliminate μ_3 and μ_4 in deriving (3.104). Eliminating μ_1 from (3.94) gives

$$\mu_2 \approx \frac{T_{41}(T_{31}+T_{32})}{T_{32}T_{41} - T_{31}T_{42}}\mu_3 - \frac{T_{31}(T_{41}+T_{42})}{T_{32}T_{41} - T_{31}T_{42}}\mu_4. \quad (3.105)$$

Likewise, eliminating μ_2 from (3.94) gives

$$\mu_1 \approx \frac{T_{42}(T_{31}+T_{32})}{T_{32}T_{41} - T_{31}T_{42}}\mu_3 + \frac{T_{32}(T_{41}+T_{42})}{T_{32}T_{41} - T_{31}T_{42}}\mu_4. \quad (3.106)$$

Substituting these two expressions into (3.104) gives

$$I = \frac{2e}{h}\left\{\left[-\frac{T_{12}T_{42}(T_{31}+T_{32})}{T_{32}T_{41} - T_{31}T_{42}} - \frac{T_{12}T_{41}(T_{31}+T_{32})}{T_{32}T_{41} - T_{31}T_{42}} - T_{13}\right]\mu_3 \\ + \left[\frac{T_{12}T_{32}(T_{41}+T_{42})}{T_{32}T_{41} - T_{31}T_{42}} + \frac{T_{12}T_{31}(T_{41}+T_{42})}{T_{32}T_{41} - T_{31}T_{42}} - T_{14}\right]\mu_4\right\}. \quad (3.107)$$

The first two terms of each of the expressions in square brackets have numerators and denominators both of order ε^2 which cancel, whereas T_{13} and T_{14} are of order ε and thus smaller compared to the first two. Neglecting terms in ε or higher then yields

$$I \approx \frac{2e}{h} T_{12}^{(0)} \frac{(T_{31} + T_{32})(T_{41} + T_{42})}{T_{31}T_{42} - T_{32}T_{41}} (\mu_3 - \mu_4), \tag{3.108}$$

which gives the four-terminal resistance

$$R_{12,34} \approx \frac{h}{2e^2} \frac{1}{T_{12}^{(0)}} \frac{T_{31}T_{42} - T_{32}T_{41}}{(T_{31} + T_{32})(T_{41} + T_{42})}. \tag{3.109}$$

In the four-terminal measurement, the measured resistance depends explicitly on the transmission coefficients into and out of the potential probes, which, depending on the relative values of the coefficients, may be positive, negative, or even zero!

To make a connection with the original Landauer formula (3.42), assume that the voltage probes 3 and 4 couple to perfect leads on either side of a scattering region or elastic scattering characterized by a simple transmission coefficient T and reflection coefficient R. The voltage probes still have tunnel barriers so that they are weakly coupled to the leads by an attenuation factor ε. The magnetic flux is assumed zero. Thus the transmission probability $T_{12} = T_{21} = T$ to lowest order in ε. On either side of the scattering region, electrons may be transmitted directly from contact to contact, or by reflecting from the scattering region; one may argue that

$$T_{31} = T_{13} = T_{42} = T_{24} = \varepsilon(1 + R), \tag{3.110}$$

whereas for carriers injected to or from one of the voltage probes through the scattering region, the transmission probabilities are proportional to T:

$$T_{32} = T_{23} = T_{14} + T_{41} = \varepsilon T. \tag{3.111}$$

Finally, carriers injected from one voltage probe to the other are attenuated twice, $T_{34} = T_{43} = \varepsilon^2 T$. Writing the four-probe resistance in this case, using (3.109), yields

$$R_{12,34} \approx \frac{h}{2e^2} \frac{1}{T} \frac{\varepsilon^2(1+R)^2 - \varepsilon^2(1-R)^2}{[\varepsilon(1+R) + \varepsilon T]^2} = \frac{h}{2e^2} \frac{R}{T}, \tag{3.112}$$

where use has been made of $R + T = 1$. Equation (3.112) is now seen to be the inverse of the Landauer conductance (3.40). Although derived with admittedly crude approximations, one sees that the factor of T/R is recovered if the voltage probes measure the potential drop across the scattering region where charge has piled up, rather than the potential at the contacts where current is injected.

3.4.4 Experimental multi-probe measurements ($B = 0$)

The most well-known measurements of the two-terminal conductance is the quantized wire results discussed in Chapter 1. Here, we briefly introduce some experimental results with zero magnetic field that demonstrate the nonlocality

Fig. 3.11 An electron micrograph of a symmetric cross structure realized using two ballistic point contact structures in series. The regions 1–4 are separately contacted regions for making multi-probe measurements (after Timp *et al.*, in *Nanostructure Physics and Fabrication*, eds. W. P. Kirk and M. Reed (New York, Academic Press, 1989), p. 331, by permission).

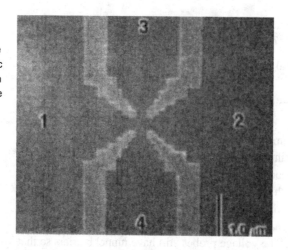

of transport in mesoscopic structures when measured with multiple probes. A widely investigated structure is the symmetric cross shown in Fig. 3.11. Experimentally, this structure is realized using split-gate electrodes above a high-mobility 2DEG heterostructure, as shown in the SEM micrograph of this figure [82]. The ports 1 through 4 are 2DEG regions that are isolated from one another and connected to Ohmic contacts for multi-probe transport measurements. The gates are individually biased so that each of the four-point contacts may be opened or closed to realize two-, three-, and four-terminal structures.

3.4.4.1 *Three-terminal measurement*

With sufficiently large negative bias applied to the bottom two electrodes, region 4 is removed from the measurement, and one has an essentially three-terminal structure. Timp and coworkers [82,83] investigated the conductance for this structure as a function of the coupling of the third electrode (3) to the system by changing the gate bias, V_{g2}, on the upper right electrode. The result is shown in Fig. 3.12 for the conductances $G_{12,12}$ and $G_{12,13}$, which theoretically are given by the inverses of (3.91) and (3.88), respectively. For $V_{g2} = 0$, there is essentially no difference between the two-terminal and three-terminal conductance because both contacts are the same, and nearly ideal quantized conductance is measured for both conductances. As the constriction is closed, the conductance $G_{12,12}$ decreases, while $G_{12,13}$ increases, particularly for high N, where N is the number of occupied subbands in the 1D channel, illustrating that the presence of the third terminal may be invasive.

3.4.4.2 *Four-terminal bend resistance*

In the previous sub-sections, the theoretical four-terminal resistance in a measurement with two current probes and two ideal voltage probes is given by Eq. (3.109). Figure 3.13 shows measured data for the four-terminal resistance

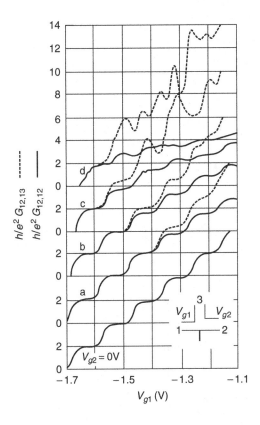

Fig. 3.12 The three-terminal conductances $G_{12,12}$ and $G_{12,13}$ for various gate biases applied to the second point contact (after Timp et al., in *Nanostructure Physics and Fabrication*, eds. W. P. Kirk and M. Reed (New York, Academic Press, 1989), p. 331, by permission).

$R_{14,32}$, measured in a symmetric cross structure similar to that shown in Fig. 3.11. As shown, the four-terminal conductance is negative and exhibits sharp minima at certain gate biases attributable to the thresholds for occupation of 1D modes in the leads. The negative four-terminal resistance follows from the four-terminal multi-probe formula as discussed above. Such structure in the resistance is sometimes referred to as the bend resistance, as the electrons are being forced to turn 90 degrees from contact 1 to contact 4. Baranger *et al.* [84] have investigated theoretically the bend resistance through such structures, both from a classical standpoint and using a four-probe recursive Green's function technique based on the theoretical approach described in the next section, including magnetic fields. The result of the calculations compares well with the bend resistance structure shown in Fig. 3.13.

3.5 Transport in quantum waveguide structures

To consider structures more complicated than a simple adiabatic point contact requires at some level a knowledge of the actual transmission characteristics of a structure in terms of its scattering matrix (e.g., the actual detailed transmission coefficient T between the leads). In split-gate structures as well as other

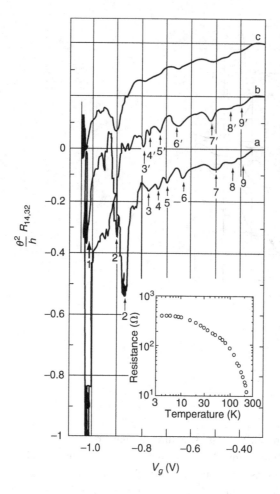

Fig. 3.13 The experimental four-terminal bend resistance, $R_{14,32}$, measured in a symmetric cross structure. The lower two traces are measured at $T = 280$ mK, after successive cycling to room temperature and back. The upper curve is the same structure at $T = 1.2$ K (after Takagaki et al., *Solid State Comm.* **71**, 809 (1989), by permission).

realizations of quantum ballistic structures, the exact treatment requires the solution of the full three-dimensional Schrödinger equation coupled with Poisson's equation for the potential (and this just within the single-particle picture!). Such calculations are quite tedious and require enormous computational resources. Thus, most of the theoretical treatment of more complicated structures are based on simplified reductions of the actual geometry into two dimensions with idealized potentials and geometries. In the first edition, we treated the mode-matching approach, which was unstable for large numbers of recursion, and the site-representation Green's function, in which a relatively large overall matrix results in the non-recursive approach. We will not treat these further. In the present section, we will consider two somewhat different approaches, both of which develop the information via a recursive approach. The first, the wave function scattering method, builds up the wave function by a forward propagation based upon the Lippmann–Schwinger equation and a subsequent back propagation to finalize the wave function itself. Later, in Section 3.5.2, we

3.5 Transport in quantum waveguide structures

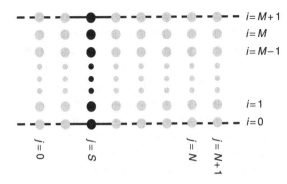

Fig. 3.14 The grid representation and spacing used in the recursive wave function and in the recursive Green's function approach.

introduce the lattice Green's function method, which is also implemented via a recursive approach from one contact to the next. In the zero-temperature format we treat in this section, these two approaches are largely equivalent, with each requiring the inversion of a Green's function matrix at each slice or step in the recursion. The former method is slightly easier to couple to the Poisson's equation in that a detailed integration of the Green's function over energy at each lattice site is not required to find the density. Rather, this integration is only required in the leads to properly normalize the various modes. Which of the two approaches is more efficient computationally is somewhat dependent upon the programmer. Both of these will be extended to the far-from-equilibrium (and finite-temperature) regime in Chapter 9.

3.5.1 The recursive scattering matrix

In wave function scattering theory, the reflected, or scattered, parts of the wave function are related to the incoming wave via the Lippmann–Schwinger equation. A useful consequence of this equation is that the scattering states satisfy precisely the same orthonormality relations as the unperturbed states [85,86,87].

The device under consideration is a normal semiconductor structure, shown schematically in Fig. 3.14, in which this structure is subject to discretization. That is, we will solve the Schrödinger equation on a grid in two dimensions, which represents the plane of a two-dimensional electron gas. As was shown in Section 2.2.5, when the Schrödinger is solved on a grid, the grid spacing introduces an artificial band structure. Hence, it is important that the grid spacing be sufficiently small that the energy range of interest is kept in the lower part of this artificial band structure so that the numerical energy dispersion is close to the actual one in the two-dimensional electron gas.

We orient the x- and y-directions in order to correspond to the length and the width of the device, respectively. In the x-direction, the input and the output contact regions are where the recursion will begin and end. In an actual device, the length of the input and the output regions would be much longer, but here using only a very few grid points still captures the important behavior. Of course, the mode structure is set by the transverse (y) direction and width. By simply defining the number of grid points in the lateral direction, the grid spacing implies a width, and the act of ending the grid at the edge grid points implements infinite potential walls at these points (this can be modified by the inclusion of a real potential at each grid point, as we will see). We implement open boundary conditions at the ends of the structure, by which we mean that the mode occupancy will be determined by the Fermi energy, which in turn is given by the 2D density of carriers. At zero temperature, the Fermi function has a sharp cutoff at the Fermi energy, and each propagating mode may be assumed to have unit amplitude at the input node.

We can now write a wave function which is described on the grid. That is, we can write the wave function as a vector

$$\psi(x,y) = \psi(n,m) = \sum_{j=1}^{N} \psi_j(m), \qquad (3.113)$$

where the subscript j refers to a slice in the transverse direction, as $x = na$, $y = ma$, and a is the grid spacing. The Schrödinger equation for the two-dimensional plane of the electrons is given as

$$\frac{-\hbar^2}{2}\left(\frac{1}{m_x}\frac{d^2}{dx^2} + \frac{1}{m_y}\frac{d^2}{dy^2}\right)\psi(x,y) + V(x,y)\psi(x,y) = E\psi(x,y). \qquad (3.114)$$

Here, we will assume that the mass is constant, in order to simplify the equations (for non-parabolic bands, the reciprocal mass enters between the partial derivatives). We have labeled the mass corresponding to the principal coordinate axes, but we will also assume that these are the same; e.g., the energy surface is isotropic. We replace the derivatives appearing in the discrete Schrödinger equation with finite-difference representations of the derivatives. The Schrödinger equation then reads

$$-t_x\left(\psi_{i+1,j} + \psi_{i-1,j}\right) - t_y\left(\psi_{i,j+1} + \psi_{i,j-1}\right) \\ + (V_{i,j} + 2t_x + 2t_y)\psi_{i,j} = E\psi_{i,j}, \qquad (3.115)$$

where t_x and t_y are the hopping energies

$$t_x = \frac{\hbar^2}{2m_x a^2}, \quad t_y = \frac{\hbar^2}{2m_y a^2}. \qquad (3.116)$$

Each hopping energy corresponds with a specific direction in the crystal. The fact that we are now dealing with an isotropic band structure and grid spacing means that we can set these two values equal to each other and to $t = t_x = t_y$.

There are other important points that relate to the hopping energy. As we mentioned at the start of this section, the discretization of the Schrödinger equation introduces an artificial band structure, due to the periodicity that this discretization introduces. As a result, the band structure in any one direction has a cosinusoidal variation with momentum eigenvalue (or mode index), and the total width of this band is $4t$ (in any one direction). Hence, if we are to properly simulate the real band behavior, which is quadratic in momentum, we need to keep the energies of interest below a value where the cosinusoidal variation deviates significantly from the parabolic behavior desired. For practical purposes, this means that $E_{\max} < t$. The smallest value of t corresponds to the effective mass, and if we desire energies of the order of the Fermi energy ~ 15 meV, then we must have $a < 6$ nm, in GaAs.

With the discrete form of the Schrödinger equation defined above, we now develop the transfer matrices relating adjacent slices in the solution space. For this, we will utilize the method in terms of slices, as indicated above, and follow a procedure first put forward by Usuki *et al.* [88,89] and used extensively by our group [90]. We begin first by noting that the transverse plane has M grid points. This produces a vector for the wave function, and it propagates via a second-rank tensor, or square matrix. Hence, the propagation is handled by a simple matrix multiplication. The vector wave function, at slice j, is

$$\Psi_j = \begin{bmatrix} \psi(j,1) \\ \psi(j,2) \\ \dots \\ \psi(j,M) \end{bmatrix}. \qquad (3.117)$$

Now, Equation (3.115) can be rewritten as a matrix equation as, with j an index of the distance along the x direction,

$$H_j \Psi_j - T_y \Psi_{j-1} - T_y \Psi_{j+1} = EI\Psi_j. \qquad (3.118)$$

Here, I is the unit matrix, E is the energy in the eigenvalue equation (and which usually represents the Fermi energy at the input lead; that is, represents the energy at which the conductance is being found), and

$$H_j = \begin{bmatrix} H_0(1,j) & t_x & \dots & 0 \\ t_x & H_0(2,j) & \dots & \dots \\ \dots & \dots & \dots & t_x \\ 0 & \dots & t_x & H_0(M,j) \end{bmatrix}, \qquad (3.119)$$

with

$$H_0(i,j) = 2(t_x + t_y) + V_{i,j}, \qquad (3.120)$$

$$T_y = t_y I. \qquad (3.121)$$

The dimension of these matrices is $M \times M$. With this setup of the matrices, the general procedure follows that laid out in the previous work [88–90]. One first solves the eigenvalue problem on slice 0 at the input end of the channel, which determines the propagating and evanescent modes for a given Fermi energy in this region. The wave function is thus written in a mode basis, but this is immediately transformed to the site basis, and one propagates toward the output end, using the scattering matrix iteration. This can be found by noting that (3.118) can be rewritten as

$$\Psi_{j+1} = T_y^{-1}(H_j - EI)\Psi_j + T_y^{-1}T_y\Psi_{j-1}. \tag{3.122}$$

If we add the identity relationship to this, we generate a super-matrix relationship for this equation, as

$$\begin{bmatrix} \Psi_j \\ \Psi_{j+1} \end{bmatrix} = \begin{bmatrix} 0 & I \\ T_y^{-1}T_y & T_y^{-1}(H_j - EI) \end{bmatrix} \begin{bmatrix} \Psi_{j-1} \\ \Psi_j \end{bmatrix}. \tag{3.123}$$

Now, as it stands, this is just a mode-matching transfer matrix. However, we interpret Ψ_{j-1} and Ψ_{j+1} as the backward- and forward-scattered wave functions at slice j. Now, we assert that the wave function has a forward and backward component which are represented by the complex amplitudes $C_j^{(1)}$ and $C_j^{(2)}$, respectively, at slice j, and postulate that the scattering matrix representation introduces two new functions $P_j^{(1)}$ and $P_j^{(2)}$, which must satisfy the relationship given by (3.123) as follows:

$$\begin{bmatrix} C_{j+1}^{(1)} & C_{j+1}^{(2)} \\ 0 & I \end{bmatrix} = \begin{bmatrix} 0 & I \\ I & T_y^{-1}(H_j - EI) \end{bmatrix} \begin{bmatrix} C_j^{(1)} & C_j^{(2)} \\ 0 & 1 \end{bmatrix} \begin{bmatrix} 1 & 0 \\ P_j^{(1)} & P_j^{(2)} \end{bmatrix}. \tag{3.124}$$

The second row of this equation sets the iteration conditions

$$\begin{aligned} C_{j+1}^{(2)} &= P_j^{(2)} = \left[C_j^{(2)} + T_y^{-1}(H_j - EI) \right]^{-1}, \\ C_{j+1}^{(1)} &= P_j^{(1)} = -P_j^{(2)} C_j^{(1)}. \end{aligned} \tag{3.125}$$

At the source end, $C_0^{(1)} = I$, and $C_0^{(2)} = 0$ are used as the initial conditions. These are now propagated to the Nth slice, which is the end of the active region, and then onto the $N + 1$ slice. At this point, the inverse of the mode-to-site transformation matrix is applied to bring the solution back to the mode representation, so that the transmission coefficients of each mode can be computed. These are then summed to give the total transmission and this is used in a version of Equation (3.61) to compute the current through the device (there is no integration over the transverse modes, only over the longitudinal density of states and energy).

If we are interested in the density at each point, or if we are to incorporate a self-consistent potential within the device, we must know the density in order to solve Poisson's equation. The density at each point in the device is determined from the wave function squared magnitude at that point. Our solution for $C_{N+2}^{(1)}$

is the wave function at this point, and this is back-propagated using the recursion algorithm

$$\Psi_{i,j}^{(\xi)} = P_j^{(1)} + P_j^{(2)} \Psi_{i,j+1}^{(\xi)}. \quad (3.126)$$

Here, we are in the mode representation, and ξ is the mode index. The density at any site (i,j) is found by taking the sum over ξ of the occupied modes at that site, as

$$n(i,j) = \sum_{\xi} \left| \Phi_{i,j}^{(\xi)} \right|^2. \quad (3.127)$$

Once the charge density is found at each site, this can be used e.g. in the Poisson's equation to find a self-consistent potential, thereby updating $V_{i,j}$ used above.

3.5.1.1 Incorporating spin in the Hamiltonian

In this section, a procedure is outlined that attempts to include the Rashba spin-orbit interaction in this formulation, particularly for ballistic electrons in quantum wires. This procedure follows the 2001 work of Mireles and Kirczenow [91] and integrates it with the method for ballistic electron transport discussed above.

We begin by taking a particular form for the Hamiltonian, and that is to incorporate the Rashba spin-orbit coupling [92]:

$$H_{SO} = \boldsymbol{\alpha} \cdot (\boldsymbol{\sigma} \times \mathbf{k}). \quad (3.128)$$

Here, $\boldsymbol{\alpha}$ is proportional to the electric field, $\boldsymbol{\sigma}$ represents the Pauli spin matrices, and $\mathbf{k}$ is the electron wave vector. If we assume a 2D electron gas in the x-y plane, an electric field applied along the z-axis, and substitute the operator form of the wave vector, then the spin-orbit Hamiltonian becomes

$$H_{SO} = i\alpha_z \hbar \left(\begin{bmatrix} 0 & -i \\ i & 0 \end{bmatrix} \frac{\partial}{\partial x} - \begin{bmatrix} 0 & 1 \\ 1 & 0 \end{bmatrix} \frac{\partial}{\partial y} \right). \quad (3.129)$$

The total Hamiltonian of the system can be written as $H = H_0 + H_{SO}$, where H_0 is given by (3.114) (the terms on the left-hand side). The Schrödinger equation is written with the wave function split into its spin-up and spin-down components as the two-vector (each term is a vector in its own right, as described previously in (3.117))

$$\hat{\psi} = \begin{bmatrix} \psi^{\uparrow} \\ \psi^{\downarrow} \end{bmatrix}. \quad (3.130)$$

Applying Equation (3.129) results in a pair of equations for the spin-up and spin-down wave functions of the electron

$$(H_0 - EI)\hat{\psi} - \alpha_z \begin{bmatrix} 0 & \left(-\frac{\partial}{\partial x} + i\frac{\partial}{\partial y}\right) \\ \left(\frac{\partial}{\partial x} + i\frac{\partial}{\partial y}\right) & 0 \end{bmatrix} \hat{\psi} = 0. \quad (3.131)$$

Here, the left-hand term is the same as in (3.114). To solve (3.131) numerically, it is discretized on a 2D grid with the index i representing sites along the x-axis and the index j representing sites along the y axis, as previously. For this problem, it is assumed that electrons are confined along the x-axis and are propagated along the y-axis. When this discretization is applied to (3.131), we obtain

$$\begin{bmatrix} (H_j - EI) & T_{SO}^{\uparrow} \\ T_{SO}^{\downarrow} & (H_j - EI) \end{bmatrix} \hat{\psi}_j - \begin{bmatrix} T_y & 0 \\ 0 & T_y \end{bmatrix} \left(\hat{\psi}_{j+1} + \hat{\psi}_{j-1}\right)$$
$$- \begin{bmatrix} 0 & it_{SO} \\ it_{SO} & 0 \end{bmatrix} \left(\hat{\psi}_{j+1} - \hat{\psi}_{j-1}\right) = 0. \quad (3.132)$$

The matrices T_{SO} are given by

$$T_{SO}^{\uparrow,\downarrow} = \begin{bmatrix} 0 & \pm t_{SO} & 0 & \ldots & 0 \\ \mp t_{SO} & 0 & \pm t_{SO} & \ldots & 0 \\ 0 & \mp t_{SO} & 0 & \ldots & \ldots \\ \ldots & \ldots & \ldots & \ldots & \pm t_{SO} \\ 0 & 0 & \ldots & \mp t_{SO} & 0 \end{bmatrix}, \quad (3.133)$$

where $t_{SO} = \alpha_{-z}/2a$. With the Schrödinger equation discretized and including the Rashba spin-orbit coupling terms, it can now be applied to the transmission method described above. To do this, Equation (3.123) above is replaced by

$$\begin{bmatrix} \hat{\Psi}_j \\ \hat{\Psi}_{j+1} \end{bmatrix} = \begin{bmatrix} 0 & I \\ K^{-1}Q & K^{-1}H'_j \end{bmatrix} \begin{bmatrix} \hat{\Psi}_{j-1} \\ \hat{\Psi}_j \end{bmatrix}. \quad (3.134)$$

The new matrices are defined by

$$K = \begin{bmatrix} T_y & it_{SO}I \\ it_{SO}I & T_y \end{bmatrix},$$
$$H'_j = \begin{bmatrix} (H_j - EI) & T_{SO}^{\uparrow} \\ T_{SO}^{\downarrow} & (H_j - EI) \end{bmatrix}, \quad (3.135)$$
$$Q = \begin{bmatrix} -T_y & it_{SO}I \\ it_{SO}I & -T_y \end{bmatrix}.$$

Using this and the stable iteration in Equation (3.124) allows us to determine the evolution of an electron wave function as it propagates through an arbitrary 2D system under the effect of the Rashba spin-orbit interaction.

To include the effects of a magnetic field in the z-direction, we normally select the Landau gauge of the vector potential, $\mathbf{A} = (0, Bx, 0)$. This results in the inclusion of the Peierl's phase factors in the definitions of the hopping matrix elements (3.116), which leads to a difference in the x- and y-directions. Additionally, the momentum term in (3.128) also contains a vector potential correction which leads to an additional potential term of the form $-\alpha_z e B x \sigma_x/\hbar$ added to Equation (3.131).

3.5.1.2 Some examples of this approach

In Fig. 3.15, we plot the carrier density for a waveguide in which a quantum point contact (QPC) has been placed. Excitation occurs only from the input end of the guide, which is at the top right of the figure. This is an important point. When we compute the conductance, we need only the transmission over the narrow range of energies represented by the small bias potential. If we want the densities, however, we need to integrate the wave function probability over the entire energy range up to the Fermi energy. As was evident in Section 3.3, this must be done for waves *initiated at both ends* of the sample. Only the "right" input is shown in Fig. 3.15, in order to better identify some aspects of the problem. Three modes propagate through the quantum point contact, whose position can be ascertained from the region where the amplitude is zero in the figure. Several things are evident in this picture. First, near the source end, the density displays rapid oscillations which are the result of the finite number of transverse modes in the narrow waveguide (200 nm width; the carrier density at the source corresponds to a Fermi energy of 14 meV). Second, there is a pileup of charge at the input side of the QPC. Such a pileup of charge is one reason that the different forms of Landauer's formula exist. Here, the distribution function is clearly out of equilibrium adjacent to the QPC. We have not solved the equations in a self-consistent manner, but if we were to do so, then the conduction band would be bent downward near the QPC to reflect this charge buildup as well as to be consistent with it. Third, the propagation of three modes through the QPC is clear in the modal structure in the region within the QPC, and this fans out once through the constriction. The transverse mode structure is again evident in the density in this region past the QPC, as well as the general expansion of the wave function in this region. The latter arises due to the quantization within the QPC. While the transverse momentum is quantized, this corresponds to two values with opposite sign, and this leads to the expansion in the output region.

Scanning gate microscopy (SGM) is a novel technique in which structure in the quantum waveguide, as seen e.g. in Fig. 3.15, may be experimentally observed. In this approach, a biased tip is used to locally modify the electron potential or density [93]. Many structures have been probed with this technique including quantum point contacts [93,94,95,96,97,98], nanowires [99,100,101], carbon nanotubes [102], quantum Hall effect edge states [103], and more recently

Fig. 3.15 The density (magnitude squared of the wave function) with a quantum point contact placed within the transport waveguide. The details are discussed in the text.

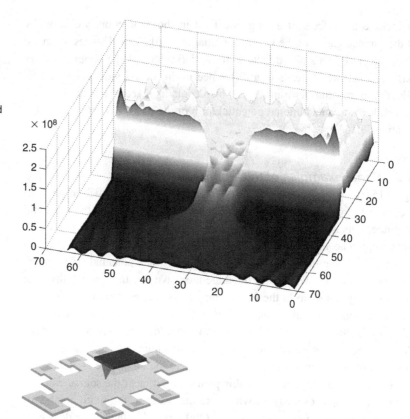

Fig. 3.16 Schematic view of an SGM. The tip is basically an AFM tip which has been metallized so that bias can be applied. This is then scanned across the device, while the conductance is measured.

quantum dots [104,105,106]. SGM allows us to measure the local potential near the pinch-off point and directly visualize the possible presence of these mechanisms. The biased tip, shown schematically in Fig. 3.16, is basically an atomic force microscope (AFM) tip that has been metallized, so that the voltage can be applied to this tip (relative to the sample ground). Then, as the tip is moved over the sample, the conductance change, induced by the tip, is monitored and this is plotted. Some of the earliest SGM measurements of the output of a quantum point contact showed the three modes as three output beams [97]. In this case, three modes were propagating through the QPC, and three relatively strong output beams were observed in the experiment.

As another example of the quantum point contact, we consider one made with in-plane gates, as shown in Fig. 3.17. Here, trenches are etched into the wafer to isolate "gate" regions from the rest of the quasi-two-dimensional electron gas in an InGaAs quantum well active layer. Then, bias applied to these regions can be used to deplete the QPC region. This allows the SGM to get into the QPC region

3.5 Transport in quantum waveguide structures

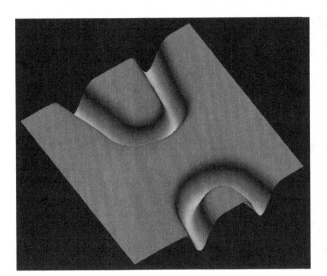

Fig. 3.17 An atomic-force microscope image of the in-plane gates formed by trench isolation [96].

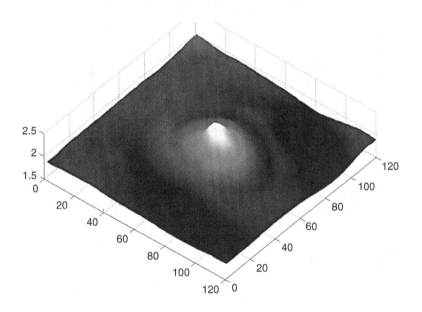

Fig. 3.18 Conductance change given by the SGM for a single mode propagating through the QPC of Fig. 3.17. The peak is located in the center of the QPC.

itself. In Fig. 3.18, we illustrate the SGM signal for the situation in which only a single mode is propagating through the QPC, fabricated in a GaAs/AlGaAs heterostructure. The peak conductance change illustrates the mode behavior. In devices fabricated with an InGaAs alloy quantum well active layer, there is considerable disorder in the sample, which causes variations in the signal. This disorder is thought to be due to the random placement of the impurities. However, it is clear that the wave function properties can be explored experimentally. In Fig. 3.19, we show an SGM signal which images the random

Fig. 3.19 Conductance image of the random potential, measured within the QPC by the SGM technique [107].

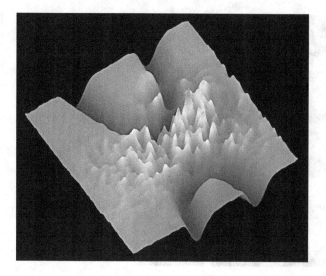

Fig. 3.20 Calculated conductance oscillations in a 0.3-μm open quantum dot. Also shown are wave function amplitudes for the scars that appear at several magnetic fields.

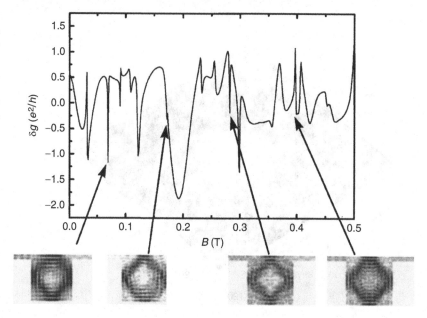

potential within the QPC itself. The image correlation function of this potential image actually shows a correlation length very close to that inferred from the measured weak localization signal in the QPC [107].

As a second example, we consider the magneto-conductance of an open quantum dot. Here, a wide waveguide (∼ 0.5–1.0 μm) is used, with local potentials defining a 0.3-μm square quantum dot, which in turn is connected to the waveguide by two QPCs. Each QPC passes one or more modes through the constriction [90]. As may be seen in Fig. 3.20, the conductance exhibits

3.5 Transport in quantum waveguide structures

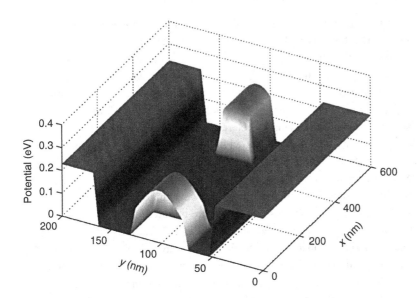

Fig. 3.21 Schematic layout of a two waveguide formulation of a waveguide coupler. The dimensions are given in the text.

significant oscillatory behavior. However, this is not universal conductance fluctuations, but a group of resonances, each of which is characterized by a Fano-like resonance shape [108]. In Fig. 3.20, the wave function amplitude is plotted for several of the resonances, showing how a particular scar is repeated at periodic values of the magnetic field (or of the gate voltage, which varies the Fermi level by moving it through the states [109]). Indeed, these resonances are remaining states of the closed quantum dot which are not "washed out" by decoherence from opening the QPCs to the environment. Such states are known as *pointer states* [110], and generally do not couple well to the environment [111,112]. They survive in the presence of coupling of the dot to the environment. These dots will be discussed more in Chapter 6.

For our next example, we consider a coupled waveguide structure which is capable of controlled transfer of density to either waveguide through the use of either a magnetic field, or by the application of an electrical bias across the device, although we consider only the former case here [113]. The structure studied here is shown in Fig. 3.21. Two parallel waveguides, separated by an electrostatic potential barrier, are coupled via a tunnel region. The input (top) waveguide has a uniform width of 35 nm from start to finish, whereas the output (bottom) waveguide is narrowed at the source end with a width of 25 nm, so that the incoming mode in the left waveguide cannot propagate in this second waveguide input, and then widens to a width of 45 nm after the coupling region in the middle of the structure. This wider output region assures that modes propagate through the coupling region and do not decay. The electrostatic potential barrier that separates the input and output waveguides begins with a width of 50 nm and then narrows to 25 nm after the coupling region.

The coupling region is 335 nm long. To achieve a more realistic potential profile for the barrier, the hardwall potential has been smoothed with a Gaussian distribution. The potential barrier, however, is still sufficiently high to prevent any leakage from the input waveguide to the output waveguide and assures all transfer of density from the input to the output occurs in the coupling region. The Fermi energy in the structure is chosen to be 2 meV, and the material is assumed to be a GaAlAs/GaAs heterojunction 2DEG. This Fermi energy is chosen so that only one mode is excited in the input waveguide of the structure. Since the input waveguide structure is wider than the output waveguide the mode that is excited at this energy will only propagate in the wider input waveguide. The particular dimensions of the waveguide structure can be easily scaled as long as the constraints mentioned are honored.

The results of this simulation are shown in Fig. 3.22. In Fig. 3.22(a), we show the density at zero applied magnetic field. The wave amplitude moves to the second guide and back again to the initial waveguide. In Fig. 3.22(b), a normal magnetic field of 0.705 T has been applied. Now, coupling is modified by the field so that the wave amplitude stays in the second waveguide. Such waveguide couplers have been suggested as being useful for perhaps configuring a qubit in quantum computation [110,114].

The study of the phase acquired by propagation in a mesoscopic device can be important for determining the presence/absence of phase interference in re-entrant geometries. In Chapter 1, we saw that the Aharonov–Bohm effect leads to phase interference in a ring structure, with the phase being modulated by a magnetic field passing through the ring. There, phase interference occurred between waves passing around the two opposite sides of the ring. As our last example of the scattering matrix formalism, we consider an equivalent phase interference effect, in which the phase variation is provided by the Rashba spin-orbit interaction [92,115]. Indeed, it only recently has been realized that an additional geometric phase can be introduced through the presence of the Rashba spin-orbit interaction in a heterostructure [116,117]. The phase shifts introduced by the spin-orbit interactions produce non-Abelian phases in the network, and the general use of non-Abelian statistics has become of interest for quantum computation [118,119]. We considered a ring structure with a radius of 100 nm and wire widths of 100 nm, with the ring in a GaAlAs/GaAs heterostructure at low temperature [120]. This structure can be seen in Fig. 3.23. In Fig. 3.24, we plot the variation of the conductance through the ring as the electric field is varied, with no magnetic field. There are resonances in the transmission due to the non-Abelian phases acquired as the electron travels around both sides of the ring. The parameter α_z is the spin-orbit parameter introduced in Eq. (3.128). The fact that the interference minima do not go to zero is a result of the small diameter of the ring relative to the width of the quantum wires. With such wide wires (relative to the diameter), the inside and outside of the wires reach resonance at slightly different values of the parameter.

(a)

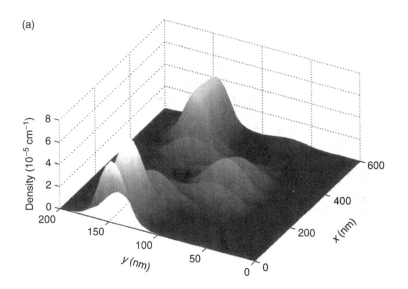

(b)

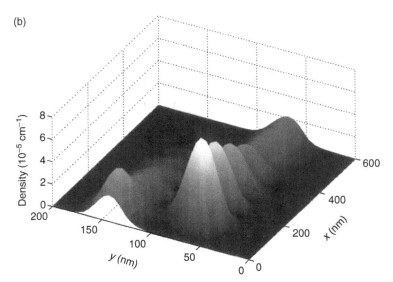

Fig. 3.22 Electron density in the guides. (a) No magnetic field is applied, and the coupling length (335 nm) is such that there is no transfer to guide 2. (b) A magnetic field of 0.705 T is applied normal to the plane, and the propagation vector is modified so that the coupling length now transfers the enter mode amplitude to the output guide.

This behavior is also the case in the Aharonov–Bohm effect. The general tuning of the transmission with the spin-orbit interaction can also be utilized to create a spin filter [121].

3.5.2 The Green's function

We now want to turn to the Green's functions. Later in this section, the concentration will be on generating a lattice representation that can be used for numerical simulations. First, however, we will introduce the analytical approach. Some

Fig. 3.23 Structure of the quantum ring structure, which is assumed to exist at the interface of a GaAs/AlGaAs heterostructure. The ring radius and wire width are both 100 nm.

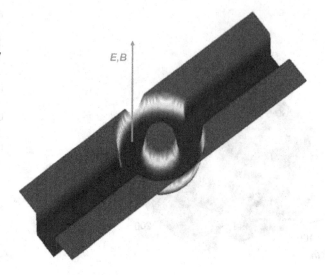

Fig. 3.24 Ring conductance as a function of the Rashba coupling term, with no magnetic field.

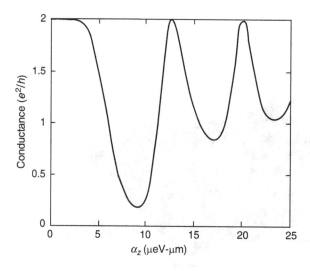

repetition of the previous material will be given in order to facilitate the understanding. In the next few sub-sections, we will basically review the use of the Green's functions in analytically calculating the conductivity. The following section will then examine the recursive Green's function and its use in numerical calculations, again primarily for the conductance.

The beginning point is the recognition that any initial wave function $\Psi(\mathbf{r}', t')$, which can be considered as an arbitrary initial condition, can be determined at any other point at a (not necessarily) later time from the relationship

$$\Psi(\mathbf{r}, t) = \int d^3\mathbf{r}' \int dt' K(\mathbf{r}, t; \mathbf{r}', t') \Psi(\mathbf{r}', t'). \tag{3.136}$$

Here, $K(\mathbf{r}, t; \mathbf{r}', t')$ is the propagator, or kernel, or Green's function $G(\mathbf{r}, t; \mathbf{r}', t') = K(\mathbf{r}, t; \mathbf{r}', t')$ for the Schrödinger equation. These quantities may be related to the actual Green's functions that one uses, the retarded and advanced Green's functions (we use the zero-temperature Green's functions in this chapter).

The kernel in (3.136) describes the general propagation of any initial wave function at time t' to any arbitrary time t (which is normally $> t'$, but not necessarily so). There are a number of methods of evaluating this kernel, either by differential equations (which will be pursued here) or by integral equations known as path integrals [122]. In general, the form shown here is developed for a system characterized by a well-defined set of basis functions, which are characteristic of the entire problem. We will see in a later chapter that one can just as easily take a thermodynamic equilibrium basis by passing to imaginary time with the substitution $t - t' \to -i\hbar\beta$, where $\beta = 1/k_B T$ is the inverse temperature. We will see still later that a further approach is to use real-time, nonequilibrium Green's functions in a manner that will require us to actually solve for the distribution function for the states in a nonequilibrium system. We work here with the first basis, in which it is assumed that the temperature $T = 0$. Thus, all states up to the Fermi energy are completely full, and all states above this energy are normally empty, in the absence of any perturbation to the equilibrium situation.

In general, one separates the kernel in the wave function (3.136) into forward and reverse time-ordering in order to have different functions for the retarded (forward in time) and advanced (backward in time, in the simplest interpretation) behavior. We do this by introducing the retarded Green's function as (for fermions)

$$G_r(\mathbf{r}, \mathbf{r}'; t, t') = -i\vartheta(t - t')\langle K(\mathbf{r}, t; \mathbf{r}', t')\rangle \\ = -i\vartheta(t - t')\langle\{\Psi(\mathbf{r}, t), \Psi^+(\mathbf{r}', t')\}\rangle, \quad (3.137)$$

where the angle brackets have been added to symbolize an ensemble average, which is also a summation over the proper basis states, and where the curly brackets indicate an anti-commutator for the anti-commuting electrons with which we are working. On the other hand, the advanced Green's function is given by

$$G_a(\mathbf{r}, \mathbf{r}'; t, t') = i\vartheta(t' - t)\langle K(\mathbf{r}, t; \mathbf{r}', t')\rangle \\ = i\vartheta(t' - t)\langle\{\Psi^+(\mathbf{r}', t'), \Psi(\mathbf{r}, t)\}\rangle, \quad (3.138)$$

and one can write the kernel itself as

$$K(\mathbf{r}, t; \mathbf{r}', t') = i[G_r(\mathbf{r}, t; \mathbf{r}', t') - G_a(\mathbf{r}, t; \mathbf{r}', t')]. \quad (3.139)$$

This particular form allows a symmetrized kernel to be used for (3.136) for which both the field operator and its adjoint satisfy this integral equation. Here, these field operators satisfy the anti-commutation relations [123]

$$\{\Psi(\mathbf{r},t), \Psi^+(\mathbf{r}',t')\} = \delta(\mathbf{r}-\mathbf{r}')\delta(t-t'). \tag{3.140}$$

The Green's functions include two time variables and two spatial variables, thus describing the propagation between two different points in the four-variable space $\mathbf{x} = (\mathbf{r}, t)$ (this four-variable notation will occasionally be used to simplify the number and complexity of integrations). Thus, temporal correlation processes can be incorporated in a fundamental manner. With this added complication – far more equations will be required to solve for the Green's functions than for the Schrödinger wave function – comes the benefit of a much more direct incorporation of dissipative processes. The equations of motion for the Green's functions can be developed for the simple propagation in the absence of any interactions from the basic Schrödinger equation. This leads to the pair of equations

$$\left(i\hbar\frac{\partial}{\partial t} - H_0(\mathbf{r}) - V(\mathbf{r})\right) G_0(\mathbf{r},\mathbf{r}';t,t') = \hbar\delta(\mathbf{r}-\mathbf{r}')\delta(t-t'), \tag{3.141}$$

$$\left(-i\hbar\frac{\partial}{\partial t'} - H_0(\mathbf{r}') - V(\mathbf{r}')\right) G_0(\mathbf{r},\mathbf{r}';t,t') = \hbar\delta(\mathbf{r}-\mathbf{r}')\delta(t-t'). \tag{3.142}$$

The Green's function here is either the retarded or the advanced function, with the difference determined by the ordering of the two time variables. In general, we can invoke the causality of the functions and make them arguments of $t - t'$, rather than of t, t' separately. Similarly, we can assert that these functions have properties that depend only on the difference in spatial variables (there is no particular value to any one point in space, and *thus the system is basically homogeneous* for such an approximation). If we then Fourier transform in space and time, we find that

$$G_0(\mathbf{k}, \omega) = \frac{\hbar}{\hbar\omega - E(\mathbf{k})}, \tag{3.143}$$

where $E(\mathbf{k})$ is the energy eigenvalue of $H_0 + V$. The latter potential is a confining potential that leads to a set of basis states so that the actual Green's function must be a sum over these states. In general, to invert the Fourier transform for the time variable, one must worry about the closure of the contour integration. Either the upper or lower half-plane must be taken in the complex ω-space. One half-plane leads to the retarded function, and the other to the advanced function. Simply speaking, these are separated by introducing an infinitesimal energy η for which

$$G_0^{r,a}(\mathbf{k}, \omega) = \frac{\hbar}{\hbar\omega - E(\mathbf{k}) \pm i\eta}, \tag{3.144}$$

where the upper sign is used for the retarded function and the lower sign is used for the advanced function. This small imaginary quantity moves the poles from the real axis into the lower half-plane for the retarded function and into the upper half-plane for the advanced function. This then gives the closure conditions to assure inclusion of the poles in the integration contour. This complexity becomes unnecessary once dissipation is incorporated in the problem.

We note that, in the case of a simple parabolic energy band for nearly free electrons, $(V \to 0)$ $E(\mathbf{k}) = \hbar^2 k^2/2m$, this leads to

$$G_0^{r,a}(\mathbf{k}, \omega) = \frac{\hbar}{\hbar\omega - \frac{\hbar^2 k^2}{2m} \pm i\eta}. \tag{3.145}$$

This is also the Fourier-space propagator for the diffusion equation if we replace $\hbar/2m$ by the diffusion constant D, so that the retarded and advanced Green's functions are often referred to as the results from *diffusion poles*. It should be noted, in particular, that the units of $\hbar/2m$ are cm^2/s, which are the same as those of the diffusion constant. The actual value has no connection to real transport numbers, but it should be recalled that this will provide a renormalization of the crucial frequencies (energies) when real numbers are used. We will use this fact later to replace the free-particle Green's function with a dressed *diffusion* particle.

3.5.2.1 Interaction and self-energies

In the classical case, and certainly before many-body theories were fully developed, the theory of electrical conduction was based upon the Boltzmann transport equation for a one-particle distribution function. Certainly this approach still is heavily used, particularly in the response of semiconductor systems in which the full quantum response is not necessary [124]; this was the case in the previous chapter. The essential assumption of this theory is the Markovian behavior of the scattering processes, that is, each scattering process is fully completed and independent of any other process. Coherence of the wave function is fully destroyed in each collision. On the other hand, here we are talking about possible coherence between a great many scattering events. Certainly, replacements for the Boltzmann equation exist for the non-Markovian world, but a more general approach is that of the Kubo formula. This linear response behavior describes transport with only an assumption that the current is linear in the applied electric field. The result, the Kubo formula, is that the current itself (or an ensemble averaged current) is given by a correlation function as

$$\langle \mathbf{J}(\mathbf{r},t)\rangle = \langle \mathbf{j}(\mathbf{r},t)\rangle + \int dt' \int d^3 r' \hat{\mathbf{G}}(\mathbf{r}-\mathbf{r}',t-t') \cdot \mathbf{A}(\mathbf{r}',t'), \tag{3.146}$$

where $\hat{\mathbf{G}}$ is the conductance tensor and $\mathbf{A}$ is the vector potential. In the diffusive limit, we can ignore the ballistic response (which leads to plasma oscillation effects in the dielectric function), and the conductance is given by the current–current correlation function

$$\hat{\mathbf{G}}(\mathbf{r}-\mathbf{r}',t-t') \cong -i\vartheta(t-t')\langle[\mathbf{j}(\mathbf{r},t),\mathbf{j}(\mathbf{r}',t')]\rangle. \tag{3.147}$$

If we recognize the current operator as

$$\mathbf{j}(\mathbf{r},t) = -i\frac{e\hbar}{2m}\{\Psi^+(\mathbf{r},t)\nabla\Psi(\mathbf{r},t) - [\nabla\Psi^+(\mathbf{r},t)]\Psi(\mathbf{r},t)\}$$
$$\rightarrow \frac{e\hbar\mathbf{k}}{m}\Psi^+(\mathbf{r},t)\Psi(\mathbf{r},t), \quad (3.148)$$

we can Fourier transform the above to arrive at

$$\sigma_{\alpha\beta} = \frac{e^2\hbar^2}{m^2}\sum_{\mathbf{k},\mathbf{k}'}k_\alpha k'_\beta G^{(2)}(\mathbf{k},\mathbf{k};\mathbf{k}',\mathbf{k}';t',t=t')\delta(E_k - E_F), \quad (3.149)$$

where $G^{(2)}(\mathbf{k},\mathbf{k};\mathbf{k}',\mathbf{k}';t',t)$ is the two-particle Green's function, and the last delta function insures that the evaluation is done at the Fermi level, since only those particles at the Fermi surface can contribute to transport at $T = 0$. The latter form is arrived at rather simply by Fourier transforming (3.146) in both space and time, noting that the commutator in (3.147) will lead to four terms, and then by using (3.148) to replace each current term. The four field operators combine to define the *two-particle Green's function*.

While the above has been written in momentum space and in time, we can examine the behavior somewhat differently in real space. The two-particle Green's function is defined in terms of four field operators. It can be written (using the time-ordering operator T) as

$$G^{(2)}(\mathbf{r},t;\mathbf{r}_1,t_1;\mathbf{r}',t';\mathbf{r}'_1,t'_1) = -\langle T[\Psi^+(\mathbf{r},t)\Psi(\mathbf{r}_1,t_1)\Psi^+(\mathbf{r}',t')\Psi(\mathbf{r}'_1,t'_1)]\rangle. \quad (3.150)$$

There are various approximations to evaluate this two-particle Green's function. The most usual are the Hartree and Hartree–Fock approximations. For the Hartree approximation we neglect possible exchange interactions, and write (3.150) as

$$G^{(2)}(\mathbf{r},t;\mathbf{r}_1,t_1;\mathbf{r}',t';\mathbf{r}'_1,t'_1) = G(\mathbf{r},t;\mathbf{r}',t')G(\mathbf{r}_1,t_1;\mathbf{r}'_1,t'_1). \quad (3.151)$$

Under the conditions for the above derivation (homogeneous conductance and the zero-frequency static conductivity), we may rewrite this in Fourier transform form as the product of two time functions, giving a convolution integral in Fourier space; since we are interested in the static result, only the frequency integral survives. The momentum integration is already contained in (3.149) and so does not appear. This result is then

$$G^{(2)}(\mathbf{k},\mathbf{k};\mathbf{k}',\mathbf{k}') = \int\frac{d\omega}{2\pi}G^r(\mathbf{k},\mathbf{k}',\omega)G^a(\mathbf{k},\mathbf{k}',\omega). \quad (3.152)$$

It is important to reiterate that this last result is a d.c. result and the factor ω in the integral is not the applied frequency, but the energy $E/\hbar$. The result is actually the d.c. (not the a.c.) conductivity, as has been stated several times. The interpretation of this term is that an electron is excited across the Fermi energy (at $T = 0$, all states are filled up to the Fermi energy), which creates an electron–hole pair in momentum state $\mathbf{k}'$. The creation is assumed to be accomplished by a non-momentum-carrying process, such as a photon (whose momentum is considerably smaller than that of the electron and/or hole). This pair then propagates, scattering from charged impurities, phonons, and other centers, to state **k**, where it recombines, again giving up the excess energy to a momentum-less particle of some type. This is then a particle–hole propagator, since the electron excited above the Fermi energy is called a quasi-particle with its characteristic energy measured from the Fermi energy itself. Similarly, the hole is a quasi-particle existing below the Fermi energy, and its energy is measured downward from the Fermi energy.

This notation arises very simply from the assertion that the absolute energy of the system is unimportant, and only the relative energies have sense. Since the total energy of these two particles remains $2E_F$, there is no confusion if we take the zero of energy at the Fermi energy. Changing the relation between the energy and momentum, which is often described by the mass in semiclassical systems, is more problematic but handled by talking about a quasi-particle mass. This treatment works well when the energy of the quasi-particles is only slightly different from the Fermi energy (i.e., in the linear response regime). Here, we retain the Fermi energy as a non-zero quantity for the moment.

Finally, let us connect the Green's functions in (3.152) with the noninteracting Green's functions of (3.145). This can be done as

$$G_0^{r,a}(\mathbf{k},\mathbf{k}',\omega) = G_0^{r,a}(\mathbf{k},\omega)\delta_{\mathbf{k}\mathbf{k}'}. \tag{3.153}$$

While there are some interactions which raise this conservation of the momentum in the Green's function of the interacting system, these will not be dealt with here. The Hartree–Fock approximation adds another term in which the two **k**-states (of the electron and hole) are interchanged due to electron exchange. This is of more interest in the electron–electron interaction and will be treated later.

3.5.2.2 Impurity scattering

As the first introduction of the methodology to be used here, we consider the contribution of impurity scattering to the resistivity (or conductivity, as the case may be). The impurity has an associated Coulomb potential that is long range in nature. This long-range interaction is usually cut off either by assuming that it is screened by the electrons (through the electron–electron interaction) or by some arbitrary distance. We will choose the latter approach as a crude approximation in that we assume the integrals (or summations) converge, but this should not be construed as any limit on the process; it is only done to avoid dealing with the divergences in some integrals that accompany the long-range interaction if it is not cut off. In general, the more advanced approach used in quantum field theory is termed *regularization*, in that additional terms are added with a variable parameter which is taken to zero after the calculation. In any real physical system, the long-range interaction is certainly cut off by at least the screening process. The impurity potential is given by

$$V_{imp} = \int d^3\mathbf{r} \sum_j U(\mathbf{r} - \mathbf{R}_j)\bar{n}(\mathbf{r})$$

$$= \frac{1}{\Omega} \sum_{\mathbf{q},j} U(-\mathbf{q})e^{i\mathbf{q}\cdot\mathbf{R}_j} \sum_{\mathbf{k},\mathbf{k}'} \Psi^+_{\mathbf{k}',s'}\Psi_{\mathbf{k},s}\delta_{\mathbf{k}',\mathbf{k}-\mathbf{q}}\delta_{ss'},$$

(3.154)

where the last Kronecker delta function conserves the spin of the particle, and Ω is the volume. Usually, the wave functions (taken here to be field operators) reduce to the creation and annihilation operators for plane-wave states. Although the primary role may well be the scattering of the electron (or hole) from one plane-wave state to another, there is also a recoil of the impurity itself, which can ultimately couple into local modes of the lattice. This latter complication will not be considered here, although it can be a source of short-wavelength phonons that can cause impurity-dominated intervalley scattering. Generating the perturbation series usually relies upon the S-matrix expansion of the unitary operator

$$\exp\left(-\frac{i}{\hbar}\int_{t'}^{t} dt'' V(t'')\right).$$

(3.155)

The treatment to be handled first is one in which we treat the two Green's functions in (3.152) independently. We will look at their interaction through the impurities later.

The expansion of the S-matrix in the scattering operator leads to an infinite series of terms, which change the equilibrium state. The higher-order terms are usually broken up by the use of Wick's theorem [125], and this leads to a diagram expansion. The formation of the Green's functions in the S-matrix expansion is accompanied by an averaging process over the equilibrium state (this is actually coupled to an average over the impurity configuration as well, which is discussed below). The equilibrium state must be renormalized in this process, and this causes a cancellation of all disconnected diagrams [126]. The result is an expansion in only the connected diagrams. Typical diagrams for the impurity scattering are shown in Fig. 3.25(a). One still must carry out an averaging process over the position of the impurities, since the end result (at least in macroscopic samples) should not depend upon the unique distribution of these impurities.

The impurity averaging may be understood by noting the summation over the impurity positions that appear in the exponential factors in (3.154). Terms like those in Fig. 3.25(a) involve the average

$$\left\langle \sum_{\mathbf{q}} e^{i\mathbf{q}\cdot\mathbf{R}_j} \right\rangle \to \delta(\mathbf{q}),$$

(3.156)

where the angular brackets denote the average over impurity positions. If the number of impurities is large, then the averaging of these positions places the important contribution of the potential from the impurities as that of a regular

(a)

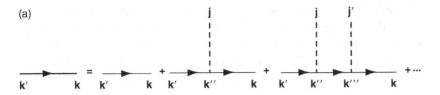

(b)

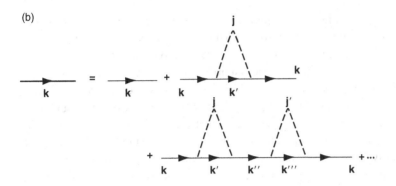

Fig. 3.25 (a) Typical impurity scattering single interactions involved in impurity averaging. (b) The second-order terms resulting in the dominant interactions after impurity averaging.

array, which can be thought of as creating a superlattice. The vectors $\mathbf{R}_j$ are then the basis vectors for this lattice, and the summation is over all such vectors. In short, the summation then represents the closure of a complete set, which yields the delta function shown on the right side of the arrow [125]. Thus, only the series of events in which the impurity imparts zero momentum to the electrons is allowed in the scattering process after the impurity averaging. Now, consider the double scattering processes shown in Fig. 3.25(b), in which a single impurity interacts to second order with the propagating electron. Now, there are two momenta imparted by the impurity, and the averaging in (3.156) becomes

$$\left\langle \sum_{\mathbf{q}} e^{i(\mathbf{q}+\mathbf{q}')\cdot\mathbf{R}_j} \right\rangle \to \delta(\mathbf{q}+\mathbf{q}'), \qquad (3.157)$$

so that one arrives at $\mathbf{q}' = -\mathbf{q}$. Thus, the interaction matrix element contained in the resultant expansion term from (3.155) for the second interaction is the complex conjugate of that for the first interaction, and the overall scattering process is proportional to the magnitude squared of the matrix element. This is the obvious result expected if one had started with the Fermi golden rule rather than Green's functions. Impurities can also interact with three lines and four lines, and more, as obvious extensions of the two situations shown in Fig. 3.25. The terms from (3.156) produce only an unimportant shift in the energy that arises from the presence of the impurities in the real crystal lattice, and the second-order interaction is the dominant scattering process. In general, the impurity interaction is sufficiently weak so that all terms with more than two coupled impurity lines usually can be safely ignored (at least in semiconductors in the absence of significant impurity-induced disorder).

It may be noticed first that, after the impurity averaging, we always have $\mathbf{k}' = \mathbf{k}$. Thus, as described above, each of the Green's functions, for which the spatial variation is in the *difference* of the two coordinates, is described by a single momentum state. This is as expected for normal Fourier transformation. Thus, we automatically recover the extension of (3.153) to

$$G^{r,a}(\mathbf{k}, \mathbf{k}', \omega) = G^{r,a}(\mathbf{k}, \omega)\delta_{\mathbf{k}\mathbf{k}'}. \tag{3.158}$$

The diagrams in Fig. 3.25(a) may be visualized as propagation, interaction, propagation, interaction, and so on. The labels indicate which impurity is involved in the process. The impurity averaging groups the terms according to the number of times that the same impurity occurs in the average. The dotted lines are connected to show this in Fig. 3.25(b), as was described just above. The light lines in each of these figures represent the unperturbed propagator G_0, and the heavy line corresponds to the actual propagator G. There is also the possibility that the intermediate state has zero momentum and is connected by impurity lines of zero momentum. This subset of diagrams is not shown, for it contributes only an arbitrary shift of the energy scale, and this can be incorporated in the overall Hartree energy. Now the third term in Fig. 3.25(b) is a replication of the second term. There could also be terms in which the two double lines overlap one another, as well as terms in which they are nested. The resummation is defined by the bringing together of all those diagrams that are topologically distinct and not simply replications of each other. These contributions are termed the self-energy. The replications are handled by replacing the output bare Green's function by the full Green's function. This leads us to be able to write the expansion as

$$G^{r,a}(\mathbf{k}, \omega) = G_0^{r,a}(\mathbf{k}, \omega) + G_0^{r,a}(\mathbf{k}, \omega)\frac{1}{\hbar}\Sigma^{r,a}(\mathbf{k}, \omega)G^{r,a}(\mathbf{k}, \omega), \tag{3.159}$$

which is *Dyson's equation* for the Green's functions. The self-energy is Σ and its expansion is shown in Fig. 3.26. Note that the internal Green's functions are the full Green's functions. In many cases, where the number of impurities is small, the scattering is weak, and the self-energy can be approximated by keeping only the first term and using the bare Green's function. Another, slightly better, approach is to keep only the lowest-order diagram but to use the full Green's function, solving the overall problem by iteration. This is often called the *self-consistent Born approximation*.

The equation for the full Green's function can be rewritten as

$$G^{r,a}(\mathbf{k}, \omega) = \frac{1}{\left[G_0^{r,a}(\mathbf{k}, \omega)\right]^{-1} - \frac{1}{\hbar}\Sigma^{r,a}(\mathbf{k}, \omega)}. \tag{3.160}$$

This form allows us to see just what the self-energy represents. In general, $\Sigma^{r,a}(\mathbf{k}, \omega)$ will have both a real and an imaginary part. If we compare the forms of (3.160) and (3.143), we note that the real part of the self-energy can be

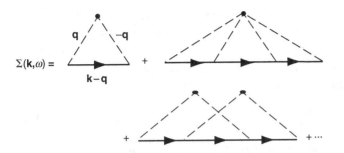

Fig. 3.26 The leading terms in the self-energy.

considered a correction to the single particle energy $E(\mathbf{k})$. This represents the *dressing* of (or change in) the energy due to the interaction with the impurities. This dressing can cause a general overall momentum-dependent shift in the energy, which also causes a change in the effective mass of the particle. On the other hand, the imaginary part of the self-energy represents the dissipative interaction that was included in an ad hoc manner by the insertion of the parameter η. Clearly the sign of the imaginary part of the self-energy is important to determine whether (3.160) represents the retarded or the advanced Green's function.

Let us now proceed to compute the scattering and conductance for the simplest case. This means that the task is really to evaluate the self-energy. However, if the full Green's function is retained in the latter term, then the process must be iterated. Here, we retain only the lowest-order term in the self-energy and treat the included Green's function with the bare equilibrium Green's function. According to (3.154), we need to sum over the momentum variable contained in the impurity interaction. For this, we will keep only the first term in Fig. 3.24, so that we assume the impurity scattering is weak. Then,

$$\Sigma^{r,a}(\mathbf{k},\omega) = \frac{N_i}{\Omega} \sum_{\mathbf{q}} |V(\mathbf{q})|^2 \frac{1}{\hbar\omega - E(\mathbf{k}-\mathbf{q}) \pm i\eta}, \quad (3.161)$$

where N_i is the total number of impurities. The impurity average has replaced the scattering by the various assortment of impurities with a single interaction, the averaged interaction, and this is multiplied by the number of impurities to arrive at the total scattering strength. At this point, it is pertinent to note that we expect the impurity potential to be reasonably screened by the free carriers, which means that the potential is very short range. It is easiest to assume a δ-function potential (in real space), which means that the Fourier-transformed potential is independent of momentum, or $V(\mathbf{q}) = V_0$. This is equivalent to assuming that the screening wavevector is much larger than any scattering wavevector of interest, or that $V_0 \sim e^2/\varepsilon_s q_s^2$. However, we will retain the form shown in (3.161) since we will need to produce a higher-order correction that modifies this term.

We are primarily interested in the imaginary parts of the self-energy, and this can be obtained by recognizing that η is small, so that we retain only the imaginary parts of the free Green's function. The real parts of the self-energy are no more than a shift of the energy scale, and we expect this to be quite small in the weak scattering limit. Then, we take the limit of small η and ignore the principal part of the resulting expansion (since it leads to the real part of the self-energy, which we have decided to ignore), so that

$$\lim_{\eta \to 0} \frac{1}{\hbar\omega - E(\mathbf{k}) \pm i\eta} \to \mp i\pi\delta(\hbar\omega - E). \qquad (3.162)$$

We now convert the summation into an integration in **k**-space, so that

$$\sum_{\mathbf{q}} \to \int\int \frac{\sin\vartheta d\vartheta d\varphi}{4\pi} \int \rho(E_q)dE_q, E_q = \frac{\hbar^2 q^2}{2m}. \qquad (3.163)$$

Here, the first two integrals represent an integration over the solid angle portions of the overall three-dimensional integral (or an equivalent in two dimensions). If the integrand is independent of these angular variables, then these integrals yield unity. It may be noted that the impurity scattering conserves energy, so that the final state energy is the same as the initial energy, or $E(\mathbf{k} - \mathbf{q}) = E(\mathbf{k})$. Thus, the energy integral can be evaluated as well, and the angular integral is at most an angular averaging of the scattering potential. This allows us to evaluate the self-energies, with $n_i = N_i/\Omega$ (Ω is the volume of the crystal) the impurity density, as

$$\Sigma^{r,a}(\mathbf{k},\omega) = \pm i\pi n_i \rho(\omega) \int\int \frac{\sin\vartheta d\vartheta d\varphi}{4\pi} |V(\mathbf{q})|^2 = \pm i\frac{\hbar}{2\tau(\omega)}. \qquad (3.164)$$

It should be noticed that, with the approximations used, the self-energies are independent of the momentum and are only functions of the energy $\hbar\omega$. We can now evaluate the conductivity as

$$\sigma_{\alpha\beta} = \frac{e^2\hbar^4}{m^2} \sum_{\mathbf{k}} k_\alpha k_\beta \int \frac{d\omega}{2\pi} \frac{1}{\hbar\omega - E(\mathbf{k}) - i\hbar/2\tau} \frac{1}{\hbar\omega - E(\mathbf{k}) + i\hbar/2\tau} \delta(E - E_F). \qquad (3.165)$$

With these approximations, and in the absence of a magnetic field, the conductivity is diagonal and isotropic (assuming that the energy band is isotropic). Either the frequency integral or the energy integral can be evaluated quickly by residues, with the other being replaced by the use of the delta function at the Fermi surface. Then, using $k_x^2 = k^2/d$ and $\rho(E_F)E_F = dn/2$, where $\rho(E)$ is the density of states (per unit volume) and d is the dimensionality, the conductivity is finally found to be

$$\sigma = \frac{e^2\hbar^4}{m^2} \sum_{\mathbf{k}} k_\alpha k_\alpha \frac{\tau}{\hbar^2} \delta(E - E_F) = \frac{e^2\hbar^2\tau}{dm^2} \rho(E_F) k_F^2 = \frac{ne^2\tau}{m}. \qquad (3.166)$$

This is the normal low-frequency result of the Drude formula, and it is the usual conductivity one arrives at in transport theory. It must be noticed, however, that most semiclassical treatments of the impurity mobility include

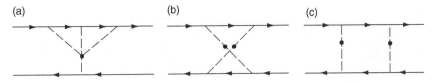

Fig. 3.27 Some typical interactions at fourth order in the impurity term. The diagrams in (c) form the so-called "ladder" diagrams which will be used to correct the conductivity.

a factor $(1 - \cos\vartheta)$, which is missing in this formulation, so that τ is a scattering time and not a relaxation time. It also must be noted that this result, which leads to a part of the normal Drude conductivity, is a result in which the two Green's functions are evaluated in isolation from one another, and there is no interaction between the two. We proceed to the more complicated case next.

3.5.2.3 Beyond the Drude result

In the previous sub-section, we treated only the interaction of the impurities with a single Green's function "line." We now consider higher-order corrections. In general, the impurity averaging still requires that the number of impurity "lines" joined by a single scattering site is an even number, with the two-line case the normal dominant term. The higher-order corrections that are most important are those in which the impurities connect the two Green's functions that appear in (3.152). The typical types of new diagrams that appear are shown in Fig. 3.27. The diagrams shown all have four interaction lines from either a single or a pair of impurities. To handle this complex situation, we will have to expand the two-particle Green's function. (As an aside, we remind ourselves that the frequencies in the two-particle Green's function are most often seen as $\omega_\pm = \omega \pm \omega_a/2$, where ω_a is the applied a.c. frequency at which the conductivity is evaluated. Here, however, we are interested in the d.c. conductivity so that $\omega_a = 0$ and we can ignore this complication.) In general, the two-particle Green's function is frequency dependent as shown, although the one we need for the static conductivity does not have any frequency dependence. We define the kernel of the conductivity (for the static conductivity) in the isotropic limit as [127,128]

$$\sigma_{\alpha\beta} = \frac{e^2\hbar^2}{dm^2} \int \frac{d\omega}{2\pi} F_{\alpha\beta}(\omega), \qquad (3.167)$$

with

$$\begin{aligned} F_{\alpha\beta}(\omega) &= \sum_{\mathbf{k},\mathbf{k}'} \mathbf{k} \cdot \mathbf{k}' G^{(2)}(\mathbf{k},\mathbf{k}',\omega;\mathbf{k}',\mathbf{k},\omega) \\ &= \sum_{\mathbf{k}} G^r(\mathbf{k},\omega) G^a(\mathbf{k},\omega) \\ &\quad \left\{ k^2 + \sum_{\mathbf{k}'} \mathbf{k} \cdot \mathbf{k}' \Lambda(\mathbf{k},\mathbf{k}',\omega;\mathbf{k}',\mathbf{k},\omega) G^r(\mathbf{k}',\omega) G^a(\mathbf{k}',\omega) \right\}. \end{aligned} \qquad (3.168)$$

In going from the tensor form of the conductivity to the form here, where the two wavevectors form a dot product, we note the previous choice of k_x^2 is replaced

Fig. 3.28 The diagram for the two-particle interacting Green's function, indicated by the symbol **L** in the figure, and the Bethe–Salpeter equation for the expansion of this function.

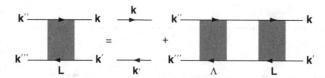

by k^2, which leads to the factor of d in the denominator. The first term in the curly brackets leads to the results obtained in the previous section. The new term is the second one in the curly brackets. Again, the various terms of Fig. 3.27(c), the ladder terms, can be summed into the form shown in Fig. 3.28, in which the last pair of Green's functions in the second term in curly brackets is an approximation to the two-particle Green's function shown in the figure. This particular form, illustrated in the figure, is known as the Bethe–Salpeter equation. In this, the quantity Λ is referred to as the irreducible scattering vertex.

The key contributions to the scattering vertex arise from the second-order interaction, which spans the two Green's functions. This is shown doubled in Fig. 3.27(c), so that the latter is a reducible term, in that it appears as a higher-order term in the series of (3.168). The replication of the lines that lead to Fig. 3.27(c) are called *ladder diagrams*. The terms of parts (a) and (b) of Fig. 3.27 are irreducible, as is the first part of (c). The impurity lines do not transport energy (the scattering from impurities is elastic), so the upper and lower Green's functions are at the same energy throughout; for the static case, these two frequencies are the same. We keep only the lowest-order correction in these ladder diagrams, which is the iterated diagrams such as those of Fig. 3.27(c). Now, we note that the kernel for the single scattering pair is

$$\Gamma_0 = \frac{n_i}{\hbar^2}|V(\mathbf{k}-\mathbf{k}')|^2, \qquad (3.169)$$

where the two momenta are those on either side of the scattering line. We want to form a series for the set of iterated iteractions and have the two integrations over **k** and **k**'. The scattering kernel (3.169) can be rewritten in terms of the scattering wavevector $\mathbf{q} = \mathbf{k} - \mathbf{k}'$, but this will make the second set of Green's functions depend upon $\mathbf{k}' + \mathbf{q}$, and the integration is then taken over **q**. However, since energy is conserved in the interaction, the argument of the potential is a function only of **k** and a scattering angle, which is not part of the integration over the last Green's functions. This means that the kernel can be treated separately from the last pair of Green's functions. We can write the series of ladder diagrams as

$$\Lambda = \Gamma_0 + \Gamma_0 \Pi \Gamma_0 + \Gamma_0 \Pi \Gamma_0 \Pi \Gamma_0 + \cdots = \frac{\Gamma_0}{1 - \Pi \Gamma_0}. \qquad (3.170)$$

The set of Green's function integrals for Π can be integrated to give

$$\Pi(\omega) = \int \rho(E) dE \frac{\hbar}{\hbar\omega - E + i\hbar/2\tau} \frac{\hbar}{\hbar\omega - E - i\hbar/2\tau} \quad (3.171)$$
$$= -2\pi\hbar\tau\rho(\hbar\omega).$$

The last pair of Green's functions in the second term of (3.168) just adds another factor of Π to the numerator of this last expression, if we take the kern in $\mathbf{k}'$ out of the integral. What remains from the dot product $\mathbf{k} \cdot \mathbf{k}' = k^2 \cos\vartheta$, where ϑ is the scattering angle. Using these last equations and properties, the kernel can be rewritten as

$$F_{\alpha\beta}(\omega) = \sum_{\mathbf{k}} G^r(\mathbf{k},\omega) G^a(\mathbf{k},\omega) k^2 \left\{ 1 + \cos\vartheta \frac{\Gamma_0 \Pi}{1 - \Pi\Gamma_0} \right\}. \quad (3.172)$$

Generally one makes the assumption here that the product $\Pi\Lambda_0$, is large compared to unity, so that one arrives at an additional term involving just the cosine of the scattering angle. Carrying out the frequency integration in (3.166) leads to the result

$$\sigma_{\alpha\beta} = \frac{e^2\hbar^2}{dm^2} \sum_k \tau k^2 (1 - \cos\vartheta) \delta(E - E_F) = \frac{ne^2\tau_m}{m}. \quad (3.173)$$

The result obtained here, with the approximations used, provides a connection with the semiclassical result for impurity scattering obtained from the Boltzmann equation discussed in Chapter 2. The complications are greater here, but this result also allows us to create a formalism that can be moved forward to treating other diagrammatic terms for new effects that are not contained in the Boltzmann equation. The angular dependence of the fraction that appears as the second term in the curly brackets of (3.172) provides the needed conversion from a simple scattering time to a momentum relaxation time. On the other hand, if we have a matrix element that really does not depend upon $\mathbf{q}$, such as a delta-function scattering potential or a heavily screened Coulomb potential where $V_0 \sim e^2/\varepsilon_s q_s^2$ (where q_s is the screening wavevector, either for Fermi–Thomas screening or for Debye screening), then the ladder correction terms do not contribute as the angular variation integrates to zero. Thus, scattering processes in which the matrix element is independent of the scattering wavevector do not contribute the angular correction. This simple fact is often overlooked in semiclassical treatments, as this latter angular variation is put in by hand, and its origin is not fully appreciated.

3.5.3 Recursive Green's functions

As we discussed at the beginning of this section, in order to consider structures more complicated than a simple adiabatic point contact requires at some level a knowledge of the actual transmission characteristics of a structure in terms of its scattering matrix (e.g., the actual detailed transmission coefficient T between the

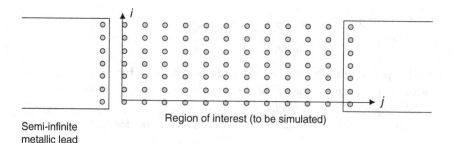

Fig. 3.29 Schematic of the active region to be simulated by a gridded structure which is connected to two semi-infinite quantum wires.

leads). Earlier in the present section, we considered a somewhat different approach, which developed the information via a recursive approach. The wave function scattering method, built up the wave function by a forward propagation based upon the Lippmann–Schwinger equation and a subsequent back propagation to finalize the wave function itself. It is also possible to do this with lattice Green's functions, which are also implemented via a recursive approach from one contact to the next [129,130,131]. In the zero-temperature format, these two approaches are largely equivalent, with each requiring the inversion of a Green's function matrix at each slice or step in the recursion. The former method is slightly easier to couple to the Poisson's equation in that a detailed integration of the Green's function over energy at each lattice site is not required to find the density. Rather, this integration is only required in the leads to properly normalize the various modes.

There is another difference, in that the wave function solution was normalized to unity amplitude of each mode at the start. For the recursive Green's function, we will assume that the active region to be simulated is coupled to two semi-infinite quantum wires, which are the input and output leads of the structure, as shown in Fig. 3.29. The Green's function for these two wires will be developed analytically, assuming that they are uncoupled to any other region. Then, the coupling to the active region, in the center of Fig. 3.29, is accomplished by perturbation theory, exactly as done in (3.122) of Section 3.5.1. That is, we assume that we can find the solution for each slice in the simulation region, and this slice is coupled to the preceding and succeeding slices by the hopping matrix terms of the Schrödinger equation, which provides the solution for G_0 in (3.141).

In the quantum wire at the input end of the structure, shown in Fig. 3.29, the solution may be written in terms of the transverse modes of the wire, given as

$$\varphi_n(x) = \sqrt{\frac{2}{W}} \sin\left(\frac{n\pi x}{W}\right), \tag{3.174}$$

where W is the width of the wire. As in Section 3.5.1, the transverse direction is taken to be the x-axis and the longitudinal direction is the y-axis. For convenience, we assume that the width of the wire is the same as that of the active region, so that $W = Ma$, where a is the grid spacing. Then, the position x is given by $x = ia$, where i is an integer, as shown in Fig. 3.14. Thus, we may rewrite (3.174) as

$$\varphi_n(i) = \sqrt{\frac{2}{Ma}} \sin\left(\frac{ni}{M}\right). \tag{3.175}$$

In the semi-infinite lead, which extends from $x = -a$ to $x = -\infty$, it is assumed that the wave function must vanish at the $j = -1$ slice, so that the wave function in the longitudinal direction is given by

$$\chi_\mu(x) = \chi_{\mu,j} = \sqrt{\frac{2}{\pi}} \sin(k_\mu a(j+1)), \tag{3.176}$$

$$E_\mu = 2t\left[1 - \cos(k_\mu a)\right].$$

In (3.176), the energy is the tight-binding band structure of the lattice and ranges from 0 to $4t$, where t is the hopping energy (3.116). We can now write down the Green's function for the wire in terms of the wave functions as

$$G(x,y;x',y') = G(i,j;i',j') \cdot$$

$$= \sum_{n,\mu} \frac{\Psi(i,j)\Psi^+(i',j')}{E - E(n,\mu) + i\eta} \tag{3.177}$$

$$= \frac{4}{\pi W} \int_0^{\pi/a} dk_\mu \sum_n \frac{\sin(k_\mu a(j+1)) \sin(k_\mu a(j'+1)) \sin\left(\frac{ni}{M}\right) \sin\left(\frac{ni'}{M}\right)}{\{E - 2t[1 - \cos(k_\mu a)] - \frac{n^2 \pi^2}{M^2} t\} + i\eta}.$$

Again, here the energy is written in terms of the hopping energy, and E represents the Fermi energy in the system. If we now evaluate this at the slice $j = j' = 0$, we have

$$G(x,0;x',0) = \frac{4}{\pi W} \int_0^{\pi/a} dk_\mu \sum_n \frac{\sin^2(k_\mu a) \sin\left(\frac{ni}{M}\right) \sin\left(\frac{ni'}{M}\right)}{\{E - 2t[1 - \cos(k_\mu a)] - \frac{n^2 \pi^2}{M^2} t\} + i\eta}. \tag{3.178}$$

In order to continue, we need to evaluate the integral over the longitudinal momentum in the quantum wire, and this becomes

$$I = \frac{1}{a} \int_0^\pi d\mu \frac{\sin^2(\mu)}{p + q \cos(\mu)} = \frac{\pi p}{qa}\left[1 - \sqrt{1 - \frac{q^2}{p^2}}\right], \tag{3.179}$$

where

$$p = E - 2t - \frac{n^2 \pi^2}{M^2} t + i\eta, \tag{3.180}$$

$$q = 2t.$$

If we write $\cos\vartheta = p/q$, then the integral becomes

$$I = \frac{\pi}{a} e^{i\vartheta} \equiv g_0. \tag{3.181}$$

Then, the Green's function at slice 0 is given as

$$G(x,0;x',0) = \frac{4g_0}{\pi W}\sum_n \sin\left(\frac{ni}{M}\right)\sin\left(\frac{ni'}{M}\right). \tag{3.182}$$

However, this Green's function is in the *mode* basis, and we need to transform it to the *site* basis. This is done with the unitary operator U, in which each column is the value of a mode at a given transverse grid point. Then, the Green's function in the site basis is given by

$$G_0 = U^+ G(x,0;x',0) U. \tag{3.183}$$

Now, given the energy structure of each mode (3.176), we can evaluate the velocity of that mode, which will be needed for later use. In the final computation, we need to propagate the value of the Green's function at slice zero to the end of the sample at slice L. In this process, we need to evaluate four Green's functions at each step of the iteration. These are G_{00}, G_{0j}, G_{j0}, and G_{jj}. This will be done for a two-dimensional system as indicated in Fig. 3.27. As in Section 3.5.1, the connection between the slices is given by the hopping interaction and this is treated as the self-energy correction to the bare Green's function of each slice. Here,

$$\begin{aligned}\Gamma \equiv H_{j,j+1} = T_y, \\ \Gamma^+ \equiv H_{j+1,j} = T_y^+,\end{aligned} \tag{3.184}$$

where T_y is given in (3.121). Since this is a diagonal matrix, the difference between this and the inverse only occurs in the presence of the magnetic field where the diagonal terms involve the Peierl's phase factors. At the first step, we initialize the four Green's functions as

$$G_{0j} = G_{j0} = G_{jj} = G_{00} = \left(E - H_j + \Gamma G_0 \Gamma^+\right)^{-1}. \tag{3.185}$$

This last is done for slice $j = 0$. Then, for $j > 0$, we generate the four Green's functions at each slice via the recursion in the following order:

$$\begin{aligned} G_{jj} &= \left[E - H_j + H_{j,j-1} G_{j-1,j-1} H_{j-1,j}\right]^{-1}, \\ G_{0,j} &= G_{0,j-1} H_{j-1,j} G_{jj}, \\ G_{j,0} &= G_{jj} H_{j,j-1} G_{j-1,0}, \\ G_{00} &= G_{00} + G_{0,j-1} H_{j-1,j} G_{j,0}. \end{aligned} \tag{3.186}$$

Finally, at the $(L+1)$st slice, the Green's function is reconnected to a semi-infinite lead through the final iteration

$$\begin{aligned} G_{L+1,L+1} &= \left[E - H_{L+1} + H_{L+1,L}(G_0 - G_{L,L}) H_{L,L+1}\right]^{-1}, \\ G_{0,L+1} &= G_{0,L} H_{L,L+1} G_{L+1,L+1}, \\ G_{L+1,0} &= G_{L+1,L+1} H_{L+1,L} G_{L,0}, \\ G_{00} &= G_{00} + G_{0,L} H_{L,L+1} G_{L+1,0}. \end{aligned} \tag{3.187}$$

3.5 Transport in quantum waveguide structures

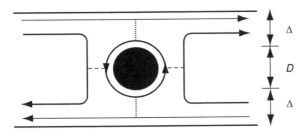

Fig. 3.30 Configuration of the edge states (thin solid lines) at the Fermi energy. The shaded area represents the antidot potential, which here is taken to be a soft wall potential. Tunnel coupling between the propagating and encircling edge states is indicated by the dotted and dashed lines [131].

The mode-to-mode transmission matrix is then determined from transforming the Green's function back to the mode representation, as

$$T_{\text{mode}} = U^+ G_{L+1,0} U, \qquad (3.188)$$

while the reflection matrix is similarly given by

$$R_{\text{mode}} = U^+ G_{00} U. \qquad (3.189)$$

Finally, the conductance is determined by summing over the propagating modes, with the magnitude squared of the transmission for each mode weighted by the ratio of the output mode velocity to the input mode velocity in the semi-infinite wire contacts, as in the normal Landauer formula. That is, the contribution from each term in the transmission matrix from mode i in the input lead to mode j in the output lead is

$$T_{ij} = \frac{v_j}{v_i} |T_{\text{mode},i,j}|^2, \qquad (3.190)$$

and the contribution of each input mode to the overall transmission is summed over all possible output modes as in (3.69). The conductance is given by (3.61).

As an example of the technique, we consider a wide quantum wire with an antidot repulsive potential located in the center of the wire. The wire size is sufficiently larger than the antidot diameter that propagation can still occur for a sufficiently high Fermi energy (carrier density). The structure is shown in Fig. 3.30 [132]. The width of the wire and the dot diameter are W and D, respectively, as shown in the figure, with $W = D + 2\Delta$. Here, the aim is to simulate the interaction of propagating edge states, in a magnetic field, with edge states trapped around the antidot [133,134,135,136]. In these studies, magneto-resistance measurements on antidot arrays have revealed a periodic oscillation superimposed on a slowly varying background due to commensurate classical orbits around the antidot. These oscillations are thought to be due to the Aharonov–Bohm effect, and disappear at both weak and strong magnetic fields. They have a maximum amplitude when the cyclotron orbit diameter is commensurate with the period of the array [132,134]. Only one antidot is used in the simulation, but an oscillation due to the Aharonov–Bohm effect is found to occur when an edge state encircling the antidot couples to the propagating edge states by tunneling

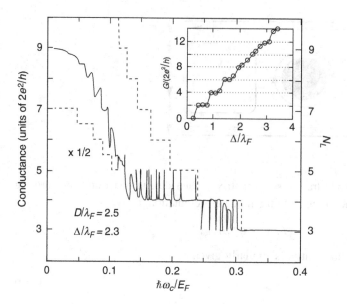

Fig. 3.31 Magnetoconductance of the wire with the antidot. The dotted line represents the number of Landau levels (N_L) in the wire away from the antidot. The inset shows the zero-field conductance as a function of the width Δ of the constrictions. The latter is quantized into steps of $4e^2/h$ [131].

through the gap between them. Such inter-edge-state tunneling has been studied by Jain and Kivelson [137]. While these latter authors found resonant peaks in the resistance, presumably due to backscattering, the antidot case also shows resonant peaks in the conductance due to forward scattering. The conductance is simulated with the recursive Green's functions discussed above, and the Fermi energy was taken such that the Fermi wavelength is $6a$, where a is the grid spacing. The magnetic field is incorporated via the Peierl's phase as discussed above. In order to have well-defined modes in the semi-infinite wires, the magnetic field is graded over a short region, from zero adjacent to these wires to its desired value closer to the antidot.

The oscillation in the conductance, shown in Fig. 3.31, is well explained by the resonant scattering between the edge states via the antidot. In addition, it was found that disorder in the local potential, due for example to randomly sited impurities, would disrupt the observation of the Aharonov–Bohm effect [138].

References

[1] R. H. Fowler and L. Nordheim, *Proc. Roy. Soc. (London)* **119**, 173 (1928).
[2] C. Zener, *Proc. Roy. Soc. (London)* **A145**, 523 (1943).
[3] L. Esaki, *Phys. Rev.* **109**, 603 (1957).
[4] I. Giaver, *Phys. Rev. Lett.* **5**, 147 (1960).
[5] I. Giaver, *Phys. Rev. Lett.* **5**, 464 (1960).
[6] J. Bardeen, *Phys. Rev. Lett.* **6**, 57 (1961).
[7] M. H. Cohen, L. M. Falicov, and J. C. Phillips, *Phys. Rev. Lett.* **8**, 316 (1962).

[8] W. A. Harrison, *Phys. Rev.* **123**, 85 (1961).
[9] E. O. Kane, *J. Appl. Phys.* **32**, 83 (1961).
[10] C. B. Duke, "Tunneling in Solids," *Sol. State Phys.* **10** (Suppl.), (New York, Academic Press, 1969).
[11] R. Tsu and L. Esaki, *Appl. Phys. Lett.* **22**, 562 (1973).
[12] L. L. Chang, L. Esaki, and R. Tsu, *Appl. Phys. Lett.* **24**, 593 (1974).
[13] T. C. L. G. Sollner, W. D. Goodhue, P. E. Tannenwald, C. D. Parker, and D. D. Peck, *Appl. Phys. Lett.* **43**, 588 (1983).
[14] E. R. Brown, T. C. L. G. Sollner, C. D. Parker, W. D. Goodhue, and C. L. Chen, *Appl. Phys. Lett.* **55**, 1777 (1989).
[15] T. J. Shewchuk, P. C. Chapin, P. D. Coleman, *et al.*, *Appl. Phys. Lett.* **46**, 508 (1985).
[16] M. Tsuchiya and H. Sakaki, *Appl. Phys. Lett.* **49**, 88 (1986).
[17] M. Tsuchiya and H. Sakaki, *Appl. Phys. Lett.* **50**, 1503 (1987).
[18] M. Tsuchiya and H. Sakaki, *Jpn. J. Appl. Phys.* **25**, L185 (1986).
[19] P. Guéret, C. Rossel, E. Marclay, and H. Meier, *J. App. Phys.* **66**, 278 (1989).
[20] H. M. Yoo, S. M. Goodnick, and J. R. Arthur, *Appl. Phys. Lett.* **56**, 84 (1990).
[21] E. R. Brown, W. D. Goodhue, and T. C. L. G. Sollner, *J. Appl. Phys.* **64**, 1519 (1988).
[22] T. Inata, S. Muto, Y. Nakata, *et al.*, *Jpn. J. Appl. Phys.* **26**, L1332 (1987).
[23] T. P. E. Broekaert, W. Lee, and C. G. Fonstad, *Appl. Phys. Lett.* **53**, 1545 (1988).
[24] L. F. Luo, R. Beresford, and W. I. Wang, *Appl. Phys. Lett.* **53**, 2320 (1988).
[25] J. R. Söderström, E. R. Brown, C. D. Parker, *et al.*, *Appl. Phys. Lett.* **58**, 275 (1991).
[26] E. R. Brown, J. R. Söderström, C. D. Parker, *et al.*, *Appl. Phys. Lett.* **58**, 2291 (1991).
[27] J. R. Söderström, D. H. Chow, and T. C. McGill, *Appl. Phys. Lett.* **55**, 1094 (1989).
[28] K. F. Longenbach, L. F. Luo, and W. I. Wang, *Appl. Phys. Lett.* **57**, 1554 (1990).
[29] S. A. Wolf, D. D. Awschalom, R. A. Buhrman, *et al.*, *Science* **294**, 1488 (2001).
[30] G. Dresselhaus, *Phys. Rev.* **100**, 580 (1955).
[31] Yu. A. Bychkov and E. I. Rashba, *J. Phys. C* **17**, 6039 (1984).
[32] A. Voshkoboynikov, S. S. Liu, and C. P. Lee, *Phys. Rev. B* **59**, 12514 (1999).
[33] E. A. de Andrade e Silva and G. C. La Rocca, *Phys. Rev. B* **59**, R15583 (1999).
[34] D. Z.-Y. Ting and X. Cartoixà, *Appl. Phys. Lett.* **81**, 4198 (2002).
[35] J. J. Zinck, D. H. Chow, K. S. Holabird, *et al.*, *Appl. Phys. Lett.* **86**, 073502 (2005).
[36] K. C. Hall, W. H. Lau, K. Gündoğdu, M. E. Flatté, and T. F. Boggess, *Appl. Phys. Lett.* **83**, 2937 (2003).
[37] D. Z.-Y. Ting, X. Cartoixà, D. H. Chow, *et al.*, *Proc. IEEE* **91**, 741 (2003).
[38] R. H. Good, Jr., and E. W. Müller, in *Handbuch der Physik,* Vol. 21, ed. S. Flugge (Berlin, Springer, 1956), p. 176.
[39] H. C. Liu, *J. Appl. Phys.* **67**, 593 (1990).
[40] T. C. McGill and D. Ting, "Fluctuations in mesoscopic systems," in the *Proceedings of the NATO ASI on Ultrasmall Devices*, eds. D. K. Ferry, H. L. Grubin, and C. Jacoboni (New York, Plenum Press, 1995).
[41] E. E. Mendez, W. I. Wang, E. Calleja, and C. E. T. Goncalves da Silva, *Appl. Phys. Lett.* **50**, 1263 (1987).
[42] A. A. Bonnefoi, T. C. McGill, and R. D. Burnham, *Phys. Rev. B* **37**, 8754 (1988).
[43] S. Luryi, *Appl. Phys. Lett.* **47**, 490 (1985).
[44] F. Capasso, K. Mohammed, and A. Y. Cho, *IEEE J. Quantum Elec.* **QE-22**, 1853 (1986).
[45] T. Weil and B. Vinter, *Appl. Phys. Lett.* **50**, 1281 (1987).

[46] M. Jonson and A. Grincwajg, *Appl. Phys. Lett.* **51**, 1729 (1987).
[47] A. D. Stone and P. A. Lee, *Phys. Rev. Lett.* **54**, 1196 (1985).
[48] L. D. Landau and E. M. Lifshitz, *Quantum Mechanics (Nonrelativistic)* (Oxford, Pergamon Press, 1977).
[49] M. Büttiker, *Phys. Rev. B* **33**, 3020 (1986).
[50] M. Büttiker, *IBM J. Res. Develop.* **32**, 63 (1988).
[51] H. C. Liu and T. C. L. G. Sollner, "High-frequency resonant-tunneling devices," in *High-Speed Heterostructure Devices*, eds. R. A. Kiehl and T. C. L. G. Sollner, Semiconductors and Semimetals, Vol. 41 (Boston, Academic Press, 1994), pp. 359–419.
[52] V. J. Goldman, D. C. Tsui, and J. E. Cunningham, *Phys. Rev. B* **35**, 9387 (1987).
[53] E. E. Mendez, L. Esaki, and W. I. Wang, *Phys. Rev. B* **33**, 2893 (1986).
[54] J. F. Young, B. M. Wood, G. C. Aers, *et al.*, *Phys. Rev. Lett.* **60**, 2085 (1988).
[55] V. J. Goldman, D. C. Tsui, and J. E. Cunningham, *Phys. Rev. Lett.* **58**, 1256 (1987).
[56] T. C. L. G. Sollner, *Phys. Rev. Lett.* **59**, 1622 (1987).
[57] F. W. Sheard and G. A. Toombs, *Appl. Phys. Lett.* **52**, 1228 (1988).
[58] A. Zaslavsky, V. J. Goldman, D. C. Tsui, and J. E. Cunningham, *Appl. Phys. Lett.* **53**, 1408 (1988).
[59] H. Ohnishi, T. Inata, S. Muto, N. Yokoyama, and A. Shibatomi, *Appl. Phys. Lett.* **49**, 1248 (1986).
[60] M. Cahay, M. McLennan, S. Datta, and M. S. Lundstrom, *Appl. Phys. Lett.* **50**, 612 (1987).
[61] K. F. Brennan, *J. Appl. Phys.* **62**, 2392 (1987).
[62] H. L. Berkowitz and R. A. Luz, *J. Vac. Sci. Technol. B* **5**, 967 (1987).
[63] W. Pötz, *J. Appl. Phys.* **66**, 2458 (1989).
[64] M. O. Vassell, J. Lee, and H. F. Lockwood, *J. App. Phys.* **54**, 5206 (1983).
[65] B. Ricco and M. Ya. Azbel, *Phys. Rev. B* **29**, 1970 (1984).
[66] L. Eaves, G. A. Toombs, F. W. Sheard, *et al.*, *Appl. Phys. Lett.* **52**, 212 (1988).
[67] P. J. Price, *Phys. Rev. B* **45**, 9042 (1992).
[68] E. Wigner, *Phys. Rev.* **40**, 749 (1932).
[69] W. R. Frensley, *Phys. Rev. B* **36**, 1570 (1987).
[70] N. C. Kluksdahl, A. M. Kriman, D. K. Ferry, and C. Ringhofer, *Phys. Rev. B* **39**, 7720 (1989).
[71] R. Landauer, *IBM J. Res. Develop.* **1**, 223 (1957).
[72] R. Landauer, *Phil. Mag.* **21**, 863 (1970).
[73] R. Büttiker, Y. Imry, R. Landauer, and S. Pinhas, *Phys. Rev. B* **31**, 6207 (1985).
[74] H. L. Engquist and P. W. Anderson, *Phys. Rev. B* **24**, 1151 (1981).
[75] Y. Imry, "Physics of mesoscopic systems," in *Directions in Condensed Matter Physics*, eds. G. Grinstein and G. Mazenko (Singapore, World Scientific Press, 1986), pp. 101–63.
[76] R. Landauer, *IBM J. Res. Develop.* **32**, 306 (1988).
[77] R. Landauer, *Physica A* **168**, 863 (1990).
[78] D. S. Fisher and P. A. Lee, *Phys. Rev. B* **23**, 6851 (1981).
[79] M. Büttiker, *Phys. Rev. Lett.* **57**, 1761 (1986).
[80] M. Büttiker, *IBM J. Res. Develop.* **32**, 317 (1988).
[81] W. Gröbner, *Matrixenrechnung* (Bibliographisches Institut AG, Mannheim, 1966), p. 47.
[82] G. Timp, R. Behringer, S. Sampere, J. E. Cunningham, and R. Howard, "When isn't the conductance of an electron waveguide quantized?" in the *Proceedings of the Int. Symp. on Nanostructure Physics and Fabrication,* ed. W. P. Kirk and M. Reed (New York, Academic Press, 1989), pp. 331–46.

[83] G. Timp, "When does a wire become an electron waveguide?" in *Nanostructure Systems*, ed. M. Reed, *Semiconductors and Semimetals* **35** (Boston, Academic Press, 1992), pp. 113–90.
[84] H. U. Baranger, D. P. DiVincenzo, R. A. Jalabert, and A. D. Stone, *Phys. Rev. B* **44**, 10637 (1991).
[85] E. Merzbacher, *Quantum Mechanics* (New York, Wiley, 1970).
[86] A. M. Kriman, N. C. Kluksdahl, and D. K. Ferry, *Phys. Rev. B* **36**, 5953 (1987).
[87] A. Szafer, A. M. Kriman, A. D. Stone, and D. K. Ferry, *unpublished*.
[88] T. Usuki, M. Takatsu, R. A. Kiehl, and N. Yokoyama, *Phys. Rev. B* **50**, 7615 (1994).
[89] T. Usuki, M. Takatsu, R. A. Kiehl, and N. Yokoyama, *Phys. Rev. B* **52**, 824 (1995).
[90] R. Akis, D. K. Ferry, and J. P. Bird, *Phys. Rev. B* **54**, 17705 (1996).
[91] F. Mireles and G. Kirczenow, *Phys. Rev. B* **64**, 024426 (2001).
[92] Y. A. Bychkov and E. I. Rashba, *J. Phys. C* **17**, 6039 (1984).
[93] M. A. Topinka, B. J. LeRoy, S. E. J Shaw, *et al.*, *Science* **289**, 2323 (2000).
[94] M. A. Topinka, B. J. LeRoy, R. M. Westervelt, *et al.*, *Nature* **410**, 183 (2001).
[95] M. A. Topinka, B. J. LeRoy, R. M. Westervelt, K. D. Maranowski, and A. D. Gossard, *Physica E* **12**, 678 (2002).
[96] N. Aoki, C. R. da Cunha, R. Akis, D. K. Ferry, and Y. Ochiai, *Appl. Phys. Lett.* **87**, 223501 (2005).
[97] M. A. Topinka, B. J. LeRoy, S. E. J. Shaw, *et al.*, *Science* **289**, 2323 (2000).
[98] B. J. LeRoy, M. A. Topinka, A. C. Bleszynski, *et al.*, *Appl. Surf. Sci.* **210**, 134 (2003).
[99] R. Crook, C. G. Smith, M. Y. Simmons, and D. A. Ritchie, *J. Phys.: Cond. Matt.* **12**, L735 (2000).
[100] T. Ihn, J. Rychen, T. Cilento, *et al.*, *Physica E* **12**, 691 (2002).
[101] R. Crook, C. G. Smith, M. Y. Simmons, and D. A. Ritchie, *Physica E* **12**, 695 (2002).
[102] S. G. Lemay, J. W. Jansen, M. van den Hout, *et al.*, *Nature* **412**, 617 (2001).
[103] N. Aoki, C. R. da Cunha, R. Akis, D. K. Ferry, and Y. Ochiai, *Phys. Rev. B* **72**, 155327 (2005).
[104] R. Crook, C. G. Smith, A. C. Graham, *et al.*, *Phys. Rev. Lett.* **91**, 246803 (2003).
[105] R. Crook, A. C. Graham, C. G. Smith, *et al.*, *Nature* **424**, 751 (2003).
[106] A. Pioda, S. Kicin, T. Ihn, *et al.*, *Phys. Rev. Lett.* **93**, 216801 (2004).
[107] C. R. da Cunha, N. Aoki, T. Morimoto, *et al.*, *Appl. Phys. Lett.* **89**, 242109 (2006).
[108] R. Akis, D. Vasileska, J. P. Bird, and D. K. Ferry, *J. Phys.: Cond. Matter* **11**, 4657 (1999).
[109] J. P. Bird, R. Akis, D. K. Ferry, *et al.*, *Phys. Rev. Lett.* **82**, 4691 (1999).
[110] W. H. Zurek, *Rev. Mod. Phys.* **75**, 715 (2003).
[111] D. K. Ferry, R. Akis, and J. P. Bird, *Phys. Rev. Lett.* **93**, 026803 (2004).
[112] D. K. Ferry, R. Akis, and J. P. Bird, *J. Phys.: Cond. Matter* **17**, S1017 (2005).
[113] M. J. Gilbert, R. Akis, and D. K. Ferry, *Appl. Phys. Lett.* **81**, 4284 (2002).
[114] A. Bertoni, P. Bordone, R. Brunetti, C. Jacoboni, and S. Reggiani, *Phys. Rev. Lett.* **84**, 5912 (2000).
[115] F. Mireles and G. Kirczenow, *Phys. Rev. B* **64**, 024426 (2001).
[116] Y. Aharonov and A. Casher, *Phys. Rev. Lett.* **53**, 319 (1984).
[117] D. Bercioux, M. Governale, V. Cataudella, and V. M. Ragaglia, *Phys. Rev. B* **72**, 075305 (2005).
[118] M. H. Freedman, M. Larsen, and Z. Wang, *Commun. Math. Phys.* **227**, 605 (2002).
[119] S. Das Sarma, M. Freedman, and C. Nayak, *Phys. Rev. Lett.* **94**, 166802 (2005).

[120] A. W. Cummings, R. Akis, and D. K. Ferry, *J. Comp. Electron.*, in press.
[121] A. W. Cummings, R. Akis, and D. K. Ferry, *Appl. Phys. Lett.* **89**, 172115 (2006).
[122] R. P. Feynman and A. R. Hibbs, *Quantum Mechanics and Path Integrals* (New York, McGraw-Hill, 1965).
[123] J. W. Negele and H. Orland, *Quantum Many-Particle Systems* (Redwood City, CA, Addison-Wesley, 1988).
[124] D. K. Ferry, *Semiconductors* (New York, Macmillan, 1991).
[125] A. L. Fetter and J. D. Walecka, *Quantum Theory of Many-Particle Systems* (New York, McGraw-Hill, 1971).
[126] G. D. Mahan, *Many-Particle Physics* (New York, Plenum Press, 1981).
[127] C. P. Enz, *A Course on Many-body Theory Applied to Solid-State Physics* (Singapore, World Scientific Press, 1992).
[128] A. A. Abrikosov, L. P. Gor'kov, and I. Ye. Dzyaloshinskii, *Quantum Field Theoretical Methods in Statistical Physics* (New York, Pergamon Press, 1965).
[129] D. S. Fisher and P. A. Lee, *Phys. Rev. B* **23**, 6851 (1981).
[130] D. J. Thouless and S. Kirkpatrick, *J. Phys. C* **14**, 235 (1981).
[131] A. MacKinnon, *Z. Physik B* **59**, 385 (1985).
[132] Y. Takagaki and D. K. Ferry, *Phys. Rev. B* **48**, 8152 (1993).
[133] C. G. Smith, M. Pepper, R. Newbury, et al., *J. Phys.: Cond. Matter* **2**, 3405 (1990).
[134] D. Weiss, M. L. Roukes, A. Menschig, et al., *Phys. Rev. Lett.* **66**, 2790 (1991).
[135] F. Nihey, S. Ishizaka, and K. Nakamura, unpublished.
[136] D. K. Ferry, *Prog. Quantum Electron.* **16**, 251 (1992).
[137] J. K. Jain and S. A. Kivelson, *Phys. Rev. Lett.* **60**, 1542 (1988).
[138] Y. Takagaki and D. K. Ferry, *Surf. Sci.* **305**, 669 (1994).

4
The quantum Hall effects

The discovery in 1980, by Klaus von Klitzing and his colleagues, of the integer quantum Hall effect (IQHE) [1] may have done more than any other single event to stimulate experimental and theoretical interest in the electrical properties of low-dimensional systems. This phenomenon has now been observed in a variety of different material systems, and is manifest as the appearance of wide and precisely quantized plateaus in the Hall resistance (R_H, or Hall resistivity ρ_{xy}), which therefore deviates strongly from the linear dependence on magnetic field that is expected classically. It is now understood that this high-magnetic-field phenomenon is associated with the formation of strongly quantized Landau levels in a two-dimensional electron gas (2DEG), under which conditions current flow is carried by ballistic edge states that are the quantum analog of classical skipping orbits (recall Section 2.5). Thus, the quantum Hall effect represents a remarkable manifestation of one-dimensional transport in a macroscopic system.

In this chapter, we begin by discussing the basic phenomenology of the (integer) quantum Hall effect, which, due to the extreme accuracy of its quantization, has now been adopted as an international standard for the definition of the ohm. We present an interpretation of this effect due to Büttiker, which begins from the concepts of the Landauer formula (Section 3.3) and explains the quantization by considering that edge states propagate ballistically, without dissipation, over the entire sample length. This remarkable property results from a strong spatial separation of forward- and backward-propagating current states when the edge states are well defined (at high magnetic fields). The microscopic structure of these edge states is strongly influenced by self-consistent screening at high magnetic fields, and a discussion of this issue is therefore also provided in this chapter.

A remarkable feature of the edge states that form at high magnetic fields is that electrons injected into them are able to propagate over extremely long distances, in some cases as much as millimeters, while preserving their initial energy distribution. It is this dissipationless aspect of transport that provides the connection to the Landauer–Büttiker picture of transport. In spite of this, however, experiments in which a locally tunable gate (potential barrier) is introduced into the system provide an ideal opportunity to test the predictions of the Landauer–Büttiker approach. In Section 4.2, we therefore discuss the details of edge state transport in nanostructures such as single barriers, quantum point contacts and quantum dots. An important concept here is that a single barrier may be used as a selective filter of edge states, giving rising to an anomalous magneto-resistance quantization that is now related to the number of edge states reflected and transmitted by the potential barrier. We shall see in our analysis that a simple extension of the Landauer–Büttiker formalism is able to account precisely for this anomalous quantization. In the case where the local barrier is actually comprised of a quantum dot, we shall also see that the dot may confine edge states completely, forming an interferometer that exhibits periodic oscillations in its magneto-conductance and providing a direct demonstration of phase-coherent propagation of electrons in edge states. These characteristics have been recently exploited to realize sensitive single-photon detectors, and to generate spin-dependent currents through quantum dots.

Then, in the last two sections, we treat the fractional quantum Hall effect (FQHE) and the many-body understanding that has arisen with regard to this effect. Here, we look first at the odd denominator fractions, then the newer even denominator fractions. This will lead us to discuss the composite fermion and to discuss some of the details of the many-body wave functions. Of more recent interest is the bilayer systems, particularly those in which each layer has a one-half filled level so that there are strong interactions between the two layers. While much of this is understood, there remains some discussion as to the exact ground state at many of the fractions, so that this remains a work in progress. Our hope here is merely to open the door to some of the ideas and theories.

4.1 The integer quantum Hall effect in two-dimensional electron systems

4.1.1 Shubnikov–de Haas effect

Prior to the discovery of the quantum Hall effect, it had long been understood that the study of the transport properties of low-dimensional systems under an applied magnetic field can provide a valuable insight into their electronic properties (as discussed in Section 2.5). The most widely studied system is that of the 2DEG with a magnetic field applied perpendicular to the plane of the gas. The first clear evidence of purely two-dimensional behavior in the Si/SiO_2 inversion layer system was reported by Fowler *et al.* [2], who studied the oscillations in the channel conductance with magnetic field in this system. Such magneto-resistance oscillations in bulk materials are often referred to as Shubnikov–de-Haas (SdH) oscillations, and provide an important tool for investigating the Fermi surface of conductors (see for example [3]). In contrast to the

4.1 The integer quantum Hall effect

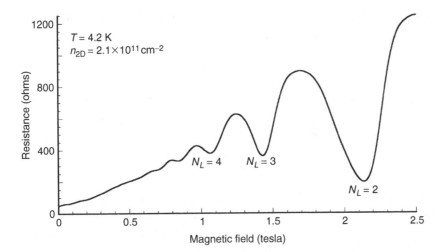

Fig. 4.1 Shubnikov–de Haas oscillations in the longitudinal magneto-resistance of the two-dimensional electron gas, formed in a GaAs/AlGaAs heterojunction quantum well. *Unpublished data.* The resistance plotted here is the longitudinal component, which is obtained by measuring the voltage dropped *parallel* to the direction of current flow.

bulk SdH effect, the observed oscillations in the Si/SiO$_2$ inversion layer depended only on the component of the magnetic field perpendicular to the 2DEG, and not on the in-plane component, demonstrating the reduced dimensionality of the electronic states. Figure 4.1 shows an experimental curve of similar SdH oscillations in the longitudinal magneto-resistance of a high mobility GaAs/AlGaAs heterostructure at 4.2 K. Due to the higher mobility and hence longer scattering times in modulation-doped heterojunction systems compared to the Si/SiO$_2$ system, features arising from the magnetic quantization are better resolved in the former system due to reduced broadening of the underlying Landau levels as discussed earlier in Section 2.5. As seen in this figure, the longitudinal resistance increases slightly with increasing magnetic field before entering a regime of sinusoidal modulation around 0.75 T. As discussed in Section 2.5, when the separation of the Landau levels exceeds the broadening of the levels due to scattering ($\omega_c \tau \sim 1$), the discrete nature of the density of states becomes observable in the transport properties, where ω is the cyclotron frequency and τ is the scattering time. It should be emphasized here that the onset field of 0.75 T is much higher than that implied by the electron mobility ($\mu = e\tau_m/m^*$), which for this sample is larger than 10^6 cm^2/Vs. As discussed already in Section 2.5, τ_m is the momentum relaxation time whereas τ is the total scattering time which may be much less in the case of anisotropic scattering mechanisms such as impurity scattering [4].

The oscillations shown in Fig. 4.1 are periodic in $1/B$ and may be related to the density of the 2DEG in a simple way. We have seen already that the quantized energy levels for electrons in a magnetic field are those of a simple harmonic oscillator, with an oscillator frequency (ω_c) that is determined by the magnetic field (see Eq. (2.45)). Since electrons follow circular paths in coordinate space due to the magnetic field, their motion in *k*-space is orbital also (for a spherical conduction band minimum such as in GaAs). By making use of the parabolic dispersion relation for free electrons, the area enclosed by such an orbit in the k_x–k_y plane can then be written as

$$A_k = \frac{2\pi eB}{\hbar}\left(n+\frac{1}{2}\right). \tag{4.1}$$

Now, for electrons at the Fermi surface, which are the ones relevant for the discussion of transport, we may use the form of the density of states at zero magnetic field to obtain

$$\pi k_F^2 = 4\pi^2 \frac{n_{2D}}{n_s n_v}, \tag{4.2}$$

where n_{2D} is the 2DEG density and n_s and n_v are the spin and valley degeneracies, respectively ($n_s = 2$ and $n_v = 1$ for GaAs, while $n_s = 2$ and $n_v = 4$ for Si). To connect these relations to the oscillatory magneto-resistance in the SdH effect, we simply equate Eqs. (4.1) & (4.2) to obtain

$$\frac{1}{B} = \frac{n_s n_v}{n_{2D}} \frac{e}{h}\left(n+\frac{1}{2}\right). \tag{4.3}$$

The physical significance of Eq. (4.3) is that it defines a series of magnetic fields for which particular Landau levels (with specific index, n) pass through the Fermi level and so depopulate. For these magnetic fields, the density of states is high since one of the broadened delta functions is coincident with the Fermi level, and the magneto-conductance consequently exhibits a peak. As the magnetic field is varied, the magneto-resistance therefore oscillates and Eq. (4.3) indicates that the resulting oscillations should be periodic in *inverse* magnetic field. More specifically, the separation of the oscillations in inverse magnetic field should be given by

$$\Delta(1/B) = \frac{n_s n_v}{n_{2D}} \frac{e}{h}. \tag{4.4}$$

It is worth pointing out here that, in contrast to the bulk SdH effect, there is no mass dependence in Eq. (4.4). For a given material, the only variable is the 2DEG density and so an analysis of the inverse-field spacing of the SdH oscillations can be used to determine this density in experiment. For the data of Fig. 4.1, for example, the periodicity of the oscillations in inverse field yields $n_{2D} = 2.1 \times 10^{11}\,\text{cm}^{-2}$, a fairly typical value for GaAs 2DEGs. While our analysis thus far has focused on the situation where just a single subband of the 2DEG is occupied, in 2DEGs with two subbands occupied different frequency oscillations are observed corresponding to the unequal densities in each subband. Magneto-resistance oscillations may also be observed in experiment by fixing the magnetic field, and so the Landau-level energy separation, but then using an external gate to change the Landau-level filling by tuning n_{2D} (see Fig. 4.3 below).

As indicated in the discussion already, the number of Landau levels occupied in a 2DEG may be tuned by variation of either the magnetic field or the electron

4.1 The integer quantum Hall effect

density. An important parameter that expresses this tunability is the Landau index (N_L), which indicates the number of occupied Landau levels under arbitrary conditions. To determine an expression for N_L, we note that at high magnetic fields all electron states in the energy range $\hbar\omega_c$ at zero magnetic field have collapsed into a given Landau level. The number of electrons per Landau level (per unit area) is therefore given by $(n_s n_v m^*/2\pi\hbar^2) \times \hbar\omega_c = (n_s n_v eB/h)$. Note that, due to the definition of the density of states, this is the number of electrons per Landau level *per unit area* of the sample. Consequently, the Landau index is given by

$$N_L = \frac{n_{2D}}{n_s n_v} \frac{h}{eB}. \tag{4.5}$$

From the density inferred above for the sample of Fig. 4.1, Eq. (4.5) can be used to determine the Landau-level filling index at any given magnetic field. A simple calculation shows that the oscillation minima near 2.3, 1.4, and 1.1 T correspond to $N_L = 2$, 3, and 4, respectively. In contrast, the oscillation maxima near 1.7, 1.2, and 0.9 T correspond to $N_L = 2.5$, 3.5, and 4.5, respectively.

The conductivity of a 2DEG in the presence of a perpendicular magnetic field was derived by Ando using the Kubo formula in linear response theory, assuming a system of short-range scatterers (see [5] for a details). For weak magnetic fields ($\omega_c \tau \leq 1$) in the extreme quantum limit, the lowest-order corrections to the longitudinal conductivity are written

$$\sigma_{xx} = \sigma_0 \frac{1}{1+\omega_c^2\tau^2} \left[1 - \frac{2\omega_c^2\tau^2}{1+\omega_c^2\tau^2} \frac{2\pi^2 k_B T_e}{\hbar\omega_c} \operatorname{cosech} \frac{2\pi^2 k_B T_e}{\hbar\omega_c} \cos \frac{2\pi^2 E_F}{\hbar\omega_c} e^{-\pi/\omega_c\tau} + \cdots \right], \tag{4.6}$$

where T_e is the electron temperature (which in general can be different from that of the lattice due to carrier heating), and $\sigma_o = n_{2D} e^2 \tau/m^*$ is the zero-field conductivity. The cosine term gives the oscillatory behavior with a period dependent on either $1/B$ or the Fermi energy (E_F, which is directly proportional to density) as discussed above. The hyperbolic cosecant term leads to a damping behavior with temperature which depends explicitly on the effective mass through the cyclotron frequency. This temperature dependence of the oscillation amplitude is sometimes used as a means of ascertaining the effective cyclotron mass in 2D systems by fitting Eq. (4.6) to experimental data on the magnetic-field and temperature-dependent amplitude of the SdH oscillations. As mentioned earlier, it is necessary to be careful regarding the interpretation of τ and its strict relation to the actual momentum relaxation time in the presence of long-range scatterers like ionized impurities.

4.1.2 The integer quantum Hall effect

As the magnetic field is increased beyond the range considered in Fig. 4.1, the sinusoidal characteristic of the magneto-resistance becomes distorted and drops to zero between the peaks. The peaks themselves split due to lifting of the spin

Fig. 4.2 The quantum Hall effect in a GaAs/AlGaAs two-dimensional electron gas. Panel (a) shows the variation of the Hall resistance of this sample while panel (b) shows the corresponding Shubnikov–de Haas oscillations in the longitudinal magneto-resistance. The measurement temperature is 50 mK and the inset to the upper panel shows a schematic of the 2DEG Hall-bar geometry with the different contacts that are used to measure the longitudinal and Hall magneto-resistance ([6], by permission).

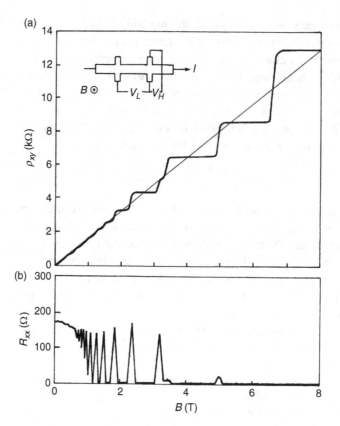

degeneracy (as well as the valley degeneracy in multi-valley semiconductors), as discussed in Section 2.5. Even more remarkable behavior is observed, however, in the Hall resistance ($R_H = V_H/I$, where V_H is the Hall voltage). In a classical derivation, it is straightforward to show that R_H should increase linearly with magnetic field. Experiments performed on two-dimensional electron gas (2DEG) systems show a marked deviation from this behavior, however, when the Landau-level quantization is well resolved. In this regime, as we illustrate in Fig. 4.2, the Hall resistance is found to saturate at certain values over a wide range of magnetic field. The saturation becomes more pronounced with increasing magnetic field and occurs for the same field range for which the longitudinal resistance vanishes (see, also, Fig. 4.2). In their landmark paper of 1980 [1], von Klitzing *et al.* pointed out that the resistance values at which these Hall plateaus occur are quantized with extremely high precision, according to the simple relation:

$$R_H = \frac{h}{ie^2}, \quad i = 1, 2, 3, \ldots \quad (4.7)$$

This quantization of R_H is universal, being independent of the material under study and of the sample quality. As we shall see below, the integer i in Eq. (4.7)

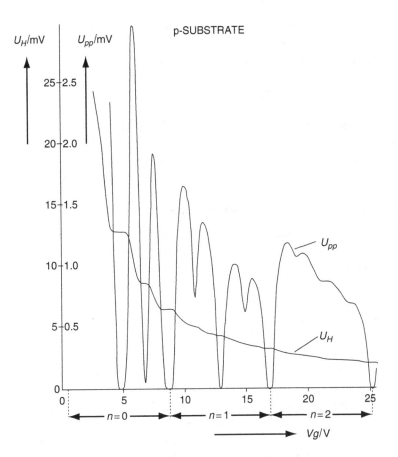

Fig. 4.3 The quantum Hall effect in an Si MOSFET two-dimensional electron gas. In contrast to the experiment of Fig. 4.2, the quantization of the Hall resistance in this experiment is observed at fixed magnetic field, while using a variation of the gate voltage to change the number of occupied Landau levels ([6], by permission).

corresponds to the number of (completely, or partially) filled (spin-resolved) Landau levels at that particular magnetic field.

In Fig. 4.2, the quantum Hall effect is observed in a GaAs/AlGaAs 2DEG with a fixed electron density [6], by using a magnetic field to vary the number of occupied Landau levels. The original experiment of von Klitzing et al. [1] was performed on Si MOSFETs, however, in which it was also possible to vary the electron density in the inversion layer, and so the number of occupied Landau levels, by means of the gate voltage. An identical quantization of the Hall resistance is found in such experiments, as we show in Fig. 4.3.

While it was apparent early on that the quantization of the Hall resistance is a universal, and extremely precise, phenomenon, subsequent studies have demonstrated the remarkable accuracy to which the quantization occurs. Since 1990, this has resulted in the quantum Hall effect being adopted by national standards laboratories over the world as a means to represent the SI-unit ohm [7]. The quantity $h/e^2 = \sim 25813\ \Omega$ is now referred to as the von Klitzing constant (R_K) and experiment has demonstrated that the accuracy of the Hall-resistance quantization to the level of 3.5 parts in 10^{10}, independent of the material system studied and the plateau index (*i*) [8]. The fact that the von Klitzing constant is determined only by the fundamental constants *h* and *e* means that it has an intimate connection to the fine structure constant (α)

$$R_K = \frac{h}{e^2} = \frac{\mu_0 c}{2\alpha}, \qquad (4.8)$$

where μ_0 is the permeability of free space and *c* is the speed of light in vacuum. Consequently, the high precision to which the Hall quantization can be determined has implications for all those areas of physics that make use of α.

4.1.3 Edge states and the integer quantum Hall effect

As a high-magnetic-field phenomenon that is observed in 2DEG systems with macroscopic dimensions, there would seem to be little connection at first between the quantum Hall effect and the earlier derivation of the Landauer formula for mesoscopic conductors (Section 3.3). In this analysis, we saw that the conductance of a perfectly transmitted (i.e. ballistic) one-dimensional channel is $2e^2/h$. As we now discuss, however, there is an intimate connection of the quantum Hall effect to the Landauer formula, which arises from the uniquely one-dimensional transport properties of the edge states that form at high magnetic fields.

In Section 2.5, we discussed already how the application of a quantizing magnetic field to a 2DEG can result in the formation of edge states, which are essentially the quantum-mechanical analog of classical skipping orbits. The edge states arise as a consequence of the Landau-level structure in real systems, which feature physical boundaries (walls) due to their finite size. Deep within the interior of such systems, the conduction band edge is essentially flat, independent of position (we return to this point shortly below), and the corresponding Landau-level energies are similarly constant. The boundaries of the sample correspond to those regions, however, where the conduction band edge rises well above the Fermi level, to confine the electron system to a finite area. Each Landau level that is occupied in the interior therefore intersects the Fermi level at two different points, located near the opposite edges of the sample, and therefore yields two counter-propagating edge states that function as one-dimensional channels.

For magnetic fields such that the Fermi energy lies well between two Landau-level energies within the bulk of the sample, the only way in which backscattering can occur is for electrons to be transferred across its entire width, a distance of around a millimeter for typical quantum Hall samples. As pointed out by Büttiker [9], in his seminal explanation of the quantum Hall effect, however, electrons propagating in edge states cannot be easily scattered over distances larger than the magnetic length, $l_B = (\hbar/eB)^{0.5}$. This can be readily understood within a picture of classical skipping orbits, since the resulting Lorentz force exerted on the electron tends to push it back towards the boundary and so

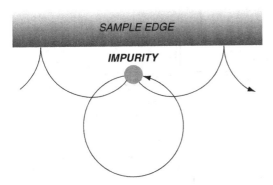

Fig. 4.4 Schematic illustration showing the suppression of backscattering for a skipping orbit in a conductor at high magnetic fields. While the impurity may momentarily disrupt the forward propagation of the electron, it is ultimately restored as a consequence of the strong Lorentz force.

maintain its skipping motion (Fig. 4.4). For a magnetic field of 4 T, l_B is little more than 10 nm, and electron backscattering is therefore strongly suppressed. In this regime, we can therefore imagine the electrical current through the sample being carried by a fixed number of edge states (corresponding to the number of occupied Landau levels) which propagate as ballistic one-dimensional channels. To generate a net current flow in such a system, it is necessary to apply a finite voltage across its current reservoirs, thereby populating edge states at opposite boundaries to different electrochemical potentials. Experiments have shown that the edge states can then propagate without dissipation over extremely long distances, as long as millimeters, while maintaining the initial populations set by their corresponding reservoirs [10,11,12].

By starting from a picture of edge states as ballistic one-dimensional channels, and making use of the Landauer formula, it is possible to account for the main features of the quantum Hall effect, as described by Büttiker [9]. We begin our discussion by assuming that the value of the magnetic field is such that N Landau levels are occupied and that, within the flat interior of the sample, the Fermi level lies well between the Nth and $(N + 1)$th Landau level. (This restriction ensures that the foregoing discussing on edge states is justified, but will be relaxed shortly when we discuss the transitions between different quantum Hall plateaus.) Now consider a conductor with a Hall bar geometry, and with several Ohmic contacts to its 2DEG, as we illustrate in Fig. 4.5. The problem of solving for the Hall resistance of this system is one that involves calculating the different electrochemical potentials that the contacts sit at, under conditions of a fixed applied current (I). We are guided in this analysis by several key assumptions, the first of which is that all edge states *leaving* any given contact are assumed to be fully equilibrated with it (that is, it is assumed that the electrochemical potential of these edge states is the same as that of the contact that they are leaving.)

Fig. 4.5 Schematic illustration of current flow via edge states in a 2DEG Hall bar under conditions of a high applied magnetic field. The gray regions are the Ohmic contacts to the 2DEG and may be used as either voltage or current probes, depending on how they are connected to an external measuring circuit. For the case shown here, $N = 3$.

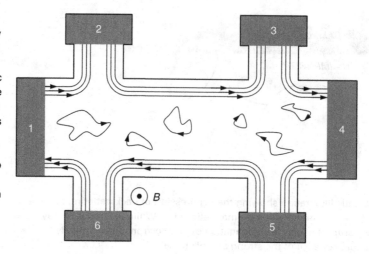

The other assumption that will be important in our analysis is that, whenever any contact of the Hall bar is configured as a voltage probe, it then draws *no* electrical current. It should be recognized here that this statement applies to the *net* current flowing through the voltage probe. This current is composed of two contributions, the first which is due to incoming edge states that are equilibrated at the electrochemical potential of the neighboring contact (downstream in the sense of the edge-state flow), while the other is carried by edge states leaving the probe. As is common in experiment, we consider that contacts 2, 3, 5, and 6 in Fig. 4.5 are to be used as voltage probes, while contacts 1 and 4 are used to source and sink current, respectively. We can then make use of the results of Section 3.3 to write the following expressions for the current drawn by the voltage probes:

$$I_2 = \frac{2Ne^2}{h}(V_2 - V_1) = 0, \tag{4.9}$$

$$I_3 = \frac{2Ne^2}{h}(V_3 - V_2) = 0, \tag{4.10}$$

$$I_5 = \frac{2Ne^2}{h}(V_5 - V_4) = 0, \tag{4.11}$$

$$I_6 = \frac{2Ne^2}{h}(V_6 - V_5) = 0. \tag{4.12}$$

In these expressions, note that we have defined current leaving any probe as positive, while current entering the probe is defined as negative. From these relations, we can therefore infer that $V_1 = V_2 = V_3$ and $V_4 = V_5 = V_6$. The current probes *do* draw net current, however, as described by the following relations

4.1 The integer quantum Hall effect

$$I_1 = \frac{2Ne^2}{h}(V_1 - V_6) = I, \qquad (4.13)$$

$$I_4 = \frac{2Ne^2}{h}(V_4 - V_3) = -I. \qquad (4.14)$$

With these results, we can now calculate the Hall resistance, which can be determined from the voltage drop between probes 2 and 6, for example

$$R_H = \frac{V_2 - V_6}{I} = \frac{h}{2Ne^2}\frac{V_2 - V_6}{V_1 - V_6} = \frac{h}{2Ne^2}, \quad N = 1, 2, 3, \ldots, \qquad (4.15)$$

where we have taken advantage of the fact that $V_1 = V_2$ in obtaining the final result. It is easy for the reader to confirm that the same result is obtained for the other Hall measurement that may be made, using voltage probes 3 and 5, instead of 2 and 6. At this point it is clear that Eq. (4.15) differs only from Eq. (4.7) by the factor of 2 in its denominator. As discussed already in the derivation of the Landauer formula (see Eq. 3.118), this factor expresses the fact that, at zero magnetic field, one-dimensional subbands are typically spin degenerate. Consequently, each subband can carry double the current expected if spin is neglected, and therefore has twice the conductance. Since the spin branches of each subband are degenerate, however, they are indistinguishable and so conductance is quantized in units of $2e^2/h$ (we return to this idea in Chapter 5). At high magnetic fields, however, spin degeneracy is lifted and the spin branches may be selectively depopulated by the magnetic field. Consequently, we should rewrite Eq. (4.15) in the form

$$R_H = \frac{h}{\nu e^2}, \quad \nu = 1, 2, 3, \ldots, \qquad (4.16)$$

where ν is the number of spin-resolved Landau levels ($\nu \equiv 2N_L$) and is referred to as the filling factor. Equation 4.16 describes precisely the quantization of the Hall effect found in experiment and reveals the physical meaning of the factor i that appears in Eq. (4.7); it is nothing more than the Landau-level filling factor.

The other unique feature of the quantum Hall effect is the vanishing of the longitudinal resistance for the same ranges of magnetic field that the quantized plateaus are observed (see Figs. 4.2 and 4.3). This feature can also be explained by the picture of non-dissipative transport via adiabatic edge states. For the geometry of Fig. 4.5, the longitudinal resistance (R_{xx}) may be determined by measuring the voltage drop between probes 2 and 3, or 5 and 6. The resistance R_{xx} vanishes for either configuration, however, just as is found in experiment

$$R_{xx} = \frac{V_2 - V_3}{I} = \frac{V_6 - V_5}{I} = 0. \qquad (4.17)$$

(In obtaining this result we have made use of Eqs. (4.9)–(4.12)). We should emphasize here that the vanishing of R_{xx} does *not* imply that the system conducts current without dissipation. Instead, it indicates that the potential applied to the sample is between its opposite edges, rather than along its length. Indeed, because of this non-uniform nature of the potential distribution, it is important that we recognize that the process of current flow through a Hall sample should be described by a tensor form, whose off-diagonal elements arise from the application of the magnetic field. Neglecting any freedom of motion in the direction perpendicular to the 2DEG plane, it is common to write the conductivity tensor for this problem in the following notation

$$\boldsymbol{\sigma} = \begin{bmatrix} \sigma_{xx} & \sigma_{xy} \\ -\sigma_{xy} & \sigma_{xx} \end{bmatrix}, \qquad (4.18)$$

where σ_{xx} is the longitudinal conductivity and σ_{xy} is the Hall conductivity. Alternatively, it is possible by inverting this matrix to obtain the resistivity tensor, which takes the same form as Eq. (4.18) but which has elements

$$\rho_{xx} = \frac{\sigma_{xx}}{\sigma_{xx}^2 + \sigma_{xy}^2}, \qquad (4.19)$$

$$\rho_{xy} = \frac{\sigma_{xy}}{\sigma_{xx}^2 + \sigma_{xy}^2}. \qquad (4.20)$$

From Eq. (4.19), we see that for conditions where the longitudinal resistivity vanishes the conductivity also vanishes! So, as mentioned already, the zero-resistance regions in the longitudinal resistance do not correspond to a superconducting state.

4.1.4 Breakdown of the quantum Hall effect

In our discussion thus far, we have focused on the situation where the current at high magnetic fields is carried by ballistic edge states that do not backscatter as they propagate through the sample. This has allowed a simple analysis in terms of the Landauer–Büttiker formalism, and has accounted for the existence of both the quantized Hall plateaus and the vanishing of the longitudinal resistivity. As such, this description is valid for magnetic fields such that, within the interior of the sample, the Fermi level lies well within the energy gap between the uppermost occupied, and lowest empty, Landau levels. As the magnetic field is increased, however, the associated increase in the Landau-level eigenenergies (Eq. (2.43)), and eventually the uppermost occupied level will approach the Fermi energy. This process is illustrated in Fig. 4.6, which shows schematically both the Landau-level distribution within the sample, and the corresponding edge states associated with these levels, for different characteristic magnetic fields. In the figure here, note that we do not draw the bottom of the potential in the interior of the sample as completely flat, but instead show it as exhibiting random fluctuations. These potential fluctuations are associated with inevitable disorder within the sample. In modulation-doped GaAs/AlGaAs 2DEGs, for

4.1 The integer quantum Hall effect 205

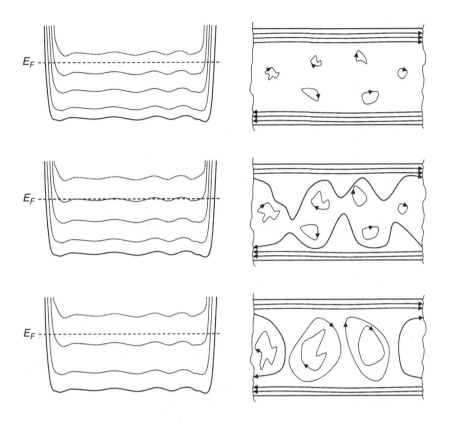

Fig. 4.6 Schematic illustration of the evolution of the Landau levels, and their associated edge states, in a magnetic field. The magnetic field is increasing from top to bottom in the figure, respectively. On the left we shown the variation of the Landau level energies (thin solid line) across the width of the sample, whose confining potential is indicated by the thick solid line. The dotted line corresponds to the lowermost empty Landau level.

example, they arise predominantly from the random distribution of ionized impurities in the donor layer of the heterostructure (as discussed already in Section 2.7). Regardless of their source, the fluctuations result in an associated broadening of the Landau levels, as illustrated in Fig. 4.7.

In Fig. 4.6, we show the evolution as the magnetic field is increased from a situation where three Landau levels are initially occupied and their associated edge states propagate without any backscattering (top). In this situation, the Fermi level lies well between the third and fourth Landau levels and electrons within the bulk of the sample are essentially pinned to local disorder by means of their cyclotron motion. In the center panels, the magnetic field has now been increased such that the third Landau level lies very close to the Fermi level. Consequently, its two counter-propagating edge states may penetrate into the interior of the sample and electrons may be backscattered between them. Since electrons injected into these edge states from their respective reservoirs now no longer maintain their initial electrochemical potentials (dissipation is occurring), the considerations leading to Eqs. (4.16) and (4.17) no longer hold. A voltage drop develops between the longitudinal probes of the device, causing ρ_{xx} to increase from zero. A the same time, ρ_{xy} increases to some non-quantized value, intermediate between $h/3e^2$ and $h/2e^2$. The exact value of the Hall resistance in

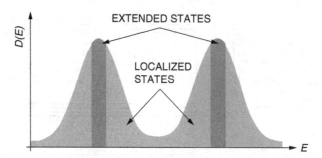

Fig. 4.7 Schematic illustration of the density of states in an infinite 2DEG. Due to the presence of disorder, which causes scattering, the Landau levels no longer correspond to delta functions in the density of states but are instead broadened as indicated. States close to the original Landau-level energies are extended while the others are localized.

this situation will be determined by the transmission coefficient of the partly transmitted edge state (T). As the magnetic field is increased in this region, T decreases smoothly from 1 to 0 and this evolution will give rise to a smooth transition from $h/3e^2$ to $h/2e^2$ in ρ_{xy}, over a narrow range of magnetic field (Fig. 4.2). The situation for even stronger magnetic field is illustrated in the lower panels of Fig. 4.6. Now, the magnetic field has increased such that the third Landau level is completely depopulated and quantized Hall transport has once again been recovered. In this situation, the longitudinal voltage drop, and so the resistivity ρ_{xx}, has returned to zero, and the Hall resistance will once again be quantized. In this situation, however, the plateau in ρ_{xy} is now at $h/2e^2$, since there are now just two edge states propagating without dissipation.

The process that we have described above is repeated cyclically with either increasing or decreasing magnetic field, which changes the number of Landau levels that contribute to transport. Finally, we note that, while for a discussion of transport in the quantized-Hall regime it is sufficient to think in terms of the edge states alone, a proper understanding of the plateau transitions, and the associated Shubnikov–de Haas oscillations, clearly requires consideration of the electronic states within the bulk of the sample. In order to explain the finite plateau width, one must invoke a picture in which *localized* states exist in the tail regions of the broadened Landau levels. In the transition from one Landau level to the next, the Fermi energy is pinned by these localized states such that for a non-zero range of magnetic fields, the transverse resistance is only determined by the occupied *extended* states below. Localized states occur naturally due to disorder in the system, and are characterized by an activated conductivity which vanishes at low temperature [5]. Models which invoke disorder-induced localization to explain the normal quantum Hall effect are sometimes referred to as "bulk" theories as the effect is a property of the 2DEG itself, and not the geometry.

4.1.5 Implications of self-consistency for the structure of edge states

In the discussion of edge states that we have given thus far, these states are associated with the intersection of different Landau levels with the Fermi energy, as the walls of the sample are approached from its interior (see Fig. 4.6, for example). According to this picture, the edge states have infinitesimally narrow width since the intersection of each Landau level with the Fermi energy occurs at a specific point in space. Actually, however, such a picture violates important principles of charge self-consistency and must therefore be modified, with dramatic implications for the physical structure of the edge states [13,14]. We have seen already (Section 4.1.1) that the number of electrons that can be accommodated in each Landau level is the *same* for each level ($n_s n_v eB/h$), regardless of its specific index. Consequently, in a model where the different Landau-level intersections with the Fermi level occur at single points in space, the depopulation of each level must be accompanied by an *instantaneous* change in the electron density, as indicated in Fig. 4.8. Such a situation is clearly unphysical, however; according to the Poisson equation, an infinite change in the electrostatic potential energy would be needed to support a discontinuity in the electron density. The solution to this problem is provided by the unique screening properties of the 2DEG in a magnetic field [15,16]. A detailed discussion may be found in Refs. [13,14], but the key point is that the self-consistent potential evolves as the magnetic field is varied, to ensure a much smoother variation of the electron density as one moves away from the sample edge and into its bulk. The smoothing of the density is achieved through a redistribution of charge near the edges of the sample at high magnetic fields. This causes the self-consistent potential to evolve from its monotonically varying form at zero field, and to develop a series of broad terraces that are separated in energy from each other by $\hbar\omega_c$ (Fig. 4.9).

An important difference with the single-particle picture (Fig. 4.8) that can be seen in Fig. 4.9 is that the occupied Landau levels are now pinned to the Fermi level over a wide range of position, instead of intersecting it at specific points. In other words, the introduction of self-consistency results in an increased edge state width (see upper part of Fig. 4.9) and an associated decrease in the inter-edge state separation. The filling of the electron states is also indicated in Fig. 4.9, from which some further comments about the nature of the edge states can be made. The edge states correspond to *compressible* regions of electron gas, in the sense that they correspond to regions where a large number of electron states are available near the Fermi energy. These regions are therefore characterized by metallic conductivity and it is their excellent screening properties that cause the self-consistent potential to be constant over their spatial range. The regions separating the edge states, on the other hand, are *incompressible*, in that they are characterized by a lack of available states close

Fig. 4.8 The formation of edge states as viewed in a single-electron description. The center shows the variation of the Landau-level energies (thin solid lines) near the boundary of a 2DEG (thick solid line). The bottom plot shows the corresponding electron-density profile expected from this figure while the upper part shows the position of the edge states relative to the 2DEG edge.

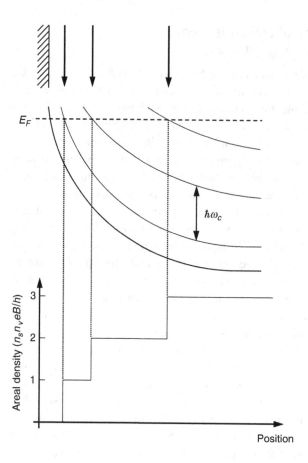

to the Fermi level and so exhibit poor screening. Consequently, these regions correspond to those points in space where the self-consistent potential transitions between different compressible plateaus. The incompressible regions are, in fact, regions of *integer* filling factor, as can be seen from inspection of Fig. 4.9. Generalizing this figure to one for a system with a total number of N occupied Landau levels ($N = 3$ in the case of Fig. 4.9), the incompressible regions are defined by the lowermost $N - 1$ levels, which are completely occupied.

A proper understanding of the microscopic structure of the edge states is clearly vital to the quantitative analysis of many experiments, for example those on inter-edge state equilibration [10,11,12]. A quantitative description of this problem has been provided by Chklovskii *et al.*, who solved the electrostatics of edge states that move along a potential boundary formed by an electrostatic gate [13,14]. They showed that the position of the kth spin-resolved edge state from the gate edge is approximately given by

4.1 The integer quantum Hall effect

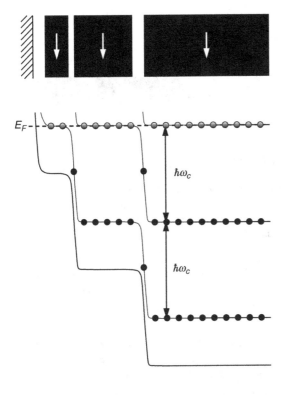

Fig. 4.9 The introduction of self-consistency to the problem of edge-state formation requires the 2DEG potential (thick solid line) to develop a step-like series of plateaus near its boundaries. The potential variation produces a similar behavior in the Landau levels (thin solid lines), which now intersect the Fermi level over a wider range to produce broad compressible edge states (upper part).

$$d + x_k = d + d\frac{\nu^2 + k^2}{\nu^2 - k^2}, \qquad (4.21)$$

where ν is the bulk filling factor, $2d$ is the depletion length around each gate at zero magnetic field ($d = V_g \varepsilon / 4\pi^2 n_{2D} e$), V_g is the gate voltage, and ε is the effective dielectric constant. Using their model, these authors were able to account for many of the characteristics of edge-state equilibration in 2DEG systems at high magnetic fields.

Over the years, numerous experimental approaches have been applied to image the edge states in 2DEG systems. Early work was based on the use of optical techniques, in which the spatial dependence of the photoconductivity [17] or the photovoltage [18] was used to map out the regions of current flow in 2DEG Hall bars. In later work, the motion of the edge states in a magnetic field was detected indirectly using the approaches of inductive imaging [19], in which a microscale pickup coil is used to detect the magnetic fields generated by the edge states, and single-electron sensing [20], in which a single-electron transistor formed on top of a Hall bar is used to monitor the evolution of the self-consistent potential. Possibly the most dramatic studies, however, are those that use a variety of different scanning probes [21,22,23,24,25,26]. By fabricating a single-electron transistor on the tip of a scanning-tunneling microscope,

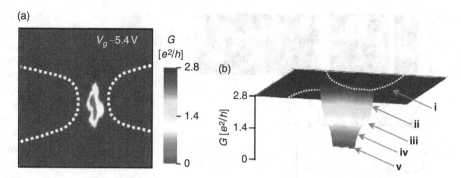

Fig. 4.10 Scanning-gate microscopy of the edge states in a quantum point contact. Panel (a) and Panel (b) are two- and three-dimensional views, respectively, showing the variation of the QPC conductance as the tip is scanned in the vicinity of the point contact ([26], by permission).

Yacoby *et al.* were able to detect the spatial distribution of the incompressible regions in a GaAs 2DEG [23]. In these experiments, the single-electron transistor current provides a sensitive probe of local charge density and, since the incompressible strips correspond to regions of constant density, they are easily identified in this approach. Most recently, Aoki *et al.* have utilized the approach developed in the Westervelt group at Harvard [27] to image the edge states in the vicinity of a quantum point contact [26]. In these experiments, changes of the quantum point contact conductance are monitored as an atomic-force-microscope tip is used to induce a local potential perturbation that is scanned over the active area of the device. For experiments in the quantum Hall regime, the presence of the point contact is critical since it brings the counter-propagating edge states in close proximity to each other and so allows the tip to induce backscattering, and thus a detectable change in conductance. The results of their study are shown in Fig. 4.10, which plots the evolution of the conductance as the microscope tip is scanned over the region of the point contact (whose geometry is indicated by the white dotted lines). The strongest edge-state scattering is obtained with the tip in the center of the point contact and a three-dimensional view of this feature indicates clear plateaus suggestive of those shown in Fig. 4.9.

4.2 Edge-state propagation in nanostructures

In the previous section, we discussed how the ballistic transport of edge states in a macroscopic 2DEG at high magnetic fields leads to the observation of the integer quantum Hall effect. We also discussed previously in Section 2.5 how, in a one-dimensional quantum wire, the presence of a static magnetic field perpendicular to the wire axis can result in the formation of hybrid magneto-electric subbands. Such subbands are a mixture of the quantization due to the lateral confining potential of the wire and the quantizing effect of the magnetic flux in

4.2 Edge-state propagation in nanostructures

terms of Landau-level formation. The symmetric dispersion relations of these magneto-electric subbands correspond to one-dimensional subbands or modes with positive and negative momentum along the wire axis.

As shown in Fig. 2.21, the dispersion relationship is distinctly nonparabolic, except in the particularly simple case of a parabolic confining potential. Although in general the dispersion relation is complicated, the current injected from a contact into one of the magneto-electric modes is independent of the magneto-electric dispersion relation, as was seen in the various formulas derived in Sections 3.4–3.9, due to cancellation of the group velocity with the density of states. Thus we may treat the magneto-electric modes in a completely analogous way to the pure magnetic edge states that were considered in our analysis of the quantum Hall effect. This important property allows us to straightforwardly extend our analysis of edge-state propagation to nanostructures with complicated geometries, as we now discuss.

4.2.1 Selective population of edge states

The edge-state picture of the quantized resistances in the quantum Hall effect (and, more generally, in the transmission picture of ballistic quantum transport developed by Landauer and Büttiker) shows that such quantization arises from the perfect transmission of carriers from one ideal contact to another due to the suppression of backscattering. The resistance is essentially given in terms of a fundamental constant times the number of transmitted edge states at the Fermi energy. A number of experiments have probed this picture of high-field magneto-transport through independent control of the number of edge states that are transmitted through a potential barrier [11,28,29,30,31,32]. Such experiments typically feature a geometry similar to that shown in Fig. 4.11, in which the key innovation is that a gate is now introduced to generate a potential barrier across the primary current path. In practice, the barrier may be realized by using a split-gate quantum point contact, as will be discussed in Chapter 5. To understand the resulting behavior in this system, it is important that we appreciate that the edge states corresponding to different Landau levels propagate along the edges of the sample while following *equipotential* paths with distinctly different guiding center energies (for nice discussions of this, see [33,34]):

$$E_G = E_F - \left(n + \frac{1}{2}\right)\hbar\omega_c, \tag{4.22}$$

where n is the Landau-level index. An important consequence of Eq. (4.22) is therefore that any tunable potential barrier may be used to selectively transmit edge states, while reflecting others, simply by adjusting the barrier height relative to the edge-state guiding-center energies. For the purpose of argument, consider the case where N spin-degenerate Landau levels are occupied in the bulk of the 2DEG, away from the barrier. The barrier itself is assumed to

Fig. 4.11 Schematic illustration of current flow via edge states in a gated 2DEG Hall bar under conditions of a high applied magnetic field. In the example shown here, two edge states are occupied in the bulk and one of these is reflected by the local potential barrier that is formed when the gates are suitably biased.

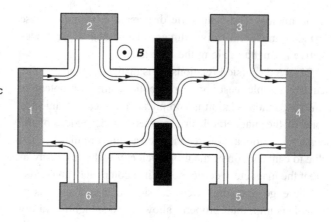

transmit only N_q of these edge states, by reflecting those $(N - N_q)$ whose guiding centers lie below the barrier. Before continuing with this analysis, a couple of further comments should be made. Firstly, this basic picture is not modified if one allows for the self-consistent structure of the compressible edge states, at least in the limit where the barrier introduces no inter-edge-state scattering [35,36]. Secondly, while it is possible in principle that the reflected edge states may tunnel through the potential barrier, this effect is insignificant for many experiments and will not be considered further here.

With the above considerations, the problem becomes one of calculating the measurable conductance of the gated device, using similar arguments to those employed in our analysis of the quantum Hall effect. To further simplify the analysis we are free to define the voltage of probe 4 to be equal to zero ($V_4 = 0$), so that the voltages of the remaining probes are then all referenced with respect to this. The most important difference with our earlier analysis is that now the edge states entering probes 3 and 6 originate from different probes, whereas, before, the absence of any barrier meant that they always came from the same probe. Taking note of this, and in analogy with Eqs. (4.9)–(4.14), we therefore write

$$I_1 = \frac{2e^2}{h}(NV_1 - (N - N_q)V_2) = I, \quad (4.23)$$

$$I_2 = \frac{2Ne^2}{h}(V_2 - V_1) = 0, \quad (4.24)$$

$$I_3 = \frac{2e^2}{h}(NV_3 - (N_q V_2 + (N - N_q)V_5)) = 0, \quad (4.25)$$

$$I_4 = -\frac{2e^2}{h}NV_3 = -I, \quad (4.26)$$

$$I_5 = \frac{2Ne^2}{h}V_5 = 0, \quad (4.27)$$

$$I_6 = \frac{2e^2}{h}(NV_6 - ((N - N_q)V_2 + N_q V_5)) = 0. \quad (4.28)$$

4.2 Edge-state propagation in nanostructures

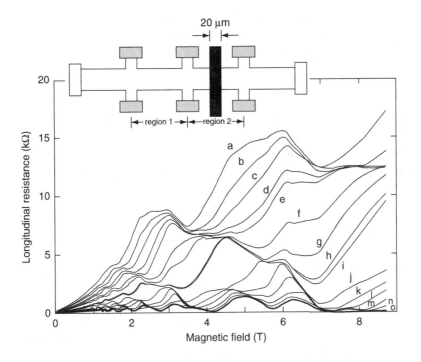

Fig. 4.12 The longitudinal magneto-resistance for a gated Hall bar at a series of different gate voltages. In this experiment, a continuous gate (inset) was used to reflect the edge states, in contrast to the split-gate considered in Fig. 4.11. Curve "o" corresponds to no applied gate voltage, while the voltage is made more negative progressing from curves "n" to "a" respectively ([37], by permission).

By simple inspection, we thus have $V_1 = V_2$, $V_3 = (N_q/N)V_2$, $V_5 = 0$, and $V_6 = (1 - (N_q/N))V_2$. With these revised expressions, the longitudinal resistance may be calculated as:

$$R_{xx} = \frac{V_2 - V_3}{I} = \frac{(N/N_q) - 1}{N} \frac{1}{2e^2/h} = \frac{h}{2e^2}\left[\frac{1}{N_q} - \frac{1}{N}\right]. \tag{4.29}$$

According to this result, the longitudinal resistance in the quantum Hall regime is zero when the barrier is such that it transmits all occupied edge states (i.e. when $N_q = N$). Whenever one or more edge states are reflected, however ($N_q < N$), the measured resistance will then be greater than zero. In fact, Eq. (4.29) predicts a new quantization of the longitudinal resistance between probes separated by the barrier. An example of this quantization is presented in Fig. 4.12 [37], which shows the measured longitudinal resistance versus magnetic field for various bias voltages applied to the gate traversing the channel. As predicted by Eq. (4.29), the zero resistances in the successive SdH minima become non-zero and approach plateau values due to the reflection of successive edge states by the potential barrier. It is easy to show, however, that the Hall resistance for this geometry

$$R_H = \frac{V_3 - V_5}{I} = \frac{h}{2Ne^2}\frac{1}{N}, \tag{4.30}$$

which is obviously unchanged from its original form (Eq. (4.15)). Another possibility that exists is to measure the "Hall resistance" using voltage probes on opposite sides of the barrier

$$\frac{V_2 - V_5}{I} = \frac{h}{2Ne^2} \frac{1}{N_q}. \tag{4.31}$$

This result is clearly different from Eq. (4.30) and can be seen instead to provide a direct determination of the number of edge states that are actually transmitted through the barrier. It is this configuration that is therefore generally used in experiments that measure the edge-state transmission through nanostructures such as quantum wires and dots.

4.2.2 Edge-state confinement in quantum dots

A natural extension of the discussion above is to consider the modifications to edge-state transport that arise when two barriers such as the one in Fig. 4.11 are arranged in series with each other. Of particular interest is the situation that arises in most lateral quantum dots, which typically consist of two quantum point contacts that are separated from each other by means of a submicron-sized cavity. As is discussed in more detail in Chapter 6, the conductance of such structures fluctuates reproducibly at low magnetic fields, reflecting the fact that transport through them is directly influenced by the bound states that form inside the cavity. At high magnetic fields, however, the nature of these bound states changes dramatically, since the possibility now exists to completely trap certain edge states within the quantum dot (Fig. 4.13). Dependent

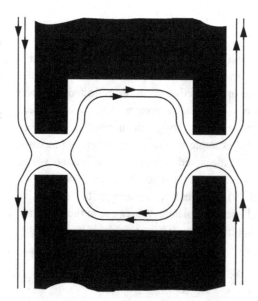

Fig. 4.13 Edge-state confinement in a quantum dot at high magnetic fields. By suitable configuration of the potential barriers that form the entrance and exit of the dot, it is possible to confine the uppermost Landau levels while allowing the others to pass freely through the dot.

upon the nature of the interaction between the propagating and trapped edge states, a rich variety of phenomena can be observed in the resulting magneto-transport.

In Fig. 4.13, we show a situation where two Landau levels are occupied in the 2DEG, and the tunable barriers of the quantum dot are configured to allow the propagation of just one of these (the lowest Landau level, with $n = 0$). The edge states associated with the uppermost Landau level are reflected at the input and exit barriers of the dot, since these have a smaller guiding center energy and so follow a lower-lying equipotential line (Eq. (4.22)). Assuming that the electron density within the dot is the same as that of the 2DEG (which should be reasonable in many situations), the filling of the upper Landau level in this dot should result in the population of a completely trapped edge state. When there is no scattering between the propagating and confined edge states of the dot, however, the latter will have no influence on the conductance, which will instead be determined by the number of transmitted edge states only, in accordance with Eq. (4.31). A number of things may happen to disrupt this situation, however, producing significant modifications in the magneto-conductance. Tunneling of electrons between the reflected edge states in the source and drain is a resonant process that is mediated by the confined edge state. The eigenspectrum of this edge state is different to that of the propagating ones, consisting of a series of discrete levels whose quantization is a consequence of the confinement of the edge state in the dot. (In fact, the spectrum of these states closely resembles the Darwin–Fock states that will be discussed in Chapter 6.) Tunneling via these discrete states is maximal whenever one of them is coincident with the Fermi energy. As we discuss in connection to the Darwin–Fock spectrum in Chapter 6, variation of the magnetic field causes different eigenstates to pass through the Fermi level, at a rate of one state for each flux quantum that is added to the cross-sectional area of the dot [38,39]. Consequently, the conductance of the dot should exhibit Aharonov–Bohm type oscillations (see Section 1.2.2) which were first reported in a beautiful experiment by van Wees *et al.* [40]. An example of these oscillations is shown in Fig. 4.14, from which their regular character can be clearly seen. In their experiment, van Wees *et al.* found that the period (ΔB) of the oscillations was roughly consistent with the field required to increase the magnetic flux threading the cross-sectional area (A_{dot}) of the dot by one quantum, i.e. $\Delta B \sim h/eA_{dot}$. For a more quantitative comparison, however, they pointed out that one should actually consider the magnetic flux enclosed by the confined edge state. Since the positions of the different edge states should evolve with magnetic field, a significant variation of the oscillation period can therefore result. Such behavior has also been reported by other authors [41,42], who have found the evolution of the oscillation period with magnetic field to be consistent with the shift in the position of the compressible edge states predicted by Refs. [13,14].

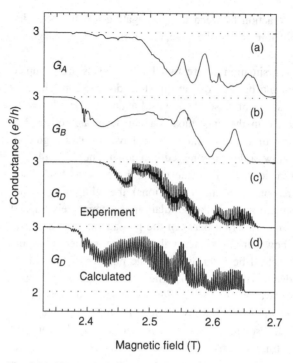

Fig. 4.14 Aharonov–Bohm type effect in the magneto-conductance of a split-gate quantum dot. The upper two curves show the variation of the conductance with magnetic field for the two barriers of the dot, which are independently tunable in this sample. The curve labeled "G_D" shows the corresponding magneto-conductance of the quantum dot, with fine, periodic, Aharonov–Bohm oscillations that are superimposed on the slower aperiodic variation due to the two barrier conductances. The bottom curve is the result of a calculation based on a model of a 1D interferometer ([40], by permission).

Subsequent to the original experiment of van Wees *et al.* [41], the properties of the Aharonov–Bohm oscillations in the edge-state regime have been investigated by numerous authors [43,44,45,46,47,48,49,50]. As mentioned already, on the basis of a noninteracting model [38,39], it is expected that the period of the oscillations should be given by $\Delta B \sim h/eA_{dot}$ (so-called *h/e periodicity*). Some experiments, however, have demonstrated the phenomenon of frequency doubling (or *h/2e periodicity*), in which the oscillations exhibit a period that is given by $h/2eA_{dot}$ over a wide range of magnetic field [46]. A similar effect has also been reported in studies of quantum *antidots* in which the antidot potential may also be used to selectively trap specific edge states [47,48,49]. An example of this behavior in the latter system is shown in Fig. 4.15, which shows results from a recent study by Kataoka *et al.* [50]. In this experiment, as indicated in the schematic, an antidot is formed at the center of a quantum point contact and may confine edge states at high fields. For the range of magnetic

4.2 Edge-state propagation in nanostructures

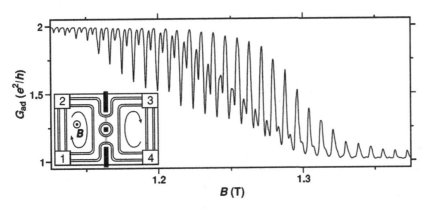

Fig. 4.15 An antidot magneto-conductance curve at 25 mK between the $v = 2$ and 1 plateaus. Bottom-left inset: schematic showing the sample geometry with four edge states (solid lines), two of each spin. Arrows indicate the direction of electron flow. Gray lines show where tunneling occurs between the extended edge states and the antidot states. The numbered rectangles on the corners represent Ohmic contacts ([50], by permission).

field considered here, the background conductance is transitioning from $2e^2h$ to e^2h, indicating that the number of spin-resolved edge states transmitted through the structure is reducing from two to one. Concomitant with this, a clear change from $h/2e$ to h/e periodicity can be observed in the data, indicating that the $h/2e$ periodicity in fact arises from the presence of two sets of oscillations with h/e periodicity that are locked in strict anti-phase with each other. In early experiments, it was suggested that the two sets of oscillations are associated with opposite spin branches of the same Landau level, and that electron–electron interactions are somehow responsible for their rigid phase locking. Recent work has suggested a different, and more complicated picture, however, in which the $h/2e$ oscillations are attributed to tunneling via a confined edge state with *specific* spin, but in which the Coulomb blockade of tunneling (Chapter 6) and Kondo physics play an important role [49,50,51].

The unique properties of confined edge states in quantum dots have also been utilized in a number of other experiments. Noteworthy among these, Komiyama *et al.* have utilized electron tunneling into confined edge states to achieve sensitive single-photon detection in the terahertz regime [52,53]. In this approach, photon absorption is utilized to induce transitions between different confined edge states and so to generate a polarization charge within the dot. This polarization charge modifies the tunneling probability through the quantum dot, under the right circumstances dramatically. In fact, in this mechanism, the absorption of just a single photon is converted into a photocurrent change of 10^6–10^{12} electrons. This should be compared with the opposite situation that arises in conventional approaches to solid-state photon detection, where each photon generates only a few electrons.

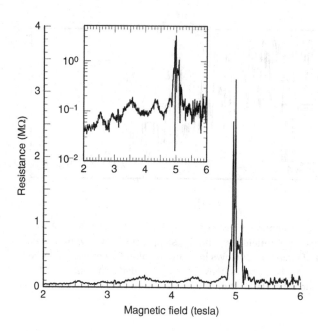

Fig. 4.16 Giant backscattering resonance in a large quantum dot. The inset shows the corresponding variation of the magneto-resistance on a logarithmic scale. While the background resistance is in the range of several tens of kilo-Ohms, it increases by almost two orders of magnitude on the resonance. For a further discussion, see Ref. [36].

In other work in this area, Sachrajda *et al.* have applied a novel approach to the study of spin injection into semiconductor nanostructures [54,55,56]. In their work, they investigated tunneling through quantum dots under conditions where the magnetic field is sufficiently strong to ensure that the spin degeneracy of the different edge states is strongly lifted. This corresponds, equivalently, to the situation where the spin branches of each Landau level are spatially separated from each other. In the tunneling regime of transport through the dot, the tunnel current will be dominated by the contribution from the edge states in the source and drain that approach the dot most closely [44,45]. When the edge states are spin polarized, this therefore allows for the injection and detection of a spin-polarized current through the quantum dot. Sachrajda *et al.* have used this technique to study spin-dependent transport through dots containing small numbers (0–45) electrons, enabling them to clarify their spin-dependent electronic structure.

There is another aspect of the magneto-transport in quantum dots that is worthy of mention, since it provides further evidence for the formation of compressible edge states through self-consistency (recall Section 4.1.4). In addition to the Aharonov–Bohm oscillations that we have been discussing, experiments on edge-state transport in quantum dots have revealed the presence of giant backscattering resonances, corresponding to a dramatic reflection of current flow in the quantum dot that is observed for a narrow range of magnetic field [35,36]. An example of such resonant behavior is shown in Fig. 4.16, which shows the magneto-resistance of a large (2-μm side) split-gate quantum dot under conditions where its quantum point contact leads have been adjusted so that they each support one mode at zero magnetic field. As the magnetic field is varied, a dramatic resonance occurs in the resistance near 5 T, where the resistance increases by nearly two orders of magnitude relative to its background value (see the inset where the same data is plotted on a log scale). The mechanism for this resonance is different from that discussed in Ref. [41] and is thought to be associated with a process in which electrons in the uppermost propagating edge state (which is closest to the trapped edge state) backscatter into their oppositely directed counterpart, in a resonant process that involves an intermediate step of scattering via the confined edge state. An important role here is played by the quantum point contact leads, in the vicinity of which the edge states rapidly change direction (see Fig. 4.13). As the voltage applied to the gates of the dot is made more negative, the point contacts narrow and scattering between the edge states becomes more effective [35,57]. It was found in experiment that the magnetic fields at which these isolated resonances occur correspond to those specific values where the uppermost Landau level depopulates completely. To properly account for this effect, it was shown [35,36] theoretically that it is necessary to formulate a model for the edge-state scattering that properly accounts for the magnetic field dependent evolution of the compressible edge states. In this way, the observed resonances were attributed to a sudden increase in backscattering at the resonant magnetic field that is mediated by edge states trapped inside the dot. More specifically, the enhanced backscattering was found to result from a swelling of the compressible strips in the dot, which occurs as a Landau level depopulates and charge is re-distributed within it. A similar treatment which ignores the evolution of the self-consistent potential profile of the quantum dot was unable to reproduce the resonances, which therefore provide an important demonstration of the role of such self-consistency for edge-state transport.

As mentioned already, the Aharonov–Bohm oscillations in quantum dots and antidots provide a dramatic demonstration of coherent electron propagation in nanostructures. Consequently, the observation of these features is typically limited to low temperatures, usually well below a degree Kelvin. An example of the temperature-dependent evolution of the quantum oscillations is shown in Fig. 4.17. It is clear that the oscillations are washed out by approximately half a degree Kelvin and there are at least two processes that can contribute to this. We have mentioned already that the oscillations may be viewed as arising from the resonant tunneling of electrons through discrete, zero-dimensional, states. These arise from the confinement of edge states within the dot and one important quantity is therefore the energy spacing of the discrete states, which is typically expected to be in the range of a few tens of micro-electron volts [38,39,40]. This corresponds to a characteristic temperature of a few tenths of a degree Kelvin, so that, when the temperature is increased beyond this, one would expect that the oscillations should no longer be observed. This is precisely the behavior that is observed in Fig. 4.17, and it was therefore suggested in Ref. [42] that thermal

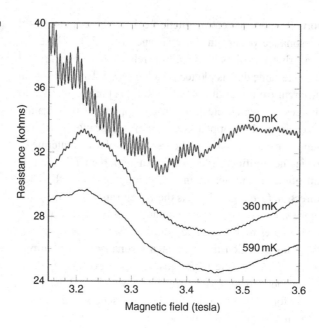

Fig. 4.17 Aharonov–Bohm oscillations in the magneto-conductance of a quantum dot at several different temperatures. For further details on the experiment, see Ref. [42].

smearing of zero-dimensional states is the mechanism responsible for the damping of the oscillations. Another mechanism that should also be considered, however, is inelastic (or, more precisely, phase-breaking) scattering of electrons (for a review of dephasing in metallic and semiconductor mesoscopic systems, see Ref. [58]).

While it is generally well understood that edge states may propagate over very long distances without undergoing equilibration of their electrochemical potentials [10,11,12], this does *not* (for a nice discussion, see [34]) necessarily mean that electrons in edge states propagate phase coherently over the same distances. While much is understood about the mechanisms for dephasing at zero magnetic field [58], the problem of phase coherence in edge states is less well understood. There are indications, however, that the phase coherence length of electrons in edge states can also be very long, and that it increases significantly with increasing magnetic field [59,60]. It therefore seems reasonable to attribute the damping of the oscillations in Fig. 4.17 to thermal smearing of the density of states.

4.3 The fractional quantum Hall effect

The remarkable results of the integer quantum Hall effect, as discussed above, are even more remarkable when it is recognized that a second quantum Hall effect also exists – the *fractional* quantum Hall effect, or FQHE. In general, one expects the Hall resistance to show the simple plateaus predicted by (4.7) and the longitudinal resistivity (or conductivity) to show a set of zeros at the plateaus. Experimentally, in high quality material such as that of a GaAs/AlGaAs heterostructure, the behavior is qualitatively different from this simple picture. The first measurements were made by Horst Störmer and Dan Tsui [61]. As can be seen from Fig. 4.18, a quite rich structure exists in the longitudinal resistance that is accompanied by the appearance of further Hall plateaus as one enters the regime of fractional filling factor. Crucially, however, these fractions are found to correspond to very specific combinations of integer numerators and denominators. As the quality of the 2DEG material has improved over time, the richness of the structure in the FQHE has also increased [62], with the appearance of stronger zeroes at more fractions.

It is clear from Fig. 4.18 that new quantized Hall states, with fractional filling factors, have appeared, and that these can lead to zeroes in the longitudinal resistance. In particular, the deep minimum at $v = 1/3$, which is much stronger than any of the other fractional states, led to the conclusion by Laughlin that the electrons had to be condensing into a new collective ground state [63,64]. That is, while the integer QHE could be understood within a single-particle (albeit single quasi-particles) picture, the fractional QHE must be a complicated many-body state. Laughlin predicted that the new collective state was a quantum fluid for which the elementary excitations, quasi-electrons and quasi-holes, were *fractionally charged* [63]. Moreover, Laughlin proposed a ground state wave function that possessed angular momentum, with the eigenvalue of $1/v$. Thus, the $v = 1/3$ state had an angular momentum of $3\hbar$. (For this discovery, Laughlin shared the 1998 Nobel prize in physics with Störmer and Tsui.)

An important consequence of the collective origin of the FQHE is the general incompressibility of the electronic state, just as in the integer QHE. While such incompressibility is a simple consequence of the Pauli principle for the integer QHE, its existence at the fractional fillings must be a result of the repulsive interactions between the electrons [65]. Haldane noted that by extending Laughlin's ideas into a spherical geometry, one could obtain a translationally invariant version and was readily extended to an entire hierarchy of fractional states [66]. An argument by Tao and Wu showed that the system remained gauge invariant, and that this led to the conclusion that the ground state had to be degenerate [67]. The entire hierarchy of fractional (and integer) states can be written as [68,69]

$$v = \frac{p}{2ps \pm 1}, \qquad (4.32)$$

although this formula is usually seen only with the positive sign. States with $s = 0$ correspond to the integer QHE, while states for $s > 0$ give rise to the set of FQHE states. It was also suggested that an excitation gap existed between the ground state of the FQHE and any excited states, and that the kinetic energy needed to bridge this gap was quenched by the high magnetic field [70].

Fig. 4.18 Plots of ρ_{xx} and ρ_{xy} on four samples at a temperature of 0.55 K. (a) and (b) are for a Si doped electron layer; (c) is for a second electron layer, while (d) is for a hole gas layer. Curve (e) is for a third electron layer. The magnetic fields have been scaled to fit a common filling factor, which is displayed across the top. (After H. L. Stormer, A. Chang, D. C. Tsui, J. C. M. Hwang, A. C. Gossard, and W. Wiegmann, *Phys. Rev. Lett.* **50**, 1953 (1983), by permission. ® American Physical Society.)

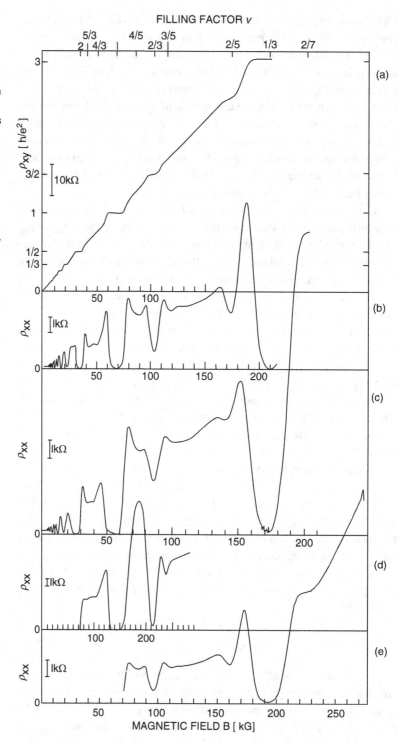

The cause of such a gap is presumably due to the many-body correlations arising from the Coulomb interaction among the electrons. An important further recognition was that the arguments above could be understood on topological grounds, in terms of the so-called first Chern class, and that the flux was exceedingly important in this context [71]. Moreover, this topological approach clearly indicated that a degenerate ground state was required to explain fractional quantization.

In deriving the expression for the integer quantized Hall resistance, we introduced the notion of the Landau filling factor in (4.16). This may be written in the form

$$\nu = \frac{N}{A}\frac{hA}{e\Phi}, \qquad (4.33)$$

where N is the total number of electrons in the sample. The important consequence of this form is that the filling factor is the number of electrons per flux quanta. Hence, $\nu = 1/3$ tells us that there are exactly three flux quanta per electron in the ground state. This leads to the conclusion that each electron is likely to be bound to three flux quanta in this state. Additionally, this filling factor is the charge-to-flux ratio. Hence, if one increases the magnetic field by a single flux quanta, while maintaining the density and area constant, then this $\nu = 1/3$ ground state is no longer full, but contains a quasi-hole. Because each particle is bonded to three flux quanta, but there is only a single extra flux quantum, the quasi-hole must have a fractional charge of $1/3$, which is the fractional charge predicted by Laughlin [64]. Similarly, reducing the magnetic field by a single flux quanta produces a quasi-electron with this same fractional charge.

At this point, ca. 1985 or so, it was thought that the key to understanding the many-body state lay in this $\nu = 1/3$ Hall plateau. But, there was more strangeness ahead, as if quasi-particles with fractional charge was not strange enough.

4.3.1 The $\nu = 1/2$ plateau

In the above discussion, it was clear that only fractional plateaus in which the filling-factor denominator was an odd integer appeared in the experimental data. However, in 1987, experimental evidence was first presented in which an even-denominator plateau was seen to begin to form [72]. In this work, in fact, evidence appeared for the $\nu = 1/2$, $3/2$, and $5/2$ plateaus (see Fig. 4.19). Observation of the latter two plateaus was particularly significant, since they correspond to the observation of fractional Hall quantization under conditions where more than one spin-resolved Landau level is occupied. Moreover, it was particularly surprising that the $\nu = 5/2$ state was the better resolved of the three, showing the clearest minima in the longitudinal resistivity. It was also found that tilting the magnetic field led to a rapid collapse of these even-denominator

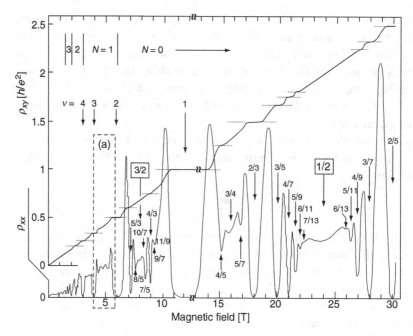

Fig. 4.19 Overview of the diagonal resistivity ρ_{xx} and Hall resistivity ρ_{xy} of a high mobility GaAs/AlGaAs heterojunction. The use of a hybrid magnet with fixed base field required composition of this figure from four different traces (breaks at 12 T). The temperature was 150 mK for the lower field traces and 85 mK for the high field trace. (After Willet et al. [72], by permission. ® American Physical Society.)

plateaus, an effect which is not seen in the odd-denominator fractional states [73]. While there is no rational reason to exclude the even-denominator fractional states, it does not sit well with the description of the fractional states introduced above. That is, the Laughlin description of the wave function fulfills the requirement for anti-symmetry (for fermions) only for odd-denominator rational filling.

Shortly after the experiments, Haldane and Rezayi [74] discovered theoretically a new incompressible quantum-liquid state of electrons which gave half-integral QHE quantization for a nonpolarized spin-singlet ground state. Moore and Read [75] then pointed out that the attachment of two quanta of a fictitious flux to each electron would lead to an acceptable order parameter for this spin-singlet state. This so-called fictitious magnetic field comes from a Chern–Simons gauge transformation, discussed more below, which introduces a gauge magnetic field [76]. The important idea here is that the electron, plus the two flux quanta, form what is known as a *composite fermion*. Moreover, the gauge field is just exactly strong enough to cancel the external field at $\nu = 1/2$. That is, when one rewrites the total Hamiltionian, with the gauge transformation, in terms of these composite particles, there is no magnetic field remaining at the value of the applied field for this filling factor. Thus, the composite fermions represent a system of spinless fermions in a (net) zero magnetic field. An additional astonishing fact is that, if the density of electrons n is held fixed, then the magnetic field corresponding to $\nu = p/(2p+1)$ ($p = 1$ for the $\nu = 1/3$ plateau) satisfies

$$\Delta B = B - B_{1/2} = \frac{h}{e}\frac{n}{p}, \quad (4.34)$$

where

$$B_{1/2} = 2\frac{h}{e}n \quad (4.35)$$

is the magnetic field corresponding to $v = 1/2$. That is, *the fractional plateaus correspond to the integer plateaus for the composite fermions* [68]! Subsequent theoretical work established that there was a band structure for the fractional states which gave a one-to-one correspondence between the strongly interacting many-body states of the fractional QHE and the weakly interacting composite fermions [77]. It was also shown that the simple trial wave functions thought of for the $v = 1/2$ plateau were, in fact, the exact ground state for a hard-core repulsive interaction model [78]. Thus, the theory was beginning to achieve closure on the composite-fermion model.

Experimental studies of the composite fermion quickly began to establish the reality of this particle by measurements of the magneto-transport, but reinterpreted in terms of the composite-fermion concept. Some of the earliest work studied the Shubnikov–de Haas oscillations for the various odd-denominator fractions near the $v = 1/2$ minima. It was found that the excitation gaps for these plateaus were equivalent to those expected for the QHE of a composite fermion in the effective magnetic field ΔB (which is measured relative to $B_{1/2}$) [79]. These studies suggested a composite-fermion mass nearly equal to the free-electron mass (as opposed to the GaAs mass of 0.067 m_0). The results are shown in Fig. 4.20. Further experiments on magnetic focusing also demonstrated the reality of the composite fermion [80]; these results are shown in Fig. 4.21. Also, studies of two composite-fermion gases separated by a tunneling barrier showed results consistent with the concept, and with the possible existence of another fractional state within the barrier [81].

Ideas about the composite-fermion mass, however, indicated a complicated behavior. That the composite fermions were *real* particles was demonstrated by geometrical resonances in an antidot superlattice [82]. Another analysis of the Shubnikov–de Haas measurements, by a British group, found somewhat different behavior than that described above. This latter group found a mass near 0.5 m_0 at zero ΔB, and this value increased only slightly with effective magnetic field, as shown in Figs. 4.21 and 4.22 [83,84]. Other measurements, however, showed a more complicated behavior, with divergence of the effective mass of the composite fermion as $\Delta B \to 0$, as shown in Fig. 4.23 [85]. Indeed, these latter results had been inferred earlier by Du *et al.* [79] and were confirmed by the Canadian group (Fig. 4.24) [86].

Measurements of the magnetic field dependence of the radiative recombination [87] in the FQHE regime suggested a composite-fermion mass perhaps four times heavier than some of the above values obtained from transport experiments.

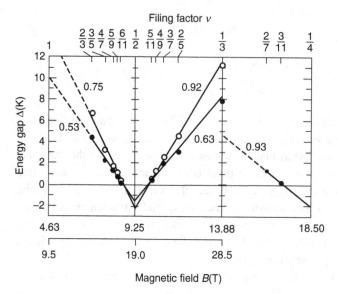

Fig. 4.20 Gap energies for various filling factors in the vicinity of $v = 1/2$ and $v = 1/4$ (right panel) in two samples differing by roughly a factor of 2 in density. The horizontal axis has been scaled so that equivalent fractions coincide (top scale). Straight lines are a guide to the eye, and the number associated with each line represents the effective mass (in units of m_0). (After Du *et al.* [79], by permission. ® American Physical Society.)

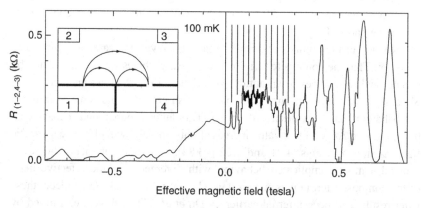

Fig. 4.21 Four-terminal focusing magneto-resistance of a sample with a constriction separation of 4.25 μm. Magnetic focusing of the composite fermions occurs for positive effective magnetic field, and the vertical lines are separated by 25 mT. The inset on the left illustrates the sample geometry. (After Goldman *et al.* [80], by permission. ® American Physical Society.)

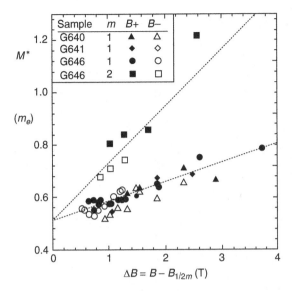

Fig. 4.22 Composite-fermion effective mass deduced by fitting maxima and minima of ρ_{xx} for several samples. (After Leadley et al. [83], by permission. ® American Physical Society.)

Studies of nuclear magnetic resonance near the half-filling plateau also confirmed the noninteracting composite fermions [88], although similar measurements, along with the Knight shift, suggested that neither a noninteracting or an interacting model of composite fermions quite fit the experiments [89]. Finally, studies of cyclotron resonance of the composite fermions indicated a composite-fermion mass more in tune with those of Fig. 4.22 [90,91]. It seems to be certain that the composite fermion is a strange beast, and there remains a question over just what the behavior of the effective mass of this composite particle really is.

Above, it was pointed out that the FQHE could be considered as the integer QHE for the composite fermion. But, this does not always follow. More recently, studies have shown the existence of fractional states at $\nu = 4/11$ and $5/13$, which do not fit this description; e.g., they are not expressible as fractional states of the composite fermion [92]. These results suggest that there are residual many-body effects between the composite fermions, and are expressed as fractional QHE states of the composite fermion [93]. Subsequent studies have suggested other fractions and the possibility of a hierarchy of composite particles (composite fermions, composite composite fermions, and so on), termed a re-entrant FQHE in the lowest Landau level (around $\nu = 1/2$) [94]. If this were the limit of the strangeness, one perhaps could be satisfied.

4.3.2 The $\nu_T = 1$ bilayer system

In early 1989, it was suggested that two layers of quasi-two-dimensional electron gas could be considered as a single quasi-two-dimensional gas with a new pseudospin index, where the up and down layers (in z) corresponds to the

Fig. 4.23 The composite-fermion mass (top) and quantum lifetime (bottom). The lines are only guides to the eye. (After Manoharan *et al.* [85], by permission. ® American Physical Society.)

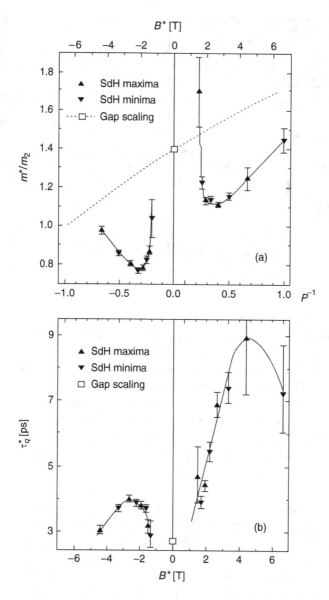

sheet index [95]. These authors considered the case where the density in each layer was the same and interaction occurred only due to the Coulomb interaction, thus neglecting tunneling between the layers. They concluded that Jastrow-like wave functions (such as those proposed by Laughlin [64]) would be appropriate but would lead to strong anomalies in the transport properties. The tunneling between the layers was considered later in fully spin-polarized layers, and it was shown that the integer QHE in each layer might be suppressed, but the fractional state at $v = 1/2$ in each layer would appear [96]. Shortly after this, experimental

4.3 The fractional quantum Hall effect

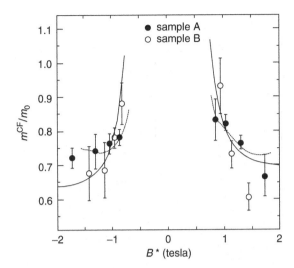

Fig. 4.24 Composite-fermion effective mass measured in two samples. The curves are estimates from two different theories. (After Coleridge et al. [86], by permission. ® American Physical Society.)

evidence was found for the new $\nu_T = 1/2$ (here ν_T refers to the *total* filling of both layers) FQHE in a wide quantum well (in which two subbands of the 2DEG are occupied) [97] and a proper double-quantum-well bilayer system [98]. Here, the interesting effects in the bilayer (or wide-quantum-well) system should occur when the interlayer separation was of the same order as the magnetic length

$$l_B = \sqrt{\frac{\hbar}{eB}}. \tag{4.36}$$

In these systems, it was recognized that the wave function of the overall system is much more spread out in the third (z) dimension, which softens the short-range part of the Coulomb interaction, which can lead to pairing correlations among the electrons of the two layers, such as p-wave pairing [99]. While the $\nu_T = 1/2$ ground-state wave function was predicted to have a single well-defined value, that for $\nu_T = 1$ belongs in some sense to two different universality classes: while the $\nu_T = 1$ is strongest when the well separation is small, and decreases as the well separation is increased, a new many-body $\nu_T = 1$ state is stabilized by a many-body gap at small interaction energy between the two layers [100]. In some sense, it is expected that the ground state would be a ferromagnetic phase of the system.

Experiments were soon undertaken to test this $\nu_T = 1$ bilayer system. Spielman et al. [101] found that the tunneling conductance between the two layers suggested a strong resonant enhancement as the density was reduced (in which the state occurred at lower values of the magnetic field), as shown in Fig. 4.25. These authors supported an interpretation that the electrons in each layer formed a Wigner crystal which filled half of the available sites, in a regular manner, but such that an electron in one layer sat opposite a hole in the adjacent layer (forming a type of exciton between the layers). However, they point out this view does not explain the resonant nature of the tunneling peak at zero interlayer

Fig. 4.25 Tunneling conductance as a function of interlayer voltage in a balanced bilayer system. N_T is the total electron concentration (in units of 10_{10} cm$_{-2}$). While the tunneling is suppressed at high density, it is greatly enhanced at low density. (After Spielman *et al.* [101], by permission. ® American Physical Society.)

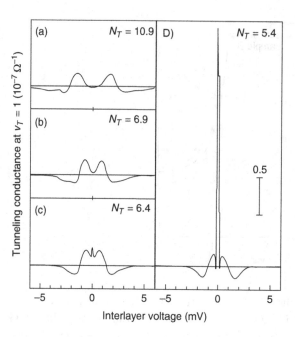

voltage. Rather, they suggest that a collective mode near zero energy which transfers charge between the two layers is required, and that such a mode may be the pseudospin Goldstone mode (a low-energy mode going to zero linear in wave number) of the structure.

It was later suggested that there should be a strong enhancement of the electron drag between the two layers at the crossover between weak and strong coupling of the two layers, due to the Coulomb interaction between the carriers [102]. Subsequent measurements indicated some novel behavior, in which the Hall resistance at $v_T = 1$ collapsed to that of $v_T = 2$ in the presence of the drag (Figs. 4.26 and 4.27) for a range of separations between the two layers [103,104]. Taking these concepts together (the electron-hole ordering between layers and the resonant enhancement of the tunneling at low densities, or strong coupling), it was suggested that the ground state of the bilayer system could well be a Bose–Einstein condensation of the electron–hole excitons [105], which is roughly equivalent to the pseudospin ferromagnet discussed above. In both cases, it is expected that a coherent phase exists between the two bilayers. Contrary to the normal QHE, in which the dissipationless transport leads to vanishing longitudinal conductivity and resistivity, the coherent interlayer phase at $v_T = 1$, one expects an infinite longitudinal conductivity in the counter-flowing current situation. This would lead to both the longitudinal and Hall resistances vanishing, an effect that is, in fact, observed in the bilayer system, as shown in Fig. 4.28 [106]. These results suggest that the bilayer system at $v_T = 1$ is a new kind of superfluid, based upon the excitonic ground state. In subsequent work, in which the densities of the two layers were varied with respect to each (thus unbalancing the system), it was shown that the

4.3 The fractional quantum Hall effect

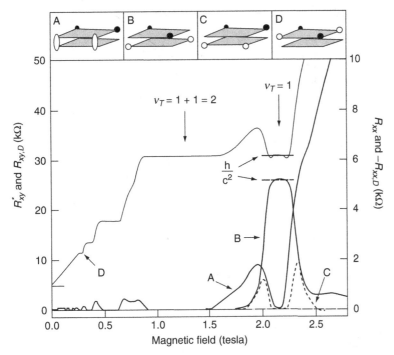

Fig. 4.26 Conventional and Coulomb drag resistances of a low-density bilayer. Panel A shows the normal case with current measured in both layers, B is the Hall drag resistance $R_{xy,D}$, C is the longitudinal drag resistance $R_{xx,D}$ (with the sign reversed for clarity), and D is the Hall resistance of the single current carrying layer. Trace B shows the quantization of the Hall drag near the $\nu_T = 1$ excitonic QHE. In the top panels, current enters and exits at the open circles and voltages are measured between the closed dots. The density of each layer is 2.6×10^{10} cm^{-2}, and $d/l = 1.6$ at the resonance (d is the interlayer separation and l is the magnetic length (4.36)). (After M. Kellogg et al. [103], by permission. ® American Physical Society.)

transition between the interlayer phase coherent state and the normal weakly coupled incompressible QHE state moved to larger interlayer separations as the electron density was unbalanced. That is, small imbalances stabilized the coherent excitonic state [107].

In later work, experiments in widely spaced quantum wells studied the so-called superfluid state of the excitonic bilayer. These studies found a thermally activated behavior in the interaction, which does not reflect the condensation energy of the superfluid $\nu_T = 1$ state [108]. With the wide separation of the two wells, no significant tunneling between the wells was observed, and it was felt that the observed activation energy was that for charge excitations in the separate layers themselves. Hence, it would seem that the excitation spectrum would be different if the well density were above or below $\nu_T = 1/2$.

Other experiments suggested that the spin degree of freedom is variable in the transition between the phases, discussed above, and that, near the critical point,

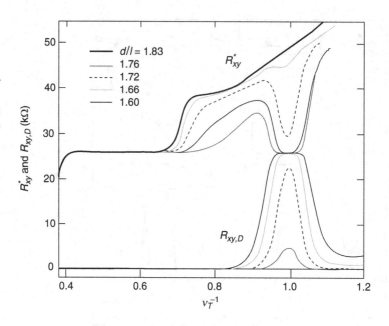

Fig. 4.27 Collapse of the $v_T = 1$ Hall drag quantization as the d/l ratio is varied. These values correspond to layer densities of 2.6, 2.8, 3.0, 3.2, and 3.4×10^{10} cm^{-2} at $v_T = 1$. The measurements were made at 30 mK. (After Kellogg *et al.* [103], by permission. ® American Physical Society.)

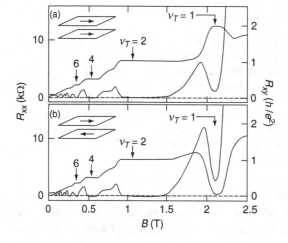

Fig. 4.28 Hall and longitudinal resistances (solid and dotted traces, respectively) in a low-density double-layer system at $T = 50$ mK. (a) Currents in parallel in the two layers give normal results, while (b) counterflow currents give a collapse of both resistances at $v_T = 1$. (After Kellogg *et al.* [106], by permission. ® American Physical Society.)

the spin polarization of the excitonic phase must exceed that of the competing non-QHE phase [109]. Here, the phase boundary is located by the appearance of the so-called Goldstone collective mode, which is characteristic of the excitonic phase. Thus, competition between the two phases also involves competition in the degree of spin polarization of the bilayer system. It was found that the nuclear spins were involved in this competition; effects which lowered the nuclear polarization, increased the Zeeman energy and favored the phase with larger electronic spin polarization – the excitonic QHE phase. Nevertheless, the experiments suggested that the spin polarization of the weakly coupled bilayer system at $v_T = 1$ cannot be complete.

This $v_T = 1$ bilayer system remains a challenge to full understanding, as its physics is quite interesting. It appears that there is a broken symmetry, in which the ground state can be either an excitonic superfluid or a pseudospin ferromagnet. Recently, a new theory of the tunneling anomaly, which highlighted the differences and similarities in these two states, was presented [110]. This theory is able to explain the exceptionally large tunnel current between the layers, which occurs at low-bias, as well as the voltage width of this transport anomaly. At this point, it seems that the data support that this plateau is a single-particle QHE state, which is stabilized by the tunneling interaction between the layers, at low density. This evolves into a many-body QHE state stabilized by strong interlayer Coulomb interactions at intermediate densities. As the density is further increased, an incompressible-to-compressible phase transition is seen with a remarkable temperature dependence [111,112], and this last phase may be related to phase transitions and superconducting-like behavior.

4.3.3 Other half-filled Landau-level states

In Fig. 4.19, the experimental studies providing evidence for the $v = 1/2$ plateau also showed the existence of $v = 3/2$ and $5/2$ states. Thus, the appearance of the even denominator plateaus is not limited to just the lowest Landau level, or even to the lowest spin-split level. Both $v = 1/2$ and $3/2$ are in the lowest Landau level, but in different Zeeman split levels (at least in the simplest theory). On the other hand, $v = 5/2$ lies in the second Landau level. Experiments around $v = 3/2$ provide a nice interpretation, from angular dependent transport studies, of a composite fermion state carrying a spin [113]. For this level, these studies give a value of the composite fermion mass of about 0.433 plus a variation with the effective magnetic field ΔB (but, with this quantity being measured from the $v = 3/2$ point). At the $v = 5/2$ plateau, however, it was found that exact quantization was achieved, with a true zero in the longitudinal resistance [114], as shown in Fig. 4.29. Strikingly different from the $v = 1/2$ state is not only the existence of the true zero at $v = 5/2$, but also the emergence of very strong maxima flanking this minima. It appears that this plateau represents a true FQHE state.

It was suggested that the $v = 5/2$ state should be spin polarized and is a true FQHE incompressible state [115]. However, theory suggested that this state should have a large overlap with states characterized by a Pfaffian wave function, and there was evidence for phase transitions to compressible states, driven by the many-body interactions [116]. The experiments, discussed above, tend to suggest that the FQHE state is the proper description. Nevertheless, further studies continued to suggest that a phase transition separating a compressible striped phase from a paired QHE state could exist [117] (we will see the stripe phase again below). However, more recent studies have shown that the Pfaffian wave

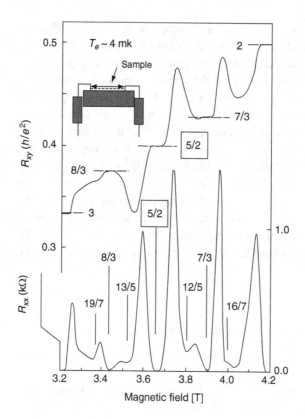

Fig. 4.29 Hall resistance and longitudinal resistance at an electron temperature of ~4 mK. Vertical lines mark the Landau-level filling factors, while the inset shows a schematic of the sample with integral heat exchangers to cool the 2DEG. (After Pan *et al.* [114], by permission. ® American Physical Society.)

function may not be correct and may not be necessary to the understanding [118,119].

In spite of the above discussion, there are indications in Fig. 4.19 that the physics of the $n = 1$ Landau level could well be somewhat different from that of the lowest Landau level. This is indicated already by the nature of ρ_{xy} in the figure. Normally, this latter quantity has a monotonic variation with magnetic field. Here, both peaks and valleys appear, which is new. Thus, it seems that the first Landau level exhibits characteristics of both the FQHE in the lowest Landau level and a charge-density wave, or liquid crystal phase, that exists in the second and higher Landau levels [120]. Indeed, studies of the $v = 9/2$ and $11/2$ states in the second Landau level showed very anisotropic transport [121] below about 150 mK, which suggested the spontaneous development of an anisotropic many-electron state. As mentioned above, it was suggested that a stripe phase could occur [115]. Studies of the stability of condensed stripe phases and bubble phases suggested that the former was more likely [122]. Subsequent experimental studies suggested that the presence of significant electron correlation, and small Zeeman splitting, in these states led to a competition between these various phases, which were thought to be nearly degenerate so that no single one could be identified as the ground state [123,124]. Subsequent calculations again showed that the anisotropic state was favorable, but that conditions between the stripe phase and the bubble phase were such that the former would dominate only on small length scales [125]. Nevertheless, current wisdom holds that the stripe phase dominates in these high Landau levels.

While we started with a general understanding of the IQHE which was the same for all Landau levels, the FQHE opened a number of doors to questions, which were exacerbated by the appearance of the even denominator states. Careful studies of the latter have given rise to an understanding that there is not a single phase throughout the range of Landau levels, and that the variations in density and magnetic field lead to a competition of many phases. Which is the proper ground state at any given density and magnetic field (any particular filling factor) seems to be highly variable with a strong competition between various possible phases.

4.4 The many-body picture

The theoretical basis for the FQHE is as old and as varied as the various experimental pictures of this effect. Thouless *et al.* were apparently the first to point out that the topology of the quasi-two-dimensional system in a magnetic field was particularly important to the understanding of its transport [126]. At the time, however, their thinking was directed toward the IQHE rather than the FQHE, as their work was contemporary with experimental efforts first reporting the latter. In this regard, their work was consistent with that of Streda [127], who pointed out that the integer QHE arose from a non-classical term in the Hall conductivity. This idea of the importance of topology was continued by Avron *et al.* [128], who showed that the integers found by Thouless *et al.* were the only possible ones associated with the energy bands as they are the only topological invariants. It was then shown that there was a connection between Thouless *et al.*'s results and the concepts of geometric phase introduced by Berry [129,130]. While the details of the importance of the topology were not crucial to the early understanding of the integer QHE, this was certainly not the case with the FQHE.

It has been known for quite some time that, in three spatial dimensions, the spin of a particle must be either integer (bosons) or half-integer (fermions). However, this holds only for systems equal to, or larger than, 3+1 dimensions (three spatial dimensions plus time). In the effectively two-dimensional situation, more possibilities exist [131], both for the spin and for the statistics. An example from Wilczek [131] considers an electron orbiting a solenoid running along the z-axis. If there is no current in the solenoid, then the total angular momentum is naturally quantized into units $l_z = $ integer. But, when a current is turned on, the charged particle will feel an electric field

$$\mathbf{E} = -\frac{\mathbf{a}_z \times \mathbf{r}}{2\pi r^2}\frac{\partial \Phi}{\partial t}, \qquad (4.37)$$

where Φ is the flux through the solenoid and $\mathbf{a}_z$ is a unit vector in the z-direction. This leads to a change in the angular momentum

$$\frac{\partial l_z}{\partial t} = [\mathbf{r} \times (e\mathbf{E})]_z = -\frac{e}{2\pi}\frac{\partial \Phi}{\partial t}. \tag{4.38}$$

The total change in angular momentum depends only upon the final flux, so that one is led to the conclusion that this angular momentum is quantized in units of

$$l_z = \text{integer} - \frac{e\Phi}{2\pi}. \tag{4.39}$$

Now, although the magnetic field is considered to vanish outside the solenoid, the vector potential does not, and in a nonsingular gauge, $A_\Phi = \Phi/2\pi$, and the azimuthal variation of the quantum wave function is just $\psi_n \sim e^{in\Phi}$, with n an integer. This leads to $l_z\psi_n = (n-e\Phi/2\pi)\psi_n$, as above. The potential outside the solenoid can be eliminated by a singular gauge transformation

$$\mathbf{A}' = \mathbf{A} - \nabla\Lambda, \tag{4.40}$$

with

$$\Lambda = \frac{\Phi}{2\pi}\varphi. \tag{4.41}$$

This gauge transformation is singular because φ is a multi-valued function. More importantly, this gauge transformation is a simple case of the more general Chern–Simons gauge transformation. The charged particle wave function now satisfies the normal Schrödinger equation, but with the unusual boundary condition

$$\psi'(\phi + 2\pi) = e^{-ie\Phi}\psi'(\varphi). \tag{4.42}$$

This boundary condition arises from the gauge transformation applied to the wave function, and requires that the wave function have the variation

$$\psi' \propto \exp\left[i\left(\text{integer} - \frac{e\Phi}{2\pi}\right)\varphi\right]. \tag{4.43}$$

Since there is no vector potential, the angular momentum is identified as usual, which just once more leads to (4.39). If one interchanges flux-tube and charge-particle composites, there will be a phase factor appearing in all gauge-invariant observables, since each of the composites must be transported covariantly in the gauge potential of the other. This resulting phase is $e^{ie\Phi}$ because of our definition of the vector potential. Now, if l_z is an integer, the phase factor is unity and the statistics are normal. That is, the composite of one flux tube and one electron obeys Fermi statistics, but if l_z is half an odd integer the normal statistics are reversed. But, in general, the angular momentum can have various values, which lead to *anyons* – particles with fractional spin and fractional statistics. It is to be expected that these composites of flux tubes and particles will have unusual statistics, which may interpolate between fermion and boson [132].

It seems to have been Halperin [133] who then recognized that the quasi-particles in the FQHE obeyed quantization rules appropriate to particles of fractional statistics, in a manner consistent with the above arguments of Wilczek, particularly in recognition of the wave functions introduced by Laughlin [64]. In particular, he pointed out that the quantization rules which determine the allowed quasi-particle spacings are just those expected for a set of identical particles having fractional statistics. From this, and the adoption of the Coulomb interaction between the particles, he found a natural set of approximations for the ground-state energies and gaps. Now, as we remarked above, the ideas of the FQHE, particularly the $v = 1/2$ state arise from coupling the particles to fluxes, which is made physical in the Hamiltonian by a Chern–Simons gauge transformation. A simple version was given above in (4.41). The more general case, for arbitrary charge density $\rho(\mathbf{r})$ is

$$\Lambda = \Phi \int d^2 r' \frac{\mathbf{a}_z \times (\mathbf{r} - \mathbf{r}')}{|\mathbf{r} - \mathbf{r}'|^2} \rho(\mathbf{r}'). \tag{4.44}$$

The simple version arises by taking a simple delta-function localized charge.

4.4.1 Fractional states

Phrases such as superfluid and incompressible have been used in the above discussions rather extensively. Basically, these are the two quantum-fluid states which are known to exist at zero temperature. The former was originally related to liquid He below the critical point, while the latter is associated with superconductivity itself. The superconductive state is incompressible if the background charge is fixed by assuming it to come from the rigid ions, and all excitations of this state have non-zero energy gaps. This incompressibility comes from the long-range Coulomb interaction. When the integer QHE was discovered [1], this led to a new class of incompressible quantum fluid, and the FQHE continued the investigation of novel properties of this fluid. It is interesting that in both high-T_c superconductivity and in the FQHE a new class of incompressible quantum spin-liquid states has been proposed – chiral spin states [134,135]. In these states, both the time-reversal symmetry and the parity symmetry are broken, and it has been shown that these states contain non-trivial topological structures [136]. In fact, the simplest FQHE states given by Laughlin [64], which are of the form

$$\psi(z_i) = \left[\prod_{j>i}(z_i - z_j)^q\right] \exp\left(-\frac{1}{4}\sum_j |z_j|^2\right) \tag{4.45}$$

where $z = x+iy$ is the in-plane coordinate, are found to be nondegenerate on a sphere [66] but q-fold degenerate on a torus (that is, in two spatial dimensions) [137]. The basic ingredients of the symmetry breaking, inherent in the chiral state, are already present in this wave function [138]. Adding one quasi-hole (increasing the magnetic field by one flux quanta, as discussed above) or two quasi-holes produces a macroscopically distinct state, but adding three quasi-holes differs from the no-quasi-hole state by only one single-particle state and these latter two are indistinct. It has been argued therefore that the ground-state degeneracy of the FQHE states is really a reflection of the topological order of the system [134].

It was quickly pointed out that Laughlin's wave function could be generalized [139]. Indeed, in this case, these more general forms became of considerable interest for the bilayer systems discussed above. If we adopt $s = 1, 2$ as the pseudospin index referring to the two layers, then the general Jastrow-type wave function could be written as

$$\Psi_{m,m,n} = \prod_{\substack{i<j \\ s=1}} (z_{i,s} - z_{j,s})^m \prod_{\substack{i<j \\ s=2}} (z_{i,s} - z_{j,s})^m \prod_{i,j} (z_{i,1} - z_{j,2})^n \exp\left[-\sum_{i,s} |z_{i,s}|^2/4l_B^2\right], \quad (4.46)$$

where l_B is the magnetic length (4.36). In particular, the so-called (3,3,1) state has been suggested as the appropriate wave function for the bilayer system at $\nu_T = 1$.

Another advance pointed out that the pairing correlations in the Hall effect suggested the use of Pfaffian wave functions [75], in their argument for the attachment of flux quanta to each particle. The term in square brackets in the equation above is closely related to the mathematical object known as a Pfaffian, which is actually expressed as

$$Pf\left(\frac{1}{z_i - z_j}\right) = A \prod_{i \text{ even}}^{N} \frac{1}{z_{i-1} - z_i}, \quad (4.47)$$

which, by example for four particles, becomes

$$Pf\left(\frac{1}{z_i - z_j}\right) = \frac{1}{z_1 - z_2}\frac{1}{z_3 - z_4} + \frac{1}{z_1 - z_3}\frac{1}{z_4 - z_2} + \frac{1}{z_1 - z_4}\frac{1}{z_2 - z_3}. \quad (4.48)$$

It was pointed out by Greiter et al. [140] that the BCS theory of paired electrons also utilized Pfaffian wave functions, from which they were drawn to the conclusion that the FQHE states might well involve paired electrons. These latter authors argue that such states are of the same universality class as those suggested by Halperin [139] for the $\nu = 1/2$ states. They suggest that this paired state might be the proper ground state in the absence of a magnetic field, and then evolve to the FQHE state, for some of the plateaus in the higher Landau levels, when the magnetic field is present. In particular, they argue that experiment results obtained for the $\nu = 5/2$ state are not inconsistent with such an interpretation.

In fact, Moore and Read [75] considered a variety of options, comparing the general Pfaffian wave functions to those of Haldane–Rezayi [74] and Halperin [139]. A key point of the argument is that the gauge transformation discussed below (4.39) is generally of the type of a Chern–Simons transformation found in quantum field theory, or more properly a conformal field theory, and these authors pointed out that, in a suitable gauge, the wave functions for the FQHE transform under interchange of the quasi-particles as a one-dimensional Abelian representation of the braid group, which becomes important later for possible applications in quantum computing. It turns out that the wave function suggested by the latter authors is the simplest paired state for spinless, or spin-polarized, particles, and they suggested further that paired states of composite particles could exist. A key point is the attachment of a pair of flux tubes to the electron to create the $1/2$ state, for which the wave function can be interpreted as a BCS paired wave function. It was subsequently shown that the quasi-holes would have properties beyond that expected from the Laughlin theory, and could thus lay the groundwork for the existence of non-Abelian statistics [141]. These ideas were further developed by Fradkin *et al.* [142], who suggested that the Pfaffian state at $1/2$ might be extended to the $5/2$ state in a single-layer system and to the new $1/2$ state in a double-layer system. Moreover, these latter authors suggested that the zero-modes, which could appear as the edge states in a finite-sized system, would be Majorana fermions – fermionic excitations whose particles are their own anti-particles. Nevertheless, the system, complicated as it appears, was not so clear, as later work showed that the non-trivial paired FQHE states are related to a weak-pairing phase with wave functions that contain the generic long-distance behavior in spinless and spin-triplet p-wave weak-pairing, but that a strong-coupling regime existed which resulted in the Halperin-like behaviour [143].

Studies of the FQHE in a harmonic confinement potential, such as that of a quantum dot, shed new light on the problem. There it was shown that the wave functions of the $1/2$ state had a relatively high overlap with the composite fermion states, but a lower overlap with the Pfaffian wave function, suggesting that the electrons might not be paired in the quantum dot [144]. It was also suggested that the $v = 5/2$ state would have a high spin polarization. Ivanov then showed that the consideration of a p-wave superconductor wave function was equivalent, in terms of statistics and vortices, to the Pfaffian Moore–Read wave function [145]. Finally, as mentioned above, it was suggested that one could understand the $5/2$ state without the Pfaffian wave function [116]. Whether this is significant or not, we would simply remark that the observation of an eigenvalue result cannot depend upon the nature of the basis set – equivalent wave functions can be obtained in many basis sets. We remark that our purpose here is not to treat the various theoretical approaches in any great detail; that would require far more space than this book is allotted. Rather, we want only to provide pointers to the wealth of modern approaches that must be considered in this regard.

4.4.2 Fractional statistics

The idea of quasi-particles with fractional statistics was intrinsic in the arguments of Wilczek [131,132], presented at the start of this section, as well as being intrinsic to the Laughlin argument [63,64]. Crucial to the subsequent discussion of quantum computing, it is important to note that when a quasi-particle adiabatically encircles another quasi-particle, an extra "statistical phase"

$$\Delta\varphi = 2\pi\nu \qquad (4.49)$$

is accumulated, where $\nu = 1/m$ is the filling factor and m is an integer [146]. For the case $\nu = 1$, $\Delta\varphi = 2\pi$, and the phase for interchanging particles is $\Delta\varphi/2 = \pi$ corresponding to Fermi statistics. However, for non-integer ν, $\Delta\varphi$ corresponds to fractional statistics, in agreement with Halperin [139]. Thus when ν is a non-integer, the change of phase when a third quasi-particle is in the vicinity will depend upon the adiabatic path taken by the quasi-particles as they are interchanged. In this situation, the pair permutation definition used for bosons or fermions no longer suffices.

Earlier in this chapter, we discussed the results of experiments in which tunneling between IQHE edge states could be induced by confining them in nanostructures such as quantum dots and antidots. In the FQHE, Goldman and Su performed measurements of resonant tunneling between edge states – they confined the edge states via a quantum point contact which was sufficiently open to allow an antidot potential in the saddle potential of the QPC – which allowed them to directly measure the charge of the quasi-particles [147]. With this technique, they measured q_{eff} to be nearly the free electron charge at $\nu = 1$ and $\nu = 2$, but $q_{\text{eff}} = 0.323e$ at $\nu = 1/3$. Subsequently, they also showed that the Aharonov–Bohm phase around the antidot is h/e even for fractionally charged quasi-particles [148], as shown in Figs. 4.30 and 4.31. This last finding can be understood by the fact that single-valuedness of the wave function requires the Berry phase difference between successive states to be an integer multiple of 2π. This condition is satisfied by fermions, but requires a fractional statistical phase contribution by the fractionally charged quasi-particles in the FQHE state.

Camino *et al.* then considered a unique type of quantum dot in which the density was non-uniform [149]. Here, edge states could enter the quantum dot and circulate around the edges, but the center had a lower density with a different set of edge states circulating this antidot. In the experiment, the outer edge states were in the $\nu = 1/3$ state, while the central antidot edge states were in the $\nu = 2/5$ state. In this experiment, quasi-particles of the 1/3 state circulated around those of the 2/5 state, and thus acquire statistical phase. Interference fringes, similar to the Aharonov–Bohm effect, were seen in which the fringes shifted by one fringe upon introduction of five magnetic flux quanta into the island, corresponding to a $2e$ charge period. These quantization periods of $\Delta\Phi = 5h/e$ and $\Delta q = 2e$ must be imposed by the symmetry properties of the two FQHE

4.4 The many-body picture

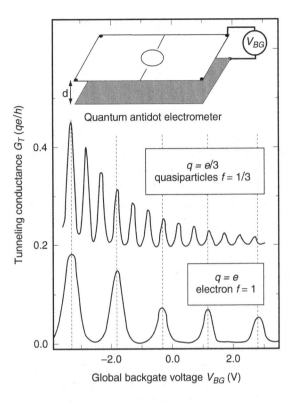

Fig. 4.30 Quantum antidot electrometer. In the experiment, a conductance peak occurs when the occupation of the antidot increments by one particle: an electron on the $v = 1$ QH plateau (the authors use the symbol f rather than v), and a quasi-particle on the fractional $v = 1/3$ plateau. (After Goldman *et al.* [148], by permission. ® American Physical Society.)

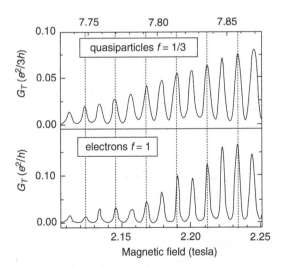

Fig. 4.31 Aharonov–Bohm periodicity. As B is varied, so is the flux Φ through the antidot area (the area encircled by the tunneling particle). Thus the period is a direct measure of the flux period for both plateaus of Fig. 4.30. Single-valuedness of the wave function requires the Berry phase difference between successive states to be an integer multiple of 2π. (After Goldman *et al.* [148], by permission. ® American Physical Society.)

fluids. If we define the relative phase shift, or relative statistics, for moving one quasi-particle around another in this experiment as $\Theta_{2/5}^{1/3}$, then we require that

$$\begin{aligned}\Delta\varphi &= \frac{q}{h}\Delta\Phi + 2\pi(\Delta N)\Theta_{2/5}^{1/3} = 2\pi \\ &= \frac{e}{3h}\left(\frac{5h}{e}\right) + 2\pi(10)\Theta_{2/5}^{1/3}\end{aligned} \quad (4.50)$$

or $\Theta_{2/5}^{1/3} = -1/15$. This idea of relative statistics should not be surprising, since all elementary charged excitations of FQHE fluids are fundamentally collective excitations of strongly interacting quasi-two-dimensional electrons. It is this property that suggests applications in quantum computing, discussed below. Subsequent calculations for a model interferometer were performed [150] and these gave good agreement with the experiments of Camion *et al.*, while a calculation of these oscillations based upon a model taken from the Haldane–Halperin fractional statistics construction also showed good agreement [151].

4.4.4 FQHE-based quantum computing

There are many reasons to believe, based upon the above discussions, that the appropriate theories to describe the incompressible QHE states are Chern–Simons gauge theories, and it is the vortex, or flux tube, excitations of the Chern–Simons theory that carry the exotic quantum statistics which characterize the long-distance interactions of the quasi-particles in the FQHE [152]. Processes which move quasi-holes around and then return them to their original positions cause the representative wave functions to be multiplied by phase factors which depend upon the linking numbers of the particle trajectories. Hence, the state furnishes a one-dimensional representation of the braid group [150]. It is this property, and the connection to the braid group for use in quantum computing, that has elicited considerable excitement.

In regard to the above situation, the movement of one set of quasi-particles around another suggests a possible implementation of a quantum bit – a qubit. One system, essentially very similar to that discussed by Camino *et al.* [149] above was suggested much earlier by Chamon *et al.* [153], in which a quantum dot, with edge states in one FQHE plateau that surround an antidot region with edge states in a different FQHE plateau. Subsequent discussion pointed out that repeated measurements, these experiments, which utilize the braid group (the braid group arises from knot theory and described the different operations that intertwining paths can perform), will lock into an eigenstate of a well-defined operator, necessary for application in quantum computation [154]. The geometric phase that results from these operations, and the resulting quantum entanglement, serve quite nicely as building blocks for quantum computation [155], a point reinforced, while discussing some limitations, by Fendley and Fradkin [156].

Contemporaneous with the Camino et al. [149] experiments, there were suggestions that the $v = 5/2$ state would be a viable candidate to probe for non-Abelian statistics [157,158]. Earlier, it was pointed out by Das Sarma et al. [159] that such states would be topologically protected in their role as qubits – they could have a relatively long coherence time. It was also pointed out that, experiments such as those of Camino et al. [149] would be complicated by tunneling between the various edge states, but that this would open another world of possible transitions in the island state which would all be relevant to fractional statistics, and therefore to quantum computing [160].

While this discussion has been overly brief, it is not our intention to discuss the entire world of quantum computing. Rather, we wish only to introduce the topic and the possible application of the extremely interesting and novel fractional statistics of the FQHE regime to this new and open field.

References

[1] K. von Klitzing, G. Dorda, and M. Pepper, *Phys. Rev. Lett.* **45**, 494 (1980).
[2] A.B. Fowler, F.F. Fang, W.E. Howard, and P.J. Stiles, *Phys. Rev. Lett.* **16**, 901 (1966).
[3] N.W. Ashcroft and N.D. Mermin, *Solid State Physics* (New York, Holt, Rinehart and Winston, 1976).
[4] J.P. Harrang, R.J. Higgins, R.K. Goodall, et al., *Phys. Rev. B* **32**, 8126 (1985).
[5] T. Ando, A.B. Fowler, and F. Stern, *Rev. Mod. Phys.* **54**, 437 (1982).
[6] M.A. Paalanen, D.C. Tsui, and A.C. Gossard, *Phys. Rev. Lett.* **25**, 5566 (1982).
[7] P.J. Mohr and B.N. Taylor, *Rev. Mod. Phys.* **72**, 351 (2000).
[8] B. Jeckelmann, B. Jeanneret, and D. Inglis, *Phys. Rev. B* **55**, 13124 (1997).
[9] M. Büttiker, *Phys. Rev. B* **38**, 9375 (1988).
[10] B.J. van Wees, E.M.M. Willems, L.P. Kouwenhoven, et al., *Phys. Rev. B* **39**, 8066 (1989).
[11] S. Komiyama, H. Hirai, S. Sasa, and S. Hiyamizu, *Phys. Rev. B* **40**, 12566 (1989).
[12] B.W. Alphenaar, P.L. McEuen, R.G. Wheeler, and R.N. Sacks, *Phys. Rev. Lett.* **64**, 677–80 (1990).
[13] D.B. Chklovskii, B.I. Shklovskii, and L.I. Glazman, *Phys. Rev. B* **46**, 4026 (1992).
[14] D.B. Chklovskii, K.A. Matveev, and B.I. Shklovskii, *Phys. Rev. B* **47**, 12605 (1993).
[15] C.W.J. Beenakker, *Phys. Rev. Lett.* **64**, 216 (1990).
[16] A.M. Chang, *Solid State Comm.* **74**, 871 (1990).
[17] R.J.F. van Haren, F.A.P. Blom, and J.H. Wolter, *Phys. Rev. Lett.* **74**, 1198 (1995).
[18] A.A. Shashkin, A.J. Kent, P.A. Harrison, L. Eaves, and M. Henini, *Phys. Rev. B* **49**, 5379 (1994).
[19] E. Yahel, A. Tsukernik, A. Palevski, and H. Shtrikman, *Phys. Rev. Lett.* **81**, 5201 (1998).
[20] Y.Y. Wei, J. Weis, K.V. Klitzing, and K. Eberl, *Phys. Rev. Lett.* **81**, 1674 (1998).
[21] S.H. Tessmer, P.I. Glicofridis, R.C. Ashoori, and L.S. Levitov, *Nature* **392**, 51 (1998).
[22] K.L. McCormick, M.T. Woodside, M. Huang, et al., *Phys. Rev. B* **59**, 4654 (1999).
[23] A. Yacoby, H. Hess, T. Fulton, L.P. Pfeiffer, and K. West, *Solid State Commun.* **1**, 111 (1999).
[24] R. Crook, C.G. Smith, M.Y. Simmons, and D.A. Ritchie, *J. Phys.: Condens. Matter* **12**, L735 (2000).

[25] S. Kičin, A. Pioda, T. Ihn, *et al.*, *Phys. Rev. B* **70**, 205302 (2004).
[26] N. Aoki, C.R. da Cunha, R. Akis, D.K. Ferry, and Y. Ochiai, *Phys. Rev. B* **72**, 155327 (2005).
[27] M.A. Topinka, B.J. LeRoy, S.E.J. Shaw, *et al.*, *Science* **289**, 2323 (2000); M.A. Topinka, B.J. LeRoy, and R.M. Westervelt, *Nature* **410**, 183 (2001).
[28] R.J. Haug, A.H. MacDonald, P. Streda, and K. von Klitzing, *Phys. Rev. Lett.* **61**, 2797 (1988).
[29] R.J. Haug, J. Kucera, P. Streda, and K. von Klitzing, *Phys. Rev. B* **39**, 10892 (1989).
[30] S. Washburn, A.B. Fowler, H. Schmid, and D. Kern, *Phys. Rev. Lett.* **61**, 2801 (1988).
[31] H. Van Houten, C.W.J. Beenakker, P.H.M. Van Loosdrecht, *et al.*, *Phys. Rev. B* **37**, 8534 (1988).
[32] B.R. Snell, P.H. Beton, P.C. Main, *et al.*, *J. Phys.: Condens. Matter* **1**, 7499 (1989).
[33] H. van Houten, C.W.J. Beenakker, and B.J. van Wees, in *Semiconductors and Semimetals*, ed. M.A. Reed (New York, Academic Press, 1992), pp. 9–112.
[34] M. Büttiker, in *Semiconductors and Semimetals*, ed. M.A. Reed (New York, Academic Press, 1992), pp. 191–277.
[35] M. Stopa, J.P. Bird, K. Ishibashi, Y. Aoyagi, and T. Sugano, *Phys. Rev. Lett.* **76**, 2145 (1996).
[36] J.P. Bird, M. Stopa, K. Connolly, *et al.*, *Phys. Rev. B* **56**, 7477 (1997).
[37] J.M. Ryan, N.F. Deutscher, and D.K. Ferry, *Phys. Rev. B* **48**, 8840 (1993).
[38] U. Sivan, Y. Imry, and C. Hartzstein, *Phys. Rev. B* **39**, 1242 (1989).
[39] U. Sivan and Y. Imry, *Phys. Rev. Lett.* **61**, 1001 (1988).
[40] B.J. van Wees, L.P. Kouwenhoven, C.J.P.M. Harmans, *et al.*, *Phys. Rev. Lett.* **62**, 2523 (1989).
[41] C.M. Marcus, R.M. Westervelt, P.F. Hopkins, and A.C. Gossard, *Surf. Sci.* **305**, 480 (1994).
[42] J.P. Bird, M. Stopa, K. Ishibashi, Y. Aoyagi, and T. Sugano, *Phys. Rev. B.* **50**, 14983 (1994).
[43] R.P. Taylor, A.S. Sachrajda, P. Zawadzki, P.T. Coleridge, and J.A. Adams, *Phys. Rev. Lett.* **69**, 1989 (1992).
[44] B.W. Alphenaar, A.A.M. Staring, H. van Houten, *et al.*, *Phys. Rev. B* **46**, 7236 (1992).
[45] A.A.M. Staring, B.W. Alphenaar, H. van Houten, *et al.*, *Phys. Rev. B* **46**, 12869 (1992).
[46] A.S. Sachrajda, R.P. Taylor, C. Dharma-Wardana, *et al.*, *Phys. Rev. B* **47**, 6811 (1993).
[47] P.J. Simpson, D.R. Mace, C.J.B. Ford, *et al.*, *Appl. Phys. Lett.* **63**, 3191 (1993).
[48] C.J.B. Ford, P.J. Simpson, I. Zailer, *et al.*, *Phys. Rev B* **49**, 17456 (1995).
[49] M. Kataoka, C.J.B. Ford, G. Faini, *et al.*, *Phys. Rev. B* **62**, R4817 (2000).
[50] M. Kataoka, C.J.B. Ford, M.Y. Simmons, and D.A. Ritchie, *Phys. Rev. Lett.* **89**, 226803 (2002).
[51] H.-S. Sim, M. Kataoka, H. Yi, *et al.*, *Phys. Rev. Lett.* **91**, 266801 (2003).
[52] S. Komiyama, O. Astafiev, V. Antonov, T. Kutsuwa, and H. Hirai, *Nature* **403**, 405 (2000).
[53] O. Astafiev and S. Komiyama, in *Electron Transport in Quantum Dots*, ed. J.P. Bird (Boston, Kluwer Academic, 2003), pp. 363–96.
[54] M. Ciorga, A. Wensauer, M. Pioro-Ladriere, *et al.*, *Phys. Rev. Lett.* **88**, 256804 (2002).
[55] M. Pioro-Ladrière, M. Ciorga, J. Lapointe, *et al.*, *Phys. Rev. Lett.* **91**, 026803 (2003).
[56] A. Sachrajda, P. Hawrylak, and M. Ciorga, in *Electron Transport in Quantum Dots*, ed. J.P. Bird (Boston, Kluwer Academic, 2003), pp. 87–122.
[57] G. Kirczenow, *Phys. Rev. B* **50**, 1649 (1994); G. Kirczenow and E. Castaño, *Phys. Rev. B* **43**, 7343 (1991).

[58] J.J. Lin and J.P. Bird, *J. Phys.: Condens. Matter* **14**, R501 (2002).
[59] J. Liu, W.X. Gao, K. Ismail, K.Y. Lee, J.M. Hong, and S. Washburn, *Phys. Rev. B* **50** 17383 (1994).
[60] T. Machida, H. Hirai, S. Komiyama, and Y. Shiraki, *Phys. Rev. B* **54**, 16860 (1996).
[61] D.C. Tsui, H.L. Störmer, and A.C. Gossard, *Phys. Rev. Lett.* **48**, 1559 (1982).
[62] J.P. Eisenstein and H.L. Störmer, *Science* **248**, 1510 (1990).
[63] R.B. Laughlin, *Phys. Rev. B* **27**, 3383 (1983).
[64] R.B. Laughlin, *Phys. Rev. Lett.* **50**, 1395 (1983).
[65] F.D.M. Haldane and E.H. Rezayi, *Phys. Rev. Lett.* **54**, 237 (1985).
[66] F.D.M. Haldane, *Phys. Rev. Lett.* **51**, 605 (1983).
[67] R. Tao and Y.-S. Wu, *Phys. Rev. B* **30**, 1097 (1984).
[68] J.K. Jain, *Phys. Rev. Lett.* **63**, 199 (1989).
[69] G. Murthy and R. Shankar, *Rev. Mod. Phys.* **75**, 1101 (2003).
[70] S.M. Girvin, A.H. MacDonald, and P.M. Platzman, *Phys. Rev. Lett.* **54**, 581 (1985).
[71] J.E. Avron and R. Seiler, *Phys. Rev. Lett.* **54**, 259 (1985).
[72] R. Willet, J.P. Eisenstein, H.L. Störmer, *et al.*, *Phys. Rev. Lett.* **59**, 1776 (1987).
[73] J.P. Eisenstein, R. Willett, H.L. Störmer, *et al.*, *Phys. Rev. Lett.* **61**, 997 (1988).
[74] F.D.M. Haldane and E.H. Rezayi, *Phys. Rev. Lett.* **60**, 956 (1988); **60**, 1886 (1988).
[75] G. Moore and N. Read, *Nucl. Phys. B* **360**, 362 (1991).
[76] B.I. Halperin, P.A. Lee, and N. Read, *Phys. Rev. B* **47**, 7312 (1993), and references therein.
[77] G. Dev and J.K. Jain, *Phys. Rev. Lett.* **69**, 2843 (1992).
[78] L. Belkhir and J.K. Jain, *Phys. Rev. Lett.* **70**, 643 (1993).
[79] R.R. Du, H.L. Störmer, D.C. Tsui, L.N. Pfeiffer, and K.W. West, *Phys. Rev. Lett.* **70**, 2944 (1993).
[80] V.J. Goldman, B. Su, and J.K. Jain, *Phys. Rev. Lett.* **72**, 2065 (1994).
[81] L. Brey and C. Tejedor, *Phys. Rev. B* **51**, 17259 (1995).
[82] W. Kang, H.L. Störmer, L.N. Pfeiffer, K.W. Baldwin, and K.W. West, *Phys. Rev. Lett.* **71**, 3850 (1993).
[83] D.R. Leadley, R.J. Nicholas, C.T. Foxon, and J.J. Harris, *Phys. Rev. Lett.* **72**, 1906 (1994).
[84] D.R. Leadley, M. van der Brugt, R.J. Nicholas, C.T. Foxon, and J.J. Harris, *Phys. Rev. B* **53**, 2057 (1996).
[85] H.C. Manoharan, M. Shayegan, and S.J. Klepper, *Phys. Rev. Lett.* **73**, 3270 (1994).
[86] P.T. Coleridge, Z.W. Wasilewski, P. Zawadski, A.S. Sachrajda, and H.A. Carmonia, *Phys. Rev. B* **52**, 11603 (1995).
[87] I.V. Kukushkin, K.V. Klitzing, and K. Eberl, *Phys. Rev. Lett.* **82**, 3665 (1999).
[88] S. Melinte, N. Freytag, M. Horvatić, *et al.*, *Phys. Rev. Lett.* **84**, 354 (2000).
[89] A.E. Dementyev, N.N. Kuzma, P. Khandelwal, *et al.*, *Phys. Rev. Lett.* **83**, 5074 (1999).
[90] I.V. Kukushkin, J.H. Smet, K. von Klitzing, and W. Wegschneider, *Nature* **415**, 409 (2002).
[91] I.V. Kukushkin, J.H. Smet, K. von Klitzing, and W. Wegschneider, *J. Supercond.* **16**, 777 (2003).
[92] W. Pan, H.L. Störmer, D.C. Tsui, *et al.*, *Phys. Rev. Lett.* **90**, 016801 (2003).
[93] C.-C. Chang and J.K. Jain, *Phys. Rev. Lett.* **92**, 196806 (2004).
[94] M.O. Goerbig, P. Lederer, and C. Morais Smith, *Phys. Rev. Lett.* **93**, 216802 (2004).

[95] D. Yoshioka, A.H. MacDonald, and S.M. Girvin, *Phys. Rev. B* **39**, 1932 (1989); these authors acknowledge an earlier effort in this direction by E.H. Rezayi and F.D.M. Haldane, *Bull. Am. Phys. Soc.* **32**, 892 (1987).
[96] S. He, X.C. Xie, S. Das Sarma, and F.C. Zhang, *Phys. Rev. B* **43**, 9339 (1991).
[97] Y.W. Suen, L.W. Engel, M.B. Santos, M. Shayegan, and D.C. Tsui, *Phys. Rev. Lett.* **68**, 1379 (1992).
[98] J.P. Eisenstein, G.S. Boebinger, L.N. Pfeiffer, K.W. West, and S. He, *Phys. Rev. Lett.* **68**, 1383 (1992).
[99] M. Greiter, X.G. Wen, and F. Wilczek, *Phys. Rev. B* **46**, 9586 (1992).
[100] S. He, S. Das Sarma, and X.C. Xie, *Phys. Rev. B* **47**, 4394 (1993).
[101] I.B. Spielman, J.P. Eisenstein, L.N. Pfeiffer, and K.W. West, *Phys. Rev. Lett.* **84**, 5808 (2000).
[102] A. Stern and B.I. Halperin, *Phys. Rev. Lett.* **88**, 106801 (2002).
[103] M. Kellogg, I.B. Spielman, J.P. Eisenstein, L.N. Pfeiffer, and K.W. West, *Phys. Rev. Lett.* **88**, 126804 (2002).
[104] M. Kellogg, J.P. Eisenstein, L.N. Pfeiffer, and K.W. West, *Phys. Rev. Lett.* **90**, 246801 (2003).
[105] J.P. Eisenstein and A.H. MacDonald, *Nature* **432**, 691 (2004).
[106] M. Kellogg, J.P. Eisenstein, L.N. Pfeiffer, and K.W. West, *Phys. Rev. Lett.* **93**, 036801 (2004).
[107] I.B. Spielman, M. Kellogg, J.P. Eisenstein, L.N. Pfeiffer, and K.W. West, *Phys. Rev. B* **70**, 081303 (2004).
[108] R.D. Wiersma, J.G.S. Lok, S. Kraus, *et al.*, *Phys. Rev. Lett.* **93**, 266805 (2004).
[109] I.B. Speilman, L.A. Tracy, J.P. Eisenstein, L.N. Pfeiffer, and K.W. West, *Phys. Rev. Lett.* **94**, 076803 (2005).
[110] E. Rossi, A.S. Núñez, and A.H. MacDonald, *Phys. Rev. Lett.* **95**, 266804 (2005).
[111] S.Q. Murphy, J.P. Eisenstein, G.S. Boebinger, L.N. Pfeiffer, and K.W. West, *Phys. Rev. Lett.* **72**, 728 (1994).
[112] T.S. Lay, Y.W. Suen, H.C. Manohara, *et al.*, *Phys. Rev. B* **50**, 17725 (1994).
[113] R.R. Du, A.S. Yeh, H.L. Störmer, *et al.*, *Phys. Rev. Lett.* **75**, 3926 (1995).
[114] W. Pan, J.-S. Xia, V. Shvarts, *et al.*, *Phys. Rev. Lett.* **83**, 3530 (1999).
[115] W. Pan, H.L. Störmer, D.C. Tsui, *et al.*, *Sol. State Commun.* **119**, 641 (2001).
[116] R.H. Morf, *Phys. Rev. Lett.* **80**, 1505 (1998).
[117] E.H. Rezayi and F.D.M. Haldane, *Phys. Rev. Lett.* **84**, 4685 (2000).
[118] C. Tőke and J.K. Jain, *Phys. Rev. Lett.* **96**, 246805 (2006).
[119] C. Tőke, N. Regnault, and J.K. Jain, *Phys. Rev. Lett.* **98**, 036806 (2007).
[120] J.P. Eisenstein, K.B. Cooper, L.N. Pfeiffer, and K.W. West, *Phys. Rev. Lett.* **88**, 076801 (2002).
[121] M.P. Lilly, K.B. Cooper, J.P. Eisenstein, L.N. Pfeiffer, and K.W. West, *Phys. Rev. Lett.* **82**, 394 (1999).
[122] M.O. Goerbig, P. Lederer, and C. Morais-Smith, *Phys. Rev. B* **69**, 115327 (2004).
[123] J.S. Xia, W. Pan, C.L. Vicente, *et al.*, *Phys. Rev. Lett.* **93**, 176809 (2004).
[124] G. Gervais, L.W. Engel, H.L. Störmer, *et al.*, *Phys. Rev. Lett.* **93**, 266804).
[125] A.M. Ettouhami, C.B. Doiron, F.D. Klironomos, R. Côté, and A.T. Dorsey, *Phys. Rev. Lett.* **96**, 196802 (2006).

[126] D.J. Thouless, M. Kohmoto, M.P. Nightingale, and M. den Nijs, *Phys. Rev. Lett.* **49**, 405 (1982).
[127] P. Středa, *J. Phys. C* **15**, L717 (1982).
[128] J.E. Avron, R. Seiler, and B. Simon, *Phys. Rev. Lett.* **51**, 51 (1983).
[129] B. Simon, *Phys. Rev. Lett.* **51**, 2167 (1983).
[130] M.V. Berry, *Proc. Roy. Soc. London A* **392**, 45 (1984).
[131] F. Wilczek, *Phys. Rev. Lett.* **48**, 1144 (1982).
[132] F. Wilczek, *Phys. Rev. Lett.* **49**, 957 (1982).
[133] B.I. Halperin, *Phys. Rev. Lett.* **52**, 1583 (1984).
[134] X.G. Wen, F. Wilczek, and A. Zee, *Phys. Rev. B* **39**, 11413 (1989).
[135] V. Kalmayer and R. Laughlin, *Phys. Rev. Lett.* **59**, 2095 (1988).
[136] X.G. Wen and Q. Niu, *Phys. Rev. B* **41**, 9377 (1990).
[137] F.D.M. Haldane and D. Rezayi, *Phys. Rev. B* **31**, 2529 (1985).
[138] P.W. Anderson, *Phys. Rev. B* **28**, 2264 (1983).
[139] B.I. Halperin, *Helv. Phys. Acta* **56**, 75 (1983).
[140] M. Greiter, X.G. Wen, and F. Wilczek, *Nucl. Phys. B* **374**, 567 (1992).
[141] N. Read and E. Rezayi, *Phys. Rev. B* **54**, 16864 (1996).
[142] E. Fradkin, C. Nayak, A. Tsvelik, and F. Wilczek, *Nucl. Phys. B* **516**, 704 (1998).
[143] N. Read and D. Green, *Phys. Rev. B* **61**, 10267 (2000).
[144] A. Harju, H. Saarikoski, and E. Räsänen, *Phys. Rev. Lett.* **96**, 126805 (2006).
[145] D.A. Ivanov, *Phys. Rev. Lett.* **86**, 268 (2001).
[146] D. Arovas, J.R. Schrieffer, and F. Wilczek, *Phys. Rev. Lett.* **53**, 722 (1984).
[147] V.J. Goldman and B. Su, *Science* **267**, 1010 (2005).
[148] V.J. Goldman, J. Liu, and A. Zaslavsky, *Phys. Rev. B* **71**, 153303 (2005).
[149] F.E. Camino, W. Zhou, and V.J. Goldman, *Phys. Rev. B* **72**, 075342 (2005).
[150] E.-A. Kim, *Phys. Rev. Lett.* **97**, 216404 (2006).
[151] V.J. Goldman, *Phys. Rev. B* **75**, 045334 (2007).
[152] C. Nayak and F. Wilczek, *Nucl. Phys. B* **479**, 529 (1996).
[153] C. de C. Chamon, D.E. Freed, S.A. Kivelson, S.L. Sondhi, and X.G. Wen, *Phys. Rev. B* **55**, 2331 (1997).
[154] B.J. Overbosch and F.A. Bais, *Phys. Rev. A* **64**, 062107 (2001).
[155] A. Stern, F. von Oppen, and E. Mariani, *Phys. Rev. B* **70**, 205338 (2004).
[156] P. Fendley and E. Fradkin, *Phys. Rev. B* **72**, 024412 (2004).
[157] P. Bonderson, A. Kitaev, and K. Shtengel, *Phys. Rev. Lett.* **96**, 016803 (2006).
[158] A. Stern and B.L. Halperin, *Phys. Rev. Lett.* **96**, 016802 (2006).
[159] S. Das Sarma, M. Freedman, and C. Nayak, *Phys. Rev. Lett.* **94**, 166802 (2005).
[160] J.K. Jain and C. Shi, *Phys. Rev. Lett.* **96**, 136802 (2006).

5
Ballistic transport in quantum wires

In this chapter we discuss a variety of issues related to the phenomenon of one-dimensional conductance quantization, probably one of the most important phenomena exhibited by mesoscopic conductors. The quantization is observed in one of the simplest of structures, namely the quantum point contact (QPC) that can be straightforwardly realized by means of the split-gate technique. The QPC is essentially a nanoscale constriction, connected at either end to macroscopic reservoirs, through which electrons may travel ballistically at low temperatures. In this chapter, we discuss how the strong lateral confinement that electrons experience as they pass through the QPC quantizes their energy into a series of discrete one-dimensional subbands. Through a simple analysis, based on a noninteracting model of transport that assumes linear response, we show that the conductance associated with these subbands takes the universal value $2e^2/h$, independent of the subband index. This results in the observation of a universal staircase structure in the conductance of QPCs, as their gate voltage is used to change the number of occupied subbands one at a time. An important requirement for the observation of this effect is that electron transport through the QPC should be ballistic, and we will see how this typically limits its observation to low temperatures ($\leq 4.2\,\text{K}$). The conductance quantization provides a striking demonstration of the validity of the Landauer approach to electrical conduction, and in this chapter we also extend the discussion to consider the influence of scattering and non-vanishing source–drain bias on the conductance. These discussions will reveal that the behavior observed under these conditions may be understood within the same framework of the Landauer approach. In spite of the successes of noninteracting models of transport in accounting for the one-dimensional conductance quantization, we will also discuss the observation of several features that *cannot* be accounted for within this approach. Prominent among these is the so-called *0.7 feature*, an additional plateau-like structure in the conductance that occurs quite ubiquitously at a value close to $0.7 \times 2e^2/h$. There is broad consensus in the community that this effect arises from a spontaneous lifting of spin degeneracy in the QPC, due to many-body interactions under conditions where its carrier density is lowered such that the conductance is about to vanish. In our discussion, we review several different models of this effect, including the results of supporting experiments. In the last sections of this chapter, we briefly discuss some of the recent work that has been done to develop novel device concepts based upon ballistic electron transport in one-dimensional conductors. We focus on experimental work on the Y-branch device, a three-terminal ballistic structure that may offer the potential of high-frequency operation. We also consider related structures that consist of a nanoscale junction with a lithographically defined scatterer located at its center. Due to the ballistic transport of carriers in the junction region, the functionality of such devices is shown to depend almost completely on the symmetry-breaking properties of the scatterer, giving rise to novel rectification effects. Finally, we review the results of experiments in which surface acoustic waves have been used to manipulate charge transport through QPCs. Such devices may be of interest in the future, for application to single-photon sources or for the implementation of flying qubits in a quantum-computing scheme.

5.1 Conductance quantization in quantum point contacts

In Chapter 3, we developed a model for transport in very small structures following the formalism developed by Landauer and Büttiker. In this picture, the conductance is determined by the number of one-dimensional channels available to carriers injected from ideal phase-randomizing contacts, and by the transmission properties of the structure for each channel. The derivation of these formulas follow from basic physical insights and certain assumptions regarding the distinction between contacts and scattering structures (if such a distinction may be made). In Chapter 4, we saw how this formalism, in combination with the concept of edge states, can be used to successfully account for the features of the integer quantum Hall effect, which may therefore be considered as a remarkable manifestation of one-dimensional transport in a macroscopic conductor. In this section, however, we discuss the application of the Landauer–Büttiker formalism to another very different system, namely the mesoscopic structures known as quantum point contacts (QPCs). In these structures we will see that a quantization of the conductance is observed that is intimately related to that which is found in the integer quantum Hall effect.

As discussed already in Section 1.2, QPCs are typically realized in a high-mobility two-dimensional electron gas (2DEG) by means of the split-gate approach [1] (shown schematically in Fig. 5.1). For zero gate bias, the 2DEG exists essentially everywhere in the space between the two Ohmic contacts. With a negative bias applied to the two Schottky contact gates, the 2DEG is depleted underneath as well as laterally from the geometric edge of the gates. In the narrow region between the two split gates, a quasi-one-dimensional electron gas (1DEG) is formed. The 1DEG density decreases as the gate bias is made more

Fig. 5.1 Schematic illustration showing the concept of using the split-gate method to form a QPC ([3], with permission).

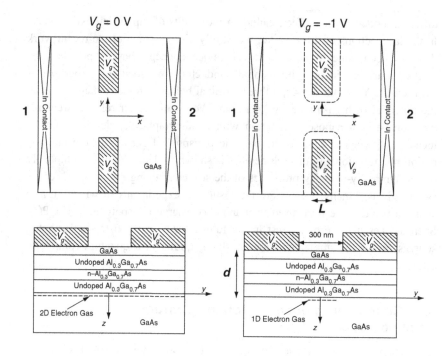

negative until it eventually vanishes between the split gates. The length of the 1DEG is defined by the width of the gate contacts and the shape of the depletion regions around the contacts. In contrast to metallic point contacts [2], which are essentially similar structures but in which the high electron density means that the electron wavelength is much smaller than the constriction size, the important feature of semiconductor QPCs is that the Fermi wavelength is comparable to that of the structure. At the same time, and in contrast to the situation in metals, the electron mean free path in semiconductors can be very long, much longer than the QPC length. Consequently, electrons travel ballistically through these structures, and it is this fact, combined with the presence of strong momentum quantization in the QPC, that gives rise to the observation of the above mentioned conductance quantization.

Fig. 5.2 shows an example of the conductance quantization that is exhibited by QPCs. In this measurement, the conductance of the QPC is measured (in the absence of a magnetic field) while its gate voltage is made more negative. Referring to the behavior shown in the inset, as the gate voltage is made more negative than ~ -0.5 V a sudden drop in the conductance is observed, indicating the full depletion of the 2DEG directly underneath the gates and, thus, the formation of the QPC. As the gate voltage is made further negative a slower decrease of the conductance occurs and it is clear from the behavior in the main panel that it develops a steplike variation. In fact, the conductance in this figure is plotted in units of $2e^2/h$ and it is clear that each step in the conductance

5.1 Conductance quantization in quantum point contacts

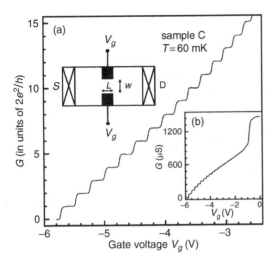

Fig. 5.2 A beautiful example of conductance quantization in a split-gate QPC ([6], with permission).

corresponds to a change by this amount. This remarkable behavior was first observed, independently, by Wharam et al. [4] and van Wees et al. [5] in 1988 and has since been confirmed in numerous experiments.

For an intuitive discussion of the origins of the conductance quantization, it is helpful to start from a discussion of the form of the self-consistent potential that arises for electrons when a gate bias is applied to the QPC. The actual form of this potential due to the space charge region and the gates is complicated and requires the 3D solution of Poisson's equation. A simple analytical model of the potential in a QPC may be obtained, however, if one neglects the effects due to space charge layer formation and simply treats the GaAs/AlGaAs heterostructure as a dielectric [7]. For two electrodes held at a constant voltage V_g forming a narrow constriction, the confining potential may be expressed as

$$V(x,y) = f\left[\frac{2x-l}{2z_0}, \frac{2y+w}{2z_0}\right] - f\left[\frac{2x+l}{2z_0}, \frac{2y+w}{2z_0}\right] + f\left[\frac{2x-l}{2z_0}, \frac{-2y+w}{2z_0}\right] - f\left[\frac{2x+l}{2z_0}, \frac{-2y+w}{2z_0}\right], \quad (5.1)$$

where

$$f(u,v) = \frac{eV_g}{2\pi}\left[\frac{\pi}{2} - \tan^{-1} u - \tan^{-1} v + \tan^{-1}\frac{uv}{\sqrt{1+u^2+v^2}}\right]. \quad (5.2)$$

Here, l and w are the lithographic width and gap between the electrodes, respectively, and z is the vertical distance between the 2DEG and the gate (the coordinates of the x–y plane are indicated in Fig. 5.1). An example of a potential calculated using the formulas above is shown in Fig. 5.3. From this figure it can be clearly seen that application of appropriate bias to the gates results in the formation of a saddle potential. This not only confines electrons in the (y) direction transverse to the wire axis, but also presents a potential barrier (formed at the saddle minimum) along the (x) direction of current flow. The form of this

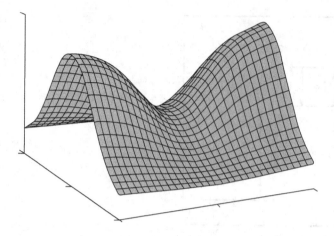

Fig. 5.3 Calculated QPC potential from Eqs. (5.1) and (5.2). Figure courtesy of Dr. A. Ramamoorthy.

potential evolves smoothly with change of gate voltage, with the saddle minimum rising in energy and the transverse width of the QPC shrinking, as V_g is made more negative. Typically, it is the combination of these two effects that eventually causes the QPC to pinch off (near -5.8 V in the case of Fig. 5.2) [8,9].

Reasonably close to the bottom of the saddle in Fig. 5.3, the variation of the electron potential energy can be well described by a parabolic form:

$$V(x, y) = V_0 - \frac{1}{2}m^*\omega_x^2 x^2 + \frac{1}{2}m^*\omega_0^2 y^2, \tag{5.3}$$

where V_0 is the height of the saddle-barrier, m^* is the electron effective mass, and ω_x and ω_0 are characteristic oscillator frequencies. ω_0 determines the energy splitting of the one-dimensional subbands in the QPC, while ω_x essentially dictates how sharply its transmission drops to zero when the Fermi level falls below the saddle minimum. For a harmonic-oscillator potential of the form of Eq. (5.3), one expects that the energy for motion along the direction of confinement should be quantized into a set of equally spaced energies. The resulting electron dispersion relation (with energy measured relative to the conduction-band edge) is then given by

$$E_n = \left[n + \frac{1}{2}\right]\hbar\omega_0 + \frac{\hbar^2 k_x^2}{2m^*}, \quad n = 1, 2, 3, \ldots \tag{5.4}$$

This relation defines a series of one-dimensional modes (subbands, or channels), each of which is characterized by a unique value of the index, n. Similar to the discussion of edge states in Chapter 4, in a system with a fixed Fermi energy, only those subbands whose energy threshold (at $k_x = 0$) lies below the Fermi energy will be populated by electrons at low temperatures (Fig. 5.4) and so contribute to current through the QPC. To obtain an expression for the number of occupied subbands (N) in the QPC [10], we note that its effective width (W) at the Fermi energy (E_F) is

5.1 Conductance quantization in quantum point contacts

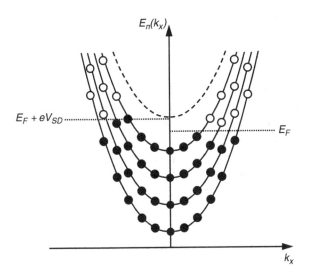

Fig. 5.4 Electron dispersion, and filling of electron states for a small applied voltage, in a QPC with a harmonic lateral-confinement potential.

$$E_F = \frac{1}{2}m^*\omega_0^2 \frac{W^2}{4}, \qquad (5.5)$$

which yields $W = 2\hbar k_F/m^*\omega_0$. The value of N is such that $E_F \geq (N-\frac{1}{2})\hbar\omega_0$ (the minus sign here since N counts the *number* of subbands while n in Eq. (5.4) is the subband index), which can therefore be rewritten as

$$N = \text{Int}\left[\frac{1}{2} + \frac{E_F}{\hbar\omega_0}\right] \approx \frac{E_F}{\hbar\omega_0} = \frac{k_F W}{4} = \frac{\pi W}{2\lambda_F}. \qquad (5.6)$$

In making this approximation, we have obviously assumed that the subband spacing is small compared to the Fermi energy. For a fairly typical GaAs 2DEG, with an electron density of 3×10^{11} cm^{-2}, $E_F = \sim 10$ meV. The subband separation on the other hand varies with gate voltage, increasing in size as the QPC width is reduced. For most split-gate samples, a reasonable estimate might be 1–2 meV for the last few subbands [10], so that the approximation in Eq. (5.6) should be reasonable. It is worth commenting here that, in the case where the confining potential is of a square-well, rather than a parabolic, form, it is straightforward to show that $N = k_F W/\pi = 2W/\lambda_F$. In other words, a new subband is populated in such a QPC each time its width is increased by $\lambda_F/2$, where λ_F is the Fermi wavelength.

In Fig. 5.4, we illustrate schematically the filling of the electron states in the different subbands of a QPC, under conditions where a *small* (compared to E_F/e) voltage (V_{SD}) is applied across its ends. For convenience, we have assumed a two-probe geometry, as is effectively the case in transport experiments where split-gate QPCs are connected to the outside measurement circuitry via ungated regions of 2DEG (see Fig. 5.1, for example). In drawing the filling of the electron states in the QPC, we have also made an implicit assumption of ballistic transport, since we have indicated that electrons traveling in opposite directions (with positive and negative k_x) preserve the electrochemical potential of the voltage probe from which they originate. From this figure we therefore see that, due to the absence of energy relaxation in the QPC, the net effect of applying the voltage is to cause an imbalance in the filling of electron states with opposite momentum and, therefore, the flow of a net electrical current. In the figure here, electron states moving to the right (with negative k_y) are filled to a higher energy than those moving to the left, indicating that a negative voltage has been applied to the right probe with respect to the left one (which may be viewed as the grounded source). For *each* one-dimensional subband, and assuming a small applied voltage, the *excess* charge due to the applied voltage may be written as:

$$\delta Q = e \frac{D_{1D}(E_F)}{2} eV_{SD} = \frac{e^2 V_{SD}}{2} \left[\frac{m^*}{2\pi^2 \hbar^2 E_F} \right]^{1/2}. \tag{5.7}$$

Focusing on the first term on the right-hand side of this equation, D_{1D} is the 1D density of states (Eq. (2.34)) and the factor of 1/2 appears since we are only interested in the filling of 1D states for one direction of motion (not both). Due to the definition of the density of states in Eq. (2.34), E_F is measured relative to the bottom of the 1D subband in question, and δQ is the excess charge per subband in the QPC *per unit length*. Consequently, the amount of charge occupying these states that passes through the QPC per unit time is given by

$$\delta Q v_F = \frac{e^2 V_{SD}}{2} \left[\frac{m^*}{2\pi^2 \hbar^2 E_F} \right]^{1/2} \left[\frac{2E_F}{m^*} \right]^{1/2} = \frac{2e^2}{h} V_{SD}. \tag{5.8}$$

This is nothing more than the current carried by the subband and the critical feature of this relationship to note is that it contains no information specific to any given subband. That is, the same current is carried by each subband, and this phenomenon of *equipartition of current* is a unique consequence of the cancellation of energy terms in the 1D density of states and the electron group velocity (Eq. 5.8). Due to the equipartition of current, we may therefore simply write the total current flowing through the QPC as

$$I = N \frac{2e^2}{h} V_{SD}, \tag{5.9}$$

which means that the corresponding conductance is given by

$$G = \frac{I}{V_{SD}} = N \frac{2e^2}{h} \equiv N G_0. \tag{5.10}$$

In other words, at low temperatures and under conditions of ballistic transport, the linear (i.e. small signal) conductance of a QPC should be quantized in units of $2e^2/h$, which is exactly the behavior observed in experiment (Fig. 5.2).

5.1 Conductance quantization in quantum point contacts

For the reader acquainted with the results of Chapter 3, it will not require much to understand that the foregoing intuitive derivation of the conductance quantization in QPCs is based entirely on the transmission picture of conduction as developed by Landauer and Büttiker. Indeed, the final result that we have ended up with (Eq. (5.10)) is nothing more than the two-probe Landauer formula (Eq. (3.61)) for the special case of perfect transmission of all channels. It is also clear that there is a strong connection between the QPC conductance quantization and that which occurs in the integer quantum Hall effect (Chapter 4). In the latter phenomenon, however, the conductance quantization is in units of e^2/h. The additional factor of two in Eq. (5.10), in contrast, arises since its derivation assumed unbroken spin degeneracy at zero magnetic field, while this degeneracy is broken by the magnetic field in the quantum Hall effect. Another important difference between these two effects is the accuracy of their conductance quantization. While we have seen already (Chapter 4) that the quantization of the Hall resistance is accurate to of order 3.5 parts in 10^{10}, in the case of QPCs it is typically much less, of order just a few percent. Part of the experimental uncertainty arises from the non-zero resistance of the Ohmic contacts and 2DEG between the source–drain and the QPC that includes the spreading resistance around the QPC itself in the 2DEG. Some of this series resistance may be eliminated by performing four-terminal measurements. However, if the probes are within approximately a phase coherence length from the QPC itself, they themselves contribute to the transmission through the structure. Therefore, much of the series resistance effect cannot be eliminated other than to measure this quantity separately and to subtract its effect from the measured results, a practice which is often employed. Another contribution that degrades the ideal conductance quantization is random inhomogeneities such as impurities and boundary roughness. Backscattering by such inhomogeneities causes the transmission coefficient to be less than unity, thus degrading the ideal quantization of the conductance.

5.1.1 Adiabatic transport model

In the previous section, we made some hand-waving arguments for the existence of quantized conductance in QPCs, by assuming perfect transmission of each occupied subband. In this section, we consider just what is required in order to achieve such perfect transmission in practice. As we have mentioned several times already, when realizing QPCs by the split-gate method, one essentially realizes a system in which the current flows into and out of the QPC via macroscopic 2DEG regions. In the sense of Eq. (5.6), we can view these 2D regions as extremely wide 1D conductors in which many 1D subbands are occupied. As current flows between the reservoirs and the QPC it is therefore redistributed among different modes, with the redistribution taking place at the transition regions between the QPC and the reservoirs. As pointed out by Imry [11], it is this redistribution that gives rise to an unavoidable contact resistance, even though electrons are transmitted ballistically through the QPC itself without any scattering. It should be pointed out that the quantized conductance of Eq. (5.10) is the *maximum* possible conductance of a QPC with N occupied modes and that observation of this conductance requires a smooth, or *adiabatic*, connection between the QPC and its reservoirs. As can be seen from the potential profile of Fig. 5.3, a smooth transition is actually typically achieved in split-gate QPCs. In the absence of such a smooth potential, however, an abrupt transition between the QPC and the reservoirs may give rise to intersubband scattering by itself and so disrupt the conductance quantization [12,13].

The formal description of transport through a smooth QPC is given by the so-called *local adiabatic model*, which we discuss in terms of the idealized QPC potential of Fig. 5.5 [12]. Here it is assumed that the confining potential in the normal (growth) direction is sufficiently narrow that the width in that direction is negligibly small compared to the lateral dimensions, which is generally a good approximation in split-gate structures. It is therefore sufficient to consider the two-dimensional stationary Schrodinger equation:

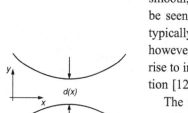

Fig. 5.5 Two-dimensional approximation for a QPC that couples smoothly to its 2DEG reservoirs.

$$-\frac{\hbar^2}{2m^*}\left[\frac{\partial^2}{\partial x^2}+\frac{\partial^2}{\partial y^2}\right]\psi(x,y)+V(x,y)\psi(x,y)=E\psi(x,y). \quad (5.11)$$

The adiabatic approximation amounts to assuming that the spatial variation of the potential in the x-direction is much slower than in the transverse y-direction. Thus, the second derivative with respect to x may be neglected in Eq. (5.11) to give the one-dimensional Schrödinger equation at a fixed x as

$$-\frac{\hbar^2}{2m^*}\frac{\partial^2}{\partial y^2}\chi_n(x,y)+V(x,y)\chi_n(x,y)=E_n\chi_n(x,y), \quad (5.12)$$

where $\chi_n(x,y)$ is the transverse eigenfunction and $E_n(x)$ is the position-dependent energy eigenvalue corresponding to the transverse potential at point x. Since the χ_n form a complete set, the solution of Eq. (5.11) may be constructed from the solutions of Eq. (5.12) as

$$\psi(x,y)=\sum_n \phi_n(x)\chi_n(x,y). \quad (5.13)$$

By substituting Eq. (5.13) into Eq. (5.11), multiplying by χ_m^* and integrating over all values of y we thus obtain

$$\left[-\frac{\hbar^2}{2m^*}\frac{\partial^2}{\partial x^2}+E_m(x)-E\right]\phi_m(x)=\sum_n A_{nm}\,\phi_n(x), \quad (5.14)$$

where A_{mn} is the operator:

$$A_{mn}=\frac{\hbar^2}{m^*}\int dy\,\chi_m^*(x,y)\frac{\partial}{\partial x}\chi_n(x,y)\frac{\partial}{\partial x}+\frac{\hbar^2}{2m^*}\int dy\,\chi_m^*(x,y)\frac{\partial^2}{\partial x^2}\chi_n(x,y). \quad (5.15)$$

5.1 Conductance quantization in quantum point contacts

The right side of Eq. (5.14) gives rise to terms coupling mode index m to n, which corresponds to intersubband scattering. However, these terms are proportional to spatial gradients in the longitudinal direction, which according to the adiabatic approximation are small. Therefore, to lowest order, the right side is zero ($A_{nm} = 0$), giving the simple one-dimensional equation:

$$\left[-\frac{\hbar^2}{2m^*}\frac{\partial^2}{\partial x^2} + E_n(x)\right]\phi_n(x) = E\,\phi_n(x), \tag{5.16}$$

This equation shows that for a slowly varying longitudinal potential, the envelope function for the motion in the x-direction satisfies the one-dimensional Schrödinger equation in an effective potential, $E_n(x)$, determined by the spatial variation of the nth energy eigenvalue of the transverse solution. The degree that this is true may be checked quantitatively by substituting back into Eq. (5.15) to determine the actual magnitude of the coefficients A_{nm}.

As an illustrative example, consider a simple case assuming hard-walled confinement. In Fig. 5.5, the potential is assumed to be infinite outside of the constriction so that the solution for $\chi_n(x, y)$ at point x is

$$\chi_n(x, y) = \sqrt{\frac{2}{d(x)}}\sin\frac{n\pi(2y+d(x))}{d(x)}, \tag{5.17}$$

where the index n takes values $n = 1, 2, 3, \ldots$ and, assuming a symmetric geometry for simplicity, the value of y is defined to be zero at the midpoint of the constriction. The corresponding effective potential is

$$E_n(x) = \frac{n^2\pi^2\hbar^2}{2m^*d(x)^2}, \tag{5.18}$$

which is just the usual expression for the energy eigenvalues of a particle in a box, although now the effective size of this box ($d(x)$) is position dependent in accordance with Fig. 5.5.

When the variation of the effective potential close to the constriction has a quadratic form

$$E_n(x) = V_0 - \frac{1}{2}m^*\omega_x^2 x^2, \tag{5.19}$$

the transmission coefficient of the nth subband may be expressed analytically as [14]

$$T_n(E) = \frac{1}{1+\exp(-2(E-V_0)/\hbar\omega_x)}. \tag{5.20}$$

Thus, expanding Eq. (5.18) into quadratic form we have

$$E_n(x) \approx E_n(1 - \alpha x^2), \tag{5.21}$$

where α characterizes the variation of the confining potential, and $E_n = n^2\pi^2\hbar^2/2m^*d^2$ with d the minimum width of the QPC (which is defined to occur at

$x = 0$). Large values of α correspond to a sharp point contact while smaller values correspond to longer structures. The resulting transmission coefficient is thus

$$T_n(E) = \frac{1}{1 + \exp(-\beta_n(E - E_n))}, \qquad (5.22)$$

where

$$\beta_n(E) = \sqrt{\frac{2m^*}{\alpha \hbar^2 E_n}}. \qquad (5.23)$$

Finally, the conductance at low temperatures is given by the two-probe multi-channel formula (Eq. (3.61)) with Eq. (5.20) above:

$$G = \frac{2e^2}{h} \sum_n \frac{1}{1 + \exp(-\beta_n(E_F - E_n))}. \qquad (5.24)$$

Equation (5.24) predicts, as the gate voltage is changed and the width of the QPC is reduced, successive energy levels, E_n, pass through the Fermi energy giving rise to plateau-like structures. For small values of α sharp steps are found, whereas for large values the steps are rounded, in accordance with experiment. The rounding is due to tunneling through the constriction for a Fermi energy just below the next unoccupied level, and then approaches unity as the Fermi energy moves above the subband minimum. Note, however, that a well-defined 1DEG in a long quantum wire is not necessary for achieving conductance quantization. The important point is that the conductance is controlled by the mode spacing at the narrowest point of the channel. Thus either a QPC or a longer wire may exhibit conductance quantization, providing that the electron transport remains ballistic (so ensuring perfect transmission of all occupied modes).

The simple example above assumed hard wall boundaries for the potential, but one could equally well use some other function. Büttiker has used the two-dimensional quadratic potential of Eq. (5.3) [8] for the potential variation around the saddle point forming the constriction, which directly gives the transmission coefficient in the form of Eq. (5.22) without inter-mode scattering and without the necessity of expanding the potential as we did earlier. In any case, the form of the transmission coefficient (3.150) represents a convenient empirical form for use in comparing to experiment treating β_n simply as an *adiabatic* parameter. Given that the form of the transmission coefficient is a Fermi function, it is not surprising that the effect of finite temperature is very similar to the effect of tunneling in the adiabatic model as discussed below.

5.1.2 Conductance quantization in the quantum Hall regime

The effect of both parallel and perpendicular magnetic field on the quantized conductance was first reported by Wharam *et al.* [15]. In Fig. 5.6, we show data

5.1 Conductance quantization in quantum point contacts

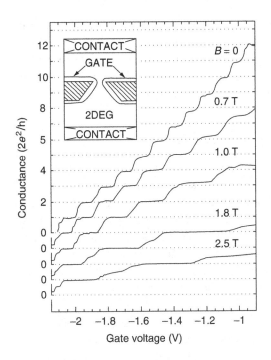

Fig. 5.6 QPC conductance as a function of gate voltage for several magnetic field values, illustrating the transition from zero-field quantization to quantum Hall effect ([16], with permission).

of van Wees *et al.* of the evolution of the conductance plateaus for various perpendicular magnetic field intensities [16]. The plateaus are observed to broaden and flatten as a function of increasing magnetic field. Additional plateaus are seen to arise at the highest fields, but now at half steps of $2e^2/h$. For a given gate bias, as a function of increasing magnetic field, the quantized conductance remains in a given integer conductance plateau until a certain critical magnetic field intensity is reached, and then it jumps down by an integer multiple (or half integer multiple at high fields) of $2e^2/h$.

A simple explanation for this behavior may be constructed by considering a QPC with a parabolic lateral potential in the presence of a perpendicular magnetic field (as introduced already in Section 2.5.2). Considering the quantum point contact to be a waveguide, the energy eigenvalues in the constriction are given by Eq. (2.37) as

$$E_n(k_y) = \left[n + \frac{1}{2}\right]\hbar\omega + \frac{\hbar^2 k_y^2}{2M}, \tag{5.25}$$

where

$$M = m^* \frac{\omega_0^2 + \omega_c^2}{\omega_0^2} \quad \text{and} \quad \omega = \sqrt{\omega_0^2 + \omega_c^2}. \tag{5.26}$$

As discussed already, ω_0 characterizes the strength of the lateral confinement, and $\omega_c = eB/m^*$ is the cyclotron frequency. In the adiabatic limit, the conductance is given by the number of occupied modes (N) below the Fermi energy in the wide 2DEG, times $2e^2/h$. If we assume for simplicity that the Fermi energy in the 2DEG is almost constant with magnetic field (it in fact oscillates due to Landau-level formation in the bulk region), the number of occupied subbands below the Fermi energy is given by

$$N = \text{Int}\left[\frac{E_F}{\hbar\sqrt{\omega_0^2 + \omega_c^2}} + \frac{1}{2}\right]. \qquad (5.27)$$

It is clear that for low fields, $\omega_c \ll \omega_0$ and N remains constant. As ω_c approaches ω_0, increasing the magnetic field causes N to decrease by successive integer values. This decrease in N with increasing magnetic field in a quantum waveguide is referred to as the *magnetic depopulation of subbands* [17]. Thus the widening of the plateaus from the zero field conductance is simply a manifestation of the increased energy separation of the subbands as the magnetic field increases, causing fewer and fewer subbands to reside below the Fermi energy, with a corresponding reduction in the conductance.

Büttiker has shown that it is not necessary to assume a waveguide geometry in explaining the transition from zero to high field magneto-conductance [18]. As discussed earlier, Eq. (5.3) describes a parabolic saddle-point potential, which gives an exact form for the transmission coefficient. This potential may easily be combined with the harmonic potential due to a perpendicular magnetic field to find the transmission coefficient using the form given by Fertig and Halperin [19]:

$$T_{nm} = \delta_{nm}\frac{1}{1 + e^{-\pi\varepsilon_n}}, \qquad (5.28)$$

where the saddle-point energies are given by

$$\varepsilon_n = \frac{E_1 - E_2(n + 1/2) - V_0}{E_1}, \qquad (5.29)$$

$$E_1 = \frac{\hbar}{2\sqrt{2}}((\Omega^4 + 4\omega_x^2\omega_y^2)^{0.5} - \Omega^2)^{0.5}, \qquad (5.30)$$

$$E_2 = \frac{\hbar}{2\sqrt{2}}((\Omega^4 + 4\omega_x^2\omega_y^2)^{0.5} + \Omega^2)^{0.5}. \qquad (5.31)$$

The term $\Omega^2 = \omega_c^2 + \omega_y^2 - \omega_x^2$ defines an effective energy, and all other parameters are defined in Eq. (5.3). Figure 5.7 shows the calculated conductance using $T_{nm}(E_F)$ above in the two-terminal multi-channel formula. For increasing B, the curves evolve from narrow, rounded conductance plateaus to sharp, broad plateaus in qualitative agreement with the experimental results shown in Fig. 5.6.

Other features of the experimental results shown in Fig. 5.6 are of interest as well. One is the flattening of the plateaus with increasing accuracy about $2Ne^2/h$. It was argued earlier in Chapter 4 that backscattering is decreased as the

5.1 Conductance quantization in quantum point contacts

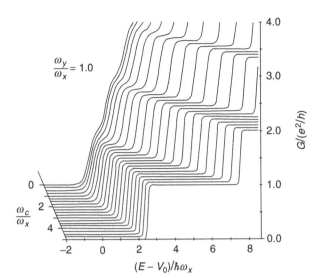

Fig. 5.7 Calculated two-terminal conductance of a QPC in a magnetic field as a function of Fermi energy. The ratio ω_c/ω_x is increased in steps of 0.25 from 0 to 5 ([18], with permission).

magnetic field increases due to an increasing amount of current being carried by edge states. As we discuss shortly below, the conductance is degraded from the ideal integer multiples of the fundamental conductance due to impurities and boundary roughness which give rise to backscattering, and thus less than unity transmission. The flattening of the experimental plateaus with increasing field is thus taken as a sign of increased transmission and reduced backscattering due to these effects.

Another feature of the conductance at high fields is the appearance of half integer plateaus. This phenomenon can be straightforwardly explained if we relax the assumption of spin degeneracy in the presence of the magnetic field. As discussed already in Section 2.5, the initially (at $B = 0$) two-fold spin-degenerate magneto-subbands are split by an amount $g^*\mu_B B$ in a magnetic field, forming distinct spin-up and spin-down states. These spin-split subbands constitute separate modes that contribute to conduction. However, they contribute only e^2/h to the conductance, since the density of states is now reduced by a factor of two. As the magnetic field is increased, the splitting of the magneto-electric subbands due to breaking of the spin-degeneracy increases in energy to the point that the individual contributions of spin-up and spin-down levels is resolvable in the conductance versus gate bias curve, as seen in Fig. 5.6.

5.1.3 Conductance quantization at higher temperatures

At non-zero temperature, it is qualitatively expected that the conductance plateaus should persist until the thermal broadening, $k_B T$, becomes comparable to

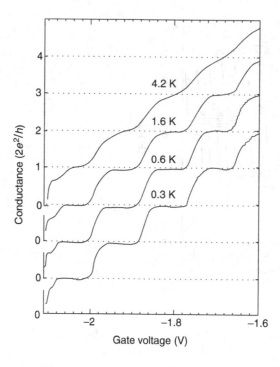

Fig. 5.8 Experimental temperature dependence of the quantized conductance in a QPC ([20], with permission).

the spacing of the 1D subbands. Figure 5.8 shows the measured conductance plateaus for several different temperatures [20]. At 4.2 K, the conductance quantization is almost completely washed out.

In order to extend the multi-channel formula to non-zero temperature, the energy dependence of the transmission coefficient must be explicitly taken into account. Since carriers are now injected at significantly different energies, we can think of each incremental energy range dE around E as a separate channel into which charge is injected. Thus the incremental charge injected into channel i on the right side from channel j on the left becomes

$$\frac{2e}{h} T_{ij}(E)(f(E - \mu_1)dE - f(E - \mu_2)dE), \quad (5.32)$$

where $f(E - \mu_i)$ is the quasi-equilibrium distribution function of the ith contact. The total current then summed over all channels and integrated over energy becomes

$$I = \frac{2e}{h} \int_{-\infty}^{\infty} dE \sum_i T_i(f(E - \mu_1) - (f(E - \mu_2))). \quad (5.33)$$

This equation also represents a starting point for considering the nonlinear response in a QPC, which we will return to shortly in this chapter. For now, however, we are interested in the influence of temperature in the linear-response regime, so that the difference in chemical potentials may be assumed to be small.

5.1 Conductance quantization in quantum point contacts

Assuming $f(E)$ to be the Fermi–Dirac distribution, the current in Eq. (5.33) may be rewritten as

$$I = -\frac{2e}{h} \int_{-\infty}^{\infty} dE \sum_i T_i(E) \frac{df}{dE} (\mu_1 - \mu_2), \qquad (5.34)$$

where

$$-\frac{df}{dE} = \frac{df}{d\mu} = \lim_{\mu_1 \to \mu_2} \frac{f(E-\mu_1) - f(E-\mu_2)}{(\mu_1 - \mu_2)}. \qquad (5.35)$$

Consider now the simple case that the transmission coefficient into the ith mode is unity for energies greater than the subband minimum and zero below. Then, the two-terminal conductance becomes simply

$$G = \frac{I}{(\mu_1 - \mu_2)} = \frac{2e}{h} \sum_i \frac{1}{1 + e^{-(E_F - E_i)/k_B T}}, \qquad (5.36)$$

which is identical in form with Eq. (5.24), with $\beta_n = 1/k_B T$. The effect on the calculated conductance versus width is identical to that due to adiabatic tunneling. As T increases, the effective β decreases, causing increased rounding of the plateaus until they eventually disappear at sufficiently high temperature.

5.1.4 Influence of disorder on the transmission properties of QPCs

In the description so far of the quantized conductance in QPCs, we have neglected the effect of unintentional inhomogeneities. Real split-gate structures are formed on modulation-doped, heterojunction layers with the narrow channel formed using high-resolution lithography. The exact distribution of the impurities in the barrier region (as well as unintentional background doping) give rise to random fluctuations of the potential in addition to that defined to lowest order by the split gate. Since the ideal quantization of conductance requires unity transmission for each mode in the narrow region, a potential fluctuation in the ballistic region of the QPC contributes to backscattering which degrades the conductance below $2Ne^2/h$. Boundary roughness may play a dramatic role as well. The interface between lattice-matched GaAs and AlGaAs may be nearly flat on an atomic scale. However, the lateral fluctuations of the metal gates used to form the QPC may easily be on the order of several nanometers due to the lithographic process required to form narrow regions as discussed in Chapter 2. Such boundary roughness also results in backscattering and a reduction in conductance. It is important to realize that in small nanostructures such as QPCs, the total scattering matrix through the structure, including impurities and roughness, depends on the exact location of the impurities and boundary fluctuations. If such inhomogeneous effects are important, the conductance in a sample with

Fig. 5.9 Measured conductance in a quantum point contact structure with a 600-nm gate length ([7], with permission).

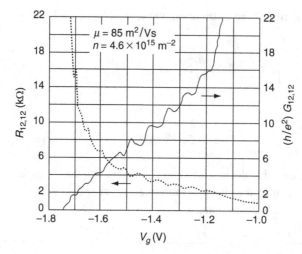

a slightly different impurity configuration will be completely different. This behavior contrasts from that of diffusive transport over very long length scales, in which the contribution of many impurities averages to a constant contribution to the conductance that depends only on their density and not their exact location.

Given that such inhomogeneities in present-day technology exist, their role should become increasingly important as the length of the QPC or waveguide is increased. Figure 5.9 shows the conductance-gate voltage characteristic of a 600-nm long point contact structure reported by Timp *et al.* [3,7]. It is clear that the conductance of this device is far from the ideal behavior shown previously in Fig 5.2. For all plateaus in this long waveguide, the conductance is degraded below $2Ne^2/h$, and some plateaus are even missing altogether. Similar behavior is observed in general by different groups, while the particular length scale where quantized conductance is degraded depends on the exact fabrication technology used.

Nixon *et al.* have calculated the explicit contribution to the conductance in a point contact due to various arrangements of random impurities in the AlGaAs doping layer [21]. They use a semiclassical, self-consistent potential model of the heterostructure including that due to discrete random charges projected onto the plane of the 2DEG. The transmission coefficients are calculated numerically using a coupled mode theory, extended to an arbitrary change of potential. Figure 5.10 shows the metal gate pattern and the calculated potential distribution around the split gates. Case (a) corresponds to the ideal potential in which a uniform positive charge for the ionized donors is assumed with a density of 2.5×10^{12} cm^{-2} located in a delta layer 42 nm above the channel. The ideal potential appears like the smooth saddle-point potential of Fig. 5.3. Cases (b) and (d) show the potential for two different realizations of the random-impurity configuration in the delta-doping plane for a 200-nm long QPC, while (c) is the case of a 600-nm QPC.

5.1 Conductance quantization in quantum point contacts

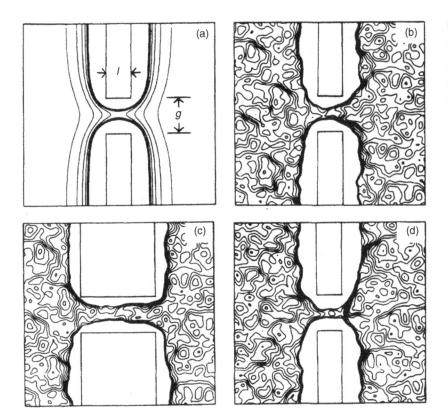

Fig. 5.10 Gate pattern and contour maps of the 2DEG electron density for 300-nm-wide split-gate constrictions with lengths of 200 and 600 nm. (a) A QPC with a smooth positive background impurity distribution; (b)–(d) contours with random arrangements of impurities with an areal density of $2.5 \times 10^{11}/cm^2$ ([21], with permission).

The calculated conductance for the 200-nm case for uniform charge (top curve) and three different impurity arrangements is shown in Fig. 5.11(a). The ideal case shows smooth conductance plateaus without resonances. With the discrete charge potential included, degradation of the plateaus is evident, even in the 200-nm length case. Certain impurity configurations may even give rise to resonant peaks in the conductance. For the 600-nm QPC, the conductance is even more degraded. The calculated results shown in Fig. 5.11(b) predict much more variation from sample to sample than is seen experimentally, which suggests some additional role of screening in the QPC, or three-dimensional effects in reducing the fluctuations. Nevertheless, the calculations demonstrate the important role that impurities play disrupting ballistic transport and the conductance quantization in QPCs.

5.1.5 Nonlinear transport in QPCs

In deriving the quantized conductance of Eq. (5.10), we assumed that the applied source–drain bias was small such that the energy dependence of the density of states could be neglected. When the source–drain bias becomes sufficiently large, however, this assumption is no longer valid and the conductance is expected to

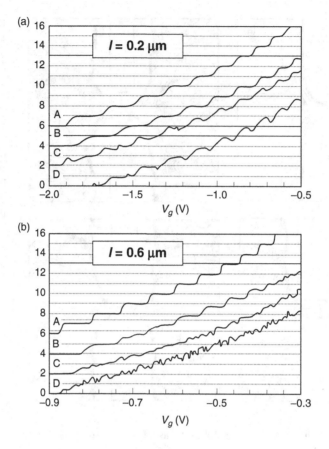

Fig. 5.11 Calculated conductance plateaus for various impurity arrangements for (a) the 200-nm, and (b) 600-nm QPC of Fig. 5.10. The upper curve in each panel corresponds to the smooth impurity configuration ([21], with permission).

deviate from the simple form of Eq. (5.10). In particular, when the voltage becomes greater than the spacing of the quasi-1D subbands of the QPC, different numbers of subbands become available for transport in the forward and reverse directions, giving rise to nonlinear conductance. [22,23].

To extend the Landauer–Büttiker model to the nonlinear regime, we may make use of Eq. (5.33). We should keep in mind that we do not take into account the self-consistent pileup of charge treated in the linear response regime in the Landauer derivation. In the nonlinear regime, we cannot simply say that Eq. (5.33) is the two-terminal simplification of a full self-consistent treatment. Rather, Eq. (5.33) in the nonlinear regime really has more relation to the Tsu–Esaki formula (Eq. (3.85)) in the limit that the transverse directions of motion are fully quantized.

To arrive at an analytical expression for the current–voltage characteristic of a ballistic QPC, consider the 1D model for the potential landscape of the point contact shown in Fig. 5.12. For small V_{SD}, the Fermi energies from the left and right reservoirs inject carriers into the same number of 1D subbands as shown in Fig. 5.12(a). Due to the applied gate bias, electrons transmitted from the left to the right experience an electrostatic barrier ($e\phi_0$). The barrier can be raised or

5.1 Conductance quantization in quantum point contacts

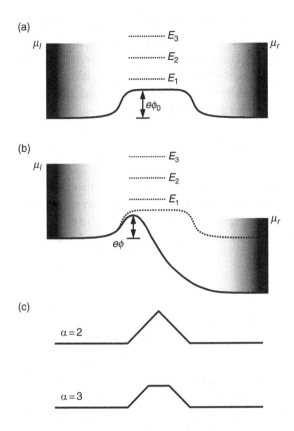

Fig. 5.12 Energy diagram of a quantum point contact with non-zero source–drain bias. (a) The point contact with zero bias. (b) The point contact with non-zero bias. (c) The change is barrier height (ϕ) with applied bias for two simple geometrical cases.

lowered by variation of the gate bias, pushing the 1D levels up through the Fermi energy, which results in conductance plateaus as discussed already. For non-zero V_{SD}, however, the Fermi energy on the right is pulled down with respect to that on the left so that electrons are injected into higher subbands by the left reservoir only (Fig. 5.12(b)). Under these conditions, the barrier height and shape changes with respect to its initial value. We assume here that this change can be characterized by a *barrier-shape parameter* (α). The barrier varies as $\phi = \phi_0 - V_{SD}/\alpha$. Figure 5.12(c) indicates that for a symmetrical triangular barrier as shown, the barrier height changes as $V_{SD}/2$ if one pulls the right horizontal region down by V_{SD} with respect to the left horizontal region. Similarly, for the lower potential profile, the barrier height changes as $V_{SD}/3$.

If we assume low temperatures, such that the Fermi functions may be regarded as step functions, and let the transmission coefficient be given by Eq. (5.22) for a parabolic saddle potential, then in the limit of zero temperature:

$$I = \frac{2e}{h}\sum_n \left[\int_{-\infty}^{\mu_l} dE T_n(E) - \int_{-\infty}^{\mu_r} dE T_n(E) \right], \quad T_n(E) = \frac{1}{1+e^{-\beta_n(E-E_n)}}. \quad (5.37)$$

Here, β_n is the adiabatic parameter of the nth 1D subband in the constriction. We may write E_n as the sum of the electrostatic potential in the constriction plus the confinement energy relative to this maximum, $E_n = e\phi + \varepsilon_n$. We assume that ε_n is measured relative to the maximum of the electrostatic potential (ϕ) and is fixed with respect to this reference under bias arising from either the gate or the source–drain. The Fermi-like functions are easily integrated to give

$$I = \frac{2e}{h}\sum_n \frac{1}{\beta_n}\ln\left[\frac{1+e^{\beta_n(\mu_l - e(\varphi_0 - V_{SD}/\alpha) - \varepsilon_n)}}{1+e^{\beta_n(\mu_r - e(\varphi_0 - V_{SD}/\alpha) - \varepsilon_n)}}\right]. \quad (5.38)$$

If we assume that the transmission coefficient is sharp (i.e. that $T_n(E) = \theta(E - E_n)$, where $\theta(x)$ is the unit step function) but also assume finite temperature, the same result is obtained with $\beta_n = 1/k_B T$. If we take the limit of β_n large (i.e., low temperature and a sharp transmission probability around the subband minimum), Eq. (5.38) simplifies to:

$$I = \frac{2e}{h}\sum_n [(\mu_l - e\varphi_0 - \varepsilon_n + eV_{SD}/\alpha)\vartheta(\mu_l - e\varphi_0 - \varepsilon_n + eV_{SD}/\alpha) - \\ (\mu_r - e\varphi_0 - \varepsilon_n + eV_{SD}/\alpha)\vartheta(\mu_r - e\varphi_0 - \varepsilon_n + eV_{SD}/\alpha)]. \quad (5.39)$$

For subbands in which the Fermi energies of both the left and right reservoirs inject carriers, the step function is non-zero and the terms involving ϕ_0, ε_n, and V_{SD}/α cancel, leaving $\mu_l - \mu_r = eV_{SD}$ and thus a contribution of G_0 to the conduction. However, under non-zero source–drain bias, the upper subbands are unequally populated from the left and right (Fig. 5.12), and the current depends directly on the shape parameter, α. As an example, assume the Fermi level on the left injects into the first subband only, and V_{SD} is such that μ_r is below the subband minimum and does not contribute in Eq. (5.39). Then the current is given by:

$$I = \frac{2e}{h}(\mu_l - e\varphi_0 - \varepsilon_1 + eV_{SD}/\alpha). \quad (5.40)$$

The corresponding conductance is given by

$$G = \frac{2e^2}{h\alpha}. \quad (5.41)$$

Since $\alpha > 1$ by definition, Eq. (5.41) predicts the quantization of the conductance at a value below the fundamental conductance, G_0.

Nonlinear transport in QPCs was first studied by Kouwenhoven *et al.* [23]. Under conditions where the applied voltage is dropped symmetrically across either side of the QPC ($\alpha = 2$), Eq. (5.41) predicts the appearance of so-called "half-plateaus" in the conductance, at e^2/h, $3e^2/h$, etc. The transition from integer to half-integer plateaus was first clearly demonstrated by Patel *et al.* [24,25] and is illustrated in Fig. 5.13. This shows the results of measuring the conductance of a QPC for a series of increasing biases applied across it. Note a systematic

5.1 Conductance quantization in quantum point contacts

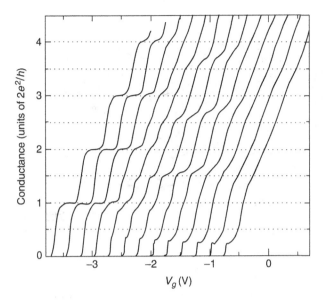

Fig. 5.13 Evolution from integer to half-integer plateaus with increasing d.c. bias in a QPC. The d.c. bias is incremented in steps of 0.5 mV from 0 mV (left) to 6 mV (right). Curves for increasing bias are shifted right in increments of 0.2 V ([25], with permission).

evolution of the conductance, from plateaus quantized in integer units of $2e/h$ to the half plateaus, as the bias is increased. Also note that, as the DC bias is increased, eventually almost all of the structure in the conductance curve is washed out. This behavior can likely be attributed to the increased importance of energy relaxation of hot electrons injected into the QPC.

While the development of the half-plateaus is clearly seen in Fig. 5.13, where $G(V_g)|_{V_{SD}}$ is plotted for a series of different values of V_{SD}, more quantitative information is obtained by constructing a figure in which $G(V_{SD})|_{V_g}$ – the *differential conductance* of the QPC – is instead plotted for a series of closely spaced values of V_g. A fairly typical result of such a measurement is shown in Fig. 5.14, for a QPC that is actually realized by a somewhat different method to the split-gate technique (Fig. 5.1). In this approach, the QPC is first defined by etching a narrow channel in a GaAs 2DEG, after which a global gate is deposited that allows the density and width of the QPC to be varied [26]. In spite of these differences, however, the variation of the differential conductance shown in Fig. 5.14 is quite representative of that found in most experiments. Note in this plot the presence of distinct regions, where the differential conductance curves for different gate voltages are closely bunched. These indicate ranges of quantized conductance, where the conductance does not vary significantly as the gate voltage is changed. Near $V_{SD} = 0$, the bunching is centered on integer values of $2e^2/h$, as we would expect from our discussion of linear transport in these structures. Clear half-plateaus are observed, however, at ∼1.5, ∼2.5, and ∼3.5 × $2e^2/h$, when $1 < |V_{SD}| < 5$ mV. There are some additional features of this graph that are less well understood, however. Note, for example, how instead of a half-plateau at $0.5 \times 2e^2/h$ the feature is instead closer to $0.2 \times 2e^2/h$, and there

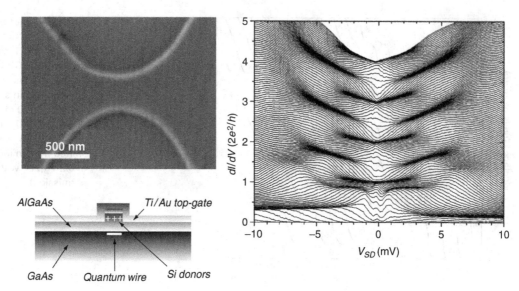

Fig. 5.14 Differential-conductance variation in a QPC with hybrid confinement. The images on the left show the realization of the device while the one on the right plots differential conductance at a series of different (fixed) gate voltages ([26], with permission).

is considerable asymmetry in the curves on reversing the bias polarity. The authors have attributed this effect to a "self-gating" of the device by the applied source–drain bias, which becomes important near pinch-off when the electron density in the QPC is reduced. This effect seems to be quite generic, however, and can also be seen in Fig. 5.13, where there is a clear plateau-like structure that shifts to ∼0.25 × $2e^2/h$ with increasing bias. Another interesting feature of Fig. 5.14, for relatively modest bias ($1 < |V_{SD}| < 3$ mV) is an unexpected bunching of curves near $0.8 \times 2e^2/h$. We shall return to a discussion of the possible origin of this feature shortly below.

An important application of nonlinear transport in QPCs is to the so-called *bias spectroscopy* [22,25] that is widely used as a means to determine the 1D subband separation in these structures. In these measurements, one is interested in the *transconductance* of the QPC – the variation of its conductance as a function of gate voltage – and on the dependence of the transconductance on the source–drain voltage. At $V_{SD} = 0$, the form of the transconductance is easy to understand. Its value is equal to zero over wide ranges of gate voltage that correspond to the plateaus in the linear conductance, and these regions are separated by a series of peaks, each of which correspond to the transition between a specific pair of plateaus. When a non-zero V_{SD} is applied, however, each of these peaks is found to split into two components of similar magnitude. This can basically be understood as arising from the fact that each integer plateau at $V_{SD} = 0$ evolves into two half plateaus, as can be seen by inspection

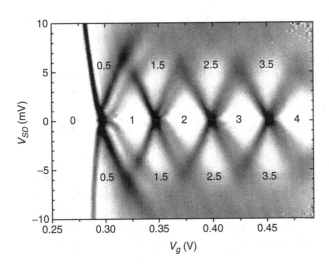

Fig. 5.15 Grayscale plot showing the variation of the transconductance of the device of Fig. 5.14 ([26], with permission).

of Fig. 5.13. The gate-voltage positions of the two peaks are typically found to shift linearly as a function of the source–drain bias, which is shown nicely in the results of Fig. 5.15. This plots transconductance as a grayscale contour, with bright regions corresponding to zero transconductance while the dark regions indicate the peaks associated with the transitions between conductance plateaus. Important to note in the figure is the existence of specific points at which a peak moving up from one integer plateau intersects that moving down from the next integer plateau above it. These points correspond to very specific conditions where the value of the applied source–drain voltage matches the energy spacing between the two uppermost subbands involved in transport. In Fig 5.15, for example, there is a crossing at a gate voltage of 0.32 V, when $V_{SD} = 6$ mV. From this it can be inferred that the energy separation of the first and second 1D subbands of the QPC is therefore 6 meV. Similar estimates may be made from the other crossings in the figure, yielding ∼5 meV, ∼4.5 meV and ∼4 meV, for the separations of the higher subbands, respectively. Another important quantity that may be determined from plots of this type is the gate-voltage *lever arm*, the factor that connects a change of the gate voltage to a corresponding shift of the subband energies within the QPC.

5.2 Non-integer conductance quantization in quantum point contacts

In the preceding section, we mentioned that the conductance quantization of QPCs and the quantization of the Hall resistance in the quantum Hall effect have essentially the same physical origins, since both effects involve the flow of electrical current via one-dimensional channels (or modes). In the case of the quantum Hall effect, the one-dimensional channels are magnetic edge states that are located near the opposite boundaries of the sample, allowing them to travel over macroscopic distances while maintaining their initial electrochemical potential. The one-dimensional modes of QPCs, on the other hand, are purely electrostatic in origin and have a much stronger spatial overlap than the magnetic edge states. The resulting conductance quantization is consequently not as accurate in such structures, typically only one part in 10^2, in comparison to the quantum Hall effect. Nonetheless, it is important to emphasize that the simple noninteracting picture of electron transport that led to Eq. (5.10) is able to describe the main features of the conductance quantization in QPCs. In this section, however, we discuss the results of several important experiments, which present evidence for quantization of the one-dimensional conductance in non-integer units of $2e^2/h$. These experiments have generated a lively debate over the past decade, and seem to demonstrate that a remarkable variety of exotic *correlated* electron behavior may arise in QPCs. This behavior is observed in spite of the fact that QPCs should correspond to one of the simplest of mesoscopic systems.

5.2.1 Spontaneous spin polarization and the 0.7 feature in QPCs

The simple prediction of Eq. (5.10) is that the low-temperature conductance of QPCs should be quantized in integer units of $2e^2/h$ and determined solely by the number of occupied one-dimensional subbands in the QPC. As the gate voltage is made more negative, and the saddle potential at the QPC center rises to approach the Fermi level, the number of occupied subbands should decrease one at a time, giving rise to a corresponding step-like decrease of the conductance. Barring any possible complications, this picture should continue to hold even as the gate voltage causes the last subband to depopulate, at which point we would expect a smooth transition from a conductance of $2e^2/h$ to zero. Numerous experiments have now shown, however, that the behavior found in this pinch-off regime is actually quite different. An unexpected (on the basis of Eq. (5.10)), but nonetheless distinct, additional plateau is typically observed in the conductance, at a value that ranges from ~ 0.5–$0.75 \times 2e^2/h$ (Fig. 5.16). This so-called *0.7 feature* was actually apparent in the earliest QPC studies of van Wees et al. [5], although these authors did not comment on its significance in that work. The pioneering experiments providing insight into the physical significance of this feature were instead performed some time later, by Thomas et al. [6,27]. These authors showed that the 0.7 feature exhibits several unusual characteristics that indicate it may be associated with the enhanced role of electron–electron interactions in the QPCs. By using a back-gate to vary the density of the 2DEG in which the QPCs were realized, for example, they were able to show that the 0.7 feature became more pronounced as the 2DEG density

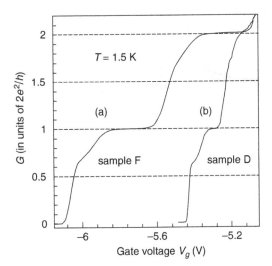

Fig. 5.16 Observation of the 0.7 feature in two different split-gate QPCs ([27], with permission).

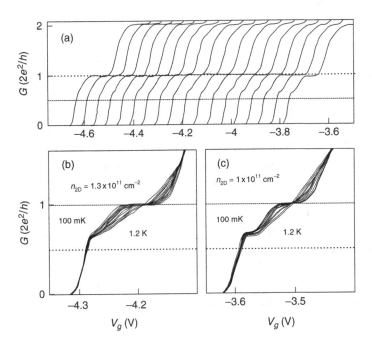

Fig. 5.17 (a) Observation of 0.7 feature as the 2DEG density is varied. In the upper panel, the 2DEG density is reduced from 1.4×10^{11} cm^{-2} to 1.1×10^{11} cm^{-2}, in steps of 1.8×10^{9} cm^{-2}, from left to right respectively. (b) and (c) show in greater detail how the 0.7 feature becomes more clearly resolved with decreasing 2DEG density ([27], with permission).

was decreased (Fig. 5.17). Quite remarkably, they also found that this feature becomes *more* pronounced as the temperature is increased, at least in the range up to ∼10 K, behavior that is opposite to typical mesoscopic phenomena (quantum interference, Coulomb blockade, size effects, etc.) Note, for example, in Figs. 5.17(b) and (c) how the 0.7 feature becomes more clearly resolved by 1.2 K, while the $2e^2/h$ plateau is washed out completely.

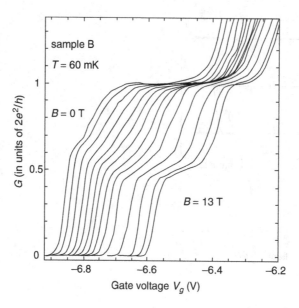

Fig. 5.18 Influence of a large in-plane magnetic field on the 0.7 feature. Magnetic field is stepped from 0 to 13 T in 1-T increments, from left to right, respectively ([6], with permission).

The observation that the 0.7 feature becomes more clearly resolved as the 2DEG density decreases is suggestive of a many-body effect, since one expects quite generally that carrier interactions should become more pronounced under such conditions. The fact that the 0.7 feature is not expected from the simple, noninteracting, derivation that leads to the one-dimensional conductance quantization (Eq. (5.10)), is furthermore supportive that some kind of carrier interaction is needed to account for it. From the early experiments of Thomas *et al.*, it was suggested that the relevant mechanism should somehow involve the electron spin. This conclusion was reached from studies of the influence of a strong magnetic field on the 0.7 feature, which was found to cause the unexpected plateau to evolve smoothly towards $0.5 \times 2e^2/h$ (Fig. 5.18). Since such a conductance value is actually expected for a single one-dimensional subband whose spin degeneracy is broken, this observation lead the authors to suggest that the 0.7 feature may actually be due to some remnant (or spontaneous) spin polarization of the electron gas in the QPC that persists *even at zero magnetic field*.

5.2.1.1 The 0.7 feature – further characteristics

There have been numerous experimental reports on the properties of the 0.7 feature, subsequent to the original work of Thomas *et al.* [28,29,30,31,32, 33,34,35,36,37,38,39,40,41,42]. These works have shown that the 0.7 feature appears to be a generic feature of transport in one-dimensional systems, being manifest in structures realized by physical etching [26], as opposed to electrostatic gating, in 2DEGs without any modulation doping [29,34,37], and in

5.2 Non-integer conductance quantization

material systems other than GaAs [28,40,41,42]. While the original experiments on this phenomenon were performed on electron systems, recent investigations have shown that a similar feature is also found in the conductance of QPCs implemented in two-dimensional hole systems.

The anomalous temperature scaling of the 0.7 feature, first revealed in the work of Thomas et al., was further confirmed in the work of Kristensen et al. [26,31], who undertook a detailed study of this issue. In their work they showed that the deviation of the QPC conductance from the integer conductance quantization follows an activated behavior, consistent with the excitation of carriers across some characteristic energy gap. The magnitude of this gap was shown to be of order 2 K and to evolve as the density in the QPC was varied, a crucial idea that remains central to many of the attempts to explain the 0.7 feature. On the basis of their results, these authors suggested that the 0.7 feature was associated with the scattering of electrons from confined 1D plasmons in the QPC. In later work [26], however, the authors emphasize the connection of the inferred energy gap to a possible breaking of spin degeneracy in the QPC.

Building on the idea that a density-dependent spin gap may open in QPCs, Reilly et al. have developed a phenomenological model that relates the 0.7 feature to the manner in which this gap opens as the QPC gate voltage is varied [37,43]. The relatively simple model that they have developed appears able to account for many of the experimental characteristics of the 0.7 feature, by introducing just one model parameter that determines the rate at which the spin gap opens (linearly) as the gate voltage is varied. The conductance is then calculated by assuming a simple step function of transmission for the two spin subbands as a function of energy and including the effects of non-zero temperature via the Fermi–Dirac distribution function. Numerically calculated results then reveal that a variety of different features may be obtained in the conductance, dependent upon the rate at which the spin gap opens. When the spin gap opens slowly, for example, a clear 0.7 feature is observed, while for a faster rate of opening the feature is closer to $0.5 \times 2e^2/h$. The latter observation appears to be in agreement with experimental results by the same authors from longer QPCs, which have shown a 0.7 feature that actually shifts towards $0.5 \times 2e^2/h$ with increasing density (Fig. 5.19) [34].

As an alternative means to shed light on the origins of the 0.7 feature, Graham et al. have used an innovative approach in which they monitor the evolution of the QPC conductance at very high in-plane magnetic fields, sufficient to cause expected crossings of the opposite Zeeman branches of neighboring one-dimensional subbands [38]. Over those regions of magnetic field where such crossings were expected, the authors in fact found a clear avoided crossing, corresponding to a spontaneous spin splitting, and the appearance of correlated "0.7 analogs" in the conductance. The latter correspond to quasi-plateau features at conductance values near 1.7 and $2.7 \times 2e^2/h$ and also show the unique temperature evolution that is characteristic of the 0.7 feature. The

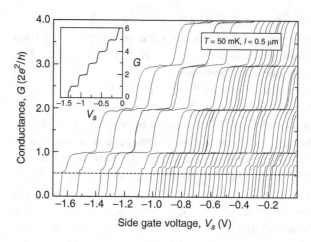

Fig. 5.19 Conductance of a 0.5-μm ballistic quantum wire for increasing density (from right to left, respectively). Note how the 0.7 feature shifts towards $0.5 \times 2e^2/h$ as the density is increased ([34], with permission).

suggestion, therefore, of this work is that the true 0.7 feature is also a manifestation of a spontaneous lifting of spin degeneracy, leading to a true spin gap in the density of states.

A related, but nonetheless distinct, picture of the physical origins of the 0.7 feature has been suggested by the experimental work of Cronenwett *et al.* [36]. These authors argued that the 0.7 feature exhibits a number of characteristics in common with the Kondo effect that occurs in quantum dots [44] (the Kondo effect in dots will be discussed in Chapter 6). This result is at first surprising, since the Kondo effect is well known to involve the interaction of conduction electrons with a localized *spin* (or magnetic moment). For Coulomb-blockaded quantum dots, the source of this localized spin is easy to explain in terms of the net spin of a (typically) odd number of electrons localized on the dot [44]. In QPCs, however, it is not immediately obvious how spin localization can arise (we return to discuss this issue shortly below). Nonetheless, in their experiment, Cronenwett *et al.* showed many features characteristic of the Kondo effect (Fig. 5.20). In the region where the 0.7 feature occurs in the linear conductance, for example, they showed the presence of a zero-bias peak in the differential conductance, as expected for a Kondo system out of equilibrium [45,46]. This zero-bias peak also developed a Zeeman splitting in the presence of an in-plane magnetic field, in good correspondence with the behavior exhibited by the Kondo effect in quantum dots. Finally, it was shown by the authors that the variation of the QPC linear conductance with temperature could be scaled to the universal Kondo form

$$g = \frac{2e^2}{h} \frac{1}{[1/2f(T/T_K) + 1/2]}, \quad (5.42)$$

simply by introducing the appropriate Kondo scaling temperature (T_K). The universal function $f(T/T_K)$ that appears here is given by [47]

5.2 Non-integer conductance quantization

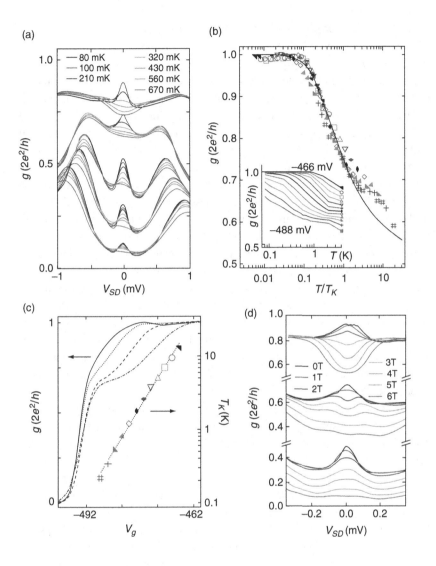

Fig. 5.20 Kondo-like behavior in QPCs. (a) Temperature dependence of the zero-bias anomaly observed near the 0.7 feature. (b) Linear conductance as a function of scaled temperature, T/T_K. (c) Variation of Kondo temperature with gate voltage. (d) Evolution of the zero-bias anomaly with in-plane magnetic field, for different conductance, set by the split-gate voltage ([36], with permission).

$$f \sim \left[1 + (2^{1/s} - 1)\left(\frac{T}{T_K}\right)^2\right]^{-s}, \quad s = 0.22. \quad (5.43)$$

In Fig. 5.20(b), we show data for QPC conductance for a series of temperatures and gate voltages and show how they are collapsed onto a universal curve. Fig. 5.20(c) shows an exponential scaling of the Kondo temperature with gate voltage. Similar behavior has been found for the Kondo effect in quantum dots, and was suggested in Ref. [36] to reflect the ability of the gate voltage to shift the energy of a bound state on the QPC with respect to the Fermi energy in the reservoirs.

5.2.1.2 Microscopic models for the 0.7 feature

The discovery of the 0.7 feature in QPCs invites comparison to discussions of the integer and fractional quantum Hall effects. While the integer QHE was explained fairly quickly after its discovery, in terms of a noninteracting model of independent 1D edge states, the many-body nature of the FQHE state rendered this phenomenon more resistant to clarification. Similarly, for the case of QPCs, the integer conductance quantization has long been well understood, while the 0.7 feature remains the topic of ongoing theoretical investigations. The many-body nature of this effect was first suggested by Thomas *et al.* [27] and remains the central component of all attempts to explain it. By far, the vast majority of these theories ascribe the 0.7 feature to the role of electron–electron interactions, which give rise to spin-dependent transport through the QPC. The details of these models can differ significantly, however, involving: spontaneous lifting of spin degeneracy at low densities in 1D [48]; the formation of two-electron bound states in the QPC [49,50]; a phenomenological assumption of spin-resolved subbands [51]; the formation of charge density waves as a precursor to Wigner crystallization [52]; and scattering from localized spins [53] and from spin fluctuations [54] inside the QPC. A notable exception to these models is provided by the work of Seelig and Matveev, who ascribe the 0.7 feature to the role of electron backscattering from acoustic phonons in the QPC, instead of any spin-dependent interaction [55].

Current consensus regarding the origins of the 0.7 feature appears to be converging on a spin-dependent phenomenon, with two distinct microscopic scenarios of *static ferromagnetism* [56,57,58,59,60] and *dynamic spin polarization* [61,62,63,64]. (A diversion here is provided by the well known Lieb–Mattis theorem, which forbids the formation of a ferromagnetic ground state for a true 1D system [65]. However, the relevance of this mathematical theorem to the *quasi*-1D structures of experiment is far from clear.) Berggren and his colleagues have worked extensively on this problem within the framework of the local spin-density approximation (LSDA) [56–58]. Their work points to the formation of a static local moment in the QPC, which arises when the exchange energy of the carriers due to their Coulomb interaction exceeds their kinetic energy. As these authors have pointed out, the results of the LSDA method are sensitive to the values of the exchange and correlation potentials. While the former is much stronger than the latter, calculations based on exchange alone are found to overestimate the spin splitting and, thus, to give rise to unrealistic conductance variations. The inclusion of correlation into the model weakens the spin polarization resulting from exchange alone, but, at the same time, gives a conductance variation that is more typical of that found in experiment. In the region where the 0.7 feature occurs, these authors found a bifurcation of the LSDA solutions into ground-state and metastable solutions. The conductance of the metastable solutions was found to be lower than that of the ground state, and this fact

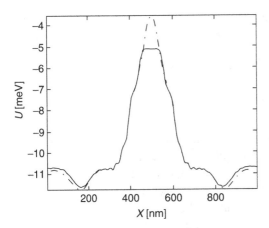

Fig. 5.21 Spin-polarized self-consistent total potentials for spin-up (solid line) and spin-down (dashed line) electrons along the transport axis in the middle of a QPC ([58], with permission).

allowed for a possible explanation of the unusual temperature dependence of the 0.7 feature [58]. In further analysis, these effects on transport were shown to be associated with strong spin-dependent modifications to the self-consistent potential profile of the QPC. This is illustrated in Fig. 5.21, which shows the results of a calculation of the different potentials seen by spin-up and spin-down electrons passing through a QPC. Note for both curves that the basic structure of the potential barrier is what one would expect to observe on moving through the saddle potential of Fig. 5.3. Figure 5.21 indicates, however, that the QPC should now function as a spin filter, with a much stronger transmission probability for spin-up electrons than spin-down. The results of Fig. 5.21 imply that a *net* spin polarization can develop in the QPC, since the local electron density is now spin dependent. It is this spin-density difference that in turn is the source of a static magnetic moment. From a knowledge of the spin-dependent potentials shown in Fig. 5.21, it is possible to estimate the net number of spins that contribute to this magnetic moment. Typically, this number is found to be just a little bit smaller than one (∼0.6).

An alternative picture [61–64] of local-moment formation in QPCs was motivated by the work of Cronenwett *et al.* [36], which revealed a number of similarities between the 0.7 feature and the Kondo effect (see Fig. 5.20). Meir *et al.* [61] provided an explanation of this effect by considering that a correlated many-body state is formed between a localized electron in the QPC and the reservoirs connected to it, resulting in the appearance of a *dynamic* magnetic moment. Although electrons constantly tunnel back and forth between the QPC and the reservoirs, the equivalence of the number of spin-up and spin-down electrons visiting the contact is broken, resulting in a non-zero net spin polarization in the QPC. As such, the key assumption is that a self-consistent modification of the QPC potential leads to the formation of a bound state, making this scenario roughly equivalent to the conventional Kondo effect for a spin localized on an impurity, or in a quantum dot [47]. To describe this problem, the authors used a modified form of the Anderson Hamiltonian [66]:

$$H = \sum_{\sigma;k \in L,R} \varepsilon k_\sigma c^\dagger_{k\sigma} c_{k\sigma} + \sum_\sigma \varepsilon_\sigma \mathbf{d}^\dagger_\sigma \mathbf{d}_\sigma + U \mathbf{n}_\uparrow \mathbf{n}_\downarrow$$
$$+ \sum_{\sigma;k \in L,R} [V^{(1)}_{k\sigma}(1 - \mathbf{n}_{\bar{\sigma}})c^\dagger_{k\sigma}\mathbf{d}_\sigma + V^{(2)}_{k\sigma}\mathbf{n}_{\bar{\sigma}}c^\dagger_{k\sigma}\mathbf{d}_\sigma + \text{H.c.}], \quad (5.44)$$

where $c^\dagger_{k\sigma}$ ($c_{k\sigma}$) creates (destroys) an electron with momentum k and spin σ in the left (L) or right (R) leads of the QPC, $\mathbf{d}^\dagger_\sigma$ ($\mathbf{d}_\sigma$) creates (destroys) a spin-σ electron on the bound state of the QPC ($\mathbf{n}_\sigma \equiv \mathbf{d}^\dagger_\sigma \mathbf{d}_\sigma$). The term U is the on-site Coulomb energy associated with the addition of a second electron to the bound state when a first is already present, while $V^{(1)}_{k\sigma}$ and $V^{(2)}_{k\sigma}$ are hybridization matrix elements for transitions between 0 and 1 electrons, and 1 and 2 electrons, on the site, respectively. Since the presence of one electron that already occupies the site should suppress the tunneling of a second, it is expected that $V^{(2)}_{k\sigma}$ should be less than $V^{(1)}_{k\sigma}$.

To determine the influence of the localized spin on the conductance of the QPC, the authors transformed the Hamiltonian of Eq. (5.44) into a Kondo form and then solved this perturbatively to determine the conductance. In this way they were able to obtain characteristics similar to those found in experiment (see Fig. 5.22), namely an anomaly in the linear conductance near $0.7 \times 2e^2/h$, which moves to $2e^2/h$ with decrease of temperature, a zero-bias anomaly in the differential conductance, and a decrease of the linear conductance towards $0.5 \times 2e^2/h$ with increasing magnetic field. According to this picture, the 0.7 plateau arises from a situation where the $0 \to 1$ valence fluctuations yield a maximum contribution of $0.5 \times 2e^2/h$, while the contribution from the $1 \to 2$ valence fluctuations is affected by the Kondo effect. At higher temperatures, the Kondo correlation is weak and so a feature close to $0.7 \times 2e^2/h$ is obtained. As the temperature is lowered, however, Kondo correlations enhance the transmission associated with the $1 \to 2$ fluctuations so that the conductance eventually rises towards $2e^2/h$, much as is found in experiment.

While the Kondo model appears able to explain many features of the experiments of Cronenwett *et al.*, this model is based on an assumption of localized-state formation in QPCs. Since these are essentially open systems, it is not immediately obvious how such a state might form in them. Light on this issue has been cast be the results of fully self-consistent spin-density-functional calculations, which reveal the formation, under quite general conditions, of an electronic state with a spin-1/2 magnetic moment in QPCs [62,64]. According to these calculations, the calculated QPC potential for one of the spin components can exhibit a double-barrier form, with an associated bound state. This unique form to the spin-dependent potential is basically a manifestation of Friedel oscillations, and is predicted to become more pronounced as the length of the QPC increases.

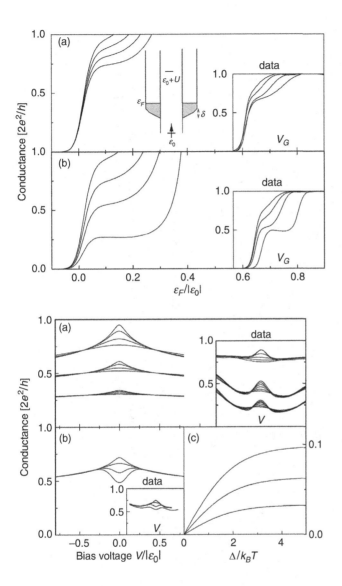

Fig. 5.22 Top: (a) Calculated QPC conductance at different temperatures (increasing from top to bottom curves) as a function of Fermi energy (ε_F). Right inset: experimental conductance of QPC at different temperatures. (b) Conductance at different magnetic fields (increasing from top to bottom curves). Inset: experimental conductance of QPC at different magnetic fields. Bottom: QPC differential conductance from the Kondo model. (a) Differential conductance versus bias for three Fermi energies. Curves for four temperatures are shown at each Fermi energy. Inset: experimental differential conductance. (b) Differential conductance at different magnetic fields (increasing from top to bottom). Inset: experimental differential conductance at different magnetic fields. (c) Spin conductance as a function of magnetic fields, for several values of Fermi energy ([61], with permission).

This behavior is shown in Fig. 5.23, in which it is clear that the QPC serves as a spin filter, much as discussed already in the context of Fig. 5.21.

The existence of self-consistently formed bound states in QPCs remains a subject of controversy in the community. Starikov *et al.* [58] have noted specifically that they do not obtain such bound states in their self-consistent calculations, while Ihnatsenka and Zozoulenko have suggested that the observed features in Refs. [62,62,64] may actually represent some spurious artifact associated with implementing the local spin-density approximation. This controversy only highlights the need for new experimental approaches that probe the

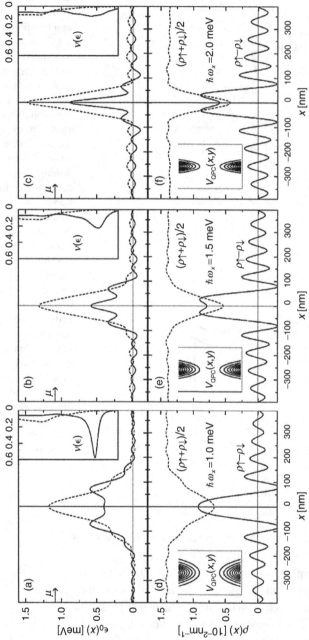

Fig. 5.23 Calculated QPC potentials of increasing sharpness (i.e. decreasing length) from left to right). (a)–(c) Self-consistent barrier (that is the energy of the bottom of the lowest 1D subband) as a function of position in the direction of current flow through the QPC. The chemical potential is indicated by an arrow on the left. Solid lines are for spin-up and dashed lines are for spin-down electrons. Insets: local density of states at the center of the QPC. (d)–(f) 1D electron density in QPC. The solid line gives the net spin-up density and the dashed line gives the spin-averaged density. Insets: contour plot of the QPC potential ([61], with permission).

microscopic origins of the 0.7 feature in other ways than through a simple conductance measurement. In the section that follows, we highlight the results of a number of such recent studies.

5.2.1.3 Alternative experimental approaches to the study of the 0.7 feature

Thus far, we have focused our discussion on the results of experiments that use direct measurements of the linear, or nonlinear, conductance of QPCs to explore the characteristics of the 0.7 feature. Although providing a valuable probe of the electronic system, the information that can be obtained from experiments of this type is limited. To obtain further information on the characteristic properties of the 0.7 feature, different approaches to this problem are instead required. Motivated by this idea, recent studies have used noise measurements [67,68], electron focusing geometries [69], studies of the correlation between the electrical and thermal conductance of QPCs [70], and nonlocal probing with coupled QPCs [71], as a means to further explore the 0.7 feature.

Measurements of shot noise in QPCs have long been used as a means to explore the implications of the Landauer–Büttiker picture of transport [72,73,74]. Shot noise corresponds to time-dependent fluctuations of the electrical current in a conductor, and results from the granular nature of the current. In mesoscopic conductors, the phase coherent nature of electron transport introduces non-classical correlations that cause the value of the shot noise to be suppressed below that expected for a classical conductor [75,76]. In a *ballistic* conductor, in which all 1D subbands are perfectly transmitting, the shot noise is expected to vanish completely, and this has been demonstrated in measurements at the integer conductance plateaus of QPCs [74]. More recently, noise measurements have been extended to investigate the 0.7 feature [67,68]. These studies have revealed a suppression of shot noise, below the value expected for spin-degenerate transport, that occurs in the region near the 0.7 feature. These results have been interpreted as providing support for the presence of two distinct channels for transmission in this regime. Presumably, the two channels arise from a lifting of the spin degeneracy of the lowest 1D subband, a conclusion supported by the finding that, at high magnetic fields, the 0.7 structure evolves towards $0.5 \times 2e^2/h$ and the shot noise vanishes.

In other work, Rokhinson *et al.* [69] have used a focusing geometry to obtain evidence for the spatial separation of the opposite spin branches of a current, injected into a two-dimensional hole gas (2DHG) near the 0.7 feature. Their results are shown in Fig. 5.24, in which the image on the left shows the coupled-QPC geometry that is used to perform the focusing experiment. It is well known that a magnetic field may be applied normal to the plane of such a device to induce curvature of the electron trajectories, and to direct electrons emitted from one QPC into the other. At this condition, the collector current (or voltage) shows a resonant peak, the first of which occurs at ~ 0.2 T in Fig. 5.24. In their

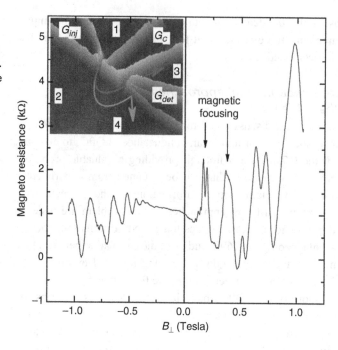

Fig. 5.24 The main panel shows the magnetoresistance of the focusing device shown in the inset. The basic geometry of the focusing experiment is illustrated in the inset. Magnetic focusing peaks are marked with arrows in the main figure. QPC conductance is varied by means of the voltages applied to the various gates (G_{det}, G_{inj}, and G_C) ([77], with permission).

experiment, Rokhinson *et al.* used a two-dimensional hole gas to implement their coupled-QPC system, and showed a complicated evolution of the primary focusing peak as a function of the conductance of the QPC that served as the hole emitter. This behavior is shown in Fig. 5.24, where, as the emitter conductance is reduced from $2e^2/h$ to $\sim 0.7 \times 2e^2/h$, the focusing peak appears to develop a splitting. In earlier work, the authors had suggested that these peaks represent spin-dependent focusing conditions for the holes in the 2DHG, which couple adiabatically to spin-resolved subbands in the injector QPC [77]. Due to the strong spin-orbit coupling in the GaAs 2DHG, these holes with opposite spins have different momenta, and so exhibit focusing at different magnetic fields. In their later work [69], these authors were able to use these ideas to relate the relative amplitude of the spin-dependent focusing peaks to extract a spin polarization for electrons leaving the injector QPC. They found values as large as 40% when the QPC conductance was below the first integer plateau. An important point emphasized by these authors is that their method is sensitive to *static* polarization, suggesting that, in contrast to dynamic polarization, a true ferromagnetic ordering may occur within the QPC.

In a recent striking experiment, Chiatti *et al.* [70] performed correlated measurements of the thermal and electrical conductance of QPCs and showed an unexpected departure from the Wiedemann–Franz law in the region where the 0.7 feature occurs. In these experiments, the current flow through a QPC was used to heat a relatively small number ($\sim$200,000) of electrons, whose resulting

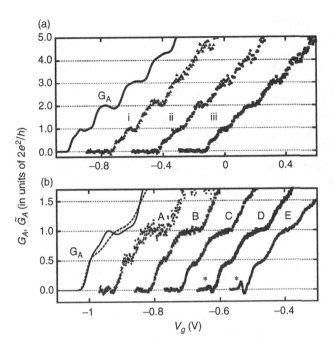

Fig. 5.25 A comparison of the linear conductance of a QPC (solid lines) with the conductance inferred from the results of thermal conductance measurements (discrete data points). Panel (a) shows a wide conductance range, to emphasize the agreement obtained between the two measurements above $2e^2/h$. In panel (b), however, the thermally inferred conductance clearly shows a plateau at $0.5 \times 2e^2/h$, while the directly measured conductance shows the 0.7 feature ([70], with permission).

temperature rise was then detected. In previous studies of this type [78,79,80], effort had focused on understanding the behavior observed under conditions where one or more propagating 1D subbands were occupied in the QPC. In the early experiment of Molenkamp *et al.* [78], the thermal conductance of such QPCs was determined and was found to exhibit a scaling with conductance in accordance with the Wiedemann–Franz law:

$$\frac{\kappa}{GT} = \frac{\pi k_B^2}{3e^2} \equiv L_0, \quad (5.45)$$

where κ is the thermal conductivity of the QPC, G is it conductance, T is the temperature, and L_0 is a constant referred to as the Lorenz number. For QPCs, the implication of this result is that κ should also exhibit a quantized staircase as a function of gate voltage, reflecting the associated quantization of the conductance.

In the recent experiment of Chiatti *et al.* [70], measurements of the QPC thermal conductance were extended to the 0.7 regime where an unexpected breakdown of the Wiedemann–Franz law was found. This behavior is illustrated in Fig. 5.25, which compares the results of direct measurements of the QPC conductance, with that extracted from the results of the thermal conductivity measurements (based on an assumption of the Wiedemann–Franz law). While these measurements show a close correspondence with one or more modes transmitted through the QPC, a clear disagreement was found below the last conductance plateau. Here, the thermally inferred conductance showed a plateau at $\sim 0.5 \times 2e^2/h$, under conditions where the directly measured conductance

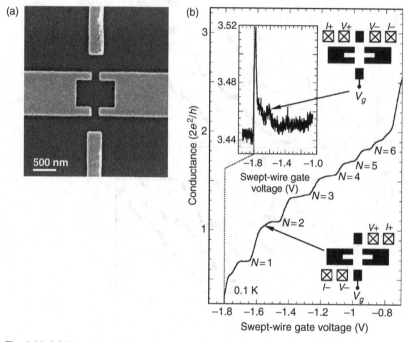

Fig. 5.26 (a) An electron micrograph of a device consisting of two QPCs that are connected via a quantum dot. Panel (b) shows the resonance phenomenon observed in this structure. In this figure, the main panel shows the measured variation of the swept-QPC conductance at 0.1 K, performed under conditions where a fixed voltage is applied to the upper three gates and the voltage applied to the lower gate is varied (as indicated by the device schematic). The inset of the right panel shows the variation of the conductance of the detector QPC for the same gate-voltage conditions as in the main panel. The measurement configuration for this experiment is also indicated schematically (for further details, see [71]).

showed the 0.7 feature (Fig. 5.25). In their discussions of the possible origins of this unexpected result, the authors pointed to a recent theoretical study by Matveev [81], who considered a model of a 1D Wigner crystal with separate charge and spin degrees of freedom. According to this model, a quantized thermal conductance of $\kappa/L_0 T = e^2/h$ is expected, just as is found in the experiment. As such, these experiments should therefore provide a valuable framework for the future development of a successful theory of the 0.7 feature.

In yet another approach to the study of the 0.7 feature, Morimoto *et al.* used a device consisting of a pair of coupled QPCs to provide evidence for electrical detection of local-magnetic-moment formation (Fig. 5.26) [71]. In these experiments, one QPC (the "detector") was used to monitor the pinch-off of the other (the "swept QPC"), and a resonance was systematically observed in the conductance of the detector as the swept QPC pinched off (Fig. 5.26). Puller *et al.* provided an interpretation of these results, in the form of a phenomenological model that starts from an a priori assumption that the resonance is associated with the formation of some self-consistently driven bound state on the swept QPC [82]. By modeling this state with the Anderson Hamiltonian, and including terms representing the coupling between the bound state and the detector QPC, these authors were able to calculate a resonantly enhanced correction to the detector conductance, reminiscent of that found in experiment. In subsequent experimental work, coupled QPCs were used to demonstrate that the localized spin that occupies the bound state may be robustly (several meV) confined, and clear evidence was presented for its Zeeman shift in a magnetic field [83]. Evidence for coherently interacting spins localized on separate QPCs has even been obtained in most recent experiments [84].

5.2.2 Non-integer conductance plateaus in ballistic transport

The 0.7 feature is not the only fascinating conductance anomaly exhibited by high-quality 1D conductors. There have been numerous reports of conductance plateaus at non-integer values of $2e^2/h$, which have, in turn, been ascribed to a variety of different mechanisms. Tarucha et al. studied the conductance characteristics of long ballistic quantum wires and found a renormalization of the lowest conductance plateau to a value below $2e^2/h$ at low (mK) temperatures [85]. These authors interpreted their data in terms of theoretical predictions for the formation of a Tomonaga–Luttinger liquid, a novel many-body state unique to one-dimensional systems. A related phenomenon was observed by Yacoby et al. [86], who measured the conductance of atomically perfect quantum wires of various length, realized by the technique of cleaved-edge overgrowth (which was discussed already in Chapter 2). Their experiments revealed a remarkable phenomenon, namely a *systematic* rescaling of *all* conductance plateaus by a factor ($f < 1$) whose value decreased with increasing wire length. In the longest wires (20 μm) studied, this factor approached a value $f \sim 0.5$, while conventional integer quantization was recovered on increasing the temperature to above a degree Kelvin. These results, also, were discussed in terms of Luttinger-liquid formation, although inconsistencies with such an interpretation were also noted.

Ogata and Fukuyama were the first to calculate the corrections to the conductance of a single-mode quantum wire due to Luttinger-liquid formation [87]. They calculated these corrections in the presence of both mutual Coulomb interaction among electrons and impurity scattering (the so-called dirty Luttinger liquid). In wires much longer than the mean free path, they predicted a reduction of the quantized conductance below the value observed in experiment, much like that subsequently observed by Tarucha et al. [85]. Later theoretical work by Maslov [88] and Oreg and Finkel'stein [89] provided further clarification of this issue. The latter authors demonstrated, in particular, that electron–electron interactions do *not* modify the conductance of a *clean* 1D wire. This result was explained by the fact that, while electron–electron interactions *do* renormalize the current in such a conductor, they also induce a modification of the electric field, and these effects cancel each other exactly when calculating the resulting conductance. In a system that is subject also to impurity scattering, however, the situation becomes more complicated and a renormalization of the conductance is expected. Recently, Levy *et al.* [90] studied the temperature dependence of the conductance of clean quantum wires formed by epitaxial growth on V-groove substrates (Chapter 2) and demonstrated behavior consistent with the predictions of Refs. [88,89]. As the temperature was lowered below $\sim$4.2 K, these authors found a suppression of the lowest conductance plateau below $2e^2/h$, with a variation that could be described as

$$G(T) = \frac{2gG'(T)}{(h/e^2)(g-1)G'(T) + 2g}, \quad G'(T) \equiv \frac{2e^2}{h} g \left[1 - \left[\frac{T}{T_0}\right]^{g-1} \right], \quad (5.46)$$

where $g < 1$ is a dimensionless parameter that is a measure of the strength of the electron interactions and T_0 is a parameter describing the strength of the disorder. Good agreement with this form was obtained for values of $g \sim 0.6$, consistent with the results of experiments that have used cleaved-edge wires to perform tunneling spectroscopy of the Luttinger state [91], and for values of T_0 less than 2 mK. The results of these experiments suggested a strong sensitivity to disorder, however, with wires with T_0 larger than $\sim$2 mK showing a temperature variation poorly described by Eq. (5.46). Levy *et al.* have suggested that this failure might be due to a breakdown of the validity of the perturbative methods used to calculate Eq. (5.46), in the presence of stronger disorder [90].

While we have thus far focused on many-body interactions as the source of departures from the integer conductance quantization in 1D systems, other mechanisms should also be considered. In our original discussion of the quantization in QPCs, we emphasized that an adiabatic transition between the constriction and the reservoirs is needed to ensure that inter-subband scattering does not arise from the coupling between these different regions. In QPCs realized by the split-gate technique, the soft confining potential induced by the application of the gate bias typically ensures that this coupling is achieved. In wires formed by epitaxial-growth techniques, however, the much more abrupt transitions between the 1D and 2D regions allow the effects of non-adiabatic coupling to be explored. The most detailed work on this issue has been performed by Kaufman *et al.* [92,93], who measured the conductance of narrow GaAs/AlGaAs quantum wires grown on V-groove substrates. By gating a portion of the wires, the authors were able to observe a staircase-like structure to the conductance, with broad plateaus that became increasingly suppressed below the integer values with increasing gate length (Fig. 5.27). Similar to the results of Yacoby *et al.* [86], for any given conductance curve in this figure, all plateaus appear to be shifted by roughly the same factor. They proposed that this result could be explained in terms of additional scattering of electrons at the

5.2 Non-integer conductance quantization

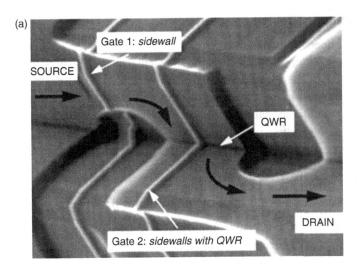

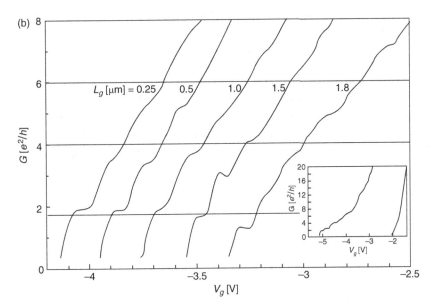

Fig. 5.27 (a) A scanning-electron micrograph of one of the gated V-groove wires studied by Kaufman et al. The wire is formed at the corner of the two sidewalls and the gate is also indicated. Panel (b) shows measured conductance characteristics at 4.2 K for wires with various gate lengths (L_g, indicated) ([92], with permission).

transition between the 1D and 2D regions of the device. Although this transition is an abrupt one structurally, the authors suggested the existence of a non-zero-length transition region over which inter-mode scattering takes place. Since this scattering represents an additional contribution to the resistance, beyond the quantum contact resistance responsible for the integer conductance quantization, it results in a suppression of the conductance steps below $2e^2/h$. Kaufman et al. studied the factors that govern the size of this transition region, which they estimated to be as much as several microns long, and found that its length increased with the strength of the confining potential of the wire.

5.3 Some ballistic device concepts

As semiconductor dimensions continue to shrink in size, an eventual end to the well-known semiconductor roadmap is expected that will prevent further scaling of CMOS technology. Alternative technologies are therefore desired that will allow continued increase in the density of memory and logic into the terabit regime. Devices that utilize the ballistic transport of carriers may be attractive for this purpose, and in this section we therefore briefly discuss some approaches to functional electronic devices that are based on ballistic transport phenomena.

5.3.1 The Y-branch switch and other ballistic junctions

The Y-branch device is a three-terminal structure, which is formed by the intersection of three narrow quantum wires at a common, nanoscale, junction (Fig. 5.28). [94,95,96]. The basic idea of this device is to deflect ballistically moving electrons incident from the center wire (the *stem*), into either the left or right wire, via the application of a suitable electric field. An important advantage of the Y-branch is that, since its switching is achieved by deflecting electrons from one branch into another, it is not necessary to completely stop the incident carriers at some barrier, as is more typical of conventional devices. In principle, at least, it should therefore be possible to use this device to achieve high-frequency switching operation. Indeed, several theoretical studies [97,98] have predicted a number of novel applications that could arise by exploiting ballistic transport in these structures. Since the ballistic transport is only required over

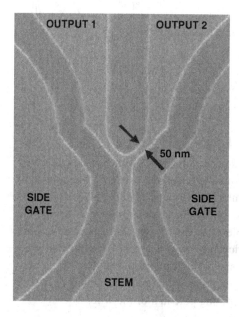

Fig. 5.28 Scanning electron image of a Y-branch device formed by etching of a GaAs/AlGaAs heterostructure (adapted from [99], with permission).

the length of the nanoscale junction in order to achieve these applications, the Y-branch can operate at much higher temperatures than other mesoscopic devices. This topic has recently been nicely reviewed in an extensive report by Worschech *et al.* [99].

Current switching in the Y-branch device may be achieved via two different approaches, the first of which involves making use of external gates in close proximity to the device, and applying a push-pull potential to these to switch current injected from the stem into the left or right outputs. While this allows current to be switched between the two waveguides [95,100], of more interest is the *self-gating* effect that can occur in these devices when applying a nonlinear source–drain voltage, instead of a gate voltage [96,97]. In this approach, the voltage of the floating stem is measured, while fixed voltages ($V_R = -V_L$) are applied to the left and right terminals. For a Y-branch in which electron motion is diffusive, the stem voltage V_s should simply correspond to the *average* of the terminal voltages V_R and V_L. By modeling the Y-branch as a ballistic cavity that is adiabatically coupled to three waveguides [97], however, it may be shown that the stem tends to follow the higher electrochemical potential of the two reservoirs. With push-pull voltages ($V_R = -V_L$) applied to the left and right waveguides, the voltage of the stem will therefore always be *negative*, rather than taking the classically expected value of zero. This is a consequence of the ballistic motion of carriers in the junction region, in the presence of which it may be shown that [97]

$$V_s = -\frac{1}{2}\alpha V^2 + O(V^4), \quad V = -V_R = V_L. \quad (5.47)$$

In this equation, α is a positive constant that varies with temperature, the reservoir chemical potential, and the center-waveguide conductance. The parabolic variation of V_s predicted by Eq. (5.47), has been observed experimentally by several different groups [101,102,103]. As we noted above, the only requirement for the observation of this effect is that electron transport in the junction be ballistic, as a result of which clear signatures of the non-classical voltage variation, predicted in Eq. (5.47), have been observed at room temperature [101,103,104] (Fig. 5.29). The non-classical response of the Y-branch switch has also been predicted [97,98] to allow for rectification and basic transistor action, second-harmonic generation, and logic operation. In particular, with biases applied to left and right reservoirs, the output voltage of the center waveguide will only be positive when a positive voltage is applied to *both* the left and right reservoirs, indicating that the Y-branch may be used as a compact AND gate. With this motivation, there has been much interest in the development of novel circuit architectures, based upon the properties of the Y-branch [105].

Nonlinear, ballistic electron transport can also give rise to usable effects in devices that incorporate deliberately introduced asymmetries in their structure. One of the most widely studied examples of such a device consists of a junction

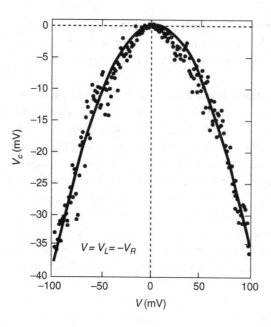

Fig. 5.29 The self-gating action of the Y-branch at room temperature. In this figure, V_c refers to the stem voltage, the sign of which is always negative, regardless of the sign of the net voltage applied between the two inputs (adapted from [103], with permission).

formed from two electron waveguides, which features a triangular antidot at its center [106] (Fig. 5.30). In order to discuss the characteristics of this device, we recall that, for current flow in the linear regime, the Landauer–Büttiker formalism predicts that the measured resistances should obey the following reciprocity relation [107] (recall Chapter 3.5)

$$R_{SD,LU}(I_{SD}) = R_{SD,LU}(-I_{SD}). \quad (5.48)$$

In this equation, $R_{SD,LU}$ corresponds to the value of the resistance measured by passing current between probes S and D and measuring voltage between probes U and L. In the results of Fig. 5.30, however, regardless of *the sign of the current*, the reciprocity relation of Eq. (5.48) is violated since the measured voltage is always negative. This behavior can easily be accounted for by considering the symmetry-breaking properties of the scatterer located at the junction center. Regardless of the direction in which carriers are incident on this (from the source or the drain), the resulting scattering that they undergo directs them into the lower contact, causing V_{LU} to always be negative. While it is surprising that such behavior is not expected from Eq. (5.48), the reason for this is that the Landauer–Büttiker approach neglects the current dependence of the subband transmission probabilities. Song has shown, however, how to extend this formalism to the nonlinear regime, and has demonstrated that it yields predictions that are consistent with the results of Fig. 5.30 [108]. (The other noteworthy feature of Fig. 5.30 is the asymmetry of the curves around $I_{SD} = 0$. This has been attributed [106] to irregularities in the antidot geometry that are introduced unintentionally during device fabrication.)

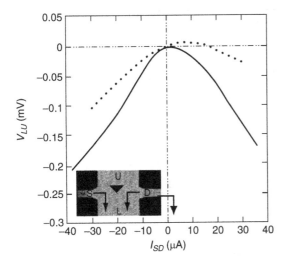

Fig. 5.30 Rectification in a cross junction that features a triangular antidot at its center (inset) (adapted from [106], with permission).

The behavior shown in Fig. 5.30 is reminiscent of that exhibited by a bridge rectifier. Because of the ballistic nature of the phenomenon that gives rise to this behavior, it has been suggested [109] that the rectification should persist to very high frequencies, and operation at 50 GHz has actually been demonstrated in a modified structure consisting of two-dimensional planar arrays of triangular antidots [110]. For the same reasons, this rectifier action can also be seen at room temperature [110].

5.3.2 Interaction of surface acoustic waves with QPCs

Although we have focused thus far on the quantization of the conductance that can be observed in the linear conductance of QPCs, in response to the application of a d.c. source–drain bias, further functionality may be obtained in these structures by utilizing their interaction with surface acoustic waves (SAWs). SAWs are easily excited in piezoelectric crystals such as GaAs, typically by placing pairs of interdigitated gates, with a geometry comprised of interlocking fingers, on top of its surface. When a microwave signal is applied between the transducers gates, SAWs can be resonantly excited at a well-defined frequency, where the SAW wavelength is commensurate with the spacing between the transducer fingers. When the finger separation is in the micron range, the resonant frequency is typically of order a few GHz. By placing pairs of such transducers at the opposite ends of a sample, SAWs may be induced to propagate through it. The strong piezoelectric fields that accompany the propagating wave induce a traveling modification of the conduction band edge, which has previously been used to induce the dissociation of optically generated electron–hole pairs (excitons), thereby allowing for acoustically controlled storage of light [111].

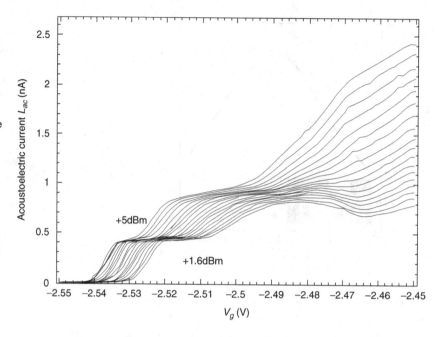

Fig. 5.31 Quantization of the acoustoelectric current in a QPC excited by a SAW at 2.7284 GHz. The different curves correspond to increasing transducer power and so increasing SAW amplitude (from [112], with permission).

When a traveling SAW passes through a QPC, its resulting lateral confinement results in the formation of a moving quantum dot. Due to the Coulomb-blockade effect (see Chapter 6), only a small number of electrons occupy this dot, and this number (N_{SAW}) may be controlled by variation of the QPC gate voltage and the transducer power [112,113]. Since each such dot passes through the QPC in a time $1/f_{SAW}$, where f_{SAW} is the transducer frequency, the possibility should exist to realize a current turnstile, with a quantized current $I = eN_{SAW}f_{SAW}$. The earliest demonstration of this effect was provided by Shilton et al. [112], who demonstrated the observation of a quantized *acoustoelectric current* in QPCs subject to SAW excitation (Fig. 5.31). In this experiment, the (effectively) d.c. current flowing through the QPC is measured as a function of the QPC gate voltage and discrete current steps of ∼0.4 nA can be seen. This quantized increment corresponds well to $ef_{SAW} = 0.44$ nA, implying it is associated with the pumping of single electrons through the QPC by the SAW.

The current-pumping action illustrated in Fig. 5.31 has subsequently been confirmed in several different experiments [114,115,116,117,118]. Much of the interest in this approach has been motivated by theoretical proposals to use SAW pumping of QPCs as a route to the realization of a single-photon source [119], and of a qubit for use in quantum computing [120]. SAW pumping has also been used to control the current flow through other nanostructures than QPCs. There have been several reports that have investigated SAW pumping of split-gate quantum dots, in which the gating ensures that a quantum dot is present even in the absence of SAW excitation [121,122,123]. Through appropriate design of the SAW transducers, the authors excited the quantum dot with SAWs with a wavelength twice the size of the quantum dot [112]. In this way, they were able to achieve a turnstile-like pumping of charge through the quantum dot, with N_{SAW} being determined by the number of discrete quantized states of the dot that are brought into resonance with the electrochemical potential of the reservoirs over the pumping cycle. SAWs have also been used to modify the operation of Y-branch structures [124]. In such an approach, the SAW is used to transport few-electron puddles from the stem of the device to its branching point, where the charge is then divided between the two branches. By combining the charge quantization of the SAW method with the high-speed operation of the ballistic Y-branch, such structures should, in principle, allow for the rapid manipulation of single electrons. It has therefore been suggested that these structures could find application as a fast electrometer with sensitivity at the sub-electron level [124].

References

[1] T.J. Thornton, M. Pepper, H. Ahmed, D. Andrews, and G.J. Davies, *Phys. Rev. Lett.* **56**, 1198 (1986).

[2] Yu.V. Sharvin, *Sov. Phys. JETP* **21**, 655 (1965); Yu.V. Sharvin and N.I. Bogatina, *Sov. Phys. JETP* **29**, 419 (1969).

[3] G. Timp, in *Nanostructure Physics and Fabrication*, eds. W.P. Kirk and M. Reed (New York, Academic Press, 1989), pp. 331–46.

[4] D.A. Wharam, T.J. Thornton, R. Newbury, *et al.*, *J. Phys. C: Solid State Physics* **21**, L209 (1988).

[5] B.J. van Wees, H. van Houten, C.W.J. Beenakker, *et al.*, *Phys. Rev. Lett.* **60**, 848 (1988).

[6] K.J. Thomas, J.T. Nicholls, M.Y. Simmons, *et al.*, *Phys. Rev. Lett.* **77**, 135 (1996).

[7] G. Timp, in *Semiconductors and Semimetals*, eds. M.A. Reed (New York, Academic Press, 1992), pp. 113–90.

[8] M. Büttiker, *Phys. Rev. B* **41**, 7906 (1990).

[9] H. van Houten, C.W.J. Beenakker, and B.J. van Wees, in *Semiconductors and Semimetals*, ed. M.A. Reed (New York, Academic Press, 1992), pp. 9–112.

[10] C.W.J. Beenakker and H. van Houten, *Solid State Physics* **44**, 1 (1992).

[11] Y. Imry, in *Directions in Condensed Matter Physics*, eds. G. Grinstein and G. Mazenko (Singapore, World Scientific Press, 1986).

[12] L.I. Glazman, G.B. Lesovik, D.E. Khmelnitsky, and R.I. Shekhter, *Pis'ma Z. Eksp. Teor. Fiz.* **48**, 218 (1988) [*JETP Lett.* **48**, 238 (1988)].

[13] A. Szafer and A.D. Stone, *Phys. Rev. Lett.* **62**, 300 (1989).

[14] H.A. Fertig and B.I. Halperin, *Phys. Rev. B* **36**, 7969 (1987).

[15] D.A. Wharam, T.J. Thornton, R. Newbury, *et al.*, *J. Phys. C* **21**, L209 (1988).

[16] B.J. van Wees, L.P. Kouwenhoven, H. van Houten, *et al.*, *Phys. Rev. B* **38**, 3625 (1988).

[17] K.-F. Berggren, T.J. Thornton, D.J. Newson, and M. Pepper, *Phys. Rev. Lett.* **57**, 1769 (1986).

[18] M. Büttiker, *Phys. Rev. B* **41**, 7906 (1990).

[19] H.A. Fertig and B.I. Halperin, *Phys. Rev.* **36**, 7969 (1987).

[20] B.J. van Wees, L.P. Kouwenhoven, E.M.M. Willems, et al., Phys. Rev. B **43**, 12431 (1991).
[21] J.A. Nixon, J.H. Davies, and H.U. Baranger, Superlatt. Microstruct. **9**, 187 (1991).
[22] L.I. Glazman and A.V. Khaetskii, Europhys. Lett. **9**, 263 (1989).
[23] L.P. Kouwenhoven, B.J. van Wees, C.J.P.M. Harmans, et al., Phys. Rev. B **39**, 8040 (1989).
[24] N.K. Patel, L. Martin-Moreno, M. Pepper, et al., J. Phys.: Condens. Matter **2** 7247 (1990).
[25] N.K. Patel, J.T. Nicholls, L. Martin-Moreno, et al., Phys. Rev. B **44**, 13549 (1991).
[26] A. Kristensen, H. Bruus, A.E. Hansen, et al., Phys. Rev. B **62**, 10950 (2000).
[27] K.J. Thomas, J.T. Nicholls, N.J. Appleyard, et al., Phys. Rev. B **58**, 4846 (1998).
[28] P. Ramvall, N. Carlsson, I. Maximov, et al., Appl. Phys. Lett. **71**, 918 (1997).
[29] B.E. Kane, G.R. Facer, A.S. Dzurak, et al., Appl. Phys. Lett. **72**, 3506 (1998).
[30] C.T. Liang, M.Y. Simmons, C.G. Smith, et al., Phys. Rev. B **60**, 10687 (1999).
[31] A. Kristensen, P.E. Lindelof, J.B. Jensen, et al., Physica B **249–251**, 180 (2000).
[32] K.J. Thomas, J.T. Nicholls, M. Pepper, et al., Phys. Rev. B **61**, R13365 (2000).
[33] K.S. Pyshkin, C.J.B. Ford, R.H. Harrell, et al., Phys. Rev. B **62**, 15842 (2000).
[34] D.J. Reilly, G.R. Facer, A.S. Dzurak, et al., Phys. Rev. B **63**, R121311 (2001).
[35] R. Wirtz, R. Newbury, J.T. Nicholls, et al., Phys. Rev. B **65**, 233316 (2002).
[36] S.M. Cronenwett, H.J. Lynch, D. Goldhaber-Gordon, et al., Phys. Rev. Lett. **88**, 226805 (2002).
[37] D.J. Reilly, T.M. Buehler, J.L. O'Brien, et al., Phys. Rev. Lett. **89**, 246801 (2002).
[38] A.C. Graham, K.J. Thomas, M. Pepper, et al., Phys. Rev. Lett. **91**, 136404 (2003).
[39] R. Danneau, O. Klochan, W.R. Clarke, et al., Phys. Rev. Lett. **97**, 026403 (2006).
[40] H.T. Chou, S. Lüscher, D. Goldhaber-Gordon, et al., Appl. Phys. Lett. **86**, 073108 (2005).
[41] G. Scappucci, L. Di Gaspare, E. Giovine, et al., Phys. Rev. B **74**, 035321 (2006).
[42] O. Gunawan, B. Habib, E.B. de Poortere, and M. Shayegan, Phys. Rev. B **74**, 155436 (2006).
[43] D.J. Reilly, Phys. Rev. B **72**, 033309 (2005).
[44] D. Goldhaber-Gordon, H. Shtrikman, D. Abush-Magder, U. Meirav, and M.A. Kastner, Nature **391** 156 (1998).
[45] Y. Meir, N.S. Wingreen, and P.A. Lee, Phys. Rev. Lett. **70**, 2601 (1993).
[46] Y. Meir and N.S. Wingreen, Phys. Rev. B **49**, 11040 (1994).
[47] D. Goldhaber-Gordon, J. Göres, M.A. Kastner, et al., Phys. Rev. Lett. **81**, 5225 (1998).
[48] A. Gold and L. Calmels, Philos. Mag. Lett. **74**, 33 (1996).
[49] V.V. Flambaum and M.Yu. Kuchiev, Phys. Rev. B **61**, R7869 (2000).
[50] T. Rejec, A. Ramsak, and J.H. Jefferson, J. Phys.: Condens. Matter **12**, L233 (2000); T. Rejec, A. Ramsak, and J.H. Jefferson, Phys. Rev. B **62**, 12985 (2001).
[51] H. Bruus, V.V. Cheianov, and K. Flensberg, Physica E **10**, 97 (2001).
[52] O.P. Sushkov, Phys. Rev. B **64**, 155319 (2003).
[53] Y. Tokura and A. Khaetskii, Physica E **12**, 711 (2002).
[54] I.A. Shelykh, N.G. Galkin, and N.T. Bagraev, Phys. Rev. B **74**, 085322 (2006).
[55] G. Seelig and K.A. Matveev, Phys. Rev. Lett. **90**, 176804 (2003).
[56] C.-K. Wang and K.-F. Berggren, Phys. Rev. B **57**, 4552 (1998).
[57] K.-F. Berggren and I.I. Yakimenko, Phys. Rev. B **66**, 085323 (2002).
[58] A.A. Starikov, I.I. Yakimenko, and K.-F. Berggren, Phys. Rev. B **67**, 235319 (2003); P. Jaksch, I. Yakimenko, and K.-F. Berggren, Phys. Rev. B **74**, 235320 (2006).

[59] B. Spivak and F. Zhou, *Phys. Rev. B* **61**, 16730 (2000).
[60] A.D. Klironomos, J.S. Meyer, and K.A. Matveev, *Europhys. Lett.* **74**, 679 (2006).
[61] Y. Meir, K. Hirose, and N.S. Wingreen, *Phys. Rev. Lett.* **89**, 196802 (2002).
[62] K. Hirose, Y. Meir, and N.S. Wingreen, *Phys. Rev. Lett.* **90**, 026804 (2003).
[63] P.S. Cornaglia and C.A. Balseiro, *Europhys. Lett.* **67**, 634 (2004).
[64] T. Rejec and Y. Meir, *Nature* **442**, 900 (2006).
[65] E. Lieb and D. Mattis, *Phys. Rev.* **125**, 164 (1962).
[66] P.W. Anderson, *Phys. Rev.* **124**, 41 (1961).
[67] P. Roche, J. Ségala, D.C. Glattli, *et al.*, *Phys. Rev. Lett.* **93**, 116602 (2004).
[68] L. DiCarlo, Y. Zhang, D.T. McClure, *et al.*, *Phys. Rev. Lett.* **97**, 036810 (2006).
[69] L.P. Rokhinson, L.N. Pfeiffer, and K.W. West, *Phys. Rev. Lett.* **96**, 156602 (2006).
[70] O. Chiatti, J.T. Nicholls, Y.Y. Proskuryakov, *et al.*, *Phys. Rev. Lett.* **97**, 056601 (2006).
[71] T. Morimoto, Y. Iwase, N. Aoki, *et al.*, *Appl. Phys. Lett.* **82**, 3952 (2003).
[72] M. Reznikov, M. Heiblum, H. Shtrikman, and D. Mahalu, *Phys. Rev. Lett.* **75**, 3340 (1995).
[73] A. Kumar, L. Saminadayar, D.C. Glattli, Y. Jin, and B. Etienne, *Phys. Rev. Lett.* **76**, 2778 (1996).
[74] R.C. Liu, B. Odom, Y. Yamamoto, and S. Tarucha, *Nature* **391**, 263 (1998).
[75] G.B. Lesovik, *JETP Lett.* **49**, 592 (1989).
[76] M. Büttiker, *Phys. Rev. Lett.* **65**, 2901 (1990).
[77] L.P. Rokhinson, V. Larkina, Y.B. Lyanda-Geller, L.N. Pfeiffer, and K.W. West, *Phys. Rev. Lett.* **93**, 146601 (2004).
[78] L.W. Molenkamp, H. van Houten, C.W. Beenakker, R. Eppenga, and C.T. Foxon, *Phys. Rev. Lett.* **65**, 1052 (1990).
[79] A.S. Dzurak, C.G. Smith, L. Martin-Moreno, *et al.*, *J. Phys.: Condens. Matter* **5**, 8055 (1993).
[80] N.J. Appleyard, J.T. Nicholls, M.Y. Simmons, W.R. Tribe, and M. Pepper, *Phys. Rev. Lett.* **81**, 3491 (1998).
[81] K.A. Matveev, *Phys. Rev. Lett.* **92**, 106801 (2004).
[82] V.I. Puller, L.G. Mourokh, A. Shailos, and J.P. Bird, *Phys. Rev. Lett.* **92**, 96802 (2004).
[83] Y. Yoon, L. Mourokh, T. Morimoto, *et al.*, *Phys. Rev. Lett.* **99**,136805 (2007).
[84] Y. Yoon, T. Morimoto, L. Mourokh, *et al.*, *J. Phys.: Cond. Matter* **20**, 164216 (2008).
[85] S. Tarucha, T. Honda, and T. Saku, *Sol. St. Comm.* **94**, 413 (1995).
[86] A. Yacoby, H.L. Störmer, N.S. Wingreen, *et al.*, *Phys. Rev. Lett.* **77**, 4612 (1996).
[87] M. Ogata and H. Fukuyama, *Phys. Rev. Lett.* **73**, 468 (1994).
[88] D.L. Maslov, *Phys. Rev. B* **52**, R14368 (1995).
[89] Y. Oreg and A.M. Finkel'stein, *Phys. Rev. B* **54**, R14265 (1996).
[90] E. Levy, A. Tsukernik, M. Karpovski, *et al.*, *Phys. Rev. Lett.* **97**, 196802 (2006).
[91] O.M. Auslaender, A. Yacoby, R. de Picciotto, *et al.*, *Phys. Rev. Lett.* **84**, 1764 (2000).
[92] D. Kaufman, Y. Berk, B. Dwir, *et al.*, *Phys. Rev. B* **69**, R10433 (1999).
[93] D. Kaufman, B. Dwir, A. Rudra, *et al.*, *Physica E* **7**, 756 (2000).
[94] T. Palm and L. Thylén, *Appl. Phys. Lett.* **60**, 237 (1992).
[95] T. Palm, *Phys. Rev. B* **52**, 13773 (1995).
[96] J.-O. J. Wesström, *Phys. Rev. Lett.* **82**, 2564 (1999).
[97] H.Q. Xu, *Appl. Phys. Lett.* **78**, 2064 (2001).
[98] H.Q. Xu, *Appl. Phys. Lett.* **80**, 853 (2002).

[99] L. Worschech, D. Hartmann, S. Reitzenstein, and A. Forchel, *J. Phys.: Condens. Matter* **17**, R775 (2005).

[100] L. Worschech, B. Weidner, S. Reitzenstein, and A. Forchel, *Appl. Phys. Lett.* **78**, 3325 (2001).

[101] K. Hieke and M. Ulfward, *Phys. Rev. B* **62**, 16727 (2000).

[102] L. Worschech, H.Q. Xu, A. Forchel, and L. Samuelson, *Appl. Phys. Lett.* **79**, 3287 (2002).

[103] I. Shorubalko, H.Q. Xu, I. Maximov, *et al.*, *Appl. Phys. Lett.* **79**, 1384 (2001).

[104] D. Hartmann, L. Worschech, S. Höfling, A. Forchel, and J.P. Reithmaier, *Appl. Phys. Lett.* **89**, 122109 (2006).

[105] S. Kasai and H. Hasegawa, *IEEE Electron Device Lett.* **23**, 446 (2002).

[106] A.M. Song, A. Lorke, A. Kriele, *et al.*, *Phys. Rev. Lett.* **80**, 3831 (1998).

[107] M. Büttiker, *Phys. Rev. Lett.* **57**, 1761 (1986).

[108] A.M. Song, *Phys. Rev. B* **59**, 9806 (1999).

[109] A.M. Song, S. Manus, M. Streibl, A. Lorke, and J.P. Kotthaus, *Superlatt. Microstruct.* **25**, 269 (1999).

[110] A.M. Song, P. Omling, L. Samuelson, W. Seifert, and I. Shorubalko, *Appl. Phys. Lett.* **79**, 1357 (2001).

[111] C. Rocke, S. Zimmermann, A. Wixforth, *et al.*, *Phys. Rev. Lett.* **78**, 4099 (1997).

[112] J.M. Shilton, V.I. Talyanskii, M. Pepper, *et al.*, *J. Phys.: Condens. Matter* **8**, L531 (1996).

[113] V.I. Talyanskii, J.M. Shilton, M. Pepper, *et al.*, *Phys. Rev. B* **56**, 15180 (1997).

[114] J. Cunningham, V.I. Talyanskii, J.M. Shilton, *et al.*, *Phys. Rev. B* **60**, 4850 (1999).

[115] J. Cunningham, V.I. Talyanskii, J.M. Shilton, *et al.*, *Physica B* **62**, 1564 (2000).

[116] A. Robinson, V.I. Talyanskii, M. Pepper, J. Cunningham, and E.H. Linfield, *Phys. Rev. B* **65**, 045313 (2002).

[117] K. Gloos, P. Utko, J. Bindslev Hansen, and P.E. Lindelof, *Phys. Rev. B* **70**, 235345 (2004).

[118] M. Kataoka, C.H.W. Barnes, H.E. Beere, D.A. Ritchie, and M. Pepper, *Phys. Rev. B* **74**, 085302 (2006).

[119] C.L. Foden, V.I. Talyanskii, G.J. Milburn, M.L. Leadbeater, and M. Pepper, *Phys. Rev. A* **62**, 011803 (2000).

[120] C.H.W. Barnes, J.M. Shilton, and A.M. Robinson, *Phys. Rev. B* **62**, 8410 (2000).

[121] N.E. Fletcher, J. Ebbecke, T.J.B.M. Janssen, *et al.*, *Phys. Rev. B* **68**, 245310 (2003).

[122] J. Ebbecke, N.E. Fletcher, T.J.B.M. Janssen, *et al.*, *Appl. Phys. Lett.* **84**, 4319 (2004).

[123] J. Ebbecke, N.E. Fletcher, T.J.B.M. Janssen, *et al.*, *Phys. Rev. B* **72**, 121311(R) (2005).

[124] V.I. Talyanskii, M.R. Graham, and H.E. Beere, *Appl. Phys. Lett.* **88**, 083501 (2006).

6
Quantum dots

The focus of this chapter is a discussion of transport in quantum dots, which are quasi-zero-dimensional nanostructure systems whose electronic states are completely quantized. The confinement of carrier motion in these structures is imposed in all three spatial directions, resulting in a discrete spectrum of energy levels much the same as in an atom or molecule. We can therefore think of quantum dots as artificial atoms, which in principle can be engineered to have a particular energy level spectrum. As in atomic systems, the electronic states in quantum dots are sensitive to the presence of multiple electrons due to the Coulomb interaction between electrons. Rich transport phenomena are therefore observed in these structures, not only because of quantum confinement and the resonant structure associated with this confinement, but also due to the granular nature of electric charge.

In contrast to quantum wells and wires, quantum dots can be sufficiently small that the introduction of even a single electron is sufficient to dramatically change the transport properties due to the charging energy associated with this extra electron. One of the main consequences of this charging energy is to give rise to a Coulomb blockade of transport, where conductance oscillations are observed with the addition or subtraction of a single electron from a quantum dot, which we discuss in detail in Sections 6.1 and 6.2. In metal quantum dots, in which quantization of the electron energy can essentially be neglected, we will discuss how the resulting transport behavior can be analyzed purely in terms of classical charging. In semiconductor dots, however, the interplay of effects due to Coulomb charging and quantum confinement can lead to the observation of a rich variety of phenomena. As we discuss in Section 6.2, these include the observation of a novel Kondo effect in which the quantum dot plays the role of a tunable magnetic impurity. The control offered by quantum dots is further demonstrated in Section 6.3, in which we discuss the results of experiments in which semiconductor quantum dots are coupled together to form artificial molecules, with the potential of application to quantum computing. Experiments on such structures, in particular, reveal remarkable coherent control of their spin states. Finally, in Section 6.4, we discuss the results of experiments that probe quantum interference and wave function coherence in quantum dots. A focus here is on discussion of *open* quantum dots, which are strongly coupled to their external reservoirs so that the Coulomb blockade is suppressed. In these structures, the observation of regular oscillations in the conductance as a function of magnetic field or energy provides a means to experimentally probe the transition from quantum mechanics to classical physics.

6.1 Fundamentals of single-electron tunneling

A quantum dot is essentially a microscopic puddle of charge that is connected to charge reservoirs that allow for the flow of current through the quantum dot. In any discussion of the transport through quantum dots, we must implicitly consider the manner in which the dot is coupled to its external environment. Under conditions where the quantum dot is isolated from its surroundings, with only weak tunneling allowed between the dot and its leads, the phenomenon of single-electron tunneling can be observed. In our discussions of this phenomenon, it will be helpful to distinguish between the situations where the quantum dot (or Coulomb island, as it is often referred to) is realized in a metal or a semiconductor, the importance difference between these two situations being the typical number of electrons that is stored on the island, and the extent to which quantum confinement influences the allowed electron states. Confinement effects in semiconductor quantum dots may be quite large (as will be discussed later in this chapter), leading to structures that justifiably may be considered artificial atoms consisting of just a few particles. Metallic systems, on the other hand, have much larger electron densities, and mean free paths at the Fermi energy of only a few nanometers. Therefore, metallic islands behave more or less as small, bulk-like, systems. However, both systems share a common feature, that the discrete nature of the electron charge becomes strongly evident when particles tunnel into and out of the structure.

When an electron is transferred from one of the reservoirs into a nearly isolated dot, there is a rearrangement of charge in the electrode, resulting in a change in the electrostatic potential of the dot. In a large system, this change in potential due to the injection of electrons via tunneling or over the barrier by thermionic emission is hardly noticeable. In sufficiently small quantum dots, however, the potential change may be greater than the thermal energy, $k_B T$, particularly at low temperatures. Such large changes in the electrostatic energy due to the transfer of a single charge may result in a gap in the energy spectrum at the Fermi energy, leading to the phenomenon of *Coulomb blockade*, in which the tunneling of electrons is inhibited until this charging energy is overcome through an applied bias.

In the theory of Coulomb blockade, the basic experimental results are conveniently discussed in terms of a macroscopic capacitance associated with the system. The change in electrostatic potential due to a change in the charge on an ideal conductor is associated with the linear relationship:

6.1 Fundamentals of single-electron tunneling

$$Q = CV, \quad (6.1)$$

where C is the capacitance, Q is the charge on the conductor, and V the electrostatic potential relative to some chosen reference (such as ground). Since we are considering an ideal conductor, any charge added to the conductor rearranges itself such that the electric field inside vanishes, and the surface of the conductor becomes an equipotential surface. Therefore, the electrostatic potential associated with the conductor relative to its reference is uniquely defined. If we consider two conductors connected by a d.c. voltage source, a charge $+Q$ builds up on one conductor and a charge $-Q$ on the other. The capacitance of the two conductor system is then defined as $C = Q/V_{12}$. The electrostatic energy stored in the two-conductor system is the work done in building up the charge Q on the two conductors and is given by

$$E = \frac{Q^2}{2C}. \quad (6.2)$$

For a system of N conductors, the charge on conductor i may be written as

$$Q_i = \sum_{j=1}^{N} C_{ij} V_j, \quad (6.3)$$

where the diagonal values C_{ii} are the capacitance of conductor i if all other conductors are grounded. The diagonal elements are commonly referred to as the *coefficients of capacitance*, the off-diagonal elements are called the *coefficients of induction*. The total electrostatic energy stored in a multi-conductor system is given by the generalization of Eq. (6.2) as

$$E = \frac{1}{2} \sum_i \sum_j (C^{-1})_{ij} Q_i Q_j, \quad (6.4)$$

where C^{-1} is the inverse capacitance matrix.

It is important to note that the polarization charge on the capacitor, Q, does *not* necessarily have to be associated with a discrete number of electrons, N. This charge is essentially due to a rearrangement of the electron gas with respect to the positive background of ions, and as such it may effectively take on a continuous range of values. It is only when we consider changes in this charge due to the tunneling of a single electron between the conductors that the discrete nature becomes apparent.

In systems of very small conductors, the capacitances approach values sufficiently small that the charging energy given by Eq. (6.2) due to a single electron, $e^2/2C$, becomes comparable to the thermal energy, $k_B T$. The transfer of a single electron between conductors therefore results in a voltage change that is significant compared to the thermal voltage fluctuations and creates an energy barrier to the further transfer of electrons. This barrier remains until the charging energy is overcome by sufficient bias. How small must the quantum dot be for such effects to become important? As a simple example, consider the case of a conducting sphere above a grounded conducting plane. This example approximates a metal cluster imbedded in an insulator above a conducting substrate, which is a commonly realized structure that has been extensively studied experimentally. The exact solution may be found using the method of images, which gives the capacitance of the sphere as [1]

$$C = 4\pi\varepsilon a\left[1 + \alpha + \frac{\alpha^2}{1-\alpha^2} + \cdots\right], \quad \alpha = \frac{a}{2l}, \quad (6.5)$$

where a is the radius of the sphere and l is the distance above the conducting substrate, As the radius of the sphere becomes small compared to l, the capacitance becomes independent of the distance of the cluster from the substrate. An alternate example is that of a flat circular disk located parallel to and a distance d above a ground plane. This example is more closely analogous to the semiconductor quantum dots fabricated by lateral confinement of a 2DEG. In the limit $d \gg R$, the capacitance of such a disk is given as [2]

$$C = 8\varepsilon R, \quad (6.6)$$

where R is the radius of the disk. Equating the charging energy with the thermal energy, we can obtain that at room temperature, $C = 3 \times 10^{-18}$ F. The corresponding radius of the sphere that yields this capacitance is $a = 28$ nm (assuming a relative dielectric constant of 1), and somewhat larger for the disk. Since $\varepsilon > \varepsilon_0$ in typical structures, and since the charging energy should be several times larger than the thermal energy this condition implies that sub-10-nm structures need to be fabricated in order to see clear single-electron charging effects at room temperature. Although it is still somewhat challenging with today's lithographic techniques to nanoengineer such structures, it is not difficult to grow insulating films with random metallic clusters on this order in which Coulomb blockade effects are readily observed, even at room temperature. Furthermore, if we perform measurements at cryogenic temperatures, then the size scale becomes comparably larger, allowing single-electron effects to be observed in nanofabricated quantum-dot structures.

We should keep in mind that in such small systems as the ones discussed above, the concept of using a "lumped capacitance" to characterize the distributed rearrangement of charges in the system may have limited validity. In semiconductor systems in particular, the capacitance in general is not linear but depends on the operating voltage. Thus the capacitance is more generally defined as the differential change in charge with voltage,

$$C(V') = \left.\frac{\partial Q}{\partial V}\right|_{V'}. \quad (6.7)$$

The capacitance associated with a depletion region (such as those used to define split-gate point contacts) is a good example of a capacitance which depends strongly on voltage due to the change of width of the depletion region with gate bias. Furthermore, for ultra-small nanostructures, the nonlocality of charge itself on the quantum level introduces its own contribution to the effective capacitance of the system [3]. Despite these caveats, the simple idea of single-electron charging associated with a macroscopic quantity such as the capacitance seems to work surprisingly well in describing a host of experimental results, some of which we review briefly below.

6.1.1 Coulomb blockade in normal metal tunnel junctions

Historically, Coulomb blockade effects were first predicted and observed in small metallic tunnel junction systems. As mentioned already, the conditions in metallic systems of high electron density, large effective mass, and short phase coherence length (compared to semiconductor systems) usually allow us to neglect size quantization effects. The dominant single-electron effect for small metal tunnel junctions is therefore the charging energy due to the transfer of individual electrons, $e^2/2C$. The effects of single-electron charging in the conductance properties of very thin metallic films was recognized in the early 1950s by Gorter [4] and Darmois [5]. Thin metal films tend to form planar arrays of small islands due to surface tension, and conduction occurs due to tunneling between these islands. Since the island size is small, the tunneling electron has to overcome an additional barrier due to the charging energy, which leads to an increase in resistance at low temperature. Such discontinuous metal films show an activated conductance, $\sigma \sim \exp(-E_c/k_B T)$, similar to an intrinsic semiconductor. Neugebauer and Webb [6] developed a theory of activated tunneling in which this activation energy was the electrostatic energy required to tunnel electrons in and out of the metal islands. In analogy to the semiconductor case, this activation energy resembles an energy gap and is therefore referred to as a *Coulomb gap*.

A number of studies have been conducted concerning the transport properties of metal clusters or islands imbedded in an insulator that are then contacted by conducting electrodes. A schematic of such a structure for Au particles imbedded in native or evaporated Al_2O_3 is shown in Fig. 6.1. Each metal cluster represents a Coulomb island. Giaever and Zeller [7] investigated the differential resistance of oxidized Sn islands sandwiched between two Al electrodes forming Coulomb islands down to 2.5 nm diameter depending upon the evaporation conditions. They measured one of the telltale signs of Coulomb blockade in these samples, that is, a region of high resistance for small bias voltages about the origin followed by a strong decrease in resistance past a voltage of approximately 1 mV. Shortly thereafter, Lambe and Jaklevic [8] performed capacitance–voltage measurements on structures similar to those of Giaever and Zeller. The structures were designed with thick oxides between the islands and the substrate so that tunneling of electrons occurred only through the top contact. Oscillatory

Quantum dots

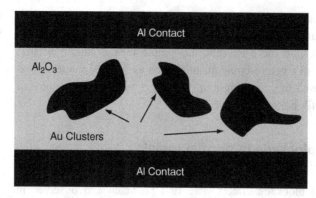

Fig. 6.1 A cross section of Coulomb island structures realized by embedding metal clusters such as Au in a dielectric medium.

behavior in the differential capacitance was interpreted in terms of the addition of charges one by one to the islands as the bias voltage increases in integral multiples of e/C_I, where here C_I is the substrate to island capacitance. Cavicchi and Silsbee [9] later investigated the frequency and temperature dependence of the capacitance and associated charge transfer in asymmetric structures similar to those of Lambe and Jaklevic.

Meanwhile, a rigorous theoretical treatment of single-electron effects during tunneling was introduced by Kulik and Shekhter [10] based on the tunneling Hamiltonian method (discussed below) in order to derive a kinetic equation for charge transport. This kinetic equation approach was later improved by Averin and Likharev [11] to derive the so-called *Orthodox* model of single-charge tunneling. A similar master equation approach was contemporaneously introduced by Ben-Jacob and coworkers [12] based on a semiclassical analysis. One prediction of this new theory was the occurrence of single-charge effects in transport through single-tunnel junctions rather than islands. Subsequent attempts to observe such effects in single-tunnel junctions have not been particularly successful due to environmental effects such as parasitic capacitances. However, the renewed interest in single-charge effects also led to new predictions for transport in Coulomb island structures. One effect in particular is the *Coulomb staircase* in the current–voltage characteristics, first observed by Kuzmin and Likharev [13] and Barner and Ruggiero [14] in metallic island structures of the same general structure as in Fig. 6.1. The difficulty of measuring single-electron effects in metal films is that transport involves many islands, and therefore fluctuations in the size, capacitance, and tunnel resistance of individual islands tend to average out effects due to single-charge tunneling. The use of a scanning tunneling microscopy (STM) tip to contact individual islands alleviates this problem and allows clean measurement of the Coulomb staircase (see Fig. 6.2) [15,16]. The current is essentially zero about the origin, evidencing the Coulomb blockade effect, and then rises in jumps, giving a staircase-like appearance. The subsequent jumps in the $I-V$ characteristics correspond to the stable voltage regimes in which one more electron is added to or subtracted from

6.1 Fundamentals of single-electron tunneling

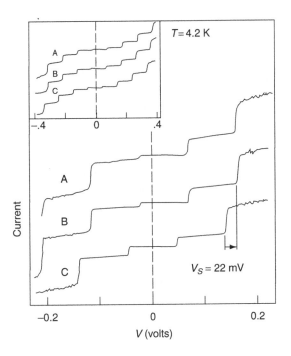

Fig. 6.2 Experimental (A) and theoretical (B and C) I-V characteristics from an STM-contacted 10-nm diameter In droplet illustrating the Coulomb staircase in a double-junction system. The peak-to-peak current is 1.8 nA. The curves are offset from one another along the current axis, with the intercept corresponding to zero current ([16], with permission).

the island. Each plateau essentially corresponds to a stable regime with a fixed integer number of electrons on the island. It is interesting to note that the $I-V$ characteristics in Fig. 6.2 are not symmetric with respect to the origin. This offset in the $I-V$ curve is due to the presence of unintentional background charges which contribute an additional charging energy to the Coulomb island. Such random charges are very difficult to eliminate experimentally and represent a severe problem in realizing practical device technologies based on single-electron devices.

To understand the Coulomb staircase effect somewhat more quantitatively, we consider the equivalent circuit of the Coulomb island shown in Fig. 6.3 in which the "island" consists of a small metallic cluster coupled weakly through thin insulators to metal leads as shown schematically in Fig. 6.4. We have introduced a circuit element representing the tunnel junction as a parallel combination of the tunneling resistance R_t and the capacitance C. In metal tunnel junctions, the tunneling barrier is typically very high and thin, while the density of states at the Fermi energy is very high. The tunneling resistance is therefore almost independent of the voltage drop across the junction. In the analysis that follows, we implicitly assume a sequential tunneling model. Electrons that tunnel through one junction or the other are therefore assumed to immediately relax due to carrier–carrier scattering so that resonant tunneling through both barriers is simultaneously neglected. One has to be careful at this point in distinguishing the tunnel resistance from an ordinary Ohmic resistance.

Island

Fig. 6.3 A quantum dot coupled to two leads connected to an external circuit.

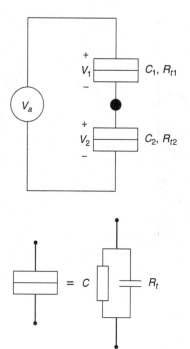

Fig. 6.4 Equivalent circuit of a metallic island weakly coupled to a voltage source through two tunnel junctions with capacitances C_1 and C_2. R_t is the tunnel resistance, n_1 is the number electrons that have tunneled *into* the island through junction 1, and n_2 is the number of electrons that have tunneled *out of* the island through junction 2.

In an ordinary resistor, charge flow is quasi-continuous and changes almost instantaneously in response to a change in electric field (at least down to time scales on the order of the collision times, or picoseconds). Tunneling represents the injections of single particles, which involve several characteristic time scales. The tunneling time (the time to tunnel from one side of the barrier to the other) is the shortest time (on the order of 10^{-14} s), whereas the actual time between tunneling events themselves is on the order of the current divided by e, which for typical currents in the nA range implies a mean time of several hundred picoseconds between events. The time for charge to rearrange itself on the electrodes due to the tunneling of a single electron will be something on the order of the dielectric relaxation time, which is also very short. Therefore, for purposes of analysis, we can consider that the junctions in the regime of interest behave as ideal capacitors through which charge is slowly leaked.

For the circuit shown in Fig. 6.4, the capacitor charges are given by

$$Q_1 = C_1 V_1, \quad Q_2 = C_2 V_2. \tag{6.8}$$

The net charge Q on the island is the difference of these two charges. In the absence of tunneling, the difference in charge would be zero and the island neutral. Tunneling allows an integer number of excess electrons to accumulate on the island so that

$$Q = Q_2 - Q_1 = -ne, \tag{6.9}$$

where $n = n_1 - n_2$ is the net number of excess electrons on the island (which can be positive or negative), with n_1 and n_2 defined as in Fig. 6.4. This convention is chosen such that an increase in either n_1 or n_2 corresponds to increasing either the junction charge Q_1 or Q_2, respectively, in Eq. (6.8). The sum of the junction voltages is just the applied voltage, V_a, so that using Eqs. (6.8) and (6.9) we may write the voltage drops across the two tunnel junctions as

$$V_1 = \frac{1}{C_{eq}}(C_2 V_a + ne),$$
$$V_2 = \frac{1}{C_{eq}}(C_1 V_a - ne),$$
(6.10)

where $C_{eq} = C_1 + C_2$ is the capacitance of the island. The electrostatic energy stored in the capacitors is given by

$$E_s = \frac{Q_1^2}{2C_1} + \frac{Q_2^2}{2C_2},$$
(6.11)

which using Eqs. (6.8) and (6.9) gives

$$E_s = \frac{1}{2C_{eq}}(C_1 C_2 V_a^2 + Q^2).$$
(6.12)

In addition, we must consider the work done by the voltage source in transferring charge in and out of the island via tunneling. The work done by the voltage source may be considered as the time integral over the power delivered to the tunnel junctions by this source:

$$W_s = \int dt V_a I(t) = V_a \Delta Q,$$
(6.13)

where ΔQ is the total charge transferred from the voltage source, including the integer number of electrons that tunnel into the island and the continuous polarization charge that builds up in response to the change of electrostatic potential on the island. A change in the charge on the island due to one electron tunneling through tunnel barrier 2 (so that $n_2' = n_2 + 1$) changes the charge on the island to $Q' = Q + e$, and $n' = n - 1$. According to Eq. (6.10), the voltage across junction 1 changes as $V_1' = V_1 - e/C_{eq}$. Therefore, from Eq. (6.8) a polarization charge flows in from the voltage source $\Delta Q = -eC_1/C_{eq}$ to compensate. The total work done to pass in n_2 charges through junction 2 is therefore

$$W_s(n_2) = -n_2 e V_a \frac{C_1}{C_{eq}},$$
(6.14)

By a similar analysis, the work done in transferring n_1 charges through junction 1 is given by

$$W_s(n_1) = -n_1 e V_a \frac{C_2}{C_{eq}},$$
(6.15)

We may therefore write the total energy of the complete circuit including the voltage source as

$$E(n_1, n_2) = E_s - W_s = \frac{1}{2C_{eq}}(C_1 C_2 V_a^2 + Q^2) + \frac{eV_a}{C_{eq}}(C_1 n_2 + C_2 n_1). \quad (6.16)$$

We may now look at the condition for Coulomb blockade based on the change in this electrostatic energy with the tunneling of a particle through either junction. At zero temperature, the system has to evolve from a state of higher energy to one of lower energy. Therefore, tunneling transitions that take the system to a state of higher energy are not allowed, at least at zero temperature (at higher temperature, thermal fluctuations in energy on the order of $k_B T$ weaken this condition; see discussion below). The change in energy of the system with a particle tunneling through the second junction is

$$\Delta E_2^\pm = E(n_1, n_2) - E(n_1, n_2 \pm 1)$$
$$= \frac{Q^2}{2C_{eq}} - \frac{(Q \pm e)^2}{2C_{eq}} \mp \frac{eV_a C_1}{C_{eq}} = \frac{e}{C_{eq}}\left[-\frac{e}{2} \pm (ne - V_a C_1)\right]. \quad (6.17)$$

Similarly, the change in energy of the system with a particle tunneling through junction 1 is given by

$$\Delta E_1^\pm = E(n_1, n_2) - E(n_1 \pm 1, n_2) = \frac{e}{C_{eq}}\left[-\frac{e}{2} \mp (ne + V_a C_2)\right]. \quad (6.18)$$

According to our previous assertion, only transitions for which $\Delta E_j > 0$ are allowed at zero temperature.

Consider now a system where the island is initially neutral, so that $n = 0$. Equations (6.17) and (6.18) reduce to

$$\Delta E_{1,2}^\pm = -\frac{e^2}{2C_{eq}} \mp \frac{eV_a C_{2,1}}{C_{eq}} > 0. \quad (6.19)$$

For all possible transitions into and out of the island, the leading term involving the Coulomb energy of the island causes ΔE to be negative until the magnitude of V_a exceeds a threshold that depends on the lesser of the two capacitances. For $C_1 = C_2 = C$, the requirement becomes simply $|V_a| > e/C_{eq}$. Tunneling is prohibited and no current flows below this threshold, as evident in the I–V characteristics shown in Fig. 6.2. This region of *Coulomb blockade* is a direct result of the additional Coulomb energy, $e^2/2C_{eq}$, which must be expended by an electron in order to tunnel into or out of the island. The effect on the current voltage characteristics is a region of very low conductance around the origin, as shown in Fig. 6.2. For large-area junctions where C_{eq} is large, no regime of Coulomb blockade is observed, and current flows according to the tunnel resistance R_t.

Figure 6.5(a) shows the equilibrium band diagram for a double-tunnel-junction system, illustrating the Coulomb blockade effect for equal capacitances.

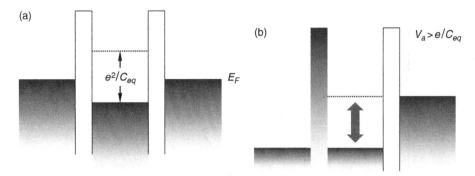

Fig. 6.5 Band diagram of a double-junction structure (a) in equilibrium and (b) under an applied bias. A gap exists in the density of states of the island system due to the Coulomb charging energy which prohibits tunneling into and out of the island below the threshold voltage.

A Coulomb gap of width e^2/C_{eq} has opened at the Fermi energy of the metal island, half of which appears above and half below the original Fermi energy, so that no states are available for electrons to tunnel into from the left and right electrodes. Likewise, electrons in the island have no empty states to tunnel to either until the blockade region is overcome by sufficient bias as shown in Fig. 6.5(b).

Now consider what happens when the double-junction structure is biased above the threshold voltage to overcome Coulomb blockade. Assume for the sake of illustration that the capacitance of the two barriers is the same, $C_1 = C_2 = C$. Suppose the threshold for tunneling, $V_a > e/2C$, has been reached so that one electron has already tunneled into the island and $n = 1$ as shown in Fig. 6.5(b). The Fermi energy in the dot is raised by e^2/C_{eq} and a gap appears that prohibits a second electron from tunneling into the island from the right electrode until a new voltage, $V_a > 3e/2C$, is reached, as is apparent from Eq. (6.17). Within this range of V_a, no further charge flows until the extra electron on the island tunnels into the left electrode, taking the dot back to the $n = 0$ state. This transfer lowers the Fermi energy in the dot and allows another electron to tunnel from the right electrode, and the process repeats itself. Thus, a correlated set of tunneling processes of tunneling into and out of the dot for the $n = 1$ configuration occur, giving rise to a non-zero current. In order to observe the Coulomb staircase of Fig. 6.2, the junctions should be asymmetric such that either the capacitances or the tunneling resistances are quite different. Assume in our present case that the capacitances are still equal but that the tunneling resistances are quite different, with $R_{t_1} > R_{t_2}$. For this situation, the limiting rate is tunneling through the first barrier, so that the island remains essentially in a charge state corresponding to the voltage range defined by Eq. (6.17) for positive voltages. As soon as an electron tunnels out of the dot through junction 1, it is immediately

Fig. 6.6 Ideal current–voltage characteristics for an asymmetric double-junction system with and without consideration of Coulomb charging effects. For this system, $C_1 = C_2 = C$ and $R_t = R_{t_1} \gg R_{t_2}$.

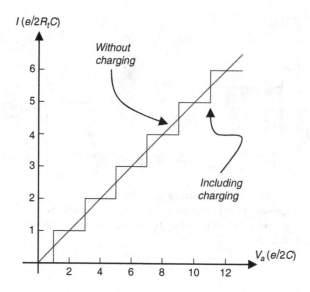

replenished by junction 2. Under these conditions, the current is approximately controlled by the voltage drop across junction 1, which is given by Eq. (6.10) for the equal capacitance case as $V_1 = V_a/2 + ne/C_{eq}$. The voltage across the first barrier therefore jumps by an amount e/C_1 whenever the threshold for increasing n is reached for junction 2. The current correspondingly jumps by an amount given by

$$\Delta I \approx \frac{\Delta V_1}{R_{t_1}} = \frac{e}{C_{eq} R_{t_1}} = \frac{e}{2CR_{t_1}}. \qquad (6.20)$$

Assuming the current does not vary much between jumps, the *I–V* characteristics exhibit the staircase structure shown in Fig. 6.6, which qualitatively explains the staircase structure in the experimental data of Fig. 6.2.

The existence of well-defined structure in the transport properties of a multi-junction system due to Coulomb charging depends on the magnitude of the Coulomb gap, e^2/C_{eq}, compared to the thermal energy. Qualitatively it is clear that this gap must greatly exceed the thermal energy, $e^2/C_{eq} > k_B T$, to observe well-defined Coulomb blockade effects at a given temperature. A further constraint is that the quantum fluctuations in the particle number, n, be sufficiently small that the charge is well localized on the island. A hand-waving argument is to consider the energy uncertainty relationship

$$\Delta E \Delta t \geq \hbar, \qquad (6.21)$$

where $\Delta E \sim e^2/C_{eq}$ and the time to transfer charge into and out of the island given by $\Delta t \approx R_t C_{eq}$ where R_t is the smaller of the two tunnel resistances. Combining these two expressions together gives

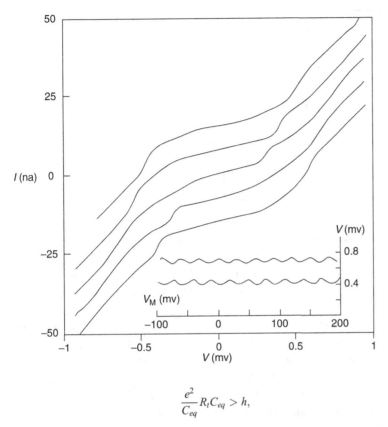

Fig. 6.7 I–V curves for a substrate-biased double-junction system at 1.1 K for different substrate biases covering one cycle. Curves are offset by increments of 7.5 nA. The inset shows Coulomb oscillations in the current as a function of the gate voltage ([17], with permission).

$$\frac{e^2}{C_{eq}} R_t C_{eq} > h,$$

so that the requirement for clear Coulomb charging effects is that the tunnel resistance be sufficiently large (otherwise it would not be a tunnel junction)

$$R_t > \frac{h}{e^2} = 25,813 \, \Omega. \quad (6.22)$$

Up to this point, we have discussed experimental results in Coulomb islands that arise due to clustering and island formation during the growth of very thin metal films. Except for the STM experiments mentioned earlier, single-electron effects were measured in an averaged way. For practical applications it is much more desirable to have the capability to fabricate nanostructure dots in an intentional fashion rather than relying on the random occurrence of such structures. Improvements in nanolithography technologies, such as direct write e-beam lithography and atomic force microscopy, now allow the routine fabrication of single island nanostructures exhibiting single-electron phenomena.

One of the first realizations of a nanofabricated single-electron structure was reported by Fulton and Dolan [17], where a shadow-mask technique using electron beam lithography was used to fabricate 30 nm × 30 nm Al–Al$_2$O$_3$–Al tunnel junctions. The device showed clear evidence of Coulomb blockade and structure due to the Coulomb staircase (see Fig. 6.7). Besides realizing

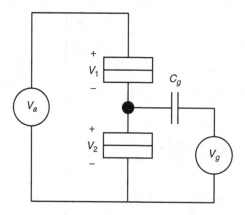

Fig. 6.8 Equivalent circuit for a single-electron transistor.

a double junction using nanolithography, the work of Fulton and Dolan was seminal in that additional gate contacts to the island were fabricated to realize a single-electron transistor. As seen in Fig. 6.7, the Coulomb staircase structure in their experiments changes as the bias applied to a substrate contact changes (the substrate contact was capacitively coupled to the island and acts as the gate in this case). The change in the I–V curves with gate bias was periodic with period approximately given by e/C_s, where C_s is the island-to-substrate capacitance.

To understand this behavior, we consider the topologically equivalent single-electron transistor circuit shown in Fig. 6.8. There, a separate voltage source, V_g, is coupled to the island through an ideal (infinite tunnel resistance) capacitor, C_g. This additional voltage modifies the charge balance on the island so that Eq. (6.8) requires an additional polarization charge:

$$Q_g = C_g(V_g - V_2). \tag{6.23}$$

The island charge becomes

$$Q = Q_2 - Q_1 - Q_g = -ne + Q_p. \tag{6.24}$$

Here Q_p has been added to represent both the unintentional background polarization charge that usually exists in real structures due to workfunction differences and random charges trapped near the junctions. The existence of such random charge was necessary in explaining the asymmetry of the experimental I–V characteristics about the origin shown in Fig. 6.2. Equations (6.8), (6.23), and (6.25) together yield the voltages across the two tunnel junctions:

$$\begin{aligned} V_1 &= \frac{1}{C_{eq}}((C_g + C_2)V_a - C_g V_g + ne - Q_p), \\ V_2 &= \frac{1}{C_{eq}}(C_1 V_a + C_g V_g - ne + Q_p). \end{aligned} \tag{6.25}$$

The equivalent capacitance of the island is that obtained by grounding the independent voltage sources:

6.1 Fundamentals of single-electron tunneling

$$C_{eq} = C_1 + C_2 + C_g. \tag{6.26}$$

The electrostatic energy is given by Eqs. (6.11) and (6.12) and now includes the energy of the gate capacitor, $e^2/2C_g$, as

$$E_s = \frac{1}{2C_{eq}}(C_g C_1 (V_a - V_g)^2 + C_1 C_2 V_a^2 + C_g C_2 V_g^2 + Q^2). \tag{6.27}$$

The work performed by the voltage sources during the tunneling through junctions 1 and 2 now includes both the work done by the gate voltage and the additional charge flowing onto the gate capacitor electrodes. Equations (6.14) and (6.15) are now generalized to

$$W_s(n_2) = -n_2 \left[\frac{C_1}{C_{eq}} eV_a + \frac{C_g}{C_{eq}} eV_g \right],$$

$$W_s(n_1) = -n_1 \left[\frac{C_2}{C_{eq}} eV_a + \frac{C_g}{C_{eq}} e(V_a - V_g) \right]. \tag{6.28}$$

The total energy for a charge state characterized by n_1 and n_2 is given by Eq. (6.16). For tunnel events across junction 1, the change in energy of the system is now given by

$$\Delta E_1^\pm = \frac{Q^2}{2C_{eq}} - \frac{(Q \pm e)^2}{2C_{eq}} \mp \frac{e}{C_{eq}}((C_g + C_2)V_a - C_g V_g)$$

$$= \frac{e}{C_{eq}} \left[-\frac{e}{2} \mp (en - Q_p + (C_g + C_2)V_a - C_g V_g) \right]. \tag{6.29}$$

and for tunnel events across junction 2, the change in energy is given by

$$\Delta E_2^\pm = \frac{Q^2}{2C_{eq}} - \frac{(Q \pm e)^2}{2C_{eq}} \mp \frac{e}{C_{eq}}(C_1 V_a + C_g V_g)$$

$$= \frac{e}{C_{eq}} \left[-\frac{e}{2} \pm (en - Q_p - C_1 V_a - C_g V_g) \right]. \tag{6.30}$$

In comparison to Eqs. (6.18) and (6.19), the gate bias allows us to change the effective charge on the island, and therefore to shift the region of Coulomb blockade with V_g. Thus, a stable region of Coulomb blockade may be realized for $n \neq 0$. As before, the condition for tunneling at low temperature is that $\Delta E_{1,2} > 0$ such that the system goes to a state of lower energy after tunneling. The random polarization charge acts as an effective offset in the gate voltage, so we may define a new voltage, $V'_g = V_g + Q_p/C_g$. The conditions for forward and backward tunneling then become

$$-\frac{e}{2} \mp (en + (C_g + C_2)V_a - C_g V'_g) > 0$$

$$-\frac{e}{2} \pm (en - C_1 V_a - C_g V'_g) > 0. \tag{6.31}$$

Fig. 6.9 A stability diagram for the single-electron transistor for the case $C_2 = C_g = C$, $C_1 = 2C$, illustrating the regions of energy where tunneling is prohibited at 0 K for various numbers of electrons on the island. The shaded areas correspond to regions where no tunneling through either junction may occur, and thus they represent regions of fixed electron number.

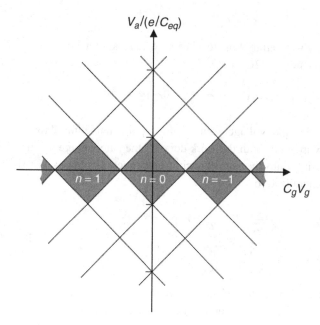

These four equations for each value of n may be used to generate a stability plot in the V_a–V_g plane, which shows stable regions corresponding to each n for which no tunneling may occur. Such a diagram is shown in Fig. 6.9 for the case of $C_g = C_2 = C$, $C_1 = 2C$. The lines represent the boundaries for the onset of tunneling given by Eq. (6.31) for different values of n. The trapezoidal shaded areas correspond to regions where no solution satisfies Eq. (6.31), and hence where Coulomb blockade exists. Each of the regions corresponds to a different integer number of electrons on the island, which is "stable" in the sense that this charge state cannot change, at least at low temperature when thermal fluctuations are negligible. The gate voltage then allows us to tune between stable regimes, essentially adding or subtracting one electron at a time to the island.

The "Coulomb diamond" structure described above, and shown in Fig. 6.9, has now been observed in a wide variety of experiments on single-electron transistors realized by a broad range of approaches. As just one example of this behavior, in Fig. 6.10 we show the Coulomb diamond characteristic of an InAs-nanowire single-electron transistor [18]. In this device, the nanowire was realized using the techniques of vapor-phase synthesis discussed in Chapter 2. The Coulomb island was comprised of a region of InAs, sandwiched between two tunnel barriers of InP, a wider-bandgap material than InAs which therefore serves as an insulating barrier. The diameter of the nanowire was 55 nm, while the separation between the 5-nm thick InP barriers, which sets the size of the Coulomb island, was of order 100 nm. The Coulomb diamond shown in the upper panel of Fig. 6.10 was obtained by measuring the variation of the transistor channel current, as a function of both the source–drain voltage

6.1 Fundamentals of single-electron tunneling 315

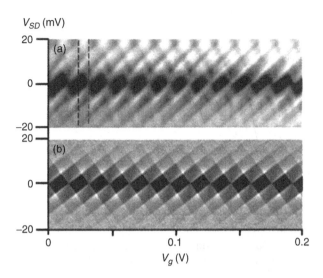

Fig. 6.10 The source–drain conductance dI/dV_{SD} plotted as a function of gate voltage (V_G) and source–drain voltage (V_{SD}). Bright areas correspond to a higher value of dI/dV_{SD}. (a) Measurement at 4.2 K for a 55-nm diameter nanowire device. (b) Theoretical fit at 0 K.

and the gate voltage. (As is common for nanowire-based implementations of transistors, the gate was formed by a conductive Si substrate, separated from the nanowire by an insulating layer of SiO_2.) This figure reveals a clear, and periodically recurring, diamond structure, just as expected from Fig. 6.9. By analyzing the slopes of this diamond pattern, the authors were able to determine the relevant capacitances of the transistor. From the width (∼4 mV) of the Coulomb diamonds as a function of source–drain voltage, they determined a total capacitance of ∼40 aF. From the period of the diamonds as a function of gate voltage, the gate capacitance of 10 aF can be determined. The lower panel of Fig. 6.10 shows the result of a model calculation, based on a commercial single-electron device simulation package (SIMON) [19], which uses realistic capacitance parameters to reproduce closely the results of experiment.

A striking implication of the Coulomb-diamond pattern of Figs. 6.9 and 6.10 is the prediction that the single-electron transistor should exhibit "Coulomb oscillations" in its conductance, when its gate voltage is varied (in the limit of small source–drain bias). To understand the origin of these oscillations, we note that, for a given gate bias, the range of V_a over which Coulomb blockade occurs is given by the vertical extent of the shaded region in Fig. 6.9. For the case shown in Fig. 6.9, the maximum blockade occurs when $C_g V'_g = me$, with $m = 0$, ± 1, ± 2, . . . As $C_g V'_g$ approaches half-integer values of the charge of a single electron, the width of the Coulomb blockade region vanishes and tunneling may occur. Therefore, for a small source–drain bias V_a across the double junction, a measurement of the current versus gate bias will exhibit peaks in the current for a narrow range of gate bias around half-integer values of the gate charge, as illustrated in Fig. 6.11. The distance between peaks is given by $\Delta V_g = e/C_g$. Between peaks, the number of electrons on the dot remains a stable integer

Fig. 6.11 Calculated conductance versus gate voltage in the linear response regime of a double-junction single-electron transistor for $k_BT = 0.05e^2/C_{eq}$.

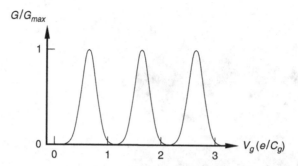

value. As long as $eV_a < k_BT$ the width of these peaks will essentially be limited by thermal broadening and therefore smear out at temperatures greater than the energy width of the Coulomb blockade regime. The conductance linewidth has been calculated in the limit of $e^2/C \gg k_BT$ by Beenakker as [20]

$$\frac{G}{G_{\max}} = \cosh^{-2}\left[\frac{e(C_g/C_{eq})(V_g^{res} - V_g)}{2.5k_BT}\right], \tag{6.32}$$

where V_g^{res} corresponds to a resonant value such that C_gV_g' is a half-integer multiple of the fundamental charge. Results using Eq. (6.32) for $k_BT = 0.05e^2/C_{eq}$ are shown in Fig. 6.11, illustrating the expected lineshape. The existence of conductance peaks may be more clearly understood from the energy band diagram of the system shown in Fig. 6.12.

The measurement of current or conductance peaks with gate bias in the linear response regime is typical of experimental results reported in semiconductor quantum dot structures, as we illustrate in Fig. 6.13. The first evidence of single-electron charging in such structures was provided by Scott-Thomas et al., who observed conductance oscillations versus gate voltage in the low temperature I–V characteristics of side gated Si MOSFET [21]. The oscillations observed in these structures were subsequently explained as arising from Coulomb charging of "accidental" quantum dots, formed by impurities along the 1D inversion layer channel [22,23]. Following on from this, the first definitive demonstration of Coulomb charging in deliberately defined quantum dots was reported by Meirav et al. using quantum point contacts to define the input and output barriers of a double-junction quantum dot structure shown in Fig. 6.13 [24,25]. The overall heterostructure was grown on a conducting, GaAs substrate, which was biased to change the electron density in the 2DEG existing at the GaAs/AlGaAs interface. The structure was biased such that the Fermi energy in the 2DEG outside of the dot was below the lowest conducting channel in the constriction, i.e., in the tunneling regime. The resistance was thus much greater than h/e^2. Figure 6.13 shows the measured conductance versus gate voltage for two different length constrictions. The oscillations are periodic in terms of $\Delta V_g = e/C_g$, which is evidenced by the increase in spacing of the oscillations when the dot area is

6.1 Fundamentals of single-electron tunneling   317

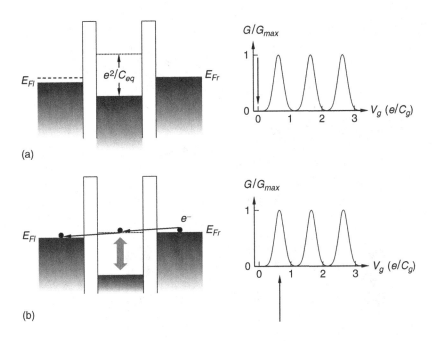

Fig. 6.12 Energy-band diagram of a double-junction system, illustrating the occurrence of conductance peaks due to Coulomb blockade. (a) The system is biased off resonance. (b) Resonance condition when the excited state due to Coulomb charging lies between the Fermi energies on the left and right in the linear response regime.

reduced. Although the periodic oscillations with gate bias are understood within the context of the double-junction model described for metallic systems, the random modulation of the peak amplitude is not. This modulation arises from the discrete nature of the quantum dot states compared to a metal, where the tunneling probability varies strongly with energy. With increasing temperature, the peak-to-valley amplitude of the conductance peaks decreases rapidly, as we illustrate in Fig. 6.14.

Building on the results of studies of Coulomb blockade in single Coulomb islands, subsequent investigations of one- and two-dimensional arrays of small tunnel junctions have demonstrated a rich field of new physical phenomena due to the coupled motion of single charges, which can exhibit nonlinear soliton-like behavior [26,27]. Just one important device is the *single-electron turnstile* [28], which clocks electrons one by one through a Coulomb island (or a series of islands) using a radio-frequency a.c. gate bias. The basic idea of the turnstile is to move an electron onto the central island from one reservoir during the first half of the a.c. cycle, and then move it out from the island to the other reservoir during the second half, thus clocking one electron per cycle through the turnstile. Under such conditions, the resulting current is therefore accurately given by $I = ef$, where f is the frequency of the a.c. source. Figure 6.15 shows the semiconductor realization of the single-electron turnstile implemented by Kouwenhoven *et al.* [29]. The inset shows the metal gate pattern above a 2DEG layer. In the actual experiments, gates 3 and 4 were grounded and therefore not utilized so that the system was basically a single quantum dot structure with

Fig. 6.13 Part (a) shows a schematic drawing of a double-constriction split-gate structure exhibiting single-electron conductance oscillations. Part (b) shows conductance as a function of V_g, for two samples with different Coulomb-island lengths. Sample 2 has length 0.8 μm, sample 3 has length 0.6 μm. Note how the oscillation period is longer for the smaller island, consistent with a smaller gate capacitance ([25], with permission).

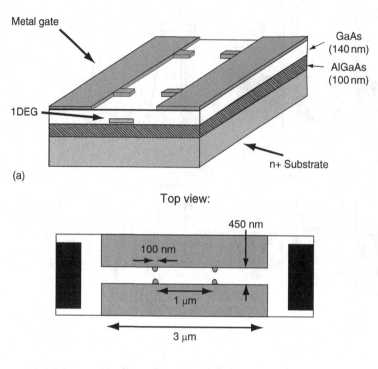

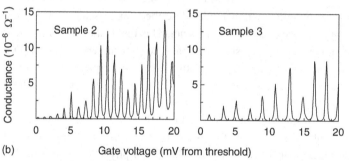

independently biased input and output QPCs. The d.c. I–V characteristics show clear evidence of the Coulomb staircase, which is controlled by the gate contact V_C, where contact C is the plunger shown in Fig. 6.15. With phase-shifted a.c. biases applied to gates 1 and 2, electrons could be clocked through the dot one electron at a time as illustrated by the main I–V characteristics shown in the figure. Several different curves, corresponding to different values of V_C, are shown for each frequency, and these all show a clear Coulomb staircase as a function of the source–drain voltage. The staircase actually arises as the number of electrons that can be clocked through the turnstile increases sequentially, one at a time, as the source–drain voltage is increased. It is also clear from this figure that, for fixed source–drain voltage, the current scales in proportion to the clocking frequency, just as expected for a single-electron turnstile. Figure 6.16

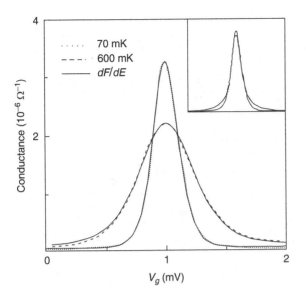

Fig. 6.14 Coulomb oscillation at two different temperatures, showing that the oscillation amplitude decreases, while the width increases, with increase of temperature. This behavior is well described by the evolution with temperature of the Fermi function, whose derivative is plotted as the solid lines ([24], with permission).

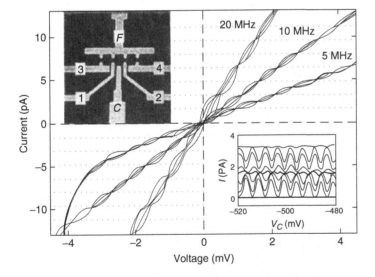

Fig. 6.15 Semiconductor turnstile device realized with a split-gate geometry on a 2DEG substrate. The main figure shows the curves for a.c. signals of various frequency applied to the barrier gates and different values of V_C. The upper inset shows the metal gate pattern, while the lower inset shows Coulomb oscillations of the source–drain current for various source–drain voltages ([29], with permission).

illustrates this operation in more detail. As the left barrier is pulled down in (b), an electron may enter the dot from the left, but not tunnel out to the right. During the opposite cycle (d), the right barrier is lowered and the electron tunnels out on the right, resulting in the transfer of a single electron per clock cycle.

The transfer of electrons through the structure of the single-electron turnstile is dissipative in that the tunneling electron gives up its excess kinetic energy when tunneling into the island. The power dissipation of this device therefore increases as the frequency increases. An alternate single-charge transfer device

Fig. 6.16 Schematic band diagram of the quantum-dot turnstile of Fig. 6.15 during various stages of its RF turnstile cycle ([29], with permission).

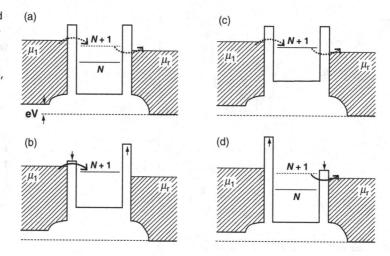

Fig. 6.17 (a) Single-electron pump circuit schematic; n_1 and n_2 are the extra numbers of electrons on the two islands. (b) Stability diagram in the domain of the two gate biases for zero bias voltage across the junctions. One turn around the point P transfers one electron across the circuit, in a direction determined by the sense of rotation. (c) I-V characteristic of the pump with and without 4-MHz gate modulation ([30], with permission).

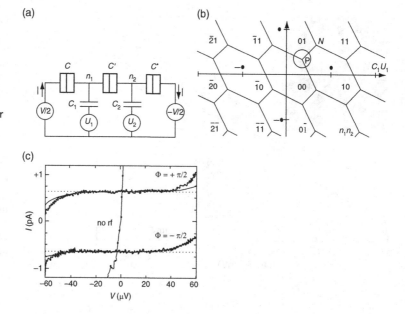

referred to as the *single-electron pump* was reported by Pothier *et al.* in which charge transfer is quasi-adiabatic (i.e., with little energy dissipation during the transfer process) [30]. Figure 6.17 shows a circuit schematic and stability diagram for the pump structure. It consists of three tunnel junctions, with separate gate biases that are capacitively coupled to each of the two Coulomb islands. For small bias voltages across the junctions ($V \approx 0$), stable configurations exist for the excess electron numbers on each island, n_1 and n_2, for various combinations of the gate bias voltages U_1 and U_2. The stability diagram showing these stable configurations has the honeycomb pattern shown in Fig. 6.17 (the structure of this diagram is discussed further in Section 6.3). By adjusting the d.c. bias to be close to a "P-type" triple point, such as point P in Fig. 6.17, one can traverse the region around the triple point in a clockwise or counterclockwise direction by using phase-shifted a.c. biases, u_1 and u_2, applied on top of the d.c. gate biases. If u_2 lags behind u_1 by a factor of $\pi/2$, the system moves from state 00 to state 10 as u_1 increases, crossing a domain boundary, meaning that an electron moves from the left electrode into island 1 during the first positive part of the AC cycle. Then as u_2 increases, a second domain is crossed, taking the system from 10 to 01 (i.e., the electron moves from island 1 to island 2), Finally, during the negative-going part of the u_1 cycle, the system crosses back down to the original domain 00, and one electron has tunneled through the whole system from left to right. The net effect is a negative current (in terms of the sign convention in Fig. 6.17) with a magnitude given by ef. If the phase of the two a.c. voltages is reversed such that u_1 lags behind u_2, the path around the triple point is traversed in the clockwise direction, causing an electron to transit the junctions from right to left and giving a positive current. If the system is closer to the "N-type" triple point labeled N, a counterclockwise rotation produces the opposite current as an electron tunnels first into island 2 and then into island 1.

Figure 6.17(c) shows the experimental I–V characteristics for a pump operated close to a "P-type" triple point. Close to the origin, the current is given by a value $I = \pm ef$ depending on whether the phase difference is positive or negative, independent of the bias voltage across the junction. The quantization of the current, and its polarity dependence on the phase difference of the two a.c. sources, illustrates nicely the pump principle. The operation of the single-electron pump is a single-electron analog of charge-coupled or "bucket brigade" devices (CCDs), which are used extensively in memory and image processing applications.

6.1.2 Orthodox theory of single-electron tunneling

We now want to introduce in somewhat more detail the theoretical models used for analyzing single-charge transport. The most widely invoked theory is the so-called "Orthodox" theory of single-electron tunneling (SET) of Averin and Likharev [11] in which a kinetic equation is derived for the distribution function describing the charge state of a junction or system of junctions. In the semiclassical limit, this theory has proved extremely valuable in analyzing the transport properties of metallic tunnel junctions where size quantization effects are negligible. The method may also be extended to the semiconductor quantum dot case by introducing quantization of the dot states in addition to Coulomb charging effects [31].

In order to establish a kinetic equation based on tunneling into and out of the Coulomb island or quantum dot, it is generally more convenient to describe tunneling in terms of transition rates using perturbation theory rather than in terms of tunneling probabilities (as we did in Chapter 3). Using the transition rates for tunneling allows one to write a detailed balance for tunneling into and out of the dot, and thus an equation of motion describing the evolution of the charge with time. For this reason, the *transfer Hamiltonian method of tunneling* is employed in the Orthodox model, described in more detail in the following section.

6.1.2.1 Transfer Hamiltonian method of tunneling

In Chapter 3 we introduced tunneling using the scattering matrix formalism. Historically, an alternative method has also been successfully employed in tunneling problems based on the transfer or tunneling Hamiltonian approach [32,33]. The transfer Hamiltonian method has been reviewed thoroughly by Duke [34]. This technique was used extensively in describing transport in superconducting tunnel junctions, and has been the basis for most models describing tunneling in small tunnel junctions including Coulomb blockade effects.

In the tunneling Hamiltonian approach, the tunneling barrier is treated as a perturbation to the (much larger) systems forming the left and right sides. The current may be investigated by calculating the rate of transfer of particles from left to right (and right to left) using time-dependent perturbation theory. As such, the applicability of the model is valid only when the perturbation is sufficiently small, which in the tunneling case usually means that the transmission coefficient is small, $T \ll 1$. However, the advantage of the method is that within perturbation theory, the powerful diagrammatic techniques of quantum field theory and many-body theory may be utilized. This allows many-body effects such as quasi-particle tunneling or phonon-assisted tunneling to be treated.

The model system for the transfer Hamiltonian approach is shown schematically in Fig. 6.18 for a simple planar rectangular barrier. The two systems, l and r, represent unperturbed systems on the left and right that are independent of one another except for the perturbation. For one particular choice of subsystems shown in Fig. 6.18, the unperturbed system on the left represents the half-space $x < d$, while the system on the right is the half-space $x > 0$; both systems contain the tunnel barrier itself. The total Hamiltonian is written:

$$H = H_l + H_r + H_t, \tag{6.33}$$

where the Hamiltonians on the left and right, H_l and H_r, presumably are known with eigenvectors and eigenvalues:

$$\begin{aligned} H_l \psi_l &= E_l \psi_l \\ H_r \psi_r &= E_r \psi_r. \end{aligned} \tag{6.34}$$

The significance of the tunneling Hamiltonian becomes particularly transparent if we write the Hamiltonian of Eq. (6.33) explicitly in second quantized form:

6.1 Fundamentals of single-electron tunneling

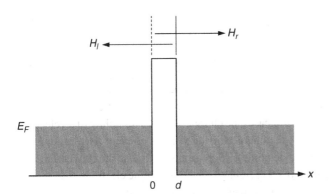

Fig. 6.18 Single barrier between two isolated systems illustrating the decomposition of the system into two independent systems in the tunnel Hamiltonian approach.

$$H_0 = H_l + H_r = \sum_{k_l} E_{k_l} c^*_{k_l} c_{k_l} + \sum_{k_r} E_{k_r} c^*_{k_r} c_{k_r}, \tag{6.35}$$

$$H_t = \sum_{k_l k_r} T_{k_l k_r} c^*_{k_r} c_{k_l} + \sum_{k_l k_r} T_{k_l k_r} c^*_{k_l} c_{k_r}, \tag{6.36}$$

where $c^*_{l,r}$ and $c_{l,r}$ are the Fermion creation and annihilation operators of the independent many-body state of the left and right sides of the barrier, respectively. In Eq. (6.35), the pairs of operators represent the occupation or number operators

$$\begin{aligned} N_{k_l} &= c^*_{k_l} c_{k_l} \\ N_{k_r} &= c^*_{k_r} c_{k_r}. \end{aligned} \tag{6.37}$$

For particles obeying Fermi–Dirac statistics, the result of such an operation can give only one or zero. For finite temperature, the expectation value averaged over the equilibrium ground state gives the Fermi–Dirac distribution:

$$<N_{k_{l,r}}> = f(E_{k_{l,r}}) = \frac{1}{1 + \exp[(E_{k_{l,r}} - E_F^{l,r})/k_B T_L]}. \tag{6.38}$$

Equation (6.36) is the tunnel Hamiltonian, and its first term annihilates a particle of wavevector k_l on the left side, and creates it in k_r on the right side. This process therefore corresponds to tunneling from left to right; the second term corresponds to the reverse process. The tunneling rate is calculated using time-dependent perturbation theory for transitions from an initial state on one side of the barrier to a final state on the other side using Fermi's golden rule, in which the tunnel Hamiltonian of Eq. (6.36) is the perturbation.

We now consider tunneling for a metallic tunnel junction within the single-particle picture using the transfer Hamiltonian method. To begin, a constant voltage V is applied to the left electrode relative to the right. Assume that the electrodes remain approximately in thermal equilibrium, so that the one-particle distribution functions are still of the form of Eq. (6.38). A positive potential applied to the left side with respect to the right lowers the Fermi energy on that side according to

$$E_F^r - E_F^l = eV. \tag{6.39}$$

Due to the capacitance of the junction, there is a certain charge buildup Q on the left electrode, and charge $-Q$ on the right electrode, associated with $CV = Q$. Tunneling from left to right increases the charge, whereas tunneling of an electron from right to left decreases the charge for the voltage convention we are using (positive voltage on the left electrode). Therefore, we will use the convention that Γ^+ is the tunneling rate from left to right which increases the electrode charge, and that Γ^- is the rate from right to left which decreases the charge.

The transition rate from an initial state $\mathbf{k}_l$ to a final state $\mathbf{k}_r$ is treated as a scattering process using Fermi's golden rule

$$\Gamma^+_{k_l \to k_r} = \frac{2\pi}{\hbar} |T_{k_l,k_r}|^2 [1 - f(E_r)] \delta(E_l - E_r), \quad T_{k_l,k_r} = <k_r|H_t|k_l>, \tag{6.40}$$

where the tunnel Hamiltonian is the perturbation and the probability that the state is unoccupied has been included. Duke has discussed in detail how to define the tunnel Hamiltonian and to apply it to the simple single barrier case [34]. The results are approximately the same as those derived using the scattering matrix approach, particularly in the limit of thick barriers. The total rate from occupied states on the left to unoccupied states on the right is given by

$$\Gamma^+(V) = \frac{2\pi}{\hbar} \sum_{\mathbf{k}_l,\mathbf{k}_r} |T_{k_l,k_r}|^2 f(E_l)[1 - f(E_r)] \delta(E_l - E_r). \tag{6.41}$$

For a typical metal tunnel junction, the barrier consists of a thin native oxide with a relatively high barrier height. It is usually a reasonable approximation for such cases to neglect the variation of the tunnel matrix element with energy and momentum so that the matrix element is treated as a constant which may be taken outside of the summation. The sums over momentum are converted to sums over energy in the usual fashion to obtain

$$\Gamma^+(V) = \frac{2\pi}{\hbar} |T|^2 \int_{E_{c_l}}^{\infty} dE_l \int_{E_{c_r}}^{\infty} dE_r D_l(E_l) D_r(E_r) f(E_l)[1 - f(E_r)] \delta(E_l - E_r), \tag{6.42}$$

where $D_{l,r}(E)$ is the density of states in energy on the left and right sides of the barrier. Since the main contribution from the integral is for a narrow range of energies (depending on the applied voltage) around the Fermi energies on the left and right, the densities of states appearing in the integral may be taken constant as well, $D_{l,r}(E) = D_{lo,ro}(E)$. The delta function then reduces one of the integrations such that

$$\Gamma^+(V) = \frac{2\pi}{\hbar} |T|^2 D_{lo} D_{ro} \int_{E_{c_m}}^{\infty} dE f(E - E_F^l)[1 - f(E - E_F^r)], \tag{6.43}$$

where the lower limit is the higher of the two conduction band minima on the left and right. The difference between the Fermi energies is given by Eq. (6.39). The same calculation for the charge flow the other way gives by symmetry:

$$\Gamma^-(V) = \frac{2\pi}{\hbar}|T|^2 D_{lo} D_{ro} \int_{E_{cm}}^{\infty} dE f(E - E_F^r)[1 - f(E - E_F^l)]. \tag{6.44}$$

The total current is written

$$I = e[\Gamma^-(V) - \Gamma^+(V)] = \frac{2\pi e}{\hbar}|T|^2 D_{lo} D_{ro} \int_{E_{cm}}^{\infty} dE[f(E - E_F^r) - f(E - E_F^l)]. \tag{6.45}$$

Choosing the Fermi energy on the right as the zero of energy, then $E_F^l = -eV$. The integral over the Fermi functions may be performed analytically in the limit that the minimum energy is far below the Fermi energy to give

$$\lim_{E_{cm} \to -\infty} \int_{E_{cm}}^{\infty} dE[f(E - E_F^r) - f(E - E_F^l)] = eV. \tag{6.46}$$

Hence from (4.93), the I–V characteristic of the metal tunnel junction is found to be Ohmic:

$$V = IR_t, \quad R_t = \frac{\hbar}{2\pi e^2 |T|^2 D_{lo} D_{ro}}, \tag{6.47}$$

where R_t defines the *tunneling resistance* invoked in our earlier analysis of single-electron circuits. The resistance decreases as the density of states and the transition probability increase. Note that the density of states factors and the transition matrix element contain the cross-sectional area of the junction itself.

For a semiconductor system, the typical Fermi energies are quite small compared to metallic systems, so the limit taken in Eq. (6.46) is not strictly valid, particularly at higher temperature. Further, the approximations leading to Eq. (6.47) in terms of the neglect of the energy dependencies of the transmission matrix element and densities of states are not well satisfied. In Section 5.1.5 we showed that for a saddle-point QPC, the conductance in the tunneling regime was in fact strongly nonlinear due to the exponential dependence of the transmission coefficient with energy below the lowest subband minimum, given by

$$R_{QPC} = \alpha \beta_n \frac{h}{2e^2}[1 + \exp(-\beta_n(\mu_l - e\varphi_0 + eV_{SD}/\alpha - \varepsilon_n))], \tag{6.48}$$

where the constants are defined as before. We note from this expression that the gate bias defining the point contact directly modulates the barrier height, φ_0. This fact allows one to sensitively control the tunneling resistance of the tunnel junctions forming the input and output of a QPC quantum dot. This control was utilized, for example, in the turnstile device of Kouwenhoven *et al.* [29] discussed earlier.

We now want to consider the modification of the transition rates for nanostructure systems in which the charging energy is no longer a negligible contribution. Consider a system of N tunnel junctions that are coupled together and characterized by the number of electrons that have passed through the junction, $\{n\} \equiv E\{n_1, n_2, \ldots, n_j, \ldots, n_N\}$. Previously, we derived the change in energy through a double junction system (including the voltage sources) due to electrons tunneling through either the first or second junction, Eqs. (6.17) and (6.18). Through a generalization of such analysis for the N-junction system, we may write the change in energy of the jth junction as

$$\Delta E_j^\pm = E\{n_1, n_2, \ldots, n_j \pm 1, \ldots, n_N\} - E\{n_1, n_2, \ldots, n_j, \ldots, n_N\}, \qquad (6.49)$$

where the $\pm$ sign refers to the forward or reverse tunneling process across the junctions. Using a "golden rule" approximation [35], the rate of tunneling of electrons back and forth through the jth junction is

$$\Gamma_j^\pm(V) = \frac{2\pi}{\hbar} \sum_{\mathbf{k}_i, \mathbf{k}_f} |T_{if}|^2 f(E_i)[1 - f(E - E_f)] \delta(E_i - E_f + \Delta E_j^\pm), \qquad (6.50)$$

where i and f refer to the initial and final states in the forward or reverse directions. The delta function now includes the change in the total energy of the multi-junction system due to a single electron tunneling across the jth junction. Again assuming weak energy dependencies for the transfer matrix element and the densities of states, the tunneling rate may be written

$$\Gamma_j^\pm(V) = \frac{1}{e^2 R_{t_j}} \int_{E_{cm}}^{\infty} dE f(E)[1 - f(E + \Delta E_j^\pm)], \qquad (6.51)$$

where R_{t_j} is given by Eq. (6.47) for junction j. Using the property of the Fermi function

$$f(E)[1 - f(E + \Delta E_j^\pm)] = \frac{f(E) - f(E + \Delta E_j^\pm)}{1 - \exp(-\Delta E_j^\pm / k_B T)}, \qquad (6.52)$$

we may perform the integration in Eq. (6.51) using Eq. (6.46) to obtain

$$\Gamma_j^\pm(V) = \frac{1}{eR_{t_j}} \frac{\Delta E_j^\pm / e}{1 - \exp(-\Delta E_j^\pm / k_B T)}. \qquad (6.53)$$

The term in the numerator looks like the Ohmic current, which normally would flow for an effective bias $\Delta E_j^\pm / e$ across the jth tunnel junction. The total current flowing across the junction is still given by Eq. (6.45) as the difference of the left and right tunneling rates. The result of Eq. (6.53) is central to the theoretical treatment of Coulomb charging in the Orthodox model discussed in the next section. The energetic arguments made earlier in our discussion of single-electron circuits now may be stated more quantitatively. We see that in the limit that $\Delta E_j^\pm$ is positive and much greater than the thermal energy

$$\Gamma_j^\pm(V) = \frac{1}{eR_{t_j}} \Delta E_j^\pm, \quad \Delta E_j^\pm \gg k_B T, \qquad (6.54)$$

so that tunneling is thus "allowed." On the other hand, when $\Delta E_j^\pm$ is large and negative, Eq. (6.53) shows that tunneling is "forbidden"

$$\Gamma_j^\pm(V) \approx 0, \quad -\Delta E_j^\pm \gg k_B T. \tag{6.55}$$

Thus the energetic arguments leading to the qualitative explanation for Coulomb blockade discussed previously are strictly valid in the limit that $|\Delta E_j^\pm| \gg k_B T$.

Considering the double-junction system of Fig. 6.8 as an example, the change in energy associated with forward and backward tunneling across the second junction was given by Eq. (6.17) as

$$\Delta E_2^\pm = \frac{e}{C_{eq}}\left[-\frac{e}{2} \mp (ne - V_a C_1)\right]. \tag{6.56}$$

For zero applied bias and an initially charge-neutral island ($n = 0$), $\Delta E_j^\pm = -e^2/2C_{eq}$. From Eq. (6.55), the tunneling current is approximately zero as long as $e^2/2C_{eq} \gg k_B T$, which sets the temperature limits for observing Coulomb blockade. In the limit that the capacitances are large so that the charging energy is small compared to the applied bias, the change in energy is just proportional to the voltage drop across that junction

$$\Delta E_2^\pm \approx eV_a \frac{C_1}{C_{eq}} = \pm eV_2. \tag{6.57}$$

The net current across the second junction is given by Eqs. (6.45) and (6.53) as

$$\begin{aligned} I &= e\left[\Gamma_2^-(V) - \Gamma_2^+(V)\right] \\ &= \frac{V_2}{R_{12}}\left[\frac{1}{1 - \exp(eV_1/k_B T)} - \frac{1}{1 - \exp(eV_2/k_B T)}\right] = \frac{V_2}{R_{12}}. \end{aligned} \tag{6.58}$$

We see that that in the limit of negligible charging effects, the simple Ohmic relation of Eq. (6.47) is recovered.

Before going on to discuss the kinetic equation governing the charge state of the junctions based on the tunneling model derived above, it is important to note that we have implicitly assumed time scales for the problem such that the charge distribution fully relaxes during the time it takes to tunnel through the junction, which is subsequently assumed to be much shorter than the time between tunnel events. We have not accounted for the coupling of the system to the electromagnetic environment surrounding the circuit, which may alter the nature of the charge relaxation and the energetics of the tunneling electron. Detailed consideration of such environmental effects have been given by Ingold and Nazarov [36] using a model system. The tunneling rates corresponding to Eq. (6.53) represent the limiting case of a low-impedance environment coupling the voltage source to the tunnel junctions.

6.1.2.2 Equation of motion for charge in a single-tunnel junction

To begin, we first consider the dynamics of a single-tunnel junction with capacitance C that is biased by an ideal current source with current I shown

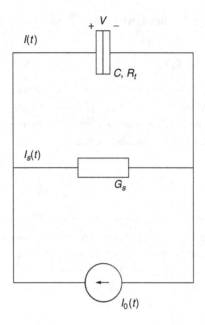

Fig. 6.19 A single-tunnel junction of capacitance and tunnel resistance biased by an ideal current source. G_s is the shunt resistance of the external circuit.

in Fig. 6.19. The basic approach in the Orthodox model is to characterize the state of the junction in terms of its charge, $Q(t) = CV(t)$. This quantity may be treated as a quantum-mechanical variable which is conjugate to the "phase" of the junction

$$\varphi \approx \int_{-\infty}^{t} V(t')dt'. \tag{6.59}$$

Hence, the charge and the phase are conjugate variables, which ultimately may be quantized for a fully quantum-mechanical behavior. Here, however, we will work with the semiclassical approach utilizing these variables.

In order to study Coulomb blockade, the equation of motion for the probability distribution governing the charge on the junction, $f(Q,t)$, must be derived. In order to do this, it is assumed that we can combine a picture of discrete random tunneling events with the continuous flow of charge associated with metallic conductors and the circuit elements characterizing the driving source and environment. For this framework to be valid, the tunneling time itself is required to be almost instantaneous, and the thermalization time of the metallic junctions themselves is assumed to be much shorter than the time between tunneling events so that we can characterize the distribution functions in the leads by quasi-equilibrium (i.e., Fermi–Dirac) functions. The major point in this approach is that tunneling of an electron from left to right or right to left changes the charge Q on the junction as $Q + e$ or $Q - e$, respectively. Further, tunneling of electrons within the same electrode is assumed to be uncorrelated.

Averin and Likharev first derived the master equation governing the evolution of the charge probability distribution in small tunnel junctions using a density matrix approach in their pioneering work [11,31,35,37]. Here, we take a simpler semiclassical kinetic equation approach to arrive at the same equation, in which Q is treated as a classical variable. The justification for this latter approach is that most of the assumptions discussed above are analogous to the same assumptions made in going from the equation of motion for the density matrix (the Liouville equation) to the Boltzmann transport equation, the semiclassical equation governing particle transport in solids [38]. A similar semiclassical approach based on the Kolmogorov master equation has been given by Ben-Jacob *et al.* [39]. For this approach, we first develop the driving force terms and then turn our attention to the "scattering" terms that represent the tunneling of the individual electrons.

In the simple semiclassical picture, we can decompose the system into classical trajectories in phase space, terminated by instantaneous scattering events, in this case tunneling, which instantaneously changes the charge on the tunnel junction by an amount e. In the simple current biased circuit of Fig. 6.19, if we neglect for the moment the effect of the shunt conductance, the change in charge between tunneling events after a short time dt is given by

$$dQ(t) = I(t)dt, \qquad (6.60)$$

where in general $I(t)$ may have some time variation. The charge distribution function at a time $t + dt$ due to this "ballistic" motion becomes $f(Q + I(t)dt, t + dt)$. Expanding to first order, we may write this latter expression as

$$f(Q + I(t)dt, t + dt) = f(Q,t) + I(t)dt \frac{\partial f}{\partial Q} + dt \frac{\partial f}{\partial t} + \cdots, \qquad (6.61)$$

so that the total rate of change of f therefore becomes

$$\frac{df}{dt} = \frac{f(Q + I(t)dt, t + dt) - f(Q,t)}{dt} = \frac{\partial f}{\partial t} + I(t)\frac{\partial f}{\partial Q} = \frac{\partial f}{\partial t}\bigg|_{tunn}, \qquad (6.62)$$

where it has been assumed that the total change in the distribution is balanced by the change induced by tunneling events.

The nonideality of the current source is accounted for with the shunt conductance, G_s, in parallel with the tunnel junction as shown in Fig. 6.19 as considered by Averin and Likharev. We need to be a little more careful on the left side of Eq. (6.62) in that we need to account for the existence of the shunt conductance (we want to write everything in terms of the *external* current) and also for fluctuations in the charge on the capacitor that arise from thermal fluctuations in the shunt conductance. Let $I_0(t)$ be the current provided by the current source. A certain fraction of this current is shunted through the conductance, I_s, which depends on the voltage drop across the tunnel junction. Therefore

$$I_s = G_s V = G_s \frac{Q}{C}, \quad I(t) = I_0(t) - G_s \frac{Q}{C}, \qquad (6.63)$$

where $I(t)$ is, as before, the current through the tunnel junction. In addition, we need to account for the contribution due to Nyquist noise in the shunt conductance. As is well known, the thermal noise in a conductor contributes a fluctuating current given by $<i_{th}^2> = 4k_B T G_s B$, where B is the bandwidth of the noise [40]. This contribution to the junction charge distribution function gives an additional term that we have to consider. Therefore, the distribution function at time $t + dt$ is $f(Q + (I_0(t) - G_s Q/C)dt, t + dt)$, which may be expanded to higher order to give

$$f(Q + (I_0(t) - I_s)dt, t + dt) = f(Q,t) + (I_0(t) - I_s)dt\frac{\partial f}{\partial Q} - fdt\frac{\partial I_s}{\partial Q} + dt\frac{\partial f}{\partial t} \\ + \frac{1}{2}(I_0(t) - I_s)^2 dt^2 \frac{\partial^2 f}{\partial Q^2} - dt^2 \frac{\partial I_s}{\partial Q}\frac{\partial f}{\partial Q} + \cdots \quad (6.64)$$

Now, in general, all the second-order terms vanish in the limit $dt \to 0$, except for the term involving the square of the shunt current. In this case, we make the connections $I_s^2 \to <i_{th}^2>$ and use the Nyquist theorem to relate the bandwidth to the sample time interval $Bdt \to 1/2$. Following the procedure leading to Eq. (6.62), this then leads to the master equation [10]:

$$\frac{df}{dt} = -I_0(t)\frac{\partial f}{\partial Q} + \frac{G_s}{C}\frac{\partial(Qf)}{\partial Q} + k_B T G_s \frac{\partial^2 f}{\partial Q^2} + \left.\frac{\partial f}{\partial t}\right|_{tunn}, \quad (6.65)$$

where the last term is given below. This equation is identical to that derived by Averin and Likharev. It should be pointed out that some quantum-mechanical features are not recovered from this semiclassical derivation although they do arise in the full derivation. In particular, a fuller treatment also defines the limits of applicability of Eq. (6.65) to the limit of $R_t \gg h/2e^2$, which in the handwaving argument given earlier arises from consideration of the fact that the leakage is actually due to tunneling!

To derive the rate of change due to tunneling, we look at a detailed balance between in-scattering and out-scattering events that modify the charge Q. To do this, consider a hypothetical ensemble of N tunnel junctions, $n(Q)$ of which are in charge state Q. The term $\Gamma^+(Q)dt$ corresponds to the number of systems that change from state Q to $Q + e$ due to tunneling of an electron from left to right in Fig. 6.19. Likewise, $\Gamma^-(Q)dt$ is the number of systems that go from Q to $Q - e$. Thus, within a small time interval dt, the change in the number of systems in state Q is given by

$$n(Q, t + dt) = n(Q,t) - \Gamma^+(Q)n(Q,t)dt - \Gamma^-(Q)n(Q,t)dt \\ + \Gamma^+(Q-e)n(Q-e,t)dt + \Gamma^-(Q+e)n(Q+e,t)dt. \quad (6.66)$$

Letting $dt \to 0$ and $f(Q,t) = n(Q,t)/N$, the rate of change due to tunneling is written:

$$\left.\frac{\partial f}{\partial t}\right|_{tunn} = -\Gamma^+(Q)f(Q,t) - \Gamma^-(Q)f(Q,t) \\ + \Gamma^+(Q-e)f(Q-e,t) + \Gamma^-(Q+e)f(Q+e,t). \quad (6.67)$$

The first two terms on the right-hand side represent outscattering, while the last two correspond to inscattering. In each case, it must be recalled that it is assumed that the time scale involved for the above master equation is such that both the tunneling time through the insulator and the equilibration time (such as plasmon decay) within the electrodes are small, so that these events can be assumed to occur instantaneously:

$$V(t) = \frac{1}{C}\int dQ Q f(Q,t) = \frac{\bar{Q}(t)}{C}, \qquad (6.68)$$

assuming $f(Q,t)$ is properly normalized. The DC I–V characteristics may be calculated using Eq. (6.68) from the stationary solutions of Eq. (6.65), $f(Q)$, for $\partial f/\partial t = 0$.

In the previous section, we showed that the tunneling rate itself depends on the change in energy of the system before and after charge transfer, $\Delta E^\pm(Q)$, given by Eq. (6.53). The change in electrostatic energy due to forward and backward tunneling for the current-biased single junction considered in Fig. 6.19 is simply the change in electrostatic stored energy (the constant current source performs negligible work in the incremental tunneling time interval dt):

$$\Delta E^\pm = \frac{Q^2}{2C} - \frac{(Q \pm e)^2}{2C} = -\frac{e}{C}\left[\frac{e}{2} \pm Q\right]. \qquad (6.69)$$

One observation from this equation is that for $|Q| < e/2$, the energy change associated with tunneling is negative which implies from the tunneling rate of Eq. (6.68) that tunneling is prohibited at zero temperature. Therefore, even in the single-junction case, a region of Coulomb blockade is expected in the range of voltages $-e/2C < V < e/2C$ for the d.c. current–voltage characteristics, similar to the experimental results shown in Fig. 6.2 for the double-junction case.

An interesting prediction of the master equation of Eq. (6.65) is that of periodic single-electron tunneling oscillations in the voltage across the junction. To understand this phenomena, consider the time evolution of the junction charge. In the range $-e/2 < Q < e/2$, tunneling is suppressed, and the charge varies linearly as $Q = Q_0 + It$, where I is the current (ignoring the shunt resistance for the moment) and t is the time from the last tunneling event. When the charge reaches $Q = e/2$, the sign of Eq. (6.69) changes and tunneling is allowed. The first electron to tunnel then drives the junction charge back to $Q = Q_0 = -e/2$, and the cycle repeats itself. The Coulomb blockade results in a correlated tunneling of single electrons with a frequency cycle of $f = e/I$, similar to the correlated tunneling in the single-electron turnstile discussed earlier.

To make this slightly more quantitative, if one looks at the regime $-e/2 < Q < e/2$, the tunneling contributions may be neglected, and one just has a diffusion-like equation for the probability distribution, $f(Q,t)$. If we can neglect the shunt current contribution corresponding to the second term on the right side

Fig. 6.20 Schematic depiction of the dynamics of the probability density (left column) and the corresponding change in the junction state (right column) in a small current-biased tunnel junction illustrating single-electron tunneling oscillations (D. V. Averin and K. K. Likharev, *J. Low Temp. Phys.* **62**, 345 (1986), with permission).

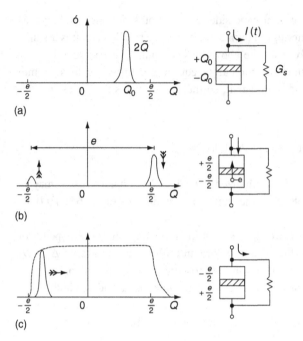

of Eq. (6.65), this equation has precisely the form of a forced diffusion equation. For the system initially in a definite charge state $Q = -e/2$ at $t = 0$ (i.e., $f(Q,0)$ is a delta function at $Q = -e/2$), the solution for subsequent times is

$$f(Q,t) = \frac{1}{\sqrt{4\pi k_B T G_s t}} \exp\left[-\frac{(Q + e/2 - I_0 t)^2}{4\pi k_B T G_s t}\right]. \qquad (6.70)$$

This solution represents a propagating Gaussian pulse centered at $Q = -e/2 + I_0 t$ which spreads in time due to the Nyquist noise contribution. Figure 6.20 illustrates the solution to the master equation showing the cyclical motion of the probability density. As the average of the probability density approaches $e/2$, the tunnel rate for $\Gamma^-(Q)$ becomes large and scatters the system back to the $-e/2$ state.

Calculated results for the d.c. *I–V* characteristics and the time-dependent voltage across a single-tunnel junction were reported by Averin and Likharev [11] under the simplifying assumption of zero temperature and $G_s = 0$ as shown in Fig. 6.21. The d.c. *I–V* characteristic in Fig. 6.21(a) exhibits a region of Coulomb blockade for small $\bar{V} < e/2C$. Single-electron tunneling oscillations are evident in the calculated voltage versus time in Fig. 6.21(b) for small current values in the Coulomb blockade regime. As the current bias increases such that the average voltage exceeds the critical voltage, the oscillations disappear as illustrated in the successive curves.

Clear experimental evidence of SET oscillations in single junctions has remained elusive. The principal problem is the realization of the ideal current

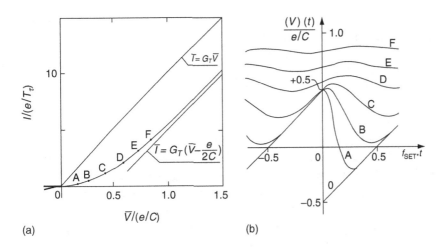

Fig. 6.21 (a) Calculated DC I-V characteristics for a single current-biased tunnel junction for $G_s = 0$ and $T = 0$. (b) Time-dependent voltage calculated for the different current bias points shown in part (a). The quantity f_{SET} is I/e and is the natural Bloch frequency of the SET itself. (D. V. Averin and K. K. Likharev, *J. Low Temp. Phys.* **62**, 345 (1986), with permission.)

source-driven junction illustrated in Fig. 6.19. In nanostructures, the parasitic capacitances of the leads often far exceed that of the junction itself. An external current source charges up the lead capacitance, which tends to behave as a voltage source with respect to the tunnel junction rather than an ideal current source. This argument again illustrates a nontrivial problem with nanostructure systems, that of isolating the behavior of the nanostructure from its surrounding environment (see for example [36]).

6.1.2.3 Multiple junctions

We saw earlier that double-junction structures show clear evidence of single electron-like phenomena such as the Coulomb staircase and periodic Coulomb oscillations. In the double junction, single-electron effects are observed with voltage rather than current bias, which circumvents the problems of observing single-electron tunneling oscillations in a single junction.

The master equation approach of the preceding section is easily extended to the case of multiple junctions. The state of an N-junction array is characterized by the set of junction charges, $\{Q_1, Q_2, \ldots, Q_N\}$. Due to the capacitance–voltage relationships of each of these junctions, we could alternately characterize the state of such a system by the excess number of electrons in each *node* between junctions, $\{n_1, n_2, \ldots, n_{N-1}\}$. The distribution function $f(n_1, n_2, \ldots, n_{N-1})$ is then the multidimensional probability of finding the system in a state characterized by this set of node occupancies. This distribution function evolves continuously under the time-dependent influence of the external sources (voltage or current), as well as through instantaneous tunneling events which change the charge state of each node. For the single-electron transistor structure of Fig. 6.8, there are two tunnel junctions and one node, driven by a constant voltage source. The following equation then describes the evolution of the one-dimensional distribution function:

$$\frac{\partial f(n,t)}{\partial t} = \sum_{j=1,2} [\Gamma_j^+(n-1)f(n-1,t) + \Gamma_j^-(n+1)f(n+1,t) \qquad (6.71)$$
$$- (\Gamma_j^+(n) + \Gamma_j^-(n))f(n,t)],$$

where j is a sum over the tunnel junctions, the tunneling rates are given by Eq. (6.53), and the change in energy during tunneling is given by Eqs. (6.29) and (6.30). The right side of Eq. (6.71) is nothing more than the balance of inscattering and outscattering terms to the distribution function that we wrote in Eq. (6.66), now generalized to include both junctions. The effects of the junction and gate biases are contained in the energy changes, $\Delta E_j^\pm$, given by Eqs. (6.29) and (6.30), which in turn determine the tunneling rates $\Gamma_j^\pm(n)$. Once $f(n)$ is determined, the average current through junction j is written in terms of the net rate of forward and backward tunneling:

$$<I_j(t)> = e \sum_n f(n,t)[\Gamma_j^-(n) - \Gamma_j^+(n)], \qquad (6.72)$$

where the sum is over all the possible node occupancies, n.

Figure 6.22 shows the calculated DC I–V characteristics due to Likharev [37] of a similar SET circuit to that of Fig. 6.8 at a temperature $k_B T = 0.1 e^2/2C_{eq}$. The circuit topology used is slightly different, such that the parameter Q_0 represents the total charge injected into the central node by the gate. For the sake of discussion, we can essentially take this charge to be $C_g V'_g$ as defined in Eq. (6.31), which was plotted in the stability diagram of Fig. 6.9. The plotted curves in Fig. 6.22 represent one complete cycle of gate charge in the stability diagram (Fig. 6.9) going from $Q_0 = 0$ to e. For this asymmetric example, the calculated I–V characteristic shows distinctly the Coulomb staircase discussed earlier, with a region of Coulomb blockade about the origin. As the gate charge approaches $e/2$, the width of the Coulomb blockade region vanishes as qualitatively predicted by the stability diagram of Fig. 6.9, and the conductance becomes non-zero. For small bias voltages, the range of gate voltages where the conductance is non-zero is small, giving rise to conductance peaks as illustrated in Fig. 6.11. Solutions to the master equation for a two-junction system using stochastic methods were also given by Ben-Jacob et al. [12]. Comparison of this calculation with the Coulomb staircase measured in STM studies by Wilkins et al. [16] was shown in Fig. 6.2.

6.1.3 Cotunneling of electrons

In the prior analysis of single-electron tunneling, we have treated the tunneling to lowest order in perturbation theory. However, in the regime of Coulomb

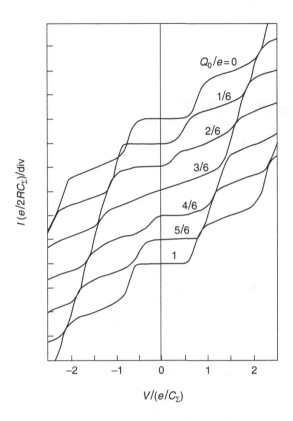

Fig. 6.22 DC I–V characteristics of a single-electron transistor circuit for several values of gate charge ($R_1 \gg R_2$, $C_1 = 2C_2$) ([37], with permission).

blockade, when the tunnel current is small, higher-order processes may become important. These higher-order processes become increasingly important when the resistances of the tunnel junction begins to approach e^2/h, such that quantum fluctuations broaden the energy levels, allowing more channels for charge transfer. In the first-order theory, an electron could not tunnel from the leads to the dot due to conservation of energy when biased in the Coulomb blockade regime. However, in a higher-order process, the electron can transfer from the left lead to the right lead (or vice versa) via a virtual state in the island, conserving energy for the entire process even if tunneling into a virtual state does not. The consideration of such higher-order effects is important for the proposed operation of single-electron transistors and metrological structures such as the turnstile, since these effects ultimately determine the accuracy of single-charge transfer.

The theoretical treatment of higher-order tunneling processes is referred to as the *macroscopic quantum tunneling of charge* (abbreviated as *q*-mqt [41,42]). This process may be either elastic or inelastic. The elastic process corresponds essentially to the same electron tunneling into and out of the virtual state, and thus is a coherent process. Inelastic tunneling, or *cotunneling*, involves one electron tunneling in from a state below the Fermi energy of the lead into the

Fig. 6.23 Illustration of the inelastic macroscopic quantum tunneling of charge, or cotunneling.

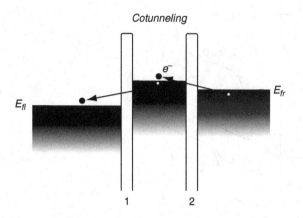

dot, and a second electron leaving from a different state in the dot into the other lead at an energy above the Fermi energy. This latter process is illustrated schematically in Fig. 6.23. As shown, the process of tunneling of an electron from the right electrode to the left creates an electron–hole excitation in the center electrode, so that the electron appearing on the left does not necessarily have to have the same energy as the initial electron on the right. This extra energy eventually is dissipated through carrier–carrier interactions in the island. Since this process results in the creation of an electron–hole pair (with respect to the Fermi energy of the dot), it is referred to as *inelastic q-mqt*. It does not involve phase coherence as the two electrons involved are different. Since this is essentially a two-electron process, it is more popularly known as cotunneling. The cotunneling process usually is dominant in comparison to the elastic q-mqt, except at very small bias voltages and temperatures [42].

Following the Fermi golden rule treatment of cotunneling by Averin and Nazarov [42], the matrix element for tunneling via an intermediate state may be written

$$<i|M|f> = T^{(1)}T^{(2)}\left[\frac{1}{\Delta E_1} + \frac{1}{\Delta E_2}\right], \tag{6.73}$$

where $T^{(i)}$ represents the tunneling amplitude through barrier i, and ΔE_i is the energy barrier difference including the charging energy between the initial state and the virtual state in the island for tunneling through the ith barrier. These states are different due to the different energies in the dot of the two tunneling electrons. The two terms represent the fact that the process could either occur due first to the electron tunneling out of the island into the left electrode and then filled from the right, or due to the first electron tunneling into the island from the right electrode and then the second one tunneling out. The transition rate is given by the usual expression

$$\Gamma = \frac{2\pi}{\hbar}|<i|M|f>|^2\delta(E_i - E_f), \tag{6.74}$$

where E_i and E_f are the initial and final energies of the system after tunneling, which says that the energy lost during the tunneling process by the electron motion from right to left electrode in Fig. 6.23 is equal to that given up in creating the electron–hole excitation in the island. The squares of the tunneling amplitudes are inversely proportional to the respective tunnel resistances, as shown in Eq. (6.47). The total rate of cotunneling is found by summing Eq. (6.74) over all possible initial and final states for the two processes indicated in Fig. 6.23. The interested reader is referred to [42] for more detail of the general relationship. Under the limiting case that the applied voltage is much less than the charging energy of the island, the I–V relationship resulting from this tunneling rate is given by

$$I_c(V) = \frac{\hbar}{12\pi e^2 R_{t_1} R_{t_2}} \left[\frac{1}{|\Delta E_1^\pm|} + \frac{1}{|\Delta E_2^\pm|} \right]^2 (e^2 V + (2\pi k_B T)^2) V, \qquad (6.75)$$

where V is the voltage across the double junction, and R_{t_i} is the tunnel resistance of junction i. The energies $\Delta E_i^\pm$ are the change in energies due to forward and backward tunneling across the ith tunnel junction given by Eqs (6.29) and (6.30) in the Coulomb blockade regime (i.e., $\Delta E_i^\pm < 0$). This equation shows that the cotunneling current goes as the inverse of the square of the tunneling resistance due to the second-order nature of the process. It becomes significant when the tunnel resistance approaches the fundamental resistance h/e^2. The current for low temperature has a characteristic power-law dependence, V^3, and a quadratic temperature dependence. In contrast, we may argue from the form of the first-order tunnel rate of Eq. (6.53) that the current decreases exponentially as the voltage decreases in the Coulomb blockade regime, and has an activated temperature dependence. Thus, while the first-order tunnel rate is usually dominant, in the Coulomb blockade regime the first-order current goes rapidly to zero, and the higher-order process given by Eq. (6.75) may dominate.

Experimental evidence for the V^3 dependence of the current measured in the Coulomb blockade regime of double-junction structures was reported by Geerligs *et al.* using metal tunnel junctions [43]. The temperature dependence evident in Eq. (6.75) was measured in metal double junctions [44]. Investigation of the dependence of the cotunneling current on the tunnel resistance was measured in a double-barrier semiconductor quantum dot [45]. In these dots, the tunnel resistance was independently controlled by the gate potential, which allowed the observation of Coulomb blockade oscillations from the strong tunneling regime $R_t \gg h/e^2$ to the regime where the resistance was less than h/e^2.

Because the q-mqt or cotunneling represents a current that flows in a regime in which to first order it should be suppressed, it is viewed as a parasitic or undesirable effect in proposals for utilizing single-electron charging for practical applications such as logic or memory elements. Understanding how to control this effect is important. If one extends the calculation outlined above for the double junction to include multiple junctions, the tunneling current at $T = 0$ is given by [42]

$$I_c(V) = \frac{2\pi e}{\hbar} \left[\prod_{i=1}^{N} \frac{h/e^2}{4\pi^2 h/e^2} \right] \frac{S^2}{(2N-1)!} (eV)^{2N-1}, \qquad (6.76)$$

where N is the number of junctions and S is the generalized sum of the energy denominators. For identical capacitances, with no stray or self-capacitance on the islands, this term may be written

$$S = N! \left[\prod_{i=1}^{N-1} |\Delta E_i| \right]^{-1}. \qquad (6.77)$$

For $N = 2$, Eq. (6.76) reduces to Eq. (6.75) in the limit of $T = 0$. As the number of junctions increases, the power-law dependence becomes stronger, and the magnitude of the current decreases as roughly α^N, where $\alpha = (h/e^2)/R_t$ (keeping the tunnel resistances the same for simplicity). For $R_t < h/e^2$, the current due to inelastic q-mqt is increasingly suppressed as the number of junctions increases. This fact implies that the stability of single-electron circuits can be increased by increasing the number of junctions per element.

6.2 Single-electron tunneling in semiconductor quantum dots

In Section 6.1, we introduced the basic principles that underlie the phenomenon of single-electron tunneling, and saw how this led to a discussion of the Coulomb-staircase and of Coulomb oscillations. We also briefly mentioned that our analysis was appropriate to Coulomb islands realized in metallic systems, containing a large number of electrons. In such systems, the Fermi wavelength of the carriers is typically much smaller than the island size, so that the influence of quantum confinement on the electronic energy spectrum can be safely neglected. In such a situation, changing the electron number on the dot by a single electron has an essentially negligible effect on both the total charge stored on the dot, and, consequently, on its self-consistent potential (and so its capacitance). Consequently, in an experiment where the Coulomb oscillations (Figs. 6.11 and 6.13) reflecting the electrochemical potential of the dot are measured, the oscillations are found to be periodic in gate voltage, reflecting the fact that the addition energy for the quantum dot is determined by the capacitance alone, independent of gate voltage. (This is even true for the results of Fig. 6.13, which were actually obtained in measurements of relatively large dots realized in heavily doped GaAs [25]. Consequently, the energy spacing of the discrete quantum states of the dot was much smaller than its charging energy, allowing almost periodic Coulomb oscillations to be obtained for these "quasi-metallic" dots.)

The study of single-electron tunneling in semiconductor quantum dots allows the observation of a vast array of new phenomenon, not observable in their metallic counterparts, which arise from the fact that strong quantum confinement in the low-density limit leads to quantized energy levels whose energy scales can be at least comparable to the Coulomb charging energy. As we discuss shortly below, the electron number in the dot can even be tuned all the way down to zero, so that when subsequent recharging of the dot is performed the capacitance that determines the charging energy must actually be considered as an electron-number-dependent quantity. The strong connection of single-electron charging to the eigenstates of semiconductor quantum dots makes the charging process sensitive to the application of a magnetic field, allowing new opportunities to discuss how the population of electron states in these dots is influenced by many-body and spin-dependent interactions. Moreover, in quantum dots realized by an extension of the split-gate method, their tunnel barriers can be implemented with quantum point contacts, allowing the tunnel resistance (R_t) to be varied over a wide range. As we discuss below, this has led to the observation of a tunable Kondo effect in these systems, a discovery that has advanced the basic understanding of this important many-body phenomenon.

The observation of single-electron tunneling effects in semiconductor quantum dots has provided researchers with a powerful means to study their discrete energy spectra with unprecedented accuracy. To appreciate how this connection is made, we can begin our analysis by using the equivalent circuit model of Fig. 6.8 to represent the basic structure of the relevant experiments in semiconductor quantum dots. As illustrated in Fig. 6.12 for a metal system, in the linear response regime in which the applied voltage is small, the condition for lifting the Coulomb blockade is to bias the gate voltage such that the Fermi energy of the dot corresponding to the addition of one electron, $n + 1$, lies between the reservoir energies on the left and right. This condition corresponds to a conductance peak in a measurement of conductance versus gate bias. In the semiconductor system, however, we must additionally account for the energy spacing of the discrete states comprising the allowed states in the dot, as shown in Fig. 6.24. We may write the total ground-state energy in a dot of n electrons as the sum of the filled single-particle energy states plus the electrostatic energy due to the filling of the dot with electrons:

$$E_t(n) = \sum_{i=1}^{n} E_i + E(n_1, n_2), \qquad (6.78)$$

where E_i are the discrete single-particle energies of the quantum dot and $E(n_1, n_2) = E_s - W_s$ is the total charging energy of the system due to tunneling through the left and right barriers given by Eqs. (6.27) and (6.28). We may define the electrochemical potential in the dot as the difference $\mu_d(n) = E_t(n) - E_t(n-1)$, that is, the energy difference associated with the removal of one electron from the dot. In the linear response regime, the applied voltage V_a is small, and therefore the charging energy does not depend on which barrier the electron tunnels through.

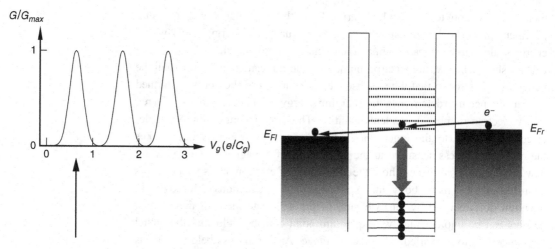

Fig. 6.24 Illustration of the condition for a conductance peak in a semiconductor double-barrier island structure where discrete states co-exist with a Coulomb gap.

Using Eqs. (6.29) and (6.30), the change in charging energy from $n \to n-1$ corresponds to $n_1 \to n_1 - 1$ and $n_2 \to n_2 - 1$, which gives for either case:

$$\mu_d(n) = E_n + \frac{e^2(n-1/2)}{C_{eq}} - e\frac{C_g V'_g}{C_{eq}}, \qquad (6.79)$$

where E_n represents the energy of the highest filled state (at zero temperature), and the prime on V_g signifies that this term may contain the random polarization charge potential Q_p/C_g if necessary. For a given gate bias V_g, if the chemical potential $\mu_d(n)$, corresponding to the addition of an electron to the dot with $n-1$ electrons, lies between μ_l and μ_r of the reservoirs, tunneling may occur. The system goes from $n-1 \to n \to n-1$ alternately, giving rise to current. An increase in gate bias causes the state with n electrons to be stable so that the dot then goes off resonance. The conductance vanishes until the level $\mu_d(n+1)$ lies between the Fermi energies of the reservoirs, and a new conductance peak appears. From Eq. (6.79), we see that the gate period corresponding to these two successive conductance peaks is

$$\Delta V_g = \frac{C_{eq}}{C_g}\frac{E_{n+1} - E_n}{e} + \frac{e}{C_g}, \qquad (6.80)$$

The second term on the right-hand side of Eq. 6.80 corresponds to the constant period of the Coulomb oscillations that we obtained previously, in our discussion of the metallic single-electron transistor (see Fig. 6.11). For semiconductor dots, however, Eq. (6.80) shows that there is an additional contribution to the Coulomb-oscillation peak spacing, which arises from the non-zero energy separation of their nth and $(n+1)$th quantized levels (see Fig. 6.24).

6.2.1 Semiconductor quantum dots as artificial atoms

The concepts that lead to the derivation of Eq. (6.80) are collectively referred to as the *constant-interaction* model [20,46,47] of single-electron tunneling, since they rely on an implicit assumption that the energy separation of the discrete states of the quantum dot remains unchanged as these states are populated or depopulated. (It is also assumed that the Coulomb interactions of carriers on the dot may be expressed in terms of an appropriate capacitance that is also independent of the charge stored on the dot.) The results of early experiments on the Coulomb blockade in semiconductor dots have been extensively reviewed by Kouwenhoven *et al.* [48]. The use of single-electron tunneling to study the level spectrum of quantum dots was initially pioneered by McEuen *et al.* [49,50,51] and by Ashoori *et al.* [52,53,54]. The full range of this approach was really demonstrated, however, in later joint studies performed by the Kouwenhoven and Tarucha groups. These authors investigated single-electron tunneling in quantum dots of the type shown in Fig. 6.25. These dots were formed in a double-barrier heterostructure, by etching excess material to leave a sub-micron-diameter column with an isolated (InGaAs) quantum dot in its middle. By forming electrical contacts to the top and bottom of this column (indicated), an electrical current could be driven through it. At the same time, a novel wraparound gate (also

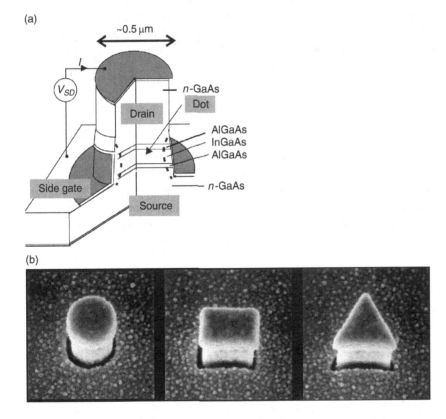

Fig. 6.25 Double-barrier quantum-dot structures used to investigate single-electron tunneling and artificial-atom behavior in few electron quantum dots. (a) shows the layer structure of these devices, while (b) shows actual devices implemented with different geometries ([56], with permission).

indicated) was used to control the number of electrons in the dot. Through suitable control, this number could be reduced to zero, and subsequently increased one at a time. A major achievement of this work was to demonstrate true "artificial-atom" behavior in the filling of electron states, in a manner directly analogous to that which occurs in natural atoms [55,56].

In natural atoms, it is well known that the filling of electronic orbitals results in the existence of particular "magic numbers" (2, 10, 18, 36, . . .) for which the resulting atoms have a stable electronic configuration. As is now well understood, these correspond to values of the electron number for which all occupied energy shells are completely filled. For atoms with electron numbers intermediate between these unique values, it is also known that Hund's rule drives the filling of states in a manner that maximizes total spin, thereby lowering the Coulomb repulsion among the different orbitals. Prior to the experiments of Ref. [55], the existence of a similar shell structure for electrons in quantum dots was anticipated [57,58,59] in a number of theoretical studies. Due to the different nature of the confining potential for electrons in an atom and a quantum dot, however, it is necessary to start from an understanding of the form of the self-consistent potential in quantum dots, to achieve a proper understanding of their magic numbers and filling characteristics. The self-consistent potential of such dots has been calculated by numerous authors [60,61,62,63,64,65,66,67]. For the case of interest here, of a circular pillar fully depleted of electrons, Fermi-level pinning at the etched surface gives rise to depletion regions that extend smoothly into the interior of the dot, so that a parabolic approximation may reasonably be used [55]:

$$V(x,y) = \frac{1}{2}m^*\omega_0^2(x^2+y^2), \tag{6.81}$$

where ω_0 is the oscillator frequency that characterizes the potential and we have assumed that the quantum dot lies in the x–y plane. An advantage of assuming this specific form for the potential is that, by introducing it into the Schrödinger equation, the resulting differential equation may be solved analytically, even for an arbitrary magnetic field applied perpendicular to the plane of the dot. In this case, one obtains the well-known Darwin–Fock spectrum [68,69], whose eigenstates are described by the form:

$$E_{n,m} = (n+1)\hbar\Omega + \frac{m}{2}\hbar\omega_c \pm \frac{1}{2}g^*\mu_B B, \tag{6.82}$$

where ω_c is the cyclotron frequency and Ω is a hybrid frequency, given by $\Omega^2 = \omega_c^2 + \omega_0^2$. The quantum number n and m satisfy the following relations:

$$n = 0, 1, 2, \ldots, \tag{6.83}$$

$$m = n, n-2, n-4, \ldots, -n+2, -n. \tag{6.84}$$

6.2 SET in semiconductor dots

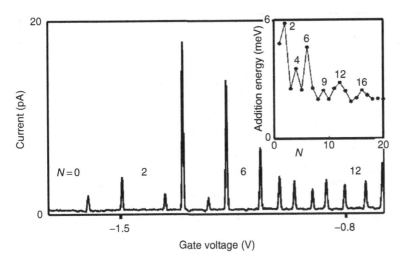

Fig. 6.26 Gate-voltage-dependent variation of current through a circular quantum dot such as that of Fig. 6.25. The first peak on the left-hand side indicates the first electron entering the dot, while successive peaks denote the addition of subsequent electrons one at a time. The gate-voltage separation of successive peaks can be used to determine the variation of the dot addition energy as a function of gate voltage, as plotted in the inset ([56], with permission).

From inspection of Eq. (6.82), it is clear that, at zero magnetic field, the quantum number n defines a series of degenerate energy shells with degeneracy factors (accounting for spin) of 2, 4, 6, 8, ... for the $n = 0, 1, 2, 3, ...$ shells, respectively. Consequently, one expects magic numbers of 2, 6, 12, 20, ... electrons for such two-dimensional artificial atoms, in contrast to the series exhibited by natural atoms. A clear example of this shell structure is shown in Fig. 6.26, in which we show the Coulomb oscillations measured at $B = 0$ in a circular dot like the one shown in the bottom-left image of Fig. 6.25. For gate voltages corresponding to the leftmost end of the figure the dot is empty of electrons ($N = 0$, as indicated in the figure). As one moves along this axis towards less-negative gate voltage, successive electrons are added to the dot, as indicated by the sequence of sharp Coulomb oscillations. It is immediately clear from this figure that these oscillations are *not* periodically spaced in gate voltage. Of particular importance, much larger spacings separate the 2nd and 3rd, the 6th and 7th, and the 12th and 13th peaks, compared to the neighboring oscillations, consistent with the expected magic numbers derived from Eqs. (6.82) and (6.83). This is illustrated quantitatively in the inset to the figure, which plots the voltage separation of successive peaks and shows local maxima for adding the 3rd, 7th, and 13th electrons. It is furthermore clear from this figure that the energy needed to add the 5th, 9th, and 17th electrons is also enhanced, although not to the same extent as for the magic numbers. Such behavior is actually a manifestation of Hund's rule, according to which it is

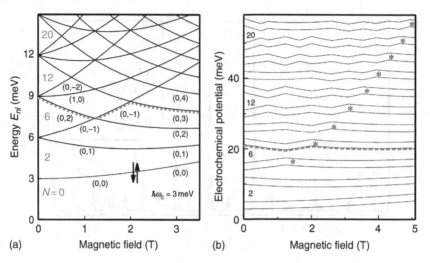

Fig. 6.27 (a) The single-particle states, or Darwin–Fock spectrum, for a parabolic dot in a transverse magnetic field. Spin degeneracy of these states is assumed here, meaning that any Zeeman term is neglected (see corresponding discussion in the text). (b) In this plot the Darwin–Fock spectrum of (a) is replotted to show the variation of the corresponding electrochemical potential, the energy needed to add successive electrons to the dot ([56], with permission).

preferred to first half fill a given energy shell with electrons with the same spin projection, thereby minimizing the exchange energy.

Richer behavior yet is revealed by studying the evolution of the Coulomb oscillations in the presence of a perpendicular magnetic field. At zero magnetic field, the energy shells of the two-dimensional oscillator are $2(n + 1)$-fold degenerate (accounting for spin), with states that are defined by unique values of the quantum number m. Physically, the significance of this quantum number is that it defines states of different orbital angular momentum:

$$L_z = m\hbar, \tag{6.85}$$

where L_z is the z-component of the angular momentum. As indicated by Eq. (6.82), the application of a magnetic field lifts the degeneracy of the orbital states, causing a dispersion that is proportional to the magnitude of the angular momentum.

Combined with the breaking of spin degeneracy, which is also described by Eq. (6.82), the lifting of orbital degeneracy in a magnetic field leads to a characteristic evolution of the Coulomb oscillations. Many (although not all) of the features of this evolution can be understood within the context of the constant-interaction model. To see how this model applies to charging of quantum dots in a magnetic field, we begin by considering Fig. 6.27(a), which shows the calculated evolution of the one-electron energy states of Eq. (6.82) (spin is neglected here). In an experiment in which a gate is used to populate

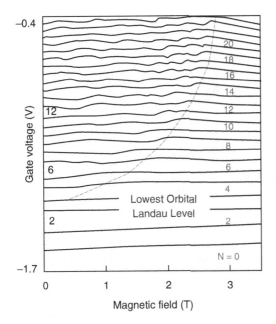

Fig. 6.28 Measurements of the electrochemical potential of a circular quantum dot, showing the correspondence to Fig. 6.27(b) ([56], with permission).

such states, we might expect that they should be filled sequentially with spin-up and spin-down electrons. In this case, we would expect from Eq. (6.80) that the gate voltage separation between Coulomb peaks associated with adding spin-up and spin-down electrons to the same dot level should be just e/C_g (since $E_{n+1} = E_n$), while populating a higher level should lead to a larger Coulomb-peak separation (proportional to $E_{n+1} - E_n$). This concept is illustrated in Fig. 6.27(b), which plots the evolution of the electrochemical potential as a function of electron number and magnetic field. (Note that in these calculations the Zeeman term in Eq. (6.82) is neglected, since in the GaAs dots of interest here it is generally much smaller than the orbital energy [55]). It is clear from this figure that one expects to observe Coulomb peaks that evolve in *pairs* with magnetic field and gate voltage, due to the addition of spin-up and spin-down electrons to each discrete state. As indicated in Fig. 6.27(a), the kinks that occur in each of the curves are associated with the avoided crossings of states from different energy shells, and with different orbital quantum numbers, which approach each other in energy with increasing magnetic field. The dashed lines in Figs. 6.27(a) and (b) denote, for a given electron (the 7th), the kinks in the Coulomb-peak evolution which correspond to different avoided crossings. The experimental demonstration of this effect is shown in Fig. 6.28, which shows behavior that is strikingly reminiscent of that of Fig. 6.27(b). Each of these curves shows a sequence of kinks up to some characteristic magnetic field, beyond which these features abruptly terminate. As denoted in the figure, the termination of the kinks demarcates the situation where all electrons have dropped to occupy the lowest Landau level, in which case no further crossings of levels are expected. Instead, all states of the quantum dot simply increase in energy with increasing magnetic field in this regime [49–51]. What is most important for the discussion here, however, is that clear grouping of pairs of curves can be seen, denoting the sequential filling of successive quantum states by spin-up and spin-down electrons. Indeed, at such high magnetic fields, all signatures of the shell spectrum are essentially lost, consistent with the fact that in this regime the eigenspectrum of the dot consists of densely spaced levels.

The results of Figs. 6.27 and 6.28 suggests that the constant-interaction model can be very successful in accounting for filling of electron states in artificial-atom quantum dots. Nonetheless, important deviations from its predictions have been found in experiment. At low magnetic fields, in particular, the pair-like bunching of Coulomb peaks departs from a simple spin-up/spin-down sequence, indicating the role of Hund's rule. This situation is illustrated in Fig. 6.29(a), which shows the Coulomb peaks associated with the addition of the 3rd–6th electrons, and their evolution with magnetic field. Beyond ∼0.5 T, it is clear that the peaks for the 3rd and 4th, and for the 5th and 6th, electrons follow the same evolution, indicating that these are associated with the population of the same dot state by a spin-up and a spin-down electron. Interestingly, however, the pairing of the curves below 0.4 T seems to involve the 3rd and 5th, and the 4th and 6th, electrons. In other words, in this regime it is preferable for successive electrons to populate *different* dot states while keeping their spins unpaired, just as expected from Hund's rule. The sequence of filling that occurs in this case is indicated schematically in Fig. 6.29(b), which shows how the 3rd and 4th electrons occupy different orbitals but have parallel spins.

With increasing magnetic field in Fig. 6.29, the pairwise grouping of Coulomb peaks noted already in Fig. 6.28 is once again recovered, indicating a transition to the level filling that is also indicated in Fig. 6.29(b). To understand the origin of this transition, it is necessary to consider the competition between the orbital energy cost ($\hbar\omega_c$, see Eqs. 6.82–6.84) in adding electrons to different orbital states, to the exchange-energy saving (K) that can be achieved by doing this. At zero magnetic field, the orbital states are degenerate and the exchange-energy consideration is dominant. As the magnetic field is increased, however, $\hbar\omega_c$ eventually becomes larger than K and, at this point, the pairwise grouping of Coulomb peaks should again resume. Based on these arguments, the transition at 0.4 T should correspond to the condition $\hbar\omega_c = K$, yielding $K = 0.7$ meV for these dots. Using this value, the authors of Ref. [55] were able to calculate the expected evolution of the Coulomb oscillations as a function of magnetic field and the results of these calculations in Fig. 6.29(b) show excellent agreement with the experiment in Fig. 6.29(a). This experiment therefore provides an excellent demonstration of the need to go beyond the constant-interaction model, to include mechanisms such as exchange when discussing the filling of states in quantum dots.

Important deviations from the constant-interaction model have also been found in other experiments. We recall that one assumption of this model is that

6.2 SET in semiconductor dots

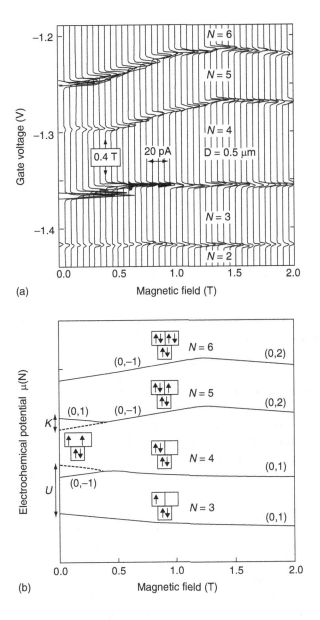

Fig. 6.29 The variation of the Coulomb peaks for the addition of the 3rd through 6th electrons of a circular quantum dot, as a function of magnetic field. The upper panel (a) shows the experimental data, while the lower panel (b) shows the corresponding evolution of the spin configuration as determined from theory ([56], with permission).

the discrete eigenspectrum of the dot is not affected by populating it with successive electrons. Patel *et al.* studied the statistics of the Coulomb-oscillation spacings for dots containing large numbers of electrons (of order several hundred) and showed evidence for "scrambling" of the eigenspectrum after the addition of only a small number of electrons to the dot [70]. Stewart *et al.* performed excitation spectroscopy of similar quantum dots and found evidence for a single-particle-like picture in which successive levels of a fixed spectrum are successively filled [71]. At the same time, however, they also found evidence for an absence of spin degeneracy, which they attributed to the presence of

Fig. 6.30 Experimental charge-stability diagram for a quantum dot, showing excited states outside of the main Coulomb diamonds (which are centered on $V_{SD} = 0$) ([56], with permission).

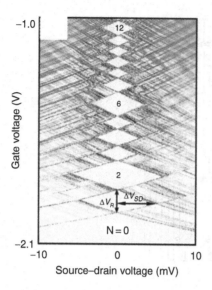

electron interactions in the dot. In other work yet, Potok *et al.* investigated the spin transitions associated with the charging of many-electron dots and also found evidence of high-spin ground states [72].

6.2.1.1 The excitation spectrum of semiconductor quantum dots

In our discussion of semiconductor quantum dots thus far, we have focused on the properties of the Coulomb oscillations that are observed in their *linear* conductance (i.e. in the limit of small source–drain bias, V_{SD}). As we saw in our previous discussion of the single-electron transistor, simultaneous variation of the transistor gate and source–drain voltages allows their *Coulomb diamond* to be determined (Figs. 6.9 and 6.10). In semiconductor quantum dots, the determination of this Coulomb diamond is also of particular value, since it provides a useful means to perform a spectroscopy of the excited states of the quantum dot. The basic principle of such measurements is to use the source–drain voltage to probe the characteristic energy splitting of the quantum states, in the limit where these are thermally resolved. In Fig. 6.30, we show the result of measuring the Coulomb diamonds for a dot similar to those of Fig. 6.25. Note how in contrast to Fig. 6.9, the size of successive diamonds (the white regions of low differential conductance) is quite different, consistent with the shell structure that we have already explained. Outside of these regions, however, clear sets of parallel lines can be seen to run parallel to the edges of the diamonds. These lines are associated with transport through excited states of the quantum dot, which become accessible as the source–drain voltage is increased. When the source–drain voltage is small, and the gate voltage is adjusted to a Coulomb peak, tunneling through the dot involves only the lowest discrete state

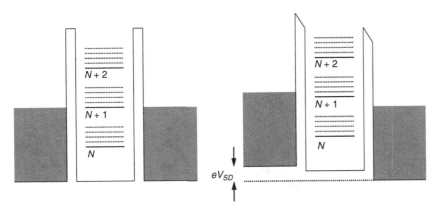

Fig. 6.31 Schematic diagram indicating the ground-state levels for successive electrons in a quantum dot (solid lines), as well as possible excited states of these electrons (dotted lines). Note in this case how the spacing of successive dot states is significantly smaller than the charging energy.

of the dot. As the source–drain voltage becomes larger than the separation of the dot states, however, the current flow can occur through multiple discrete states of the quantum dot (Fig. 6.31; this picture is therefore most appropriate when the spacing of dot states is significantly smaller than the charging energy). As the gate voltage is varied at fixed source–drain bias, the number of dot states that can contribute to transport changes [73], and it is this process that leads to the appearance of the fine lines in the Coulomb diamond.

Some further comments should be made about the additional structure that appears in the excitation spectroscopy for semiconductor dots. As we have discussed previously, the main diamonds, centered about $V_{SD} = 0$, correspond to situations for which the electron number on the dot is fixed and transport is correspondingly blockaded. Transport through the dot is allowed, however, for combinations of the gate voltage (V_g) and V_{SD} for which one moves outside of the diamonds. If we now consider the diamond that corresponds to the occupation of the Nth electron on the dot, the upper (lower) edges of this diamond then indicate the transition to a new ground state of $N + 1$ ($N - 1$) electrons. Now we consider how this behavior is modified as a result of transport through excited states of the dot, using the charge stability diagram shown in Fig. 6.32. In this figure, which resembles that of Fig. 6.9, regions denoted "CB," "SET," and "2ET," correspond to regions of fixed electron number on the dot (Coulomb blockade), single-electron tunneling, and two-electron tunneling, respectively. (Tunneling for even larger numbers will be possible for diamonds at even larger V_{SD} than shown here.) Also shown in the figure are dotted (dashed) lines that run parallel to the upper (lower) edges of each diamond. These denote the condition for accessing excited states of the dot and for clarity here we only consider the case of tunneling via the first excited state (the actual situation in experiment is more complicated, as indicated in Fig. 6.30). For the Coulomb diamond corresponding to the Nth electron, the dotted (dashed lines) therefore indicate tunneling via the first excited state of the $N + 1$ ($N - 1$) electron dot. Note that both sets of lines are assumed to be parallel to the edges

Fig. 6.32 Schematic diagram indicating the expected form of the charge-stability diagram for a semiconductor quantum dot with discrete levels. Regions of Coulomb blockade (CB), single-electron tunneling (SET) and two-electron tunneling (2ET) are indicated. Dotted and dashed lines indicate tunneling via excited states. See text for a more detailed discussion.

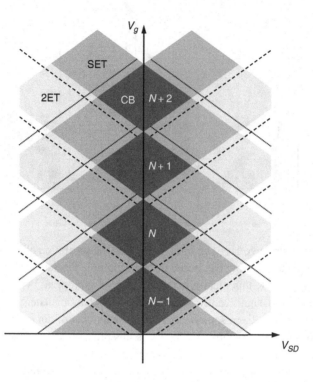

of the Coulomb diamonds, which is actually only valid within the constant-interaction approximation.

An interesting feature of transport via excited dot states, in contrast to the situation for single-electron tunneling via the ground state, is that accessing additional states for transport does *not* necessarily result in an increase of conductance, and may even lead to a decrease. This behavior was pointed out by Weis *et al.* [74], who discussed it in terms of the manner in which the source–drain voltage influences the accessibility of dot states to the two reservoirs. They pointed out that, in the regime where single-electron tunneling occurs, the onset of tunneling via an excited state of the dot must be accompanied by a simultaneous suppression of tunneling via the ground state. Due to different spatial properties of the wave function in the ground and excited states, which will determine the effective time that the electron spends trapped on the dot, the conductance may therefore actually be decreased by accessing the excited states. In their measurements of the excitation spectroscopy of their dot, these authors actually found regions of negative differential conductance, which were later explained in terms of a *spin-blockade* effect by Weinmann *et al.* [75]. The idea here is that, when considering transport through discrete states, in addition to considerations related to the Coulomb energy, one must also take spin selection rules into account. In particular, if the excited state is associated with a large total spin, which cannot be realized by adding a single electron to the dot, then the

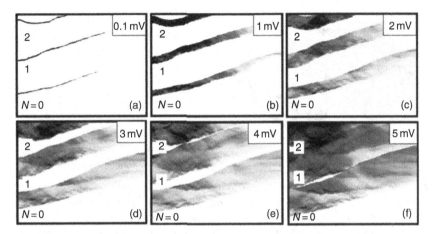

Fig. 6.33 Measurements of the electrochemical potential variation of a quantum dot as a function of magnetic field, for several distinct values of the source–drain bias voltage. Plots are obtained by sweeping gate voltage and measuring the resulting variation of conductance, as the magnetic field is incremented. Dark lines in (a) denote the usual peaks of Coulomb oscillations as successive electrons are added. Note how these lines broaden into *stripes* as the source–drain voltage is increased ([56], with permission).

current will be suppressed by this state. Probably some of the nicest experimental demonstrations of spin-blockade effects has been provided in the subsequent work by Sachrajda and co-workers [76,77,78,79,80]. These authors investigate single-electron tunneling in single and coupled lateral quantum dots at high magnetic fields, where electrons in the reservoirs all occupy the lowest Landau level. As discussed already in Chapter 4, the current in these regions is carried by spin-polarized edge states which are located at different distances from the confining boundaries of the sample. Consequently, at the points where electrons tunnel between these edge states and the quantum dot, the tunneling will be dominated by the edge state located closest to the dot. In practice, this means that it is possible to use the reservoirs as a means to inject spin-polarized carriers into and out of the quantum dot. These authors have made extensive use of this approach to study the spin transitions in quantum dots at high magnetic fields, including singlet–triplet transitions, and, more recently, cotunneling through coupled quantum dots.

With a non-zero V_{SD} applied, the curves denoting the evolution of specific Coulomb peaks in a magnetic field (Fig. 6.28), broaden to form stripes of width eV_{SD}/α on the gate voltage scale, where α is the "lever arm" that relates a change of gate voltage to a change of electrochemical potential of the dot. Such behavior is clearly shown in Fig. 6.33, which shows that, for sufficiently large V_{SD} (in this case 5 mV), there can be significant overlap of different strips, indicating the possibility of simultaneous transport involving ground and excited states for

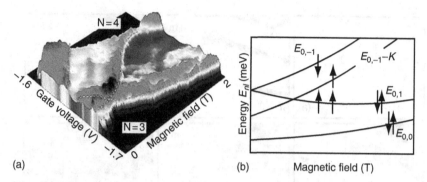

Fig. 6.34 (a) Variation of the stripe due to the addition of the 4th electron, as a function of magnetic field. The source–drain bias applied here is 1.6 meV. (b) Schematic diagram illustrating the relation between the features in the experimental data of (a) and different spin configurations for the four electrons ([56], with permission).

different electron numbers. Kouwenhoven *et al.* have used this technique to investigate various transitions between the ground and excited states of quantum dots [81]. In Fig. 6.34 we show their results for a four-electron dot, in which spin-dependent transitions for the 3rd and 4th electrons are induced with increasing magnetic field.

In the discussion thus far, we have seen how the magneto-tunneling spectra of isolated quantum dots can be remarkably well explained in terms of the properties of the noninteracting Darwin–Fock states. While these states experience a dispersion in a magnetic field, the details of this are assumed to be well described by Eq. (6.82) for a fixed confining potential. In reality, however, the application of a magnetic field distorts the electron wave functions from their original forms, essentially compressing the electron probability distribution, behavior that is neglected in the constant-interaction model. At the same time, the wave functions are also distorted by electron interactions, which will tend to drive electrons apart to minimize their repulsion. A proper treatment of these effects requires a fully self-consistent analysis, particularly in the regime of small electron number (where we have already seen pronounced shell structure.)

An important example of effects that go beyond the constant-interaction model is provided by the singlet–triplet transition that has been found to occur in two-electron dots in a magnetic field. According to the noninteracting Darwin–Fock spectrum, a transition from a (spin $S = 0$) singlet to a ($S = 1$) triplet state is expected due to a crossing of the electron orbitals, at a magnetic field in excess of 25 T. By taking account of the magnetic-field-induced shrinkage of the electron wave functions, however, this transition is reduced to as small as a few Tesla. Tarucha *et al.* have studied this transition in detail and have found that it is replicated when the electron number stored on the dot is even [82]. They have also used these investigations to fully parameterize the electron–electron interactions, quantitatively determining the magnitude of the exchange and Coulomb interactions, as well as their dependence on the dot-electron number. The exchange energy was found to decrease with increasing electron number, due to the increased importance of screening. The agreement with theory obtained in these studies is therefore truly impressive.

Finally in this section, we note that in the discussion thus far we have been concerned with inferring information on level transitions in quantum dots, from studies of their d.c. conductance. Further important information on these transitions can be obtained, however, from measurements of the transient conductance of such dots [83,84,85]. Measurements of this type allow the relaxation time of electrons in specific spin states to be studied, revealing the role of transition rules in such processes. A typical approach in experiments of this type is to apply a (repetitive) pulse to the gate of the device to first induce tunneling into an excited state of the dot, and to then remove this pulse and to study how the electron relaxes to the ground state. When the transition is one between states with the same spin, but different orbital quantum number (m), the relaxation process is found to be fast, on the scale of a few ns. The relaxation time is also found to decrease with increasing magnetic field, which has been well explained in terms of the efficient relaxation of the excess energy via spontaneous phonon emission. In marked contrast to this, relaxation times that are longer by several orders of magnitude (approaching 10^{-3} s!) are obtained for transitions between excited and ground states that require a spin flip. By studying the dependence of the relaxation process on the initial excitation state, Fujisawa *et al.* were able to show that the main mechanism for the "forbidden" transitions is actually a cotunneling process, in which the electron spin in the dot is flipped by carriers tunneling from the leads, after which the relaxation can proceed rapidly [85].

6.2.2 The Kondo effect in semiconductor quantum dots

In the preceding sections, we have seen how quantum dots may behave as artificial atoms, with discrete energy levels that may be tuned by variation of a magnetic field, or the electron number. In this section, we discuss another important phenomenon exhibited by quantum dots, in which they essentially function as an "artificial impurity," more specifically a magnetic impurity that gives rise to unique manifestations of the Kondo effect.

Before proceeding with a discussion of the features of the Kondo effect in quantum dots, let us first recall the main features of the Kondo effect as exhibited by dilute magnetic alloys. While the resistance of pure metals is well known to decrease with decreasing temperature, due to the associated suppression of electron–phonon scattering, it was first shown more than seventy years ago [86] that doping such metals with a very dilute concentration of magnetic impurities (such that the interaction among these impurities can be neglected) causes a change in the sign of the temperature coefficient of the resistance, which now *increases* with decrease of temperature. This behavior is only observed at low temperatures, of order a few degrees Kelvin, and its origins remained a mystery for some thirty years, until the pioneering work of Kondo [87].

The key advance in understanding made by Kondo was the realization that, to properly describe the scattering of conduction electrons in the host metal by magnetic impurities, it is necessary to go beyond usual first-order perturbation treatments of scattering. This is because of the importance of virtual *exchange-scattering* processes, in which the conduction electrons scatter between some initial and final state, via an intermediate state in which their spin is flipped. This particular type of scattering is possible since the magnetic impurity can serve as a spin buffer to conserve total spin momentum, flipping its spin simultaneously with that of the conduction electron. While this Kondo effect is therefore a many-body phenomenon, it is worth mentioning that it is not one that involves a direct electron–electron interaction. Rather, the interaction between electrons is mediated by the spin of the magnetic impurity, with a first electron flipping this spin, which then interacts with a second electron (and so on) [88]. A first-order perturbation treatment of this scattering yields a resistivity contribution that does not differ quantitatively from that for scattering from non-magnetic impurities, and so is unable to account for the behavior found in experiment. Kondo showed, however, that the higher-order terms of this perturbation, which would normally be expected to be small, actually yield a divergent resistivity that should become infinite at higher temperatures. While this treatment captures the essence of the unusual resistivity variation found in experiment, the divergence itself is unphysical and is a consequence of the breakdown of the perturbation approximation in the low-temperature limit. In a later approach, Wilson studied this problem using non-perturbative numerical renormalization [89]. He showed that the divergence is indeed an artifact and that the resistance due to exchange scattering should saturate in the low-temperature limit. In this regime, the localized magnetic moment of the magnetic impurity is essentially completely screened by the spin exchange with the conduction electrons.

The Kondo effect causes a distinctive modification of the density states of the non-magnetic metal, with the net effect of repeated exchange scattering being the formation of a narrow resonance that is situated at the Fermi level. The energy width of this resonance is nothing more than the Kondo temperature, $k_B T_K$, which can be expressed in turn in terms of the parameters of the so-called Anderson model [90]. This treats the magnetic impurity as a discrete quantum level that may be occupied by a spin-1/2 electron, while an on-site energy U must be overcome to add an additional electron to the level. The coupling of this level to the host metal causes it to be broadened by an amount Γ. Subsequent to the work of Anderson and Kondo, Haldane [91] later showed that the Kondo temperature is related directly to the parameters of the Anderson Hamiltonian, according to

$$T_K = \frac{1}{2}(\Gamma U)^{1/2} \exp\left[\pi\varepsilon \frac{\varepsilon + U}{\Gamma U}\right], \qquad (6.86)$$

where ε is the energy of the discrete level that supports the local magnetic moment and the other parameters have already been defined.

Subsequent to the original work on dilute magnetic alloys, Glazman and Raikh [92] and Ng and Lee [93] predicted that, by functioning as an Anderson-type system, it should be possible for quantum dots to also exhibit Kondo behavior. The first experimental verification of this was provided ten years later by Goldhaber-Gordon and colleagues [94,95], and many demonstrations have subsequently followed. In order to observe the Kondo behavior, it is necessary to study the behavior in the regime where the barriers of the dot are close to conducting (i.e. their individual conductances are approaching $2e^2/h$). In this limit, an unusual temperature-dependent behavior is observed in the conductance, at the minima of the Coulomb oscillations that correspond to the occupation of an odd number of electrons (and so a net spin of $1/2$). As described by Eq. (6.32), it is normally expected that a decrease of temperature should induce a decrease of the conductance in between the Coulomb-oscillation peaks, behavior that is indeed found in usual experiments. The characteristic signature of the Kondo effect, on the other hand, is an *increase* of the conductance with *decreasing* temperature in the odd-electron valleys, while the even valleys show the usual behavior of Eq. (6.32). An example of this behavior is provided in Fig. 6.35, which shows results of the study of van der Wiel *et al.* [96]. As indicated in Fig. 6.35(a), the even valleys indeed show a decrease of conductance with decreasing temperature, while the odd valleys show the opposite behavior.

The unexpected increase of the conductance with decreasing temperature, noted for the odd-electron Coulomb minima, is opposite to the behavior found in the traditional Kondo effect, in which the *resistance* instead increases with decreasing temperature. To understand the origins of the Kondo effect in quantum dots, consider the diagram shown in Fig. 6.36. The initial state corresponds to the situation at a Coulomb-oscillation minimum, where the source–drain voltage is insufficient to overcome the energy needed to add an additional electron to the dot, under which conditions we would normally expect the conductance to exhibit a minimum. For the case considered in Fig. 6.36, however, the uppermost occupied energy level of the dot is occupied by only a single spin. In this situation, aided by the relatively high transmitivity of the barriers, a cotunneling process is allowed, in which the electron on the quantum dot tunnels out to the right reservoir, momentarily leaving the initially occupied level empty (note the shift of the dot energy by the charging energy in the middle panel of Fig. 6.36). At this point, an electron may tunnel into the dot to replace the missing electron, at which point it appears that the initial state has been

Fig. 6.35 The characteristic signatures of the Kondo effect due to a localized spin in a quantum dot. (a) In linear conductance, the Kondo valleys corresponding to the localization of an odd number of spins show an unexpected trend of increasing conductance with decreasing temperature.
(b) Conductance as a function of temprature, measured for the three gate voltages indicated by arrows in (a). (c) The data of (b) can be rescaled onto a universal curve by defining an appropriate Kondo temperature (Kouwenhoven and Glazman, *Physics World*, **33** (2001), with permission).

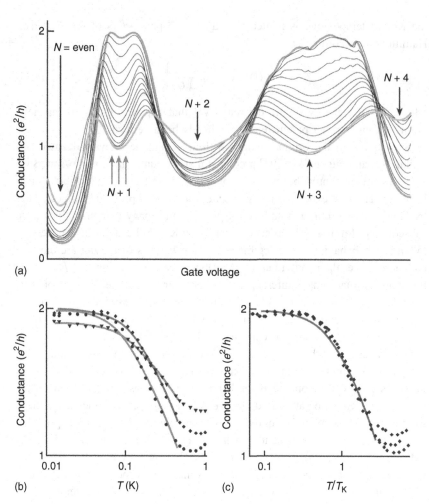

restored. Note, however, in Fig. 6.36 that the spin of the electron on the dot is now opposite to its initial state. This is therefore the equivalent of the exchange process discussed above for the Kondo effect in dilute magnetic alloys. It is important to appreciate here that the electron that tunnels into the dot to achieve the final state may have either spin. Since the cotunneling process takes place continuously, the time-averaged value of the dot spin will essentially vanish. (At the same time, the combined effect of the continued scattering is to give rise to a narrow Kondo resonance at the Fermi level, as we indicate in the left panel of Fig. 6.36, in which we superimpose the density of states of the dot on top of the energy-level diagram.) That is, by taking account of exchange scattering, the original spin on the dot is essentially screened out by the reservoirs, just as in the conventional Kondo effect for metals. An important difference, however, is that it is clear that the cotunneling in Fig. 6.36 *enhances* transmission through

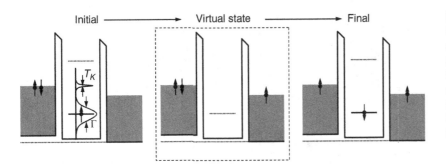

Fig. 6.36 Schematic figure illustrating the tunneling sequence that yields a spin flip on a quantum dot in the Kondo regime. The spin flip leads to a sharp Kondo resonance at the Fermi energy, as indicated in the left schematic.

the dot, so that we would expect an increase of the dot conductance as this Kondo effect grows in prominence. This is exactly what we see for the odd-electron minima in Fig. 6.35.

A remarkable feature of the Kondo effect, in either its conventional or quantum-dot forms, is the ability to perform a *universal* scaling of the conductance, in terms of the Kondo temperature. In dilute magnetic alloys, this scaling is best expressed in terms of the resistance:

$$\frac{R(T)}{R_0} = f(T/T_K), \tag{6.87a}$$

indicating that the resistance (R), when expressed in terms of the zero-temperature resistance (R_0) can be expressed purely as some function of temperature divided by the Kondo temperature ($f(T/T_K)$). Goldhaber-Gordon and colleagues showed [95] that, for the corresponding case of the Kondo effect in quantum dots, the temperature-dependent variation of the *conductance* is well described by

$$\frac{G(T)}{G_0} = \left[\frac{T_K'^2}{T^2 + T_K'^2}\right]^s, \tag{6.87b}$$

where:

$$T_K' = \frac{T_K}{\sqrt{2^{1/s} - 1}}, \tag{6.88}$$

and s is a fit parameter whose value should be close to 0.2 for a spin-1/2 impurity. This result (Eq. (6.87b)) is actually an analytical approximation to the results of the numerical renormalization for the Anderson impurity model [97]. In their work, Goldhaber-Gordon *et al.* found that the temperature dependence of the conductance in the Kondo valley could be well described by the form of Eq. (6.87b), using a value of $s = 0.20$, in agreement with expectations. At the lowest temperatures, the value of the conductance in the Kondo valley can rise to a value of $2e^2/h$ over a wide range of gate voltage. This behavior, which can be clearly seen in Fig. 6.35, was first demonstrated by van der Wiel *et al.* [96] and is referred to as the *unitarity limit* [98]. When this limit is reached, the dot becomes completely transparent since the localized spin on the dot is fully screened by

Fig. 6.37 Zero-bias anomaly in the differential conductance of a quantum dot in the Kondo regime. The anomaly corresponds to the enhanced conductance (dark) at the center of the Coulomb diamond ([103], with permission).

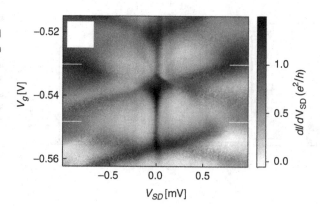

spin-flip processes involving the reservoirs. Consequently, the transmission is limited by the number of modes in the tunnel barriers. In quantum dots realized with point contact barriers, the number of modes in question is just one, yielding the saturation of the valley conductance at $2e^2/h$. Prior to reaching the Kondo limit, however, the conductance is found to increase *logarithmically* with decrease of temperature, reminiscent of the logarithmic increase of resistance found in the traditional Kondo effect.

The Kondo resonance formed at the Fermi level may also be probed by the application of a magnetic field, or non-zero source–drain bias [99,100]. To understand the influence of a finite bias voltage (V_{SD}), it is important to realize that, at thermal equilibrium, the exchange scattering process that leads to an additional resonance in the density of states of the dot also enhances the density of states at the Fermi level in both reservoirs. With a sufficient source–drain bias applied, however, electrons at the quasi-Fermi level in the higher-energy lead can no longer resonantly tunnel into the enhanced density of states of the other reservoir, and the Kondo enhancement of the conductance is suppressed. Consequently, one expects to observe a zero-bias peak (or zero-bias anomaly) in the differential conductance as a function of the source–drain bias. The width of this peak in source–drain voltage is just $k_B T_K/e$, providing a direct means to determine the Kondo temperature [94–96,101,102]. Similarly, in a magnetic field, the localized state that binds the spin on the quantum dot should develop a Zeeman splitting, so that the Kondo resonance becomes similarly split [99,100]. A measurement of the differential conductance as a function of source–drain bias should then exhibit *two* distinct peaks, one each at non-zero V_{SD} but with opposite polarity (recall that we discussed these ideas briefly already in Chapter 5, in regards to proposals for a Kondo effect in quantum point contacts). An example of the zero-bias anomaly is shown in Fig. 6.37 [103]. Here, the differential conductance is actually plotted as a gray-scale contour, as a function of gate voltage and V_{SD}, and the central white regions in this plot correspond to the Coulomb diamonds where the conductance is suppressed. Note, however, in

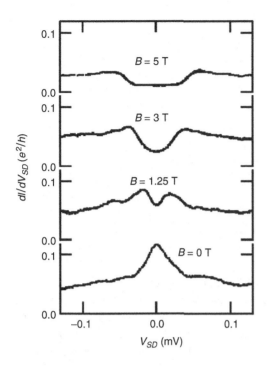

Fig. 6.38 The splitting of the zero-bias anomaly in a magnetic field. Arrows indicate the splitting of the original peak in a magnetic field. Figure reproduced with permission from A. Kogan et al., *Phys. Rev. Lett.* **93**, 166602 (2004).

this figure, the dark vertical line running upwards at $V_{SD} = 0$. This corresponds to a narrow region of enhanced conductance, and is the Kondo resonance that is suppressed by applying a source–drain bias. In Fig. 6.38, we show the splitting of the zero-bias peak in a magnetic field. Note the steady increase in the separation of the two peaks with magnetic field, which is consistent with a total Zeeman-energy splitting of $2g^*\mu_B B$.

It is clear already from the discussion above that the Kondo effect occurring in quantum dots reveals significant new features of the interaction between a localized spin and a continuum. Some other features of this Kondo effect are also worthy of discussion, however. We have noted that the Kondo effect is typically seen when the number of electrons localized on the quantum dot is odd, since in this case the corresponding net spin should be $1/2$ and the exchange-scattering process is possible. By the same token, one does not normally expect to observe a Kondo effect when the number of electrons on the dot is even, and this is indeed apparent in the data of Fig. 6.35, in which the even-electron Coulomb valleys show the usual temperature-dependent behavior expected for Coulomb oscillations. In spite of this, however, it has been found that a Kondo effect may be observed in quantum dots containing an even number of electrons, when a large magnetic field is used to induce appropriate transitions among the different orbitals of the dot [104,105]. To understand the behavior in this regime, we can consider the collective states of the two uppermost electrons

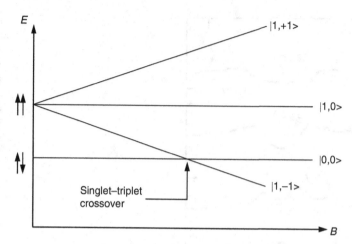

Fig. 6.39 Schematic illustration showing the evolution of the singlet and triplet states of a two-spin system in a magnetic field. Note the crossover at non-zero magnetic field from a singlet to a triplet ground state.

in the dot, and their evolution as a function of magnetic field. At zero field, we would expect these spins to form a singlet with net spin zero. A higher energy configuration for these two spins, at least at zero magnetic field, is a triplet with net spin of one. As the magnetic field is increased, however, the degeneracy of the triplet states is lifted and the lower triplet branch approaches, and eventually becomes lower in energy than, the singlet (which shows no dispersion in the magnetic field, Fig. 6.39). At the crossing point of these two branches, the singlet and triplet states of the uppermost two electrons are degenerate with each other. In this situation, the net spin on the dot may alternate between one and zero, with tunneling electrons utilizing the degenerate (singlet and triplet) dot states as intermediate states [106]. Such behavior has been demonstrated in measurements of carbon nanotubes [105], which show the singlet–triplet transition at a magnetic field around a Tesla. In all regards, the Kondo effect revealed in this case is similar to that usually found for spin-half quantum dots at zero field.

A somewhat different, yet nonetheless related, singlet–triplet Kondo effect was found in studies by Sasaki *et al.* [104]. These authors studied Kondo behavior in the pillar-type dots of Fig. 6.25, and found an unexpected behavior of the total spin with the dot occupied by a small even number (predominantly six) of electrons. Due to Hund's-rule filling, a spin triplet with net spin of one is actually the favored configuration for the six electrons at zero magnetic field. In this experiment, rather than lifting the degeneracy of the triplet states, the main effect of the magnetic field is to increase their energy collectively, due to its influence on their orbital state [104]. At a magnetic field of order a few hundred mTesla, the triplet states eventually become degenerate with the singlet state so that a triplet-to-singlet transition is expected to occur. Similar to the discussion above, for the magnetic-field-induced single-to-triplet transition, cotunneling allows the dot spin to be screened at this degeneracy point, leading to a Kondo effect. In this case, however, the singlet state is now effectively coupled to all three triplet states, rather than just one in the case where the magnetic field breaks the triplet degeneracy. This Kondo effect is therefore expected to be particularly strong, with an enhanced Kondo temperature [107,108]. Indeed, the experiments show that the zero-bias anomaly found at the triplet-to-singlet transition is much more pronounced than those obtained for the usual Kondo effect in the five and seven electron-number Coulomb valleys.

Singlet–triplet transitions can also lead to a *two-stage* Kondo effect [109,110], although the observation of this requires the use of dots whose tunnel barriers support only a single mode (unlike the vertical dots of Fig. 6.25 whose barriers allow coupling via many electron channels) [111]. In experiments of this type, a strong Kondo effect has been observed at a magnetic field where a singlet-to-triplet transition is expected to occur. Consistent with the discussion above, this resonance is found to be extremely robust, persisting to temperatures approaching 10 K. As the temperature is lowered, the conductance initially increases as expected for the Kondo effect, eventually reaching the unitary limit of $2e^2/h$ near 100 mK. With further decrease of temperature, however, the conductance was actually found to decrease. This behavior has been attributed to a two-stage Kondo process, with two Kondo temperatures, T_{K1} and T_{K2} ($T_{K1} < T_{K2}$). The first-stage Kondo effect, with temperature scale T_{K1}, reduces the net spin on the dot from one to one half. Consequently, this process is referred to as *underscreened*. In the second-stage effect, the spin is screened completely, forming a spin singlet.

In yet other experiments on the Kondo effect, Schmid *et al.* [103] also noted deviations from "even–odd filling" (i.e. filling orbitals successively with spin-up and spin-down electrons), from which it would normally be expected that the Kondo effect should be observed at every other Coulomb minimum (corresponding to an odd number of electrons on the dot). With just a few electrons occupying the dot, however, Schmid *et al.* were able to observe Kondo resonances for successive Coulomb minima, indicating that the even–odd filling breaks down in this limit, as we have discussed already. Simmel *et al.* [102] studied the Kondo effect for a dot with highly asymmetric tunnel barriers, and thus revealed an asymmetric zero-bias Kondo resonance. This was shifted away from non-zero bias, indicating the pinning of the Kondo resonance to the Fermi level of the more strongly coupled lead. In addition, while we have focused here on the behavior exhibited by semiconductor quantum dots, there have been numerous reports of similar behavior in carbon-nanotube dots [112,113,114], including a Kondo effect that arises from orbital, rather than spin, degeneracies [115].

6.3 Coupled quantum dots as artificial molecules

Having clearly demonstrated in the preceding sections a strong correspondence of Coulomb-blockaded semiconductor quantum dots to artificial atoms, in this section we discuss how such dots may be connected to each other in appropriate configurations to implement what may essentially be viewed as artificial molecules. As with the discussion of real molecules in nature, where new *molecular orbitals* are formed as a result of the wave function overlap between the component atoms, we will see here that such overlap can give rise to new electronic states in coupled quantum dots. The collective character of these states has the potential to lead to new classes of electronic devices, particularly for application to quantum computing, where one is interested in using the superposition states of quantum systems as the basis for computing. In discussions of such collective phenomena, however, there is an important need to distinguish between essentially classical collective effects that arise from Coulomb charging between otherwise isolated dots, and true quantum effects that are a consequence of controlled wave function coupling. In the following discussion, we therefore first focus on the role of charging effects in coupled quantum dots, after which we focus on hybridization effects due to wave function overlap. The topic of transport in coupled quantum dots has recently been the focus of two excellent reviews [116,117], and in our discussion of this problem we closely follow the treatment of van der Wiel *et al*. in Ref. [116].

6.3.1 Single-electron tunneling in electrostatically coupled quantum dots

In our earlier treatment of the single-electron transistor, we determined the conditions under which the Coulomb blockade may be overcome by determining the total energy of the circuit and requiring that this energy be lowered as a result of single-electron tunneling. In this problem, we treated the tunnel barriers of the quantum dot as leaky capacitors and also considered the capacitive coupling of the Coulomb island to its gate (Fig. 6.8). The extension of this concept to the problem of single-electron tunneling through a pair of sequentially coupled quantum dots (dots 1 and 2) is indicated in Fig. 6.40. In this setup, the number of electrons stored on dot 1 (2) is N_1 (N_2) (note the difference with the notation n_1 and n_2 earlier, which denoted the number of electrons to have tunneled through junctions 1 and 2, respectively, of the single-electron transistor) and the system now features three tunnel barriers (as opposed to two in the original problem). Two of these (with parameters R_L, R_R, C_L, and C_R) provide a single connection between one of the dots and either the source or drain reservoir, while the third controls the electrostatic coupling (via the capacitance C_m) between the two dots. As in the case of the single-electron transistor, the electrostatic energy of each dot is regulated via an independent gate voltage (V_{g_1} and V_{g_2}), which couples to the charge island via an associated capacitance (C_{g_1} and C_{g_2}). In order to focus on the key effects arising from the influence of the electrostatic coupling between the dots, we assume that the circuit model

6.3 Coupled quantum dots as artificial molecules

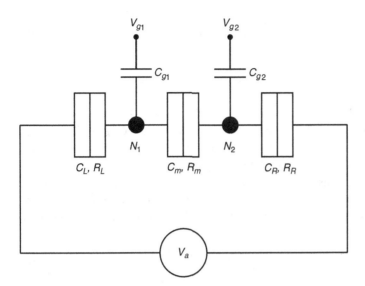

Fig. 6.40 Classical equivalent circuit to describe the charging of two quantum dots connected in series to each other. As before, the tunnel barriers are represented by a parallel combination of a capacitor and a resistor (see Fig. 6.4).

of Fig. 6.40 characterizes the system completely, and that there are therefore no other parasitic or cross capacitances relevant to the problem. (In addition to transport through series-coupled dots, a similar analysis has also been applied to the case of parallel-coupled dots, in which the net current flows through just one of the dots but is influenced by its coupling to a second dot outside of the current path [118].)

A classical analysis of the energetics of Coulomb-coupled quantum dots has been performed by several authors, by neglecting the discrete quantization of energy within the dots [119,120,121]. Following the derivation by van der Wiel *et al.* [116], we begin by considering the case of linear transport through the double-dot system, corresponding to $V_a = 0$. Under such conditions, it can be shown that the electrostatic energy of the coupled-dot system, containing N_1 and N_2 electrons on dots 1 and 2, respectively, may be written as:

$$U(N_1, N_2) = \frac{1}{2} N_1^2 E_{C_1} + \frac{1}{2} N_2^2 E_{C_2} + \frac{1}{2} N_1 N_2 E_{C_m} + f(V_{g_1}, V_{g_2}). \qquad (6.89)$$

Here, $E_{C_{1(2)}}$ is a characteristic Coulomb charging energy of dot 1 (2), while E_{C_m} can be considered as a coupling energy that determines by how much the energy of one of the dots is changed when a single electron is added to (or removed from) the other. It is the presence of this coupling term that leads to the new physics of the coupled-dot system, compared to the behavior of the single Coulomb island. These three charging terms may be written as

$$E_{C_1} = \frac{e^2}{C_1} \left[1 - \frac{C_m^2}{C_1 C_2} \right]^{-1}, \qquad (6.90)$$

$$E_{C_2} = \frac{e^2}{C_2}\left[1 - \frac{C_m^2}{C_1 C_2}\right]^{-1}, \quad (6.91)$$

$$E_{C_m} = \frac{e^2}{C_m}\left[\frac{C_1 C_2}{C_m^2} - 1\right]^{-1}. \quad (6.92)$$

In addition to the parameters introduced in Fig. 6.40, $C_{1(2)}$ corresponds to the total capacitance of dot 1(2), and may be written as $C_{1(2)} = C_{L(R)} + C_{g1(2)} + C_m$. We therefore note the similarity of Eqs. (6.90) and (6.91) with the expression derived previously (Eq. (6.19)) for the charging energy of the Coulomb island in a single-electron transistor. The major difference is that the charging energy of each dot is now modified by virtue of its coupling to the other. The other feature of Eq. 6.89, is the presence of the term $f(V_{g_1}, V_{g_2})$, which may be expressed as:

$$f(V_{g_1}, V_{g_2}) = -\frac{1}{e}(C_{g_1}V_{g_1}(N_1 E_{C_1} + N_2 E_{C_m}) + C_{g_2}V_{g_2}(N_1 E_{C_m} + N_2 E_{C_2})) \\ + \frac{1}{2e^2}(C_{g_1}^2 V_{g_1}^2 E_{C_1} + C_{g_2}^2 V_{g_2}^2 E_{C_2} + 2C_{g_1}V_{g_1}C_{g_2}V_{g_2}E_{C_m}). \quad (6.93)$$

To examine the implications of these results, we can consider two extreme cases. In the limit of zero coupling between the dots, corresponding to $C_m \to 0$, only the first two terms on the right-hand side of Eq. (6.89) survive, indicating that the total electrostatic energy of the system is simply that of the two isolated dots. The other limit arises in the situation where C_m is the largest capacitance in the system, in which case $C_{1(2)} \approx C_m$ and an analysis shows that the problem essentially reduces to one involving the charging of a large single dot formed by the two smaller ones.

To consider charging of the coupled-dot system under general conditions, we need expressions for the electrochemical potentials of the two dots ($\mu_{1(2)}(N_1, N_2)$). Within the context of this system, this potential corresponds to the energy needed to add the $N_{1(2)}$th electron to dot 1 (2), when dot 2 (1) is occupied by $N_{2(1)}$ electrons. Expressions for these potentials can be derived from the electrostatic energy defined in Eq. (6.89) (since we are considering the limit $V_a = 0$ here, we do not consider the work done by the power supply to move charge into the circuit, as was the case for the single-electron transistor):

$$\mu_1(N_1, N_2) = U(N_1, N_2) - U(N_1 - 1, N_2) = (N_1 - 1/2)E_{C_1} + N_2 E_{C_m} \\ - ((C_{g_1}V_{g_1}E_{C_1} + C_{g_2}V_{g_2}E_{C_m})/e), \quad (6.94)$$

$$\mu_2(N_1, N_2) = U(N_1, N_2) - U(N_1, N_2 - 1) = (N_2 - 1/2)E_{C_2} + N_1 E_{C_m} \\ - ((C_{g_1}V_{g_1}E_{C_{m1}} + C_{g_2}V_{g_2}E_{C_2})/e). \quad (6.95)$$

These relationships can be used to construct a charge-stability diagram that is somewhat analogous to that which we introduced previously for the single-electron transistor (Fig. 6.9). In Fig. 6.9, we show regions of stable charge

6.3 Coupled quantum dots as artificial molecules 365

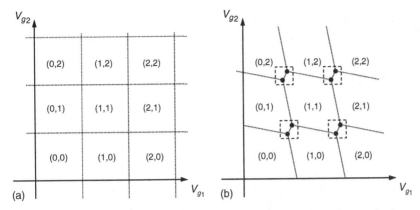

Fig. 6.41 Ground-state charge-stability diagram for the coupled-dot system of Fig. 6.40. In (a) the inter-dot capacitance (C_m) is taken to be zero, so that the two dots charge independent of each other. In (b), on the other hand, this capacitance is no longer zero and the characteristic honeycomb pattern appears. Note the characteristic triple points that are indicated by the filled symbols and enclosed by the dashed-line boxes. In both figures the stable electron numbers on the two dots are indicated as (N_1, N_2) in the different regions of the stability plots.

number on the Coulomb island as a function of the gate voltage and the source–drain bias. In the case that we consider now, however, the latter bias is zero and we can instead construct a contour showing regions of stable charge number, on the two dots, as a function of the two gate voltages. This contour indicates the charge populations of the dots at thermal equilibrium, which, for a given combination of V_{g1} and V_{g2}, are the largest allowed values of N_1 and N_2 for which *both* $\mu_1(N_1, N_2)$ *and* $\mu_2(N_1, N_2)$ are *less* than zero. In the case of completely isolated dots (i.e. $C_m = 0$), this contour simply corresponds to a series of squares (Fig. 6.41(a)), indicating that each gate voltage is applied to change the number of electrons on its associated dot, but does *not* affect the charge stored on the other dot. For non-zero coupling between the dots, however, the contour becomes distorted, and the original four-fold intersections of the different charge squares in Fig. 6.41(a) develop instead into *triple points* (Fig. 6.41(b)). The existence of these triple points is critical for transport, since they can allow for the flow of current with a small source–drain bias present, by letting the electron number on both dots fluctuate by one. The origin of these triple points is easy to explain in terms of the mutual capacitance between the two dots in the system. When this capacitance is non-zero, the charging of one of the dots by an additional single electron modifies the electrostatic energy of the other, and this effectively causes a repulsion of the resonance lines in the region where the resonance lines of both dots intersect. Consider for example, the triple point highlighted in Fig. 6.41(b), which indicates the values of the gate voltages (V_{g_1} and V_{g_2}) for which the three charge populations (N_1, N_2) = (0,0), (1,0), and (0,1)

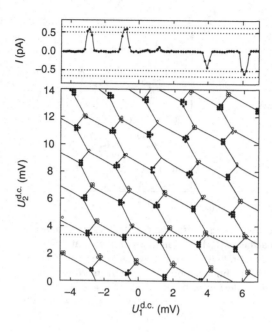

Fig. 6.42 Experimental measurement of the charge-stability diagram for a coupled-dot system ([119], with permission).

are degenerate with each other. Consequently, the flow of current via single-electron tunneling is possible at this degeneracy point, via the charge sequence $(0,0) \to (1,0) \to (0,1) \to (0,0), \ldots$ The other triple point highlighted in Fig. 6.41(b) corresponds to a degeneracy point for the three charge populations $(1,0)$, $(0,1)$, and $(1,1)$, and the energy difference of this triple point with that mentioned above is just given by the charging energy (E_{C_m}) of Eq. (6.92). For this second possible sequence, the total number of electrons on the system fluctuates between one and two, as opposed to the first sequence where it fluctuates between zero and one. While the first triple point therefore involves single-electron transfer through the double-dot system, the second can be viewed as a process in which a single hole is clocked through the system, in the opposite direction to that in which the electron is clocked at the other triple point.

Similar to the discussion of the charge diamond for the single-electron transistor, away from the triple points in Fig. 6.41(b) the charge number of the double-dot system is fixed and current through the device will therefore be blockaded at low bias and temperature. One of the earliest measurements of the charge-stability diagram for the double-dot system was performed by Pothier et al. [119], who were able to clearly show the existence of the triple points in studies of aluminum tunnel junctions (Fig. 6.42).

In our discussion thus far, we have discussed the charge-stability diagram for the double-dot system based on two key restrictions, namely an assumption of thermal equilibrium ($V_a = 0$), and a neglect of level quantization on the dots. In the discussion that now follows, we relax both of these restrictions, beginning

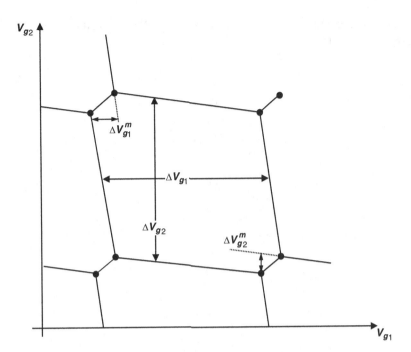

Fig. 6.43 A schematic illustration of a section of the charge-stability diagram of Fig. 6.41(a), indicating several sets of triple points (filled symbols).

by first of all considering how the charge-stability diagram is modified at equilibrium by allowing for quantization of the dot states. In this situation, just as we discussed previously for single-electron tunneling through a single Coulomb island, the non-zero energy spacing between their quantum levels must be considered when calculating the electrochemical potential of each dot. In particular, it is possible to define different *addition energies* for the population of specific states of the dots. For example, the addition energy for adding the $(N_1 + 1)$th electron to level m of dot 1, when the Nth electron occupies level n of the same dot can be written as

$$\mu_{1,m}(N_1 + 1, N_2) - \mu_{1,n}(N_1, N_2) = E_{C_1} + (E_m - E_n). \quad (6.96)$$

(Note that in the case where the quantization can be neglected this relation reduces to just the charging energy of dot 1.) Similarly, the addition energy for the other dot may be written as:

$$\mu_{2,m}(N_1, N_2 + 1) - \mu_{2,n}(N_1, N_2) = E_{C_2} + (E_m - E_n). \quad (6.97)$$

Within the context of the charging diagram that we have already introduced for the double-dot system, the consequence of including level quantization is to modify the characteristic shape of the charge-stability regions. This point is illustrated in Fig. 6.43, which indicates the form of a typical charge-stability region and its corresponding dimensions in the V_{g_1}–V_{g_2} plane. In the case where energy quantization can be neglected, the parameters identified in this figure can be expressed as [116]

$$\Delta V_{g_1} = \frac{e}{C_{g_1}}, \tag{6.98}$$

$$\Delta V_{g_2} = \frac{e}{C_{g_2}}, \tag{6.99}$$

$$\Delta V_{g_1}^m = \frac{eC_m}{C_{g_1}C_2} = \Delta V_{g_1}\frac{C_m}{C_2}, \tag{6.100}$$

$$\Delta V_{g_2}^m = \frac{eC_m}{C_{g_2}C_1} = \Delta V_{g_2}\frac{C_m}{C_1}. \tag{6.101}$$

When the discrete quantization of dot states must be taken into account, these expressions are modified as follows:

$$\Delta V_{g_1} = \frac{e}{C_{g_1}}\left[1 + \frac{E_m - E_n}{E_{C_1}}\right], \tag{6.102}$$

$$\Delta V_{g_2} = \frac{e}{C_{g_2}}\left[1 + \frac{E_m - E_n}{E_{C_2}}\right], \tag{6.103}$$

$$\Delta V_{g_1}^m = \frac{eC_m}{C_{g_1}C_2}\left[1 + \frac{E_m - E_n}{E_{C_m}}\right], \tag{6.104}$$

$$\Delta V_{g_2}^m = \frac{eC_m}{C_{g_2}C_1}\left[1 + \frac{E_m - E_n}{E_{C_m}}\right]. \tag{6.105}$$

Since we are concerned for now with the situation at thermal equilibrium, electrons populating both dots will do so in a manner that ensures that the total energy of the double-dot system takes its lowest possible value, the *ground state* of the system. While it is possible far away from equilibrium (i.e. with a large source–drain bias present) that electrons may occupy higher-energy configurations, the population of such *excited states* is not allowed here. This condition will continue to be imposed, even when we apply a small bias to drive a current through the system, providing that the value of this bias is much less than the characteristic energy-level spacing. In this situation, charge transport through the double-dot system therefore resembles that discussed already for negligible dot-level spacing, since only ground-state configurations can participate in transport. (Of course, the quantitative details are not the same, since we have seen above how the presence of the energy quantization modifies the spacing of the triple points in the charge-stability diagram.)

Finally, we must consider how transport through a double-dot system with strongly quantized energy levels is modified in the presence of a finite bias. To do this, however, it is first of all instructive to consider the role of such a bias for a system of dots whose energy quantization can be neglected. In the schematic image of Fig. 6.40, the drain is held at ground while the external bias (V_a) is applied to the source contact. The effect of this bias is to change

6.3 Coupled quantum dots as artificial molecules 369

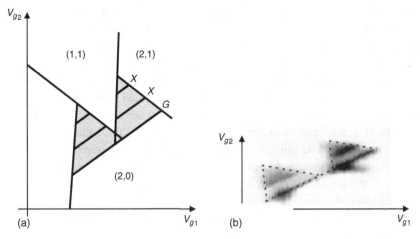

Fig. 6.44 (a) In this schematic figure, we show the charge-stability diagram for a double dot with discrete levels when it is subject to a large source–drain bias. The behavior is shown for the region corresponding to a triple point, and the dark lines correspond to resonances in the double-dot current that arises when discrete states in the separate dots align. The resonance line denoted "G" corresponds to the situation where the level alignment involves the ground-state levels of the two dots. Lines denoted "X" on the other hand involve an alignment with an excited level of at least one of the dots. (b) Experimental measurements of the triple points for a coupled-dot system ([116], with permission).

the electrostatic energy of the double-dot system, whose component dots are capacitively coupled to the biased reservoir. Consequently, it can be shown that Eq. (6.89) for the total electrostatic energy ($U(N_1, N_2)$) is modified by replacing terms of the form ($C_{g_{1(2)}}V_{g_{1(2)}}$) by ($C_{g_{1(2)}}V_{g_{1(2)}} + C_{L_{1(2)}}V_a$). Here, $C_{L_{1(2)}}$ is the capacitance of the left lead to dot 1(2). In terms of its influence on the charge-stability diagram, the applied bias causes the well-defined triple points discussed earlier to evolve into broad, triangular-shaped, regions, spanning values of $V_{g_{1(2)}}$ for which current flow now becomes possible (Fig. 6.44(a)). Essentially, the effect of the applied bias is to create an energy window of total width eV_a, so that transport becomes possible when the total energy of each charge configuration needed to shuttle electrons through the double-dot system lies within this window. In terms of the notation of Fig. 6.44(a), the dimensions of the triangles in the charge-stability diagram are [116]

$$\alpha_1 \delta V_{g_1} = \frac{C_{g_1}}{C_1} e \delta V_{g_1} = eV_a, \qquad (6.106)$$

$$\alpha_2 \delta V_{g_2} = \frac{C_{g_2}}{C_2} e \delta V_{g_2} = eV_a, \qquad (6.107)$$

where $\alpha_{1(2)}$ are conversion factors that connect a variation of gate voltage to one in energy. Now, if we consider what happens if we allow for the discrete quantization of the two sets of dot states, we note that transport through the double-dot system should ideally only occur when discrete states in the two dots align with each other. Due to the finite size of the energy window opened by the bias voltage, however, one can have a situation where different combinations of ground and excited states in the two dots yield this alignment. Consequently, the triangular region in the charge-stability region should be reduced to a series of discrete lines that correspond to tunneling via these different combinations of states (Fig. 6.44(a)). An experimental manifestation of this behavior is shown in Fig. 6.44(b). While this clearly shows the anticipated features of tunneling via different combinations of states it is also clear that the conductance does *not* vanish in the regions of the triangles that do *not* correspond to the overlap of specific pairs of states. This is because, for these particular combinations of V_{g_1} and V_{g_2}, a smaller current can flow off resonance via inelastic processes and cotunneling. In spite of this, it is still nonetheless possible to identify multiple resonances in Fig. 6.44(b) that correspond to the alignment of different excited and ground states.

Among early studies of the tunneling transport through coupled quantum dots, van der Vaart *et al.* [122] studied the linear and nonlinear transport through a system of two series-connected dots. Although they did not measure the charge-stability diagram for their system, they did show that in linear transport the linewidth of the tunneling resonances can be much smaller than the thermal energy. Such behavior arises because, if the separation of the quantized levels in each dot is much larger than the thermal energy, then linear transport through it involves only the ground state. The energy width of the resulting current resonance is then independent of the thermal smearing of occupied states in the reservoirs and is set instead by the overlap of the discrete levels in the two dots. While these states will be broadened by virtue of their coupling to the reservoirs and the other dot, this broadening will typically be much smaller than $k_B T$, allowing a sharp current resonance to be obtained. In other work, as mentioned already, Hoffman *et al.* [118] investigated the transport through a parallel-coupled dot and were also able to nicely demonstrate the charge-stability diagram, and to explain the main features of this without appeal to quantized dot states. One of the earliest studies of single-electron transport through coupled dots was performed by Molenkamp and colleagues [123,124]. They focused, however, on the so-called *stochastic Coulomb blockade* exhibited by such systems, when two dots with distinct tunneling resonances are connected to each other. In such systems, at low temperatures, the coincidence of these resonances for the two dots rarely occurs, as a result of which the dot current is found to be strongly suppressed over a broad range of gate voltage. As the temperature is increased, however, such that it becomes much larger than the level spacing (in contrast to the experiment of van der Vaart *et al.* [122]), transport via "overlapping" levels becomes possible and the amplitude of the Coulomb oscillations can actually *increase* (before decreasing later at higher temperatures as the charging energy becomes washed out).

Although we have focused here on the tunneling transport through two coupled dots, more complicated systems have been investigated by other authors. Westervelt and colleagues have studied the transport through series-connected chains of three quantum dots, and have shown clear evidence for the influence of the capacitive coupling between these [125,126,127]. Another approach worthy of highlighting is the quantum-dot-based realization of quantum cellular automata (QCA) that has been proposed and implemented by Lent *et al.* [128,129,130]. QCA is a *currentless* approach to computation, in which the binary states of a digital logic scheme are represented by means of quantum-dot cells with bistable charge configurations. No current flows into or out of these cells, and information is instead processed by means of the Coulomb interaction between different cells. Consequently, this approach may offer the potential for high-density device integration and high-speed computation with low power dissipation. The basic building block of any QCA is therefore its *cell*, which, in the quantum-dot implementation, is composed of four quantum dots which in turn hold two mobile electrons. Each electron is able to tunnel between two of the dots within its cell, and interacts electrostatically with the second electron that is able to tunnel between the other two dots. The desire to minimize total energy therefore results in two stable configurations within the cell, in each of which the electrons are localized at opposite corners, as represented in Fig. 6.45(a). By connecting series of cells in specific geometric configurations, it should therefore be possible to realize various logic gates, such as a majority gate (Fig. 6.45(b)) that can be used to achieve AND or OR operation. In their experimental work on these problems, Lent and his colleagues have implemented various components of the QCA scheme using metallic Coulomb islands realized by the shadow-evaporation technique. In this two-step fabrication process, aluminum is evaporated at two different angles through a suspended mask, with an oxidation step performed in between to form insulating Al_2O_3. In this way a reproducible tunnel barrier is formed at the regions where the two layers of metal overlap. Figure 6.45(c) shows a single QCA cell implemented by this technique [129]. Four quantum dots (labeled D_1–D_4) can be identified and two of these (D_1 and D_2, D_3 and D_4) are connected by tunnel barriers that allow for the transfer of charge between them. At the same time, the two pairs of dots are connected via a purely capacitive coupling that allows their interaction via Coulomb forces, although no charge transfer is possible between them. In the work of Ref. [129], the authors show that they could configure this cell such that the tunneling of electrons between dots in one half of the cell, could be used to induce single-electron tunneling between those in the other half, thereby mimicking the basic cell action (Fig. 6.45(d)). In later work, the same authors also demonstrated the propagation of a switching signal along a

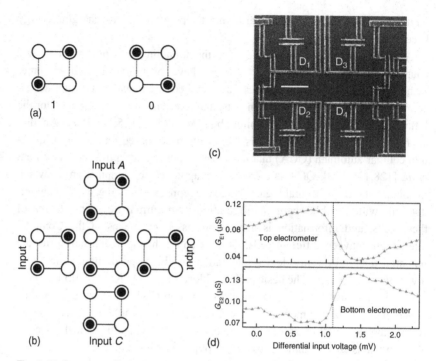

Fig. 6.45 Quantum-dot implementation of quantum cellular automata (QCA). (a) Schematic illustration of the configurations of the electrons (filled symbols) in a single QCA cell. The four dots of the cell are presented by open circles that are coupled by tunneling (solid lines) and electrostatic interaction (dotted lines). (b) Implementation of five different QCA cells to realize a so-called majority gate. The interaction between different cells is purely electrostatic. (c) Scanning electron micrograph showing an implementation of a quantum-dot QCA cell. The shadow evaporation technique is used to realize the four dots of the cell (labeled D_1–D_4). The white spacer bar denotes a distance of 1 micron. (d) Operation of the QCA cell. In the upper panel, an electron is driven from one dot to the other, causing a change of the electrometer voltage that monitors the upper part of the cell. At the same time, the anti-correlated switching of the signal in the lower panel is for an electrometer that monitors the lower part of the cell. This signal indicates that the electron in the lower half of the cell has moved from its original position to the opposite dot. Thus, switching between the two states of (a) has been achieved. For further discussion, see [129].

linear chain of three cells [131], as well as an experimental implementation of digital logic by using a six-dot QCA to mimic the behavior of the majority gate of Fig. 6.45(b). Even more recently, there have been demonstrations of clocked single-electron switching with quantum-dot QCA [132], and power gain in a QCA latch [133]. A semiconductor quantum-dot QCA cell has also been implemented by making use of the split-gate technique discussed already [134]. In spite of this progress, since these approaches to QCA rely on the Coulomb blockade to ensure the single-electron character of tunneling, the QCA effects typically wash out before temperature is increased beyond a few degrees Kelvin. One attempt to overcome this issue involves replacing the single-electron quantum dots in the quantum-dot QCA scheme, with single-domain nanomagnets that can be switched between bistable configurations, even at room temperature [135].

6.3.2 Tunneling via molecular states of coupled quantum dots

In our discussion in the preceding sections, we have focused on essentially a classical treatment of the transport in coupled quantum dots, in which the main effect of the inter-dot coupling is to introduce an electrostatic interaction of one dot with the other, thereby modifying the energetics of single-electron tunneling. By implication, we have therefore been considering the case of weakly coupled quantum dots, in which wave function leakage from either dot into the other is minimal, and, consequently, the discrete quantum states of the individual dots are not significantly affected by their coupling. As the strength of the coupling between the dots is increased, however, the resulting increase in the wave function overlap can result in the formation of *collective molecular states* that are characteristic of the coupled-dot system, rather than the individual dots. Such behavior is in direct analogy to the formation of bonding and anti-bonding orbitals that occurs when atoms are brought together to form a molecule, such as the canonical example of the hydrogen molecule. As can be demonstrated using the concepts of degenerate perturbation theory [136], when two systems with initially identical energy levels are brought together such that their wave functions overlap, these levels split to form bonding and anti-bonding states, whose associated wave functions extend over the entire system but which have different energies. The bonding state is lower in energy, by an amount $2t$ compared to the anti-bonding state. Here, t is essentially an overlap integral of the form

$$t = |<B|V|A>|, \qquad (6.108)$$

where $|A>$ and $|B>$ are the eigenfunctions of the individual levels of the two systems (A and B), while V is the perturbation that yields the coupling between the two systems. As in the discussion of molecular orbitals, in the bonding state the electron wave function is delocalized and spread over both dots, while in the anti-bonding state it is strongly localized on one dot or the other. As the coupling between the two dots is increased, the matrix element of Eq. (6.108) grows in magnitude and the energy splitting of the bonding and anti-bonding states becomes more pronounced. In terms of the charge-stability diagram for the double-dot system, the formation of molecular-like states modifies the energy separation of the triple points, and furthermore increases the number of states by which transport can occur. This behavior is illustrated in Fig. 6.46, showing the charge-stability diagram in a region close to a pair of triple points. The

Fig. 6.46 Form of the charge-stability diagram in the presence of strong tunnel coupling between two dots. The "electron" and "hole" triple points are indicated by filled and empty circles, respectively. The dotted lines indicate the form of the stability diagram for classical charging. The solid lines correspond to the boundaries taking account of tunnel splitting ($2t$).

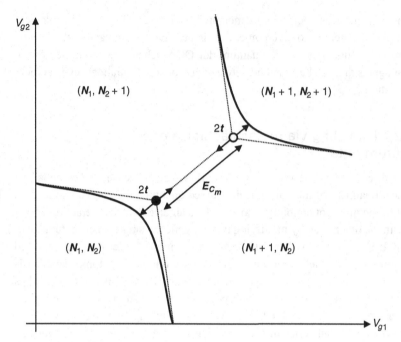

expected position of these triple points, based on a classical picture of charging, is indicated by the filled and open circles, which are located at the two intersection points of the honeycomb (the dotted lines). For each of the triple points, the tunnel splitting should now yield two distinct states for transport, so that a total of four states should mediate transport. The solid lines in the figure correspond to the bonding and anti-bonding states that are the ground states of the system near the electron and hole triple points.

Some of the earliest experimental evidence for the predicted tunnel splitting of the triple points was provided in the experiments by Blick and co-workers [137,138,139,140,141]. In their experiments, they investigated the detailed lineshape of the Coulomb oscillations in the regions of the charge-stability diagram close to its triple points [137,138]. By tuning the gate voltages that control the charge number on the dots, they demonstrated a strongly enhanced valley conductance between Coulomb peaks that they attributed to the abovementioned tunnel splitting [138]. This behavior is illustrated in Fig. 6.47, which shows the result of sweeping the voltage applied to a global top gate in the region near a triple point. The discrete points in the figure are the experimental data, while the solid lines represent fits to single resonances that are broadened by Fermi-function smearing. In the region between the two peaks, the experimental data clearly deviate strongly from the fits (note that the conductance data are plotted on a logarithmic scale). In addition, a weak shoulder can be seen on the left shoulder of the right-hand peak, which the authors attributed to one of the excited anti-bonding states. By using a controlled

6.3 Coupled quantum dots as artificial molecules

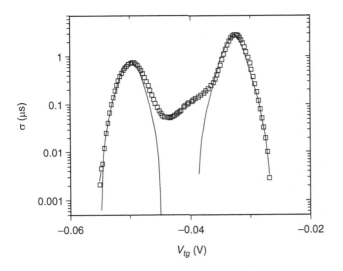

Fig. 6.47 Coulomb oscillations near the triple point of a coupled quantum dot system. Note the strong enhancement of the conductance (open squares) in the region in between the two Coulomb peaks. The solid lines are fits based on individual tunneling resonances that are Fermi-function broadened ([138], with permission).

variation of the gate voltage to move across the triple point, this shoulder was found to move systematically from the left shoulder of the right peak to the right shoulder of the left peak. The authors discussed this effect in terms of the contribution from excited anti-bonding states of the electron and hole triple points, and were able to infer an estimate for the magnitude (~ 100 μeV) of the tunnel splitting arising from the coupling between the dots. In a separate experiment, they moreover were able to demonstrate a similar effect in measurements of the Coulomb oscillations of *parallel*, rather than series-coupled, dots [140,141].

A valuable means to probe the energy structure of mesoscopic systems such as quantum dots is to investigate their response to a microwave field that is resonant with their characteristic energy splittings [48]. In the particular case of interest here, it is expected that microwaves at a frequency matching the tunnel splitting could induce *Rabi oscillations* that cause a periodic sloshing motion of charge backwards and forwards between the two dots. Blick *et al.* have attempted to explore the possibility of observing the signatures of such oscillations, by using a broadband microwave source and measuring the complex photoconductance of coupled dots as a function of frequency. These authors observed strong variations in the phase of the microwave signal at high frequencies ($> \sim 30$ GHz), which they attributed to the possible excitation of Rabi oscillations. Although this evidence is somewhat indirect, it is nonetheless consistent with the magnitude of the tunnel splitting (~ 100 μeV $= \sim 24$ GHz) inferred in the abovementioned transport experiments.

More direct evidence of coherent charge oscillations in coupled quantum dots was provided in the work of Fujisawa and co-workers, who measured the transient response of such structures to the application of controlled voltage pulses between the source and drain [117,142]. Before discussing their

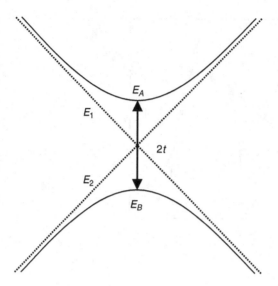

Fig. 6.48 Schematic illustration of the renormalization of the eigenstates (E_1 and E_2) of a two-level system to form bonding (B) and anti-bonding (A) levels in the presence of tunnel coupling between the states.

experimental results, however, it is worth considering the properties of a coherently coupled two-level system in further detail [116,143]. While this problem is well understood for the case of two identical systems, in which case it reduces to a textbook application of quantum mechanics, for the analysis of transport in coupled quantum dots it is important to consider how the tunnel splitting of the energy levels of the separate dots varies as these levels are *detuned* from each other. Quite generally, for two systems with energy levels E_1 and E_2 that are coupled by a matrix element t (Eq. (6.108)), their coupling results in the formation of bonding and anti-bonding states whose energies are given, respectively, by

$$E_B = \frac{E_1 + E_2}{2} - \sqrt{\frac{1}{4}(E_1 - E_2)^2 + t^2}, \qquad (6.109)$$

$$E_A = \frac{E_1 + E_2}{2} + \sqrt{\frac{1}{4}(E_1 - E_2)^2 + t^2}. \qquad (6.110)$$

The anti-crossing behavior predicted by the relations above is indicated in Fig. 6.48, which shows that the effect of the inter-dot coupling is only significantly apparent when the two energy levels, E_1 and E_2, are close to each other. With these energy levels strongly detuned ($E_1 \neq E_2$), in contrast, the bonding and anti-bonding states approximate very closely E_1 and E_2, indicating that the two systems are only weakly affected by their coupling.

In the experiments of Fujisawa and colleagues, the energy detuning of Fig. 6.48 was manipulated to induce coherent charge oscillations between a pair of coupled quantum dots. The basic setup of their experiment is shown in Fig. 6.49, which shows a coupled-dot system that was realized by patterning Schottky gates on top of an etched GaAs wire. With these dots configured to

6.3 Coupled quantum dots as artificial molecules

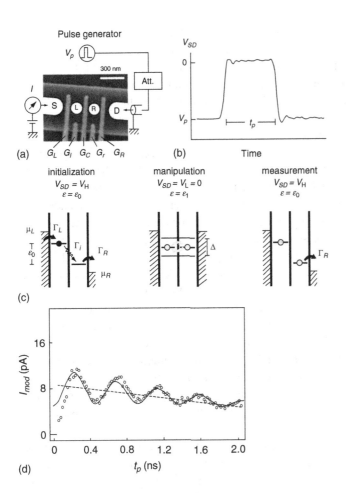

Fig. 6.49 Driven oscillations of a two-level system formed by a pair of coupled quantum dots. (a) Measurement setup showing the formation of two quantum dots by patterning Schottky gates on top of an etched quantum wire. (b) Form of the voltage pulse applied to the source and drain. (c) Initialization, manipulation, and measurement configurations. (d) Pulse-modulated current as a function of pulse length, t_p. The dots are the experimental data while the solid line is a fitting function. The electron temperature is estimated to be ~100 mK ([117], with permission).

hold a fixed number of electrons, the idea of the experiment is to apply a large voltage pulse ($V_p \sim 1$ mV) across the source and drain to modify the detuning of the two dot levels. With no pulse applied, the gate voltages are adjusted such that the level detuning is zero and the bonding and anti-bonding states are split due to the tunnel coupling alone (Eqs. (6.109) and (6.110)). At the same time, the system is configured so that the Coulomb blockade prevents the tunneling of electrons out of the dot to the reservoirs. Applying the voltage pulse affects the system in two ways. Firstly, due to the capacitive coupling of the two dots to the source, it causes a detuning of their initially quasi-degenerate energy levels. At the same time, when the magnitude of the applied voltage is large enough that the two dot states lie within the energy window, defined by the reservoir electrochemical potentials, charge may be exchanged between the dots and the reservoirs.

These concepts lead to the possibility of the initialization → manipulation → measurement sequence indicated in Fig. 6.49. In the initialization phase, the voltage pulse is applied, detuning the energy levels, and allowing an electron to tunnel from the source into the neighboring dot. After some short time, the voltage pulse is then suddenly turned off and left off for some time t_p (see Fig. 6.49). During this interval, the dots are brought back into resonance and the excess electron that tunneled into the dot should tunnel back and forth between the two dots. At the end of the manipulation phase, the voltage V_p is then again applied, at which time the excess electron may be in either the left or the right dot (dependent on the waiting time, t_p). If the electron is in the right dot, it can tunnel to the drain and so contribute to a current. If, on the other hand, the electron is located in the left dot, it will not contribute to the current. By varying the period of the pulse (t_p) and measuring the resulting current, the authors were therefore able to observe the resulting oscillations that are shown in Fig. 6.49. These oscillations represent a direct manifestation of the Rabi oscillations of the electron between the two dots. From the period of these oscillations ($\sim$2.3 GHz), an energy splitting $E_A - E_B \sim 10$ µeV could be inferred for zero detuning ($E_1 - E_2 = 0$). At the same time, it is clear that the amplitude of the oscillations decreases with increasing time, reflecting the importance of decoherence sources that disrupt the electron as it attempts to undergo its coherent sloshing motion between the dots. From the exponential decay of the oscillation envelope in Fig. 6.49, a decoherence time of $\sim$1 ns could be inferred and was attributed to cotunneling and the exchange of energy with acoustic phonons.

Another means by which the tunnel splitting of the energy levels of coupled quantum dots may be detected is from studies of their microwave-induced conductance [116,117,144]. It has long been known from studies of single-electron tunneling in *single* quantum dots that the current may be enhanced under off-resonance conditions, in an *inelastic* process in which electrons gain the additional energy needed to overcome the Coulomb blockade by absorbing a photon of appropriate frequency [145]. The tunneling process leads to the appearance of so-called photon sidebands to the Coulomb oscillations as a function of gate voltage, which shift farther away from their associated Coulomb oscillation as the microwave frequency, and so the photon energy, is increased. A similar process of photon-assisted tunneling may also be observed in coupled quantum dots, in which case the process is sensitive to the energy separation of the tunnel-split bonding and anti-bonding states. An important consideration in experiments of this type, in which one measures a net d.c. current that flows in response to the photon excitation, is the role of the level detuning in the coupled system. In the case where this detuning is set to zero, the energy separation of the bonding and anti-bonding states is just $2t$. When a photon with an energy equal to this value is incident on the system, it may excite an electron from the bonding to the anti-bonding state. This will not yield a net current, however, since the delocalized nature of the electron wave function will mean that it is equally likely to escape to either reservoir, yielding a time-averaged current of zero (Fig. 6.50(a); note that the source–drain bias is vanishingly small here). This

6.3 Coupled quantum dots as artificial molecules

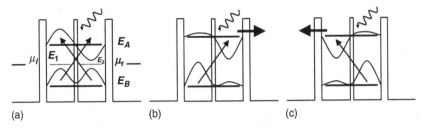

Fig. 6.50 The process of photon-assisted tunneling in coupled quantum dots for conditions of different level detuning. In (a), the level detuning is equal to zero and the bonding and anti-bonding states are split due to their tunnel coupling $2t$. In this case, the excitation of an electron from the bonding to anti-bonding state can result in it escaping to either reservoir, yielding a time-averaged current of zero. By detuning the dot levels, however, the wave function amplitude becomes localized in opposite dots for the bonding and anti-bonding states, allowing a net current to be generated as a result of photon-assisted tunneling. The two possible situations are shown in (b) and (c).

situation changes dramatically if a detuning of the levels is now introduced, by adjusting the various gate voltages that define the double-dot potential. The wave functions of the bonding and anti-bonding levels may be written, respectively, as [116]:

$$\psi_B = -\sin\frac{\vartheta}{2} e^{-i\vartheta/2}\psi_1 + \cos\frac{\vartheta}{2} e^{i\vartheta/2}\psi_2, \tag{6.111}$$

$$\psi_A = \cos\frac{\vartheta}{2} e^{-i\vartheta/2}\psi_1 + \sin\frac{\vartheta}{2} e^{i\vartheta/2}\psi_2, \tag{6.112}$$

where $\tan\vartheta = 2t/(E_1 - E_2)$ and $\psi_{1(2)}$ are the eigenfunctions of dot 1(2). We have seen already (Fig. 6.48) that, as the detuning is increased away from zero, the bonding and anti-bonding levels (E_A and E_B) approach the unperturbed levels of the isolated dots (E_1 and E_2). Consistent with this, Eqs. (6.111) and (6.112) indicate that their associated wave functions become localized on the respective dots (Figs. 6.50(b) and (c)). Under this condition, excitation of the double dot with photons with energy equal to $E_A - E_B$ can lead to a net current flow, since the excitation between the bonding and anti-bonding levels will predominantly involve a transfer from one dot to the other.

To investigate the microwave response of coupled quantum dots, Oosterkamp *et al.* made measurements of the induced photocurrent in the presence of monochromatic microwaves at frequencies up 50 GHz [144]. As indicated in Fig. 6.50, these measurements were performed in the so-called *pumping configuration*, in which the applied source–drain bias is made negligibly small. The photocurrent that flows in response to the microwave excitation is then measured as a function of the detuning of the two dot levels. Experiment shows that this current is essentially zero for all values of the detuning, except for the two situations indicated by Figs. 6.50(b) and (c). These figures indicate the

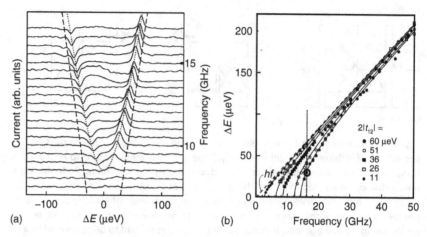

Fig. 6.51 Microwave photon-assisted tunneling in coupled quantum dots. (a) Induced photocurrent as a function of level detuning. Successive curves are shifted upwards for clarity and correspond to incrementing the microwave frequency from 7.5 to 17 GHz, in 0.5 GHz steps. The dashed lines indicate the expected positions of the resonances in the absence of tunnel splitting. At any given frequency, the fact that the current resonances occur at *smaller* frequencies than expected for isolated dots reflects the presence of the tunnel splitting, which increases as the coupling strength between the dots is increased. (b) Plot of the separation between the photon-assisted current resonances as a function of microwave frequency and inter-dot coupling strength ([116], with permission).

situation where the photon energy is resonant with the detuning, i.e. $hf = |E_A - E_B|$, so that an electron may undergo photon-assisted tunneling from one dot to the other, allowing it to contribute to the electrical current. As is clear from Figs. 6.50(b) and (c), the currents that arise in these two situations flow in opposite directions, so that the photocurrent is expected to exhibit two resonances with opposite sign as the detuning is varied. Such behavior can be seen in Fig. 6.51(a), which shows measured photocurrent at a series of frequencies between 7.5 and 17 GHz. The two current resonances due to photon-assisted tunneling can clearly be observed in each curve and the energy separation of these two peaks increases with increasing microwave frequency, as would be expected for photon-assisted tunneling. In Fig. 6.51(b), the separation of the two current resonances (more strictly, half of this separation) is plotted as a function of the microwave frequency. Several sets of data are presented and correspond to different values of the inter-dot coupling. This data clearly reveals evidence for the formation of coherent bonding and anti-bonding states, under conditions of sufficiently strong inter-dot coupling. For the smallest coupling strength ($2t = 11$ μeV), the peak separation varies linearly over almost the entire frequency range, indicating that the dots are nearly isolated in this limit and so the detuning of their levels predominantly determines the photon resonance. With increasing coupling strength (which is varied by means of a

6.3 Coupled quantum dots as artificial molecules

suitable gate), however, the peak separation deviates from this linear behavior, since the tunnel splitting now results in the formation of bonding and anti-bonding states whose energy separation is larger than the level detuning. According to Eqs. (6.109) and (6.110), the separation of the bonding and anti-bonding states is $((E_1 - E_2)^2 + (2t)^2)^{0.5}$. Consequently, the variation of the ΔE in Fig. 6.51(b) can be described as:

$$\Delta E = \sqrt{(hf)^2 - (2t)^2}. \tag{6.113}$$

The solid lines in Fig. 6.51(b) correspond to fits to the form of Eq. (6.113), using the tunnel splitting $2t$ as a parameter, and clearly reproduce the behavior found in experiment. These results therefore provide a clear demonstration of the extent to which coupled dots may be tuned continuously between the limits of weak (or *ionic*) and strong (*molecular*) coupling.

6.3.3 Single-electron quantum dots for quantum computing

One of the driving motivations for the study of coherent transport in coupled quantum dots is provided by the potential that these structures offer for application in the area of quantum computing. Although we defer from a broad discussion of this field, we simply note that there has been much theoretical interest in the possibility of using localized spins on quantum dots as the basis for realization of a quantum-mechanical bit, or *qubit*. Experimental groups in Harvard and Delft, in particular, have been vigorously exploring the key aspects of the scheme proposed by Loss and DiVincenzo, in which the control of the exchange interaction between spins on separate dots is proposed as the means to implement logic operations [146]. This scheme presents critical experimental challenges of controlled spin manipulation and spin readout, and important progress in both of these areas has been made in recent years. In this brief review, we begin by discussing work on the readout problem, after which we discuss experiments demonstrating the external control of electron coherence in coupled quantum dots. Regardless of whether this work ultimately leads to the successful implementation of a quantum computer or not, it has already greatly advanced our understanding of quantum control in the solid state.

The first challenge faced in utilizing the electron spin for computational purposes is how to detect single-electron spins in electrical measurements, and, moreover, how to distinguish between the two spin eigenstates (up and down). Work by the Delft group on this problem approaches the so-called technique of *spin-to-charge conversion*, in which one infers information about the spin state via a spin-dependent detection of the electron charge. It has been understood for more than a decade that the charge state of a quantum dot may be monitored by coupling it capacitively to a quantum point contact (QPC) and

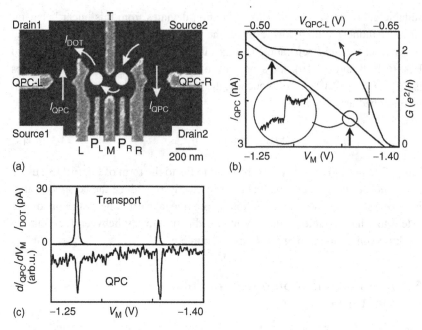

Fig. 6.52 The use of quantum point contacts as a charge sensor for quantum-dot systems. (a) In this case, the device under study consists of two coupled quantum dots, each of which has an integrated point contact (QPC-L and QPC-R) to detect charging. (b) The upper curve shows the variation of the left-QPC conductance as a function of its own gate voltage and the dotted lines indicate the point of maximum charge-detection sensitivity, where the slope of the conductance is maximal. The lower curve shows the left-QPC current induced by varying the voltage applied to gate M. The two plateau-like features that appear in this curve indicate points corresponding to single-electron charging of the left dot. (c) The upper panel shows Coulomb oscillations measured in direct transport through the left dot, while the lower panel shows the derivative of the detector current in (b), indicating the connection of the sensor-current plateaus to dot charging ([150], with permission).

measuring the current that flows through this [147,148,149]. Under appropriate conditions, the addition (or removal) of just a single electron to (or from) the dot can yield, via the capacitive coupling, a measurable change in the QPC conductance, allowing one to use this conductance to count the number of electrons stored on the dot. An example of such a device is shown in Fig. 6.52, the upper panel of which is an electron micrograph of two coupled quantum dots, realized by the split-gate technique, each of which is coupled to its own unique QPC charge detector. The lower panels of this figure show the results of measurements in which the gates of the right-hand dot and QPC are grounded, so that only the left-hand side of the device is formed. The current through the left QPC then shows clear steps (or peaks in the derivative of the current) as the voltage applied to one of the gates of the quantum dot is varied to change the electron number. These steps are clearly correlated to the Coulomb oscillations of the dot

6.3 Coupled quantum dots as artificial molecules

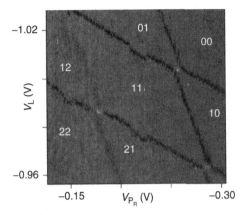

Fig. 6.53 Charge-stability diagram for the double dot shown in Fig. 6.52, measured by using the right quantum point contact as a charge sensor. Although not shown here, a high degree of correspondence was found with the charge-stability diagram obtained in direct measurements of the double-dot conductance ([150], with permission).

itself, demonstrating their connection to the dot charging [150,151]. A great advantage of this technique is that it may be used to detect changes in charge occupancy even when the transport that results from this yields only an immeasurably small current. This is shown in Fig. 6.53, in which the QPC detector is actually used to sense changes in the charge state of a coupled quantum dot (the gates of both dots are now activated and the right QPC of Fig. 6.53 is being used as the charge sensor). The well-known honeycomb pattern, with its characteristic triple points, can clearly be seen in this figure.

The behavior shown in Figs. 6.52 and 6.53 provides a means to identify when single-electron charging of a quantum dot occurs. When combined with knowledge of the spin states that are populated/depopulated by such processes, it is therefore possible to relate this information on dot charging into a corresponding measurement of the electron spin. In order to demonstrate such spin-to-charge conversion, the Delft group studied a single quantum dot with an integrated QPC detector, in the presence of a large in-plane magnetic field that was used to lift the spin degeneracy of the dot eigenstates. In one set of experiments [151], they then applied periodic pulses of predetermined amplitude to a quantum-dot gate to selectively populate specific spin states, and to measure ensuing spin-flip relaxation to the ground state. These measurements revealed an extremely long time scale for such relaxation, at least as long as 50 μs at 7.5 T and 20 mK. In these experiments, however, the authors were only able to infer a lower bound on the spin-flip time, due to the limited time window (<10 μs) of their measurements. In later experiments [152], however, the same authors applied a single-shot technique that allowed them to determine the spin-flip time more accurately and to also resolve the spin of *single* electrons in real time. Their approach

Fig. 6.54 (a) Pulse sequence used in single-shot measurements of single-electron spin on a quantum dot. (b) Corresponding change in the QPC sensor current due to charging and discharging of single electrons on the dot. As indicated by the dotted line, the sensor current is expected to show different behavior for spin-up and spin-down electrons.

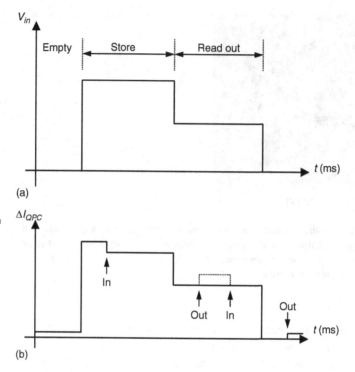

involved making use of the pulse sequence shown in Fig. 6.54, to first charge the dot with a single electron (STORE) and to then read out its spin via spin-to-charge conversion (READ OUT). In this experiment, the authors actually made use of the right-hand set of gates of the device of Fig. 6.52 and applied the pulse to gate P_R to modulate the potential of the right-hand quantum dot. The static voltages applied to the other dot gates, and the amplitude of the applied pulse, are chosen such that, initially, the dot is completely empty of electrons since its two lowest, Zeeman split, levels lie above the Fermi level in the reservoirs. During the storage phase of the pulse, the dot potential is suddenly lowered such that the initially empty spin levels are brought below the Fermi level in the reservoirs and held there for some predetermined time (~ms). During this period, an electron may tunnel into the dot after some characteristic time (which is determined in experiment by adjusting the height of the tunnel barriers), and when it does so it may occupy either of the spin states. Once this has occurred, however, the Coulomb charging energy prevents a second electron from populating the dot over the remainder of the storage phase. In the readout phase, the amplitude of the pulse voltage is again changed, this time to move the two Zeeman split levels so that the Fermi level in the reservoirs lies in between them. In this situation, two distinct tunneling scenarios are possible, dependent upon the spin of the electron localized on the dot. If the electron occupies the lower Zeeman state, it will remain trapped on the dot during the storage stage, whereas

6.3 Coupled quantum dots as artificial molecules 385

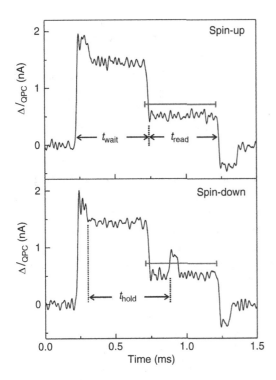

Fig. 6.55 Single-shot measurements showing the identification of spin-up and spin-down electrons in response to the application of the voltage pulse of Fig. 6.54. Figure provided courtesy of L. P. Kouwenhoven, please see [152] for further details.

if it occupies the upper state it will eventually tunnel off the dot. In doing so, however, the Coulomb blockade will be lifted so that after some characteristic time, again set by the tunnel barrier strength, a electron will tunnel back from the reservoir, in this case populating the lower Zeeman state. The electron will then remain trapped on the dot for the rest of the readout stage.

The tunneling sequences described above will lead to two highly distinctive pulse variations in a single-shot measurement where a QPC sensor is used to detect the charging and discharging of the quantum dot. These two types of pulse sequence are illustrated in Fig. 6.55, which shows results of measuring the response of the sensor to the application of a single-shot pulse. To understand the form of these pulses, we consider the schematic variation illustrated in the lower part of Fig. 6.54. This shows that, to a large part, the capacitive coupling between the QPC and the pulsing gate results in the QPC current following the variation of the pulse voltage. At the same time, however, the current changes in a discrete manner each time a single electron tunnels into or out of the dot. During the storage phase, the QPC current decreases suddenly as an electron enters the dot, regardless of its spin. In the readout phase, however, a distinct difference in the QPC current should be seen, dependent on the electron spin. In the case where the spin that tunneled into the dot in the storage phase occupied the lower Zeeman state, the QPC current will remain constant during readout, reflecting the fact that the electron is in the ground state. If, however, it occupied the upper Zeeman state, then the process by which an electron leaves the dot and then re-enters, as described above, should yield an additional pulse in the current, as indicated by the dotted line in Fig. 6.54(b). In Fig. 6.55, we show different single-shot traces that demonstrate exactly this behavior [152]. These traces clearly indicate the possibility of using spin-to-charge conversion to implement a measurement of electron spin. It is important to realize here that, due to the single-shot nature of these measurements, the tunneling processes involved occur at random times relative to the pulse sequence. By performing statistical averaging, however, the authors were able to measure the spin-flip time of the electrons directly and found values as large as 1 ms, explaining the very weak spin-flip scattering inferred earlier in Ref. [151]. In subsequent work, the same authors have refined this measurement approach [153], and have also adapted it to identify the singlet and triplet states of coupled quantum dots [154]. This work has most recently been reviewed in Ref. [155].

While we have been focusing on the use of QPCs as charge sensors to resolve electron spin, other notable work has used the charge-sensing technique to perform highly sensitive electron counting [156,157]. Ensslin and his colleagues have pointed out that it is possible to use this approach of electron counting for a quantum dot, in a situation where the actual current flowing through the dot is too small to be accurately measured by conventional approaches. Charge sensing allowed these authors to measure currents as small as just a few electrons per second, corresponding to a current of less than 1 aA. Exploiting this approach, they were able to study the distribution function of the time-dependent current fluctuations through the dot, and to accurately determine the higher-order moments of these current fluctuations, which are related to the intrinsic noise properties of the current. Fujisawa *et al*. were also able to perform somewhat similar measurements, although in their case they detected the charging of a quantum dot using another as an electrometer [158]. In their most recent work [159], Ensslin and his colleagues have used the QPC sensor to monitor charge on a *coupled* quantum dot, and have implemented this as a means to achieve single-photon detection. In this experiment, one is essentially concerned with the *back-action* of the QPC on the double-dot system, something that we have not discussed thus far. The principle of this experiment involves introducing a non-zero detuning of the double-dot levels, so that an inelastic process is required to allow an electron to tunnel from the source to the drain quantum dot. The source of the energy transfer in this system is photons that are radiated by the QPC, as a result of the shot noise in the current flowing through it.

In addition to the possibility of identifying the spin state of single electrons, for quantum-computing purposes it is also necessary to be able to perform specific operations on either a single spin or a small ensemble of spins, flipping, for example, the spin of an electron, or changing the total spin of a pair of electrons. This is another area that has witnessed significant experimental progress in the last few years [160,161,162,163,164,165,166,167], driven by

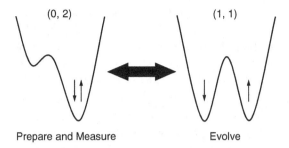

Fig. 6.56 Schematic illustration of the experiment of Petta *et al.* [165], in which they demonstrated the coherent manipulation of coupled spins in a double quantum dot. The idea of this experiment is to use gate-pulsing sequences that drive one of the electrons backwards and forwards between the two dots, while the other remains localized in the right-hand dot.

efforts by the Delft and Harvard groups. Since the work in this area has been the subject of a recent review by the original authors of this work [155], we only briefly mention the main advances here. In one experiment, Petta *et al.* used a QPC as a charge sensor to detect the response of coupled, one-electron, quantum dots to transient pulses that were used to control the exchange coupling between two electrons localized on them [165]. The basic idea of their experiment is summarized in Fig. 6.56, which shows the state of the quantum dot during successive PREPARE, EVOLVE and MEASURE phases. In terms of the charge-stability diagram discussed previously for coupled dots, in this experiment the authors used transitions between the (1, 1) and (0, 2) charge states, driving one electron back and forth between the two dots. In the PREPARE phase, the two electrons are driven into the same dot to form the (0, 2) configuration. The spins of the electrons in this situation are anti-parallel, since the opposite configuration, for which their spins would be parallel, has too high a cost in orbital energy. In the next phase of the gate-pulse sequence, the spins are driven into separate dots, forming the (1, 1) state and initially preserve their original spin-singlet alignment. In this EVOLVE phase, the (1, 1) state is maintained for some predetermined time, after which the gate bias restores the initial configuration and tries to drive the electrons back into the (0, 2) configuration. Dependent upon the length of the EVOLVE phase, it is possible that the two spins remain in their original singlet configuration, or that they evolve instead into a parallel (triplet) configuration as a result of dephasing. At the low temperatures where these experiments are performed, the main source of such dephasing is the hyperfine interaction of the localized spins with the nuclear spins of the host (GaAs) material [162]. In the case where the singlet configuration is preserved, the excess electron can easily be driven from the (1, 1) to the (0, 2) state, and this charging of the right-hand dot (in the figure) can be detected as a change in the QPC signal. If the two electrons are in a triplet state, however, a spin-blockade effect [168] prevents the additional electron from tunneling into the right dot.

In their experiment, Petta *et al.* therefore varied the time that the electrons are held in the separated state and measured the averaged QPC-sensor signal for repeated pulsing. In this way they determined the resulting probability of remaining in the singlet state. This probability was found to decay with a characteristic spin-dephasing time of order 10 ns. In a most important development, however, the same authors found that they could apply known techniques from magnetic resonance, to essentially restore the coherence of the spin pair over very much longer times ($\sim\mu$s). This was done in a spin-echo experiment, in which gate control was used to modulate the exchange splitting of the singlet and triplet states, allowing the hyperfine field of the nuclei to manipulate the coupled-spin system in a controlled manner.

In addition to controlled operations on coupled spins, driven oscillations of a single spin in a quantum dot have also been demonstrated by utilizing an on-chip micro-strip line to induce a local RF magnetic field in the vicinity of one of a pair of quantum dots [167]. The idea in this experiment involved making use of crossed magnetic fields, one static and the other time-varying, to drive electron-spin resonance at a frequency determined by that of the latter field. The static field ($B_{\parallel}$) was applied in the plane of the two-dimensional electron gas, thereby inducing a Zeeman splitting of the dot levels ($g^*\mu_B B_{\parallel}$). When the a.c. current passed through the micro-strip line at a frequency (f) that matched the Zeeman splitting ($hf = g^*\mu_B B_{\parallel}$) of the levels, it was found that the resulting electron-spin resonance could be detected via a measurement of the current through the quantum dots. The experiment makes use of a similar spin-blockade effect to that discussed above, in which, with one electron localized on one of the dots, a second can only tunnel into it from the other if its spin is opposite to that of the initial electron. The role of the a.c. field generated by the strip line is to periodically induce spin flips of the tunneling electron, allowing a current to therefore flow when it should normally be blocked.

The experiments briefly described in this section have demonstrated the exquisite control of single spins that can be achieved in semiconductor quantum dots. While there is still much work to be done to implement a practical scheme for quantum-information processing based on these advances, they nonetheless made significant advances towards this ultimate goal.

6.4 Quantum interference due to spatial wave function coherence in quantum dots

Thus far, we have discussed the results of several different experiments that have utilized the coherence of the spin-component of the electron wave function to perform various manipulations of quantum dots. A common feature of these experiments is that they rely on, or utilize, the long coherence times for electron spins in semiconductors. In materials that exhibit weak spin-orbit scattering, such as gallium arsenide or silicon, the resulting spin coherence times can be as long as milli-seconds at low temperatures, since the typical elastic or inelastic scattering processes in these materials alter carrier momentum or energy without affecting spin. In spite of this, there has been huge interest over the years in the manifestations of quantum interference of the spatial component of the wave function, in structures such as quantum wires, Aharonov–Bohm rings, and quantum dots. The dephasing time for this component of the wave function is typically many orders of magnitude shorter than that for carrier spin, due to the large number of processes that are able to disrupt the phase evolution of the spatial component of the wave function as carriers propagate through a crystal. Reviewed in detail in Ref. [169], and in Chapter 6 of the previous edition of this book, these processes include electron–electron scattering, electron–phonon scattering, and scattering from magnetic impurities. At sufficiently low temperatures, however, these processes can be suppressed sufficiently to allow the electron to propagate coherently over sufficiently long distances that a variety of quantum-interference effects can be observed. In this section, we review some of these effects, as manifested in quantum dots. We begin by discussing the observation of the Fano effect in tunnel-coupled quantum dots, which provides a sensitive probe of the quantum coherence of the electron wave function during tunneling processes.

6.4.1 The Fano effect in tunnel-coupled quantum dots

In our discussions of single-electron tunneling in semiconductor quantum dots, we have presented a picture in which each Coulomb oscillation is associated with tunneling via a specific dot eigenstate. In such situations, there is therefore only one path for electron transmission through the dot, so that interference effects can be neglected. The Fano effect, on the other hand, is a generic quantum interference effect that occurs in systems whose transmission involves the interference between two distinct channels, one of which involves direct (non-resonant) transmission while the other of which involves a resonant bound state. The Fano effect has been reported for a variety of different physical systems, including neutron scattering [170], atomic photo-ionization [171], Raman scattering [172], and optical absorption [173]. In recent years, it has also become apparent that the Fano effect can be manifested directly in the conductance of mesoscopic semiconductor devices, whenever they are configured such that their transmission is affected by the coherent coupling between resonant and non-resonant paths. Some of the most extensive studies of this phenomenon have been performed by Kobayashi and co-workers [174,175,176,177], who have made wide use of a device geometry consisting of an Aharonov–Bohm ring that features a quantum dot in one of its arms. An example of such a device is shown in Fig. 6.57, in which the ring structure has been realized in a GaAs 2DEG by a combination of electron-beam lithography and etching. Sets of gates traverse both arms of the ring, allowing, in principle, for the formation of

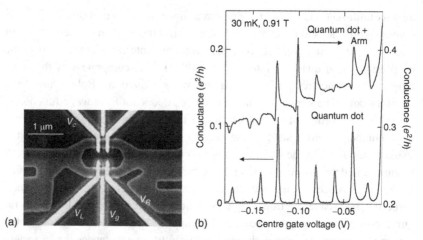

Fig. 6.57 Fano resonances in an Aharonov–Bohm ring with an embedded quantum dot. (a) Device structure. Bright regions are metal gates that can be used to modulate either arm of the ring. (b) The upper curve shows Fano resonances in the conductance when both arms of the ring conduct, while the lower curve shows Coulomb oscillations when only the dot is transmitting ([192], with permission).

a quantum dot in each arm. In the experiments of interest here, however, only one of these sets of gates is activated, forming a dot in one arm while the other is left unmodulated.

In experiments performed on the device of Fig. 6.57, the unmodulated arm represents the direct path for transmission, while the resonant path is formed by the quantum dot, the eigenstates of which can successively be brought into resonance with the Fermi level by suitable tuning of its gate voltages. The key observation of such an experiment is presented in Fig. 6.57(b), the lower curve of which shows the results of measurements where the normally unmodulated arm of the ring has been cut off completely by applying suitable voltages to its gates. In this situation, electron transmission through the ring is regulated by the quantum dot, and the ring conductance then shows a series of Coulomb oscillations as the plunger-gate of this dot is swept. As we have discussed already, these oscillations are quite symmetric. Very different behavior is obtained, however, when the originally closed arm is opened up. In this case, transmission through the ring now involves the interference between electron partial waves that pass through the unmodulated arm, and those which tunnel through the quantum dot. As indicated by the upper curve in Fig. 6.57(b), the Coulomb oscillations in this system now exhibit a highly asymmetric character that is well known from the Fano effect. In fact, it is found that the resonances that arise in this situation can be well described by a form consistent with the Fano formula [178]. By making the substitution $\varepsilon \equiv 2(V_g - V_0)/\Gamma$, where V_g is the gate voltage that is varied to induce successive resonances, V_0 is the gate voltage at which the resonance occurs, and Γ is the resonance width, this Fano form can be expressed as [174]

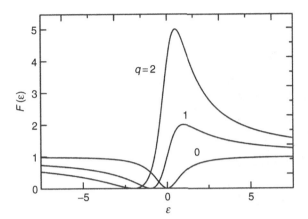

Fig. 6.58 The form of the Fano function of Eq. (6.114) for several different values of q ([192], with permission).

$$G(\varepsilon) \propto \frac{(\varepsilon+q)^2}{(\varepsilon^2+1)}. \quad (6.114)$$

In this expression, q is a parameter related to the strength of the coupling between the resonant and non-resonant paths. (More precisely, q is proportional to the ratio of matrix elements linking the input state to the resonant and non-resonant parts of the output state.) Eq. (6.114) describes a resonance whose lineshape depends sensitively on the value of q. In this case, $q = \infty$ corresponds to the situation where there is no coupling between the two paths, and yields a symmetric (so-called Breit–Wigner) resonance, due to resonant transmission via a specific bound state. A symmetric resonance is also obtained for $q = 0$, although in this case it corresponds to an anti-resonance, for which the conductance decreases to a local minimum at the gate voltage V_0. For other values of q, the resulting resonance is asymmetric, and this asymmetry is most pronounced for $q = \pm 1$. Some of these different variations are illustrated in Fig. 6.58.

In their experimental work, Kobayashi *et al.* were able to show that the resonances observed in the situation where interference between the two arms of the ring is significant can be well described by the Fano form of Eq. (6.114). An example of their fitting is shown in Fig. 6.59, which shows a Fano resonance at a series of different temperatures. The solid lines in these figures represent fits to the Fano form and it is clear that these account well for the observed experimental variations. It can also be seen from this figure that the Fano effect washes out with increase of temperature beyond 500 mK, which can be attributed to a loss of electron coherence at higher temperatures. The sources of electron decoherence in mesoscopic structures have recently been extensively reviewed in Ref. [179], and are also addressed in Section 8.3. We shall therefore not review this topic here, but note simply that typical sources of decoherence are electron–electron and electron–phonon scattering, with the former dominating at the low temperatures where these experiments are performed. In order to fully explain the results of their experiments, Kobayashi *et al.* have actually found

Fig. 6.59 Fano resonance at a series of different temperatures. Solid lines are fits to the Fano function that take account of the non-zero temperature ([192], with permission).

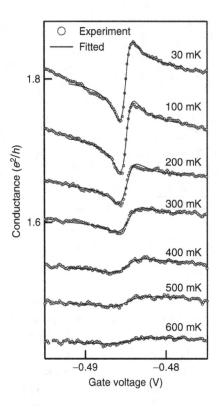

that it is necessary to assume a *complex* value for the parameter q. This is essentially a consequence of their experimental geometry, since in the original treatment by Fano the parameter q was considered to be real [178]. In his treatment, however, the resonant and non-resonant paths for transmission were assumed to be spatially coincident, so that no phase evolution was associated with the coupling between these paths (alternatively, the resonant and non-resonant paths could be assumed to enclose no effective area). This is quite different to the experiments of Kobayashi *et al.*, in which the resonant and non-resonant paths are spatially separated and enclose a significant area. By varying the magnetic flux enclosed by the ring, these authors were able to show that the magnitude of q was unchanged but that its real and imaginary components oscillate out of phase with each other, with a period given by the field needed to add a single flux quantum to the cross-sectional area of the ring [175]. This is nothing more than a manifestation of the Aharonov–Bohm effect, in which the magnetic field modulates the accumulated phase of the electron partial waves that traverse the two different arms of the ring. A similar modulation of the Fano effect was also reported by Fuhrer *et al.* [180], who studied the conductance of a ring device, one of whose arms featured a side-connection to a quantum dot.

Formally, a complex value for the parameter q can be expected in situations where time-reversal symmetry is broken, in which case the matrix elements that determine q need not be real [175]. This situation is certainly realized in the experiments of Kobayashi *et al.*, since a significant magnetic field is present. Clerk *et al.* have also pointed out, however, that a complex q can arise in situations where the interference between the resonant and non-resonant paths is modified by decoherence, providing a means, in principle, to determine the decoherence time [181]. A Fano effect has also been reported in measurements of the conductance of the junction that can be formed by crossing two carbon nanotubes so that they make a nanoscale contact [182,183]. Kim *et al.* showed that the resonance in this situation is also modified in a similar manner to that reported by Kobayashi *et al.* for their semiconductor ring structures, and that the resulting variation can again be described by assuming a complex q-parameter whose real and imaginary components are modulated by the magnetic field. In this particular experiment, the magnetic-field modulation of the Fano effect was attributed to an associated change in the magnetic flux penetrating the effective area of the nanoscale contact between the two nanotubes.

The first clear experiments to report the observation of the mesoscopic Fano effect were actually reported by Goldhaber-Gordon and colleagues [184,185, 186,187], who studied the Coulomb oscillations in single-electron transistors in the regime where their tunnel barriers were lowered to provide strong coupling of the dot to its reservoirs. In this situation, the resulting Coulomb oscillations were found to exhibit the Fano form, with a lineshape that is well described by a real value of the parameter q. In these experiments, the resonant and non-resonant paths both exist within the same quantum dot, although the source of the non-resonant path is not completely clear. In early work it was suggested that this might arise from transmission at energies higher than those of the barrier [184,185]. Later work, however, performed on dots with much higher barrier strengths, discounted this possibility, and also excluded cotunneling as the possible source of the non-resonant contribution [187]. Another interesting study by Johnson *et al.* explored the competition between quantum-interference and charging effects in small quantum dots connected to quantum point contacts [188]. These authors identified a regime of intermediate coupling between these structures where both charging and interference effects need to be considered. By taking account of both of these effects, the authors were able to explain the unusual lineshape of the Fano resonances observed in their experiments.

It is clear from the discussion above that studies of the Fano effect in mesoscopic systems can reveal a rich variety of physical behavior. This is demonstrated even more strikingly, however, by the observation of the Fano–Kondo effect, which combines the single-particle Fano effect with the many-body spin physics of the Kondo effect. Sato *et al.* observed this effect [177] by studying the device shown in Fig. 6.60, in which a one-dimensional wire is

Fig. 6.60 Experimental observation of the Fano–Kondo effect. Shown in (a) is the split-gate device used in this study, which consists of an upper quantum dot coupled to a lower quantum wire. Shown in (b) is the result of measurements of the conductance of the quantum wire as a function of the gate voltage (V_g) applied to the dot. The various curves are for different voltages (V_m) applied to the pair of gates that control the dot–wire coupling. Shown in (c) is the conductance measured at a series of temperatures for a range of V_g where the Kondo coupling is present. Kondo temperatures extracted from these measurements are also indicated ([192], with permission).

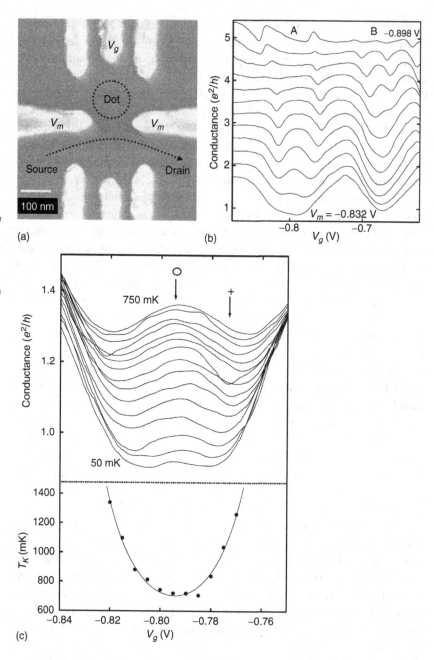

implemented with a side coupling to a quantum dot. With weak coupling between the dot and wire, the Fano effect was found to produce a series of dips in the conductance of the wire (upper curves in Fig. 6.60(b)). As the coupling to the dot was increased, however, the conductance in the regions between successive dips was found to increase and decrease alternately. This behavior was ascribed to the role of the Kondo effect, under conditions where there is an odd number of electrons localized on the quantum dot.

An important implication of the observation of the Fano effect in single-electron tunneling through semiconductor dots is that it indicates that this tunneling must be (at least partially) coherent in nature. This is not a trivial point if one considers that, in our earlier analysis, we applied a circuit model to determine the conditions for single-electron tunneling. In such a model, one assumes the presence of specific voltage drops across the tunnel barriers of the dot, which implies the presence of energy dissipation that should break quantum coherence. The Fano character of the observed resonances, however, indicates that coherent interference between resonant and non-resonant paths is possible.

The earliest studies of phase coherence in tunneling were actually performed by Yacoby, Heiblum and colleagues who used challenging nanofabrication techniques to implement the first Aharonov–Bohm ring with a quantum dot embedded in one of its arms [189,190]. In their experiments, the observation of Aharonov–Bohm oscillations in the current through the ring, in spite of the presence of the dot in one arm, provided direct evidence that coherence is at least partly preserved during tunneling [189]. In subsequent experiments, these authors we able to measure the phase change associated with tunneling through the dot, and showed that very similar variations of the phase were exhibited by successive resonances [190]. These issues have been reviewed in Refs. [191,192].

In other related work, Holleitner and colleagues studied phase coherence in tunneling by investigating the characteristics of the device shown in Fig. 6.61 [140]. This consists of two quantum dots that are connected in such a manner that current from source to drain flows through them in parallel (Fig. 6.61(a) and (b)). In this sense, the dots form a ring structure in which electrons tunnel through both dots to transfer from one reservoir to another. In Fig. 6.61(c), we show Aharonov–Bohm oscillations in the ring current as a function of the magnetic field. Oscillations with a period of ~ 17 mT are observed, consistent with the area defined by the two paths of the interferometer. Once again, such results demonstrate the preservation of phase coherence during tunneling. In a related experiment [193], Sigrist *et al.* investigated an interferometer geometry consisting of a quantum dot embedded in each arm of a ring. In this experiment, they also were able to measure the change in wave function phase as the number of electrons on each dot was changed. Their experiments revealed an evolution of the phase with magnetic field that could not be explained within existing theory, with unexpected phase lapses at certain magnetic field. It was suggested that this might reflect the additional influence of geometric resonances on the wave function. In later work [194], the same authors applied a similar method to investigate wave function coherence in the cotunneling regime. Normally, it might be expected that inelastic cotunneling should destroy phase coherence. In their experiment, however, Sigrist *et al.* were able to use an interferometer geometry to demonstrate Aharonov–Bohm oscillations, even in the presence of

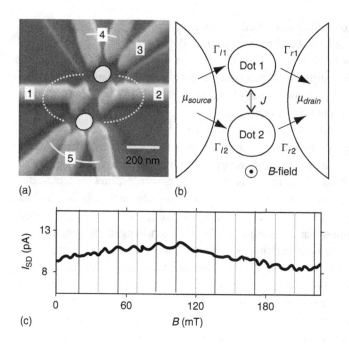

Fig. 6.61 Device comprising two coupled dots that provide parallel paths for current flow from source to drain. A scanning electron micrograph of the device is shown in (a) while in (b) the function of this device as an interferometer is indicated. In (c), the source–drain current of the double-dot system shows Aharonov–Bohm oscillations as a function of the external magnetic field ([140], with permission).

elastic and inelastic cotunneling. This remarkably surprising observation was suggested to be related to the fact that the occurrence of the cotunneling does not allow one to determine which arm of the interferometer the cotunneling electron traverses.

6.4.2 Quantum interference in open quantum dots

In discussing transport through quantum dots, it is possible to identify two distinct regimes of behavior, dependent upon the strength of the coupling between the dot and its reservoirs. Thus far we have focused on the case of *weakly coupled* quantum dots, which are connected to their reservoirs by means of tunnel barriers. At sufficiently low temperatures, we have seen that the charging energy associated with the addition of a single electron to such dots can exceed the thermal energy. In this Coulomb blockade regime, current then flows by the process of single-electron tunneling. In this section, however, we consider the problem of transport in *open* quantum dots that are much more strongly coupled to their reservoirs. This coupling is typically provided by a pair of quantum-point-contact leads, which are configured so that their saddle barriers (see Chapter 5) lie below the Fermi level (i.e. $R_T < h/e^2$). The Coulomb-blockade is therefore washed out in such dots, allowing their electrical properties to be dominated by the nature of their confinement-induced eigenstates. Consequently, these structures are ideally suited to investigate the manner in which the properties of quantum systems are modified through their coupling to the environment. The electrical properties of these dots are strongly influenced by the interference of electrons confined within them. In this sense, the dots may be viewed as the quantum analog of classical wave-scattering systems, such as microwave cavities [195,196], fluid systems [197], and semiconductor lasers [198]. At the same time, the analysis of transport through these structures is particularly amenable to semiclassical methods, in which path integrals over classical trajectories are used to calculate their conductance (for a recent review, see [199]).

As discussed in the first edition of this book, initial studies of the transport properties of open dots were performed by Marcus *et al.*, who measured the magneto-resistance of circular- and stadium-shaped dots [200]. At temperatures below a degree Kelvin, the magneto-resistance of the dots was found to exhibit reproducible fluctuations, which were ascribed to coherent interference of geometrically scattered electrons [201]. The spectral content of the fluctuations was found to be different for quantum dots fabricated with stadium and circular geometries, which are expected, classically, to give rise to chaotic and regular scattering of particles, respectively. Another feature of this experiment, that was confirmed in subsequent studies [202,203,204], was the observation of a zero-field peak in the magneto-resistance, which was later ascribed to the ballistic analog of weak localization [205]. Based on semiclassical arguments, it was proposed that the lineshape of this peak should provide a probe of electron scattering in the dots, with a Lorentzian form predicted for chaotic scattering and a linear lineshape expected for regular dynamics. These predictions appeared to be convincingly confirmed (but read on) in a later experiment by Chang *et al.*, who studied the magneto-resistance of multiply connected arrays of circular- and stadium-shaped dots [206].

While early studies of quantum transport in open dots played an important role in clarifying the connection of the classical scattering dynamics to the resulting quantum-transport properties, subsequent studies have revealed a picture that is more complicated than originally considered. In this section, we therefore review the recent progress in this area [207,208], with particular emphasis on our own work on the use of open quantum dots as a tool for the investigation of quantum decoherence theory. The manner in which the properties of quantum systems are revealed in the results of classical measurements, as well as the manner in which these quantum properties evolve into classical ones, has been the focus of investigation since the formulation of quantum theory. One interpretation, which explicitly includes the coupled systems, is that of decoherence [209]. While the description (and interpretation) of the decoherence process has varied widely, the key feature is the interaction of the system upon the environment, and that of the environment upon the system. It has been proposed by Zurek that the former interaction leads to a preferred, discrete set of quantum states (or *pointer* states), which remain robust as their superposition with other states, and among themselves, is reduced by the decoherence process [210]. This decoherence-induced selection of pointer states has been termed *einselection*.

In recent work, we have suggested that open quantum dots represent a physical system in which the properties of pointer states can be directly observed [208]. To understand the connection of these structures to the predictions of decoherence theory, it is important to consider how they are typically realized in practice. The most widely used approach makes use of the split-gate technique, in which the depletion fields generated by surface gates on a semiconductor induce the confinement of electrons to a sub-micron-sized cavity (Fig. 6.62). As pointed out by Ketzmerick [211], the inherently *soft* nature of the confining potential that arises in such structures (see Fig. 6.62) causes their associated classical dynamics to be *mixed*, exhibiting *both* chaotic *and* regular character. This point has significant implications for the study of transport in quantum dots, since many theories base their semiclassical analyses of transport through these structures on an assumption that the dynamics is purely chaotic. While this allows considerable mathematical simplification in the resulting theoretical treatments, it misses completely the exciting properties of mixed systems, whose dynamical phase space is considerably more complicated than that of chaotic systems (see below). Moreover, since mixed, rather than fully chaotic, systems actually tend to be the rule in nature, it means that theories based on assumptions of fully chaotic scattering are actually of limited practical use.

The connection of quantum-dot transport to decoherence theory is provided by studies of their conductance as a function of either magnetic field, or the confining gate potential (via gate voltage). As we illustrate in Fig. 6.63, this results in the observation of reproducible oscillations, which wash out quickly with increasing temperature (Fig. 6.64), indicating that they are related to a quantum-interference effect [207]. These oscillations become stronger in high-quality material and exhibit just a small number of dominant frequencies that are dot-size dependent. These oscillations are therefore very different from the aperiodic universal conductance fluctuations (UCF) that arise in mesoscopic systems due to impurity-induced disorder (see Section 7.4 for a discussion of these fluctuations).

An understanding of the origins of the conductance oscillations in Figs. 6.63 and 6.64 can be achieved by performing numerical calculations of quantum transport through open dots, since such calculations provide a means to connect the quantum states of these systems to their measured electrical characteristics [212,213,214]. In isolated quantum dots, containing just a few electrons, it is well known that the transport can be strongly influenced by the discrete energy spectrum that is induced by confining electrons in the dot. When the dot is opened to its external reservoirs, however, it is often (mistakenly) assumed that all details of this level spectrum are washed out completely. Numerical simulations show, however, that the coupling of such dots to their reservoirs results in a *non-uniform* broadening of their eigenstates, with certain states developing only a very small broadening as a result of this coupling, whereas others couple more strongly and so are energetically washed out. Quite intuitively, the states of the originally closed system that survive the introduction of the reservoir coupling are found to be ones that are scarred by (i.e. which have high probability density

6.4 Spatial coherence in dots 399

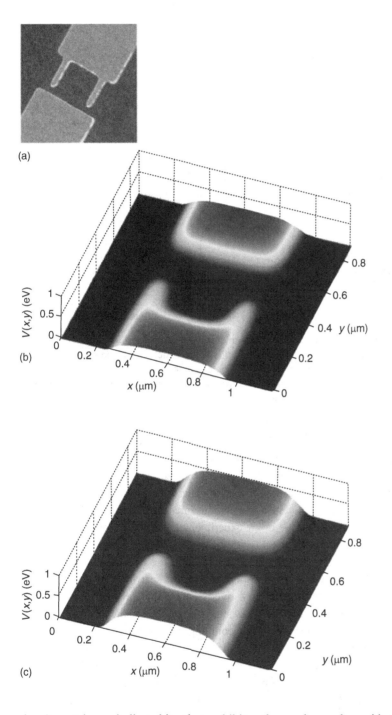

Fig. 6.62 (a) Scanning-electron micrograph of a split-gate quantum dot with a 1 micron cavity. (b) (c) Calculated potential profiles at two gate voltages. The gate voltage is more negative in the lowest panel ([207], with permission).

along) certain periodic orbits that exhibit only weak overlap with the dot leads. These orbits can therefore be viewed as weakly coupled quantum systems and the principles of semiclassical quantum mechanics imply [215] that each such orbit should give rise to a series of discrete eigenstates of varying energy.

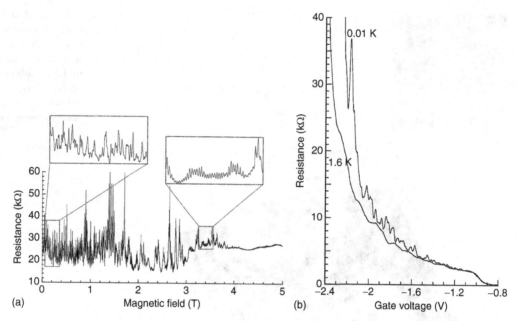

Fig. 6.63 (a) Magneto-resistance of a 1-micron split-gate quantum dot, at a fixed gate voltage, showing the detailed evolution of its structure with magnetic field at 10 mK. (b) Gate-voltage dependent variation of the resistance of a 1-micron split-gate quantum dot, at two different temperatures. The inset shows a micrograph of the dot geometry ([207], with permission).

When the magnetic field is varied in experiment, or if the potential profile of the dot is modified by means of a suitable gate voltage, these states may be swept past the Fermi level, giving rise to reproducible oscillations in the conductance that may be observed at low temperatures. The distinct frequency components that appear in the conductance oscillations are therefore related to specific scarred wave functions and, in an experimental study we were able to show evidence that, by changing the position of the dot leads relative to its perimeter, we could ensure that different orbits could be selected to induce the scarring behavior [216]. Since the orbits responsible for the scarring have weak overlap with the leads, the features that they give rise to in the wave function are quite stable as the leads are opened to allow stronger coupling to the environment, and in fact seem to be insensitive to this coupling, which agrees with experiments which have demonstrated that the dominant frequency components of the conductance oscillations are extremely stable to variation of the coupling strength over a wide range [207]. The robustness of the scarring is illustrated in Fig. 6.65, which shows the form of the wave function in the open dot for an energy where a clear scar is observed. This feature remains almost unchanged when the number of modes in the lead is increased from two to six, for which condition the vast majority of dot eigenstates are washed out completely due to their reservoir coupling.

6.4 Spatial coherence in dots 401

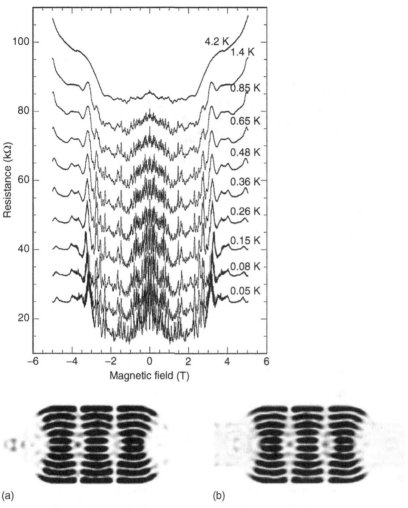

Fig. 6.64 Temperature dependence of the magneto-resistance of a 1-micron split-gate quantum dot. Successive curves are shifted upwards from each other.

Fig. 6.65 Calculations of the wave function of an open quantum dot with two (a) and six (b) propagating modes in its leads. Dark regions correspond to higher probability density ([207], with permission).

An important feature revealed in the quantum simulations of these dots is that closed-dot states very close in energy can exhibit very different behavior when the dot is opened, dependent upon the probability distribution of their eigenfunctions [208]. This, however, is exactly the property that defines the pointer states, and distinguishes these states from the original ones of the dot. This property is illustrated in Fig. 6.66, in which we show the numerically calculated variation of the conductance as a function of energy and gate voltage. This contour shows a series of resonances that move as the gate bias is varied, which changes the coupling to the environment as well as the dot size. It may be seen, however, that the linewidth does not change significantly as the

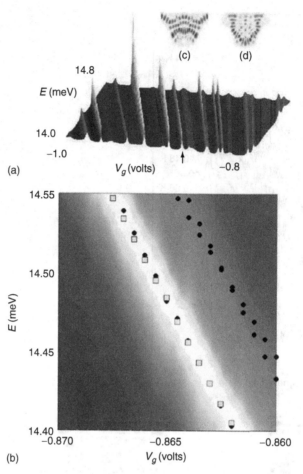

Fig. 6.66 (a) Calculated density-of-states peaks, corresponding to conductance resonances, in an open quantum dot. The peaks correspond to the stable wave functions in the dot as the gate voltage and Fermi energy are varied. As the gate voltage is varied, the dot becomes more open, but the stable resonances show no increase in width as the coupling to the environment is increased. The arrow indicates a resonance, whose wave function is indicated by inset (d), for which the state of inset (c) is quite near, but is completely damped. (b) An enlargement of panel (a) showing the lack of interaction between the stable wave functions and nearby states. The dark regions are those of lower amplitude in the density of states. The filled squares are the stable resonance indicated by the arrow in the upper panel and the wave function of inset (d) to that figure. The black circles are nearby states, which are completely damped by the interaction with the environment ([208], with permission).

environmental coupling is increased, even though the resonances shift in energy due to the change in the dot size. The resonance indicated by the arrow in this figure has another closed-dot eigenstate nearby, which is completely damped by its interaction with the environment, so no peak appears in this figure. The wave function of the resonance is shown in inset (d), while the nearby state is shown

6.4 Spatial coherence in dots   403

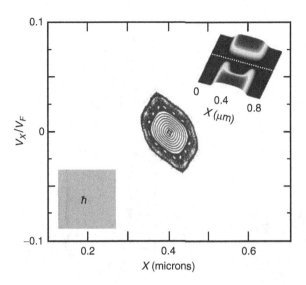

Fig. 6.67 Poincaré section of motion calculated for an open quantum dot like that shown in the inset. The section is taken along the dotted line in the upper inset. The lower inset shows an area of phase space comparable in size to $\hbar$ ([207], with permission).

in inset (c). In Fig. 6.66, we also show an expanded portion of the main contour. The filled squares correspond to the robust conduction state while the solid circles are related to nearby states that are completely damped by their interaction with the environment. Even though one of these damped states actually crosses the resonant state, there is no interaction as the two wave functions are orthogonal, as may be seen from the related wave functions in the inset.

An interesting question at this point concerns how to connect the discussion of the pointer states above to the underlying classical dynamics of these systems. The starting point for such an analysis is the classical phase space for electron motion in the dot. In the case of fully chaotic classical dynamics, all orbits are unstable and all initial conditions eventually result in escape from the dot in a finite time (except for a set of null measure). The Poincaré section of the motion in this case therefore corresponds to a dense set of points (the chaotic sea), with no evidence for any stable orbits. As we have mentioned already, however, the classical dynamics of particles in the soft potential of a quantum dot is typically not fully chaotic, but instead exhibits mixed behavior. In this situation, the section of motion exhibits well-known KAM (Kolmogorov–Arnold–Moser) islands, corresponding to isolated periodic orbits that are separated from the chaotic sea by classically inaccessible regions. The importance of this mixed phase space has been emphasized in the analysis of several experiments [217, 218,219] and in Fig. 6.67 we show a calculated Poincaré spectrum for a gated quantum dot [220]. This shows a KAM island that is centered on a period-one orbit that bounces back and forth at the center of the dot. Such an orbit is classically inaccessible to electrons injected into the dot by its connecting leads, however, and would not normally be expected to play a role in transport. In Ref. [220], however, the Gützwiler formula was used to calculate the quantized energy states arising from isolated orbits within the KAM island. By studying

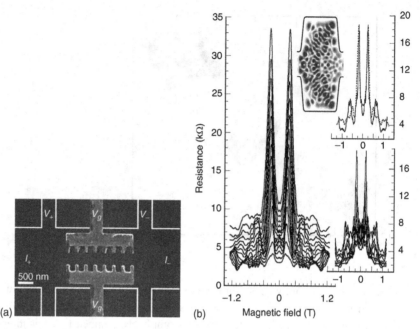

Fig. 6.68 (a) A micrograph of a seven-dot array realized by the split-gate technique. Lighter regions are the metal gates and the white lines show schematically the Hall-bar structure on top of which these are deposited. (b) Magneto-resistance of the seven-dot array at several different gate voltages and 10 mK. The lower inset shows the simulated gate-voltage-dependent magneto-resistance of the array. The upper inset shows a comparison of the experimental (dotted line) and computed (solid line) magneto-resistance ([225], with permission).

the evolution of these states as a function dot size, it was shown that they could be related to the periodic components of the conductance, found in both quantum simulations and experiment. This connection therefore provides a strong indication that the pointer states of interest are related to classically inaccessible periodic orbits of the dot, which can only be accessed by the quantum-mechanical process of phase-space tunneling (so-called dynamical tunneling [221,222,223,224]).

It is clear from the preceding discussion that there is an intimate connection between the classical dynamics of open quantum dots and their corresponding quantum characteristics. We have also emphasized the role that the isolated KAM islands can play in giving rise to phenomena beyond the usual semi-classical approximations. The connection is clearly demonstrated in more recent studies that we have performed of resonant transport through coupled quantum dots [225,226]. In these experiments, we investigate the magneto-transport of chains comprised of up to seven quantum dots, and observe a pronounced resonance in their resistance at a particular magnetic field (Fig. 6.68). The magnitude of this resonance becomes increasingly pronounced with an increasing

6.4 Spatial coherence in dots 405

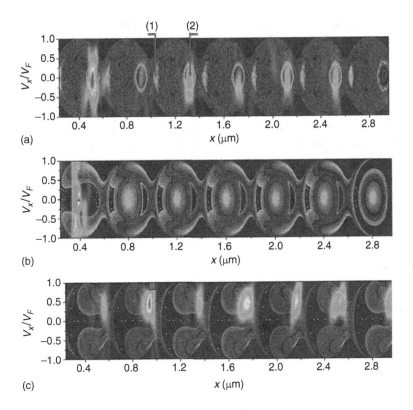

Fig. 6.69 Poincaré sections for a seven-dot quantum-dot array at three different magnetic fields, just prior to (a), at (b), and just after (c) the resonance field condition in Fig. 6.68 ([226], with permission).

number of dots in the array, while it occurs at a magnetic field for which the cyclotron orbit becomes commensurate with the dimensions of the dot. Under such conditions, enhanced backscattering strongly suppresses the forward transmission of electrons, yielding the resistance resonance. The resonant backscattering is reproduced in simulations of quantum transport through the arrays, which show an evanescently damped character to the wave function at the resonance condition [225]. Further insight into the origin of this behavior is provided by an analysis of the classical dynamics in the arrays, which shows that the magnetic field induces dramatic modifications to their classical phase space. This control of chaos appears to be a unique characteristic of the magnetic field, since it is extremely difficult in practice to realize dots whose phase space is fundamentally changed by appropriate design of their gates. Due to the remote nature of the split-gate technique, in which one applies potential to surface gates to deplete electrons in a layer below them, one almost inevitably ends up realizing a dot with soft walls and a mixed phase space. In our simulations of Ref. [226], we were able to show, however, that, without confining the quantum-dot potential, it is possible to induce a transition from a KAM to a non-KAM system by varying magnetic field. This behavior is illustrated in Fig. 6.69, which shows the calculated Poincare sections of motion at three different magnetic fields close to the resonance condition for backscattering.

The middle contour shows the case at the resonance condition and demonstrates that, under this condition, the chaotic sea surrounding the KAM islands has been largely "drained." It is also clear from a comparison with the upper and lower panels of the figure that this draining occurs over a narrow range of magnetic field. Such results furthermore demonstrate the intimate connection of quantum transport in open quantum dots to the structure of their mixed phase space.

Finally, we note now that the earlier suggestion that the nature of the magneto-resistance around zero magnetic field depends crucially upon the details of the energy spectrum of the closed dot itself. While it may seem to appear as first thought, it is more crucial to determine if the Fermi energy lies in the chaotic sea or is aligned with one of the pointer states lying within a KAM island. Whether the dot is really circular or stadium in shape plays little role, as it is the overlying mixed phase space that is most important.

References

[1] D. K. Cheng, *Field and Wave Electromagnetics*, 2nd edition (Reading, MA, Addison-Wesley Publishing, 1989), pp. 172–4.
[2] J. D. Jackson, *Classical Electrodynamics*, 2nd edition (New York, John Wiley and Sons, 1975), p. 133.
[3] M. Macucci, K. Hess, and G. Iafrate. *Phys. Rev. B* **48**, 17354 (1993).
[4] C. Gorter, *Physica* **17**, 777 (1951).
[5] E. Darmois, *J. Phys. Radium* **17**, 210 (1956).
[6] C. A. Neugebauer and M. B. Webb, *J. Appl. Phys.* **33**, 74 (1962).
[7] I. Giaever and H. R. Zeller, *Phys. Rev. Lett.* **20**, 1504 (1968).
[8] J. Lambe and R. C. Jaklevic, *Phys. Rev. Lett.* **22**, 1371 (1969).
[9] R. E. Cavicchi and R. H. Silsbee, *Phys. Rev. Lett.* **52**, 1453 (1984); *Phys. Rev. B* **37**, 706 (1987).
[10] I. O. Kulik and R. I. Shekhter, *Sov. Phys. JETP* **41**, 308 (1975).
[11] D. V. Averin and K. K. Likharev, in the *Proceedings of the Third International Conference on Superconducting Quantum Devices* (SQUID), Berlin, 1985, eds. H.-D. Hahlbohm and H. Lubbig (Berlin, W. de Gruyter, 1985), p. 197; D. V. Averin and K. K. Likharev, *J. Low Temp. Phys.* **62**, 345 (1986).
[12] E. Ben-Jacob, Y. Gefen, K. Mullen, and Z. Schuss, in the *Proceedings of the Third International Conference on Superconducting Quantum Devices* (SQUID), Berlin, 1985, eds. H.-D. Hahlbohm and H. Lubbig (Berlin, W. de Gruyter, 1985), p. 203; E. Ben-Jacob, D. L. Bergman, B. J. Matkowsky, and Z. Schuss, *Phys. Rev. B* **34**, 1572 (1986).
[13] L. S. Kuzmin and K. K. Likharev, *JETP Lett.* **45**, 495 (1987).
[14] J. B. Barner and S. T Ruggiero, *Phys. Rev. Lett.* **59**, 807 (1987).
[15] P. J. M. van Bentum, R. T. M. Smokers, and H. van Kempen, *Phys. Rev. Lett.* **60**, 2543 (1988).
[16] R. Wilkins, E. Ben-Jacob, and R. C. Jaklevic, *Phys. Rev. Lett.* **63**, 801 (1989).
[17] T. A. Fulton and G. J. Dolan, *Phys. Rev. Lett.* **59**, 109 (1987).
[18] C. Thelander, T. Martensson, M. T. Bjork, *et al.*, *Appl. Phys. Lett.* **83**, 2052 (2003).
[19] C. Wasshuber, *Computational Single-Electronics* (Berlin, Springer, 2001).
[20] C. W. J. Beenakker, *Phys. Rev. B* **44**, 1646 (1991).
[21] J. H. F. Scott-Thomas, S. B. Field, M. A. Kastner, H. I. Smith, and D. A. Antoniadis, *Phys. Rev. Lett.* **62**, 583 (1989).

[22] L. I. Glazman and R. I. Shekhter, *J. Phys.: Condens. Matter* **1**, 5811 (1989).
[23] H. van Houten and C. W. J. Beenakker, *Phys. Rev. Lett.* **63**, 1893 (1989).
[24] U. Meirav, M. A. Kastner, and S. J. Wind, *Phys. Rev. Lett.* **65**, 771 (1990).
[25] M. A. Kastner, *Rev. Mod. Phys.* **64**, 849 (1992).
[26] P. Delsing, in *Single Charge Tunneling: Coulomb Blockade Phenomena in Nanostructures*, eds. H. Grabert and Michel H. Devoret, NATO ASI Series B 294 (New York, Plenum Press, 1992), pp. 249–73.
[27] J. E. Mooij and G. Schon, in *Single Charge Tunneling: Coulomb Blockade Phenomena in Nanostructures*, eds. H. Graben and Michel H. Devoret, NATO ASI Series B 294 (New York, Plenum Press, 1992), pp. 275–310.
[28] L. J. Geerligs, V. F. Anderegg, P. A. M. Holweg, *et al.*, *Phys. Rev. Lett.* **64**, 2691 (1990); L. J. Geerligs and J. E. Mooij, in *Granular Nanoelectronics*, eds. D. K. Ferry, J. R. Barker, and C. Jacoboni, NATO ASI Series B 251 (New York, Plenum Press, 1991), pp. 393–412; L. Geerligs, in *Physics of Nanostructures*, eds. J. H. Davies and A. R. Long (London, Institute of Physics Publishing, Ltd., 1992), pp. 171–204.
[29] L. P. Kouwenhoven, A. T. Johnson, N. C. van der Vaart, C. J. P. M. Harmans, and C. T. Foxon, *Phys. Rev. Lett.* **67**, 1626 (1991).
[30] H. Pothier, P. Lafarge, C. Urbina, D. Esteve, and M. H. Devoret, *Europhys. Lett.* **17**, 249 (1992); D. Esteve, in *Single Charge Tunneling: Coulomb Blockade Phenomena in Nanostructures*, eds. H. Grabert and Michel H. Devoret, NATO ASI Series B **294** (New York, Plenum Press, 1992), pp. 109–37.
[31] D. V. Averin, A. N. Korotkov, and K. K. Likharev, *Phys. Rev. B* **44**, 6199 (1991).
[32] J. Bardeen, *Phys. Rev. Lett.* **6**, 57 (1961).
[33] M. N. Cohen, L. M. Falicov, and J. C. Phillips, *Phys. Rev. Lett.* **8**, 316 (1962).
[34] C. B. Duke, *Tunneling in Solids*, Chapter 7, *Solid State Physics* **10**, eds. F. Seitz, D. Turnbull, and H. Ehrenreich (Academic Press, New York, 1969).
[35] D. V. Averin and K. K. Likharev, "Single electronics: a correlated transfer of single electrons and cooper pairs in systems of small tunnel junctions," in *Mesoscopic Phenomena in Solids*, eds. B. L. Altshuler, P. A. Lee, and R. A. Webb (Oxford, Elsevier Science Publishers, 1991), pp. 173–271.
[36] G.-L. Ingold and Yu. V. Nazarov, "Charge tunneling rates in ultrasmall junctions," in *Single Charge Tunneling: Coulomb Blockade Phenomena in Nanostructures*, eds. H. Grabert and Michel H. Devoret, NATO ASI Series B 294 (New York, Plenum Press, 1992), pp. 21–108.
[37] K. K. Likharev, *IBM J. Res. Develop.* **32**, 144 (1988).
[38] W. Kohn and J. M. Luttinger, *Phys. Rev.* **108**, 590 (1957).
[39] E. Ben-Jacob, Y. Gefen, K. Mullen, and Z. Schuss, *Phys. Rev. B* **37**, 7400 (1988).
[40] A. Van der Ziel, *Noise in Solid State Devices and Circuits* (New York, Wiley, 1986).
[41] D. V. Averin and A. A. Odintsov, *Phys. Lett. A* **140**, 251 (1989).
[42] D. V. Averin and Yu. V. Nazarov, *Phys. Rev. Lett.* **65**, 2446 (1990); D. V. Averin and Yu. V. Nazarov, in *Single Charge Tunneling: Coulomb Blockade Phenomena in Nanostructures*, eds. H. Grabert and Michel H. Devoret, NATO ASI Series B 294 (New York, Plenum Press, 1992), pp. 217–47.
[43] L. J. Geerligs, D. V. Averin, and J. E. Mooij, *Phys. Rev. Lett.* **65**, 3037 (1990).
[44] T. M. Eiles, G. Zimmerli, H. D. Jensen, and J. M. Martinis, *Phys. Rev. Lett.* **69**, 148 (1992).
[45] D. C. Clank, C. Pasquier, U. Meirav, *et al.*, *Z. Phys. B* **85**, 375 (1991); C. Pasquier, U. Meirav, F. I. B. Williams, D. C. Glattli, Y. Jin, and B. Etienne, *Phys. Rev. Lett*, **70**, 69 (1993).

[46] L. Jacak, P. Hawrylak and A. Wojs, *Quantum Dots* (Springer, Berlin, 1998).
[47] A. N. Korotkov, D. V. Averin, and K. K. Likharev, *Physica B* **165/166**, 927 (1990).
[48] L. P. Kouwenhoven, C. M. Marcus, P. L. McEuen, *et al.*, in *Mesoscopic Electron Transport*, eds. L. L. Sohn, G. Schön, and L. P. Kouwenhoven (Dordrecht, Kluwer Series E 345, 1997) pp. 105–214.
[49] P. L. McEuen, E. B. Foxman, U. Meirav, *et al.*, *Phys. Rev. Lett.* **66**, 1926 (1991).
[50] P. L. McEuen, E. B. Foxman, J. Kinaret, *et al.*, *Phys. Rev. B* **45**, 11419 (1992).
[51] E. B. Foxman, P. L. McEuen, U. Meirav, *et al.*, *Phys. Rev. B* **47**, 10020 (1993).
[52] R. C. Ashoori, H. L. Störmer, J. S. Weiner, *et al.*, *Phys. Rev. Lett.* **68**, 3088 (1992).
[53] R. C. Ashoori, H. L. Störmer, J. S. Weiner, *et al.*, *Phys. Rev. Lett.* **71**, 613 (1993).
[54] N. B. Zhitenev, R. C. Ashoori, L. N. Pfeiffer, and K. W. West, *Phys. Rev. Lett.* **79**, 2308 (1997).
[55] S. Tarucha, D. G. Austing, T. Honda, R. J. van der Hage, and L. P. Kouwenhoven, *Phys. Rev. Lett.* **77**, 3613 (1996).
[56] *For a review, see*: L. P. Kouwenhoven, D. G. Austing, and S. Tarucha, *Rep. Prog. Phys.* **64**, 701 (2001).
[57] M. Macucci, K. Hess, and G. J. Iafrate, *Phys. Rev. B* **48** 17354 (1993); M. Macucci, K. Hess, and G. J. Iafrate, *J. Appl. Phys.* **77**, 3267 (1995).
[58] Y. H. Zeng, B. Goodman, and R. A. Serota, *Phys. Rev. B* **47**, 15660 (1993).
[59] A. Wojs and P. Hawrylak, *Phys. Rev. B* **53**, 10841 (1996).
[60] M. Stopa, *Phys. Rev. B* **54**, 13767 (1996).
[61] W. D. Heiss and R. G. Nazmitdinov, *Phys. Lett. A* **222**, 309 (1996).
[62] T. Ezaki, N. Mori, and C. Hamaguchi, *Phys. Rev. B* **56**, 6428 (1997).
[63] A. Angelucci and A. Tagliacozzo, *Phys. Rev. B* **56**, 7088 (1997).
[64] S. Nagaraja, P. Matagne, V.-Y. Thean, *et al.*, *Phys. Rev. B* **56**, 15752 (1997).
[65] O. Steffens, U. Rössler, and M. Suhrke, *Europhys. Lett.* **42**, 529 (1998).
[66] I.-H. Lee, V. Rao, R. M. Martin, and J.-P. Leburton, *Phys. Rev. B* **57** 9035 (1998).
[67] J. Harting, O. Mulken, and P. Borrmann, *Phys. Rev. B* **62**, 10207 (2000).
[68] V. Fock, *Z. Phys.* **47**, 446 (1928).
[69] C. G. Darwin, *Proc. Cambridge Philos. Soc.* **27**, 86 (1930).
[70] S. R. Patel, D. R. Stewart, C. M. Marcus, *et al.*, *Phys. Rev. Lett.* **81**, 5900 (1998).
[71] D. R. Stewart, D. Sprinzak, C. M. Marcus, C. I. Duruöz, and J. S. Harris, *Science* **278**, 1784 (1997).
[72] R. M. Potok, J. A. Folk, C. M. Marcus, *et al.*, *Phys. Rev. Lett.* **91**, 016802 (2003).
[73] A. T. Johnson, L. P. Kouwenhoven, W. de Jong, *et al.*, *Phys. Rev. Lett.* **69**, 1592 (1992).
[74] J. Weis, R. J. Haug, K. v. Klitzing, and K. Ploog, *Phys. Rev. Lett.* **71**, 4019 (1993).
[75] D. Weinmann, W. Häusler, and B. Kramer, *Phys. Rev. Lett.* **74**, 984 (1995).
[76] M. Ciorga, A. S. Sachrajda, P. Hawrylak, *et al.*, *Phys. Rev. B* **61**, R16315 (2000).
[77] M. Ciorga, A. Wensauer, M. Pioro-Ladrière, *et al.*, *Phys. Rev. Lett.* **88**, 256804 (2002).
[78] J. Kyriakidis, M. Pioro-Ladrière, M. Ciorga, A. S. Sachrajda, and P. Hawrylak, *Phys. Rev. B* **66**, 035320 (2002).
[79] M. Pioro-Ladrière, M. Ciorga, J. Lapointe, *et al.*, *Phys. Rev. Lett.* **91**, 026803 (2003).
[80] M. Korkusiński, P. Hawrylak, M. Ciorga, M. Pioro-Ladrière, and A. S. Sachrajda, *Phys. Rev. Lett.* **93**, 206806 (2004).
[81] L. P. Kouwenhoven, T. H. Oosterkamp, M. W. S. Danoesastro, *et al.*, *Science* **278**, 1788 (1997).
[82] S. Tarucha, D. G. Austing, Y. Tokura, W. van der Wiel, and L. P. Kouwenhoven, *Phys. Rev. Lett.* **84**, 2485 (2000).

[83] T. Fujisawa, Y. Tokura, and Y. Hirayama, *Phys. Rev. B* **63**, 081304 (2001).
[84] T. Fujisawa, D. G. Austing, Y. Tokura, Y. Hirayama, and S. Tarucha, *Phys. Rev. Lett.* **88**, 236802 (2002).
[85] T. Fujisawa, D. G. Austing, Y. Tokura, Y. Hirayama, and S. Tarucha, *Nature* **419**, 278 (2002).
[86] W. J. de Haas, J. H. de Boer, and G. J. van den Berg, *Physica* **1**, 1115 (1934).
[87] J. Kondo, *Prog. Theor. Phys.* **32**, 37 (1964).
[88] See Chapter 18 of R. D. Mattuck *A Guide to Feynman Diagrams in the Many-Body Problem*, 2nd edition (New York, McGraw-Hill, 1976).
[89] K. G. Wilson, *Rev. Mod. Phys.* **47**, 773 (1975).
[90] P. W. Anderson, *Phys. Rev.* **124**, 41 (1961).
[91] F. D. M. Haldane, *Phys. Rev. Lett.* **40**, 416 (1978).
[92] L. I. Glazman and M. E. Raikh, *JETP Lett.* **47**, 452 (1988).
[93] T. K. Ng and P. A. Lee, *Phys. Rev. Lett.* **61**, 1768 (1988).
[94] D. Goldhaber-Gordon, H. Shtrikman, D. Mahalu, *et al.*, *Nature* **391**, 156 (1998).
[95] D. Goldhaber-Gordon, J. Gores, M. A. Kastner, *et al.*, *Phys. Rev. Lett.* **81**, 5225 (1998).
[96] W. G. van der Wiel, S. De Franceschi, J. M. Elzerman, *et al.*, *Science* **281**, 540 (1998).
[97] T. A. Costi, A. C. Hewson, and V. Zlatic, *J. Phys.: Condens. Matter* **6**, 2519 (1994).
[98] A. Kawabata, *J. Phys. Soc. Jpn.* **60**, 3222 (1991).
[99] Y. Meir, N. S. Wingreen, and P. A. Lee, *Phys. Rev. Lett.* **70**, 2601 (1993).
[100] S. M. Cronenwett, T. H. Oosterkamp, and L. P. Kouwenhoven, *Science* **281**, 540 (1998).
[101] J. Schmid, J. Weis, K. Eberl, and K. von Klitzing, *Physica B* **182**, 256–258 (1998).
[102] F. Simmel, R. H. Blick, J. P. Kotthaus, W. Wegscheider, and M. Bichler, *Phys. Rev. Lett.* **83**, 804 (1999).
[103] J. Schmid, J. Weis, K. Eberl, and K. von Klitzing, *Phys. Rev. Lett.* **84**, 5824 (2000).
[104] S. Sasaki, S. De Franceschi, J. M. Elzerman, *et al.*, *Nature* **40F5**, 764 (2000).
[105] J. Nygard, D. H. Cobden, and P. E. Lindelof, *Nature* **408**, 342 (2000).
[106] M. Pustilnik, Y. Avishai, and K. Kikoin, *Phys. Rev. Lett.* **84**, 1756 (2000).
[107] M. Eto and Y. V. Nazarov, *Phys. Rev. Lett.* **85**, 1306 (2000).
[108] M. Eto and Y. V. Nazarov, *Phys. Rev. B* **66**, 153319 (2002).
[109] M. Pustilink and L. I. Glazman, *Phys. Rev. Lett.* **87**, 216601 (2001).
[110] W. Hofstetter and H. Schoeller, *Phys. Rev. Lett.* **88**, 016803 (2002).
[111] W. G. van der Wiel, S. De Franceschi, J. M. Elzerman, *et al.*, *Phys. Rev. Lett.* **88**, 126803 (2002).
[112] M. R. Buitelaar, A. Bachtold, T. Nussbaumer, M. Iqbal, and C. Schönenberger, *Phys. Rev. Lett.* **88**, 156801 (2002).
[113] W. Liang, M. P. Shores, M. Bockrath, J. R. Long, and H. Park, *Nature* **417**, 725 (2002).
[114] J. Park, A. N. Pasupathy, J. I. Goldsmith, *et al.*, *Nature* **417**, 722 (2002).
[115] P. Jarillo-Herrero, J. Kong, H. S. J. van der Zant, *et al.*, *Nature* **434**, 484 (2005).
[116] W. G. van der Wiel, S. De Franceschi, J. M. Elzerman, *et al.*, *Rev. Mod. Phys.* **75**, 1 (2003).
[117] T. Fujisawa, T. Hayashi, and S. Sasaki, *Rep. Prog. Phys.* **69**, 759 (2006).
[118] F. Hofmann, T. Heinzel, D. A. Wharam, *et al.*, *Phys. Rev. B* **51**, 13872 (1995).
[119] H. Pothier, P. Lafarge, C. Urbina, D. Esteve, and M. H. Devoret, *Europhys. Lett.* **17**, 249 (1992).
[120] I. M. Ruzin, V. Chandrasekhar, E. I. Levin, and L. I. Glazman, *Phys. Rev. B* **45**, 13469 (1992).
[121] D. C. Dixon, L. P. Kouwenhoven, P. L. McEuen, *et al.*, *Phys. Rev. B* **53**, 12625 (1996).

[122] N. C. van der Vaart, S. F. Godijn, Y u. V. Nazarov, C. J. P. M. Harmans, and J. E. Mooij, *Phys. Rev. Lett.* **74**, 4702 (1995).
[123] M. Kemerink and L. W. Molenkamp, *Appl. Phys. Lett.* **65**, 1012 (1994).
[124] L. W. Molenkamp, K. Flensberg, and M. Kemerink, *Phys. Rev. Lett.* **75**, 4282 (1995).
[125] F. R. Waugh, M. J. Berry, D. J. Mar, *et al.*, *Phys. Rev. Lett.* **75**, 705 (1995).
[126] F. R. Waugh, M. J. Berry, C. H. Crouch, *et al.*, *Phys. Rev. B* **53**, 1413 (1996).
[127] C. Livermore, C. H. Crouch, R. M. Westervelt, K. L. Campman, and A. C. Gossard, *Science* **274**, 1332 (1996).
[128] C. S. Lent, P. D. Tougaw, W. Porod, and G. H. Bernstein, *Nanotechnology* **4**, 49 (1993); C. S. Lent and P. D. Tougaw, *Proc. IEEE* **85**, 541 (1997).
[129] A. O. Orlov, I. Amlani, G. H. Bernstein, C. S. Lent, and G. L. Snider, *Science* **277**, 928 (1997).
[130] I. Amlani, A. O. Orlov, G. Toth, *et al.*, *Science* **284**, 289 (1999).
[131] A. O. Orlov, I. Amlani, G. Toth, *et al.*, *Appl. Phys. Lett.* **74**, 2875 (1999).
[132] A. O. Orlov, I. Amlani, R. K. Kummamuru, *et al.*, *Appl. Phys. Lett.* **77**, 295 (2000).
[133] R. K. Kummamuru, J. Timler, G. Toth, *et al.*, *Appl. Phys. Lett.* **81**, 1332 (2002).
[134] S. Gardelis, C. G. Smith, J. Cooper, *et al.*, *Phys. Rev. B* **67**, 033302 (2003).
[135] A. Imre, G. Csaba, L. Ji, *et al.*, *Science* **311**, 205 (2006).
[136] See Chapter 6 of David K. Ferry, *Quantum Mechanics: An Introduction for Device Physicists and Electrical Engineers*, 2nd Edition (Bristol, Institute of Physics Publishing, 2001).
[137] R. H. Blick, R. J. Haug, J. Weis, *et al.*, *Phys. Rev. B* **53**, 7899 (1996).
[138] R. H. Blick, D. Pfannkuche, R. J. Haug, K. V. Klitzing, and K. Eberl, *Phys. Rev. Lett.* **80**, 4032 (1998).
[139] R. H. Blick, D. W. van der Weide, R. J. Haug, and K. Eberl, *Phys. Rev. Lett.* **81**, 689 (1998).
[140] A. W. Holleitner, C. R. Decker, H. Qin, K. Eberl, and R. H. Blick, *Phys. Rev. Lett.* **87**, 256802 (2001).
[141] A. W. Holleitner, R. H. Blick, A. K. Huttel, K. Eberl, and J. P. Kotthaus, *Science* **297**, 70 (2001).
[142] T. Hayashi, T. Fujisawa, H. D. Cheong, Y. H. Jeong, and Y. Hirayama, *Phys. Rev. Lett.* **91**, 226804 (2003).
[143] C. Cohen-Tannoudji, B. Diu, and F. Laloe, *Quantum Mechanics*, Vol. 1 (New York, Wiley, 1977).
[144] T. H. Oosterkamp, T. Fujisawa, W. G. Van der Wiel, *et al.*, *Nature* **395**, 873 (1998).
[145] L. P. Kouwenhoven, S. Jauhar, J. Orenstein, *et al.*, *Phys. Rev. Lett.* **73**, 3443 (1994).
[146] D. Loss and D. P. DiVincenzo, *Phys. Rev. A* **57**, 120 (1998); D. P. DiVincenzo and D. Loss, *Superlatt. Microstruct.* **23**, 419 (1998); G. Burkard, D. Loss, and D. P. DiVincenzo, *Phys. Rev. B* **59**, 2070 (1999); and G. Burkard and D. Loss, in *Semiconductor Spintronics and Quantum Computation*, eds. D. D. Awschalom, D. Loss, and N. Samarth (Berlin Heidelberg, Springer-Verlag, 2002), pp. 229–76.
[147] M. Field, C. G. Smith, M. Pepper, *et al.*, *Phys. Rev. Lett.* **70**, 1311 (1993).
[148] M. Field, C. G. Smith, M. Pepper, *et al.*, *J. Phys.: Condens. Matter* **6**, L273 (1994).
[149] D. Sprinzak, Y. Ji, M. Heiblum, D. Mahalu, and H. Shtrikhman, *Phys. Rev. Lett.* **88**, 176805 (2002).
[150] J. M. Elzerman, R. Hanson, J. S. Greidanus, *et al.*, *Phys. Rev. B* **67**, 161308 (2003).
[151] R. Hanson, B. Witkamp, L. H. W. Von Beveren, J. M. Elzerman, and L. P. Kouwenhoven, *Phys. Rev. Lett.* **91**, 196802 (2003).

[152] J. M. Elzerman, R. Hanson, L. H. W. van Beveren, *et al.*, *Nature* **430**, 431 (2004).
[153] R. Hanson, L. H. W. van Beveren, I. T. Vink, *et al.*, *Phys. Rev. Lett.* **94**, 196802 (2005).
[154] T. Meunier, I. T. Vink, L. H. W. van Beveren, *et al.*, *Phys. Rev. B* **74**, 195303 (2006).
[155] R. Hanson, L. P. Kouwenhoven, J. R. Petta, S. Tarucha, and L. M. K. Vandersypen, *Rev. Mod. Phys.* **79**, 1217 (2007).
[156] R. Schleser, E. Ruh, T. Ihn, *et al.*, *Appl. Phys. Lett.* **85**, 2005 (2004).
[157] S. Gustavvson, R. Leturcq, B. Simovic, *et al.*, *Phys. Rev. Lett.* **96**, 076605 (2006).
[158] T. Fujisawa, T. Hayashi, Y. Hirayama, H. D. Cheong, and Y. H. Jeong, *Appl. Phys. Lett.* **84**, 2343 (2004).
[159] S. Gustavsson, M. Studer, R. Leturcq, *et al.*, *Phys. Rev. Lett.* **99**, 206804 (2007).
[160] J. R. Petta, A. C. Johnson, C. M. Marcus, M. P. Hanson, and A. C. Gossard, *Phys. Rev. Lett.* **93**, 186802 (2004).
[161] A. C. Johnson, C. M. Marcus, M. P. Hanson, and A. C. Gossard, *Phys. Rev. B* **71**, 115333 (2005).
[162] A. C. Johnson, J. R. Petta, J. M. Taylor, *et al.*, *Nature* **456**, 925 (2005).
[163] J. R. Petta, A. C. Johnson, A. Yacoby, *et al.*, *Phys. Rev. B* **72**, 161301 (2005).
[164] A. C. Johnson, J. R. Petta, C. M. Marcus, M. P. Hanson, and A. C. Gossard, *Phys. Rev. B* **72**, 165308 (2005).
[165] J. R. Petta, A. C. Johnson, J. M. Taylor, *et al.*, *Science* **309**, 2180 (2005).
[166] E. A. Laird, J. R. Petta, A. C. Johnson, *et al.*, *Phys. Rev. Lett.* **97**, 056801 (2006).
[167] F. H. L. Koppens, C. Buizert, K. J. Tielrooij, *et al.*, *Nature* **442**, 766 (2006).
[168] K. Ono, D. G. Austing, Y. Tokura, and S. Tarucha, *Science* **297**, 1313 (2002).
[169] J. J. Lin and J. P. Bird, *J. Phys.: Condens. Matter* **14**, R501 (2002).
[170] R. K. Adair, C. K. Bockelman, and R. E. Peterson, *Phys. Rev.* **76**, 308 (1949).
[171] U. Fano and A. R. P. Rau, in *Atomic Collisions and Spectra* (Orlando, Academic Press, 1986).
[172] F. Cerdeira, T. A. Fjeldy, and M. Cardona, *Phys. Rev. B* **8**, 4734 (1973).
[173] J. Faist, F. Capasso, C. Sirtori, K. W. West, and L. N. Pfeiffer, *Nature* **390**, 589 (1997).
[174] K. Kobayashi, H. Aikawa, S. Katsumoto, and Y. Iye, *Phys. Rev. Lett.* **88**, 256806 (2002).
[175] K. Kobayashi, H. Aikawa, S. Katsumoto, and Y. Iye, *Phys. Rev. B* **68**, 235304 (2003).
[176] K. Kobayashi, H. Aikawa, A. Sano, S. Katsumoto, and Y. Iye, *Phys. Rev. B* **70**, 035319 (2004).
[177] M. Sato, H. Aikawa, K. Kobayashi, S. Katsumoto, and Y. Iye, *Phys. Rev. Lett.* **95**, 066801 (2005).
[178] U. Fano, *Phys. Rev.* **124**, 1866 (1961).
[179] J. J. Lin and J. P. Bird, *J. Phys.: Condens. Matter* **14**, R501 (2002).
[180] A. Fuhrer, P. Brusheim, T. Ihn, *et al.*, *Phys. Rev. B* **73**, 205326 (2006).
[181] A. A. Clerk, X. Waintal, and P. W. Brouwer, *Phys. Rev. Lett.* **86**, 4636 (2001).
[182] J. Kim, J.-R. Kim, J.-O. Lee, *et al.*, *Phys. Rev. Lett.* **90**, 166403 (2003).
[183] M. Kida, T. Mihara, K. Miyamoto, *et al.*, *IPAP Conf. Series* **5**, 61–64 (2004).
[184] D. Goldhaber-Gordon, J. Gores, M. A. Kastner, *et al.*, *Nature* **391**, 156 (1998).
[185] D. Goldhaber-Gordon, J. Gores, M. A. Kastner, *et al.*, *Phys. Rev. Lett.* **81**, 5225 (1998).
[186] J. Gores, D. Goldhaber-Gordon, S. Heemeyer, *et al.*, *Phys. Rev. B* **62**, 2188 (2000).
[187] I. G. Zachzria, D. Goldhaber-Gordon, G. Granger, *et al.*, *Phys. Rev. B* **64**, 155311 (2001).
[188] A. C. Johnson, C. M. Marcus, M. P. Hanson, and A. C. Gossard, *Phys. Rev. Lett.* **93**, 106803 (2004).

[189] A. Yacoby, M. Heiblum, D. Mahalu, H. Shtrikman, *Phys. Rev. Lett.* **74**, 4047 (1995).
[190] R. Schuster, E. Buks, M. Heiblum, *et al.*, *Nature* **385**, 417 (1997).
[191] G. Hackenbroich, *Phys. Rep.* **343**, 463 (2001).
[192] S. Katsumoto, *J. Phys.: Condens. Matter* **19**, 233201 (2007).
[193] M. Sigrist, A. Fuhrer, T. Ihn, *et al.*, *Phys. Rev. Lett.* **93**, 066802 (2004).
[194] M. Sigrist, T. Ihn, K. Ensslin, *et al.*, *Phys. Rev. Lett.* **96**, 036804 (2006).
[195] H.-J. Stöckmann, *Quantum Chaos – an Introduction* (Cambridge, Cambridge University Press, 1999).
[196] Y.-H. Kim, M. Barth, H.-J. Stöckmann, and J. P. Bird, *Phys. Rev. B* **65**, 165317 (2002).
[197] D. V. Evans and R. Porter, *J. Eng. Math.* **35**, 149 (1999).
[198] Y. F. Chen, K. F. Huang, and Y. P. Lan, *Phys. Rev. E* **66**, 046215 (2002).
[199] J. P. Bird, *J. Phys.: Condens. Matter* **11**, R413 (1999).
[200] C. M. Marcus, A. J. Rimberg, R. M. Westervelt, P. F. Hopkins, and A. C. Gossard, *Phys. Rev. Lett.* **69**, 506 (1992).
[201] R. A. Jalabert, H. U. Baranger, and A. D. Stone, *Phys. Rev. Lett.* **65**, 2442 (1990).
[202] M. W. Keller, O. Millo, A. Mittal, D. E. Prober, and R. N. Sacks, *Surf. Sci.* **305**, 501 (1994).
[203] M. J. Berry, J. A. Katine, C. M. Marcus, R. M. Westervelt, and A. C. Gossard, *Surf. Sci.* **305**, 495 (1994).
[204] J. P. Bird, K. Ishibashi, Y. Aoyagi, T. Sugano, and Y. Ochiai, *Phys. Rev. B* **50**, 18678 (1994).
[205] H. U. Baranger, R. A. Jalabert, and A. D. Stone, *Phys. Rev. Lett.* **70**, 3876 (1993).
[206] A. M. Chang, H. U. Baranger, L. N. Pfeiffer, and K. W. West, *Phys. Rev. Lett.* **73**, 2111 (1994).
[207] J. P. Bird, R. Akis, D. K. Ferry, *et al.*, *Rep. Prog. Phys.* **66**, 583 (2003).
[208] D. K. Ferry, R. Akis, and J. P. Bird, *Phys. Rev. Lett.* **93**, 026803 (2004).
[209] W. H. Zurek, *Rev. Mod. Phys.* **75**, 715 (2003).
[210] W. H. Zurek, *Phys. Rev. D* **24**, 1516 (1981).
[211] R. Ketzmerick, *Phys. Rev. B* **54**, 10841 (1996).
[212] R. Akis, D. K. Ferry, and J. P. Bird, *Phys. Rev. B* **54**, 17705 (1996).
[213] R. Akis, D. K. Ferry, and J. P. Bird, *Phys. Rev. Lett.* **79**, 123 (1997).
[214] R. Akis, J. P. Bird, and D. K. Ferry, *Appl. Phys. Lett.* **81**, 129 (2002).
[215] M. C. Gutzwiller, *Chaos in Classical and Quantum Mechanics* (New York, Springer, 1990).
[216] J. P. Bird, R. Akis, D. K. Ferry, *et al.*, *Phys. Rev. Lett.* **82**, 4691 (1999).
[217] A. P. Micolich, R. P. Taylor, R. Newbury, *et al.*, *J. Phys. Condens.: Matter* **10**, 1339 (1998).
[218] A. S. Sachrajda, R. Ketzmerick, C. Gould, *et al.*, *Phys. Rev. Lett.* **80**, 1948 (1998).
[219] Y. Takagaki, M. Elhassan, A. Shailos, *et al.*, *Phys. Rev. B* **62**, 10255 (2000).
[220] A. P. S. de Moura, Y.-C. Lai, R. Akis, J. P. Bird, and D. K. Ferry, *Phys. Rev. Lett.* **88**, 236804 (2002).
[221] M. J. Davis and E. J. Heller, *J. Chem. Phys.* **75**, 246–254 (1981).
[222] O. Bohigas, S. Tomsovic, and D. Ullmo, *Phys. Rep.* **223**, 43 (1993).
[223] S. Tomsovic, ed., *Tunneling in Complex Systems* (Singapore, World Scientific, 1998).
[224] S. Tomsovic, *Phys. Scr.* **T90**, 162 (2001).
[225] M. Elhassan, R. Akis, J. P. Bird, *et al.*, *Phys. Rev. B* **70**, 205341 (2004).
[226] R. Brunner, R. Meisels, F. Kuchar, *et al.*, *Phys. Rev. Lett.* **98**, 204101 (2007).

7
Weakly disordered systems

In the preceding chapters, and indeed in the subsequent chapters, most of the discussion is on semiconductors in which the Bloch theory of extended states prevails. There is another class of semiconductors that has received considerable attention over the past several decades, and that is disordered (or amorphous) semiconductors. Here, in the realm of nanostructures, we really do not want to discuss the entire field of amorphous semiconductors, and would generally ignore strongly disordered materials as well. However, recent experiments have shown the presence of a metal–insulator transition in quasi-two-dimensional systems. Consequently, one needs to understand the difference between localized (disordered) systems, weakly disordered systems, and the normal Bloch band picture of conductance.

Generally, in disordered (or, strongly localized) systems, the Boltzmann equation fails to describe transport adequately except under very special circumstances. Disordered materials can stem from several sources, ranging from amorphous materials to relatively good single crystals with very high doping concentrations. In particular, the latter exhibit a form of impurity-induced disorder when the concentration of the impurity reaches a significant fraction of the atomic concentration of the host lattice. This, in turns connects to weak localization which can also arise from impurity-induced coherence effects even when the concentration is not too high.

The central idea of impurity-induced disorder is that the perturbation to the lattice introduces variations in the crystal potential in which the spectrum of allowed energy levels is significantly modified. In some sense, nearly all of the present understanding evolves from a seminal paper by P. W. Anderson in 1958 [1] and by extensive subsequent work by Sir Nevill Mott. Of course, there has been an extensive range of experimental work during this period, and this work has provided important clues to the current level of understanding. However, even this early date is misleading, as this paper remained relatively poorly understood for several years; only in the past few decades has real understanding of the topic begun to emerge. In this regard, a scaling theory [2] was put forward which suggested that all states in one and two dimensions would ultimately be localized as the system size increased. That is, only three-dimensional systems would show extended state conductivity. We now know that this is wrong, and even quasi-two-dimensional systems *can* show a metal–insulator transformation. But, we know from e.g. the quantum Hall effect that exceedingly high levels of conductivity can be achieved in quasi-two-dimensional systems, and very good levels of conductivity are being found in quasi-one-dimensional systems. In the next section, we will review the understanding of (strongly) localized systems and where the current understanding really lies today.

Another localization effect, due to phase coherence of the propagating electron waves on the nanoscale, can also exist. This was discussed already in Chapter 1. In this case, the interference of the wave with itself, after successive backscattering from a number of impurity atoms, leads to a resistance enhancement, termed *weak* localization [3]. The fact that it is a coherent phenomenon is demonstrated by the effect caused by the presence of a magnetic field. Since the path around a loop of scattering centers, which returns to the origin, is a time-reversed copy of the path traversing in the opposite direction, these two paths interfere with each other to give the resistance enhancement. However, the magnetic field breaks time-reversal symmetry, and the two paths no longer interfere. Hence, a magnetic field (normal to the quasi-two-dimensional plane) breaks up the weak localization enhancement of the resistance. We will treat this topic, as well as the closely allied concept of universal conductance fluctuations in the later sections of this chapter.

7.1 Disordered semiconductors

The fundamental problem in disordered semiconductors is how the spectrum of allowed energy levels, and the corresponding nearest-neighbor coupling energies, are modified in going from a crystalline state to the disordered state. Even in this simple statement, however, one is assuming that there is a basic similarity between the two types of materials, which is not always true. Of course, the ultimate aim is to understand the conduction properties of the disordered materials, and for this to be possible, it is necessary to have some idea of the statistical properties and the energy "band" structure. It is known that most disordered semiconductors of interest here apparently form covalent bonds, and the bond order will be retained at least on a short-range basis. It is presumably only the long-range order that is destroyed. This implies that the distance between neighboring atoms, and the bond angles of the covalent bonds, will be changed only slightly from those in the crystalline state. Indeed, this is the case in even silicon dioxide, where the bond angles in the amorphous crystal are distributed in a relatively narrow range around those of crystalline quartz [4]. Even the concept of the wave vector is likely to carry forward as it was predicated upon the periodicity of the entire crystal and the periodic boundary conditions. However, within the tight-binding band structure, it is the nearest-neighbor interactions, and the Bloch sums over nearest neighbors, that provide the details of the band structure, so that it is expected that some (many) features of the band structure should be retained due to their connection to short-range order.

Since the long-range order is broken, one can think of the disordered material as being composed of many small regions, with the band properties varying between the regions but more or less uniform within a region. This leads to the concept of energy bands, but bands with very fuzzy edges and regions of states existing throughout the so-called bandgaps. Extensive studies have been performed to study the conductance properties of most disordered semiconductors derived from tetrahedrally bonded solids. These studies indicate that there is an excitation gap, such as the bandgap between conduction and valence bands in crystalline solids, although the nature of the gap is quite different.

The electrical conductivity of disordered semiconductors, whether it is measured by electrical properties or by optical absorption, is primarily intrinsic in nature – it displays an activation energy E_a so that the temperature variation is given by

$$\sigma = \sigma_0 \exp\left(-\frac{E_a}{k_B T}\right). \tag{7.1}$$

The general form of this equation is, of course, exactly what would be observed for intrinsic conduction in a semiconductor, where the activation energy is just that of the intrinsic concentration – one half of the bandgap. In disordered materials, however, this connection is not found, but the behavior of (7.1) is observed over many orders of magnitude, and from conductivities that are almost free-electron-like to conductivities that are orders of magnitude smaller than normal semiconductors. The pre-factor is usually an order of magnitude smaller than that found in an intrinsic semiconductor, but the activation energy is generally somewhat larger than that found in the crystalline material. This suggests that the excitation gap in the disordered material is larger than the bandgap in the corresponding crystalline material.

7.1.1 Localized and extended states

The general features of the disordered materials can be expressed in the presence of almost a continuum of allowed states, of which certain ranges are localized (such as those arising from impurity levels in the bandgap region of crystalline material) and other ranges are extended (almost free-electron-like). As will be seen below, this viewpoint can be traced to the Anderson treatment and has been well summarized in a model developed subsequently by Mott – the ideal covalent glass [5]. In this model, the ideal covalent glass is a semiconductor, which may be elemental, such as Ge or Si, or a compound, such as arsenic selenide. It may even be a multi-component alloy. In general, the glass possesses a one-, two-, or three-dimensional random network with excellent short-range order. The major features of the glass are governed by the covalent bonding requirements, so that there are, in fact, no dangling bonds, no essential structural defects, and the bonding requirements of each atom are satisfied. This model supports a short-range order comparable to that of crystalline solids, and there are general band features that can be identified with the conduction and valence bands of the perfectly ordered semiconductor. Since most semiconductors do form structures similar to these ideal glasses in the weakly disordered state, one expects that the disordered materials will have some universal features in their electronic band structure that correlate with our normal understanding.

In keeping with the model above, it is therefore expected that there will be "bands" that extend over a wide range of extended electronic states, whose properties will be very much like the electronic states in the conduction band of a perfectly ordered material. These extended states are characterized by wave functions that possess long-range order only in their amplitudes, but phase coherence exists only over short distances. At the edges of the bands of extended states are tails of localized states, which have only short-range order, in both phase and amplitude. The transition from extended states to localized states is quite sharp in energy, and it is these transitions that are connected with the excitation energies observed in optical absorption and transport. A simple "conceptual picture" of the transition is of a very rugged terrain in which are hundreds of small "lakes." As long as the water level remains low, the lakes are isolated. As the level rises, however, the lakes spread and begin to coalesce. Finally, there is only one well-connected "sea," which is the Fermi sea. The water level corresponds to the energy of the carriers (for the case of the bottom of the conduction band), and as the energy is raised, the localized states transition to extended states.

The localized states extend out from both the conduction band and the valence band. Indeed, they are actually states that are removed from these bands and whose density is sufficiently low that the wave functions no longer overlap with neighboring atoms. This is shown in Fig. 7.1, where the transition energies are termed the conduction and valence band "edges." The width of the localized state density tail depends on the degree of disorder in the solid. In most disordered semiconductors, the width of the disordered, localized state region is sufficiently great that those from the conduction band overlap those from the valence band.

In crystalline (ordered) structures, the activation energy for the conductivity arises from the presence of the bandgap; that is,

$$n_i = \sqrt{N_C N_V} \exp\left(-\frac{E_{gap}}{2k_B T}\right), \tag{7.2}$$

where the various parameters are usually defined (N_C and N_V are the effective density of states for the conduction and valence bands, respectively). The bandgap variation arises because the density of states in the ordered material has a sharp cutoff at the band edges. However, in disordered material, the

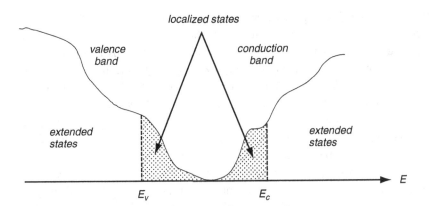

Fig. 7.1 The Anderson–Mott model of the bands in disordered semiconductors. The bands are viewed as being made of mostly extended states with tails of localized states in the regions normally occupied by the energy gap.

existence of the tails of localized states below E_c and above E_v would appear to eliminate the possibility of a sharp activation energy. This is not really the case, because of the basic difference in the nature of the extended and localized states. In the extended states, the wave functions have some coherence (at least in amplitude) over many "unit cells" of the material, so that one might expect some of the concepts of effective mass and mobility to be applicable. In this sense, the mobility exists and may be not too much smaller than in the ordered state. In the localized states, on the other hand, the wave functions do not have any long-range coherence and do not overlap the neighboring atoms, or "unit cells," to any great degree. Therefore, the concept of nearly free motion is just not applicable, and one must think of the electron moving from one cell to the next by a tunneling process, or "hopping" process, in which the carrier is thermally activated over the potential barrier between the atoms, only to be retrapped at the next atom. There is thus a large barrier preventing motion among the atoms in the localized states, and the mobility is essentially zero (very small) on the scale of the extended states.

Thus, at the transition energy between the localized and the extended states, it is to be expected that the mobility will fall by several orders of magnitude. In the disordered semiconductors, the boundaries between these two types of states are therefore the boundaries between states for which the carriers are free to move and those in which the movement is largely forbidden. The disordered semiconductor therefore has regions of allowed and forbidden *mobility*, as opposed to allowed and forbidden states, and the activation energy is that required to take the electron from an extended state in the valence band to an extended state in the conduction band. Consequently, the activation energy will usually be larger than the bandgap, as part of the otherwise allowed states in the conduction and valence bands will in fact have forbidden mobilities due to the disorder-induced localization.

When the energy is in the range of the localized states, the conductivity proceeds mainly by a mechanism of hopping, whereby an electron jumps from one site to a neighboring site. Repeated hopping leads to the possibility of the carrier transiting through the entire sample, but the conductivity is reduced by the need to be excited over the intersite barrier (which leads to an activation barrier). Since the wave functions are localized on a single site, the probability of the jump, either by a phonon-assisted transition over the barrier or tunneling through the barrier, is proportional to the overlap of the wave function on the two neighboring sites, which falls off as $\exp(-\alpha R)$, where R is the hopping distance and α is a decay constant (which for tunneling can be calculated if the details of the potential barrier are known). Such hopping is known as *nearest-neighbor hopping* [6] and is often found in the case of impurity conduction in highly doped semiconductors. The conductivity is proportional to the density of states at the Fermi level and the width of the Fermi–Dirac distribution, the difference in the probabilities for forward and backward hopping when a field is present, and an effective *velocity* that is approximately the distance times a "hop frequency." The latter is related to the phonon frequency in phonon-assisted hopping. Thus, the current density is given by [7]

$$J \sim e k_B T \rho(E_F) R v_{ph} \exp\left(-2\alpha R - \frac{\Delta E \pm eFR}{k_B T}\right)$$
$$= 2 e k_B T \rho(E_F) R v_{ph} \exp\left(-2\alpha R - \frac{\Delta E}{k_B T}\right) \sinh\left(\frac{eFR}{k_B T}\right), \quad (7.3)$$

where ΔE is the (effective) barrier height. For weak fields F, the hyperbolic sine function can be expanded, and the conductivity is just

$$\sigma = 2 e^2 \rho(E_F) R^2 v_{ph} \exp\left(-2\alpha R - \frac{\Delta E}{k_B T}\right). \quad (7.4)$$

Nearest-neighbor hopping is expected mainly when $\alpha R_0 \gg 1$, where R_0 is the average distance of the hops. In this case, the conductance is greatly reduced by the exponential factor. In this limit, the energy levels that are allowed are expected to be rather widely spaced, and an estimate of the barrier may be obtained from the density of states $\rho(E_F)$, which is the number of states per unit volume (in d-dimensional space). Then, the number of states per unit energy is just $R_0^d \rho(E_F)$, and the average energy spacing per state is $\Delta E \sim [R_0^d \rho(E_F)]^{-1}$. But, if this spacing is that observed at a single site, it is comparable to the separation of the atomic levels at that site and hence is of the order of the bandwidth. This leads us to the conclusion that such nearest-neighbor hopping is expected only in the case for which all levels in the conduction band (or the impurity band) are localized, so that any mobility edge lies in a higher-lying band [7].

On the other hand, it is often the case that $\alpha R_0 \leq 1$, so that the hopping range may extend well beyond that of the nearest neighbor. This is the regime of *variable-range* hopping. Here, the distance over which the hop is expected to occur increases with increasing temperature, and the exponential argument is a very low power of the inverse temperature [8,9]. The argument for this result is that, at a given temperature, the electron will "hop" to a site that lies somewhere inside a radius of order R, which gives $\gamma_d(R/R_0)^d$ available sites for the hop, where γ_d is a numerical factor that depends upon the dimensionality (this factor is $4\pi/3$ in three dimensions, π in two dimensions, and 2 in one dimension). The hop generally occurs to a site for which the activation energy is the lowest. The latter is given by the same argument of the last paragraph, and is $\Delta E \sim [\gamma_d R_0^d \rho(E_F)]^{-1}$. Thus, the effective barrier depends upon the distance over which the hop can occur. The probability of a hop is proportional to

$$v_{ph} \exp\left(-2\alpha R - \frac{\Delta E}{k_B T}\right). \tag{7.5}$$

The phonon frequency is generally constant over the range of temperatures of interest, so that the maximum value of the exponential occurs when

$$\alpha = \frac{1}{2\gamma_d R^{d+1} \rho(E_F) k_B T}. \tag{7.6}$$

In computing the conductivity, the prefactor term in R needs to be the average value of the hop distance over the radius, and this is just

$$\langle R \rangle = \frac{d}{d+1} R. \tag{7.7}$$

Finally, the conductivity is given by

$$\sigma = 2e^2 \rho(E_F) \langle R \rangle^2 v_{ph} \exp\left(-2\alpha R - \frac{B}{T^{1/(d+1)}}\right), \tag{7.8}$$

where

$$B = \frac{4\alpha^{d/(d+1)}}{[2\gamma_d \rho(E_F) k_B]^{1/(d+1)}}. \tag{7.9}$$

Equation (7.9) leads to the famous $T^{-1/4}$ behavior of the conductivity that is often found in disordered material. It should be noted that, in two dimensions, the expected behavior is $T^{-1/3}$, and this has been found in activated transport conductivity in the inversion layer of silicon MOSFETs at low temperature [10]. These temperature dependencies are widely viewed as indicative of variable-range hopping.

7.1.2 Localization of electronic states

In the foregoing discussion of localized and extended states, the concept of a rapid transition, at a single critical energy, from a set of localized states with no long-range order to a set of extended states was remarkable. This is a somewhat unusual concept and it is necessary to search further for understanding of the localized state, which can be achieved with some rather straightforward techniques. This approach is due to Anderson [1], who was one of the first to treat localized states in disordered materials. He considered the existence of localized states in a system for which the energies of the lattice sites are randomly distributed over an energy range, and he considered a number of different distributions of these energies. For example, one possibility is that these site energies are uniformly distributed over a range of width $W (= \Delta E)$. That is, the probability density for any energy is $1/W$ so long as the actual energy lies in the range of e.g. 0 to W (a shift of energy would yield the equivalent $-W/2$ to $W/2$ range). As will be seen, variations in both the definition of the probability density and the range of the energy can be introduced for different conditions. The principal fact is the energy of a given lattice site E_s is a random variable and has a given probability density for which there is a given characteristic width W that is centered about the mean value.

Anderson's paper has been recognized for its importance and applicability to a variety of problems. The model clearly shows that there is a critical value of the width W_c such that, if $W > W_c$, the states at the middle of the band (and therefore all of the states) are localized. The transition at $W = W_c$ (the mobility edge), for which the existence of extended states vanishes, is usually termed the Anderson transition. Its importance is quite clear. For a degree of disorder such that $W > W_c$, one cannot expect conductivity to occur with any reasonable value. The essence of the model is that, on each atomic site (which may not lie on a regular lattice), there is a "site" energy, which corresponds to the atomic energy level of the isolated atom, and an "overlap" energy describing the interaction between the wave function on the atom and that of its neighbors. This is just an adaptation of a bond orbital or tight-binding model for band structure. The difference here is that the atomic energy will be a random variable. Normally, one thinks of the overlap energies as being random, but here it is the site energies that will be taken as random. Thus, the Schrödinger equation may be written as

$$i\hbar \frac{\partial \psi_s}{\partial t} = E\psi_s = E_{0,s}\psi_s + \sum_{s'} V_{ss'}\psi_{s'}. \qquad (7.10)$$

Here, $E_{0,s}$ varies from site to site as a random variable according to an assumed probability density function. The overlap energy $V_{ss'}$ is taken to be a relative constant value, which is non-zero only for nearest neighbors. The right-hand side of (7.10) is just the perturbation series expansion of the wave function at a single site in terms of the interactions with its neighbors. Normally, (7.10) is just the result of applying a tight-binding calculation, in which the sum runs over a set of neighbors weighted by the Bloch terms. Here, however, it is assumed that the average over the neighbors yields a set of energies centered about $E = 0$.

Equation (7.10) is an infinite set of equations for the wave functions ψ_s. It very closely resembles a normal perturbative series and this is actually the way in which a solution will be developed [11,12]. It is the singularities of the matrix of energies that gives the allowed values of the energy for the overall system. If one ignores the off-diagonal terms, the energies are just the site energies, which may be interpreted as a set of localized states. It is the overlap of the wave functions that leads to the extended states that are of interest, and will be investigated with a perturbation series. In such a series, the site energy is the zero-order first term in the final energy, and this final energy can be written in perturbation theory as

$$E_s = E_{0,s} + \sum_{s' \neq s} \frac{|V_{ss'}|^2}{E_s - E_{0,s}} + \sum_{s',s'' \neq s} \frac{V_{ss'} V_{s's''} V_{s''s}}{(E_s - E_{0,s'})(E_s - E''_{0,s})} + \cdots, \qquad (7.11)$$

where one must be careful not to include the terms for which the denominator vanishes. Unless E_s is real, the amplitude of the wave function decays with time, since this quantity appears in the time variation as $\exp(iE_s t/\hbar)$. Thus, the nature of the states will be investigated by examining the convergence properties of (7.11). If the series converges to a well-defined E_s, then the state is localized at that site. On the other hand, if the series does not converge, it must be assumed that the energy lies within a band of energies which correspond to wave functions that are extended over the entire crystal and are therefore extended states.

7.1.2.1 Uniform distribution of energies

The expression for the energy eigenvalues E_s given above is a stochastic series, since the site energies are a random variable, by our construction. Each term in the series contains $V^{(L+1)}$ as a factor, where L is an integer expressing the order of the particular term. Here, it may be assumed that the overlap integral is constant, as discussed in the preceding section. In a general lattice, each atomic site (recall that all the disorder is contained in the site energies and the atoms are assumed to lie on a regular lattice) has Z neighbors, so that there are Z^L contributions to the Lth term, each of which is a product of the form

$$T_{s'} = \frac{V}{E_s - E_{0,s'}}. \qquad (7.12)$$

If the values of the site energies on the primed subscripted sites are statistically uncorrelated, the magnitude of the contribution of such a product term can be estimated by taking the average of its logarithm, as

$$\langle \ln|T_s T_{s'} \ldots T_{s''}|\rangle \sim L\langle \ln|T|\rangle. \qquad (7.13)$$

The form of the terms, expressed in (7.13), indicates that the perturbation series (7.11) is to be interpreted as a geometric series, at least in the sense of the average term. A geometric series is a special form of a power series which will converge provided that the ratio of subsequent terms approaches a limit, as one takes the highest-order terms, and the limit is less than unity. Thus, if the coefficients of the power series are A_n, the series will converge if

$$\lim_{n \to \infty} \left| \frac{A_{n+1}}{A_n} \right| x < 1, \qquad (7.14)$$

where x is the argument of the series $f(x)$ (i.e., the general term is $A_n x^n$). It is this convergence criteria that will be applied to the perturbation series to determine the range of allowed bandwidth in the disordered energy spectrum. This leads us to the conclusion, from (7.13) and (7.14), that the series will converge if

$$Z \exp(\langle \ln|T| \rangle) < 1. \qquad (7.15)$$

In the standard case, which is the one discussed by Anderson [1], one can take the site energies to be uniformly distributed over a width W. The probability distribution function is then $P(E_{0,s}) = 1/W$ for $-W/2 < E_{0,s} < W/2$, and zero otherwise. It is now possible to evaluate the average of the logarithm term by including the form (7.12) with the probability distribution, as

$$\langle \ln|T| \rangle = \frac{1}{W} \int_{-W/2}^{W/2} \ln \left| \frac{V}{E_s - E_{0,s'}} \right| dE_{0,s'} \qquad (7.16)$$

$$= 1 - \frac{1}{2} \left[\ln \left| \frac{4E_s^2 - W^2}{4V^2} \right| + \frac{2E_s}{W} \ln \left| \frac{2E_s + W}{2E_s - W} \right| \right].$$

The value of W required for localization evidently depends upon the energy of the site at which one is looking. However, the energy in (7.16) is the averaged energy and should now be identified with E_s, which is a smooth variable. It is clear that for $E_s = 0$ (i.e., at the center of the band), there is a minimum value of W for which the series does not converge. For this condition, convergence occurs for

$$Z \exp\left(1 - \ln \frac{W}{2V}\right) < 1 \qquad (7.17)$$

or

$$\frac{W}{2V} > e^1 Z \sim 2.7 Z = 10.8, \qquad (7.18)$$

where the last form is for the tetrahedrally coordinated semiconductors. Equation (7.18) indicates that if the disorder is sufficiently large (large W), the series converges even at the center of the band $E_s = 0$. This means that, if the series converges, the site is a localized site in that a localized atomic-like energy state exists at that site. Thus, for extended states, one does not want the series to

7.1 Disordered semiconductors

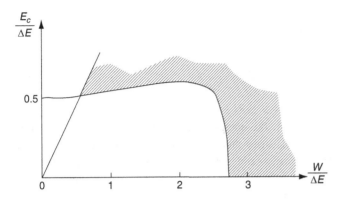

Fig. 7.2 The critical energy found from the discussion in the text for localization in a case where the probability for the local site energy is uniformly distributed. The normalization is $\Delta E = 2VZ$, where V is the overlap energy between neighboring sites.

converge – a localized state is not wanted. For a site energy away from the center of the band ($|E_s| > 0$), the logarithmic terms in (7.16) increase in amplitude (the second term increases much faster than the first term decreases), and a lower value of disorder is required for the series to converge and the states to be localized. In Fig. 7.2, the critical transition energy E_c is plotted as a function of the disorder. Here, the energies are normalized to $\Delta E = 2VZ$. For a case in which the disorder W is greater than the critical energy, one expects to have band tails of localized states. These localized state regions begin at the edges of the bands and move toward the band center as the disorder increases. It is also apparent that there is a critical disorder W below which no localized states exist, which explains why the small disorder introduced by impurities does not produce band tails of localized states in normal semiconductors.

7.1.2.2 Binary alloys

Another type of disorder, other than bond-energy disorder, has to do with alloys in which the disorder is in the arrangement of the atoms. Materials such as GaAlAs are random alloys for which one normally introduces a virtual-crystal approximation. Here, the interest is now in looking at the nature of the localized states and disorder that can arise from such a crystal. It will be assumed that the site energies $E_{0,s}$ can take on the values E_1 or E_2 with relative probabilities c and $1 - c$. Thus, compound 1 occurs with relative probability c and compound 2 occurs with relative probability $1 - c$. The site energy is still distributed randomly in *space*. The expression that arises for the average (7.15) is even simpler than in the previous case, and for the alloy,

$$\langle \ln|T| \rangle = c \ln \left| \frac{V}{E_s - E_1} \right| + (1 - c) \ln \left| \frac{V}{E_s - E_2} \right|. \tag{7.19}$$

If one chooses to take $E_2 = 0$ and $c \ll 1$, only the second term of (7.19) needs to be retained (it should be recalled that the band is centered about E_2 and has a width of $4V$ in a one-dimensional solid composed solely of atom type 2). The condition for localized states then becomes just

$$\frac{ZV}{|E_s|} < 1, \tag{7.20}$$

which is just the quantity $|E_s| > \Delta E/2$ in terms used for Fig. 7.2. This implies that for localization to occur, the energy must lie outside the energy band, a somewhat comforting result, since this is the approximation that was used in the virtual-crystal approximation and explains just why the random alloys work so well and why low concentrations of impurity atoms (e.g., dopants) do not disorder the crystal to any great extent.

In general, if we have $E_1 \neq E_2$, the introduction of a small fraction of impurity atoms leads to localized states lying just outside the normal energy band – these are the impurity levels corresponding to donors and acceptors. As the difference between the atomic energies is increased, these states begin to "nibble" into the energy band and eventually lead to a splitting of the band into two branches. In essence, as the concentration increases, or the energy difference increases, or both, the localized state regions begin to spread in energy, drawing states out of the energy band, which pushes the critical mobility edge into the band. At a sufficiently high degree of disorder, the band splits into two bands, one centered about each of the two atomic energies, with localized states between the two and on the "outside" of each branch. Consider, for example, the case for which $c = 0.5$ and $E_1 = -E_2 = W/2$ (which centers the two branches about the average virtual crystal band center $E_s = 0$). Then

$$\langle \ln|T| \rangle = \frac{1}{2} \ln \frac{4V^2}{4E_x^2 - W^2}, \tag{7.21}$$

and

$$E_s > \frac{1}{2}\sqrt{W^2 + (\Delta E)^2} \quad \text{or} \quad E_s < \frac{1}{2}\sqrt{W^2 - (\Delta E)^2} \tag{7.22}$$

defines the regions of localized states. Here, ΔE has the same definition as used previously, and the regions of localized states are shown in Fig. 7.3.

7.1.2.3 Lorentzian distribution of energies

Let us now turn our attention to a somewhat more realistic probability distribution function for the random site energy $E_{0,s}$. For this, a distribution that is Lorentzian in shape will be assumed to be

$$P(E_{0,s}) = \frac{1}{\pi} \frac{\Gamma}{E_{0,s}^2 + \Gamma^2}. \tag{7.23}$$

7.1 Disordered semiconductors

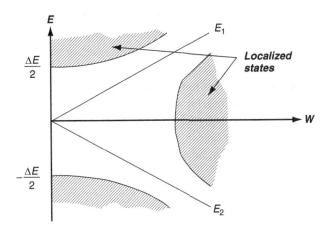

Fig. 7.3 Ranges of localized states for the two-component alloy are shown here for the case of equal concentrations. The borders between the localized and extended states are given by (7.22).

This can now be introduced into the averaging function as

$$\langle \ln|T| \rangle = \frac{1}{\pi} \int_{-\infty}^{\infty} \frac{\Gamma}{E_{0,s}^2 + \Gamma^2} \ln\left|\frac{V}{E_s - E_{0,s'}}\right| dE_{0,s'}. \tag{7.24}$$

Here, the quantity Γ is the half-width of the distribution, so that the mean width of the distribution W is just 2Γ. Equation (7.24) can be evaluated by complex integration. However, the essential singularity in the logarithm function is not of interest, and a branch cut must be chosen so that this singularity is excluded from the integration. Thus, the integral may be written as

$$\langle \ln|T| \rangle = \frac{1}{2\pi} \lim_{\delta \to 0} \left[\int_{C_1} \frac{\Gamma}{(E_{0,s} + i\delta)^2 + \Gamma^2} \ln\left|\frac{V}{E_s - E_{0,s'} - i\delta}\right| dE_{0,s'} \right.$$
$$\left. + \int_{C_2} \frac{\Gamma}{(E_{0,s} - i\delta)^2 + \Gamma^2} \ln\left|\frac{V}{E_s - E_{0,s'} + i\delta}\right| dE_{0,s'} \right]. \tag{7.25}$$

In this integral, the contour C_1 is taken to lie just above the real axis and is closed in the upper half plane, while C_2 is taken to lie just below the real axis and is closed in the lower half plane. This means that the branch cut is taken along the real axis from E_s to ∞. The integration can then be carried out by residues and yields the straight forward result

$$\langle \ln|T| \rangle = \frac{1}{2} \ln \frac{V^2}{E_s^2 + \Gamma^2}, \tag{7.26}$$

for which the condition for the existence of localized states becomes

$$\frac{ZV^2}{\sqrt{E_s^2 + \Gamma^2}} < 1. \tag{7.27}$$

Here, again it is found that for $E_s = 0$, one requires $\Gamma > \Delta E/2$ for localizing the entire band, and for $\Gamma = 0$, one requires $E_s > \Delta E/2$, which means that the localized states lie outside the normal conduction band. This result is the same result as that obtained in the preceding two sections, in which localized states exist only outside the energy band. The first result reinforces the conclusion that there is also a critical value of disorder for which all states are localized. Between these two extremes, one generally has a range of localized states in the band tails and a range of extended states in the center of the band, with a transition energy, the mobility edge, given by

$$E_c = \pm\sqrt{\left(\frac{\Delta E}{2}\right)^2 - \Gamma^2}. \qquad (7.28)$$

Several examples of the band nature are given in Fig. 7.4.

The comparison between the mobility edge given by (7.28) and that obtained in earlier examples demonstrates that there is very little sensitivity to the exact model used for the probability distribution. Rather, drastic changes have been made in the details of the model, and in the corresponding probability distribution function, with relatively little change in the degree of disorder required to totally localize all states and with relatively little change in the energy at which the mobility edge occurs. It may be concluded from this behavior that the

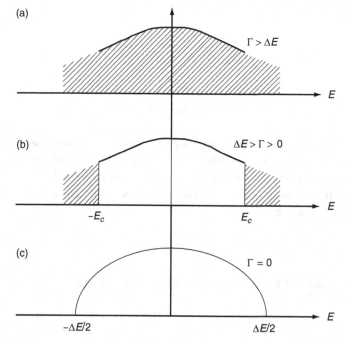

Fig. 7.4 The solutions to (7.28) for three conditions are shown here for the Lorentzian model of the disorder distribution. The curve in (a) shows the case for all states localized, while the curve in (b) shows the presence of a mobility edge. In (c), all states are extended.

relationship $W > \Delta E = 2ZV$ is a fundamental quantity that defines the onset of the Anderson transition in disordered material.

7.2 Conductivity

A very simple, hand-waving argument was presented above to arrive at the expression for variable-range hopping in the localized states. Here, it is desired to look at more fundamental and basic approaches to the conductivity in disordered material. The theory of transport properties in disordered materials is one of the areas that is presently under rather intense investigation. One reason for this is the concept of the sharp mobility edge separating localized states with very low mobility and extended states with nearly free-electron-like mobility. More interestingly, experiments have shown the presence of an equivalent metal–insulator transition in the quasi-two-dimensional electron (or hole) gas, and this is presumably the latest incarnation of the localized–extended state transition. We return to this below, after discussing the scaling of the conductivity.

It is difficult to explain the sharp drop in mobility by several orders of magnitude at the mobility edge just by the change in the wave functions, as one really needs to have another physical mechanism that relates the onset of localization to the onset of rather strong scattering or trapping. Essentially, when the state becomes localized, it is necessary to understand that the transport ceases to be drift-like and must become hopping in nature, with excitation between neighboring sites by e.g. variable-range hopping, field-assisted tunneling, or some other mechanism. The full understanding of these mechanisms is still not available.

The ease with which crystalline semiconductors carry current and the high mobility of the carriers is thought to be associated with the regular periodicity and long-range order of the lattice as well as with the relative weakness of the scattering processes. The problem in transport in the disordered materials is to explain the mechanisms by which the current is carried. In some sense, one would like to know exactly the variation in the mobility at the mobility edge, but this becomes very model dependent. Here, the aim is to focus more on the universal properties of the transport. Some of these factors are the sensitivity to boundary conditions, the scaling of the conductivity, and some examples of transport in the Anderson model.

7.2.1 Minimum metallic conductivity

When the system is composed of entirely localized states, or when the available states are filled completely up to the Fermi level, the conductivity vanishes at $T = 0$. This can easily be shown with the Green–Kubo formula for conductivity, which will be used in the following. In general, the conductivity is related to the current–current correlation function, since the mobility involves the number of carriers while the current involves both the number of carriers and the velocity. One can quite generally write the Kubo formula for the conductivity as [13]

$$\sigma_{ij} = V \int_0^\infty dt \int_0^\beta d\beta' Tr\{\rho_0 j_j(-t - i\hbar\beta') j_i\}, \quad (7.29)$$

where V is the sample volume and $\beta = 1/k_B T$. The inverse temperature that appears as an imaginary time contribution signifies that a finite-temperature Green's function is being used (we deal with these more in the next chapter). The imaginary time builds in the Boltzmann factor from a Maxwellian distribution function in the exponential time factor arising from the Schrödinger equation via

$$e^{iHt/\hbar} \to e^{i(H/\hbar)(-t-i\hbar\beta)} = e^{iHt/\hbar} e^{-\beta H} = e^{-i\omega t - E/k_B T}. \quad (7.30)$$

The presence of the Maxwellian distribution function indicates that there must be a basic connection with normal Boltzmann transport in the extended state regime, and this will be illustrated below. The trace operation (a sum over the energy, which represents a sum over the diagonal terms of the density matrix ρ_0) is achieved by inserting a complete set of energy eigenfunctions $\{\psi_n\}$, which will actually be denoted in the Dirac notation, and summing over the trace operation for the resulting matrices. The Pauli exclusion principle will also be incorporated along the way, and the expansion of (7.29) is given by

$$\sigma_{ij} = \frac{1}{V} \int_0^\infty dt \int_0^\beta d\beta' \sum_{n,m} e^{-\beta' H} \langle n|j_j(-t - i\hbar\beta')|m\rangle \left[1 - e^{-\beta' E_m}\right] \langle m|j_i|n\rangle. \quad (7.31)$$

The first matrix element may be expanded as

$$\langle n|j_j(-t - i\hbar\beta')|m\rangle = \langle n|e^{-iH(t+i\hbar\beta')} j_j e^{iH(t+i\hbar\beta')}|m\rangle$$
$$= \langle n|j_j|m\rangle e^{-it(E_n - E_m)/\hbar} e^{\beta'(E_n - E_m)}. \quad (7.32)$$

The first exponential on the right-hand side can be integrated over t and yields the energy-conserving delta-function factor $\hbar\pi\delta(E_n - E_m)$. The second exponential can be integrated over β' to yield just a factor β. This combines with the $f(1-f)$ to give the negative of the derivative of the distribution function, just as in the Boltzmann equation case. The diagonal terms in the conductivity are then given by

$$\sigma = \frac{\pi\hbar}{V} \sum_{n,m} \left(-\frac{\partial f}{\partial E}\right) |\langle n|j|m\rangle|^2 \delta(E_n - E_m)$$
$$= \frac{\pi\hbar}{V} \int dE' \int dE'' \rho(E') \rho(E'') \left(-\frac{\partial f}{\partial E'}\right) |\langle n|j|m\rangle|^2 \delta(E' - E''), \quad (7.33)$$

where a continuous energy variable has been introduced in the last line. This is very like the result obtained in normal Boltzmann transport if it is recognized that one factor of the density of states and the current matrix elements go into the velocity and relaxation terms. The energy derivative of the distribution function just gives an additional delta function (at low temperatures) at the Fermi energy for degenerate semiconductors, and after introducing the current operators, we may write this as [14]

$$\sigma = \frac{\pi e^2 \hbar^3}{Vm^2}[\rho(E_F)]^2|D_{mn}|^2, \tag{7.34}$$

where

$$D_{mn} = \int d^3\mathbf{r}\, \psi_n^* \nabla \psi_m \tag{7.35}$$

is the reduced momentum matrix element.

In three dimensions, the density of states appearing in (7.34) is easily evaluated to be just $Vmk/\pi^2\hbar^2$, as was found earlier. The evaluation of the matrix element is a little more complicated. For this, an argument due to Mott [15] will be followed. An effective volume Ω, which is defined by the bulk mean free path of the electron as

$$\Omega = \frac{4\pi}{3}\lambda^3 \tag{7.36}$$

is introduced. This volume differs from that of the integrations above, but describes essentially the sphere (in three dimensions) that is obtainable by a carrier before it undergoes scattering. In a sense, this volume is more closely related to that of the hop used in the previous section. For plane waves, the integral in (7.35) produces a term that varies as

$$e^{-i\mathbf{k}\cdot\mathbf{r}}\nabla e^{i\mathbf{k}\cdot\mathbf{r}} = i\mathbf{k}e^{i(\mathbf{k}-\mathbf{k}')\cdot\mathbf{r}}. \tag{7.37}$$

If the exponential varies only little over the volume, then the result is just $i\mathbf{k}$, but if the exponential varies significantly, the integral averages to zero. Thus, the magnitude is just

$$R = \begin{cases} \frac{k\Omega}{V}, & |\mathbf{k}-\mathbf{k}'|\lambda < 1, \\ 0, & \text{otherwise}. \end{cases} \tag{7.38}$$

Then, it is found that (7.35) leads to

$$|D_{mn}|^2 = \frac{N}{\Omega}R^2\xi = \frac{V}{\Omega}\left(\frac{k\Omega}{V}\right)^2\xi, \tag{7.39}$$

where ξ is a factor that arises from the angular averaging

$$\xi \sim \frac{1}{2}\int_0^{1/k\lambda}\sin\vartheta\, d\vartheta = \frac{1}{4k^2\lambda^2}. \tag{7.40}$$

Finally, we can combine the equations above to yield the longitudinal conductivity as

$$\sigma = \frac{e^2 k^2 \lambda}{3\pi^2 \hbar} = \frac{ne^2 \tau}{m}. \tag{7.41}$$

In the last form, $\lambda = v\tau = \hbar k \tau/m$, with τ evaluated at the Fermi surface, has been used and the relationship between the density n and the Fermi wavevector has been introduced as $n = k^3/3\pi^2$. The last form in (7.41) is readily recognized as the standard conductivity in a metal or a semiconductor, which reinforces the conclusion that the approach has some validity. In particular, the method has been shown to have strong similarities to the Boltzmann equation approach and yields the normally expected conductivity in the metallic (extended state) regime.

We now want to use this approach, specifically (7.33), for the localized disordered system. Here, the goal is to establish just what the minimum value of the conductivity can be in a system that is not completely localized. This quantity is called the *minimum metallic conductivity*. As indicated above, values of the mean free path that are smaller than $1/k$ cannot be accepted. Indeed, if the wave functions are extended, they must lead to a mean free path that is larger than the shortest spatial variation of the momentum. A random change of the phase of the wave function from one site to the next actually corresponds to a mean free path of $1/k$ and is the shortest mean free path that can be achieved for an extended state. It is this case, for which $k\lambda \sim 1$, that is of interest to evaluate now. If there are N potential wells (N atoms) in the solid, then the effective momentum operator can be written as

$$D \sim N^{1/2} R, \tag{7.42}$$

where the integration over the matrix element in (7.35) is now limited to just the volume of a single well. For this it may be assumed that D is dominated by nearest neighbor hops, so that (7.38) leads to

$$R \sim \frac{ka^3}{V}, \tag{7.43}$$

where a^3 is the elemental volume per atom. In the middle of the energy band, which is the location of the last states to localize, $k = \pi/a$. Since $N = V/a^3$, this leads to

$$D = \frac{\pi}{2}\sqrt{\frac{a}{V}}. \tag{7.44}$$

Using this result in (7.34) and noting that the density of states will be reduced at the center of the band by the spread W of the energy probability function by the amount $1/W$ (in keeping with the preceding section, $\Delta E = W$), so that we write the density of states as $2V/a^3 W$, and

$$\sigma = \frac{\pi^3 e^2 \hbar^3}{m^2 a^5 W^2}. \tag{7.45}$$

The energy at the center of the band can be found from the assumed band shape discussed above, $E = E_0 - 2V\cos(ka)$, which leads to the effective mass at the bottom of the band as $m^{-1} = 2a^2V/\hbar^2$. Then, relative to the bottom of the band ($E_0 - 2V$), the center of the band is just $2V$, or

$$E_0 = \frac{\hbar^2}{ma^2}, \tag{7.46}$$

and this may be used in (7.45) to give the conductivity in these reduced potentials as

$$\sigma = \frac{\pi^3 e^2}{\hbar a}\left(\frac{E_0}{W}\right)^2. \tag{7.47}$$

This equation gives the conductivity in the center of the band when the states there are nonlocalized [14]. From the preceding section we know that when the disorder reaches $W = 2.7\Delta E = 5.4VZ$, in terms of the number of nearest neighbors and the overlap energy, all of the states are localized. Here, the band width ΔE is just $4V$ in the zero-disorder limit. For this limit, the energy at the center of the band is just one-half of the bandwidth, so that

$$\sigma \sim \frac{\pi^3}{(5.4)^2}\frac{e^2}{\hbar a} \sim 1.3 \times 10^4 / \Omega\text{cm} \tag{7.48}$$

for the parameters of silicon. In fact, there is a great range of conductivities that is observed in disordered systems, and these conductivities are often many orders of magnitude smaller than this value without being thermally activated. The catch here is the use of the inter-atomic spacing for a. In many cases, one should instead use a correlation length over which short-range order is maintained. As this length can be orders of magnitude (but usually is not) larger than the inter-atomic spacing, it is not unusual to expect a value of the minimum metallic conductivity smaller than (7.48) by one or two orders of magnitude. The important point is that as the last states are localized, the conductivity will drop by a significant number of orders of magnitude.

7.2.2 Scaling the conductivity

A somewhat different approach to obtaining the so-called minimum metallic conductivity has been pioneered by Thouless and co-workers [16,17,18,19], which has led to the ability to set up a scaling theory. Such scaling theories allow us to examine the dimensionality and limits as T goes to zero. The arguments for this approach are quite simple in spirit. One assumes that the semiconductor sample is such that the electrons move in a potential that is uniform on a macroscopic scale, but is disordered on a microscopic scale. Nevertheless, it is assumed that the entire band is not localized, but retains a region in the center for which extended states are dominant and yield a conductivity that is non-zero as the temperature goes to zero. For this material, the density of electronic states per unit energy per unit volume is just dn/dE. Since the sample has a finite volume, the electronic states are discrete levels determined by the size of this volume. These individual energy levels are sensitive to the boundary conditions applied to the ends of the sample and can be shifted small amounts, on the order of $\hbar/t$, where t is the time required for an electron to diffuse to the end of the sample. In essence, one is defining here a broadening of the levels that is due to the finite lifetime of the electrons in the sample, which in turn defines a *coherence length* in terms of the length of the sample. This coherence length is just the sample length that is inherently assumed also to be the inelastic mean free path, the distance over which the electrons lose phase memory, which is normally longer than the length of the sample. The time required to diffuse to the end of the sample (or from one end to the other) is L^2/D, where D is the diffusion constant (for the electron or hole, as the case may be). The conductivity is related to the diffusion constant, in a degenerate electron gas (the current argument is being made for the case $T \to 0$), by

$$\sigma = \frac{e^2 D}{2} \frac{dn}{dE}. \qquad (7.49)$$

If L is now introduced as the effective length and t the time for diffusion, both from D, one finds that

$$\frac{\hbar}{t} \sim \frac{2\hbar}{e^2} \frac{dE}{dn} \frac{\sigma}{L^2}. \qquad (7.50)$$

the quantity on the left-hand side will be defined to be an average broadening of the energy levels ΔE_a, and the dimensionless ratio of this width to the average spacing of the energy levels may be defined as (here d is still the dimensionality of the sample)

$$\frac{\Delta E_a}{dE/dn} = \frac{2\hbar}{e^2} \sigma L^{d-2}. \qquad (7.51)$$

This equation can be derived in another manner. If the sample is connected between two metallic reservoirs, which serve as the contacts, the current will flow by virtue of the separation between the Fermi levels in these two reservoirs (this argument will be revisited again below). Since the Fermi levels at the two ends are separated by an energy difference eV, there are $eV(dn/dE)$ electron states contributing to the current, and each of these carries a current of e/t. Combining these two relations again gives (7.50).

It is the quantity on the left-hand side of (7.51) that is of interest in setting the minimum metallic conductivity. In a disordered material, the quantity V/W is important. The factor V is the overlap energy between neighboring sites and is therefore related to the width of the energy levels or bands, while the factor W

is the disorder-induced broadening of the site energies. In essence, if V/W is small, it is hard to match the width of the energy level on one site with that on the neighboring sites, so that the allowed energies do not overlap and there is no appreciable conductance through the sample. On the other hand, if V/W is large, the energy levels easily overlap, giving rise to extended wave functions and large conductance through the sample. (It may be noted here that the conductance is just σL^{d-2}.) The quantity $e^2/2\hbar$ is related to the fundamental unit of conductance, and is just 1.21×10^{-4} S (the inverse is 8.24 kΩ).

It is now possible to define a dimensionless conductance, which has been called the "Thouless number" by Anderson and co-workers [20], in terms of the dimensionless conductance as

$$g(L) = \frac{2\hbar}{e^2} G(L), \qquad (7.52)$$

where $G = \sigma L^{d-2}$ is the actual conductance. The latter authors have given a scaling theory based on renormalization group theory, which gives us the dependence on the scale length L and the dimensionality of the system. They consider small hypercubes of size L on an edge and dimension d. In the case where L is large compared to the mean free path, it is expected that the bulk conductivity will dominate the behavior. In this case, one may compute a critical exponent factor for the conductance $g(L)$, as a function of L, through the scaling function

$$\beta_d \equiv \frac{d[\ln g(L)]}{d(\ln L)} \underset{g \to \infty}{\to} d - 2. \qquad (7.53)$$

Since $G = \sigma L^{d-2}$, then $\ln g(L) = \ln(2\hbar\sigma/e^2) + (d-2)\ln L$, and the above equation follows immediately. The above is an interesting result, and suggests that a one-dimensional system will always be localized, for any degree of disorder, as the conductance will vanish exponentially with the sample length. On the other hand, the conductance would grow with length in three dimensions, which means that the conductance is limited by actual scattering processes rather than by sample length. In two dimensions, the exponent vanishes, which means that the conductance is independent of length for lengths smaller than the inelastic mean free path. Disorder in the latter case is thought to lead to power-law localization rather than exponential localization.

When L is much smaller than the mean free path, there is phase coherence on the scale of L only and g is no longer described well by (7.52). On the other hand, it can be shown that g is still describable in terms of the Thouless argument, but only on general grounds. In this case, it may be assumed that the desired conductivity is that of the hypercube embedded in a perfect crystal. With this in mind, it is clear that the dimensionless ratio $g(L)$, which is obtained from the conductance by dividing out the factor $2e^2/\hbar$, is the proper quantity to use to describe the overlap of the energy levels when two hypercubes are brought together. This is inherent in Thouless' argument, and can be extended from the original Anderson model merely by considering the size of the unit cell, in which the atomic energy levels are calculated, as L. If the atomic size is really of the order of L_0, one adopts an Anderson model with $(L/L_0)^d$ energy levels and a spectral width

$$W = \frac{dE}{dn}\left(\frac{L}{L_0}\right)^d. \tag{7.54}$$

This is based on the assumption that it is only the granularity of the energy levels that is important in deciding whether or not wave functions can overlap and produce extended states in the crystal.

The crystal now may be considered to be composed of b^d cubes, which are arranged into blocks bL on a side, which will produce a new L' length of the structure. It is now necessary to examine just how the concepts of granularity of the energy levels go over to the new length, which is the manner in which renormalization groups are usually evaluated. Thus, interest lies in the new $\Delta E'/(dn/dE)'$ at the new length scale, particularly in describing this new quantity in terms of the old. In general, it is possible to write $g(bL)$ as a function $f(b,g(L))$, which should produce the left-hand terms of (7.53) with $\beta = \beta[g(L)]$. In this sense, the scaling trajectory has only the single parameter $g(L)$, which is the normalized conductance.

At large values of g, one can use the general asymptotics described above, where $G = \sigma L^{d-2}$, so that the right-hand side of (7.53) is recovered as mentioned. For small g, where $V/W \ll 1$, exponential localization is surely to be the case; that is, we expect the wave function to fall off exponentially with distance, so that

$$g = g_0 e^{-\alpha L}. \tag{7.55}$$

Thus, one finds that

$$\lim_{g \to 0} \beta_d = \frac{d(\ln g)}{d(\ln L)} = \frac{d(\ln g_0 - \alpha L)}{d(\ln g)} = \frac{d(\ln g_0 - \alpha e^{\ln L})}{d(\ln g)}$$
$$= -\alpha e^{\ln L} = \ln\left(\frac{g}{g_0}\right). \tag{7.56}$$

In these expressions, g_0 is a dimensionless constant of order unity.

Since β represents the joining of fundamental blocks together, and the size of these groups is finite, there is no singularity in this quantity. This means that it must have a smooth variation from the negative quantity for $g < g_0$ up to $d - 2$ for large g. The significance of this lies in the fact that if the left-hand terms of (7.53) are integrated, it is found that $g \sim L^\beta$, which means that there is no sharp discontinuous behavior in g as the length scale is varied. Thus, in moving from a total localization to free-electron-like behavior, g varies in a smooth manner. For three dimensions, the exponent β changes from negative for small g to positive for large g, so that the idea of a sharp mobility edge must be interpreted as the point at which $\beta = 0$. Again, this is expected to occur near $g = g_c \sim 1$, so that the conductance itself is on the order of the fundamental conductance $e^2/\pi\hbar$. In two dimensions, Abrahams et al. [20] point out that there is no critical g_c, where $\beta = 0$. Rather, the value of β is negative for small conductance and approaches zero for large conductance. Instead of a sharp transition to a minimum metallic conductivity, there is a universal crossover from logarithmic localization at large conductance to exponential localization at small conductance. We will see in the next section, that this view is over-simplistic and ignores the possibility that the curve can indeed cross zero and then approach zero from the *positive* side. We plot these trends in Fig. 7.5, with the solid curves for the predictions of this scaling theory, and a dashed curve for a possible variation in two dimensions. Similar behavior occurs in one dimension, except that the scaling trajectory never approaches the $\beta = 0$ crossover. Thus, one might expect that all states will be localized for $d \leq 2$ if there is any disorder at all. Yet, the validity of this claim has not been demonstrated in real systems. As mentioned, it is quite likely that the trajectory for $d = 2$ actually crosses the $\beta = 0$ line, and current experiments in quantum wires show good extended conductance, with little evidence for real localization, as discussed at several points elsewhere in this book.

7.2.3 The metal–insulator transition in two dimensions

According to the above discussion, and the scaling theory of Abraham et al. [20], there can be no metallic state, that is no state with extended wave functions,

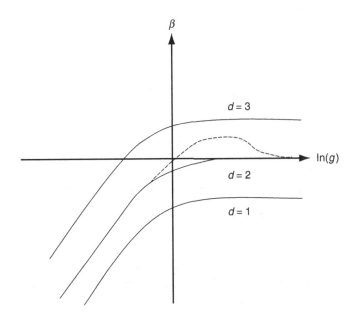

Fig. 7.5 Scaling trajectories according to (7.53) and (7.56). The solid curves correspond to these estimates. The dashed curve corresponds to an alternative trajectory in two dimensions that is supported by the observation of a metal–insulator transition.

in two dimensions in zero magnetic field. With reducing temperature, the resistance is expected to grow dramatically. With the strong localization discussed above, this growth should be exponential. In fact, in high-mobility quasi-two-dimensional systems, such as those which have been discussed in the earlier chapters of this book, this is just not the case. Instead, there appears to be a metal–insulator transition, which is a function of the two-dimensional carrier density. That is, at high density (or high conductance), the semiconductor appears to be metallic, while at low density (or low conductance), the semiconductor appears to be localized at low temperature. This would follow the dashed curve for $d = 2$ in Fig. 7.5.

Now, there are a great many types of metal–insulator transitions, and these appear in a great many different materials, some of which are semiconductors. It is not our aim here to review the entire field, as this would fill a book larger than the current one. Rather, we want to discuss the narrow range of interest, and that is the transition that appears to occur in quasi-two-dimensional systems at a heterojunction interface, whether this is a GaAs/AlGaAs heterojunction or a Si/SiO$_2$ interface. To be sure, the transition itself depends upon the degree of disorder, and in (relatively) highly disordered systems, the metallic phase did not seem to appear. Thus, in the early 1980s, it was found that transport in Si MOSFETs really seemed to fit the localization picture discussed above – no metallic phase was considered to exist [21,22,23]. Yet, with the development of high-quality Si MOSFETs (those with very high mobility at low temperature), convincing evidence for the existence of a two-dimensional metallic state at zero magnetic field was found [24,25]. The conclusion is that a one-parameter scaling theory, as discussed in the last section, is inadequate to explain the experimental findings [26]. It is more likely that the two-dimensional curve follows the dashed one in Fig. 7.5, which shows a metal–insulator transition at the point where it crosses the line $\beta = 0$. Following Pudalov et al. [26], we may summarize the observations as:

(i) The resistivity drops exponentially fast as the temperature decreases below a critical temperature $T_0 \sim 2$ K, and may be expressed as

$$\rho(T) = \rho_0 + \rho_1 \exp[-(T_0/T)^p], \tag{7.57}$$

where the parameter $p \sim 1$. The resistivity drop is more pronounced in high-mobility material [27,28]. The critical temperature T_0 is sample dependent and increases with the carrier density, with a form that varies something like

$$T_0 \propto |n - n_c|^q, \tag{7.58}$$

where n_c is a critical transition density and $q \sim 1$.

(ii) The resistivity for each particular sample may be scaled into two generic branches, using a scaling parameter T/T_0. This critical temperature demonstrates a critical behavior around a critical density n_c [29,30].

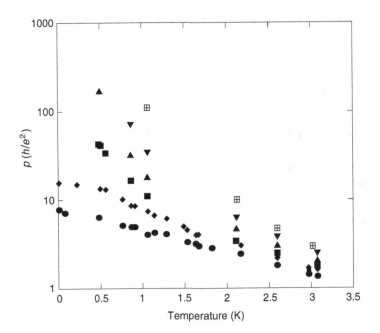

Fig. 7.6 Resistivity versus temperature for a Si MOS two-dimensional electron gas for various inversion densities (described in the text in detail). The data has been taken from Ref. [26], and replotted to show more clearly the behavior for the insulating phase at lower densities.

(iii) A regular negative magneto-resistance is found in a weak perpendicular magnetic field, which indicates the presence of weak localization (discussed in the next section below). This is felt to indicate the presence of quantum interference as a contribution to the conductivity of the two-dimensional metallic state. In addition, there are effects that arise from the presence of an in-plane magnetic field, but it is not clear if these are fundamental to the process.

The presence of the metal–insulator transition in high-quality material has led to its observation in the transport of holes in a SiGe quantum well, placed between two Si layers [31], and in GaAs/AlGaAs heterostructures [32,33,34]. Generally, the transition density n_c is at a very low value; for example, it is estimated that in undoped GaAs/AlGaAs heterostructures, n_c is somewhere in the range 2.3–2.6 × 10^9 cm^{-2} for electrons [35]. For high-mobility strained Si quantum wells, this number seems to be about an order of magnitude larger, 3.2 × 10^{10} cm^{-2} for electrons [36]. To demonstrate the observable behavior, we will illustrate the data for a Si MOS structure, for a particular sample which showed a mobility of 51,000 cm^2/Vs at 0.3 K [26]. In Figs. 7.6 and 7.7, we show the corresponding set of the temperature dependence for a range of inversion densities. These various inversion densities exhibit a difference of behavior, and vary from 4.49 × 10^{10} cm^{-2} to 4.985 × 10^{11} cm^{-2}. The set of curves are determined by: the lowest 11 curves are for values ranging from 4.49 to 9.89 (in units of 10^{10} cm^{-2}) in steps of 0.54, the subsequent curves (again moving upward) are for 10.97, 12.05, 14.21, 16.4, 17.45, 28.2, 39.0, and 49.85. The lowest six densities are shown in Fig. 7.6 and exhibit exponentially increasing resistivities as the temperature is lowered. The highest five densities are shown

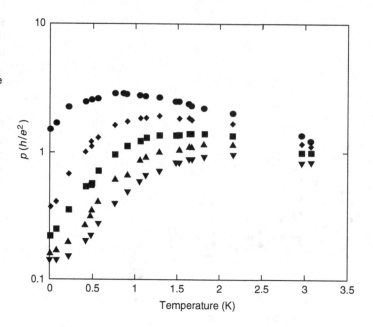

Fig. 7.7 Resistivity versus temperature for a Si MOS two-dimensional electron gas for various inversion densities (described in the text in detail). The data has been taken from Ref. [26], and replotted to show more clearly the behavior for the metallic phase at higher densities.

in Fig. 7.7 and decreasing resistivities as the temperature is lowered. This different behavior is thought to be a metal–insulator transition. The critical density appears to be somewhere between 1.2 and 1.4 × 10^{11} cm^{-2}. The dashed horizontal lines suggest a separation between the metallic and the insulating phases. As the temperature decreases, both the characteristic magnitude and steepness of the drop in resistivity decrease, and the drop shifts to higher densities.

Abrahams [37] has raised the question as to what differentiates the high-quality samples, which exhibit the two-dimensional metal–insulator transition, from the older samples which do not. The quantities of interest are the Coulomb energy of the carriers, the Fermi energy and the disorder. A measure of the Coulomb energy is the quantity (in two dimensions)

$$V_c = \frac{e^2}{2\pi\varepsilon_s r_0}, \quad (7.59)$$

where r_0 is the average interparticle spacing, and is related to the density by $r_0 = 1/\sqrt{\pi n_s}$. The Fermi energy is given by $E_F = \hbar^2 k_F^2/2m = 2\pi n_s/ms$, where s is a factor for spin and valley degeneracy. In Si, with only the lowest subband occupied, $s = 4$ (2 for spin and 2 for the doubly degenerate valleys). In $d = 2$, the resistivity is proportional to the disorder in the sense that (for weak disorder) it is approximately related to $1/(k_F l)$, where l is a mean free path. Now, the ratio of the Coulomb energy to the Fermi energy produces a critical parameter

$$\frac{V_c}{E_F} = \frac{s}{a_B\sqrt{\pi n_s}} = r_s, \quad (7.60)$$

where $a_B = 4\pi\hbar^2\varepsilon_s/me^2$ is the Bohr radius in the semiconductor (about 2.8 nm in Si). What distinguishes the new devices from the old is the fact that the densities are low, below 10^{11} cm^{-2} in nearly all cases, and the mobility is high. The low density gives values for $r_s > 10$, so that the Coulomb energy is much larger, relatively speaking. In Abrahams' view [37], this led to a conclusion that these devices were clearly exhibiting a strongly coupled, many-body behavior, a point which he reinforced later [38]. Nevertheless, he points out that the theoretical methods for such a regime are still poorly developed.

In general, the view can be made that the transition is a percolation transition [39,40]. As the density is lowered, screening becomes progressively weaker and strongly nonlinear. A small decrease in the density leads to a much larger decrease in the screening, and this eventually leads to a highly inhomogeneous electron gas, which is unable to screen the disorder potential. This gives rise to a set of valleys in which electrons can be localized, which means that both the Coulomb energy and the disorder strength are going to be important. The percolation transition occurs when carriers can begin to successfully negotiate through the sample by successive transfer through the valleys. More recently, Kastrinakis [41] has suggested that strong spin-density correlations are important in the metallic phase below the critical temperature, so that the final understanding is probably not fully at hand at present.

7.3 Weak localization

Above we talked about *strong* localization, which is a total lack of coherence on any reasonable scale. Here, we want to talk about *weak* localization which requires some coherence. The presence of weak localization results in a reduction in the actual conductivity of the sample, that is, weak localization is a negative contribution to the conductivity. The basic idea was illustrated briefly in Chapter 1 (and discussed above) as a result of interference around a diffusive loop between two trajectories. Let us repeat part of that discussion for convenience. What is crucial here is that there is a set of elastic scatterers, for which a particle undergoes multiple scattering and returns essentially to its original position (and momentum). Since time-reversal symmetry is not broken, the particle may follow either the original path among the scatterers or its time-reversed image. The interference between these paths leads to the enhanced resistance known as weak localization. A typical device is shown in Fig. 7.8. Here, a low-mobility GaAs MESFET structure (a heavily doped epitaxial layer on top of a semi-insulating substrate) has been biased nearly to depletion, and the resulting strong impurity scattering leads to the weak localization in the sample. (Also shown is the Hall voltage, but this is not of interest to us here.) The normal magneto-resistance has not been subtracted out of this plot. The weak localization contribution here is about 30 pS (the structure observed will be discussed in a later section).

Fig. 7.8 Weak localization in a heavily doped epitaxial layer that has been biased by a surface gate to near pinch-off, giving large impurity scattering.

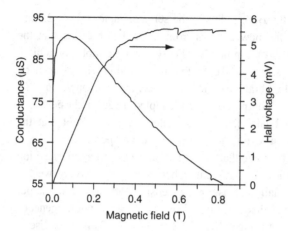

The weak localization correction is damped by the application of a magnetic field. This follows as the magnetic field breaks time-reversal symmetry, and causes the two reverse paths to diverge, hence reducing the quantum interference. We will see below that a critical magnetic field can be defined as that at which the weak localization correction has been reduced to half of its peak value, and this defines a critical magnetic field. It will be argued that this value of critical magnetic field is just enough to enclose one flux quanta *(h/e)* within the area of the phase-coherent loop. In this case, this field value is about 0.01 T for the MESFET structure of Fig. 7.8, which is quite small. (Here, as in most cases, one can only make a rough guess of the actual value due to the conductance fluctuations present in most samples.) These values will be used later in the semiclassical theory discussed below.

Before beginning to talk about the quantum treatment of conductance fluctuations and weak localization, we need to refresh our memory of the semiclassical transport theory in terms of correlation functions. The reason for this is that the primary measurement on the fluctuations is to determine their correlation function as the field or density is varied. Thus, we need to understand that the correlation functions are a general property of transport and not a new phenomenon for describing these effects.

7.3.1 Semiclassical treatment of the conductance

The conductivity can be generally related to the current–current correlation function, a result that arises from the Kubo formula, but it also is dependent upon an approach involving retarded Langevin equations [42]. In general, the conductivity can be expressed simply as

$$\sigma = \frac{ne^2}{m\langle v^2(0)\rangle}\int_0^\infty \langle v(t)v(0)\rangle dt \sim \frac{ne^2}{m}\int_0^\infty e^{-t/\tau}dt = \frac{ne^2\tau}{m} \qquad (7.61)$$

where it has been assumed that the velocity decays with a simple exponential, and where the conductivity is measured at the Fermi surface (although this is not strictly required for the definitions used in this equation). The exponential in (7.61) is related to the probability that a particle diffuses, without scattering, for a time τ, which is taken as the mean time between collisions. Now, in one dimension, we can use the facts that $D = v_F^2\tau$ (or more generally, $D = v_F^2\tau/d$, where d is the dimensionality of the system), $l_e = v_F\tau$, and $n_1 = k_F/\pi$ to write (7.61) as

$$\sigma_1 = \frac{e^2}{\pi\hbar}2D\int_0^\infty \frac{1}{2l_e}e^{-t/\tau}dt. \qquad (7.62)$$

In two dimensions, we use the facts that $D = v_F^2\tau/2$ and $n_1 = k_F^2/2\pi$ to give

$$\sigma_2 = \frac{e^2}{\pi\hbar}2D\int_0^\infty \frac{k_F}{2l_e}e^{-t/\tau}dt. \qquad (7.63)$$

We note that the integrand now has the units L^{-d} in both cases. It is naturally expected that this scaling will continue to higher dimensions. We note, however, that the quantity inside the integral is no longer just the simple probability that a particle has escaped scattering. Instead, it now has a prefactor that arises from critical lengths in the problem. To continue, one could convert each of these to a *conductance* by multiplying by L^{d-2}. This dimensionality couples with the diffusion constant and time integration to produce the proper units of conductance. To be consistent with the remaining discussion, however, this will not be done.

In weak localization, we will seek the correlation function that is related to the *probability of return* to the initial position. We define the integral analogously to the above as a *correlator*, specifically known as the *particle–particle correlator*. Hence, we define

$$C(\tau) = \int_0^\infty W(t)dt \sim \int_0^\infty \frac{1}{2L^d}e^{-t/\tau}dt, \qquad (7.64)$$

where $W(t)$ is the time-dependent correlation function describing this return, and the overall expression has exact similarities to the above equations (in fact, the correlator in the first two equations is just the integral). What is of importance in the case of weak localization is that we are not interested in the drift time of the free carriers. Rather, we are interested in the diffusive transport of the carriers and in their probability of return to the original position. Thus, we will calculate the weak localization by replacing C or W by the appropriate quantity defined by a diffusion equation for the strongly scattering regime. Hence, we must find the expression for the correlation function in a different manner than simply the exponential decay due to scattering.

Following this train of thought, we can define the weak localization correction factor through the probability that a particle diffuses some distance and *returns to the original position*. If we define this latter probability as $W(T)$, where T is the time required to diffuse around the loop, we can then define the conductivity correction in analogy with (7.63) to be

$$\Delta\sigma = -\frac{2e^2}{h} 2D \int_0^\infty W(t)dt, \, x(t) \to x(0), \qquad (7.65)$$

and the prefactor of the integral is precisely that occurring in the first two equations of this section. (The negative sign is chosen because the phase interference *reduces* the conductance.) In fact, most particles will not return to the original position. Only a small fraction will do so, and the correction to the conductance is in general small. It is just that small fraction of particles that actually does undergo backscattering (and reversal of momentum) after several scattering events that is of interest. In the above equations, the correlation function describes the decay of "knowledge" of the initial state. Here, however, we use the "probability" that particles can diffuse for a time T and return to the initial position while retaining some "knowledge" of that initial condition (and, more precisely, the phase of the particle at that initial position). Only in this case can there be interference between the initial wave and the returning wave, where the "knowledge" is by necessity defined as the retention of phase coherence in the quantum sense. Now, while we have defined $W(t)$ as a probability, it is not a true probability since it has the units L^{-d}, characteristic of the conductivity in some dimension *(DWt* is dimensionless). In going over to the proper conductance, this dimensionality is correctly treated, and in choosing a properly normalized probability function, no further problems will arise. This discussion has begun with the semiclassical case, but now we are seeking a quantum-mechanical memory term, exemplifying the problems in connecting the classical world to the quantum-mechanical world. There are certainly other approaches to get to this latter equation, but the approach used here is chosen to enhance its connection with classical transport; what we have to do is describe the phase memory in such an approach, and this is traditionally ascribed to the WKB method in quantum mechanics.

It has been assumed so far that the transport of the carriers is diffusive, that is, that the motion moves between a great many scattering centers so that the net drift is one characterized well by Brownian motion. By this, we assume that quantum effects cause the interference that leads to (7.65), and that the motion may be dominantly described by the classical motion (the quantum treatment is taken up below). This means that $kl_e < 1$, where k is the carrier's wavevector (usually the Fermi wavevector) and l_e is the mean free path between collisions, which is normally the elastic mean free path (which is usually shorter than the inelastic mean free path). This means that the probability function will be

7.3 Weak localization

Gaussian (characteristic of diffusion), and this is relatively easily established by the fact that $W(t)$ should satisfy the diffusion equation for motion away from a point source (at time $t = 0$), since the transport is diffusive. This means that [43]

$$\left(\frac{\partial}{\partial t} - D\nabla^2\right) W(t) = \delta(\mathbf{r})\delta(t), \tag{7.66}$$

which has the general solution

$$W(\mathbf{r}, t) = \frac{1}{(4\pi Dt)^{d/2}} \exp\left(-\frac{r^2}{4Dt}\right). \tag{7.67}$$

In fact, this solution is for unconstrained motion (motion that arises in an infinite d-dimensional system). If the system is bounded, as in a two-dimensional quantum well or in a quantum wire, then the modal solution must be found. At this point we will not worry about this, but we will return to this in the formal theory below. Our interest is in the probability of return, so we set $\mathbf{r} = 0$. There is one more factor that has been omitted so far, and that is the likelihood that the particle can diffuse through these multiple collisions without losing phase memory. Thus, we must add this simple probability, which is an exponential. This leads us to the probability of return after a time T, without loss of phase, being

$$W(t) = \frac{1}{(4\pi Dt)^{d/2}} e^{-t/\tau_\varphi}. \tag{7.68}$$

Here, the phase-breaking time τ_ϕ has been introduced to characterize the phase-breaking process. We note at this point that the dimensionality of $W(t)$ is L^{-d} (Dt has the dimensions of L^2), which is the dimensionality of the integrand for the conductivity, not the conductance. Thus, this fits in with the discussion above.

One further modification of this simple semiclassical treatment has been suggested by Beenakker and van Houten [44]. This has to do with the fact that we do not expect to find these diffusive effects in ballistic transport regimes. Thus, it can be expected that on the short-time basis, these effects go away. Here, "short time" is appropriate in that collisions must occur before diffusive transport can take place. If there are no collisions, there is little chance for the particle to be backscattered and to return to the original position. Thus, these authors suggest modifying (7.68) to account for this process. This gives the new form for the probability of return to be

$$W(t) = \frac{1}{(4\pi Dt)^{d/2}} e^{-t/\tau_\varphi}\left(1 - e^{-t/\tau}\right). \tag{7.69}$$

At this point, we have slipped in the only quantum mechanics in the current approach. This quantum mechanics is connected with the phase of the electrons and is described by the phenomenological phase relaxation time τ_ϕ, which has been introduced. We have not actually carried out a quantum-mechanical calculation, yet we have introduced all the necessary phase interference through this phenomenological term. The actual quantum calculations are buried at this point, but it is important to recognize where they have entered in the discussion.

The dimensionality correction to the probability of return will go away if we work with the total conductance, rather than the conductivity. However, as above, we will not make this change. We can use (7.69) in (7.65). This gives the conductivity corrections for weak localization to be

$$\Delta\sigma = -\frac{e^2}{\pi\hbar} \begin{cases} \frac{1}{2\pi l_\varphi}\left(\sqrt{1+\tau_\varphi/\tau} - 1\right), & d = 3, \\ \frac{1}{2\pi}\ln(1+\tau_\varphi/\tau), & d = 2, \\ l_\varphi\left(1 - \sqrt{\frac{\tau}{\tau+\tau_\varphi}}\right), & d = 1. \end{cases} \quad (7.70)$$

It is clear that the important length in this diffusive regime is the phase coherence length $l_\varphi = \sqrt{D\tau_\varphi}$. To be sure, this result is the most simple one that can be obtained within reasonable constraints. Nevertheless, the results are quite useful to point out that the weak localization reduction of conductance is relatively universal in its amplitude, but also that it has an adjustment depending upon the ratio of the important time scales in the transport problem. Nevertheless, the quantum mechanics is buried in the ad hoc introduction of the phase coherence time τ_ϕ. Without this introduction, none of the above formulas would be meaningful. The detailed quantum treatment, described in a later section, is more formal, but it will basically return these same values for the weak localization corrections to the conductance. Let us now compare with the experiments presented above. In the quasi-two-dimensional GaAs MESFET device, we had a density of 4×10^{11} cm^{-2}, and the mobility was only 2.0×10^4 cm^2/Vs. Using these values, it is estimated that the value for the phase coherence length is only about 0.16 μm.

7.3.2 Effect of a magnetic field

In the presence of a magnetic field, the diffusive paths that return to their initial coordinates will enclose magnetic flux. As discussed above, the situation is slightly different than that of the Aharonov–Bohm effect, in that each of the two interfering paths cycles completely around the loop, doubling the enclosed flux. Thus the expected periodicity is in $h/2e$ (rather than h/e). To examine this behavior is only slightly more complicated than that of the above treatment, and we will take the magnetic field in the z-direction and in the Landau gauge $\mathbf{A} = (0, Bx, 0)$. We will also consider only a thin two-dimensional slab with no z variation at present, so that (7.66) can be written as

$$\left[\frac{\partial}{\partial t} - D\left(\nabla - i\frac{2e\mathbf{A}}{\hbar}\right)^2 + \frac{1}{\tau_\varphi}\right] W(\mathbf{r},t) = \delta(\mathbf{r})\delta(t), \quad (7.71)$$

or

$$\left[\frac{\partial}{\partial t} - D\frac{\partial^2}{\partial x^2} - D\left(\frac{\partial}{\partial y} - i\frac{2eBx}{\hbar}\right)^2 + \frac{1}{\tau_\varphi}\right] W(\mathbf{r},t) = \delta(\mathbf{r})\delta(t). \quad (7.72)$$

Before proceeding, the two equations above should be justified. The earlier approach was contained in (7.66). In the equations here, an extra term appears related to the phase coherence time, and this is just the expected term that will give rise to the exponential decay introduced earlier. However, the gradient operator has also been replaced by a term combining the gradient and the vector potential (divided by $\hbar$). It is this replacement, which seems somewhat questionable in the classical approach, that has been followed so far. Here, we are introducing another aspect of the quantum behavior, first studied for band electrons in terms of the susceptibility. The connection is that the ∇^2 term is related to the square of the momentum in the total energy operator (as related to the Schrödinger equation in Chapter 1). Classically, when one introduces a magnetic field described by the vector potential $\mathbf{A}$, the proper conjugate "momentum" becomes $\mathbf{p} + e\mathbf{A}$ [45]. Quantum mechanically, we expect the vector potential, which is the source of the magnetic field, to modify the momentum as $\hbar\mathbf{k} \to \mathbf{p} + e\mathbf{A} \to -i\hbar\nabla + e\mathbf{A}$, where it is the relation to quantum mechanics that allows us to replace the total conjugate momentum $\mathbf{p}$ by $-i\hbar\nabla$, and the replacement of the square of the simple momentum in the diffusion equation that has been used above. In addition, the vector potential is modified by the factor of 2 for the special circumstances of the problem of interest. This is the second adjustment that arises from quantum mechanics. We have used the connection between the diffusion equation and the Schrödinger equation discussed in Chapter 1 to guide this replacement, so now our quantum treatment lies in the phase-breaking process and in the modification of the diffusion equation itself. In the following, we go essentially completely to the quantum treatment but leave the proper derivation of the above equation to a later section.

This approach is of course quite similar to that for Landau levels. The major difference is that we are developing the response with the diffusion equation, and the decay due to the inelastic (phase-breaking) time represented by (7.68) has been included explicitly. However, we will proceed in exactly the same manner as in the case of Landau levels by assuming that the y variation is characterized by a free momentum plane wave in this direction with wavevector k, so that the last equation can be rewritten as

$$\left[\frac{\partial}{\partial t} - D\frac{\partial^2}{\partial x^2} - D\left(ik - i\frac{2eBx}{\hbar}\right)^2 + \frac{1}{\tau_\varphi}\right] W(\mathbf{r},t) = \delta(\mathbf{r})\delta(t). \quad (7.73)$$

At this point, we shall diverge from a Landau level treatment and recognize that in the final process the limit $x = 0$ will be introduced. Moving directly to this, it may be recognized that the inelastic decay time of (7.68) may be replaced as

$$\frac{1}{\tau_\varphi} \to \frac{1}{\tau_\varphi} + Dk^2 = \frac{1}{\tau_\varphi} + (2n+1)\frac{eBD}{\hbar}, \qquad (7.74)$$

where it has been assumed in the last expression that the relevant momentum is the Fermi momentum with $E_F = (n + 1/2)\hbar\omega_c$. The simplest assumption is to replace the phase-breaking time in (7.70) by the combination of the phase-breaking and magnetic times (with $n = 0$) defined in (7.74). Of course, this treatment is oversimplified, and a more extensive exact treatment is required. However, the important point is that the role of the magnetic field is to dramatically increase the phase-breaking *rate*, thus reducing the *effective* phase-breaking time. As a consequence, the magnetic field breaks up the correlation of the time-reversed paths that led to the weak localization correction, thus reducing this term. (On a formal basis, the magnetic field breaks the time-reversal symmetry of the two counter-propagating waves in the ring. Thus they are no longer equivalent to one another, and the weak localization interference is destroyed.) By estimating the magnetic field at which the weak localization correction is reduced by a factor of two, one has an estimate of the value of the phase-breaking time, since it will be equal to the magnetic time defined in the second term of (7.74)

$$B_c \sim \frac{h}{el_\varphi^2}, \qquad (7.75)$$

where it is assumed that D is known. This leads us to conclude that the critical magnetic field is that necessary to couple one flux quantum (h/e) through a phase-coherent area defined by the diffusive coherence length. Hence, one can thus determine the phase coherence length by two measurements: the amplitude of the weak localization correction and the magnetic field at which this correction has decayed by a factor of 2. In actual fact, the equations (7.70) give a slightly different value for the effective phase-breaking time to reduce the correction by a factor of 2, but this is a detail correction, not a fundamental variation.

To compare, we note that, in the quasi-two-dimensional GaAs MESFET structure, the reduction of the weak localization by a factor of two at 0.01 T (which is a crude estimate given the scales of the data in the figure), suggests a value of 0.11 µm, which compares reasonably well with the above estimate of the phase coherence length. As will be seen below, there are a number of correction terms that need to be included before these estimates can be taken to be accurate.

The one point left out of the latter development, but alluded to earlier, is the fact that the particle may return to its initial point many times due to the Landau quantization of the orbits. This is an equivalent effect to the Aharonov–Bohm oscillations. The integration in (7.65) does not include the possibility of a periodicity in the probability for return. This fact must be taken out by hand at this point

7.3 Weak localization

by noting that the time of interest should be reduced by the period of the orbit around the closed path. This orbit may be due to the Landau quantization, or just due to the fact that the circumference of the path is smaller than the coherence length. It is the latter that produces the competition to the Aharonov–Bohm effect. Moving completely around the loop produces a phase correction of twice that of the Aharonov–Bohm effect, so that there is a correction to the probability of return in the form of

$$\sum_{n'=-\infty}^{\infty} \exp\left(-2i\pi \frac{2eBa^2}{h} n'\right), \qquad (7.76)$$

where a^2 is the enclosed area of the loop. Thus, this factor (using only the case of $n' = \pm 1$) produces a periodicity in $h/2e$. This is shown in Fig. 7.9 for metal

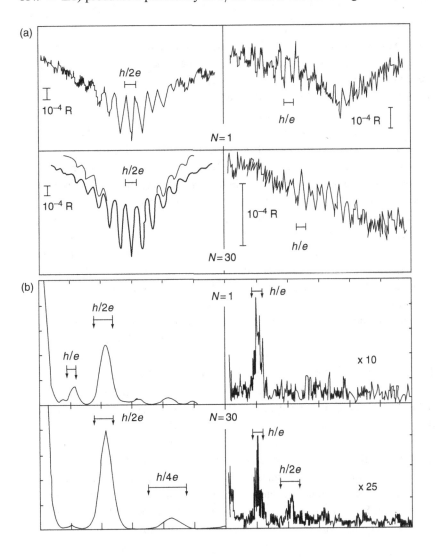

Fig. 7.9 The oscillations in (a) show the resistance oscillations, and (b) shows the Fourier transform, for arrays of 1 and 30 rings, which are discussed in the text. (After Umbach *et al.* [46], by permission.)

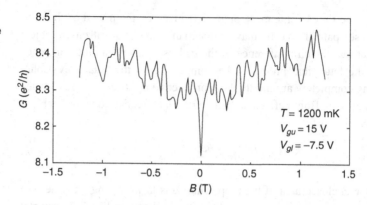

Fig. 7.10 Typical magneto-resistance curve for a short quantum wire in Si. The negative going peak near $B = 0$ is the weak localization. [After de Graaf et al. [47], by permission.]

rings (similar effects are also observed in semiconductor samples) [46]. For small magnetic fields, the $h/2e$ oscillations dominate the conductance correction, whereas for large magnetic fields, where the weak localization effects are damped, the Aharonov–Bohm oscillations dominate the conductance. On the other hand, the latter are heavily damped (actually exponentially attenuated) as the number of rings increases, while the weak-localization $h/2e$ oscillations are attenuated only as $1/N$, where N is the number of rings in the structure. Equivalently, N is the number of equivalent regions of size dimension comparable to the inelastic mean free path. Because these are well-formed rings, the oscillations remain coherent. In a bulk region, where the "rings" are phase-coherent regions of random size, the oscillations are expected not to be coherent and to produce interfering effects leading to universal conductance fluctuations.

In Fig. 7.10, the weak localization measurements for a short "wire" formed in a Si MOSFET are shown [47]. The devices used are dual-gate Si MOSFETs, in which a short one-dimensional channel is formed by a split pair of lower gates (biased negatively), while the overall inversion layer density is controlled by a large upper gate (biased positively to attract carriers into the inversion channel). The width of the gap in the split lower gates is 0.35 μm, but the width of the conducting channel is considerably smaller due to the fringing fields of the gate. The length of the split-gate defined wire is about 0.73 μm. The real one-dimensional channel will be somewhat longer than this, again due to the fringing fields around the actual metal gates. The low-field magneto-resistance of this device is given in the figure. The strong dip in the conductance around zero magnetic field is due to weak localization of the carriers. In addition, universal conductance fluctuations are seen in the conductance at higher fields. First, in these one-dimensional wires, surface scattering from the side walls of the wire strongly affect the conductance and actually increase the amplitude of the weak localization. Second, the shortness of this wire means that the two-dimensional areas to which the wire is attached (the areas that are not covered by the lower

7.3 Weak localization

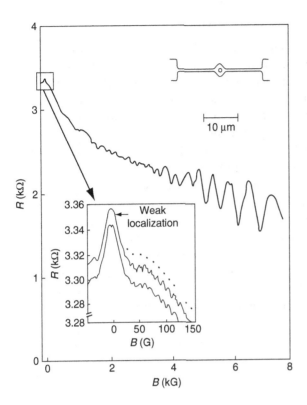

Fig. 7.11 Magneto-resistance of a typical bridge sample (inset) in which the probe lengths and width are less than the elastic mean free path. The Hall voltage is also shown in the inset. [After J. Simmons et al. [48], by permission.]

split gates) play an important role in the size and field dependence of the weak localization. We will turn to these effects below, but the analysis presented there suggests that the phase coherence lengths for this device (at the measuring temperature of 1.2 K) are about 0.7 µm in the two-dimensional regions and about 0.2 µm in the wire itself.

The results of the weak localization in a wire formed in a GaAs/AlGaAs heterostructure are shown in Fig. 7.11 [48]. The structure of the wire is shown in the inset to this figure. The magneto-resistance for this structure is for a wire segment of length $L = 2$ µm at 0.3 K. The wires, in this case, are formed by etching them to a width of 1 µm. Universal conductance fluctuations and the onset of the quantum Hall effect are shown at higher fields for this sample. The large resistance peak near $B = 0$ is the weak localization. The phase coherence length in this sample has been estimated from the universal conductance fluctuations to be about 12 µm. It should be noted, however, that the basic heterostructure has very high mobility so that both the length and width of the wire segments are less than the *elastic* scattering mean free path. The mobility (10^6 cm^2/Vs) was such that the latter was on the order of 6 µm in this wire structure.

7.3.3 Size effects in quantum wires

In general, weak localization seems to be stronger in quantum wires. This is because the longitudinal resistivity (along the magnetic field) in a quasi-two-dimensional system remains independent of the magnetic field, at least within the semiclassical treatment. However, in a high-mobility quantum wire, the transverse motion causes the carriers to interact strongly with the "walls" of the quantum wire. If the transport is primarily diffusive in moderate mobility material, this is not an important effect. On the other hand, in high-mobility wires, where the transport can be quasi-ballistic, this effect is a natural extension of the properties of boundary scattering in the wire itself.

Earlier, it was shown how boundary scattering changes the basic resistivity of the region contained within the quantum wire. Let us review that discussion. We consider that the channel (or wire) is defined by the presence of hard walls at $x = \pm W/2$, at which point the carriers are diffusely scattered. By diffusive scattering we mean that the carriers lose all information of the incoming angle of propagation and are reflected uniformly in an angle of π radians. If they have some memory of their initial momentum direction, they are then said to have been scattered specularly [49,50,51]. The stationary distribution function for carriers in the wire satisfies the simple Boltzmann equation, within the relaxation time approximation [52],

$$\mathbf{v} \cdot \nabla f(\mathbf{r}, \alpha) = -\frac{f(\mathbf{r}, \alpha)}{\tau} + \frac{1}{\tau} \int_0^{2\pi} \frac{f(\mathbf{r}, \alpha')}{2\pi} d\alpha', \qquad (7.77)$$

where $\mathbf{r}$ is the two-dimensional position vector in the quantum wire and α is the angle that the velocity makes with the transverse x-axis. The second term on the right is a balancing term so that integrating over all angles α' produces no net flow term on the left side (as required by current continuity along the wire). With diffuse scattering, the distribution function should be independent of the direction of the velocity for those velocities directed *away* from the boundary. Current continuity then leads to the boundary conditions

$$f(\mathbf{r}, \alpha) = \begin{cases} \dfrac{1}{2} \displaystyle\int_{-\pi/2}^{\pi/2} f(\mathbf{r}, \alpha') \cos \alpha' d\alpha', & x = \dfrac{W}{2}, \quad \dfrac{\pi}{2} < \alpha < \dfrac{3\pi}{2}, \\ \dfrac{1}{2} \displaystyle\int_{\pi/2}^{3\pi/2} f(\mathbf{r}, \alpha') \cos \alpha' d\alpha', & x = -\dfrac{W}{2}, \quad -\dfrac{\pi}{2} < \alpha < \dfrac{\pi}{2}. \end{cases} \qquad (7.78)$$

Since there is no magnetic field, it may be assumed that the density is uniform in α', so that the integral over the angle in (7.77) vanishes. Beenakker and van Houten [44] show that the solution is given by (except for a normalization constant)

$$f(\mathbf{r},\alpha) = -y + l_e \sin\alpha \left[1 - \exp\left(-\frac{W}{2l_e|\cos\alpha|} - \frac{x}{l_e \sin\alpha}\right)\right], \quad (7.79)$$

where $l_e = v_F \tau$ is the elastic mean free path at the Fermi surface. For $W \ll l_e$, this leads to

$$\rho = \rho_0 \frac{\pi l_e}{2W \ln(l_e/W)} = \frac{m v_F}{n e^2 W} \frac{\pi}{2 \ln(l_e/W)}. \quad (7.80)$$

In a magnetic field, it is possible for skipping orbits to form. Even when the magnetic field is directed along the axis of the wire, it is possible for particles, whose motion would normally be along the wire, to be deflected into the surface and to undergo backscattering. Our major interest, however, is for the case in which the magnetic field is normal to the quasi-two-dimensional layer. Then the motion is much clearer. In the presence of this z-directed magnetic field, (7.77) becomes

$$\mathbf{v} \cdot \nabla f(\mathbf{r},\alpha) + \omega_c \frac{\partial f(\mathbf{r},\alpha)}{\nabla \alpha} = -\frac{f(\mathbf{r},\alpha)}{\tau} + \frac{1}{\tau} \int_0^{2\pi} \frac{f(\mathbf{r},\alpha')}{2\pi} d\alpha', \quad (7.81)$$

where $\omega_c = eB/m$ is the cyclotron frequency. If we have *specular* scattering, then we require $f(\mathbf{r},\alpha) = f(\mathbf{r},\pi-\alpha)$ at $x = \pm W/2$. This leads to the result [52]

$$f(\mathbf{r},\alpha) = -y + \omega_c \tau x + l_e \sin\alpha. \quad (7.82)$$

The interesting aspect is that the diffusive current along the wire $I_y = \pi W v_F l_e$, is the same whether or not there is a magnetic field. This implies that, in the presence of specular scattering at the surface, the motion along the wire is not affected by the magnetic field. Longitudinal motion is converted to edge-state skipping orbits, as discussed in previous chapters, but the overall effect of the magnetic field on the resistance is a zero effect. In the presence of diffusive scattering at the boundaries, however, this is no longer the case. As we have seen above, diffusive scattering increases the resistivity. This is further increased in a small magnetic field, in which trajectories that would normally move directly along the wire are diverted into the surface. In a high magnetic field, the cyclotron motion inhibits the backscattering that can occur from the diffuse scattering by bending the orbits back into the forward direction. Thus, when $W > 2l_{cycl}$, where $l_{cycl} = r_c$ is the cyclotron orbit at the Fermi surface, the resistivity increase due to diffusive backscattering is dramatically reduced. This is apparent in Fig. 7.12. where experiments from Thornton *et al.* [53] are shown. Data for several different widths are shown for comparison. First, it is clear that the resistance increase at zero magnetic field is much larger than that which arises from the simple reduction of the wire width (decreasing the width by a

Fig. 7.12 The role of boundary scattering in thin wire structures is indicated here for a set of wires of varying width. (After Thornton *et al.* [53], with permission.)

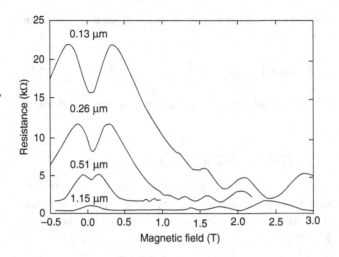

Fig. 7.13 The cyclotron radius at the peak position of the resistance, plotted versus wire width. The inset shows the dependence upon the carrier density. (After Thornton *et al.* [55], with permission.)

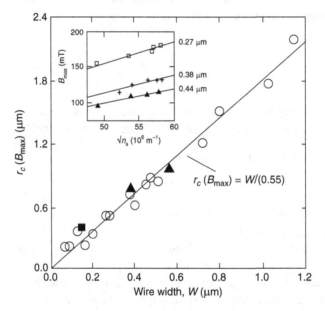

factor of two increases the resistance by a factor of 2.5–3.5). There is a rise in the resistance at small magnetic field, which is the effect of deflecting the longitudinal trajectories into the walls. Finally, at higher magnetic fields, the diffusive scattering is greatly reduced due to the formation of the edge states and their skipping orbits. Finally, at the highest magnetic fields, the normal Shubnikov–de Haas effect is seen. The position of the anomalous resistance peak, $B_{\max}$, is determined by the ratio of the width of the wire to the cyclotron radius. Calculations by Forsvoll and Holwech [54] and by Pippard [50] suggest that $W/l_{cycl} = 0.55$ at the maximum. In Fig. 7.13, the peak position (in magnetic field) of the resistance maximum is plotted for a variety of wire widths [55].

The straight line is a fit to this theory. The dependence on the cyclotron radius also suggests a connection to the carrier density and this is illustrated in the inset. In general, the fit to the theory is quite good.

That this is a high magnetic field effect in relatively high-mobility material is evident from the fact that the peak is found to occur at $B = 0.2$–0.4 T, which suggests that $\omega_c \tau = \mu B > 2$, or that $\mu > 5$–10×10^4 cm^2/Vs at low temperature. No conductance fluctuations are seen in the data, which is further indication that the impurity concentration is sufficiently low that diffusive scattering within the bulk of the wire is not significant.

It is important to note that this diffusive edge-scattering effect is different from the weak localization discussed above. Weak localization is broken up by the magnetic field so that the enhanced resistance decays already from zero magnetic field. This leads to a reduction in resistance with magnetic field. Diffusive edge scattering, however, is *enhanced* in a small magnetic field, which causes an increase in the resistance with magnetic field; this increase is much larger than normally expected from the magneto-resistance. Thus, weak localization and diffusive edge scattering show opposite behavior with magnetic field in the small field regime. However, in many cases it is not clear that the two effects can be easily separated, particularly in quantum wires. Note that for the boundary scattering, the overall change in the resistance at low magnetic fields is opposite to that of the weak localization. However, the role of boundary scattering is decreased with much stronger magnetic fields than those at which the coherence of the time-reversed orbits is broken. Consider Fig. 7.11, discussed earlier, as an example. There is a general decay of the resistance with magnetic fields in the range around 1–3 T. This decay could be due to the decay of surface scattering as well as the general onset of field quantization in the quantum wire. The weak localization, however, is observed clearly only at the lowest magnetic fields and has been totally broken up by a field value of about 1 T. Although it can not be seen clearly in this figure, one would expect the weak localization to be gone at fields of the order of 0.1 T or so.

7.3.4 The magnetic decay "time"

As discussed above, the presence of surface (or edge) scattering will modify the behavior of the magneto-resistance, and particularly the weak localization, in the presence of a magnetic field. In general, we may consider that the effectiveness of the magnetic field in suppressing the weak localization contribution to the resistance lies in the amount of magnetic flux enclosed in the loop formed by the two time-reversed trajectories. If the bulk trajectories are modified by scattering from the boundary of the quantum wire, then the amount of flux will be modified. Thus there are different regimes, characterized by the relative sizes of the important lengths: the mean free path $l_e = v_F \tau$,

Table 7.1 *Magnetic relaxation time and critical magnetic field for weak localization in a channel*

Length scales	τ_B	B_c
$l_e, l_\varphi \ll W$	$\dfrac{l_m^2}{2D}$	$\dfrac{\hbar}{2el_\varphi^2}$
$l_e \ll W \ll l_\varphi$	$\dfrac{3l_m^4}{W^2 D}$	$\dfrac{\sqrt{3}\hbar}{eWl_\varphi}$
$W \ll l_e, Wl_e \ll l_m^2$	$\dfrac{C_1 l_m^4}{W^3 v_F}$	$\dfrac{\hbar}{eW}\sqrt{\dfrac{C_1}{Wv_F \tau_\varphi}}$
$W \ll l_e, W^2 \ll l_m^2 \ll Wl_e$	$\dfrac{C_2 l_m^2 l_e}{W^2 v_F}$	$\dfrac{\hbar}{eW}\left(\dfrac{C_2}{Wv_F \tau_\varphi}\right)$

the coherence length for diffusive transport $l_\phi = (D\tau_\phi)^{1/2}$, the magnetic length $l_m = (\hbar/eB)^{1/2}$, and the width of the wire W (as well as the length of the wire L). The case for $l_e \ll W$, in which the elastic mean free path is much less than the width of the wire, corresponds to a narrow quasi-two-dimensional structure rather than to a proper quantum wire. This limit has been called the *dirty metal regime*, but it is sometimes found in Si quantum wires. In this limit, transport is mostly characterized by that of a disordered material. The opposite limit, for which $l_e \gg W$, usually called the *pure metal regime*, is one in which the differences between specular and diffusive scattering from the boundaries can be delineated. The latter regime is the one that we have called the quasi-ballistic regime.

Another relationship between the various lengths determines whether we are concerned with quasi-two-dimensional corrections to the conductivity or truly quantum wire corrections. This relationship concerns the coherence length and the width of the wire. If $l_\varphi \ll W$, it is generally felt that the weak localization corrections are those corresponding to a quasi-two-dimensional medium. On the other hand, if $l_\varphi \gg W$, it is generally felt that the system is properly a quasi-one-dimensional wire. The key factors in the magnetic field dependence of the weak localization are the magnetic "lifetime" τ_B, given for example by the last term in (7.74), and the magnetic field B_c, at which τ_B and τ_φ produce comparable effects in the weak localization. Several theories have been presented for these quantities, and the predictions are summarized in Table 7.1 [44,56,57,58]. The two constants C_1 and C_2 depend upon the nature of the boundary (surface) scattering. For specular scattering, it is generally found that $C_1 = 9.5$ and $C_2 = 4.8$, while for diffusive scattering, $C_1 = 4\pi$ and $C_2 = 3$ [44].

To compare with the results for the pseudo-wire case, one can consider a quasi-two-dimensional device for which $l_\varphi \ll W$. Then, the magnetic length is

7.3 Weak localization

given by the result of (7.74), in which $\tau_B \sim h/2eBD = l_m^2/2D$, and the critical magnetic field is found from $\tau_B \sim \tau_\varphi$, so that $B_c \sim h/2eD\tau_\varphi = h/2el_\varphi^2$. These values are for the lowest Landau level, and smaller values are expected for higher levels. We note that it is just this value that is used to estimate the quantum wire data above, and apparently this estimate was made in the wrong limits of lengths, since this wire clearly had a width smaller than the phase coherence length.

The full expression for the magneto-conductance correction due to weak localization is found by expanding the solution to (7.73) with a summation over the index n, leading to [43]

$$\delta G_{WL}^{2D}(B) - \delta G_{WL}^{2D}(0) = \frac{W}{L}\frac{ge^2}{4\pi^2\hbar}\left[\Psi\left(\frac{1}{2}+\frac{\tau_B}{2\tau_\varphi}\right) - \Psi\left(\frac{1}{2}+\frac{\tau_B}{2\tau}\right) + \ln\left(1+\frac{\tau_\varphi}{\tau}\right)\right], \quad (7.83)$$

where g is a degeneracy factor that accounts for spin and valley degeneracies, and $\Psi(x)$ is the digamma function. Normally, τ_B will be given by the value in the first row of Table 7.1. In the absence of the magnetic field, (7.70) is recovered. The weak localization will be completely destroyed by the magnetic field when the magnetic field is such that $\tau_B < \tau$, which occurs for $\hbar/2el_e^2 < 1$. These fields are still much weaker than the usual quantum limit.

In the one-dimensional case $W \ll l_\phi$, the time-reversed trajectories are squeezed by the finite width of the wire. This compression of the orbits leads to the second line of Table 7.1 [57]. The full expression for the magnetic field dependence of the conductivity correction in this limit is given by

$$\delta G_{WL}^{1D}(B) = \frac{ge^2}{hL}\left(\frac{1}{D\tau_\varphi}+\frac{1}{D\tau_B}\right)^{-1/2}. \quad (7.84)$$

It should be noted that one expects to see a crossover from quasi-one-dimensional to quasi-two-dimensional results at a magnetic field for which $l_m \sim W$. The reason is that the lateral confinement is replaced by skipping orbit formation and the dominance of edge-state conduction in the wire. We will see more of this result in the discussion of the universal conductance fluctuations below.

In semiconductor nanostructures formed in high-mobility material, the elastic mean free path is usually quite long. For example, in GaAs with a mobility of 10^5 cm^2/Vs, it was shown in Chapter 1 that the average elastic mean free path is about 1 µm. Now it is possible in heterostructures to have the mobility 10–100 times larger than this, which means that the mean free path can reach 10–100 µm. For wire widths of less than 1.0 µm, it is easy to obtain $l_e \gg W$. This is then the "pure metal" regime (or pure quasi-ballistic regime), in which scattering is relatively weak and surface scattering dominates the diffusive process. This regime seems to have been first treated by Dugaev and Khmelnitskii [59]. There

Fig. 7.14 Illustration of the manner in which diffusive scattering can cause cancellation of enclosed flux in an orbit.

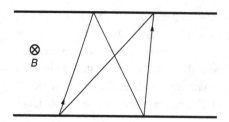

is a major new factor that becomes important in this regime, and that is the possibility of *flux cancellation* [44] by multiple reflections from the wire boundary. By this we mean that the diffusive scattering from the boundaries causes trajectories to fold back upon themselves. Consider Fig. 7.14 as an example. Here, a trajectory is indicated that actually closes upon itself. If the magnetic field is directed out of the paper, then part of the trajectory is closed in a positive manner, while the other is closed in a negative manner. Thus, these two contributions to the enclosed flux may cancel one another. Then, the consideration of weak localization is more complex than that discussed above. Now, for example, a single impurity scattering can break up the flux cancellation and lead to the onset of weak localization with a much greater effect than expected from a single scattering center. On the other hand, in trajectories that normally do not have flux cancellation, a single impurity scattering can cause the onset of weak localization by the influence of the boundaries. In this regime, one can distinguish the difference between a weak field (third row of Table 7.1) and a strong field (last row of the table) regime. In the former, many impurity scattering events are required to achieve weak localization. In the latter regime, on the other hand, a single impurity may be sufficient to cause this effect. Both of these cases are for the situation in which the magnetic length is large compared to the width, however, so that edge states and skipping orbits are not important. These effects lead to the numerical coefficients for the magnetic quantities in Table 7.1. In numerical simulations of quantum wires, it is generally found that neither limit is well formed, and an average of the two magnetic times is often preferred as [44]

$$\tau_{B,eff} = \tau_B^{weak} + \tau_B^{strong}. \tag{7.85}$$

The conduction correction for weak localization is now given for these 1D channels as

$$\delta G_{WL}^{1D}(B) = \frac{ge^2}{hL}\left[\left(\frac{1}{D\tau_\varphi}+\frac{1}{D\tau_B}\right)^{-1/2} - \left(\frac{1}{D\tau_\varphi}+\frac{1}{D\tau_B}+\frac{1}{D\tau}\right)^{-1/2}\right]. \tag{7.86}$$

This is, of course, the same result as (7.84), but with the addition of the term arising from elastic scattering, and the change in the definition of the effective magnetic relaxation time.

7.3 Weak localization

The use of these latter expressions improves the estimates of the coherence length from the magnetic decay for the short quantum wires discussed above. These now give numbers near 0.5 μm, but it is clear that these are nowhere near the value estimated from the amplitude of the weak localization correction at zero magnetic field. The shortness of the wire itself is the last culprit to be examined.

7.3.5 Extension to short wires

In the above discussions, it has been inherently assumed that the length L of the wire was large compared to any other characteristic length. In short wires, different effects may be observed. One needs to consider how the coherence may be broken within the regions at the ends of the wire. In other words, one needs to question whether the wire is terminated in two quasi-two-dimensional contact regions or by a network of other quantum wires, such as in the inset of Fig. 7.11. The question that must be addressed in the experiments is just what is being measured. When samples have characteristic lengths that are smaller than the coherence length, the amplitude of the conductance fluctuations often are found to be larger than the universal value of e^2/h [60,61]. Second, the magneto-conductance is found to be asymmetrical in magnetic field [62]. More importantly, the values of the coherence length l_ϕ are smaller than those obtained from measuring long wires, even in the same material. Thus, it seems that in measurements in these short wires, the nature of the "contacting" regions becomes very important in determining just what is being measured. This problem was examined by Chandrasekhar *et al.* [63]. In steady state, the weak localization contribution to the conductivity is given by (7.65) to be

$$\Delta\sigma(\mathbf{r}) = -\frac{2e^2 D}{\pi\hbar} C(\mathbf{r}, \mathbf{r}')\delta(\mathbf{r} - \mathbf{r}'). \tag{7.87}$$

Here, $C(\mathbf{r}, \mathbf{r}')$ is the time-integrated probability of return (e.g., when $\mathbf{r}' \to \mathbf{r}$) and therefore is given by the steady-state form of (7.71), in a magnetic field, as the solution to the equation

$$D\left[\left(-i\nabla - \frac{2e\mathbf{A}}{\hbar}\right)^2 + \frac{1}{l_\varphi^2}\right] C(\mathbf{r}, \mathbf{r}') = \delta(\mathbf{r} - \mathbf{r}'). \tag{7.88}$$

The quantity $C(\mathbf{r}, \mathbf{r}')$ is also called the particle–particle propagator because it describes the spatial correlation of the particle, and in this form will be our connection to a more basic quantum approach. On an insulating boundary, $C(\mathbf{r}, \mathbf{r}')$ is required to satisfy the condition of zero flow across the boundary, or

$$\left(-i\nabla_n - \frac{2e\mathbf{A}_n}{\hbar}\right)C(\mathbf{r},\mathbf{r}') = 0, \tag{7.89}$$

where the subscript n indicates the normal component of the vector operator. For a simple wire of length L, this leads to the weak-localization contribution to the resistance to be ($\Delta G \ll G$)

$$\left.\frac{\Delta R}{R}\right|_{WL} = -\left.\frac{\Delta G}{G}\right|_{WL} = \frac{2e^2 D}{\pi \hbar \sigma_0 L}\int_0^L C(\mathbf{r},\mathbf{r})d\mathbf{r}. \tag{7.90}$$

It is quite usual to set the particle–particle propagator path to zero, as in this equation. Yet in short wires, one cannot rule out the possibility that the coherent propagation extends well into the quasi-two-dimensional contacting regions. Here, $G = \sigma_0 L$ is the conductance for a one-dimensional wire. The above equation provides a simple approach to determine the weak-localization correction for any one-dimensional sample (one-dimensional in the fact that there is only an average of the particle–particle propagator over the length of the wire; higher dimensions require a more complete averaging process).

Chandrasekhar *et al.* [63] consider the general case in which the wire is terminated by a network of other one-dimensional wires, as well as the case in which it is terminated by two-dimensional regions. Here, we demonstrate only the latter situation. We work first in the absence of a magnetic field and count upon the substitutions developed above for the insertion of the field. In the one-dimensional wire, $C(\mathbf{r},\mathbf{r}')$ depends only on the coordinates along the wire, which is taken to be the x-axis. Following these authors, the solution to (7.89) can be formulated in terms of Green's functions (here, the solutions will be hyperbolic sines and cosines) and connected to the two-dimensional regions. This leads to

$$\left.\frac{\Delta R}{R}\right|_{WL} = \frac{e^2 l_\varphi R_{sq}}{\pi \hbar W}\left[\frac{(\eta^2 + \alpha^2)\coth(L/l_\varphi) - (l_\varphi/L)(\eta^2 - \alpha^2) + 2\alpha\eta}{(\eta^2 + \alpha^2) + 2\alpha\eta\coth(L/l_\varphi)}\right], \tag{7.91}$$

where $\alpha = W/l_\varphi$, and

$$\eta = \pi \ln\left(\frac{2l_\varphi^{2D}}{l_e}\right) \tag{7.92}$$

describes the effects of the two-dimensional boundary layers. Here, $R_{sq} = 1/\sigma_{sq}$ is the sheet resistance of the two-dimensional layer. It is clear that in the case for which $L \gg l_\phi$, the result reduces to the normal quantum wire result

$$\left.\frac{\Delta R}{R}\right|_{WL} = \frac{e^2 l_\varphi R_{sq}}{\pi \hbar W}. \tag{7.93}$$

In the opposite limit, $L \ll l_\phi$, the fluctuations are dominated by the two-dimensional regions, and

$$\left.\frac{\Delta R}{R}\right|_{WL} = \frac{e^2 R_{sq}}{\pi \hbar \eta}. \tag{7.94}$$

It should be mentioned that, if the wire is measured in a four-probe configuration with narrow probes attached at the junction of the two dimensional region and the wire of interest, then $\eta = \eta_1 + \eta_2$, where η_1 corresponds to (7.92) with the angle π replaced by the appropriate acceptance angle of the two-dimensional region, and $\eta_2 = W_p/l_\varphi^p$, which describes the nature of the voltage probe wires. In the presence of the magnetic field, the magnetic relaxation time is introduced through the replacement of the appropriate τ_ϕ according to the same procedure as above,

$$\frac{1}{\tau_\varphi} \to \frac{1}{\tau_\varphi} + \frac{1}{\tau_B}, \tag{7.95}$$

with the appropriate τ_B introduced into each of the coherence lengths used in (7.91).

The major effect of the results obtained in this section is that the propagator in the wire does not end (is not terminated to a zero value) at the contacts but rather transitions smoothly into the appropriate one- or two-dimensional propagators for the "contacting regions." These differ by having either one- or two-dimensional values for the coherence lengths. The latter differ in the wire and in the two-dimensional regions due to the additional effect of the side-wall scattering as discussed above. Careful analysis of the data discussed above, with the formulas of this section, now leads to an estimate for the coherence length of 2.5 μm in the wire section and about 5.0 μm in the two-dimensional sections to which the wire is connected. The actual fit is not particularly sensitive to the latter quantity, however, so the value for this quantity is not well determined by such an analysis. The value for the wire is now quite close to that estimated from the amplitude of the weak-localization correction at zero magnetic field.

7.4 Universal conductance fluctuations

It is clear from the above discussion for weak localization that the existence of time-reversed paths can lead to quantum interference that causes a correction to the conductivity. These paths are modified by the change in the electrochemical potential (changes in the wavephase velocity) and in the magnetic field (similar changes in the momentum occur through the vector potential). If a sample is composed of a great many such loops (and not necessarily time-reversed loops), with each contributing a phase-dependent correction to the conductivity, then the summation over these loops may or may not ensemble average to zero, depending upon the number of such loops contained within the sample (and, hence, upon the size of the sample). On the other hand, in mesoscopic systems, where the number of such loops is relatively small, this is not the case, and these loops lead to the presence of universal conductance fluctuations. Since the impurity distribution is reasonably fixed for a given sample configuration, the particular oscillatory pattern is often thought of as a fingerprint of an individual sample. That is, the impurity distribution is different from one sample to another, and therefore the detailed nature of the interference pattern seen in any one sample is a characterization of its unique impurity distribution. Perhaps the most remarkable feature of these oscillations is that, for $L \gg l_e$, the amplitude of these oscillations seems to satisfy a universal scaling with a nominal value of $\delta G = e^2/h$, regardless of the size of the sample (but notice the discussion of this below) and the degree of disorder [64], [65]. One view of this universality in amplitude arises from the Landauer formula, in which the conductance is quantized for each possible channel through the sample. Here, the conductivity is given by (see Chapter 3) [66]

$$G = \frac{e^2}{h} \sum_{i,j=1}^{N} |t_{ij}|^2, \qquad (7.96)$$

where t_{ij} is the transmission from incoming channel i to outgoing channel j. In this regard, one conclusion that may be drawn from the above is that the fluctuation in conductance arises from the random turning on and/or off of a single channel of the many channels that are occupied. Generally, one would now construct the variance of G, but the problem lies in the fact that there will be correlation between the individual t_{ii}, especially for similar pairs of ingoing and outgoing waves. However, since we are considering channels that are primarily transmitting, the correlation between the reflection coefficients can be much smaller (and therefore more easily averaged) [67]. Thus, we rewrite (7.96) as

$$G = \frac{e^2}{h} \left[N - \sum_{i,j=1}^{N} |r_{ij}|^2 \right], \qquad (7.97)$$

where N is the number of channels, and the variance of the conductance is given simply by

$$\text{var}(G) = \left(\frac{e^2}{h}\right)^2 \text{var}\left[\sum_{i,j=1}^{N} |r_{ij}|^2\right] = \left(\frac{e^2}{h}\right)^2 N^2 \text{var}\left(|r_{ij}|^2\right), \qquad (7.98)$$

with uncorrelated reflection coefficients. The exact evaluation of the variance is quite difficult and will be deferred until later. However, one simple approximation is to assume that Wick's theorem carries over to this problem, by which we mean that there are no high-order correlation functions. Thus we can use the simple approximation

$$\langle |r_{ij}|4 \rangle = 2\langle |r_{ij}|^2 \rangle^2. \qquad (7.99)$$

7.4 Universal conductance fluctuations

To lowest order, the last term in the angular brackets is simply $1/N$, so that

$$\text{var}(G) \sim \left(\frac{e^2}{h}\right)^2 N^2 \frac{2}{N^2} \sim 2\left(\frac{e^2}{h}\right)^2. \quad (7.100)$$

The numerical prefactors are more problematic and have to be treated carefully [68], [69]. However, the general result can be written in the form [52], [65]

$$\delta G = [\text{var}(G)]^{1/2} = \frac{g}{2}\frac{C}{\sqrt{\beta}}\frac{e^2}{h} \quad (7.101)$$

where g is a factor for spin and valley degeneracies, C is a constant of order unity (C is about 0.73 for a narrow channel with $L \gg W$, but on the order of $\sqrt{W/L}$ in the opposite limit of a wide channel), and β is unity in zero magnetic field but takes a value of 2 when the magnetic field lifts the time-reversal symmetry of the system. (It should also be noted that the lifting of the spin degeneracy by the magnetic field leads to this contribution to the degeneracy factor to be reduced from 2 to $\sqrt{2}$.)

The simplest experiments were discussed in Fig. 1.7, which was measured for a wide wire formed in a Si MOSFET. These measurements have been studied for a large range of effective wire lengths L. If we calculate the variance of this data, where the amplitude is relatively constant down to zero field, we arrive at a value of C of about 0.4.

In general, most measurements are made under constant current conditions (as was the latter data), so that fluctuations in the voltage couple to resistance fluctuations (a similar effect appeared in the discussion of the previous section). If the resistance fluctuations are independent of the sample length, then it is expected that $\delta G \approx \delta R/R^2 \sim L^{-2}$, since the resistance is expected to scale linearly with the length of the wire. The results of an extensive study of Si wires are shown in Fig. 7.15, where the reduced conductance fluctuation (in units of e^2/h) is plotted as a function of L/l_φ [70]. For $L < l_\varphi$, this behavior is clearly observed; that is, for lengths smaller than the coherence length, the resistance fluctuation is independent of the length of the sample. This result indicates that the voltages on the individual probes are fluctuating with some coherence to one another (the measuring length is still l_φ, which can reach around from one voltage probe to another). The net voltage fluctuation is then independent of L, since the critical length for coherence is l_φ. On the other hand, when $L > l_\varphi$, the measurement is over many phase-coherent regions (approximately L/l_φ, such regions), so that the fluctuation in the resistance is proportional to the number of these resistances. In this case, the two voltage probes are considered to be oscillating independently from one another, with the amplitude of the variation proportional to the square of the number of such phase-independent cells. Thus, $\delta R \sim \sqrt{N} \sim L^{1/2}$, which leads to a variation in the conductance fluctuation as $L^{-3/2}$, and this behavior is also clearly seen in the figure. Finally, we note that the estimate of the data from Fig. 5.1 lies within the spread of

Fig. 7.15 Variation of the conductance fluctuation with the wire length, demonstrating the scaling behavior. (After Skocpol et al., Phys. Scripta **T19**, 95 (1987), with permission.)

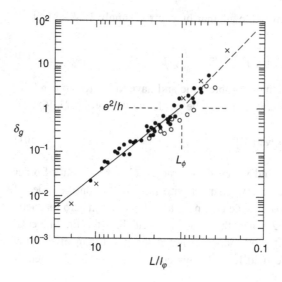

Fig. 7.16 (a) The measured voltage fluctuations, and (b) the conductance fluctuations as a function of wire length. (After Benoit et al., Phys. Rev. Lett. **58**, 2343 (1987), with permission.)

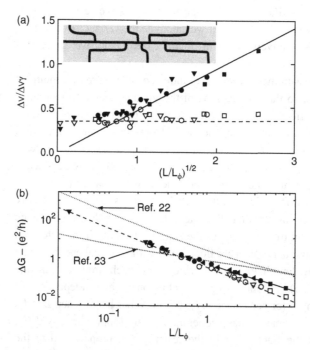

experimental points in Fig. 7.15 when we consider that the value of the phase coherence length determined previously is slightly larger than the length of the wire itself.

The behavior of the scaling of the voltage (or the resistance) with the length is also illustrated in Fig. 7.16 [60]. Here, it is important to note that under the

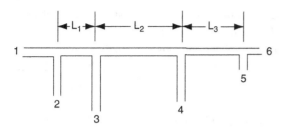

Fig. 7.17 A typical multi-lead sample geometry with the leads numbered for reference.

application of a magnetic field, there is a part of the fluctuation that is symmetric in magnetic field and a part that is asymmetric. Only the symmetric part satisfies this behavior of relative independence for $L > l_\varphi$. In Fig. 7.16(a), the voltage fluctuation itself is plotted as a function of the relative length of the sample. Here, the symmetric and asymmetric parts are separated and plotted. In Fig. 7.16(b), the corresponding conductance fluctuation is plotted. Although one might think that there should not be a part of the voltage fluctuation that is asymmetric in magnetic field (most of us are ittaught that magnetic field effects should he symmetric in the field), it should be remembered that these mesoscopic devices behave like small circuits rather than as discrete devices. As circuits, they exhibit nonlocal properties, whether simple circuit theory or the equivalent microwave modal theory (see Chapter 3) is introduced. Consider, for example, the device of Fig. 7.17. If we feed the current through the contacts labeled 1 and 4, the voltage measured between contacts 2 and 3 will depend upon the manner in which leads 5 and 6 are terminated. If these latter two leads are grounded, they act as current sinks. If they are left open-circuited, then they can have floating voltages. The principle that this will strongly affect the measurements is well known from introductory circuit theory through Thevenin's equivalent circuit principles. In fact, if an Aharonov–Bohm ring is attached between lead 5 and ground, the voltage measurements between leads 2 and 3 will show oscillations that arise from the varying conductance in lead 5 to ground. (This is true even if the end of the ring is left "open," since there really is no such thing in the semiconductor circuits due to parasitic resistances and capacitances.) These effects are seen in mesoscopic devices [71]. The reversal of the magnetic field, and the consequent asymmetry, brings another basic circuit principle into play, the principle of reciprocity, in which measurements are required to be equal only when the current and voltage leads are interchanged in a multi-loop/node circuit (this was discussed in Chapter 3). In reversing the magnetic field, the symmetry of the problem requires the invoking of reciprocity, so that [72]

$$V_{ij,kl}(B) = V_{kl,ij}(-B), \qquad (7.102)$$

where the first set of indices refers to the current probes and the second set of indices refers to the voltage probes. The full importance of this relation was not appreciated when mesoscopic systems were first studied, and only after several experiments and the explanation of Büttiker [72] was it realized that (7.102) was the proper form in which the Onsager relation would appear. Since the current is a constant value in the experiments, the voltages indicated are precisely just a constant value different from the resistances, so that $R_{ij,kl} = R_{kl,ij}$.

It was indicated in the earlier treatment that the magnetic field would cause the weak-localization correction to decay. But, we have pointed out that the universal conductance fluctuation arises from similar considerations, and that any variation in the position of the electrochemical potential in the density of states, whether caused by a gate bias or by a magnetic field, would also cause the correlation in the loops to decay (or at least to change). This is treated in the universal conductance fluctuation by the presence of a correlation function

$$C(\Delta E, \Delta B) = \langle G(E + \Delta E, B + \Delta B) G(E, B) \rangle - \langle G(E, B) \rangle^2. \tag{7.103}$$

It is clear that the magnetic field variation causes a change in the correlation function. The energy variation arises from the facts that the impurities cause an inhomogeneous energy surface and that different current paths will see different local variations in the chemical potential. This means that small changes in the overall chemical potential, due to an applied gate voltage, for example, will cause larger variations in the local chemical potential. This will lead to an equivalent correlation "length" in the energy, or in the gate voltage. In Fig. 7.18, the correlation function is plotted for several values of ΔV_g (which corresponds to ΔE) as a function of the magnetic field for the measurements shown in Fig. 1.5. The correlation function has a normal decay and can be characterized by correlation energies ΔE_c, and magnetic fields ΔB_c. (The tendency is to use a different term than the correlation field B_c, which has been used previously to define the value of the field at which the magnetic decay time is comparable to the phase-breaking time in weak localization, although the two

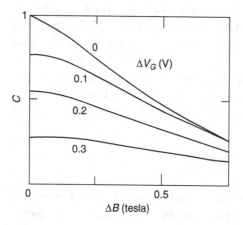

Fig. 7.18 The correlation function for the fluctuations measured in Fig. 1.5 are plotted as a function of the change in magnetic field, with the change in gate voltage as a parameter. (After Skocpol *et al.*, *Phys. Rev. Lett.* **56**, 2865 (1986), with permission.)

7.4 Universal conductance fluctuations

terms arise from the same physical processes and are both used often with the same symbolism.) At least for low magnetic fields, these correlation fields and energies are universal quantities related to the size of phase-coherent regions.

The correlation "lengths" that can be defined from the correlation function (7.103) and measurements such as those of Fig. 7.18, are in terms of magnetic fields $\Delta B = \Delta B_c$ and energies $\Delta E = \Delta E_c$, which may formally be defined theoretically, as will be done later in this chapter. The magnetic field correlation length ΔB_c, the correlation field, is the critical field that characterizes the magnetic-field-induced dephasing of the interference phenomena. The correlation field is defined as that field at which C is reduced to its half-height, so that the correlation field is the half-width at half-height of the correlation function. As in weak localization, the correlation function is made up of three factors [3]: (1) the probability of return of the trajectory from **r** to **r'**, in a time T, independent of the magnetic field, (2) a phase relaxation factor, and (3) the average phase factor itself. The averages in the correlation function are taken over all classical trajectories that can contribute to the diffusive transport being measured. In fact, therefore, the critical correlation held corresponds to coupling flux quanta through the area of a so-called phase-coherent region corresponding to some product of length and width within the sample. One can expect that these length scales will then be the smaller of the characteristic length of the sample (width or length) and the phase coherence length. Also, as in the case of weak localization, the dephasing is caused by the magnetic field in a manner in which the effective inelastic scattering process is given by (we use a common notation here, which will be explained further below)

$$\frac{1}{\tau_\varphi} \to \frac{1}{\tau_\varphi} + \frac{1}{\tau_{\Delta B/2}}, \quad (7.104)$$

so that the magnetic field actually does produce dephasing of the correlation function in the same manner as it reduces the weak-localization correction to the conductance. It is for this reason that these phenomena are termed *universal*. The magnetic field at which the correlation function is reduced to half-height is the same quantity as given in Table 7.1, but with the substitution $B \to \Delta B/2$. The factor of 2 that enters this latter substitution has to do with the nature of the averaging process that appeared in (7.99), in that we are no longer dealing with two time-reversed interfering paths. Since we are not using two time-reversed paths, but only a *single net loop* around the enclosed area, the magnetic field coupled to the loop is reduced by a factor of 2 from that in the weak-localization correction. Thus, for UCF, we replace $2B$ in the magnetic decay time by ΔB (or B by $\Delta B/2$), where ΔB is the shift in the magnetic field used in the correlation function (7.103). This means that for a wire in the dirty metal regime

$$\tau_{\Delta B/2} = \frac{12 l_m^4}{W^2 D} = 12\left(\frac{\hbar}{e\Delta B}\right)^2 \frac{1}{W^2 D}, \quad l_e \ll W \ll l_\varphi, \quad (7.105)$$

and this leads to the correlation field

$$\Delta B_c = 0.55 \frac{h}{e} \frac{1}{W l_\varphi}, \quad l_e \ll W \ll l_\varphi. \tag{7.106}$$

In fact, the numerical pre-factor can vary somewhat with the measurements. It should be noted that this value differs from the equivalent value obtained from the weak localization. For a wire in the pure metal regime,

$$\tau_{\Delta B/2} = 4C_1 \left(\frac{\hbar}{e\Delta B}\right)^2 \frac{1}{W^3 \nu_F} + 2C_2 \left(\frac{\hbar}{e\Delta B}\right) \frac{l_e}{W^2 \nu_F}, \quad W \ll l_e, \tag{7.107}$$

where the weak and strong magnetic field results have been combined. The quantities C_1 and C_2 were discussed below in Table 7.1. This now leads to

$$\Delta B_c = \frac{C_2 h}{4\pi e} \frac{l_e^2}{W^2 l_\varphi^2} \left[1 + \sqrt{1 + \frac{8C_1}{C_2} + \frac{W l_\varphi^2}{l_e^3}}\right], \quad W \ll l_e. \tag{7.108}$$

It should be remarked, however, that this version of the correlation magnetic field differs from that obtained in the earlier detailed theories, where it was found that [68]

$$\Delta B_c = \beta \frac{h}{e} \frac{1}{W l_\varphi}, \quad W \ll l_\varphi, l_e. \tag{7.109}$$

Here, $\beta \sim 0.25$ is found for a variety of long wires in high-mobility material [73]. A similar behavior can be found in the low-magnetic-field limit of (7.108) only if $W \sim l_e$ is invoked and much smaller values of the constants are used. In this situation, the elastic mean free path is that determined from the surface scattering, rather than from the basic mobility of the material. Thus, one should use some care in actually using this great variety of formulas for computing any one quantity.

While it can not easily be seen from Fig. 5.1, the correlation field has been determined to be about 0.8 mT at a width of 0.5 μm. If we use the previously determined values for l_φ, and this correlation field, it is found that $\beta \sim 0.4$ for this wire. On the other hand, if we use (7.106), then a good fit to the earlier value of coherence length is found. This suggests that the constants C_1 and C_2 are somewhat high for this experimental wire. It should be apparent that obtaining a unique value for the coherence length is difficult, and only when a number of determinations from different sets of measurements are achieved with some consistency is the end result believable.

The values of the constants depend on the nature of the surface scattering at the edge of the wire. In actual samples being measured, it may not be clear just which regime is appropriate for using a formula. This has complicated the evaluation quite often, and sometimes it is simpler to use numerical coefficients in these expressions in a manner that allows them to be adjusted for a particular sample. However, the above discussion leads to the conclusion that, once the regime of operation is known and the values for the width, length, elastic scattering length, and coherence length are known, the amplitude of the fluctuations and the correlation functions are universal for any value of the magnetic field and bias. This is what is meant by universal behavior. This is found to be the case so long as the magnetic field remains relatively small, that is, so long as the motion remains diffusive in the bulk of the sample. We discuss the high-magnetic-field behavior in a later section.

7.5 Green's functions in disordered materials

At this point, we want to begin to try to understand the standard methods by which weak localization and universal conductance fluctuations are treated by the theoretical world in a quantum-mechanical manner. This involves some rather complicated mathematical approaches, because the long-range correlations are important for phase interference, and these are usually difficult to evaluate. The approach we follow here is based on the Green's functions introduced in Chapter 3. The following sections will examine weak localization and universal conductance fluctuations.

7.5.1 Weak localization

The important phenomenon that we have used in the introductory sections above to discuss weak localization is that of phase coherence, that is, the phase memory is maintained through several scattering events in the time-reversed paths. This requires the scattering to be elastic, in which only the direction of the momentum is changed. Moreover, we require that this coherence be maintained through a series of scattering events by which the particle returns (is scattered back) to its original position, or to a momentum directly opposite to its original momentum state (in momentum space). Weak localization itself is the interference between the two time-reversed paths around the scattering ring. The resistivity corrections that were calculated above assume that there is no coherence through the scattering process, so that the impurity scattering introduces disorder into the system. Strong disorder, of course, will localize part or all of the states in the system [74]. As discussed above with the experimental studies, the most remarkable sign of weak localization is a backscattering peak at zero magnetic field. This backscattering is a representation of multiple elastic scattering. Consider the construction of Fig. 7.19. A momentum vector is gradually rotated in momentum space by elastic scattering until it points in the opposite direction. In the figure, the various momentum vectors have been translated to a common origin, so that the momentum imparted by the impurity ($\mathbf{q}_i$) is clearly seen. There are, of course, two directions in which the momentum vector can be rotated, and these are the two time-reversed paths. We note that the momentum,

Fig. 7.19 The multiple scattering can be linked with this ring for time-reversed paths.

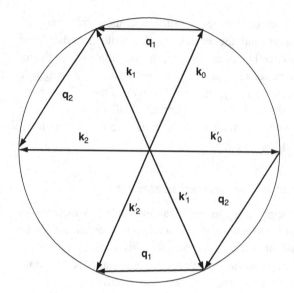

which starts out as $\mathbf{k}_0$, is rotated to $\mathbf{k}_1$ by $\mathbf{q}_1$ and then to $\mathbf{k}_2$ by $\mathbf{q}_2$, and so on. The primed vectors denote the opposite sense of rotation. The important point is that the set of momentum vectors can be made precisely the same for each sense of rotation, and it is this coherence that leads to the interference of the time-reversed paths; the wavevectors are matched by their complements, which makes the motion difficult. The coherent backscattering is the result of this constructive interference in the set of multiple scatterings. If we consider that each process of scattering involves an amplitude then we note that each a_i is matched by the equivalent a'_i. Thus, the coherence that this entails leads to [75]

$$\frac{|a_i + a'_i|}{|a_i|^2 + |a'_i|^2} \sim 2 \tag{7.110}$$

if there is perfect coherence between the two scattering amplitudes. If there were no coherence, the ratio in (7.110) would be unity. This leads to scattering almost a factor of two stronger, and hence to a significant reduction in the conductivity. From this consideration, the quantity $\langle a_i a'^*_i \rangle_{imp}$, where the brackets denote the impurity averaging, gives us the degree of coherence in this process and represents the coupled scattering processes. This is just the effect that we discussed in the previous section, since this coupled impurity (two interaction lines) spans across propagators that are rotating in different directions. We note that the energy-conserving delta function is not in the basic impurity interaction Hamiltonian, but occurs in the transition probability when we compute the matrix elements and the Green's functions. (The delta function arises from the Green's function itself in the absence of interactions; it is replaced here by the broadened spectral density.) Thus, one can evaluate $\langle a_i a'^*_i \rangle_{imp}$, without concern about this, showing that there is a singularity in this quantity when the backscattering condition is fulfilled. Since the propagators are rotating around the loop in opposite directions, the interactions in which we are interested are the set of maximally crossed diagrams first studied by Langer and Neal [76].

At the end of the previous section we considered those interactions that spanned the two Green's functions but did not cross. These formed the ladder diagrams. As mentioned here, our interest is now in that set of diagrams that are maximally crossed. This raises an important general point about the perturbation series. In two cases above, only a particular set of diagrams – certainly not all possible diagrams that can be drawn – were chosen for consideration. Here, still a third set of possible diagrams is being considered. In truth, one decides a priori just what the nature of the physics will require and tries to find a subset of all possible diagrams that first fits the physics and second can be conveniently resummed into a simple analytic expression. Only experience can guide the practitioner as to the manner in which the terms are chosen, and hopefully other series contributions are small enough that the chosen diagrams are the dominant ones.

As mentioned, we now want to evaluate the maximally crossed diagrams. These are shown in Fig. 7.20. The first diagram was used earlier as part of the ladder diagram series; the next two are the next-higher orders of the interaction, which are the terms of interest. The approach that we follow is mainly due to Bergmann [77] although it builds on earlier work of Altshuler *et al.* [56] and others [58], [78].

7.5.1.1 The cooperon correction

The value of this group of electron–hole propagators is an additional term to Λ of Eq. (5.81). Consider the third diagram of Fig. 7.20. The arrangement of the Green's functions and matrix elements in the diagram is usually changed by reversing the direction of the Green's functions on one of the horizontal lines, usually the hole line, so that it becomes an electron–electron propagator and the interactions create a new ladder diagram. This changes $\mathbf{k}' \to -\mathbf{k}'$ and changes the integrals that result from the new ladder diagram. The maximally crossed diagrams now become the normal set of ladder diagrams in this new description. Such a diagram reversal was first used in discussing superconductivity, so that the set of ladder diagrams with the electron–electron propagators is often called a *cooperon*. Rather than the entire two-particle Green's function, we are interested only in the correction term. Previously we have let the applied frequency

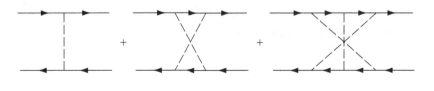

Fig. 7.20 The set of maximally crossed diagrams that lead to weak localization.

$\omega_a \to 0$, for the static conductance, but here we will find a divergence as $q, \omega_a \to 0$. Normally, one enters the two frequencies through $\omega \pm \omega_a/2$, as discussed before (5.81). Here, we will continue to let $\omega_a \to 0$ but will introduce an arbitrary decay function for the phase as the phase-breaking self-energy $i\hbar/2\tau_\varphi$ ($\omega_\pm = \omega \pm i\hbar/2\tau_\varphi$) in the critical polarization terms of the ladder diagram series. Following Bergmann, we will adopt an approach suitable for a d-dimensional system. First, it should be noted that Λ depends only on $\mathbf{k} + \mathbf{k}' = \mathbf{q}$, which is different from the result in the preceding section (where we were interested in $\mathbf{k} - \mathbf{k}' = \mathbf{q}$), since the direction of $\mathbf{k}'$ has been reversed but has little internal structure apart from the divergence already mentioned. Therefore, we can write this term as $\Lambda(\mathbf{q}, \omega)$. The two integrations (summations) that are indicated in Eq. (5.83) become integrations over $\mathbf{q}$ and $\mathbf{k}$. We further note that since we are interested in the low q limit, we can make the approximation (which is on the Fermi surface, so that $k \to k_F = m v_F/\hbar$)

$$E(k') = \frac{\hbar^2}{2m}(\pm \mathbf{k} + \mathbf{q})^2 \simeq E(k) \pm \hbar \mathbf{v}_F(k) \cdot \mathbf{q}. \tag{7.111}$$

Now, the integration over $\mathbf{k}$ in the Green's functions can be performed as

$$I = \int \frac{d\omega}{2\pi} \int \frac{d^d \mathbf{k}}{(2\pi)^d} k_x^2 G^r(\mathbf{k}, \omega_-) G^a(\mathbf{k}, \omega_-) G^r(-\mathbf{k}+\mathbf{q}, \omega_-) G^a(-\mathbf{k}+\mathbf{q}, \omega_-) \delta(E - F_F)$$

$$= \int \frac{d\omega}{2\pi} \int \rho(E) \frac{k^2}{d} dE \frac{1}{\hbar\omega - E - (i\hbar/2\tau)} \frac{1}{\hbar\omega - E + (i\hbar/2\tau)} \delta(E - E_F)$$

$$\times \frac{1}{\hbar\omega - E + \hbar \mathbf{v}_F(k) \cdot \mathbf{q} + (i\hbar/2\tau)} \frac{1}{\hbar\omega - E + \hbar \mathbf{v}_F(k) \cdot \mathbf{q} - (i\hbar/2\tau)}. \tag{7.112}$$

The contour of integration (the energy runs from $-\infty$ to ∞ to account for both electrons and holes in the diagrams) is closed in the upper half-plane, giving two poles at

$$\hbar\omega = E + (i\hbar/2\tau),$$
$$\hbar\omega = E + \hbar \mathbf{v}_F(k) \cdot \mathbf{q} + (i\hbar/2\tau), \tag{7.113}$$

which leads to

$$I = -i\frac{k_F^2}{d}\rho(E_F) \frac{1}{i/\tau} \frac{1}{\hbar \mathbf{v}_F \cdot \mathbf{q}} \frac{2\mathbf{v}_F \cdot \mathbf{q}}{(i/\tau)^2 - (\mathbf{v}_F \cdot \mathbf{q})^2}. \tag{7.114}$$

For small q, this becomes

$$I \cong 2\frac{k_F^2}{d}\rho(E_F)\tau^3. \tag{7.115}$$

This now leads to the conductivity correction

$$\delta\sigma = -\frac{2e^2\hbar^2 k_F^2 \rho(E_F)\tau^3}{dm^2} \int \frac{d^d\mathbf{q}}{(2\pi)^d} \Lambda(\mathbf{q}, \omega) \tag{7.116}$$

where we have used the coefficient from earlier results.

Now, we must calculate the kernel $\Lambda(\mathbf{q},\omega)$. Neglecting the terminal Green's functions which have been included already in the above calculation, we can write the third-order ladder interaction as

$$\frac{1}{\hbar^6}\sum_{\mathbf{r},\mathbf{s}} V_{\mathbf{r}}G^r(\omega_+,\mathbf{k}+\mathbf{r})V_{\mathbf{s}}G^r(\omega_+,\mathbf{k}+\mathbf{r}+\mathbf{s})V_{\mathbf{s}'}^*V_{\mathbf{r}}^*G^a(\omega_-,\mathbf{k}'-\mathbf{r})$$
$$V_{\mathbf{s}}^*G^a(\omega_-,\mathbf{k}'-\mathbf{r}-\mathbf{s})V_{\mathbf{s}'}^*. \tag{7.117}$$

With the reversal of one of the particle lines, the crossed diagrams are converted to a simple ladder, so that a series is more easily generated. For simplicity, we assume that the scattering is isotropic (independent of the scattering wavevector, as discussed earlier), since the angular dependence of the potential is not crucial to the presence of these terms in the series. This leads to

$$\frac{n_i}{\hbar^2}|V_{\mathbf{r}}|^2 = \cdots = \frac{n_i}{\hbar^2}|V_0|^2 = \frac{1}{2\pi\hbar\rho(E_F)\tau} \equiv \Gamma_0 \tag{7.118}$$

where we have used the definitions of the scattering time from the previous section. Further, we make the definition

$$\Pi(\mathbf{k}+\mathbf{k}',\omega,\omega') = \sum_{\mathbf{g}} G^r(\omega_+,\mathbf{k}+\mathbf{g})G^a(\omega_-,\mathbf{k}'+\mathbf{g}) \tag{7.119}$$

which depends on only the sum of the two wavevectors, as discussed previously. Then, the term (7.117) can be written as

$$\Gamma_0\Pi\Gamma_0\Pi\Gamma_0. \tag{7.120}$$

Each of the terms in the series can be written in this way, and the series can be resummed as before, which gives

$$\Lambda = \Gamma_0 + \Gamma_0\Pi\Gamma_0 + \Gamma_0\Pi\Gamma_0\Pi\Gamma_0 + \cdots = \frac{\Gamma_0}{1-\Pi\Gamma_0}. \tag{7.121}$$

Because there are no internal integrations (the terms Π are functions only of the scattering wavevector), the result (7.121) is equivalent to a Dyson's equation for the two-particle propagator. Thus, we need only to evaluate the vertex functions, and for this we will set $\mathbf{k}'' = \mathbf{k}+\mathbf{g}$ and $\mathbf{k}+\mathbf{k}' = \mathbf{q}$ in (7.121). Then, the reversal of the hole line takes $\mathbf{k}'' \to -\mathbf{k}''$, and the polarization function between the elements of the ladder, when we use (7.111) for small q, becomes

$$\Pi(\mathbf{q};\omega) = \int\frac{d^d\mathbf{k}''}{(2\pi)^d}G^r(\omega_+,\mathbf{k}'')G^a(\omega_-,-\mathbf{k}''+\mathbf{q})$$
$$= \rho(E_F)\int\frac{dS_{k''}}{S_{k''}}\int dE\left\{\frac{\hbar}{\hbar\omega - E - (i\hbar/2\tau) + (i\hbar/2\tau_\varphi)}\right. \tag{7.122}$$
$$\left.\times\frac{\hbar}{\hbar\omega - E + \hbar\mathbf{v}_F(k'')\cdot\mathbf{q} - (i\hbar/2\tau) - (i\hbar/2\tau_\varphi)}\right\},$$

where we have introduced the additional phase-breaking term to the self-energy. The integration over the energy can be performed as

$$\Pi(\mathbf{q};\omega) = -\rho(E_F)\int \frac{dS_{k''}}{S_{k''}} 2\pi i \frac{\hbar^2}{\hbar\mathbf{v}''\cdot\mathbf{q} - i\hbar/\tau - i\hbar/2\tau_\varphi}$$
$$\cong 2\pi\hbar\rho(E_F)\tau\left[1 - \frac{\tau}{\tau_\varphi} - Dq^2\tau - \cdots\right] \quad (7.123)$$

where $D = v_F^2\tau/d$. We note that the density of states is inside the above integral over frequency and energy, so that it is actually evaluated at the Fermi surface. This can then be used with (7.118) to give the kernel

$$\Lambda(\mathbf{q};\omega') = \frac{1}{2\pi\hbar\rho(E_F)\tau}\frac{1}{Dq^2\tau + \frac{\tau}{\tau_\varphi}} \quad (7.124)$$

and the conductivity correction is simply

$$\delta\sigma = -\frac{e^2}{\pi\hbar}(D\tau)\int\frac{d^d\mathbf{q}}{(2\pi)^d}\frac{1}{Dq^2\tau + \tau/\tau_\varphi}. \quad (7.125)$$

We see that it does indeed diverge as $q \to 0$ and for $\tau_\varphi \to \infty$. This must be examined with some care. Normally, one ignores the last factor in the denominator, but we shall carry it along here. We note here that the first fraction in the integral looks like a Green's function but involves the diffusion constant D. This replacement for the Green's function, which has naturally arisen, was discussed much earlier in this chapter and is termed the *diffusion pole*.

The main contribution to the integral (7.125) arises from small values of the momentum, hence the finite size of the system will appear here. For example, in a film of thickness t and a mean free path l_e, which should be less than t, the first integration over the energy arose from a three-dimensional integration. However, in the integration left in (7.125), the finite thickness of the film can limit the integration to just two dimensions if there is quantization in the direction normal to the film. Here, we will explicitly deal with only the two-dimensional case, which is the one normally encountered when studying mesoscopic effects in semiconductors. Then, we will further limit the integration to values of q below those defined by the length given by the diffusion process (the diffusion length), so that $q < 1/\sqrt{D\tau}$. In essence, this limit says that we are interested only in the small q divergence of the integral, and

$$\delta\sigma = -\frac{e^2}{4\pi\hbar}(D\tau)\int_0^{1/\sqrt{D\tau}} Dq^2 \frac{1}{Dq^2\tau + \tau/\tau_\varphi} = -\frac{e^2}{4\pi\hbar}\ln\left[\frac{1+\tau/\tau_\varphi}{\tau/\tau_\varphi}\right]. \quad (7.126)$$

Thus, we finally arrive at the two-dimensional correction to the conductivity due to weak localization as

$$\delta\sigma = -\frac{e^2}{4\pi\hbar}\ln\left(1 + \frac{\tau}{\tau_\varphi}\right) \quad (7.127)$$

where $\tau < \tau_\varphi$ is needed for the entire approach of this section to be fully valid. If $\tau = \tau_\varphi/10$, then the logarithm term is about 2.4.

The integrand of Eq. (7.125) is often called the diffusion propagator. It has a form quite similar to that of the Green's functions themselves, even though it is real. It often appears with the fraction in the denominator expressed as $i\tau\omega$, which introduces the Fourier representation. In this form, it appears more evenly in the form of the Green's function. We note that this diffusion propagator did not appear in the treatment of the particle–hole ladder of the previous section, since the small terms are irrelevant to that development, whereas here they are critical to the development. Following the same line of argument, it can be shown that the q-dependent diffusion treated in the previous section gives the same polarization (7.124) as obtained here, provided that there is no magnetic field. This is a result of the time-reversal symmetry of the diagrams used for this purpose. Consequently, the diffusion propagator is often used for both the particle–hole and the particle–particle ladders, and these are termed the *diffuson* and the *cooperon*, respectively. It may also be noted that the leading Green's functions in each of the forms of the conductivity used in these sections form a basic polarization term but are weighted by a term in k^2. The fact that the two Green's functions give essentially the same weight as a polarization term is important, and the overall diagram of the left side of Fig. 5.16 can be redrawn as in Fig. 7.21. Here, the two Green's function lines on the left side have been pulled together, since they represent the same point **r**. Similarly, the two Green's function lines on the right side have been drawn together since they represent the same point **r**′. The two squiggly lines represent the current carried by the factors k (one for each side of the diagram which gives the k^2), as **k** is the Fourier transform variable for **r** − **r**′. The overall diagram of Fig. 7.21 is termed the "current bubble." In the absence of either the diffuson or the cooperon contributions, the interactions spanning the two Green's function branches (top of the diagram and bottom of the diagram) are absent, and one has just the two simple Green's functions. These are of course the full Green's functions with the interaction important to the single lines, but with no interactions spanning the individual electron and hole (or electron line in the cooperon) lines.

7.5.1.2 Role of a magnetic field

If we now apply a magnetic field normal to the two-dimensional electron (and hole) gas, this has a nonnegligible effect on the conductance correction, as was discussed in the opening sections of this chapter. The vector potential of the magnetic field modifies the phase of the wave functions. In the treatment here, we must assume that the elastic mean free path is much smaller than the cyclotron radius, since we cannot have closed orbits for our diffusive motion. The main effect of the vector potential, and the magnetic field, is then to change the relative phase between any two points on the path more than would normally arise from

Fig. 7.21 Bubble form of the current arising from the Green's functions on the left side of the general conductivity.

the propagation. Hence, the phase of one wave function relative to the other one in the Green's function is shifted by the vector potential according to

$$G(\mathbf{r},\mathbf{r}',B) = G(\mathbf{r},\mathbf{r}',0)\exp\left[\frac{ie}{\hbar}\int_\mathbf{r}^{\mathbf{r}'} \mathbf{A}(\mathbf{r}'') \cdot \mathbf{dr}''\right]. \qquad (7.128)$$

The form of the additional term is easiest to conceptualize when we think of plane waves. Here, one replaces the momentum by the proper conjugate momentum discussed earlier in this chapter, so that $\mathbf{k} \to \mathbf{k} + e\mathbf{A}/\hbar$, which is then integrated over the path. This vector potential breaks the translational invariance of the Green's function but does not change the basic approach that we have used. The major change will be in the summation of the maximally crossed diagrams. For the moment, this calculation is done in real space, but it will be transferred to the Fourier space after some preliminary considerations. The polarization (5.98) is now given by

$$\Pi(\mathbf{r},\mathbf{r}';\omega;B) = G^r(\mathbf{r},\mathbf{r}';\omega;B)G^a(\mathbf{r},\mathbf{r}';\omega;B)$$

$$= G^r(\mathbf{r},\mathbf{r}';\omega;B)G^a(\mathbf{r},\mathbf{r}';\omega;B)\exp\left[\frac{i2e}{\hbar}\int_\mathbf{r}^{\mathbf{r}'} \mathbf{A}(\mathbf{r}'') \cdot \mathbf{dr}''\right] \qquad (7.129)$$

$$= \Pi(\mathbf{r},\mathbf{r}';\omega;0)\exp\left[\frac{i2e}{\hbar}\int_r^{r'} \mathbf{A}(\mathbf{r}'') \cdot \mathbf{dr}''\right].$$

The factor of 2 in the phase arises from the fact that each of the two lines, the electron line and the hole line, contribute a factor equivalent to that in Eq. (7.128). This is quite similar to the Aharonov–Bohm effect. The path indicated in the integral is just one-half of the overall loop, and each side of the loop contributes one-half of the total phase integral leading to the Aharonov–Bohm phase. (We will see later that the universal conductance fluctuations do not constitute a complete loop, but that the two sides of the loop tend to cancel one another, so that it is only the difference in the coupled field from the two sides that will remain in the phase integral.) Another way of seeing this is that while the first Green's function is a function of $\mathbf{r} - \mathbf{r}'$, the second is a function of the reverse of this quantity, and is a complex conjugate, so that it leads to a doubling of the phase shift. Now, we need to show that this is equivalent to the Peierl's substitution in the scattering wave vector $\mathbf{q} \to \mathbf{q} + 2e\mathbf{A}/\hbar$, with the factor of 2 coming from the phase-doubling inherent in the last equation. Moreover, it is important to show that it arrives as an eigenvalue of the diffusive operator for the polarization.

To show the important property of the eigenvalues, we note that the polarization is an operator that can operate on an arbitrary wave function and produce an eigenvalue equation according to

$$\int d^3 r' P(\mathbf{r},\mathbf{r}';\omega;B)\psi_i(\mathbf{r}') = \lambda_i \psi_i(\mathbf{r}), \qquad (7.130)$$

where P is an operator expression that will be determined below. To proceed, we introduce (7.129) into the integral and expand the exponential function and the wave function up to second order in a Taylor series about **r**. These terms are then integrated. The expansion gives

$$\int d^3\mathbf{r}' \left\{ P(\mathbf{r}-\mathbf{r}';\omega;0) - \frac{(\mathbf{r}-\mathbf{r}')^2}{2} P(\mathbf{r}-\mathbf{r}';\omega;0)\left(-i\nabla+\frac{2eA}{\hbar}\right)^2 \right\} \psi_i(\mathbf{r}) = \lambda_i \psi_i(\mathbf{r}). \quad (7.131)$$

The wave function can he brought outside the integration (but carefully, as it is still subject to the operators). Then, the first term in the integral is recognized as the Fourier transform if we let $\mathbf{q} \to 0$. Similarly, if we use this Fourier transform connection, the second term becomes the second derivative with respect to the Fourier variable, and we may rewrite the integral in transformed variables as

$$\left[P(\mathbf{q}=0;\omega;0) + \frac{1}{2}\frac{\partial^2 P(\mathbf{q};\omega;0)}{\partial q^2}\bigg|_{\mathbf{q}=0} \left(-i\nabla+\frac{2eA}{\hbar}\right)^2 \right] \psi_i(\mathbf{r}) = \lambda_i \psi_i(\mathbf{r}). \quad (7.132)$$

The terms in the square brackets should be recognized as the first two terms in a Taylor series for the polarization, but expanded around zero momentum. Then, the second term tells us that we connect the momentum with $(-i\nabla + 2eA/\hbar)$, and it is clear that the eigenfunctions of the operator in momentum space are identical with wave functions for particles of charge $2e$ in a magnetic field (which is another connection with the cooperon). Comparison with the development much earlier in this chapter relates the operator P to the diffusion operator, which is just the polarization above that gives rise to the diffusion pole type of "Green's function." Hence, in the magnetic field, we can replace the momentum by a Peierl's substitution in vector potential for a doubly charged particle, but this leads to the replacement

$$q^2 \to q_n^2 = \frac{4eB}{\hbar}\left(n+\frac{1}{2}\right), \quad (7.133)$$

where we have used the cyclotron energy relation as $\hbar^2 q_n^2/2m = \hbar\omega_c(n+1/2)$ and doubled the charge in the cyclotron frequency. Now, using Eq. (7.123), the eigenvalue can be written as

$$\lambda_i = 2\pi\hbar\rho(E_F)\tau\left[1 - \frac{\tau}{\tau_\varphi} - D\frac{4eB}{\hbar}\left(n+\frac{1}{2}\right) - \cdots\right]. \quad (7.134)$$

One could now proceed to compute the actual value of the polarization in real space and Fourier-transform it, but the important point is that the only change in the summation for the kernel is in the quantization of the momentum. Thus, Eq. (7.124) becomes

$$\Lambda(\mathbf{q}_n;\omega) = \frac{1}{2\pi\hbar\rho(E_F)\tau}\frac{1}{Dq_n^2\tau + \tau/\tau_\varphi} \quad (7.135)$$

and the conductivity may be found from the first line of (7.126) as

$$\delta\sigma = -\frac{e^2}{\pi\hbar}(D\tau)\frac{eB}{\pi\hbar}\sum_{n=0}^{\hbar/4eDB\tau}\frac{1}{D\tau(4eB/\hbar)(n+1/2)+\tau/\tau_\varphi}$$
$$= -\frac{e^2}{4\pi^2\hbar}\left[\Psi\left(\frac{1}{2}+\frac{\hbar}{4eDB\tau}\right)-\Psi\left(\frac{1}{2}+\frac{\hbar}{4eDB\tau_\varphi}\right)\right] \quad (7.136)$$

where a different normalization has been used to take care of the degeneracy of the quantized state that appears in the polarization, and $\Psi(x)$ is the digamma function. Quite often the first digamma function is replaced by its large argument limit, which is the natural logarithm of the second term in the argument.

The first line of (7.136) carries an important message. The denominator in the summation is basically a sum over $(\lambda'_i - 1)$, where λ'_i is λ_i reduced by the prefactor on the right side of (7.134). That is, the conductivity is given basically by a summation over the reciprocal of the eigenvalues of the diffusion equation. This same behavior will arise in a later section for the universal conductance fluctuations. Clearly, the reduction of the conductivity correction that causes weak localization is because of an increase in the eigenvalues with magnetic field. This increase is explicit in Eq. (7.134) through the third term in the large brackets. Understanding this behavior, in which the weak-localization amplitude depends upon the eigenvalues of the diffusion equation describing the electron–electron correlation, is crucial and will open the door to some interesting effects in the next section.

7.5.1.3 Periodic eigenvalues for the magnetic effects

It is important to note that each term in the series of (7.136) decays as the magnetic field is increased. This is the rationale for why the weak-localization correction decays with the magnetic field. Recall, however, that the terms that arise come from the eigenvalues of the basic diffusion equation for the polarization operator. If there is some part of the system that makes these eigenvalues periodic, then some unusual effects can occur. One such unusual behavior is found in periodic superlattices. Among the earliest to study the magnetic field and periodic potential in two dimensions was Harper [79]. Azbel [80] subsequently studied the problem and found that the magneto-transport could show oscillations periodic in the magnetic field, and also periodic in the reciprocal of the magnetic field [81], depending on the relative strengths of the periodic potential and the Landau quantization energy $\hbar\omega_c$. The most extensive studies, at least up to a few years ago, were these of Rauh et al. [82] and Hofstadter [83]. One major problem is the need to have the various lengths in the problem (the magnetic length, the periodicity of the lattice, etc.) rationally related. Through the use of Harper's equation, Hofstadter developed the discrete energy levels that arise when the ratio of the lengths are rationally related;

$$\phi(m+1) + \phi(m-1) + 2\cos[2\pi m\alpha - \nu]\phi(m) = \varepsilon\varphi(m) \quad (7.137)$$

7.5 Green's functions in disordered materials

where

$$\alpha = \frac{ea^2 B}{\hbar} \tag{7.138}$$

is the normalized flux coupled through each unit cell of the periodic potential of basis vector a, and $\psi(x,y) = \phi(ma)e^{iky}$, with $v = k_y a$. Here, ϕ is the portion of the wave function. It is clear from Eq. (7.137) that the energy eigenvalues are periodic in the magnetic field, a result that is induced by the periodic potential. While Hofstadter's result is reasonable for small magnetic fields, it is only valid for a single Landau level in high magnetic fields. Geisel has given a more complete derivation for the high-magnetic-field case that accounts for coupling between the Landau levels [84], but here we are primarily interested in the low-magnetic-field regime, where the motion has not been quantized into Landau levels. In this regime, the periodicity in magnetic fields of the energy "bands" formed from the above equation is important. In the normal case of semiconductor materials, the small lattice constants force this periodicity to occur for megagauss magnetic fields, a clearly untenable and unreachable level of field. On the other hand, in artificial superlattices, for which the lattice constant can be relatively large, the magnetic fields are readily attainable.

Of interest in this discussion is the possibility of really connecting the weak-localization behavior with the eigenvalues of the diffusive nature of the polarization function. For this, the first requirement is sufficient impurity scattering so that the transport is diffusive. A structured gate electrode written by electron beam lithography is shown in Fig. 7.22. Here, the gate forms a two-dimensional periodic potential that is induced upon the electron gas under it. The potential is created by the depletion under the actual gate lines, leaving higher carrier density in the open regions. Just before pinch-off of the entire conducting region, this potential is induced in an effective two-dimensional electron gas, even in a normal MESFET structure. The structures in which the best results are seen have epitaxial layers only 50 nm thick grown on semi-insulating GaAs. The grown layers are also GaAs, but doped to 1.5×10^{18} cm^{-3}. Near pinch-off, most of the dopants are ionized and lead to significant impurity scattering within the electron gas. The resultant conductance curves (actually current through the device) are shown in Fig. 7.23 for a variety of source–drain biases [85]. Each of the curves shows a series of dips in conductance, with the dips essentially equally spaced in magnetic field. (A 2-mV source–drain potential corresponds to essentially 11.4 µV across each cell of the superlattice, a relatively small potential drop that is orders of magnitude smaller than the estimate of the minibands resulting from the superlattice.) The onset of the magnetic-field-induced dips, interpreted as replicas of the weak localization, generally does not occur at zero magnetic field, except for the lowest values of the drain bias (i.e., at and below 2 mV). In Fig. 7.24, the locations of these dips in conductance are plotted versus the magnetic field. (Each dip is assigned an arbitrary index, and the magnetic field at

Fig. 7.22 A metallized gate that produces a periodic potential on a two-dimensional electron gas by depletion under the gate fingers. Here the lines are 40 nm wide with a periodicity of 165 nm.

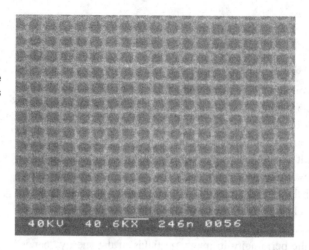

Fig. 7.23 Measured current through a device in which a superlattice potential has been applied. The parameter is the drain bias. (After Ma *et al.*, *Surf. Sci.* **229**, 341 (1991).)

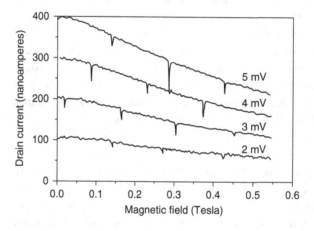

which the dip is seen is plotted as a function of this index number.) The various data lie on a single curve. The required shift of the data for each bias is defined as an index shift. It may be seen from this latter figure that all conductance dips essentially line up with a single periodicity in magnetic field. The slope of the curve, if we are to believe the periodicity in Eq. (7.137), should correspond to the coupling of integer numbers of flux quanta per unit cell. The slope of this curve indicates a periodicity of the superlattice of about 170 nm, which is to be compared with the estimate from the electron micrographs of 165 nm.

Using the theory for weak localization around zero magnetic field, the inelastic mean free path can be estimated for each of the dips that occur in Fig. 7.23. The value obtained in this manner is relatively independent of the magnetic

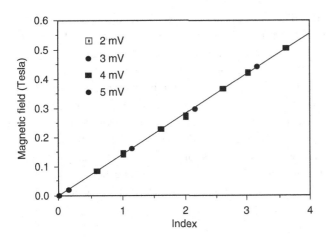

Fig. 7.24 The magnetic field at which dips in conductance occur can be plotted against an arbitrary index, which relates to the flux coupled in each unit cell.

field and bias applied and has a value near 0.55 ± 0.1 μm. The conductance of the sample, at zero magnetic field, rises from 40 μS at 2 mV bias to 80 μS at 5 mV bias. The conductance expected from a fully conducting channel with the Landauer formula is about 77 μS so it is clear that the conductance is not a free electron channel but rather a diffusive (hopping) transport through the array of quantum dots. The data given in the figures were measured at 5.7 K, and the effects persist up to about 15 K. It is also interesting that the increase of the drain potential from 2 mV to 5 mV has shifted the spectrum almost exactly one unit cell, bringing the spectra back into commensurability (with a weak localization dip at zero magnetic field).

The shifts in the dips induced by the source–drain potential can be understood within the context of Harper's equation. With a source–drain potential applied to the device, the energy levels are not only shifted along the channel, in reference to their values at the source end, but also distorted slightly within each unit cell. It is this latter effect that can give rise to the measured shifts, but this is within the context of the derivations provided by Hofstadter. Here, both the electric field and the magnetic field must be treated in the vector potential, so that

$$A_y = Bx + e \int_0^\infty e^{-t/\tau} E \, dt \qquad (7.139)$$

where the elastic scattering time has been introduced to limit the time range of the integral. The addition of the second term in the vector potential modifies (7.137) to the form

$$\varphi(m+1) + \varphi(m-1) + 2\cos[2\pi m\alpha - \omega_B \tau - \nu]\varphi(m) = \varepsilon\varphi(m) \qquad (7.140)$$

where

$$\omega_B = \frac{eEa}{\hbar} \qquad (7.141)$$

is the Bloch frequency. The presence of the electric field, and the resulting change in the drift velocity shifts the v-momentum that appears in v. This shifts the zero of the cosine function and hence the value of the magnetic field that corresponds to "zero." However, we would not expect any shift until the drift momentum becomes comparable with the Fermi momentum, and for these samples this means biases above 2 mV. In fact, the total shift of the spectrum by one index unit in moving from 2 mV bias to 5 mV bias leads to an estimate of the mobility within 25% of the measured value of 2×10^4 cm^2/Vs. These measurements, and the above discussion, strongly suggest an interpretation of the dips as replicas of the zero-magnetic-field weak localization, an interpretation in keeping with the dependence of weak localization upon the eigenvalues, as found in the basic theory above.

7.5.2 Conductance fluctuations

In this section, we want to extend the Green's function approach to the calculation of the correlation function for the universal conductance fluctuations. This will give us both the amplitude of these fluctuations and the manner in which they decay with magnetic field. As above, we will build this up in parts, by discussing the basic approach, which is similar to that above, and then look at the energy variation of the correlation function. We then look at the inclusion of the magnetic field. Most of this work is built around the electron–hole propagator, which has become known as the diffuson, and the cooperon or electron–electron propagator. Also, as above, the approach that we follow is to work with the impurity-averaged diagrams, which is a method of constructing ensemble averages for the Green's functions, and the approach to be adopted here closely follows the seminal paper of Lee, Stone, and Fukuyama [68]. The crucial assumption is that the ergodic hypothesis adopted here implies that rms$(g) = [\text{var}(g)]^{1/2}$ is a good measure of the typical amplitude of the fluctuations that are found in the conductance.

In this approach, we want to work with real-space Green's functions rather than momentum-space Green's functions. That is, we want to work with $G(\mathbf{r},\mathbf{r}',\omega)$, as originally described above. The diagram for the electron–hole propagator is shown in Fig. 7.25. This should be compared with Fig. 5.13a for the momentum-space representation and with Fig. 7.21. The diagram in Fig. 7.25 represents the creation of an electron-hole pair (excitation of an electron above the Fermi energy leaving a hole below the Fermi energy) at $\mathbf{r}$ and the subsequent propagation of this pair to $\mathbf{r}'$, where it recombines. The dashed lines are the impurity interactions, and the "wiggly" lines are the position connections for the diagram (which couple in the k^2 term in the momentum representation). Impurity averaging connects the impurity lines, just as discussed in the treatment of conductance above. For the full Green's functions, the impurity lines connect to other impurities on the same propagator, while for

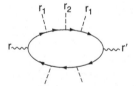

Fig. 7.25 The real-space electron–hole propagator with unconnected impurities.

higher-order terms (beyond the Drude approximation for the diffuson and the cooperon) the connections span across the two Green's functions. Here, however, we are interested in the quantity $\langle \delta g(E,B) \delta g(E+\Delta E, B+\Delta B)\rangle$, which is a conductance–conductance correlation function, so that we want to consider diagrams in which there are two loops, each of which is like that of Fig. 7.25. That is, we shall make a "double bubble" diagram, and our interest lies in those interaction terms in which the impurity lines span *between these two bubbles*, thus building in correlations between the two δg terms in the correlation function. This basic premise lies in the Kubo formula (3.149) introduced in Chapter 3 (which is in momentum space), in which the two-particle Green's function is represented by the electron–hole propagator discussed here (the lowest-order term in the electron–hole propagator). It is assumed that each of the Green's functions in this figure is already the total Green's function including the self-energy corrections due to impurities that average together for these terms. Thus, only those new interactions that span the two pairs of propagators are of interest in this treatment and give rise to the fluctuations that are of interest.

One reason for this approach is that, if we consider only the terms that lie on one of the two bubbles and do not couple the two bubbles, then these terms are already included in the proper calculation of $g(E,B)$ and cancel when we compute the correlation of the *fluctuations* in the conductivity. Thus, the correlation function describes small changes in the conductivity at one energy due to interactions with a second conductivity bubble. This means that such an interaction must arise from terms that span the two bubbles in our double-bubble diagram.

7.5.2.1 The correlation function in energy
Let us now turn to the evaluation of the correlation function for the fluctuations in the conductance. For simplicity in this section, the magnetic field effects will be ignored, and we will worry only about the variation with the energy. This variation is introduced by varying, for example, the Fermi energy of the two-dimensional electron gas. The diagrams are two nested polarization loops, as discussed above; each pair of loops represents one of the two conductivity bubbles given in the product of the fluctuations in $\langle \delta g(E) \delta g(E+\Delta E)\rangle$. One loop is at the energy E, and the second one is at the energy $E+\Delta E$. Since the quantity of interest is the correlation of the *fluctuation* in the conductance, the first term in the coupling of the two loops, which corresponds to no coupling between the loops (the isolated loops), cancels with the term in $\langle g(E)\rangle\langle g(E+\Delta E)\rangle$; that is, $\langle \delta g(E)\delta g(E+\Delta E)\rangle = \langle g(E)g(E+\delta E)\rangle - \langle g(E)\rangle\langle g(E+\Delta E)\rangle$. When there are no impurity lines coupling the two loops, there is no correlation between them, and the contribution cancels the last term. Thus, after impurity averaging, the only diagrams of interest in computing the correlation in the fluctuations are those in which the impurity lines span the two loops, as discussed above. The most important diagrams are those in which the impurity lines do not cross,

such as those diagrams that would normally contribute to ladder diagrams. There are a variety of ways in which these two nested loops can be arranged. For example, one loop consists of the position vectors **r** and **r**′. The second loop consists of position vectors $\mathbf{r}_1$ and $\mathbf{r}_1'$. In the most logical case, **r** and $\mathbf{r}_1$ are coincident while **r**′ and $\mathbf{r}_1'$ are coincident (the electron–hole pairs propagate close to one another). In another case, **r** and **r**′ are coincident, while $\mathbf{r}_1$ and $\mathbf{r}_1'$ are at different locations (the electron–hole pairs diverge from one another in their propagation). A third possibility is that all four positions are unrelated. A fourth possibility is that each loop comes back upon itself so that **r** and **r**′ are coincident. Still there is another possibility in which the latter case is complicated by impurity lines spanning the bubble. The latter two, however, are unimportant because they tend to be canceled by higher-order terms. Only the first three terms in which two, three, and four diffusons are involved need to be considered, since they tend to produce contributions of the same order. The first three terms are shown in Fig. 7.26(a)–(c), respectively.

The above diagrams are in fact generated by the adoption of the two current loops (the two polarization bubbles), with the two irreducible vertices (the factors Λ, which represent the scattering plus polarization) inserted in all possible manners. Another set of diagrams arises from the maximally crossed impurity lines, in which the rotation of the "hole" propagator is reversed, giving a set of electron–electron (cooperon) bubbles. In the presence of normal impurity scattering (no spin-dependent scattering), these latter diagrams give exactly the same contribution to the correlation function. Thus, a factor of two will be added to the final result. The feature that leads to the recognition that these diagrams are the important contribution is tied to the basic singularity at small **q** and the frequency of the diffusion pole itself, as discussed in the previous section. When the arrows on the Green's functions are oppositely directed (this assumes a particular orientation of the two bubbles), as is normal in the particle–hole two-particle Green's function, the diagrams represent density fluctuations and have the characteristic diffusion pole. The set of diagrams shown in Fig. 7.26 are the only ones in which the diffusion poles all occur with the same value of momentum transfer.

Each of the bubble-pair combinations has two end groups corresponding to the connections to the currents, and a central group corresponding to the

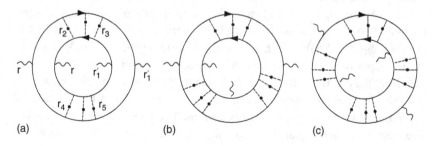

Fig. 7.26 The three diagrams that correspond to the first three contributions to the conductance correlation function.

7.5 Green's functions in disordered materials

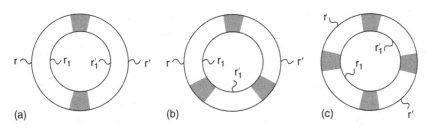

Fig. 7.27 A rearrangement of Fig. 7.26(a,b) showing the positions of the polarizations.

impurity-polarization combination, just as in the previous sections. (For example, the central impurity-polarization part led to Λ, and the end groups added an additional pair of Green's functions to the overall two-particle Green's functions.) Consider Fig. 7.26(a). Each bubble has both an electron and a hole propagator (and the equivalent cooperon propagator), leading to four possible combinations of Green's functions in which the upper part of the pair of rings is represented by $G^r G^r$, $G^a G^r$, $G^r G^a$, and $G^a G^a$ respectively. Coupled to this are two current connections, each of which has a pair of Green's functions. Now consider one of these, as shown in Fig. 7.27. In (a), the pair of rings is shown with a set of coordinates, while (b) indicates the current connections. In essence, the current connections span the coordinates of the conductivity two-particle Green's function and couple to the diffusion coordinates, and the overall conductance fluctuation contribution can be written in the form of the conductance found earlier, which we now write

$$F_a = \frac{e^4 \hbar^4}{m^4} \int d^d \mathbf{r}_2 \int d^d \mathbf{r}_3 \int d^d \mathbf{r}_4 \int d^d \mathbf{r}_5 \Lambda(\mathbf{r}_2, \mathbf{r}_3) \Lambda(\mathbf{r}_4, \mathbf{r}_5) j(\mathbf{r}_2, \mathbf{r}_4) j(\mathbf{r}_3, \mathbf{r}_5). \quad (7.142)$$

The terms Λ are the diffuson (or cooperon) contributions and are given by (7.112), and the terms j are the current connections. We have previously found the cooperon, or polarization, terms including the inelastic decay rate, so that

$$\Lambda(\mathbf{q}; \omega) = \frac{1}{2\pi \hbar \rho(E_F) \tau} \frac{1}{Dq_n^2 \tau + \frac{\tau}{\tau_\varphi}} \quad (7.143)$$

which is in Fourier transform mode. The current kernel is given equivalently by (7.112) modified for the present purpose. That is, we need to pull Green's function lines together to make a current vertex as done in the calculation of the conductivity above. This may be accomplished by letting the current operator be

$$j(\mathbf{r}_3, \mathbf{r}_5) = j_0 \delta(\mathbf{r}_3 - \mathbf{r}_5) \quad (7.144)$$

and

$$j_0 = \int \frac{d^d \mathbf{k}}{(2\pi)^d} k_d^2 [G^r(\mathbf{k}) G^a(\mathbf{k})]^2 = 2\rho(E_F) \tau^3 \frac{k_F^2}{d}. \quad (7.145)$$

This is well and good for infinite samples. However, when we deal with finite samples, the polarization functions must he modified to account for the boundary conditions. We work with the reduced polarization functions, represented by the terms in the last fraction in (7.143). When there is no current flow through a lateral boundary, then one has the normal derivative of Π equal to zero. On the other hand, when there are good conducting metallic contacts at the boundary, the flow is ballistic (nondiffusive) and the boundary condition must be to set Π equal to zero. In general, the polarization can be found from the differential equation in (7.132), but in real space as

$$\Pi(\mathbf{r},\mathbf{r}',\Delta E) = \sum_m \frac{Q_m^*(\mathbf{r})Q_m(\mathbf{r}')}{\lambda_m} \tag{7.146}$$

where $Q_m(\mathbf{r})$ and λ_m are the eigenfunctions and eigenvalues of

$$\tau\left[-D\nabla^2 - i\frac{\Delta E}{\hbar} + \frac{1}{\tau_\varphi}\right]Q_m(\mathbf{r}) = \lambda_m Q_m(\mathbf{r}) \tag{7.147}$$

subject to the boundary conditions discussed above, and where an effective diffusion constant has been introduced in place of the bare $\hbar/2m$. For a sample of cross section $L_x \times L_y$, and length (to the metallic contacts) L_z, in three dimensions,

$$Q_m(\mathbf{r}) = \sqrt{\frac{8}{L_x L_y L_z}} \sin\left(\frac{m_z \pi z}{L_z}\right)\cos\left(\frac{m_y \pi y}{L_y}\right)\cos\left(\frac{m_x \pi x}{L_x}\right) \tag{7.148}$$

and

$$\lambda_m = \tau D \pi^2 \left[\left(\frac{m_z}{L_z}\right)^2 + \left(\frac{m_y}{L_y}\right)^2 + \left(\frac{m_x}{L_x}\right)^2\right] - i\frac{\tau\Delta E}{\hbar} + \frac{\tau}{\tau_\varphi}. \tag{7.149}$$

Thus, we can now write, after integration in the transverse directions,

$$\begin{aligned}F_a &= \frac{\hbar^4 e^4}{m^4}\left[2\rho(E_F)\tau^3\frac{k_F^2}{d}\right]^2\left[\frac{1}{2\pi\hbar\rho(E_F)\tau}\frac{4}{L_z}\right]^2\sum_m\frac{1}{\lambda_m^2}\\ &= \left(\frac{e^2}{\hbar\pi}\right)^2\left(\frac{4}{\pi}\right)^2\sum_m\frac{1}{\tilde{\lambda}_m^2}\end{aligned} \tag{7.150}$$

where $D = v_F^2\tau/d$. and where

$$\tilde{\lambda}_m = \lambda_m \frac{1}{D\tau}\left(\frac{L_z}{\pi}\right)^2 \tag{7.151}$$

produces a dimensionless eigenvalue. The other three possible arrangements of the advanced and retarded Green's functions for this simple diagram lead to changes in the sign of the wavevector (for the cooperon terms), and the three taken together lead to a factor of two multiplier on Eq. (7.150).

7.5 Green's functions in disordered materials

The terms that arise from Fig. 7.26(b),(c) can be computed in the same manner, but they add extra summations over the eigenvalues since they have more diffusons (or cooperons) in the overall diagram. We will not go through the full derivation, as it follows clearly the outline above. The end result is that

$$F(\Delta E) = \left(\frac{e^2}{\hbar\pi}\right)^2 \left(\frac{4}{\pi}\right)^2 \sum_{m_x,m_y=0} \sum_{m_z=1,3,5...} \left\{ 2\,\text{Re}\left[\frac{1}{\tilde{\lambda}_m}\right]^2 \right.$$

$$- 8\text{Re} \sum_{n_z=2,4,6...} \frac{f_{mn}^2}{\tilde{\lambda}_m \tilde{\lambda}_n} \left[\frac{1}{\tilde{\lambda}_m} + \frac{1}{\tilde{\lambda}_n}\right] \quad (7.152)$$

$$\left. + 24\text{Re} \sum_{p_z=1,3,5...} \sum_{n_z q_z=2,4...} \frac{f_{mn}f_{np}f_{pq}f_{qm}}{\tilde{\lambda}_m \tilde{\lambda}_n \tilde{\lambda}_p \tilde{\lambda}_q} \right\}$$

where

$$f_{mn} = \frac{4m_z n_z}{\pi(m_z^2 - n_z^2)}. \quad (7.153)$$

Clearly, the amplitude of the universal conductance fluctuation is given by the square root of F and has a value of the order of $e^2/\pi\hbar$, with a numerical factor on the order of unity depending on the number of modes excited in the system and the extent to which the lateral dimensions are meaningful. These numerical factors have been computed to be [68] 0.729 in 1D, 0.862 in 2D, and 1.088 in 3D, when the current flows in the z-direction.

Just as in the previous section for weak localization, the key factor in the dependence of the correlation function on the energy difference is the reduced eigenvalue of the diffusion propagator, which is given by (7.149) and (7.151) as

$$\lambda_m = \left[m_z^2 + m_y^2\left(\frac{L_z}{L_y}\right)^2 + m_x^2\left(\frac{L_z}{L_x}\right)^2\right] - i\frac{\Delta E L_z^2}{\pi^2 \hbar D} + \frac{L_z^2}{\pi^2 \tau_\varphi D}. \quad (7.154)$$

For a two-dimensional system, in which the current path is short compared to the lateral dimensions but large compared to the coherence length, the second and third terms in the square brackets are negligible, and the eigenvalue is dominated by the last two terms. Then, the critical value of the energy change required to reduce the correlation function to one-half its peak value is just

$$\Delta E_{C,2D} = \hbar/\tau_\varphi, L_x, L_y \gg L_z \gg L_\varphi. \quad (7.155)$$

On the other hand, for a truly short current path, in which the length is much smaller than the phase coherence length $(D\tau_\varphi)^{1/2}$, the last term is also negligible, and the critical correlation energy is when the imaginary term is equal to m_z^2, which we take to be unity for the lowest eigenstate, and

$$\Delta E_{C,2D} = \frac{\pi^2 \hbar D}{L_z^2} L_z \ll L_\varphi, L_x, L_y. \quad (7.156)$$

If we are dealing with a quantum wire, in which $L_z \gg L_x, L_y$, then the transverse modes dominate the eigenvalue. The correlation energy is then defined by the condition

$$\frac{\Delta E_{C,1D} L_z^2}{\pi^2 \hbar D} = \left[m_y^2 \left(\frac{L_z}{L_y} \right)^2 + m_x^2 \left(\frac{L_z}{L_x} \right)^2 \right] + \frac{L_z^2}{\pi^2 \tau_\varphi D} \tag{7.157}$$

or

$$\Delta E_{C,1D} = \frac{\pi^2 \hbar D}{L_x L_y} \left[\frac{m_y^2 L_x^2 + m_x^2 L_y^2}{L_x L_y} + \frac{L_x L_y}{\pi^2 L_\varphi^2} \right] \sim \frac{\pi^2 \hbar D}{L_x L_y}, L_x, L_y \ll L_\varphi^2. \tag{7.158}$$

7.1.5.2 Correlation function in a magnetic field

When we now turn to the magnetic field variation, the same diagrams will come into play. Now, however, the Green's functions that contribute to these diagrams are modified by the magnetic field (actually, by the vector potential) according to Eq. (7.128). This leads to the general changes as discussed in the previous section on weak localization. There are some differences, however. There, we were discussing the cooperon, or particle–particle channel, for which the actual value of the magnetic field appears, since the contribution of interest is of order $G(\mathbf{r}, \mathbf{r}')^2$. Thus, for the particle–particle terms, the important equation of motion (7.147) is changed by

$$-i\nabla \rightarrow -i\nabla - e(2A + \nabla A). \tag{7.159}$$

The eigenvalues for the particle–particle channel will decay away very quickly with magnetic field, as does the weak localization (unless there is a periodicity in the eigenvalues, as discussed in the previous section). For all practical purposes, the particle–particle channel can be ignored except at the lowest magnetic fields. Thus, the behavior of the correlation function near $B = 0$ will be somewhat different than at higher magnetic fields, where the particle–particle channel has been damped out. Nevertheless, where the magnetic field clearly is not large, the particle–particle channel must still be evaluated but its contribution is no longer equal to that of the particle–hole channel.

In distinction to the particle–particle channel, the particle–hole channel relies upon the Green's function product $G(\mathbf{r}, \mathbf{r}')G(\mathbf{r}', \mathbf{r})$, so that it is dependent only on the *difference* in the vector potential over the two paths. As discussed earlier in this chapter, the particle–particle channel is similar to the Aharonov–Bohm effect in that each propagator corresponds to one branch and the two add their effects together to measure the coupled flux. With the particle–hole channel, however, one propagator *cancels* the phase of the other, so that only the *difference* in phase between the two paths is important. Thus, in this case the equation of motion (7.147) is modified to depend only on the difference in the magnetic field between the two Green's functions, or

7.5 Green's functions in disordered materials

$$-i\nabla \to -i\nabla - e\nabla A. \tag{7.160}$$

Here, only the difference in the magnetic field appears in the eigenvalues for the diffuson. The equation of motion must still be solved, and this is considerably more difficult now since the magnetic field, taken to be in the x-direction, couples the y- and z-motions. This leads to a complicated coupled-mode solution for the wave function.

In the high-magnetic-field limit, and for a mainly two-dimensional system, where the Landau levels are fully formed, we have $BL_zL_y \gg \Phi_0 = h/e$. In this case, the eigenvalues for the particle–particle channel are just those of the Landau levels formed in the two-dimensional system, because of the role played by the absolute value of the magnetic field. This leads to [68]

$$\tilde{\lambda}_m^{pp}(B, \nabla B) = \frac{4}{\pi}\left(n + \frac{1}{2}\right)\frac{(2B + \Delta B)L_z^2}{\varphi_0}. \tag{7.161}$$

Hence, the lowest value of this eigenvalue is still much greater than unity, regardless of the value of ΔB. Certainly, this will not be the case for the diffuson contributions, which depend on only this increment in the field, and the particle–particle channel contribution will be negligible. Thus, the amplitudes of the fluctuations are reduced by a factor of two over the values quoted above.

The eigenvalues for the particle–hole channel can be computed in two limits. The first is where the difference magnetic field can be treated as a small perturbation on the normal eigenvalues (7.154). (We generally take $\Delta E = 0$ in this discussion.) If we take the vector potential as $\mathbf{A} = \Delta By\mathbf{a}_z$, where $\mathbf{a}_z$ is a unit vector pointing in the z-direction, the perturbation has two terms:

$$V' = -2ie\Delta By\frac{\partial}{\partial z} + (e\Delta By)^2. \tag{7.162}$$

For consistency, one needs to calculate the effect of the first term to second order and the effect of the second term to first order (thus giving a correction to order B^2 in both cases). This is much easier for the case of a quantum wire, in which $L_z \gg L_x, L_y$. In the case in which only the lowest lateral mode is chosen, the first term has no diagonal corrections due to the symmetry of the system, and the second term leads to the correction

$$\tilde{\lambda}_m(\Delta B) \simeq m_z^2 + \frac{1}{3}\left(\frac{\Delta B_y L_y L_z}{\varphi_0}\right)^2 + \frac{L_z^2}{\pi^2\tau_\varphi D}. \tag{7.163}$$

If the last term is negligible, this leads to a correlation magnetic field (for $m_z = 1$)

$$\Delta B_C \sim \sqrt{3}\frac{\varphi_0}{L_y L_z}, \quad L_z \ll l_\varphi. \tag{7.164}$$

It must be recalled that the magnetic field is in the x-direction, so that the area in this equation is the sample area normal to the magnetic field. In the other limit, in which the length is much larger than the coherence length, the correlation magnetic field is given by

$$\Delta B_C \sim \frac{\sqrt{3}}{\pi} \frac{\varphi_0}{L_y L_z}, \quad L_z \gg l_\varphi. \tag{7.165}$$

A full perturbation treatment, with evaluation of all the diagrams (and the variations in f_{mn}) suggests that the factor of $\sqrt{3} \to 1.2$ for the case of (7.164) [68]. As the sample dimensions change from a long quantum wire to a square two-dimensional region, the perturbation contribution from the first term of Eq. (7.162) no longer vanishes but instead begins to mix states with different m_y. This contribution is negative and therefore reduces the coefficient of the ΔB^2 term in the eigenvalue. Thus, higher dimensionality will reduce the coefficient and increase the value of the correlation magnetic field. This is expected to have less of a shape dependence, and the product $L_y l_\varphi \to l_\varphi^2$ for $L_z, L_y \gg L_\varphi$.

References

[1] P. W. Anderson, *Phys. Rev.* **109**, 1492 (1958).
[2] E. Abrahams, P. W. Anderson, D. C. Licciardello, and T. V. Ramakrishnan, *Phys. Rev. Lett.* **42**, 673 (1979).
[3] See, e.g., S. Chakravarty and A. Schmid, *Phys. Rept.* **140**, 193 (1986), and references contained therein.
[4] R. L. Mozzi and B. E. Warren, *J. Appl. Crystalography* **2**, 164 (1969).
[5] N. F. Mott, *Adv. Phys.* **16**, 49 (1967).
[6] A. Miller and S. Abrahams, *Phys. Rev.* **120**, 745 (1960).
[7] N. F. Mott and E. A. Davis, *Electronic Processes in Non-Crystalline Materials* (Oxford, Clarendon Press, 1979).
[8] N. F. Mott, *J. Noncrystal. Sol.* **1**, 1 (1968).
[9] N. F. Mott, *Philos. Mag.* **19**, 835 (1969).
[10] T. Ando, A. B. Fowler, and F. Stern, *Rev. Mod. Phys.* **54**, 437 (1982), and references therein.
[11] J. M. Ziman, *J. Phys. C* **2**, 1230 (1960).
[12] E. N. Economou and M. H. Cohen, *Mater. Res. Bull.* **5**, 577 (1970).
[13] R. Kubo, *J. Phys. Soc. Jpn.* **12**, 570 (1957).
[14] S. E. Edwards, *Proc. Roy. Soc.* London **A267**, 578 (1958).
[15] N. F. Mott, *Philos. Mag.* **22**, 7 (1970).
[16] J. T. Edwards and D. J. Thouless, *J. Phys. C* **5**, 807 (1972).
[17] D. C. Licciardello and D. J. Thouless, *J. Phys. C* **8**, 4157 (1975).
[18] D. J. Thouless, *Phys. Rept.* **13C**, 93 (1974).
[19] D. J. Thouless, *Phys. Rev. Lett.* **39**, 1167 (1977).
[20] E. Abrahams, P. W. Anderson, D. C. Licciardello, and T. V. Ramakrishnan, *Phys. Rev. Lett.* **42**, 673 (1979).
[21] G. J. Dolan and D. D. Osheroff, *Phys. Rev. Lett.* **43**, 721 (1979).
[22] D. J. Bishop, D. C. Tsui, and R. C. Dynes, *Phys. Rev. Lett.* **44**, 1153 (1980).

[23] M. J. Uren, R. A. Davies, and M. Pepper, *J. Phys. C* **13**, L985 (1980).
[24] M. D'Iorio, V. M. Pudalov, and S. G. Semenchinsky, *Phys. Rev. B* **46**, 15992 (1992).
[25] S. V. Kravchenko, W. Mason, J. E. Furneaux, and V. M. Pudalov, *Phys. Rev. Lett.* **75**, 910 (1995).
[26] V. M. Pudalov, G. Brunthaler, A. Prinz, and G. Bauer, *Physica E* **3**, 79 (1998).
[27] S. V. Kravchenko, W. E. Mason, G. E. Bowker, et al., *Phys. Rev. B* **51**, 7038 (1995).
[28] Y. Hanein, U. Meirav, D. Shahar, et al., *Phys. Rev. Lett.* **80**, 1288 (1998).
[29] D. Popović, A. B. Fowler, and Washburn, *Phys. Rev. Lett.* **79**, 1543 (1997).
[30] M. Y. Simmons, A. R. Hamilton, M. Pepper, et al., *Phys. Rev. Lett.* **80**, 1292 (1998).
[31] M. D'Iorio, D. Brown, J. Lam, et al., *Superlatt. Microstruc.* **23**, 55 (1998).
[32] A. Gold, *Phys. Rev. B* **44**, 8818 (1991).
[33] H. W. Jiang, C. E. Johnson, K. L. Wang, and S. T. Hannahs, *Phys. Rev. Lett.* **71**, 1439 (1993).
[34] Y. Hanein, D. Shahar, J. Yoon, et al., *Phys. Rev. B* **58**, 13338 (1998).
[35] M. J. Lilly, J. L. Reno, J. A. Simmons, et al., *Phys. Rev. Lett.* **90**, 056806 (2003).
[36] K. Lai, W. Pan, D. C. Tsui, et al., *Phys. Rev. B* **72**, 081313 (2005).
[37] E. Abrahams, *Physica E* **3**, 69 (1998).
[38] E. Abrahams, S. V. Kravchenko, and M. P. Sarachik, *Rev. Mod. Phys.* **73**, 251 (2001).
[39] S. Das Sarma and E. H. Hwang, *Sol. State Commun.* **135**, 579 (2005).
[40] S. Das Sarma, M. P. Lilly, E. H. Hwang, et al., *Phys. Rev. Lett.* **94**, 136401 (2005).
[41] G. Kastrinakis, *Physica B* **387**, 109 (2007).
[42] D. K. Ferry, *Semiconductors* (New York, Macmillan, 1991).
[43] S. Chakravarty and A. Schmid, *Phys. Rept.* **140**, 193 (1986).
[44] C. W. J. Beenakker and H. van Houten, *Phys. Rev. B* **38**, 3232 (1988).
[45] L. D. Landau and E. M. Lifshitz, *The Classical Theory of Fluids* (Reading, MA, Addison-Wesley, 1962), p. 50.
[46] C. P. Umbach, C. van Haesondonck, R. B. Laibowitz, S. Washburn, and R. A. Webb, *Phys. Rev. Lett.* **56**, 386 (1986).
[47] C. de Graaf, J. Caro, and S. Radelaar, *Phys. Rev. B* **46**, 12814 (1992).
[48] J. A. Simmons, D. S. Tsui, and G. Weimann, *Surf. Sci.* **196**, 81 (1988).
[49] E. H. Sondheimer, *Adv. Phys.* **1**, 1 (1052).
[50] A. B. Pippard, *Magnetoresistance in Metals* (Cambridge, Cambridge University Press, 1989).
[51] K. Fuchs, *Proc. Cambridge Phil. Soc.* **34**, 100 (1938).
[52] C. W. J. Beenakker and H. van Houten, in *Solid State Physics*, vol. 44, eds. H. Ehrenreich and D. Turnbull (New York, Academic Press, 1991), pp. 1–228.
[53] T. J. Thornton, M. L. Roukes, A. Scherer, and B. P. van der Gaag, *Phys. Rev. Lett.* **63**, 2128 (1989).
[54] K. Forswell and L. Holwech, *Philos. Mag.* **9**, 435 (1964).
[55] T. J. Thornton, M. L. Roukes, A. Scherer, and B. P. van der Gaag, in *Granular Nanoelectronics*, eds. D. K. Ferry, J. R. Barker, and C. Jacoboni (New York, Plenum Press, 1991) pp. 165–79.
[56] B. L. Altshuler, D. Khmelnitskii, A. I. Larkin, and P. A. Lee, *Phys. Rev. B* **22**, 5142 (1980).
[57] B. L. Altshuler and A. G. Aronov, *Pis'ma Zh. Eksp. Teor. Fiz.* **33**, 515 (1981) [*JETP Lett.* **33**, 499 (1981)].

[58] S. Hikami, A. I. Larkin, and Y. Nagaoka, *Prog. Theor. Phys.* **63**, 707 (1980).
[59] V. K. Dugaev and D. E. Khmelnitskii, *Zh. Eksp. Teor. Fiz.* **86**, 1784 (1984) [*Sov. Phys. JETP* **59**, 1038 (1981)].
[60] A. Benoit, C. P. Umbach, R. B. Laibowitz, and R. A. Webb, *Phys. Rev. Lett.* **58**, 2343 (1987).
[61] W. J. Skocpol, P. M. Mankiewich, R. E. Howard, *et al.*, *Phys. Rev. Lett.* **58**, 2347 (1987).
[62] A. Benoit, S. Washburn, C. P. Umbach, R. B. Laibowitz, and R. A. Webb, *Phys. Rev. Lett.* **57**, 1765 (1986).
[63] V. Chandrasekhar, D. E. Prober, and P. Santhanam, *Phys. Rev. Lett.* **61**, 2253 (1988); V. Chandrasekhar, P. Santhanam, and D. E. Prober, *Phys. Rev. B* **44**, 11203 (1991).
[64] B. L. Altshuler, *Pis'ma Zh. Eksp. Teor. Fiz.* **41**, 530 (1985) [*JETP Lett.* **41**, 11203 (1991)].
[65] A. Lee and A. D. Stone, *Phys. Rev. Lett.* **55**, 1622 (1985).
[66] R. Landauer, *IBM J. Res. Develop.* **1**, 223 (1957).
[67] P. A. Lee, *Physica* **140A**, 169 (1986).
[68] A. Lee, A. D. Stone, and H. Fukuyama, *Phys. Rev. B* **35**, 1039 (1987).
[69] B. L. Altshuler and D. E. Hkmelnitskii, *Pis'ma Exsp. Teor. Fiz.* **42**, 291 (1985) [*JETP Lett.* **42**, 359 (1985)].
[70] W. J. Skocpol, *Physica Scripta* **T19**, 95 (1987).
[71] S. Washburn, in *Mesoscopic Phenomena in Solids*, eds. B. L. Altshuler, P. A. Lee, and R. A. Webb (Amsterdam, North-Holland, 1991), pp. 1–36.
[72] M. Büttiker, *Phys. Rev. Lett.* **57**, 1761 (1986).
[73] H. Haucke, S. Washburn, A. D. Benoit, C. P. Umbach, and R. A. Webb, *Phys. Rev. B* **41**, 12454 (1990).
[74] P. W. Anderson, *Phys. Rev.* **109**, 1492 (1958); *Philos. Mag. B* **52**, 505 (1985).
[75] C. P. Enz, *A Course on Many-Body Theory Applied to Solid-State Physics* (Singapore, World Scientific Press, 1992).
[76] J. S. Langer and T. Neal, *Phys. Rev. Lett.* **16**, 984 (1966).
[77] G. Bergmann, *Phys. Rept.* **107**, 3 (1984).
[78] S. Maekawa and H. Fukuyama, *J. Phys. Soc. Jpn.* **50**, 2516 (1981).
[79] P. G. Harper, *Proc. Phys. Soc.* (London) **A 68**, 874 (1955).
[80] M. Ya. Azbel, *Sov. Phys. JETP* **17**, 665 (1963).
[81] M. Ya. Azbel, *Sov. Phys. JETP* **19**, 634 (1964).
[82] A. Rauh, G. H. Wannier, and G. Obermair, *Phys. Stat. Sol. (b)* **63**, 215 (1974).
[83] D. R. Hofstadter, *Phys. Rev. B* **14**, 2239 (1974).
[84] R. Fleischmann, T. Geisel, R. Ketzmerick, and G. Petschel, *Semicon. Sci. Technol. B* **9**, 1902 (1994).
[85] J. Ma, R. A. Puechner, W. P. Liu, A. M. Kriman, G. N. Maracas, and D. K. Ferry, *Surf. Sci.* **229**, 341 (1991).

8
Temperature decay of fluctuations

When the temperature is raised above absolute zero, the amplitudes of both the weak-localization, universal conductance fluctuations and the Aharonov–Bohm oscillations are reduced below the nominal value e^2/h. In fact, the amplitude of nearly all quantum phase interference phenomena is likewise weakened. There is a variety of reasons for this. One reason, perhaps the simplest to understand, is that the coherence length is reduced, but this can arise as a consequence of either a reduction in the coherence time or a reduction in the diffusion coefficient. In fact, both of these effects occur. In Chapter 2, we discussed the temperature dependence of the mobility in high-mobility modulation-doped GaAs/AlGaAs heterostructures. The decay of the mobility couples to an equivalent decay in the diffusion constant, $D = v_F^2 \tau/d$, where d is the dimensionality of the system, through both a small temperature dependence of the Fermi velocity and a much larger temperature dependence of the elastic scattering rate. The temperature dependence of the phase coherence time is less well understood but generally is thought to be limited by electron–electron scattering, particularly at low temperatures. At higher temperatures, of course, phonon scattering can introduce phase breaking.

Another interaction, though, is treated by the introduction of another characteristic length, the thermal diffusion length. The source for this lies in the thermal spreading of the energy levels or, more precisely, in thermal excitation and motion on the part of the carriers. At high temperatures, of course, the lattice interaction becomes important, and energy exchange with the phonon field will damp the phase coherence. This is introduced through the assumed thermal broadening of a particular energy level (any level of interest, that is). This broadening of the states leads to a broadening of the available range of states into which a nearly elastic collision can occur, and this leads to wave function mixing and phase information loss. Imagine two interfering electron paths from which an interference effect such as the Aharonov–Bohm effect can be observed. We let the enclosed magnetic field (or the path lengths) be such that the net phase difference is δ. At $T = 0$, all of the motion is carried out by carriers precisely at the Fermi energy. When the temperature is non-zero, however, the motion is carried out by carriers lying in an energy width of about $3.5k_B T$ (this is the full-width at half-maximum of the derivative of the Fermi–Dirac distribution), centered on the Fermi energy, where k_B is the Boltzmann constant. Thus, thermal fluctuations will excite carriers into energies near, but not equal to, the Fermi energy. The carriers with energy $E_F + k_BT$ will produce an additional phase shift $\delta\omega t$, where t is approximately the transit time along one of the interfering paths, where we assume that $\delta\omega \sim k_BT/\hbar$. This leads to a total phase difference that is roughly $k_BTt/\hbar$. In *fact,* the actual energies are distributed over a range of energies corresponding to the width mentioned above, so there will be a range of phases in the interfering electrons. The phase difference is always determined modulo 2π, so that when the extra phase factor is near unity (in magnitude), there will be a decorrelation of the phase due to the distribution of the actual phases of individual electrons, which means some phase interference is destroyed. Thus, one can think about the time over which this thermally induced phase destruction occurs as $t \sim \hbar/k_BT$. This time is a sort of thermal phase-breaking time, and so it can be used with the diffusion coefficient to define a thermal length $l_T = (D\hbar/k_BT)^{1/2}$ in analogy to the phase coherence length introduced in the previous chapters. Thus l_T is the length over which dephasing of the electrons occurs due to the thermal excitations in the system. We call this length the *thermal diffusion length*. We note that the above process can be summarized by three steps: (1) the non-zero temperature means that a spread of energies, rather than a single well-defined energy, is involved in transport at the Fermi energy; (2) this spread in energies defines a dephasing time $\hbar/k_BT$, which describes interference among the ensemble of waves corresponding to the spread in energy; and (3) the normal phase coherence time τ_φ, is replaced in the definition of the coherence length by this temperature-induced dephasing time. Thus, the thermal diffusion length describes dephasing introduced specifically by the spread in energy of contributing states at the Fermi surface.

The introduction of the thermal diffusion length means that mesoscopic devices are now characterized by a more complex set of lengths and the presence of an additional temperature-dependent quantity. We can also set a nominal temperature at which we do not need to worry about thermal effects. If $l_T > l_\varphi$, then the dominant dephasing properties are contained in the phase coherence length, and the thermal effects are minimal. On the other hand, if $l_T < l_\varphi$, then the primary dephasing interaction is due to thermal excitations in the system, and this becomes the important length defining the mesoscopic phenomena. Temperature dependence of mesoscopic phenomena are provided by both l_T and l_φ.

In this chapter, we will examine how the temperature variations arise. First, in the next section, we examine the temperature dependence of the phase coherence length, as determined by measurements of weak-localization and universal conductance fluctuations. We also look at the few measurements of the decay with temperature of the phase coherence time itself. We then turn to how the temperature spread of the distribution function can introduce the dephasing. For this we will predominantly use zero-temperature Green's functions to describe the correlation function and carry out some averages over the energy spread to achieve the desired results. Finally, we turn to the decay of the phase coherence time itself through, for example, electron–electron scattering, and this entails a development of the temperature Green's functions.

8.1 Temperature decay of coherence

In this section, we want to begin to look at the decay of weak-localization and universal conductance fluctuations with temperature, both to illustrate the role of the thermal diffusion length l_T and to try to identify the mechanisms responsible for phase breaking in these mesoscopic systems. As mentioned, the basic resistance of the mesoscopic structure changes with temperature. The mobility of the electrons themselves is reduced as the temperature is increased, as discussed in Chapter 2. This carries over to nanostructures as well. In Fig. 8.1, we show the resistance that is measured in high-mobility GaAs/AlGaAs (mobility of 6.4×10^6 cm^2/Vs at 1.5 K) quantum wires as the temperature is varied [1]. Figure 8.1 describes the temperature dependence for wires of various widths and lengths. The resistances are estimated by using a sidewall depletion of about 0.2 μm in

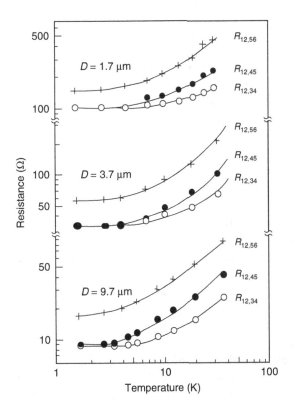

Fig. 8.1 Temperature dependence of high-mobility quantum wires. (After S. Tarucha et al., IOP Conf. Series **127**, 127 (1992), by permission.)

the calculation of total device width. At high temperature, all devices measured were in the diffusive limit, so the longitudinal resistance was proportional to the wire length. As the temperature was lowered, the length dependence of the transport became "quenched" and the transport itself was more quasi-ballistic. The width dependence clearly indicates the importance of the quasi-one-dimensional nature of the wires.

The solid lines in Fig. 8.1 are calculations based upon the multi-terminal Landauer–Büttiker formula, assuming a thermal scrambling of the various transfer coefficients $T_{ij,jk}$ for modes that differ in longitudinal and transverse quantum number. (Here, the multiple subscripts describe both the mode and terminal numbers, as discussed in Chapter 3.) This implies that the temperature effect is primarily a consequence of the thermal mixing of the modes, so one would assume that $l_T < L$ even though the wires are dominantly quasi-ballistic in nature. In these structures, the temperature dependence of the mobility within the wire itself is not as important to the overall resistance as the transmission coefficients for the various modes of the multi-mode structure. This is a clear example of temperature dependence due to the thermal spread in the energy at the Fermi surface, and the asymptotes tend toward a $T^{1/2}$ variation, characteristic of the thermal diffusion length.

8.1.1 Decay of the coherence length

As mentioned above, there are a variety of contributors to the decay of the coherence length itself with temperature. In many cases, however, these various mechanisms are not separated in an experimental measurement. Rather, only the change in the coherence length is measured by measuring, for example, the change in the behavior of the weak localization or the change in the amplitude of the fluctuations. In this section, we review some of the measurements of the temperature variation of the coherence length. It is important to remember, however, that this length is not itself a key factor, but that it summarizes a variety of temperature variations due to the diffusion "constant" and to the phase coherence time.

One of the first careful experiments concerning the temperature dependence tried to address an early discrepancy in the fit of various theories to the experiments [2]. In this set of experiments, metal loops were measured in a four-terminal configuration (the reader is referred to the discussion of the multi-terminal Landauer–Büttiker approach in Chapter 3), so that all of the various interference mechanisms were present. These authors then measured the voltage fluctuation under constant current conditions and determined the resistance fluctuations (and hence the conductance fluctuations). Figure 8.2 shows the results of measurements on Sb loops; several different methods of computing the coherence length are shown for comparison. The lower curve in the figure illustrates the determination of the coherence length l_φ, from the weak localization. In addition, the correlation

8.1 Temperature decay of coherence

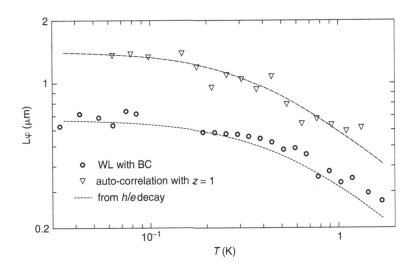

Fig. 8.2 Temperature variation of the coherence length, determined three different ways (discussed in the text). (After R. A. Webb *et al.*, in *Physics and Technology of Submicron Structures*, eds. H. Heinrich *et al.* (Berlin, Springer-Verlag, 1988), by permission.)

function for the fluctuations was determined. These latter experiments were carried out at high magnetic field so that the particle–particle contribution to the universal conductance fluctuation was absent. (As discussed in the previous chapter, the cooperon contribution from the particle–particle correlation function decays rapidly with the magnitude of the magnetic field.) The dotted curve corresponds to a determination of l_φ from the decay of the amplitude of the correlation function with temperature, where it is assumed that the fluctuation amplitude decays as $g = b\exp(-aL/l_\varphi)$, with the coefficient b found to be about 0.2 and $a = 1$. This particular functional form is not one that was discussed in the previous chapter. Nevertheless, the fit to the values obtained from the weak localization seems to be relatively good. However, the value of the coherence length found from the half-width of the correlation function is somewhat different, and this is plotted as the triangles in the figure (the dashed line is merely a guide to the eye). This value yields results that are more than a factor of 2 larger and in essence require a different value of a. This deviation has also been found in MOSFETs [3] and in wires. The asymptote, at high temperatures, is a behavior slightly slower than T^{-1}.

We next turn to measurements made in a short wire, fabricated by the split-gate technique in a Si MOSFET structure. Here, the wire was fabricated to be rather short, so that the combination formula (7.90) for quasi-two-dimensional and quasi-one-dimensional structures was used to infer the coherence lengths for both the one-dimensional and two-dimensional regions [4]. The typical drawn length of the one-dimensional channel was 0.73 μm, although the actual length was longer due to fringing depletion around the metal edges of the gates. The magnetic field dependence of the weak localization was fit to Eq. (7.90), and values for $l_{\varphi,2D}$ and $l_{\varphi,1D}$ were determined from this fit. The results are plotted in Fig. 8.3. The squares and circles plots $l_{\varphi,2D}$ and $l_{\varphi,1D}$ obtained from

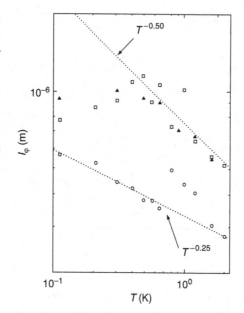

Fig. 8.3 Phase coherence lengths determined for a short wire from Eq. (7.90). The various data are discussed in the text. (After C. de Graaf et al., *Phys. Rev. B* **46**, 12814 (1992), by permission.)

this fit are shown as a function of the temperature of the sample. The full triangles plot $l_{\varphi,2D}$ is obtained from measurements in the homogeneous quasi-two-dimensional electron gas. The agreement of $l_{\varphi,2D}$ is certainly quite good between the two measurements, and the generally smaller value found for $l_{\varphi,1D}$ is consistent with the results discussed in the last chapter. It should be noted that the decays vary as $T^{-1/2}$ for the two-dimensional regions and as $T^{-1/4}$ for the one-dimensional region, shown by the dotted lines for reference, and these seem a reasonably good fit to the data. These decays seem to fit well to theoretical predictions for electron–electron scattering [5], [6] (which will be discussed later in this chapter)

$$l_{\varphi,1D} = \left[\frac{\pi W n_{2D} \hbar D}{a\sqrt{2}} l_T\right]^{1/2}, \tag{8.1}$$

$$l_{\varphi,2D} = \left[\frac{4\pi n_{2D} \hbar D}{a} l_T^2\right]^{1/2}. \tag{8.2}$$

The parameter a has a theoretical value of unity at zero temperature and decreases with the thermal length. The experimental factor for the ratio of $l_{\varphi,1D}/l_{\varphi,2D}$ from Fig. 8.3, is 0.43 ± 0.04, while the theoretical value from the above equations is 0.29 at 1 K. These latter authors consider that this fit is within the experimental error in determining the coherence lengths. Nevertheless, the dependence upon temperature and the fit to these latter two equations suggest that the dominant temperature effects are the important introduction of the thermal diffusion length and the mixing of modes near the Fermi surface.

8.1 Temperature decay of coherence

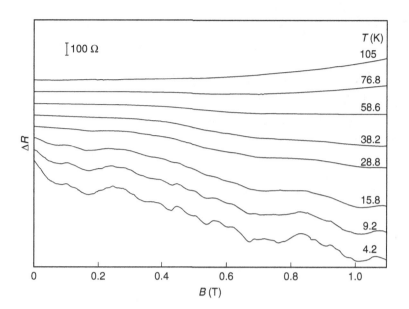

Fig. 8.4 Magnetoresistance for 0.32-μm wide etched wires in a high-mobility GaAs/AlGaAs structure. (After R. P. Taylor et al., J. Phys.: Cond. Matter **1**, 10413 (1989), by permission.)

Typical curves of resistance versus magnetic field, for high-mobility GaAs quantum wires, are shown in Fig. 8.4 [7]. Here, the data are for an etched wire of width 0.32 μm. The basic structure is a high-electron-mobility structure with a two-dimensional density of about 6×10^{11} cm^{-2} and a mobility of 4.7×10^5 cm^2/Vs. The sharp negative magneto-resistance below 0.1 T is attributed to weak localization. The universal conductance fluctuations can be quantified by the variance of the fluctuation amplitude, that is, by the correlation function for the fluctuations. The effective dimensionality of the wire is really determined by the relative magnitudes of l_φ and l_T. These give [8]

$$\text{var}(G) = \alpha \frac{e^2}{2\pi\hbar} \left(\frac{l_\varphi}{L}\right)^{3/2}, \quad l_\varphi \ll l_T, \tag{8.3}$$

and

$$\text{var}(G) = \beta \frac{e^2}{2\pi\hbar} \left(\frac{l_T^2 l_\varphi}{L^3}\right)^{1/2}, \quad l_T \ll l_\varphi. \tag{8.4}$$

Here, α and β are numerical coefficients with values near unity. Numerical simulations [9] suggest that $\alpha = \sqrt{6}$ and $\beta = \sqrt{4\pi/3}$. However, in many cases these two lengths are quite similar and one needs to interpolate between these two formulas, as

$$\text{var}(G) = \alpha \frac{e^2}{2\pi\hbar} \left(\frac{l_\varphi}{L}\right)^{3/2} \left[1 + \frac{9}{2\pi}\left(\frac{l_\varphi}{l_T}\right)^2\right]^{-1/2}. \tag{8.5}$$

Using this approach, the phase coherence length has been determined for these wires and is shown in Fig. 8.5.

Fig. 8.5 The phase coherence length determined from the etched GaAs wires shown in Fig. 8.4. (After R. P. Taylor *et al.*, *J. Phys.: Cond. Matter* **1**, 10413 (1989), by permission.)

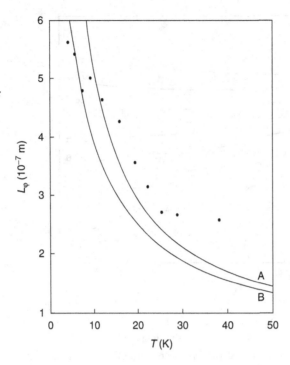

At low temperatures, generally it is felt that the main phase-breaking mechanism is carrier–carrier scattering. This rate, for a one-dimensional channel, has been suggested to be [10], [11]

$$\frac{1}{\tau_\varphi} = \frac{\pi}{2}\frac{(k_BT)^2}{\hbar E_F}\ln\left(\frac{E_F}{k_BT}\right), \quad k_BT > \frac{\hbar}{\tau_c}, \tag{8.6}$$

$$\frac{1}{\tau_\varphi} = \left(\frac{k_BT}{D^{1/2}W\rho(E_F)\hbar^2}\right)^{2/3}, \quad k_BT < \frac{\hbar}{\tau_c}. \tag{8.7}$$

Here, $\rho(E_F)$ is the density of states and τ_φ is the momentum relaxation time (a weighted average of the scattering time as discussed in the last chapter). In Fig. 8.5, the lines are for a value of the coherence length using the phase-breaking time of Eq. (8.6) (curve A) and a phase-breaking time combining the effects of both equations (curve B). In this fit, the two-dimensional density of states was used to evaluate the phase-breaking time. The derivation of the proper phase-breaking times will be discussed later in this chapter.

Another set of data for a high-mobility quantum wire is shown in Fig. 8.6. [12]. Here, the width of the wires was in the range 0.5–1.2 μm, and the basic quasi-two-dimensional gas had a mobility as high as 10^6 cm^2/Vs. The wires were measured in the low-magnetic field regime, and the universal conductance fluctuations decayed smoothly as $T^{-1/2}$. This behavior is consistent with a temperature-independent phase coherence length that varies as Eq. (8.4), and

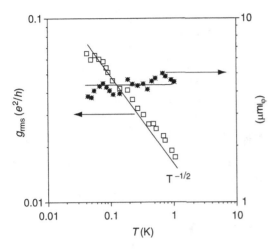

Fig. 8.6 Variance of the universal conductance fluctuations and the phase coherence length found in high-mobility wires at low magnetic field. (After J. P. Bird *et al.*, *J. Phys.: Cond. Matter* **3**, 2897 (1991), by permission.)

this is also consistent with the data for the magneto-resistance. Surprisingly, these authors found that the phase coherence length saturated at a value smaller than the probe spacing in the four-terminal measurement scheme. Although not generally expected from the theory, this was felt to be consistent with a result that the correlation magnetic field was independent of the temperature (which supports a temperature-independent phase coherence length, contrary to the earlier assumption). This result suggests that there is likely to be a mobility-dependent length scale, on the order of several microns, which limits the extent of quantum diffusion processes. While these authors do not speculate on the cause of this quantity, the result is that phase-breaking processes may well be due to processes other than inelastic electron–electron scattering.

8.1.2 Decay of the coherence time

Most of the variations described in the previous section can be attributed primarily to the introduction of the thermal diffusion length l_T into the discussion of the effective lengths and coherence length. The transport parameters themselves are then not expected to show much variation with temperature. As mentioned, the experiments are not usually analyzed to separate the variation of the transport parameters and, for example, the determination of the temperature dependence of the phase coherence time τ_φ.

However, one can separate out the phase coherence time and examine its temperature dependence in the proper circumstances. Here, we compare the phase-breaking time with a variety of other measurements. In Fig. 8.7, the phase-breaking time, determined from a variety of GaAs/AlGaAs quantum wires, is shown as a function of temperature [13]. These samples include both

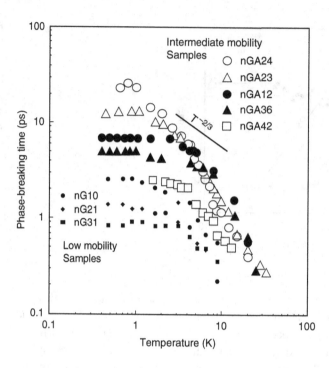

Fig. 8.7 Temperature dependence of the phase-breaking time for various samples. (After Ikoma *et al.*, IOP Conf. Series **127**, 157 (1992), by permission.)

modulation-doped single heterostructures and double heterostructures with doping in the GaAs channel itself. This allowed for sheet densities of 6.0–6.7 × 10^{11} cm^{-2} in the former case and of 0.34–1.3 × 10^{13} cm^{-2} in the latter case. The saturation of the phase-breaking time at low temperatures is interesting in that there is no acceptable theoretical basis for this. As seen from the figure, this saturation occurs in all samples; it has been reported by other authors as well [14]. The latter authors suggested that spin-orbit scattering might cause such a saturation, but this would introduce positive magneto-resistance not seen in these samples. Thus, it is thought that spin-orbit scattering does not play any role in these structures. Another possibility, which has not been thoroughly pursued, is surface-roughness scattering [15]. The diffusive nature of the surface scattering serves to mix modes of the quantum wire, and this scattering can introduce a localization length in the system, which in effect introduces a phase-breaking length [16]. Consequently, it is expected that such normally elastic scattering processes can in fact introduce a phase-breaking process. At higher temperatures, the fall-off with temperature seems to be dominated by a $T^{-3/2}$ behavior, which is different from that expected for electron–electron scattering in a one-dimensional system.

In Fig. 8.8, measurements of the phase-breaking time for a single sample with relatively high mobility (21,000 at the lowest temperature) are shown. These measurements show that the phase coherence time decays rapidly for temperatures above about 5 K in these devices. The two solid curves are various theories

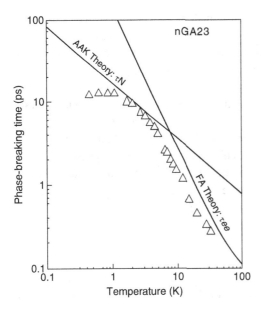

Fig. 8.8 The temperature dependence of the phase-breaking time. (After Ikoma *et al.*, IOP Conf. Series **127**, 157 (1992), by permission.)

for the breakup of phase by carrier–carrier processes and will be discussed later. We note, however, that the curve marked "AAK" decays roughly as $T^{-2/3}$, whereas the curve marked "FA" (these are authors' initials) decays roughly as $T^{-3/2}$, at least at the lower temperatures. These values correspond nicely to the temperature dependencies of Eqs. (8.7) and (8.6). Both of these theories can be thought of as forms of the electron–electron interaction, which will be discussed further below. The connection of the phase-breaking time to the phase coherence time depends on the dimensionality of the system and the temperature variation of the diffusion constant as well, so it is difficult to bring the various theories to bear on the actual temperature variation of the coherence length itself.

The general summary of these measurements, which are quite typical of those found in a variety of systems, is that the phase coherence length is relatively independent of temperature up to some critical temperature, and then decays for higher temperatures. The value of the critical temperature does not seem to be a universal quantity; rather it depends on the mobility (diffusion constant), the sample size, and the carrier density (the Fermi energy). The temperature dependence of the weak-localization and universal conductance fluctuations, in general, cannot be ascribed solely to the temperature variation of the phase coherence length, but must be studied for the effects of the thermal broadening and its introduction of the thermal length. In spite of early studies to the contrary, it appears that the temperature dependence of the coherence length obeys a power-law behavior, not an exponential behavior. This is consistent with expected thermal behavior in quantum diffusion, rather than simply an excitation argument. Moreover, it is generally thought that at these low temperatures the phase-breaking mechanism is electron–electron scattering. We turn to this in a later section. First, we discuss the general temperature dependence that is expected for the various parameters in the correlation functions developed in the last chapter. Nevertheless, it is important to note that the variety of experimental measurements provide a wide range of temperature variations, from which it is possible to conclude only that the experiments need further refinement, as does the theory. To draw any further conclusions would require a thorough investigation of the variations in the experimental conditions, and a subsequent evaluation of just what was really measured. That is, the implications concerning the temperature variation of the phase coherence length, or the phase coherence time, must he subject to de-convolution from the data that actually involve the temperature variations of a number of other possible parameters (density, surface-scattering rate, diffusion constant, etc.).

8.1.3 Summary

At this point, we should step back and review what the above experiments tell us. In the previous chapter, measurements of the phase coherence time in quantum dots led to a variation of the phase-breaking time as T^{-1}. In the experiments discussed in this section, a great variety of decay rates have been found. Webb et al. [2] find that l_φ decays slightly slower than T^{-1}, and the data suggest $T^{-2/3}$ in metal rings. De Graaf et al. [4], on the other hand, found that l_φ decays as $T^{-1/2}$ in two-dimensional structures and as $T^{-1/4}$ in one-dimensional structures of a Si MOSFET. In wires composed of high-mobility modulation-doped heterostructures, Taylor et al. [7] found that the phase-breaking length decay could be fit fairly well by the theory of [8], in which the phase-breaking rate decays slowly at $T^{-1/3}$. Conversion of these numbers to phase-breaking rates depends on knowledge of how the diffusion "constant" (or the elastic scattering rate) varies with temperature, and this is not usually given. Nevertheless, one may conclude that the phase-breaking rate decays slower with temperature than l_φ decays, if for no other reason than that it appears as a square-root factor in this last parameter. Direct measurements of the phase-breaking rate are fewer in number, as a consequence of the difficulty of the direct measurement, which is why the first data mentioned here are so important. For example, in earlier work, Bird et al. [12] found in long wires fabricated in high-mobility material that the only temperature variation arose from l_T and not from l_φ, mentioned above. Ikoma et al. [13] were able to separate out the phase-breaking rate and found that it decayed with temperature approximately as $T^{-3/2}$. This decay is much faster than can be inferred from any of the previous measurements but is much closer to the measurements mentioned in the last chapter. The problem is that if one is trying to infer the value of the phase-breaking time from measurements of conductance fluctuations, then this is complicated by the fact that the effective length of the sample, the diffusion constant for samples in the diffusive regime, the elastic mean free path, and the thermal length are all varying with temperature, and all have an impact on the amplitude of the fluctuations and the correlation function. Therefore, before proceeding to a discussion of the electron–electron scattering process, which is usually inferred as the culprit for phase breaking (at low temperatures), we will review somewhat the role of temperature on the fluctuations.

8.2 The role of temperature on the fluctuations

In Chapter 7, we reviewed the correlation functions that arise from the universal conductance fluctuations (UCF) at zero temperature. We want to begin to understand how the temperature variation will affect these results. For this, we basically follow the arguments of [8]. Early work in the theory of quantum transport in metals assumed that there were two characteristic lengths: the coherence length or inelastic diffusion length $l_\varphi = \sqrt{D\tau_\varphi}$, and the length L of the sample. For the cases in which $l_\varphi > L$, pure quantum transport was to be expected, whereas in the opposite case one should expect Boltzmann transport [17]. This suggests that in the former cases, the fluctuations should be governed by the zero-temperature theory worked out in the last chapter with the only temperature variation arising from l_φ. On the other hand, we expect that the UCF will exhibit a second temperature dependence once the temperature is sufficiently high that the latter situation is realized. This would lead us to accept the fact that the fluctuations should decrease with increasing temperature with one behavior in short samples and a different behavior in relatively long samples.

This is somewhat easier to understand in considering that any sample is composed of a group of small regions, each the size of the coherence length, and these will lead to an additive behavior of the fluctuation amplitude. That is, the sample can be thought of as being composed of a number of phase-coherent regions, through which a constant current is flowing. Then, in one dimension, the var(R) is given by the summation of the voltage fluctuations across each region, as discussed in Section 7.4, since these voltage fluctuations are assumed to be independent of one another. Thus for a short sample, only a single phase-coherent region exists, and the voltage fluctuation is independent of length. However, in more than one dimension, there are a variety of parallel paths, so that var(R) varies as L^0/L^{d-1}, where $d-1$ is the number of parallel paths. Then, var(R)/$\langle R \rangle^2 \sim$ var(G)/$\langle G \rangle^2 \sim L^{d-3}$, which leads to the $\delta G \sim L^{-2}$ behavior in one dimension discussed in connection with Fig. 7.15.

For a long sample, however, there are a number of phase-coherent regions that act independently of one another, and these cause the variance of each region to be added together in an almost classical manner so that var(R) varies as L in one dimension. Since the resistances add linearly, we have var(R)/$\langle R \rangle^2 \sim 1/L$. In two dimensions (and in three dimensions), the classical conductance is essentially an additive function of the constituent conductances, and this result generalizes for the resistances. Then, var(R)/$\langle R \rangle^2 \sim$ var(G)/$\langle G \rangle^2 \sim L^{-d}$, so that the $\delta G \sim L^{-(d-4)/2}$ and the $\delta G \sim L^{-3/2}$ behavior in one dimension also discussed in connection with Fig. 7.15 arises.

In a conductor at elevated temperature, the term $\text{var}[G(l_\varphi)]$ will be modified by the spread in the Fermi distribution function, which introduces *energy averaging*. This means that several different modes, all with energies quite near the Fermi energy, will *mix* together. This will lead to a decay of the fluctuations with temperature. In general, however, this decay is only as a slow power law in T rather than an exponential as expected for excitation across a barrier or gap. The simple arguments that lead to the behavior above suggest that one can determine the length variation of the fluctuations once the latter are known within a subregion of size l_φ at a given temperature. The complication that arises in this argument is that there are two temperature-dependent length scales. The first of these is the coherence length itself, and the second is the thermal length $l_T = (\hbar D/k_B T)^{1/2}$. Both of these lengths enter the analysis of the variance of the conductivity. The major point to be shown below is that if $l_\varphi < l_T$, the $T = 0$ formulation is the proper approach. On the other hand, if $l_\varphi > l_T$, we have to address the proper role of the energy averaging that arises as a result of the spread in the Fermi function. In addition, one can conceive of an intermediate regime $l_\varphi > L > l_T$, where the temperature dependence arises only from the energy averaging and not from any change in the coherence length itself. It is generally considered sufficient (and convenient) to calculate any changes in the fluctuations from the zero-temperature result through the integration

$$G(T) = -\int dE_k \frac{\partial f_k}{\partial E_k} g(E_k), \tag{8.8}$$

which is the embodiment of a result found below in Eq. (8.61). In essence, $g(E_k)$ is the conductivity at energy E_k, and Eq. (8.8) is a restatement of the Boltzmann theory result that we must average this conductivity over the states that contribute to the conductivity. The derivative of the Fermi function f_k represents the weighting of each energy state. For example, at $T = 0$, the derivative is a delta function at the Fermi energy. Following this simple train of thought, the correlation function itself should be energy averaged in the simple process

$$F(\delta\mu, \Delta B, T) = \int dE_1 \int dE_2 \frac{\partial f(E_1, \mu)}{\partial E_1} \frac{\partial f(E_2, \mu + \delta\mu)}{\partial E_2}$$
$$\times \langle \delta g(E_1, B) \delta g(E_2, B + \Delta B) \rangle \tag{8.9}$$
$$= \int d(\Delta E) K(\Delta E, \delta\mu) F(|\Delta E|, \Delta B),$$

where μ is the Fermi energy and

$$K(\Delta E, \delta\mu) = \int dE_1 \frac{\partial f(E_1, \mu)}{\partial E_1} \frac{\partial f(E_1 + \Delta E, \mu - \delta\mu)}{\partial E_1}, \tag{8.10}$$

8.2 The role of temperature on the fluctuations

is the convolution of the two Fermi functions. In principle, a more accurate approach is to recompute the correlation diagrams using the real-time Green's functions. It is reassuring that both methods seem to give the same result. Thus a simpler approach will be followed here.

Since the derivative of the Fermi function is large only for a region of order $k_B T$ on either side of the Fermi energy, the function $K(\Delta E)$ decays exponentially for $\Delta E > k_B T$ and is

$$K(\Delta E, \delta\mu) = \beta e^\xi \int d\eta \left[\frac{e^\eta}{(1+e^\eta)(1+e^{\eta+\xi})} \right]^2, \quad (8.11)$$

where $\beta = k_B T$, $\eta = (\beta E_1 - \mu)$ and $\xi = \beta(\Delta E_1 - \delta\mu)$. For small values of ξ, this may rewritten as

$$K(\Delta E, \delta\mu) = \beta \frac{e^\xi}{(e^\xi - 1)^2} \int d\eta \left[\frac{1}{(1+e^\eta)} - \frac{1}{(1+e^{\eta+\xi})} \right]^2$$

$$\sim \beta \frac{\xi^2 e^\xi}{(e^\xi - 1)^2} \int d\eta \left[\frac{\partial f}{\partial \eta} \right]^2 \sim \beta, \quad \xi < 1, \quad (8.12)$$

since the integral is of order unity. This function is relatively flat for values $\xi < 1$; in fact, it has decreased to only about 85% of the peak value at $\xi = 1$. For large values of ξ, we can rearrange (8.11) by use of the identity

$$\frac{a}{a+e^x} \frac{e^x}{b+e^x} = \frac{a}{b-a} \frac{d}{dx} \left[\ln\left(\frac{a+e^x}{b+e^x} \right) \right] \quad (8.13)$$

to write

$$K(\Delta E, \delta\mu) = \beta \frac{e^\xi}{(e^\xi - 1)^2} \int d\eta \left[\frac{d}{d\eta} \ln\left(\frac{e^{-\xi}+e^\eta}{1+e^\eta} \right) \right]^2$$

$$\sim \beta \frac{e^\xi}{(e^\xi - 1)^2} \int d\eta [f]^2 \sim \beta e^{-\xi}, \quad \xi \gg 1. \quad (8.14)$$

Thus, for most practical purposes, the kernel can be approximated as $K(\Delta E, \delta\mu) \sim \beta \Theta(\beta^{-1} - \Delta E - \delta\mu)$.

The zero-temperature correlation function has a dependence on the two energies at which the conductance is evaluated, which are different in Eq. (8.9). Thus, the energy decay of the correlation function enters into the averaging process even though the Fermi energies are the same for the two bubbles. This is where the temperature dependence enters into the magnetic field correlation function. If the decay of $K(\Delta E)$ is faster than the asymptotic decay of the zero-temperature correlation function, then the former term can he treated as a delta function in ΔE and the zero-temperature result is valid. On the other hand, if $K(\Delta E)$ decays more slowly than the zero-temperature function, which happens when $k_B T > E_c$, the averaging integral provides only a relatively sharp cutoff of the integral in energy at $k_B T$. In fact, we can evaluate the convolution for this simple behavior, assuming that the integral is cut off at the lower end at E_c. From Eq. (7.152), the correlation function basically varies with ΔE as $(\Delta E/\Delta E_c)^{-2}$, where ΔE_c is the "correlation length" or correlation energy. Then, in the asymptotic limit $\xi > 1$, Eq. (8.9) becomes

$$F(\Delta E, T) \sim e^{-\beta E_c}(\beta \Delta E_c) \tag{8.15}$$

which clearly illustrates that the power law dominates when $k_B T > E_c$, and that the exponential dominates in the opposite situation, where the zero-temperature result will be valid. We now consider the case for which the two bubbles are at the same energy ($\Delta \mu = 0$), so that we are interested in the magnetic field-induced response.

The problem arises in the fact that the perturbative series for the correlation function in the diffusive systems, for example given by Eq. (7.152), is quite difficult to express in closed form due to the multiple contributions of comparable weight. Even when this is done, the results are expressible as derivatives of some special functions, which themselves are expressible only as power series. These are not very useful. Therefore, the simplest method of evaluating the temperature dependence is actually to evaluate the zero-temperature series, and then numerically integrate Eq. (8.9) at a given temperature. In lieu of this, we will (again following [8]) give only the asymptotic results in terms of the various parameters of the equations above. This will ignore prefactors of order unity, and so these results are really useful only in plotting the behavior of the correlation functions over a range of magnetic field or temperature to look at the trends rather than the absolute values. For this purpose, the asymptotic limiting forms will be obtained by setting $K(\Delta E) \sim \beta \vartheta (\beta^{-1} - \Delta E)$. It is also sufficient to consider just the terms arising from the diagram of Fig. 7.26(a), since all the terms have comparable asymptotic behavior. For simplicity, a number of dimensionless variables will be utilized in the intermediate results. These are: $\eta = \Delta E/\Delta E_c, \delta = 1/\beta \Delta E_c$, and $\gamma = (L/\pi l_\varphi)^2$, and initially it will be assumed that all dimensions are the same.

8.2.1 Fluctuation amplitudes

In one dimension, the sums over the lateral modes are omitted from the expression for the correlation function (7.150)–(7.153), and the averaged value becomes on the order of

$$\bar{F}_{1D}(T) \sim \frac{1}{\delta} \sum_{m=1}^{\infty} \frac{1}{(m^2+\gamma)^2} \int_0^\delta d\eta \left[1 + \frac{\eta^2}{(m^2+\gamma)^2}\right]^{-2}. \tag{8.16}$$

The integral can be evaluated exactly, but the result is not particularly useful for examining the asymptotic behavior. Instead, it should be noted that the argument of the integral drops off rapidly for $\eta^2 > (m^2+\gamma)^2$. This suggests that one cuts off the integral at this point, and the difference will be whether or not this value is beyond the upper limit. If $(m^2 + \gamma) > \delta$, then the integral is just δ. On the other hand, if $(m^2 + \gamma) < \delta$, then the integral is just $(m^2 + \gamma)$. Thus, we write the result as

$$\bar{F}_{1D}(T) \sim \frac{1}{\delta} \sum_{m=1}^{\infty} \frac{1}{(m^2+\gamma)^2} \left[(m^2+\gamma)\vartheta(\delta - m^2 - \gamma) + \delta\vartheta(m^2 + \gamma - \delta) \right]$$

$$\sim \frac{1}{\delta} \left\{ \sum_{m=1}^{M} \frac{1}{(m^2+\gamma)} + \delta \sum_{m=M+1}^{\infty} \frac{1}{(m^2+\gamma)^2} \right\}$$

(8.17)

where M is a cutoff where the theta functions transition. For $\delta \gg \gamma$, the first summation is just $\sim \gamma^{1/2}$ and the second is $\delta^{-1/2}$, so that [8]

$$\bar{F}_{1D}(T) \sim \frac{\sqrt{\gamma}}{\delta} + O\left(\delta^{-3/2}\right).$$

(8.18)

The fact that the first correction to the asymptotic behavior varies as $\sqrt{\gamma/\delta} \sim (l_T/l_\varphi)$ just points out that we are unlikely to be in any special strong limit, since the phase coherence length seldom exceeds the thermal length by more than an order of magnitude.

In two and three dimensions, we can follow the same procedure if we replace the sum over m^2 in Eq. (5.130) by one over either two or three sets of integers. If $\gamma \gg 1$, the transition in the theta functions can be set independent of the indices of the summations (in this case, $\gamma > M$ is actually required). Then, the double sums can be converted to integrals, and these are combined into two- or three-dimensional "spherical" integrals as

$$\bar{F}_d(T) \cong \frac{1}{\delta} \left[\int_0^{\sqrt{\delta-\gamma}} \frac{m^d - 1}{m^2 - \gamma} dm + \delta \int_{\sqrt{\delta-\gamma}}^{\infty} \frac{m^d - 1}{(m^2 - \gamma)^2} dm \right],$$

$$\sim \frac{1}{\delta} \times \left\{ \begin{array}{ll} \left[\ln\left(\frac{\delta}{\gamma}\right) + 1\right], d = 2 \\ \sqrt{\delta}, \quad d = 3 \end{array} \right\}.$$

(8.19)

It is now clear from these results that, up to a logarithmic correction in two dimensions, the phase coherence length has dropped out of the temperature dependence. We need to be reminded, however, that this assumed that $l_T < l_\varphi$.

Let us now consider the opposite extreme for which $l_T > l_\varphi$, but still for the high-temperature regime where $L \gg l_T > l_\varphi$. This is equivalent to $\gamma > \delta$. In this case, the zero-temperature correlation function varies as γ^{-2}, which is independent of η. In this regime, the zero-temperature correlation function can be removed from the integral, which itself just gives unity when combined with the prefactor of $1/\delta$. Hence, the results are those for the zero-temperature values.

In one dimension, this was found to be $(l_\varphi/L)^{3/2}$. In general, the results found above can be used, and $\text{rms}[G(L)] = \text{rms}[G(l_\varphi)](l_\varphi/L)^{(4-d)/2}$.

We now recall that the rms value of the universal conductance fluctuations is proportional to the square root of the correlation function. We can summarize the high-temperature behavior where the lengths are small compared in the sample size L. This leads us to (here it is also assumed that $\tau_\varphi \sim T^{-p}$,

$$\text{rms}[G(T)] \sim \frac{l_T}{L}\left(\frac{l_\varphi}{L}\right)^{1/2} \sim T^{-1/2-p/4}, \quad d=1, \quad l_T \ll l_\varphi < L, \tag{8.20}$$

$$\text{rms}[G(T)] \sim \left(\frac{l_T}{L}\right)^{(4-d)/2} \sim T^{(d-4)/2}, \quad d=2,3, \quad l_T \ll l_\varphi < L, \tag{8.21}$$

$$\text{rms}[G(T)] \sim \left(\frac{l_\varphi}{L}\right)^{(4-d)/2} \sim T^{(d-4)/2}, \quad d=1,2,3, \quad l_T \ll l_\varphi < L. \tag{8.22}$$

These results are consistent with the view that the behavior begins to become classical when the length scales become longer than l_φ. Hence, one can always write the results as

$$\text{rms}[G(T,L)] \sim \text{rms}[G(T,l_\varphi)]\left(\frac{l_\varphi}{L}\right)^{(4-d)/2}, \quad l_T \ll l_\varphi < L. \tag{8.23}$$

However, it must be pointed out that this is not equivalent to saying that one can divide the sample into small phase-coherent volumes of size l_T, since this particularly fails in the one-dimensional case given here. Care must be taken in low-dimensional systems in setting up arguments based on ensemble averaging phase-coherent regions.

8.2.2 Dimensional crossover

In the above discussion, it was assumed that all characteristic dimensions of the sample were the same; for example, $L_x = L_y = L_z = L$. This is seldom the case in real systems, and one must begin to consider the dimensions of actual samples. Thus, it is possible to have lengths large compared to the coherence (or thermal) length in one dimension but not in other dimensions. This was seen previously, where the effective dimensionality at zero temperature depended on the actual sample shape. The form that needs to be used for multiple dimensions, as found in the last chapter, is to replace m_x^2 of the last subsection with $(L_z/L_x)^2 m_x^2 = \alpha m_x^2$, and similarly for m_y^2 of the last subsection with $(L_z/L_y)^2 m_y^2 = \varsigma m_x^2$. To discuss the 1D-to-2D crossover, it is convenient to separate out the terms with $m_x^2 = m_y^2 = 0$, which then gives the proper 1D result. This allows us to write

$$\bar{F}_d(T) \cong \bar{F}_{1D}(T) + \bar{F}_C(T), \tag{8.24}$$

where

$$\bar{F}_C(T) \sim \frac{1}{\delta} \int_0^\delta d\eta \sum_{m_x,m_y,m_z=1}^\infty \frac{\left(m_z^2 + \alpha m_x^2 + \varsigma m_y^2 + \gamma\right)^2}{\left[\left(m_z^2 + \alpha m_x^2 + \varsigma m_y^2 + \gamma\right)^2 + \eta^2\right]^2}. \qquad (8.25)$$

To find the 2D-to-3D crossover, one would make a similar separation for the terms for which $m_y^2 = 0$, which would then produce the 2D result, with an additional term that creates the 3D correlation function.

Let us first consider the 2D-to-1D crossover, so we set $m_y^2 = 0$ in Eq. (8.25). This similarly omits the summation over the y variables. For this, we will convert the sums to integrals and rescale the variables with the new definitions $q_z = m_z$, $q_x = \sqrt{\alpha} m_x$. This raises the limit on the integral from unity to $\sqrt{\alpha}$ for the x-coordinate. The region of integration is not spherically symmetric, but the small region for $1 < q_z < \sqrt{\alpha}$ can be ignored for reasonably large γ. Then, the integral can be rewritten as

$$\bar{F}_C(T) \sim \frac{1}{\delta\sqrt{\alpha}} \int_0^\delta d\eta \int_{\sqrt{\alpha}}^\infty q dq \frac{(q^2+\gamma)^2}{\left[(q^2+\gamma)^2+\eta^2\right]^2}. \qquad (8.26)$$

Except for the lower limit of the revised integral/summation, this is precisely the same integral as Eq. (8.16) above. Moreover, an equivalent procedure can be followed for the 2D-to-3D crossover, and the same integral is obtained if it is assumed that $\zeta = 1$. Using the theta function separation discussed in the previous subsection, this becomes

$$\bar{F}_C(T) \sim \frac{1}{\delta\sqrt{\alpha}} \int_0^\delta d\eta \left\{ \int_{\sqrt{\alpha}}^{\sqrt{1-\gamma}} q^{d-1} dq \frac{1}{(q^2+\gamma)} + \int_{\sqrt{1-\gamma}}^\infty q^{d-1} dq \frac{1}{(q^2+\gamma)^2} \right\}. \qquad (8.27)$$

This assumes that $\delta > \gamma, \alpha$. In two dimensions, this gives a result essentially the same as previously with only the modification for the scale factor, as

$$\bar{F}_C(T) \sim \frac{1}{\delta\sqrt{\alpha}} \left[\ln\left(\frac{\delta}{\sqrt{\alpha}+\gamma}\right) + 1 \right] \sim \frac{1}{\delta\sqrt{\alpha}}, \quad d = 2. \qquad (8.28)$$

We can then write the amplitude of the fluctuations as

$$rms[G(T)] \sim \frac{l_T}{L_z} \left(\frac{l_\varphi}{L_z} + \frac{L_x}{L_z}\right)^{1/2}, \quad d = 2, \quad l_T \ll l_\varphi < L_z. \qquad (8.29)$$

The crossover from 2D to 1D occurs when $L_x < l_\varphi$. Similarly, the crossover from 3D to 2D can be expressed by the result

$$\text{rms}[G(T)] \sim \left(\frac{l_T}{L_z}\right)^{1/2} \left(\frac{l_\varphi}{L_z} + \frac{L_x}{L_z}\right)^{1/2}, \quad d = 3, \quad l_T \ll l_\varphi < L_z, L_x, \qquad (8.30)$$

and crossover occurs when $L_y < l_T$. All of this is for $l_T < l_\varphi$. When the reverse case is true, the crossover always occurs when the sample size becomes smaller than the coherence length. These latter equations explain some of the unusual temperature behaviors found in the experiments discussed at the beginning of this chapter.

We note from this crossover behavior that a sample that is much wider than it is long can have a fluctuation amplitude much larger than e^2/h, and this effect has been seen in some samples [18]. It has been pointed out previously that this is precisely the ensemble behavior one would expect from putting L_x/l_φ parallel resistors together and then placing L_z/l_φ groups of these in series, when each resistor has a fluctuation amplitude of e^2/h [19].

8.2.3 Correlation ranges

With the results of the above scaling behavior, we are now in a position to estimate the temperature dependence of the correlation "lengths" in energy and magnetic field. In the last chapter, the correlation functions themselves were calculated, and it was assumed that the magnetic field did not alter the basic eigenvalues of the diffusor equation for the correlator. As long as this remains true, then the magnetic field can be treated as an additional factor, just as the inelastic length is introduced. Hence, the dimensionless ratio $b = \Delta B/B_c(0)$ gives an additive correction to the eigenvalues in the same way that the dimensionless ratio c does. Here, we note that the zero-temperature value $B_c(0)$ is given by Eq. (7.164) as $\sqrt{3}\Phi_0/L_yL_z$, when $L_y, L_z \ll l_\varphi$, and by Eq. (7.165) as $\sqrt{3}\Phi_0/\pi L_y l_\varphi$ when $L_y \ll l_\varphi \ll L_z$. The relationship for the similar nature of b and γ leads us to simply replace γ by $\gamma + b$ in (8.16), and proceed exactly as in the above sections. We then can ask just what value of b reduces the amplitude of the fluctuation by a factor of 2. In one dimension, reference to (8.18) tells us that this occurs about where $\gamma \sim b$ when $l_T < l_\varphi$. This same result is obtained in all dimensions when $l_T > l_\varphi$. Thus, in one dimension, the temperature dependence is always determined by the phase coherence length, and Eq. (7.165) holds so long as the dimensions of the sample satisfy the designated limits. Hence, the magnetic correlation function has a temperature dependence that arises solely from the temperature dependences of the inelastic scattering time (or phase relaxation time) and the diffusion constant. This result is independent of the relative sizes of l_T and l_φ.

If we have a quasi-two-dimensional sample in which $L_y \ll l_\varphi \ll L_z, L_x$, and we are interested in a case where the field is directed longitudinally along the sample length, then the sample is still governed by the same calculation of the eigenvalues for the diffuson correlator. Since the eigenstates in the direction parallel to the field are independent of the field, we can still use the approach of replacing γ by $\gamma + b$ to obtain the correlation magnetic field $B_c(T)$. Hence, for the longitudinal magneto-conductance,

$$B_c(T) \sim \frac{\varphi_0}{L_{\min}L_y}, \quad L_{\min} = \min(l_T, l_\varphi). \tag{8.31}$$

This completely describes the case where the area normal to the field is essentially one-dimensional.

Let us turn to the case where there is a large quasi-two-dimensional system with the magnetic field normal to the plane of the sample. The eigenvalues are the Landau levels themselves. When this summation is made in the above expressions for the correlation function, the zero-temperature function decays very slowly, with respect to any appreciable variation due to the inelastic cutoff γ. This means that the latter parameter plays no part in the decay, and the magnetic correlation length arises from the value $b \sim \delta$, independent of the thickness of the sample. This is true when $l_T < l_\varphi$. When the phase coherence length is the shortest appropriate length, then it dominates everything. These results suggest that the temperature-dependent correlation magnetic field is given by

$$B_c(T) \sim \frac{\varphi_0}{L_{\min}^2}. \tag{8.32}$$

For the energy correlation length, we already have determined the dependence of the convolution function $K(\delta\mu)$ in Eq. (8.11). The behavior of this relation tells us that μ_c can never be much larger than $k_B T$ because of the exponential decay of K. Thus, we really don't have to consider the case where $l_T > l_\varphi$, because it will give essentially the zero-temperature results. We need only to consider the opposite case, and this just modifies the limits on the η integration to the range $(\bar{\mu}, \delta + \bar{\mu})$, where $\bar{\mu} = \delta\mu/E_c$. Not surprisingly, similar results are obtained as for the magnetic correlation length: for example, in one dimension, $\bar{\mu} \sim \gamma$, and in two and three dimensions, $\bar{\mu} \sim \delta$. This leads to

$$\mu_c \sim \frac{\hbar}{\tau_\varphi}, \quad l_T \ll l_\varphi < L_z \tag{8.33}$$

in one dimension, and

$$\mu_c \sim k_B T \tag{8.34}$$

for all other cases.

8.3 Electron–electron interaction effects

The study of the interacting electron gas has a long history. From the earliest days, there has been interest in computing the self-interactions that arise from the Coulomb force between individual pairs of electrons, as opposed to the interaction between the electrons and the impurities that was of primary interest in the last chapter. Early studies, using simple, straightforward perturbation theory, led to a singularity near the Fermi surface. That is, as the energy approached the Fermi energy from either above or below (in a degenerate system for which the Fermi energy lies within the conduction band), the interaction terms diverged; this divergence has been termed a Fermi-edge singularity. It was then realized that a proper calculation, taking into account the screening of any single pair-wise interaction by the remaining electrons (in a manner similar to carrying out the ladder-diagram summation in the last chapter, a calculation to which we return later in this section), removed all the singularities [20] except in one dimension, where the Fermi-edge singularity remains [21]. Following this early work, it was assumed that in general there would be no further complications arising from additional scattering by impurities [22]. However, as we have seen in the last chapter, this is certainly not the case in mesoscopic systems. Altshuler and Aronov [23] were the first to show that additional singularities arose from the presence of the impurity interaction, and that weak singularities would be seen in the tunneling density of states, particularly in Josephson junctions. In addition, it is now believed that the dominant phase-breaking interaction at low temperature in mesoscopic systems arises from the electron–electron interactions. The normal interaction gives rise to the Hartree or Hartree–Fock self-energy shifts and broadening from the self-energy. What is likely responsible for the phase-breaking is the redistribution of momentum and energy among the electrons, so that any one electron loses memory of its specific phase.

In this section, we want to examine the interaction effects, particularly as they are modified by the presence of the strong impurity scattering process for mesoscopic devices. However, these are temperature-dependent effects, so we cannot simply evaluate everything at the Fermi energy. We must begin to allow for a temperature broadening in the Fermi distribution, as discussed in the previous section, so that there is a range of energy in the vicinity of the Fermi energy that becomes involved. In order to treat this properly, we need to modify the Green's functions that we have been using, since they basically assume that $T = 0$. Our application of the temperature Green's functions will focus primarily on the electron–electron interaction and the manner in which this interaction leads to phase breaking. First, however, we review the overall properties of the electron–electron interaction, since there is considerable physics to be understood before calculations are addressed.

The simplest understanding of the electron–electron interaction can be found simply by a straightforward evaluation in terms of the Fermi golden rule. Such an evaluation will clearly point out the role played by various effects, but will keep the principal factors in sight. We may express the electron–electron scattering rate as

$$\frac{1}{\tau_{ee}} = \frac{2\pi}{\hbar}|V_q|^2 \rho(E) n_e \quad (8.35)$$

where V_q is the scattering matrix element arising from the screened Coulomb interaction, $\rho(E)$ is the density of final states, and n_e is the number of electrons with which the incident electron can scatter. Energy can be exchanged in this interaction, although the full two-particle interaction is energy-conserving. That is, the incident electron, with which we are concerned, can gain or lose energy to the scattering center, which in this case is another electron. Because energy can be exchanged, this interaction can lead to the loss of phase coherence on the part of the incident electron. The number of electrons with which the incident electron can scatter is given essentially by $\rho(E)\delta E$, where δE is the allowed range of interaction energy of the two electrons. In most cases we are interested in, $\delta E = k_B T$. If we insert this result into Eq. (8.35) we get

$$\frac{1}{\tau_{ee}} = \frac{2\pi}{\hbar}|V_q|^2 \rho^2(E)\delta E. \qquad (8.36)$$

In three dimensions, $\rho(E) \sim E^{-1/2}$, and in nondegenerate semiconductors, δE can be taken as E as well. For an energy-independent matrix element, as in the case of long-range interactions (small q) that are fully screened, this leads to $1/\tau_{ee} \sim E^2$, and this behavior is found in the scattering rate for impact ionization in bulk semiconductors [24], which leads to the so-called soft threshold for ionization processes [25]. When the energy is averaged over a Maxwellian distribution, we find that the electron–electron scattering time, proportional in many cases to the phase-breaking time, varies as $\tau_{ee} \sim 1/T^2$. If the distribution is degenerate, then the density of states functions can be approximated by taking their values at the Fermi surface, and by using the fact that the important energy spread is given by the thermal spread of the Fermi–Dirac distribution, as in (8.36). Then, it is found that the scattering time varies as $\tau_{ee} \sim 1/T$.

In two dimensions, the density of states is constant, so the only variation in the scattering rate (other than that possible from the matrix element) arises from the energy spread involved in the interaction. Using $\delta E = k_B T$, one finds then that the scattering rate varies as $\tau_{ee} \sim 1/T$. This result arises whether the distribution is degenerate or nondegenerate. In one dimension, $\rho(E) \sim E^{-1/2}$, and this behavior is seen as the source of the Fermi-edge singularity. However, if we take the interaction energy range as simply the energy, we find that the scattering rate is independent of temperature. The same result is found in zero dimensions (such as in a fully quantized quantum dot), since the density of states is a delta function, and the number of scatterers is simply the number of electrons in the isolated energy level.

These give first-order approximations to the scattering rates in order to understand the source of the various temperature dependencies. Certainly, energy and temperature variations in the matrix element can modify these simple behaviors, and it must be pointed out that we have ignored the matrix element contributions. The calculations often will be quite complicated, so the basic understanding may well be lost in plodding through the equations. Thus, it is important to keep the simple results in mind throughout. Nevertheless, the above discussion assumes a variety of things: screened interaction, loss to electron–hole pairs, and so on. Before proceeding it is useful to examine what these concepts mean in the electron–electron interaction.

8.3.1 Electron energy loss in scattering

Electrons that will lose energy to the overall electron gas usually are considered to be in *excited states*. That is, for one reason or another, the "incident" electron is in a state with an energy and momentum that place it above the Fermi energy (and Fermi momentum). This could as easily be the case for a hole lying below the Fermi energy, but here we will concentrate solely upon the excited electron because in this approach the results are most easily carried over to the non-degenerate case. In creating this excited state, the electron was excited from an initial state $\mathbf{k}_0$ lying below the Fermi energy to a state $\mathbf{k}$ lying above the Fermi energy. This leaves an empty state below the Fermi energy, which is the corresponding *hole* (or uncompensated electron at $-\mathbf{k}_0$). In order for this process to occur, one must provide a momentum $\mathbf{q} = \mathbf{k} - \mathbf{k}_0$ and energy $E(q) = \left(\hbar^2/2m^*\right)\left(k^2 - k_0^2\right)$ from an external source. In the previous chapter, electron and hole diagrams were treated in which the transferred energy and momentum were zero. This is a special case, and when the excitation comes from scattering from another electron, the momentum and energy transfer can be different from zero.

The energy and momentum of the excited electron, and the remaining hole, are not uniquely related to each other. (For a different view from that of low-temperature mesoscopic devices, one should review the arguments in cross-bandgap excitation of electron–hole pairs by hot carriers in impact ionization [24].) For a given momentum transfer $\hbar\mathbf{q}$, there is a range of allowed electron and hole momenta and energies. There are two cases to be considered. In the first case, $q < 2k_F$, only a fraction of the electrons lying within the Fermi sphere can participate in the interaction. That is because a significant fraction of the states with $\mathbf{k} = \mathbf{q} + \mathbf{k}_0$ are already occupied and therefore forbidden as possible final states. Thus, we set limits on k to be $k_F < k < k_F + q$. These limits must be carefully evaluated. When $k = k_F$, then k_0 can range from $-k_F$ to 0, and q correspondingly varies from 0 to $2k_F$. Throughout this variation, however, the excess energy of the excited particle $E_k = E(k) - E(k_F) = 0$. That is, the excited state lies right at the Fermi surface. When $k = k_F + q$, k_0 can range from $-k_F$ to k_F. Correspondingly, the transferred momentum q varies from 0 up to $2k_F$, as before. The excess energy of the excited particle now varies from $E_k = 0$ up to $8E(k_F)$. Thus, there is an entire range of energies and momenta possible for the interaction. This is why the two densities of states, mentioned in the introduction to this section, must be convolved with one another.

When $q > 2k_F$, the Pauli exclusion principle is no longer a determinant on the range of allowed energy and momentum. Rather, every possible k_0 is allowed for every value of k. In this case, the convolutions run over the full Fermi sphere. The possible final states that are allowed by the Pauli exclusion principle must lie at least an energy $\left(\hbar^2/2m^*\right)\left[(q - k_F)^2 - k_F^2\right]$ above the Fermi energy, and the maximum energy that can be transferred is, as previously,

$(\hbar^2/2m^*)\left[(q+k_F)^2 - k_F^2\right]$. This leads to the limitations on the energy of the excited state as

$$q(q - 2k_F) = k^2 < q(q + 2k_F). \tag{8.37}$$

When $q > 2k_F$, we see that there is a minimum energy, and hence a minimum momentum, in the excited state that is produced by the interaction. Thus, there is a minimum amount of energy that must be transferred to the excited electron–hole pair. This energy must come from the interacting particle that gives up the energy $E(q)$ and momentum $\hbar\mathbf{q}$.

The previous arguments have been for degenerate semiconductors, and that generally is the assumption that we will make below. It is useful to review the arguments above for the case of a nondegenerate semiconductor, for which k_F is imaginary ($E_F < 0$, where the bottom of the conduction band is taken to be the zero of energy). Obviously, then, we need only the last case discussed, since the momentum transfer cannot be less than the Fermi momentum. The maximum energy that can be transferred is obviously $\hbar^2 k^2/2m^*$ for which $\mathbf{q} = \mathbf{k}$, but we should be quick to point out that the incident particle at $\mathbf{k}$ can actually gain energy and momentum from another electron, as well as lose it to the other electron. In this case, however, the roles of the two particles are interchanged. First, consider the case for which the electrons at $\mathbf{k}$ and $\mathbf{k}_0$ are on the same energy surface, so that $k^2 = k_0^2$. Consider first the simple case in which one electron gives up all of its energy to the other. If we set $\mathbf{q} = \mathbf{k}$, we are required to have the angle between $\mathbf{k}_0$ and $\mathbf{q}$ be $\pi/2$, which specifies that the initial two particles are propagating at right angles to one another. That is, if one particle is directed along the (001) axis, the other must lie in the plane normal to this axis. In a quasi-two-dimensional system, this is a very special case in which only one possible state is allowed. Now, consider the general case, in which the particle at $\mathbf{k}$ is scattered to $\mathbf{k} - \mathbf{q}$, while the particle at $\mathbf{k}_0$ is scattered to $\mathbf{k}_0 + \mathbf{q}$, which conserves momentum. Energy conservation then gives

$$q = \frac{k^2 - k_0^2}{k \cos \vartheta + k_0 \cos \vartheta_0} \tag{8.38}$$

where ϑ is the angle between $\mathbf{k}$ and $\mathbf{q}$, and ϑ_0 is the angle between $\mathbf{k}_0$ and $\mathbf{q}$. This formula does not work for the simple case above, since setting $k = k_0$ when $\cos \vartheta = 1$ forces $\cos \vartheta_0 = -1$, and this gives $q = k + k_0$, which means that the two particles exchange momentum; this is not a real scattering event. Thus, Eq. (8.38) is good only off a common energy shell (i.e. $k \neq k_0$). Nevertheless, we can see that the maximum value of q for which the incident particle at $\mathbf{k}$ loses energy is $2k$. In fact, in order to satisfy energy conservation, the scattering momentum must lie in the range

$$-2k_0 \cos \vartheta_0 < q < 2 < 2k \cos \vartheta_0. \tag{8.39}$$

The ranges of allowed final states that arise from the energy and momentum transfer considerations define a spectrum of single-particle excitations that can result from the electron–electron interaction. Outside of the range defined by (8.37), the electron–electron interaction is not effective, although we will see below that scattering from the *collective* modes of the electron gas is possible. While the discussion of this section has been concentrated on the excited electron–hole pair, we must remember that our primary interest lies in the properties of the particle that provides the excitation. This particle has been characterized here by the momentum $\hbar\mathbf{q}$ and energy $E(q)$. We now turn our attention to the Hamiltonian for the electron gas and discuss how we want to partition the various terms.

8.3.2 Screening and plasmons

The Coulomb interaction that enters into the scattering of one electron by the other electrons is a very long range interaction. Consequently, it is usually necessary to cut off the range of this interaction, as is done for impurity scattering as well. This screening is usually introduced ad hoc, but this is not necessary. In this section, we wish to show how this can be accomplished, while at the same time introduce the collective modes, the *plasmons*, into the discussion. The approach we follow is essentially adapted from that presented by Madelung [26].

The electron gas may be considered, for the case at present, as a uniform, homogeneous quantity. When an additional electron is introduced into the gas, the Coulomb interaction causes the other electrons to be slightly repelled from it, which leaves a positive charge around the new electron. This extra positive charge in essence *screens* the electron. However, the process by which this occurs is a dynamic process involving the movement of the entire electron gas through internal interactions. This process is such that the movement of the electrons away from the extra charge is usually too great, and so the electrons must move back again, leading to oscillations of the electron gas. These are the collective oscillations, which are the plasmons. So, all of these processes must be involved in the manner in which the screening of the simple Coulomb interaction arises. In this subsection we pursue a classical understanding of the effects, leaving the quantum-mechanical approach to later subsections, where we can deal with it properly in terms of Green's functions. Our starting point is the simple Hamiltonian

$$H = \sum_i \frac{p_i^2}{2m^*} + \frac{e^2}{8\pi\varepsilon_0} \sum_{i,j \neq 1} \frac{1}{|r_i - r_j|}, \tag{8.40}$$

where the extra factor of two in the Coulomb term arises because of the double counting in summing over both subscripts completely. An ad hoc introduction of

8.3 Electron–electron interaction effects

screening is often made by adding an exponential decay term within the second term of Eq. (8.40), such as $\exp(-\lambda|\mathbf{r}_i - \mathbf{r}_j|)$, where λ is an inverse screening length, such as the Debye length or the Fermi–Thomas length. However, this would eliminate the long-range part of the Coulomb interaction, which would have to be examined separately. Almost the same effect can be achieved by limiting the range of k in any Fourier transform of the Coulomb interaction. Consequently, with the Fourier transformation of the second term, defined by

$$\frac{e^2}{8\pi\varepsilon_0} \sum_{i,j \neq 1} \frac{1}{|\mathbf{r}_i - \mathbf{r}_j|} = \frac{e^2}{2V_0\varepsilon_0} \sum_{i,j \neq 1} \sum_{\mathbf{k}} \frac{1}{k^2} e^{i\mathbf{k}\cdot|\mathbf{r}_i-\mathbf{r}_j|}, \qquad (8.41)$$

where a separation will be introduced between long-range effects and short-range effects through the use of a cutoff wavevector $k_c = \lambda$. With this separation, Eq. (8.40) can be rewritten as

$$H = \sum_i \frac{p_i^2}{2m^*} + \frac{e^2}{2V_0\varepsilon_0} \sum_{i,j \neq 1} \left(\sum_{k<\lambda} + \sum_{k>\lambda} \right) \frac{1}{k^2} e^{i\mathbf{k}\cdot|\mathbf{r}_i-\mathbf{r}_j|}. \qquad (8.42)$$

Along with the screening of the individual electron by the rest of the electron gas, the collective oscillations act back upon the original charge through a self-consistent field produced by their own Coulomb potential. This may be described by the irrotational vector potential $\mathbf{A}(\mathbf{r}_i)$. This vector potential can he expressed through its own Fourier transform as

$$\mathbf{A}(\mathbf{r}_i) = \frac{1}{\sqrt{V_0\varepsilon_0}} \sum_{k>0} \frac{\mathbf{k}}{k} Q_{\mathbf{k}} e^{i\mathbf{k}\cdot\mathbf{r}_i}. \qquad (8.43)$$

The unit vector inside the summation assures that the vector potential is irrotational. The requirement that the vector potential be real leads to $\mathbf{k}Q_{\mathbf{k}}^* = -\mathbf{k}Q_{-\mathbf{k}}$, so that $Q_{\mathbf{k}}^* = -Q_{-\mathbf{k}}$. The electric field that arises from the vector potential may be written as

$$\mathbf{E} = -\frac{\partial \mathbf{A}}{\partial t} = -\frac{1}{\sqrt{V_0\varepsilon_0}} \sum_{k>0} \frac{\mathbf{k}}{k} \frac{\partial Q_{\mathbf{k}}}{\partial t} e^{i\mathbf{k}\cdot\mathbf{r}_i} = \frac{1}{\sqrt{V_0\varepsilon_0}} \sum_{k>0} \frac{\mathbf{k}}{k} P_{\mathbf{k}}^* e^{i\mathbf{k}\cdot\mathbf{r}_i} \qquad (8.44)$$

where the momentum $P_{\mathbf{k}}^*$ conjugate to $Q_{\mathbf{k}}$ has been introduced in the last term. The $Q_{\mathbf{k}}$ and $P_{\mathbf{k}}$ can he taken to be the collective coordinates of the field that describes the collective motion of the electron gas as a whole. Thus, we introduce these coordinates into (8.42) in the following manner. First, the leading term which describes the kinetic energy of the electrons is modified by the replacement $\mathbf{p}_i \rightarrow \mathbf{p}_i + e\mathbf{A}(\mathbf{r}_i)$, so that the momentum responds to the collective self-consistent field of the electron gas as a whole. Second, the part of the summation in the second term, which represents the long-range part of the Coulomb interaction ($k < \lambda$), is replaced by the energy of the self-consistent electric field interactions among the electrons. Thus, we may write (8.42) as

$$H = \sum_i \frac{1}{2m^*}\left(\mathbf{p}_i + \frac{e}{\sqrt{V_0\varepsilon_0}}\sum_{k>0}\frac{\mathbf{k}}{k}Q_\mathbf{k}e^{i\mathbf{k}\cdot\mathbf{r}_i}\right)^2 + \frac{e^2}{2V_0\varepsilon_0}\sum_{i,j\neq 1}\sum_{k>\lambda}\frac{1}{k^2}e^{i\mathbf{k}\cdot|\mathbf{r}_i-\mathbf{r}_j|}$$
$$+ \frac{1}{2V_0}\sum_{k,k'<\lambda} P_\mathbf{k} P_{\mathbf{k}'} \frac{\mathbf{k}\cdot\mathbf{k}'}{k\cdot k'}\int e^{j(\mathbf{k}+\mathbf{k}')\cdot\mathbf{r}_i}dV.$$
(8.45)

The integration of the last term produces a delta function, which allows us to write the last term as

$$\frac{1}{2}\sum_{k<\lambda} P_\mathbf{k}^* P_\mathbf{k}. \qquad (8.46)$$

At this point, we depart from the classical approach and go over to the quantum-mechanical approach. This is achieved by replacing the various variables with noncommuting operators, which satisfy the commutation relationships

$$[p_{iv}, r_{j\xi}] = -i\hbar\delta_{ij}\delta_{v\xi} \qquad (8.47)$$

$$[P_\mathbf{k}, Q_{\mathbf{k}'}] = i\hbar\delta_{\mathbf{k},\mathbf{k}'} \qquad (8.48)$$

where v and ξ refer to the x-, y-, and z-components of the position and momentum vectors. The first term in the brackets of Eq. (8.45) then contains the term

$$\mathbf{p}_i e^{i\mathbf{k}\cdot\mathbf{r}} + e^{i\mathbf{k}\cdot\mathbf{r}}\mathbf{p}_i = 2e^{i\mathbf{k}\cdot\mathbf{r}}\mathbf{p}_i - \hbar\mathbf{k}e^{i\mathbf{k}\cdot\mathbf{r}}. \qquad (8.49)$$

Using this expression, (8.45) can be arranged to yield

$$H = \sum_i \frac{p_i^2}{2m^*} + \frac{e^2}{2V_0\varepsilon_0}\sum_{i,j\neq 1}\sum_{k>\lambda}\frac{1}{k^2}e^{i\mathbf{k}\cdot|\mathbf{r}_i-\mathbf{r}_j|} + \frac{1}{2}\sum_{k<\lambda}\left(P_\mathbf{k}^* P_\mathbf{k} + \omega_p^2 Q_\mathbf{k} Q_\mathbf{k}^*\right)$$
$$+ \frac{e}{\sqrt{V_0\varepsilon_0}m^*}\sum_{k>0} Q_\mathbf{k} e^{i\mathbf{k}\cdot\mathbf{r}_i}\frac{\mathbf{k}}{k}\sum_i (\mathbf{p}_i - \hbar\mathbf{k})$$
$$+ \frac{e^2}{2V_0 m^*\varepsilon_0}\sum_{k,k'<\lambda}\left[1-\delta_{\mathbf{k},-\mathbf{k}'}\right]\frac{\mathbf{k}\cdot\mathbf{k}'}{kk'}\sum_i e^{j(\mathbf{k}+\mathbf{k}')\cdot\mathbf{r}_i}$$
(8.50)

where $\omega_p^2 = ne^2/m^*\varepsilon_0$ is the low-frequency plasma frequency (the plasmons have this characteristic frequency for frequencies well below the polar optical phonon frequency), and $n = N/V_0$ is the electron density.

In introducing the collective coordinates $P_\mathbf{k}$ and $Q_\mathbf{k}$ for the plasma oscillations, we have raised the number of degrees of freedom, so there must be some additional constraints connecting these two degrees of freedom. These extra constraints come from the Poisson equation, in which we require that $\nabla\cdot\mathbf{E} = -\rho/\varepsilon_0$. Expanding the charge density into Fourier coefficients, and using the above relationships for the electric field, the added constraints may be expressed as

8.3 Electron–electron interaction effects

$$P_{\mathbf{k}} - i\sqrt{\frac{e^2}{V_0\varepsilon_0 k^2}} \sum_j e^{i\mathbf{k}\cdot\mathbf{r}_j} = 0. \tag{8.51}$$

In the transition to quantum mechanics, this becomes an operator equation, and the constraint is applied by requiring that this operator, when applied to the wave equation, yields a zero result.

Now, let us turn to the meaning of Eq. (8.50). The first line corresponds to a gas of nearly free electrons interacting with one another through a *screened* interaction, where the screening is introduced by the wavevector cutoff. The second line of the equation describes the harmonic-oscillator-like vibrations and energy levels of the collective plasma oscillations of the electron gas. The last two lines correspond to the interaction of the nearly free electrons with the collective modes of the electron gas. The third line is clearly the scattering of the nearly free electron through the emission or absorption of plasmons. This can be seen clearly by going over to the number representation for the electrons and plasmons, but this will not be pursued at this point. The last term is usually ignored by adopting what is called the *random phase approximation*, in which it is assumed that the phase in the exponential term varies so rapidly that no terms for which $\mathbf{k} \neq -\mathbf{k}'$ can really contribute to the energy. A careful expansion of the operators in this latter term in the number representation shows that it corresponds to the processes in which a momentum $\mathbf{k} + \mathbf{k}'$ is transferred to the emission or absorption of a *pair* of plasmons, or to the simultaneous emission and absorption of a plasmon. Use of the random phase approximation in this context implies ignoring terms in which more than a single electron interacts with more than a single plasmon.

What has been achieved by the introduction of the cutoff wavevector is that the Hamiltonian is now expressible in terms of the nearly free energy of the electrons and the plasmons separately, and two interaction terms in which the nearly free electrons interact with each other through a screened Coulomb interaction and in which an electron can emit or absorb plasmons. The crucial step was the introduction of the cutoff wavevector λ, which corresponds to the screening length. Since momentum is transferred to the collective modes, one must expand the plasmon energy to higher orders, and this leads to $E_{pl} \cong \hbar\omega_p + \alpha\lambda^2$. Now, since pair excitations cannot excite or absorb plasmons, the plasmon interaction must come from the single particle spectrum, so we require that $E = \hbar^2 k^2/2m^* > E_{pl}$, which leads to a cutoff wavelength $\lambda \sim \omega_p/v_F \sim 0.7 k_{FT}$, where k_{FT} is the Fermi–Thomas screening wavevector (hence the usual argument that the latter overestimates screening). In a non-degenerate semiconductor, the Fermi velocity is replaced by the thermal velocity, which leads to the Debye screening wavevector, a result found in careful studies of plasmon scattering in hot electron systems [27].

In most systems, however, the screening is not given by a simple cutoff wavevector but is described by a dynamic polarization function for the electron gas. In order to more fully treat this dynamic screening, and to properly compute the scattering self-energy for the electron–electron single-particle scattering, we will move to the Green's function treatment. However, since the system is usually at a non-zero temperature, we must first introduce the appropriate Green's function formulation, to which we now turn.

8.3.3 Temperature Green's functions

At temperatures above absolute zero, there is an analogous Green's function to those that have been introduced in Chapter 3. However, this function is more complicated, and the calculation of the equilibrium properties and excitation spectrum also is more complicated. The first part, the determination of the equilibrium properties, is handled by the introduction of a *temperature Green's function*, often called the Matsubara Green's function, while the second step requires the computation of a time-dependent Green's function that describes the linear response of the system [28]. In principle, the definition of the temperature Green's function is made simply by taking account of the fact that there exist both a distribution function in the system and a number of both full and empty states. The principal change is that the average over the basis states represented by $\langle ..f(A)..\rangle$ goes into $\mathrm{Tr}\{\hat{\rho}[...f(A)...]\}$, where $\hat{\rho} = \exp[\beta(\Omega - K)]$ is the system density matrix, $\beta = 1/k_B T$, Ω is the free energy (defined through the partition function), and $K = H - \mu N$ represents the grand canonical ensemble, which includes the chemical potential μ (usually equal to the Fermi energy) and total number of electrons N. With this new Hamiltonian, we introduce the modified Heisenberg picture

$$A(\mathbf{x}, \tau) = e^{K\tau/\hbar} A(\mathbf{x}) e^{-K\tau/\hbar}, \tag{8.52}$$

where a new complex variable τ has been introduced as a replacement for the time, but which can be analytically continued to *it*. That is, we have replaced the variable *it* with τ primarily to make the connection with the thermal distribution $e^{-\beta H}$ (although we shall use the grand canonical ensemble mentioned above), and we will have to be quite careful about various trajectories and paths that are taken, particularly for factors such as time ordering operators. This then leads to the single-particle Green's function

$$G_{rs}(\mathbf{x}, \tau; \mathbf{x}', \tau') = -\mathrm{Tr}\{\hat{\rho} T_\tau [\Psi_r(\mathbf{x}, \tau) \Psi_s^+(\mathbf{x}', \tau')]\}, \tag{8.53}$$

where r and s are spin indices (spin will now be important), T_τ is the "time ordering" operator in the imaginary τ domain, and the field operators are

$$\Psi_r(\mathbf{x}, \tau) = e^{K\tau/\hbar} \Psi_r(\mathbf{x}) e^{-K\tau/\hbar},$$
$$\Psi_s^+(\mathbf{x}, \tau) = e^{K\tau/\hbar} \Psi_s^+(\mathbf{x}) e^{-K\tau/\hbar}. \tag{8.54}$$

The temperature Green's function is useful because it will allow us to calculate the thermodynamic behavior of the system. If the Hamiltonian is time independent, then the Green's function depends only on the difference $\mathbf{x} - \mathbf{x}'$, as for the previous case. Similarly, a homogeneous system will lead to the Green's function being a function of $\mathbf{x} - \mathbf{x}'$. One useful property is shown by

$$\sum_s G_{rs}(\mathbf{x}, \tau; \mathbf{x}', \tau')|_{\tau' \to \tau^+} = \sum_s \text{Tr}\{\hat{\rho} T_\tau [\Psi_s^+(\mathbf{x}', \tau) \Psi_r(\mathbf{x}, \tau)]\}$$
$$= e^{\beta\Omega} \sum_s Tr\{e^{-\beta K} e^{K\tau/\hbar} \Psi_s^+(\mathbf{x}') \Psi_r(\mathbf{x}) e^{-K\tau/\hbar}\} \qquad (8.55)$$
$$= e^{\beta\Omega} \sum_s Tr\{e^{-\beta K} \Psi_s^+(\mathbf{x}') \Psi_r(\mathbf{x})\} = \langle n(\mathbf{x}) \rangle,$$

where we have used the anti-commutation of the fermion field operators and the cyclic property of the trace. The total density is found by integrating this over all space. Here, it becomes obvious why the complex time variable τ has been introduced. It allows the temporal evolution operators to be cast into the same exponential form as the statistical density matrix itself, and allows directly for the commutation of these different functions of the Hamiltonian.

Let us consider a simple noninteracting system, in which the creation and annihilation operators $a_{\mathbf{k}\lambda}^+$ and $a_{\mathbf{k}\lambda}$ create and destroy an electron in momentum state $\mathbf{k}$ with spin λ. Then, the field operators can be written as

$$\Psi(\mathbf{x}) = \frac{1}{\sqrt{V}} \sum_{\mathbf{k}\lambda} e^{i\mathbf{k}\cdot\mathbf{x}} \eta_\lambda a_{\mathbf{k}\lambda}, \qquad (8.56)$$

$$\Psi^+(\mathbf{x}) = \frac{1}{\sqrt{V}} \sum_{\mathbf{k}\lambda} e^{-i\mathbf{k}\cdot\mathbf{x}} \eta_\lambda^+ a_{\mathbf{k}\lambda}^+, \qquad (8.57)$$

where η_λ is a spin wave function and V is the volume (this usage differs from previous chapters). The equation of motion for the creation and annihilation operators is easily found, for example, by using

$$\hbar \frac{\partial a_{\mathbf{k}\lambda}}{\partial \tau} = e^{K\tau/\hbar} \{K, a_{\mathbf{k}\lambda}\} e^{-K\tau/\hbar} = -(E_k - \mu) a_{\mathbf{k}\lambda}, \ E_k = \frac{\hbar^2 k^2}{2m}, \qquad (8.58)$$

or

$$a_{\mathbf{k}\lambda}(\tau) = a_{\mathbf{k}\lambda} e^{-(E_k-\mu)\tau/\hbar}, \ a_{\mathbf{k}\lambda}^+(\tau) = a_{\mathbf{k}\lambda}^+ e^{(E_k-\mu)\tau/\hbar}. \qquad (8.59)$$

Then, the Green's function can be written as

$$G_{rs}(\mathbf{x}, \tau; \mathbf{x}', \tau') = -e^{\beta\Omega} \text{Tr}\{e^{-\beta K} T_\tau [\Psi_r(\mathbf{x}, \tau) \Psi_s^+(\mathbf{x}', \tau')]\}$$
$$= -\frac{1}{V} \sum_{\mathbf{k},\mathbf{k}'} \sum_{\lambda,\lambda'} e^{i\mathbf{k}\cdot\mathbf{x} - i\mathbf{k}'\cdot\mathbf{x}'} (\eta_\lambda)_r (\eta_{\lambda'})_s \qquad (8.60)$$
$$\times e^{-(E_k-\mu)\tau/\hbar + (E_{k'}-\mu)\tau'/\hbar} \langle a_{\mathbf{k}\lambda} a_{\mathbf{k}'\lambda'}^+ \rangle.$$

The spin wave functions are orthonormal, so that $\lambda' = \lambda$. Similarly, the last average requires that the momenta states are the same, and $\langle a_{\mathbf{k}\lambda} a_{\mathbf{k}\lambda}^+ \rangle = 1 - \langle a_{\mathbf{k}\lambda}^+ a_{\mathbf{k}\lambda} \rangle = 1 - f_k$, where

$$f_k = \frac{1}{1 + e^{\beta(E_k-\mu)}} \tag{8.61}$$

is the Fermi–Dirac distribution function, with the chemical potential being exactly the Fermi energy in this case. Then,

$$G_{rs}^0(\mathbf{x},\tau;\mathbf{x}',\tau') = \begin{cases} -\dfrac{\delta_{rs}}{V}\sum_{\mathbf{k}} e^{i\mathbf{k}\cdot(\mathbf{x}-\mathbf{x}')} e^{-(E_k-\mu)(\tau-\tau')/\hbar}(1-f_k), & \tau > \tau', \\ \dfrac{\delta_{rs}}{V}\sum_{\mathbf{k}} e^{i\mathbf{k}\cdot(\mathbf{x}-\mathbf{x}')} e^{-(E_k-\mu)(\tau'-\tau)/\hbar} f_k, & \tau < \tau'. \end{cases} \tag{8.62}$$

One important point of these Green's functions is the periodicity in complex time of the functions themselves. One obvious reason for using the complex time version of the operators is that the Heisenberg temporal propagator is in a form that commutes with the density matrix itself. However, this practice creates some unusual behaviors. Consider the case in which, for convenience, we take $\tau = 0$, $\tau' > 0$, so that we can write (8.53) as (we take only the fermion case here)

$$\begin{aligned} G_{rs}(\mathbf{x},0;\mathbf{x}',\tau') &= e^{\beta\Omega}\mathrm{Tr}\{e^{-\beta K}\Psi_s^+(\mathbf{x}',\tau')\Psi_r(\mathbf{x},0)\} \\ &= e^{\beta\Omega}\mathrm{Tr}\{\Psi_r(\mathbf{x},0)e^{-\beta K}\Psi_s^+(\mathbf{x}',\tau')\} \\ &= e^{\beta\Omega}\mathrm{Tr}\{\Psi_r(\mathbf{x},0)e^{-\beta K}\Psi_s^+(\mathbf{x}',\tau')e^{-\beta K}e^{\beta K}\} \\ &= e^{\beta\Omega}\mathrm{Tr}\{e^{-\beta K}\Psi_r(\mathbf{x},\beta\hbar)\Psi_s^+(\mathbf{x}',\tau')\} \\ &= -G_{rs}(\mathbf{x},\beta\hbar;\mathbf{x}',\tau'). \end{aligned} \tag{8.63}$$

Here, because of the periodicity, we have to assume that $0 < \tau' < \beta\hbar$. Thus, we find that the temperature Green's function is *antiperiodic* in $\beta\hbar$. This can easily be shown to be true in the second time variable as well. (The boson version is periodic in $\beta\hbar$.) This result is very important and builds the properties of the density matrix into the results for the Green's function. The importance of this is that the perturbation series is now integrated only over the range $(0, \beta\hbar)$ instead of $(0, \infty)$. In the usual situation in which the Hamiltonian is independent of time, the Green's function depends only on the difference in time coordinates, and the time is readily shifted so that

$$G_{rs}(\mathbf{x},\mathbf{x}';\tau-\tau'<0) = -G_{rs}(\mathbf{x},\mathbf{x}';\tau-\tau'+\beta\hbar>0). \tag{8.64}$$

For the noninteracting Green's function (8.62), this leads to the important result

$$f_k e^{\beta(E_k-\mu)} = 1 - f_k, \tag{8.65}$$

which insures that the equilibrium distribution function is a Fermi–Dirac function.

With this periodicity in mind, we can now introduce the Fourier transform representation (in time) for the temperature Green's function. We note that both the boson and fermion temperature Green's functions are fully periodic in $2\beta\hbar$. We let $\tau'' = \tau - \tau'$, and define the Fourier transform from

8.3 Electron–electron interaction effects

$$G_{rs}(\mathbf{x}, \mathbf{x}'; \tau'') = \frac{1}{\beta\hbar} \sum_n e^{-i\omega_n \tau''} G_{rs}(\mathbf{x}, \mathbf{x}'; \omega_n), \tag{8.66}$$

where, for the moment,

$$\omega_n = \frac{n\pi}{\hbar\beta}. \tag{8.67}$$

We remark here that τ'' in this equation is restricted to lie within the principal periodic region (e.g., $-\hbar\beta < \tau'' < \hbar\beta$) since it is a complex quantity, and we must assure that the summation does not diverge because of this. Principally, this means that the real time is restricted to a finite range. This now leads to the transform itself,

$$G_{rs}(\mathbf{x}, \mathbf{x}'; \omega_n) = \frac{1}{2} \int_{-\hbar\beta}^{\hbar\beta} d\tau'' e^{i\omega_n \tau''} G_{rs}(\mathbf{x}, \mathbf{x}'; \tau''). \tag{8.68}$$

It is convenient to separate this integral into two parts, the positive and negative time segments, as

$$\begin{aligned}
G_{rs}(\mathbf{x}, \mathbf{x}'; \omega_n) &= \frac{1}{2} \int_{-\hbar\beta}^{0} d\tau'' e^{i\omega_n \tau''} G_{rs}(\mathbf{x}, \mathbf{x}'; \tau'') + \frac{1}{2} \int_{0}^{\hbar\beta} d\tau'' e^{i\omega_n \tau''} G_{rs}(\mathbf{x}, \mathbf{x}'; \tau'') \\
&= -\frac{1}{2} \int_{-\hbar\beta}^{0} d\tau'' e^{i\omega_n \tau''} G_{rs}(\mathbf{x}, \mathbf{x}'; \tau'' + \beta\hbar) \\
&\quad + \frac{1}{2} \int_{0}^{\hbar\beta} d\tau'' e^{i\omega_n \tau''} G_{rs}(\mathbf{x}, \mathbf{x}'; \tau'') \\
&= \frac{1}{2}\left(1 - e^{-i\omega_n \beta\hbar}\right) \int_{0}^{\hbar\beta} d\tau'' e^{i\omega_n \tau''} G_{rs}(\mathbf{x}, \mathbf{x}'; \tau'').
\end{aligned} \tag{8.69}$$

The prefactor vanishes for n even. A similar approach shows that the prefactor will vanish for the bosonic form for n odd. For these cases, the prefactor is unity. This further implies that for fermions we use only the odd frequencies in (8.67), whereas for bosons we use only the even frequencies. Then, for fermions (8.17) reduces to

$$G_{rs}(\mathbf{x}, \mathbf{x}'; \omega_n) = \frac{1}{2} \int_{-\hbar\beta}^{\hbar\beta} d\tau'' e^{i\omega_n \tau''} G_{rs}(\mathbf{x}, \mathbf{x}'; \tau''), \quad \omega_n = \frac{(2n+1)\pi}{\hbar\beta}. \tag{8.70}$$

Since this Green's function is often called a *Matsubara Green's function*, the frequencies are referred to as the *Matsubara frequencies*.

To examine the nature of this function, let us expand the field operators in a set of basis functions $\{\varphi_m\}$, so that in Dirac notation we can write (8.70) using (8.60) as

$$G_{rs}(i\omega_n) = -e^{\beta\Omega}\sum_{m,m'}|\langle m|\Psi(\mathbf{x})|m'\rangle|^2 e^{-\beta E_m}\int_0^{\beta\hbar} d\tau'' e^{i\omega_n\tau''} e^{\tau''(E_m-E_{m'})/\hbar}$$
$$= e^{\beta\Omega}\sum_{m,m'}|\langle m|\Psi(\mathbf{x})|m'\rangle|^2 \frac{e^{-\beta E_m}+e^{-\beta E_{m'}}}{i\omega_n+(E_m-E_{m'})/\hbar}. \tag{8.71}$$

In particular, if the basis set is that for which the creation and annihilation operators are defined for the field operator, then the matrix elements are zero except for particular connections between the two coefficients; that is, $m' = m \pm 1$, with the sign determined by which of the two operators is used to define the matrix elements. However, there is no requirement that this basis set be the same as that used in the field operator definition. On the other hand, if we use momentum wave functions (the plane waves), then the difference in energies produces just the single momentum E_k, where $\mathbf{k}$ is the Fourier transform for the difference in position $\mathbf{x} - \mathbf{x}'$.

In direct analogy with the results of a previous chapter, it is now possible to write the retarded and advanced temperature Green's functions as

$$G_{rs}^r(\mathbf{x},\tau;\mathbf{x}',\tau') = -\vartheta(\tau-\tau')\text{Tr}\left\{\hat{\rho}T_\tau\left[\Psi_r(\mathbf{x},\tau),\Psi_s^+(\mathbf{x}',\tau')\right]_+\right\}, \tag{8.72}$$

$$G_{rs}^a(\mathbf{x},\tau;\mathbf{x}',\tau') = \vartheta(\tau'-\tau)\text{Tr}\left\{\hat{\rho}T_\tau\left[\Psi_r(\mathbf{x},\tau),\Psi_s^+(\mathbf{x}',\tau')\right]_+\right\}. \tag{8.73}$$

(Note that since we are using the curly braces to represent the quantity over which the trace is performed, we can no longer use these to indicate the anti-commutator; therefore we will use the normal commutator with the subscript "+" to indicate that the anti-commutator relationship is used for the fermions.) For now, let us consider just the retarded function and take the Fourier transform of this quantity. First, we make the same basis function expansion as above, and this leads to

$$\begin{aligned}G_{rs}^r(\tau'') &= -\vartheta(\tau'')e^{\beta\Omega}\sum_{m,m'}e^{-\beta E_m}\left\{|\langle m|\Psi(\mathbf{x})|m'\rangle|^2 e^{i\tau''(E_m-E_{m'})/\hbar}\right.\\ &\quad\left. + |\langle m'|\Psi(\mathbf{x})|m\rangle|^2 e^{-i\tau''(E_m-E_{m'})/\hbar}\right\}\\ &= -\vartheta(\tau'')e^{\beta\Omega}\sum_{m,m'}\left\{|\langle m|\Psi(\mathbf{x})|m'\rangle|^2 e^{i\tau''(E_m-E_{m'})/\hbar}\right.\\ &\quad\left.\times\left(e^{-\beta E_m}+e^{-\beta E_{m'}}\right)\right\},\end{aligned} \tag{8.74}$$

where we have interchanged the two dummy indices in the second term of the first line. It is important to note that the normal Heisenberg time variation has been used, and not the limited imaginary time form. We now Fourier-transform this quantity over the entire frequency range, so that

$$G_{rs}^r(\omega) = e^{\beta\Omega} \sum_{m,m'} |\langle m|\Psi(\mathbf{x})|m'\rangle|^2 \frac{e^{-\beta E_m} + e^{-\beta E_{m'}}}{\omega + i\eta + (E_m - E_{m'})/\hbar}, \quad (8.75)$$

with the normal small quantity η required for convergence of the integration. Comparing this result with that of (8.71) leads to an informative result, that the Matsubara Green's function can be transformed into the retarded temperature Green's function through the limiting process

$$\lim_{i\omega_n \to \omega + i\eta} G_{rs}(i\omega_n) = G_{rs}^r(\omega). \quad (8.76)$$

This limiting process is referred to as analytic continuation [29]. A similar result, with the sign of the small quantity reversed, leads to the advanced temperature Green's function.

Another quantity of interest is the spectral density function $A(\mathbf{k}, \omega)$, which was defined in a previous chapter. This quantity is defined from the retarded or advanced Green's functions through

$$A(\mathbf{k}, \omega) = -2\text{Im}\{G_{rs}^r(\mathbf{k}, \omega)\} = 2\text{Im}\{G_{rs}^a(\mathbf{k}, \omega)\}. \quad (8.77)$$

From the form of the retarded function (8.75), the only imaginary part arises from the denominator of the last fraction. This term can be expanded as

$$\frac{1}{\omega + (E_m - E_{m'})/\hbar + i\eta} = P\frac{1}{\omega + (E_m - E_{m'})/\hbar} - i\pi\delta(\omega + (E_m - E_{m'})/\hbar). \quad (8.78)$$

Using this in the definition of the retarded function, we arrive at

$$\begin{aligned}A(\mathbf{k}, \omega) &= 2\pi e^{\beta\Omega} \sum_{m,m'} |\langle m|\Psi(\mathbf{x})|m'\rangle|^2 \left(e^{-\beta E_m} + e^{-\beta E_{m'}}\right)\delta(\omega + (E_m - E_{m'})/\hbar) \\ &= 2\pi e^{\beta\Omega}\left(1 + e^{-\hbar\beta\omega}\right)\sum_{m,m'} |\langle m|\Psi(\mathbf{x})|m'\rangle|^2 e^{-\beta E_m}.\end{aligned} \quad (8.79)$$

We note that the right side is positive definite, so the spectral density can be interpreted as a probability density function. This leads to the important result that

$$\begin{aligned}1 &= \int \frac{d\omega}{2\pi} A(\mathbf{k}, \omega) = e^{\beta\Omega} \sum_{m,m'} |\langle m|\Psi(\mathbf{x})|m'\rangle|^2 \left(e^{-\beta E_m} + e^{-\beta E_{m'}}\right) \\ &= e^{\beta\Omega} \sum_m e^{-\beta E_m}\{\Psi(\mathbf{x})\Psi^+(\mathbf{x}) + \Psi^+(\mathbf{x})\Psi(\mathbf{x})\} \\ &= e^{\beta\Omega} Tr\{e^{-\beta K}\} = 1.\end{aligned} \quad (8.80)$$

In the noninteracting case, we expect the spectral density to be a delta function relating the energy to the momentum. The broadening of this function by the interaction eliminates the simple relationship between energy and momentum that exists in classical mechanics, and thus is the major introduction of quantum effects in this regard. Finally, we note that by comparing (8.71), (8.75), and (8.79), we can write the Green's functions in terms of the spectral density as

$$G_{rs}(i\omega_n) = \int \frac{d\omega'}{2\pi} \frac{A(\mathbf{k},\omega')}{i\omega_n - \omega'}, \quad (8.81)$$

$$G^r_{rs}(\omega) = \int \frac{d\omega'}{2\pi} \frac{A(\mathbf{k},\omega')}{\omega - \omega' + i\eta}. \quad (8.82)$$

The connection between these two equations is an example of the process of analytic continuation. Here we take the complex frequency $i\omega_n$ and analytically continue it to the proper real frequency (with an imaginary convergence factor, as has been used in the previous chapters), $\omega + i\eta$.

8.3.4 One-particle density of states

In Chapter 2, we developed the concept of the density of states (per unit energy per unit volume). There, for example, it was shown that the density of states could be written as

$$\rho_d(\omega) = \frac{2}{\hbar} \int \frac{d^d\mathbf{k}}{(2\pi)^d} \delta(\omega - E_k/\hbar), \quad (8.83)$$

where d is the dimensionality of the system, and the factor of $s=2$ is for electrons, where the opposite spins do not raise the spin degeneracy (or represents a summation over the diagonal spin indices of the Green's function). For three dimensions, one recovers the familiar

$$\rho_3(\omega) = \frac{1}{2\pi^2} \left(\frac{2m}{\hbar^2}\right)^{3/2} (\hbar\omega)^{1/2}. \quad (8.84)$$

If we write this in the noninteracting Green's function form (with the latter described by plane-wave states and putting in the values of the creation and annihilation operators), this can be more easily related to the noninteracting Green's function as

$$\rho_d(\omega) = -\frac{1}{\pi\hbar} \sum_{r,s} \int \frac{d^d\mathbf{k}}{(2\pi)^d} \mathrm{Im}\{G^r_{rs}(\mathbf{k},\omega)\}\delta_{rs}, \quad (8.85)$$

and hence the density of states is related to the spectral density. This relationship can be expressed as

$$\rho_d(\omega) = -\frac{1}{2\pi\hbar} \sum_{r,s} \int \frac{d^d\mathbf{k}}{(2\pi)^d} A(\mathbf{k},\omega)\delta_{rs}. \quad (8.86)$$

In general, the decay of excitations is much stronger in the disordered (large impurity scattering and diffusive transport) system than it is in the pure system. This is of concern, since most mesoscopic systems were treated early on as disordered systems because of the randomness introduced by the impurity scattering. As we saw in previous chapters, this is now not often the case. One of the important properties in disordered material, therefore, is the one-particle density of states. Now, by one-particle density of states we don't mean the quantity for a single electron, but rather the density of states that is appropriate for a single electron in the sea of other electrons and impurities. The temperature Green's function can be recognized through the reverse of the analytic continuation procedures, which was

$$\rho_d(\omega) = -\frac{1}{\pi\hbar} \sum_{r,s} \int \frac{d^d\mathbf{k}}{(2\pi)^d} \text{Im}\{G_{rs}(\mathbf{k}, i\omega \to \omega + i\eta)\}\delta_{rs}, \quad (8.87)$$

and this may be related to the noninteracting Green's function and the self-energy by Dyson's equation as

$$\rho_d(\omega) = -\frac{1}{\pi\hbar} \sum_{r,s} \int \frac{d^d\mathbf{k}}{(2\pi)^d} \text{Im}\left\{\frac{1}{\left[G_{rs}^0(\mathbf{k},\omega)\right]^{-1} - \Sigma(\mathbf{k},\omega)}\right\}\delta_{rs}, \quad (8.88)$$

where the self-energy includes those parts from impurities and the electron–electron interaction, which itself may be mediated by the impurity scattering in the diffusive limit, as we shall see below. The result for the interacting Green's function is essentially the same as that of the last chapter for zero-temperature Green's functions, and arises from the properties of the linear perturbation expansion and resummation that leads to Dyson's equation itself, and not from any particular property of any one type of Green's function.

Thus, the spectral density in the interacting system represents the entire density of states, including both the single-particle properties and their modification that arises from the self-energy corrections that come from the presence of the interactions. In the next few sections, we will examine the electron–electron interaction and its modification in the presence of strong impurity scattering. This will entail an approach that is quite different from the one usually found in high-mobility materials, where the electron–electron interaction is thought to be the dominant interaction process. Let us now turn to the self-energy for the electron–electron interaction.

8.3.5 The effective interaction potential

In a semiconductor, the interactions of the electrons can be with a variety of scattering centers, whether from impurities, from other electrons, or from phonons. Generally, one can write the diagrammatic expansion of the perturbation series as shown in Fig. 8.9. Only the impurity and the electron–electron interactions are indicated in this figure. It also is important to note that only the lowest-order terms have been retained. The first term in the electron–electron diagram series is valid generally only for small momentum exchange, and therefore it is sensitive to the bare interaction potential. There are higher-order terms, on the other hand, which tend to involve larger momentum exchanges, and one needs to consider the screening of the interaction potential by the other

Fig. 8.9 Contributions to impurity (a) and electron–electron (b) self-energy.

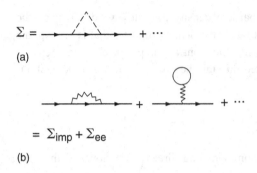

electrons. The Coulomb interaction between individual electrons is a long-range interaction, and never should it be taken into account only in first-order perturbation theory. Rather, one should use an effective interaction potential that takes into account the screening, as discussed above. This was the procedure used in the earlier chapters, and it leads to the need to account for the potential and its self-screening. This interaction potential may be found to be

$$U_{eff} = \frac{U}{1 - U\Pi(\omega)},\tag{8.89}$$

where $\Pi(\omega)$ is a *polarization* resummation, as discussed in connection with other diagrammatic expansions. However, as will be explained below, this polarization is slightly different in that the terms included in the summation are a different set of diagrams. We want to examine how this resummation of the terms occurs in this section. The interaction potential itself is the bare Coulomb interaction $e^2/4\pi\varepsilon r$, which leads to the Fourier transformations

$$U(\mathbf{q}) = \begin{cases} \dfrac{e^2}{\varepsilon q^2}, & d = 3, \\ \dfrac{e^2}{2\varepsilon q}, & d = 2, \\ \dfrac{e^2}{4\pi\varepsilon}\ln\left(1 + \dfrac{q_0^2}{q^2}\right), & d = 1, \end{cases}\tag{8.90}$$

where q_0 is an artificial cutoff in momentum for the Coulomb potential (which was discussed in a previous section) [30]. It is important to point out that the forms that appear in (8.90) are for true dimensionality d, since integration over the other directions in a quasi-d-dimensional system produces some modifications. These will not be dealt with here, but they can be quite important in some cases.

The summation represented by (8.89) is different than the earlier summations in some crucial details. Here, we are replacing the bare interaction line (the squiggly line in Fig. 8.9) with a summation of an infinite number of terms, which leads to a renormalized, dressed interaction. The set of diagrams that are most important in this interaction is shown in Fig. 8.10. Here, the polarization $\Pi(\omega)$ is composed of the two Green's functions that form the ring in a summation of *ring* diagrams. (We emphasize that, as in all such cases, experience tells

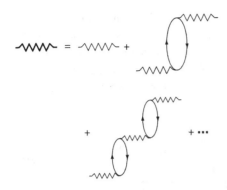

Fig. 8.10 The expansion of the interaction into ring diagrams.

us just which set of diagrams should be resummed to give the dominant interactions.) That is, the polarization function that is important here is the closed pair of Green's functions, in which the expansion is a summation of such Green's function rings, connected at each end by an interaction line. Each ring corresponds to a density fluctuation, and the long-range correlations of these disparate density fluctuations are coupled through the carrier–carrier interaction into the effective interaction. The interaction lines at the input and output all carry momentum **q**. However, the polarization is different here from that treated in the previous chapter in that the two Green's functions not only have different momentum **k** and **k − q**, but also different frequencies $\omega_n + \omega_1$ and ω_1. This is because it is a summation of ring diagrams rather than ladder diagrams. Thus, we denote this polarization $\Pi(\omega)$. However, it is also complicated by the fact that the two Green's functions are also representative of the diffusive nature of the transport in diffusive systems, so there can be a variety of corrections to the ring diagrams themselves. We deal with one such set of corrections in the next section. The expansion shown in Fig. 8.10 can be written as

$$U_{\it eff} = U + U\Pi(\omega)U + U\Pi(\omega)U\Pi(\omega)U + \cdots = \frac{U}{1 - U\Pi(\omega)}. \tag{8.91}$$

This is just the result of (8.89). In many cases, this is written as

$$U_{\it eff} = \frac{U_0}{\varepsilon_{\it eff}(\omega)}, \tag{8.92}$$

where U_0 is the appropriate form from (8.90) with the dielectric constant removed. This leads to the general definition of the dielectric function as

$$\varepsilon_{\it eff}(\omega) = \varepsilon[1 - U\Pi(\omega)]. \tag{8.93}$$

In some cases, other contributions to the overall polarization, such as that due to the lattice in polar materials, can be added to the dielectric function.

The lowest-order polarization, composed of the ring diagrams, is still written as a summation over a pair of Green's functions, but now a summation over the new frequency ω_1 has been added, as well as over the momentum **k**. The end result, of course, modifies the interaction potential, which is a function of **q**, so that these new variables are integrated. It is this integration over interior variables in the pair of Green's functions that distinguishes this polarization, which is related to the density–density correlation function. We can now write the polarization as

$$\Pi(\mathbf{q},\omega_n) = \frac{2}{\beta\hbar^2}\int \frac{d^d\mathbf{k}}{(2\pi)^d}\sum_{\omega_1}\frac{1}{i\omega_1 - (E_k-\mu)/\hbar - i/2\tau} \\ \times \frac{1}{i\omega_1 + i\omega_n - (E_{k+q}-\mu)/\hbar + i/2\tau}, \quad (8.94)$$

and the role of the impurities has been included here through a lifetime in the Green's functions as a replacement for the normal factor η. The momentum integration is over a single set of spin states. A typical term in the frequency summation is of order $1/|\omega_1|^2$ for large values of the frequency, and the summation therefore converges absolutely [28]. Although one could do the contour integration, a more useful result is obtained by introducing a convergence factor (which must be handled carefully due to the complex frequencies) and rewriting the two sums as

$$\Pi(\mathbf{q},\omega_n) = \frac{2}{\beta\hbar^2}\int \frac{d^d\mathbf{k}}{(2\pi)^d}\frac{1}{i\omega_n - (E_{k+q}-E_k)/\hbar - i/\tau} \\ \times \sum_{\omega_1}\left[\frac{1}{i\omega_1 - (E_k-\mu)/\hbar - i/2\tau}\right. \\ \left. - \frac{1}{i\omega_1 + i\omega_n - (E_{k+q}-\mu)/\hbar + i/2\tau}\right]. \quad (8.95)$$

At this point, we want to develop a useful identity that represents the frequency sums needed to evaluate (8.95). The summation that we would like to consider is the fermion sum

$$\sum_n e^{i\omega_n\eta}\frac{1}{i\omega_n - x}, \quad n \text{ odd}. \quad (8.96)$$

This sum would diverge (we ignore the lifetime effects of τ in this argument) without the convergence factor, so η must remain positive until after the sum is evaluated (which implies directly that the separation done above is valid only when the convergence factor is included). The most direct approach is to use contour integration of a suitable function. The function $-\beta\hbar(e^{\beta\hbar z}+1)^{-1}$ has simple poles at the values $z = i(2n+1)\pi/\beta\hbar = i\omega_n$. Each pole has unit residue with the chosen normalization. Thus, we consider the contour integral

$$-\frac{\beta\hbar}{2\pi i}\oint_C \frac{dz}{e^{\beta\hbar z}+1}\frac{e^{\eta z}}{z-x}. \quad (8.97)$$

8.3 Electron–electron interaction effects

The choice now is the contour that we use. We pick a simple contour that encloses the imaginary axis (and includes the point $z = x$) in the positive sense (counterclockwise); that is, we use a circle of radius R with $R \to \infty$. The poles are then the set of fermion Matsubara frequencies $z = i(2n + 1)\pi/\beta\hbar = i\omega_n$ and the pole $z = x$. Now, as $R \to \infty$, the argument of the integral vanishes (for $\beta\hbar > \eta > 0$), so that the integral around the contour itself vanishes. Thus,

$$0 = \sum_n e^{i\omega_n \eta} \frac{1}{i\omega_n - x} - \beta\hbar \frac{e^{\eta x}}{e^{\beta\hbar x} + 1}. \tag{8.98}$$

After letting the convergence factor vanish, this gives [28,30],

$$\sum_n e^{i\omega_n \eta} \frac{1}{i\omega_n - x} = \beta\hbar \frac{1}{e^{\beta\hbar x} + 1} = \beta\hbar f(\hbar x). \tag{8.99}$$

The result for the frequency summation yields the Fermi–Dirac distribution (by construction). This result can now be used in (8.95) to give the polarization as

$$\Pi(\mathbf{q}, \omega_n) = -\frac{2}{\hbar} \int \frac{d^d\mathbf{k}}{(2\pi)^d} \frac{f_{k+q} - f_k}{i\omega_n - (E_{k+q} - E_k)/\hbar - i/\tau}. \tag{8.100}$$

Adding and subtracting a term $f_{k+q}f_k$ in the numerator allows us to rearrange this last equation as

$$\Pi(\mathbf{q}, \omega_n) = -\frac{2}{\hbar} \int \frac{d^d\mathbf{k}}{(2\pi)^d} \frac{f_{k+q}(1 - f_k) - f_k(1 - f_{k+q})}{i\omega_n - (E_{k+q} - E_k)/\hbar - i/\tau}. \tag{8.101}$$

Making the change of variables $\mathbf{k} + \mathbf{q} \to \mathbf{k}$, and using the spherical symmetry of the energy bands, this equation can be rearranged into

$$\Pi(\mathbf{q}, \omega_n) = -\frac{2}{\hbar} \int \frac{d^d\mathbf{k}}{(2\pi)^d} f_{k+q}(1 - f_k) \frac{(E_{k+q} - E_k)/\hbar}{(\omega_n + 1/\tau)^2 - [(E_{k+q} - E_k)/\hbar]^2}. \tag{8.102}$$

There are a great variety of approximations that can be developed for the polarization. We discuss a few of these.

8.3.5.1 Static screening

In the case of static screening, we will essentially take the small-momentum, low-frequency limit of (8.100). In this approach, we ignore the frequency and the scattering rate as both being small. Then, the numerator and denominator are expanded as

$$f_{k+q} - f_k \approx \mathbf{q} \cdot \frac{\partial f_k}{\partial \mathbf{k}} \approx \left[\mathbf{q} \cdot \frac{\partial E}{\partial \mathbf{k}}\right] \frac{\partial f_k}{\partial E} \tag{8.103}$$

and

$$E_{k+q} - E_k \approx \mathbf{q} \cdot \frac{\partial E}{\partial \mathbf{k}}. \tag{8.104}$$

For nondegenerate semiconductors, we can also write

$$\frac{\partial f_k}{\partial E} = -\beta f_k. \tag{8.105}$$

Now, using

$$2 \int \frac{d^d \mathbf{k}}{(2\pi)^d} f_k = n, \tag{8.106}$$

we can rewrite (8.100) as

$$\Pi(\mathbf{q}, 0) = -\frac{n}{k_B T}. \tag{8.107}$$

The dielectric function may then be written as

$$\varepsilon(0) = \varepsilon \left[1 + \frac{q_D^2}{q^2} \right], \quad q_D^2 = \frac{ne^2}{\varepsilon k_B T}, \tag{8.108}$$

where the latter quantity is the square of the reciprocal Debye screening length. This latter quantity provides exactly the cutoff behavior discussed in the above sections. For degenerate material, the average energy $k_B T$ is replaced by the Fermi energy E_F, and the quantity becomes the reciprocal of the squared Fermi–Thomas screening length.

8.3.5.2 Plasmon-pole approximation

Still another approximation occurs when we assume that the frequency is large compared to the energy exchange in the denominator of (8.100). This equation is first rewritten as

$$\Pi(\mathbf{q}, \omega_n) = -\frac{2}{\hbar} \int \frac{d^d \mathbf{k}}{(2\pi)^d} f_k \left[\frac{1}{i\omega_n - (E_k - E_{k-q})/\hbar - i/\tau} \right.
\left. - \frac{1}{i\omega_n - (E_{k+q} - E_k)/\hbar - i/\tau} \right] \tag{8.109}$$

where a change of variables has been made in the first term. This can now be combined as

$$\Pi(\mathbf{q}, \omega_n) = -\frac{2}{\hbar} \int \frac{d^d \mathbf{k}}{(2\pi)^d} f_k$$
$$\times \frac{2E_k - E_{k+q} - E_{k-q}}{[i\omega_n - (E_k - E_{k-q})/\hbar - i/\tau][i\omega_n - (E_{k+q} - E_k)/\hbar - i/\tau]}. \tag{8.110}$$

The numerator of the fraction can be written

$$2E_k - E_{k+q} - E_{k-q} \sim -\frac{\hbar^2 q^2}{m}, \tag{8.111}$$

and the denominator is approximately $(i\omega_n)^2$. Using the summation (8.106), the dielectric function can now be written (after analytically continuing the frequency into the real domain)

$$\varepsilon(0) = \varepsilon \left[1 - \frac{\omega_p^2}{\omega^2} \right], \qquad (8.112)$$

where

$$\omega_p^2 = \frac{ne^2}{m\varepsilon} \qquad (8.113)$$

is the free-carrier plasma frequency.

In many cases, one wants to add the lattice contributions to the dielectric function, in which case the polarization in polar semiconductors may be added to (8.113) as [24]

$$\varepsilon(0) = \varepsilon \left[1 + \frac{\omega_{LO}^2 - \omega_{TO}^2}{\omega_{TO}^2 - \omega^2} - \frac{\omega_p^2}{\omega^2} \right], \qquad (8.114)$$

where ω_{TO} and ω_{LO} are the transverse and longitudinal optical phonon frequencies, respectively. When $\omega_p < \omega_{LO}$, there are two zeros and two poles of the dielectric function. The poles are at zero frequency and at the transverse optical phonon frequency. The zeroes are at a down-shifted plasma frequency (evaluated at the low-frequency dielectric constant) and at the longitudinal phonon frequency. Here, the dielectric constant ε is inferred to be the high-frequency dielectric constant. In the opposite case, where $\omega_p > \omega_{LO}$, a zero at the transverse optical frequency cancels the pole at this frequency, so there remains only the poles at zero frequency and at the plasma frequency. This latter is the hybridized polar phonon–plasmon frequency, and corresponds to what Ridley [25] has called the *descreened* frequency.

In essence, the interaction between the single carrier and the background electron gas is evaluated at the zeros of the dielectric function, which means that in this approximation the dominant scattering is from a single particle to either the longitudinal optical phonon or to a collective oscillation (the plasmons). In the simple approximation that has been used so far in this section, the only scattering is to the lattice or to the collective modes. This approximation has been termed the plasmon-pole approximation [31,32], but it ignores the important single-particle scattering interactions. This is especially significant, since the single-particle interactions are thought to be dominant in phase-breaking processes. That is, in the last subsection for low frequencies (low energies), we predominantly found a screened interaction representing the scattering of an incident electron by the gas of nearly free electrons – single-particle scattering. Here, in the high-frequency (high-energy) situation, we predominantly find an interaction representing the scattering of an incident electron by the collective modes of the electron gas (and the lattice polar modes if that term is included). In the next subsection, we will probe between these two limits.

8.3.5.3 Momentum-dependent screening

Although the low-frequency approximation will be used here, the desire is to evaluate carefully the summation over the free carriers that is involved in the summation (8.109), and to remove the approximation $q \ll k$. While we can continue to use the low-frequency approximation, we now want to examine a range of values for the momentum exchange q. Thus, the dielectric function may be rewritten as

$$\varepsilon(\mathbf{q}, 0) = \varepsilon + \frac{e^2}{q^2} \int \frac{d^d \mathbf{k}}{(2\pi)^d} f_k \left[\frac{1}{E_{k+q} - E_k} - \frac{1}{E_k - E_{k-q}} \right]. \quad (8.115)$$

As in the previous sections, the energy denominators can be expanded as

$$E_{k \pm q} - E_k = \frac{\hbar^2 q^2}{2m} \pm \frac{\hbar^2 kq}{m} \cos \vartheta, \quad (8.116)$$

where the mass is, of course, the effective mass in the semiconductor of interest. The integration will be carried out for three dimensions but is readily extendible to lower dimensionality. The integration over the angles is straightforward, and this leads to

$$\varepsilon(\mathbf{q}, 0) = \varepsilon + \frac{2e^2 m}{q^2 \hbar^2} \int_0^\infty \int_0^\pi \frac{k^2 dk \sin \vartheta d\vartheta}{2\pi^2} f_k \left[\frac{1}{q^2 + 2kq \cos \vartheta} \right.$$

$$\left. + \frac{1}{q^2 - 2kq \cos \vartheta} \right] \quad (8.117)$$

$$= \varepsilon + \frac{e^2 m}{\pi^2 q^2 \hbar^2} \int_0^\infty f_k k \ln \left| \frac{k + 2q}{k - 2q} \right| dk.$$

The form of the argument of the logarithm arises from having factored $\hbar^2 q / 2m$ out of each term in the numerator and the denominator, so the magnitude sign is required to assure that the argument is positive definite. To proceed further, the following normalized variables are now introduced:

$$\xi^2 = \frac{\hbar^2 q^2}{8mk_B T}, \quad x^2 = \frac{\hbar^2 k^2}{2mk_B T}, \quad \mu = \frac{E_F}{k_B T}. \quad (8.118)$$

It may be noted that the temperature here is that of the distribution function and represents the electron temperature, not the lattice temperature, so that this formulation may well be used in nonequilibrium situations such as those of the next chapter. Although this has not been noted by any subscript on the temperature, it should not be confusing since this is the only temperature in the problem. By incorporating these normalizing factors into the expression (8.117), we obtain

8.3 Electron–electron interaction effects

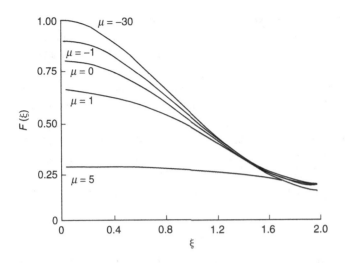

Fig. 8.11 Momentum dependence of the screening function.

$$\varepsilon(\mathbf{q}, 0) = \varepsilon\left[1 + \frac{q_D^2}{q^2} F(\xi, \mu)\right], \qquad (8.119)$$

where

$$F(\xi, \mu) = \frac{1}{\sqrt{\pi}\xi F_{1/2}(\mu)} \int_0^\infty \frac{x\,dx}{1 + e^{x^2 - \mu}} \ln\left|\frac{x + \xi}{x - \xi}\right|, \qquad (8.120)$$

and $F_{1/2}(\eta)$ is a standard Fermi–Dirac integral

$$F_\eta(\mu) = \int_0^\infty \frac{x^{3/2 - \eta} dx}{1 + e^{x^2 - \mu}}. \qquad (8.121)$$

In the case of a nondegenerate semiconductor (μ large and negative), $F(\xi,\mu)$ becomes Dawson's integral and is a tabulated function. In either case, however, best results are obtained from numerical simulation. In Fig. 8.11, the overall behavior of this function is plotted, and the topmost curve ($\mu = -30$) corresponds to the nondegenerate limit. The behavior is not very dramatic. As $q \to 0$ (i.e., $\xi \to 0$), $F(\xi,\mu) \to 1$ for nondegenerate material, and the usual Debye screening behavior is recovered. On the other hand, as $q \to \infty$, $F(\xi,\mu) \to 0$, and the screening is broken up completely. Thus, for high-momentum transfer in the scattering process, the scattering potential is completely descreened $q \gg q_D$. The upshot of this is that the nonlinearity in Coulomb scattering appears once again as a scattering cross section that depends upon itself through the momentum transfer $\hbar q$ (and hence through the scattering angle θ). $F(\xi,\mu)$ has decreased to a value of one-half its maximum already at a value of $\xi = 1.07$, for which $q = 1.75 q_T$, where $q_T = mv_T/\hbar$ is the thermal wavevector, corresponding to the thermal velocity of a carrier. For GaAs at room temperature, $q_T = 2.61 \times 10^6$ cm^{-1}. This value is only 3.5 times larger than the Debye wavevector at a carrier density of 10^{17} cm^{-3} and becomes smaller than the Debye screening wavevector at higher carrier densities. Although screening is not totally eliminated, it is greatly reduced, and this can lead to more effective single-particle scattering than expected.

On the other hand, as the carrier density is increased, the material becomes degenerate, and the Maxwellian approximation cannot be used. In the case of the Fermi–Dirac distribution above, the integrals are more complicated but still readily evaluated, as evident in Fig. 8.11. It is clear that the screening is strongly reduced at all wavevectors as the carrier density is increased, but the variation with the wavevector is also strongly reduced. The equivalent results for two dimensions have been evaluated by Ando *et al.* [33], and a figure equivalent to Fig. 6.11 appears in their review.

8.3.5.4 Dynamic screening

As a further consideration, we consider a number of factors. First, the product of the Fermi–Dirac functions for small $q \to 0$ give essentially the derivative of the function, which is equivalent to a delta function at the Fermi energy (at sufficiently low temperature). But, for this term the angular average vanishes. If we wish to keep the frequency variables in the dielectric function, so that we have two approaches to consider. In one, we expand the term in f_{k+q} and expand the energy differences, keeping only the lowest-order terms in q. The expansion of the energy around the momentum k yields a function of the angle between the two vectors, which is involved in the averaging process of the d-dimensional integration. In the second approximation, we use the expansion already found in (8.110). In the former case, we arrive at the derivative of the distribution function, which is useful for nondegenerate materials. However, in the case of degenerate materials, we must use the second approach. Let us examine how these differ.

We begin with the form (8.100) for the dielectric function (ignoring the phonon contribution), which we write as

$$\varepsilon(\mathbf{q},\omega) = \varepsilon + \frac{2e^2}{\hbar q^2} \int \frac{d^d\mathbf{k}}{(2\pi)^d} \frac{f_k - f_{k+q}}{\omega + (E_{k+q} - E_k)/\hbar - i/\tau} . \tag{8.122}$$

Then, the numerator is evaluated by (8.113), and the denominator is expanded into a power series

$$E_{k+q} - E_k \approx \mathbf{q} \cdot \frac{\partial E}{\partial \mathbf{k}} + \cdots = \hbar \mathbf{q} \cdot \mathbf{v} + \cdots \tag{8.123}$$

This allows us to write the dielectric function as

$$\varepsilon(\mathbf{q},\omega) = \varepsilon + \frac{2e^2}{\hbar q^2} \int \frac{d^d\mathbf{k}}{(2\pi)^d} \frac{\partial f_k}{\partial E} \frac{-\hbar \mathbf{q} \cdot \mathbf{v}}{\omega + \mathbf{q} \cdot \mathbf{v} - i/\tau} . \tag{8.124}$$

This equation can now be rewritten for a nondegenerate distribution, using (8.115) as

8.3 Electron–electron interaction effects

$$\varepsilon(\mathbf{q},\omega) = \varepsilon + \frac{2e^2}{q^2 k_B T}\int \frac{d^d\mathbf{k}}{(2\pi)^d} f_k \frac{\mathbf{q}\cdot\mathbf{v}}{\omega + \mathbf{q}\cdot\mathbf{v} - i/\tau}. \tag{8.125}$$

The dot product involves the polar angle in the integration, and this makes it quite complicated. However, we can expand the fraction as

$$F = \frac{\mathbf{q}\cdot\mathbf{v}}{\omega + \mathbf{q}\cdot\mathbf{v} - i/\tau} = 1 - \frac{\omega - i/\tau}{\omega + \mathbf{q}\cdot\mathbf{v} - i/\tau}$$

$$= 1 - \left[1 - \frac{\mathbf{q}\cdot\mathbf{v}}{\omega - i/\tau} + \frac{(\mathbf{q}\cdot\mathbf{v})^2}{(\omega - i/\tau)^2} + \cdots\right]. \tag{8.126}$$

In carrying out the angular average, the second term in the square brackets will always vanish. If we are in three dimensions, then the integral involves a $\sin\theta\, d\theta$ factor, which couples with the $\cos\theta$ term to yield zero. A similar effect occurs in both two dimensions and one dimension. On the other hand, the third term will give a factor of $1/d$, where d is the dimensionality. Hence, we may carry out the angular average, noting that the rest of the integration over momentum gives the density, to rewrite (8.125) as

$$\varepsilon(\mathbf{q},\omega) = \varepsilon + \frac{2e^2}{q^2 k_B T}\int \frac{d^d\mathbf{k}}{(2\pi)^d} f_k \left\{1 - \left[1 + \frac{q^2 D/\tau}{(\omega - i/\tau)^2}\right]\right\}$$

$$= \varepsilon + \frac{ne^2}{q^2 k_B T}\left[1 - \frac{(\omega - i/\tau)^2}{(\omega - i/\tau)^2 - q^2 D/\tau}\right] \tag{8.127}$$

$$= \varepsilon - \frac{ne^2}{k_B T}\frac{D/\tau}{(\omega - i/\tau)^2 - q^2 D/\tau}.$$

Here, we have used the generalized diffusion constant as $D = v^2\tau/d$, although it must be identified with an average velocity in bringing this quantity outside of the integration. We note that if $\omega \to 0$, and we neglect the scattering, we recover the Debye screening result (8.108).

A somewhat different approach must be followed for degenerate situations, such as at low temperature, since we cannot simply relate the derivative of the distribution function to the distribution itself as was done in (8.103). Instead, we begin with the expansion (8.109), which we rewrite as

$$\varepsilon(\mathbf{q},\omega) = \varepsilon - \frac{2e^2}{\hbar q^2}\int \frac{d^d\mathbf{k}}{(2\pi)^d} f_k$$

$$\times \frac{2E_k - E_{k+q} - E_{k-q}}{\left[\omega - (E_k - E_{k-q})/\hbar - i/\tau\right]\left[\omega - (E_{k+q} - E_k)/\hbar - i/\tau\right]}.$$

$$\approx \varepsilon - \frac{2e^2}{m}\int \frac{d^d\mathbf{k}}{(2\pi)^d} f_k \frac{1}{[\omega + \mathbf{q}\cdot\mathbf{v} - i/\tau]^2} \tag{8.128}$$

$$= \varepsilon + \frac{2e^2\tau}{m}\int \frac{d^d\mathbf{k}}{(2\pi)^d} f_k \frac{1}{1 + [i\omega\tau + i\mathbf{q}\cdot\mathbf{v}\tau]^2}.$$

The fraction term can now be written, by expanding, and performing the angular integration as done previously, as

$$F = \frac{1}{(1+i\omega\tau + i\mathbf{q}\cdot\mathbf{v}\tau)^2}$$
$$= \frac{1}{(1+i\omega\tau)^2}\left[1 - 2i\frac{\mathbf{q}\cdot\mathbf{v}\tau}{1+i\omega\tau} - 3\frac{(\mathbf{q}\cdot\mathbf{v}\tau)^2}{(1+i\omega\tau)^2} - \cdots\right] \quad (8.129)$$
$$= \frac{1}{(1+i\omega\tau)^2 + Dq^2\tau}.$$

Here, we have used the same arguments on the angular integration as above, so that the dielectric function now becomes

$$\varepsilon(\mathbf{q},\omega) = \varepsilon + \frac{ne^2}{m}\frac{\tau^2}{(1+i\omega\tau)^2 + Dq^2\tau}. \quad (8.130)$$

Again, this satisfies the proper limits.

The difference in the two approaches lies in the portion of the spectrum in which the effects will occur. First, we note that in each case the Coulomb interaction that appears as a product with the Lindhard potential is that appropriate for a three-dimensional system. We will address the changes appropriate for reduced dimensions in the next section. Here, however, we note that (8.127) may be rewritten, using (8.108) as

$$\varepsilon(\mathbf{q},\omega) = \varepsilon - \frac{ne^2}{k_B T}\frac{D/\tau}{(\omega - i/\tau)^2 - q^2 D/\tau}$$
$$= \varepsilon\left[1 + \frac{q_D^2 D\tau}{(1+i\omega\tau)^2 + Dq^2\tau}\right], \quad (8.131)$$

which should be compared with the rewritten form of (8.130). If we use (8.113), we can rewrite this as

$$\varepsilon(\mathbf{q},\omega) = \varepsilon + \frac{ne^2}{m}\frac{\tau^2}{(1+i\omega\tau)^2 + Dq^2\tau}$$
$$= \varepsilon\left[1 + \frac{\omega_p^2 \tau^2}{(1+i\omega\tau)^2 + Dq^2\tau}\right]. \quad (8.132)$$

These two forms are essentially the same, once it is recognized that there is a close relationship between D and $k_B T$ in the nondegenerate case (the Einstein relationship). Thus, (8.131) is the proper form for nondegenerate material, and (8.132) is the proper form for degenerate material.

8.3.6 Screening in low-dimensional situations

The approach that was followed above was focused upon three-dimensional systems. This dimensionality appeared in at least two different places. First,

8.3 Electron–electron interaction effects

the Fourier transform of the induced and applied potentials assumed a three-dimensional form (e.g., **q** was a three-dimensional variable). Secondly, the wave function was taken to be a plane wave in three dimensions. The corrections for these two assumptions, in the case of a reduced dimensionality, are actually closely related. Let us consider, as an example, the case of a quasi-two-dimensional system arising in the inversion layer of a Si-SiO$_2$ or GaAs-AlGaAs heterostructure interface. In the first case, the Fourier transform is changed to a two-dimensional variation in the plane of the interface and an integration over the z-component (normal to the interface) of the wave functions in this direction. In the second case, the correction accounts for momentum plane waves in two dimensions and the localized wave function in the z-direction normal to the interface. While we will pursue the corrections for the two-dimensional case, the results are easily extended to even lower dimensions.

We basically start by rewriting the wave function that appears in the Lindhard potential as

$$|\mathbf{k}\rangle = \varphi_n(z)e^{i\mathbf{k}\cdot\mathbf{r}} = |\mathbf{k}, n\rangle, \tag{8.133}$$

except that both **k** and **r** are now two-dimensional vectors lying in the plane of the interface. The heart of the change lies in the two-dimensional Fourier transform of the potential matrix element, as

$$V_{nn'}(\mathbf{q}) = \langle \mathbf{k}, n|V(\mathbf{r},z)|\mathbf{k}+\mathbf{q}, n'\rangle$$
$$= \int d^2\mathbf{r} \int_{-\infty}^{\infty} dz \varphi_{n'}^*(z)\varphi_n(z) V(\mathbf{r},z) e^{i\mathbf{q}\cdot\mathbf{r}} e^{-qz}. \tag{8.134}$$

Within the linear response and random-phase approximations that we have been using, the z-variation of the potential is ignored as being small. Normally, the z-integration yields a δ-function on the indices, but the e^{-qz} term spoils this orthogonality, even when the z-variation of the potential is ignored. The result is

$$V_{nn'}(\mathbf{q}) = \frac{e^2}{2\varepsilon q} \int_{-\infty}^{\infty} dz \varphi_{n'}^*(z)\varphi_n(z) e^{-qz}. \tag{8.135}$$

However, when this term is included in the summation appearing in the Lindhard potential, one has to account for contributions both to the local potential and to the distribution function from the envelope functions in the z-direction. Hence, in this case, the dielectric function becomes

$$\varepsilon_{nn'}(\mathbf{q},\omega) = \varepsilon\left[\delta_{nn'} + \frac{e^2}{2\varepsilon q}\sum_{mm'} F_{mm'}^{nn'}(q) L_{mm'}(\mathbf{q},\omega)\right], \tag{8.136}$$

where

$$F_{mm'}^{nn'}(q) = \int_{-\infty}^{\infty} dz \int_{-\infty}^{\infty} dz' \varphi_{m'}^*(z)\varphi_m(z)\varphi_{n'}^*(z')\varphi_n(z') e^{-q|z-z'|} \tag{8.137}$$

and

$$L_{mm'}(\mathbf{q}, \omega) = \sum_{\mathbf{k}} \frac{f_m(\mathbf{k}) - f_{m'}(\mathbf{k+q})}{E_{k+q,m'} - E_{k,m} + \hbar\omega - i\hbar/\tau}. \quad (8.138)$$

Finally, the two-dimensional density is given by

$$n_s = 2\sum_{\mathbf{k},n} f_{kn}, \quad f_{kn} = \left[1 + \exp\left(\frac{E_k - E_n - E_F}{k_B T}\right)\right]^{-1}, \quad (8.139)$$

where E_n is the subband energy of the nth subband.

In the special case of a quasi-two-dimensional system in the quantum limit with only one subband occupied, (8.136) reduces to

$$\varepsilon_{00}(\mathbf{q}, \omega) = \varepsilon\left[\delta_{nn'} + \frac{e^2}{2\varepsilon q} F_0(q) L_{00}(\mathbf{q}, \omega)\right]. \quad (8.140)$$

To evaluate F_0, we need to use some assumed form for the envelope function. A common choice is the variational wave function [35]

$$\varphi_0(z) = \sqrt{\frac{b^3}{2}} e^{-bz/2}, \quad (8.141)$$

for which

$$F_0(q) = \frac{b^6}{(b^2 + q^2)^3}. \quad (8.142)$$

In general, $b > q$, and is sufficiently larger that $F_0 \sim 1$. In this situation, the major change to the previous sections is only the Fourier transform of the potential, the leading term in (8.135). With this change, the results of the previous sections can be used in the low-dimensional case as well.

8.3.7 The self-energy

We would now like to calculate the self-energy for the single-particle electron–electron interaction to see how it yields an insight into the scattering dynamics. For this, we will use the dynamic screening approximation (8.131) to the polarizability, and the effective interaction can then be written as

$$U_{eff} = \frac{e^2/2\varepsilon q}{1 - \frac{e^2}{2\varepsilon q} \frac{n}{k_B T} \frac{Dq^2/\tau}{(\omega - i/\tau)^2 - q^2 D/\tau}}$$

$$= \frac{e^2}{2\varepsilon q \left[1 - \frac{e^2}{2\varepsilon} \frac{n}{k_B T} \frac{Dq/\tau}{(\omega - i/\tau)^2 - q^2 D/\tau}\right]}, \quad (8.143)$$

for the two-dimensional case (in the quantum limit) pursued here. For relatively low frequencies, such as those involved in the single-particle scattering, this can be rearranged to yield (in lowest order for which $Dq^2 \gg \tau, \omega\tau$)

8.3 Electron–electron interaction effects

$$U_{\text{eff}} = \frac{Dq^2 e^2}{2\varepsilon q[Dq^2 + \chi_2 Dq - \tau(\omega_n + 1/\tau)^2]}, \quad (8.144)$$

where

$$\chi_2 = \frac{n_s e^2}{2\varepsilon k_B T}. \quad (8.145)$$

Since both the Green's function and the interaction strength are functions of frequency, and since these would be a product in real space, it is necessary to convolve them in frequency space, so the self-energy can be written as

$$\Sigma_{ee}^{r,a}(\mathbf{k}, i\omega_n) = -\sum_m \frac{1}{\beta\hbar^2} \int \frac{d^2\mathbf{q}}{4\pi^2} G^{r,a}(\mathbf{k} - \mathbf{q}, i\omega_n + i\omega_m) \\ \times \frac{Dq^2 e^2}{2\varepsilon q[Dq^2 + \chi_2 Dq - \tau(\omega_n + 1/\tau)^2]}. \quad (8.146)$$

Fukuyama [34] argues that we are really interested in cases in which the momentum change is quite small, so that the energy E_k in the Green's function differs little from the Fermi energy. Fukuyama makes the additional observation that since we are interested in a small frequency change, the Green's functions can be approximated as

$$G^{r,a}(\mathbf{k}, i\omega) = \frac{1}{i\omega - (E_k - \mu)/\hbar - i/2\tau} \sim i2\tau, \quad (8.147)$$

where τ is the broadening in the Green's function that arises from other scattering, such as from impurity scattering. With this approximation, the self-energy is momentum independent, since the last momentum variables will be integrated, as will the frequency dependence. However, this is not quite true if we take the zero of frequency at the Fermi energy for convenience; the range of the convolution integral is limited to the singular case for which one frequency is below the Fermi energy and the other is above the Fermi energy, or $\omega_n(\omega_n + \omega_m) < 0$. This limits the summation over frequency to those values for which $\omega_m < -\omega_n$. We can now rewrite the self-energy as

$$\Sigma_{ee}^{r,a}(\mathbf{k}, i\omega_n) = -\sum_{\omega_m < -\omega_n} \frac{2i\tau}{\beta\hbar^2} \int \frac{d^2\mathbf{q}}{4\pi^2} \frac{Dq^2 e^2}{2\varepsilon q[Dq^2 + \chi_2 Dq - \tau(\omega_n + 1/\tau)^2]}. \quad (8.148)$$

It is clear that this formula has singularities. The self-energy is singular in two dimensions, in the limit of $q, \omega_n \to 0$, with a variation as $\ln(\omega_n) \sim \ln(T)$. Nevertheless, this integral can be rewritten as

$$\Sigma_{ee}^{r,a}(\mathbf{k}, i\omega_n) = -\sum_{\omega_m < -\omega_n} \frac{i\tau e^2}{2\pi\varepsilon\beta\hbar^2} \int_0^\infty \frac{q^2 dq}{2\varepsilon q[q^2 + \chi_2 q - \tau(\omega_n + 1/\tau)^2/D]}. \quad (8.149)$$

There is a general problem with which integration should be pursued first; that over q or that from the summation over ω_m. Either leads to the need to introduce a cutoff into the actual integration, either a cutoff on the largest value for q

(which was discussed already in the leading sections of this chapter) or in the lower frequency limit. This is slightly easier in the case of the momentum, so we can rewrite the last equation in the leading terms as

$$\Sigma_{ee}^{r,a}(\mathbf{k}, i\omega_n) = -\sum_{\omega_m < -\omega_n} \frac{i\tau e^2}{2\pi\varepsilon\beta\hbar^2} \frac{\chi_2 q_{max}(q_{max} + \chi_2)v^2}{4(\omega_m + 1/\tau)^2}. \tag{8.150}$$

In this last form, a logarithmic term has been expanded under the assumption that the last fraction is small, and the diffusion constant has been expanded in terms of the velocity found earlier. The value of q_{max}, is an upper cutoff on the momentum, and this was found earlier to be approximately $2k_F$. The summation over the frequencies can now be converted to an integral and

$$\Sigma_{ee}^{r,a}(\mathbf{k}, i\omega_n) = \frac{i\tau e^2}{2\pi\varepsilon\hbar} \frac{\chi_2 q_{max}(q_{max} + \chi_2)v^2}{4(\omega_m + 1/\tau)}. \tag{8.151}$$

The factor in the denominator is predominantly related to the Fermi energy, whereas the term in the velocity squared leads to a variation as T when the thermal averaging is done.

Chaplik [35] and Giuliani and Quinn [36] have carried out evaluations of the "lifetime" of an electron due to single-particle scattering from the Fermi sea. The latter is perhaps the most often cited for these calculations, and their result is

$$\frac{1}{\tau_{ee}} \sim \frac{E_F}{4\pi\hbar} \left(\frac{E_k - E_F}{E_F}\right)^2 \left[\ln\left(\frac{E_k - E_F}{E_F}\right) - \frac{1}{2} - \ln\left(\frac{2q_{FT}}{\hbar k_F}\right)\right]. \tag{8.152}$$

Here, the leading term can be rearranged as

$$\frac{E_k - E_F}{E_F} \sim \frac{p^2 - p_F^2}{p_F^2} = \frac{(p + p_F)(p - p_F)}{p_F^2} \sim \frac{2(p - p_F)}{p_F}. \tag{8.153}$$

Since the temperature in the strongly degenerate limit describes the fluctuations around the Fermi energy, the leading behavior in the electron–electron self-energy varies as $(p - p_F)^2 \sim 2mk_BT$, so the same linear behavior in temperature is found (Giuliani and Quinn, however, seem to have missed this point and claim a T^2 behavior), with the overall temperature behavior being $T \ln(T)$, a result found in some of the experiments cited at the start of this chapter. In (8.151), a similar behavior is found if we relate the velocity to the quasi-particle velocity $[v^2 \sim (p - p_F)^2/m^2]$: the cutoff relates to the Fermi momentum $q_{max}^2 \to E_F$, and $\hbar\omega_n \to E_F$, the same leading-order behavior is recovered. We will return to this discussion later in a more explicit discussion of the appropriate diagram terms, but this basic result (with some slight modification) will be recovered.

8.3.8 Self-energy in the presence of disorder

It is now clear that the nature of the temperature variations will be found to some extent in the self-energy, since it describes the important broadening of the single-particle density of states. This, in turn, relates to the mixing of phase-coherent states due to broadening of the Fermi distribution function. Thus, the task *in disordered systems* is to define the self-energy in the system in which both impurities and electron–electron interactions (and possibly others as well) exist. In treating the self-energy, it is necessary to decide how one wants to resum the various diagrams for the interactions. For example, in disordered systems it was assumed in the last chapter that the disorder-inducing impurity interactions are the dominant interactions, and the electron–electron interactions were ignored. Above, however, we dealt with screening of an interaction by the electrons, so that the electron–electron interaction was assumed to be the dominant interaction. In this section, we will continue to assume that the disorder-inducing impurity interactions are the dominant scattering process, and that the electron–electron interactions are a perturbation of this process. Now, clearly, the entire set of Feynman diagrams that relate to scattering by impurities and by the electron–electron Coulomb interaction can be separated into sets of terms which lead to a sum of two distinct self-energies. Thus, the impurity scattering processes are treated as previously, and we add a new self-energy term Σ_{ee}, which accounts for the interaction effects that are themselves mediated by the impurities. This was shown in Fig. 8.9, where the jagged line represents the electron–electron interaction and the dashed line represents the impurity (averaged) interaction. If these were the only diagrams in the perturbation expansion, life would indeed be quite easy. However, we must recall that, for the impurities alone, we wound up with a great many different types of diagrams. There was the simple summation of independent Coulomb scattering events which were impurity averaged and which led to the simple impurity self-energy

$$\Sigma_{imp}^{r,a}(i\omega_n) = \mp i \frac{\hbar}{2\tau(i\omega_n)}, \tag{8.154}$$

found in the last chapter. In addition, however, the two-particle Green's function contribution to the polarization in the Kubo formula led to a set of ladder diagrams which led to the diffuson and a set of maximally crossed diagrams which led to the cooperon. These additional contributions were important for weak localization and for universal conductance fluctuations.

The problem with two perturbing species at hand is what to do with diagrams like Fig. 8.12, where the two perturbations interfere with one another. The question is really to which of the self-energies will we ascribe such terms (or should they be ignored), and then how they are to be included in the selected self-energy. The normal approach (normal in the sense of semiclassical transport with the Boltzmann equation) is to assume that the electron–electron interaction is the dominant scattering process. This usually results in some sort of assumption of a drifted Maxwellian (or Fermi–Dirac) distribution, with the interaction being subsequently ignored except for its role in screening other scattering processes. This screening arises just from the diagrams of the type

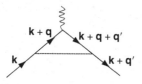

Fig. 8.12 Diagram with both impurity and electron–electron interactions.

in Fig. 8.12, although this is not often recognized. This leads to a philosophy of the following strategic approach for when the various perturbing interactions interfere with one another: (1) an assumption is made regarding the dominant perturbation interaction, and this self-energy is treated as if the other processes are absent; and (2) the other interactions are treated as if they are *screened* by the dominant perturbation through the introduction of *vertex corrections*. Normal screening by the electrons is such a vertex correction.

In disordered materials, where the disorder is induced by the heavy impurity scattering, the impurity scattering is the dominant perturbation. Thus, we will subsequently screen the electron–electron interaction by the impurity interaction, with the latter leading to a vertex correction. That is, we calculate the interaction effects not between free electrons (which would be simple plane-wave states) but between the diffusive electrons. This is handled by the means of a vertex correction for the Coulomb interaction between the diffusing electrons. The impurity self-energy is still assumed to be given by its previous form, but with the temperature dependence included. This temperature dependence does not occur within the self-energy but arises from the energy (frequency) dependence of $\tau(\omega)$, which is coupled to an energy dependence when the integration over frequency is performed in the calculation of the conductivity (and therefore of an average relaxation time). Thus, the sign on (8.154) is such that we no longer have to worry about the small quantity η (as we did in the last chapter). Here, we are interested in investigating the self-energy term arising from the interaction between diffusing electrons in a sea of impurity scattering.

It was stated above that the electron–electron interaction would be subject to a vertex correction. Let us examine just what this means. Consider Fig. 8.13. In the figure, we draw four possible electron–electron interactions within the sea of impurity scattering. Diagrams (a) and (b) refer to the exchange interaction, and diagrams (c) and (d) refer to the Hartree correction. Both pairs are first-order corrections to the energy of the "free" particles. In the diagrams, however, impurity interactions have been drawn around the points (the *vertices*) at which the interaction line connects to the Green's function lines. In diagrams (a) and (c), the single dashed line represents the diffuson interaction, whereas in diagrams (b) and (d) the double-dashed line refers to the cooperon contribution. But the diffuson and the cooperon are two-particle Green's function interactions. How do we describe these *dressings* of the vertex that are produced by the two-particle Green's function ladders (note that they actually connect with four Green's functions) that were treated in the last chapter? Consider the two-particle Green's function and the notation of momentum in Fig. 8.12. The two momenta $\mathbf{k}$ and $\mathbf{k}'=\mathbf{k}+\mathbf{q}'$ on the left (the right side) become the momenta $\mathbf{k}+\mathbf{q}$ and $\mathbf{k}+\mathbf{q}+\mathbf{q}'$, respectively, in Fig. 8.12, where $\mathbf{q}'$ is the momentum transferred via the electron–electron interaction. Thus, the two Green's function lines on the right of the two-particle structure are pulled together to meet at the vertex of the

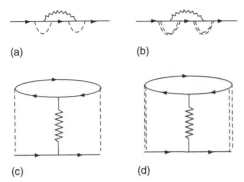

Fig. 8.13 The first-order exchange interaction, (a) and (b), and Hartree interaction, (c) and (d), dressed by the diffusons in (a) and (c) and by cooperons in (b) and (d).

electron–electron interaction. Now, **q** is the total momentum transferred to the impurities in the ladder diagram summation, so **k″** becomes the momentum **k** in Fig. 8.12. Similarly, the momentum **k‴** is associated with **k** + **q′** in Fig. 8.12. Thus, the single impurity line in Fig. 8.12 becomes the full two-particle Green's function represented by the ladder diagrams that lead to the diffuson. Similarly, the set of impurity lines that cross one another from one side of Fig. 8.12 to the other lead to the set of maximally crossed diagrams, giving rise to the two-particle Green's function representation of the cooperon. The vertex corrections then are the resummations of the ladder diagrams and/or maximally crossed diagrams that represent the two-particle Green's function dressing of the vertex of the electron–electron interaction.

As a consequence, the vertex corrections lead to a Coulomb interaction not between two free electrons, but between two dressed particles described either by the diffuson or by the cooperon. These interactions are appropriate only for diffusive transport when the energy exchange between the carriers is quite small; that is, the diffusive interactions with the impurities involves very small energy exchange (it was assumed to be zero in the last chapter). The cooperon contribution to the vertex correction is given by the ladder summation of the (reversed) maximally crossed interactions $\Lambda = (1 - \Pi V_0)^{-1}$ of the last chapter, or by reinserting the new frequency,

$$\Lambda_C(q, \omega_n) = \frac{1}{2\pi\hbar\rho(\mu)\tau^2} \frac{1}{Dq^2 + |\omega_n| + 1/\tau_\varphi}. \tag{8.155}$$

On the other hand, the diffuson contribution to the vertex correction is given by the resummation of the ladder diagrams themselves,

$$\Lambda_D(q, \omega_n) = \frac{1}{2\pi\hbar\rho(\mu)\tau^2} \frac{1}{Dq^2 + |\omega_n|}. \tag{8.156}$$

There is an important difference in these two equations, over and above the elimination of the phase-breaking time in the latter expression. For the diffuson, **q** = **k** − **k′**, where the latter two momenta are those of the Green's function ladder that gives rise to the polarization Π in the diffuson, while $\mathbf{q} = \mathbf{k} + \mathbf{k}'$ in the cooperon. This difference is related to the conservation laws on the total number of particles. The diffuson represents the modification of the matrix elements by the dressed diffusing particles, while the cooperon relates directly to the phase interference properties of these diffusing electrons [36].

Altshuler and Aronov first studied the role of interaction effects in a three-dimensional disordered material [23]. Later, Altshuler *et al.* extended this treatment to the two-dimensional situation [37]. In the first case, the authors considered only the diagram Fig. 8.13(a), while in the latter work, they included this diagram plus the equivalent Hartree term from Fig. 8.13(c). In these two cases, they found that the effect of these two terms led to a $\ln(T)$ dependence in the density of states, which is a noticeable renormalization of the energy structure around the Fermi energy at low temperatures. Fukuyama [38] later treated all four diagrams of Fig. 8.13 and found that they all contribute comparably in the lowest order of the interaction. Here, we want to illustrate the results but will treat only the first process, given by Fig. 8.13(a). This diagram is redrawn in Fig. 8.14, where the curly line is the full electron–electron interaction, and the shaded triangles are the vertex correction, represented by the diffuson propagator.

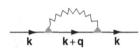

Fig. 8.14 The lowest-order Hartree diagram for the electron–electron self-energy.

By looking at the diagram in Fig. 8.14, it is clear that the self-energy for this term involves the product of one Green's function line with momentum $\mathbf{k} - \mathbf{q}$, and an interaction line at wavevector $\mathbf{q}$. In the simplest case, this would lead to the product of just the interaction potential and the Green's function, integrated over the scattering wavevector $\mathbf{q}$. However, as we have discussed, the simple interaction is modified by the vertex correction of the strong background of impurity scattering. In essence, we seek a correction that is one end of a ladder diagram in which there are two Green's functions at momentum $\mathbf{k}$ and $\mathbf{k} - \mathbf{q}$, which meet with the interaction line at momentum $\mathbf{q}$. In the presence of strong impurity scattering, one can consider that there may be an intermediate state connecting these lines, which represents a set of impurity scattering events organized so that the end momenta stay the same. This is just the set of ladder diagrams that we described above. Here, these modify the bare interaction potential in a way to allow for the weak interaction scattering in the "sea of impurity scattering." The ladder we need involves impurity-averaged scattering events between two Green's functions which always differ by the momentum $\mathbf{q}$. This is just the ladder sum we used in the diffuson correction for weak localization. In short, we need the sum that appears to modify the bare impurity interaction in (8.156). Now, in this particular case, there are vertex corrections at each end of the interaction line, so the net interaction line carries a modification of the strength by $[\Lambda_D(q,\omega_n)/\Gamma_0]^2 = [2\pi\hbar\rho(\mu)\tau\Lambda_D(q,\omega_n)]^2$ (since we want only the summation and not the leading interaction term in the ladder diagram of the last chapter, which really introduces the number of impurities to the summation). The net interaction is now

8.3 Electron–electron interaction effects

$$\frac{U(\mathbf{q}, i\omega_n)}{[Dq^2 + |\omega_n|]^2 \tau^2}. \tag{8.157}$$

Since both the Green's function and the interaction strength are functions of frequency, and since these would be a product in real space, it is necessary to convolve them in frequency space, and the self-energy can be written (just as in the last section) as

$$\Sigma_{ee}^{r,a}(\mathbf{k}, i\omega_n) = -\sum_m \frac{1}{\beta\hbar^2\tau^2} \int \frac{d^d\mathbf{q}}{(2\pi)^d} G^{r,a}(\mathbf{k}-\mathbf{q}, i\omega_n + i\omega_m) \frac{U(\mathbf{q}, i\omega_m)}{[Dq^2 + |\omega_m|]^2 \tau^2}. \tag{8.158}$$

Fukuyama [38] argues that we are really interested in cases in which the momentum change is quite small, so that the energy E_k differs little from the Fermi energy. Since we are interested in a small frequency change, Fukayama also observes that the Green's functions can be approximated as

$$G^{r,a}(\mathbf{k}-\mathbf{q}, i\omega_n + i\omega_m) = \frac{1}{i\omega_m + i\omega_n - (E_{k-q} - \mu)/\hbar - i/2\tau} \sim 2i\tau. \tag{8.159}$$

With this approximation, the self-energy is momentum-independent, since the last momentum variables will be integrated, as will the frequency dependence. However, this is not quite true if we take the zero of frequency at the Fermi energy for convenience; the range of the convolution integral is limited to the singular case for which one frequency is below the Fermi energy and the other is above the Fermi energy, or $\omega_n(\omega_m + \omega_n) < 0$. This limits the summation over frequency to those values for which $\omega_m < -\omega_n$. We can now rewrite the self-energy as

$$\Sigma_{ee}^{r,a}(\mathbf{k}, i\omega_n) = -\sum_{\omega_m < -\omega_n} \frac{2i}{\beta\hbar^2\tau} \int \frac{d^d\mathbf{q}}{(2\pi)^d} \frac{U(\mathbf{q}, i\omega_m)}{[Dq^2 + |\omega_m|]^2 \tau^2}. \tag{8.160}$$

It is clear that this formula has singularities. Even if we ignore the variables in the interaction potential and treat the potential as constant, the self-energy is singular in two dimensions with a variation as $\ln(\omega_n) \sim \ln(T)$. The other diagrams in Fig. 8.13 have similar results and have to be taken into account on an equal footing.

If we take the interaction as constant within the germane region of momentum and frequency, $U(\mathbf{q}, i\omega_m) = U_0$, then the momentum integration can be carried out quickly. In two dimensions

$$\begin{aligned}\Sigma_{ee}^{r,a}(\mathbf{k}, i\omega_n) &= -\sum_{\omega_m < -\omega_n} \frac{iU_0}{\beta\hbar^2 D\tau} \frac{1}{2\pi} \int \frac{Dd(q^2)}{[Dq^2 + |\omega_m|]^2 \tau^2} \\ &= -\frac{iU_0}{2\pi\beta\hbar^2 D\tau} \sum_{\omega_m < -\omega_n} \frac{1}{|\omega_m|}.\end{aligned} \tag{8.161}$$

Thus, if we add this self-energy to the self-energy due to impurity scattering, or

$$\Sigma^{r,a}(\mathbf{k}, i\omega_n) = \Sigma^{r,a}_{imp}(\mathbf{k}, i\omega_n) + \Sigma^{r,a}_{ee}(\mathbf{k}, i\omega_n), \tag{8.162}$$

we find that the total lifetime is modified to be

$$\frac{1}{\tau} \to \frac{1}{\tau}[1 + \lambda g G_2(\omega_n T)], \tag{8.163}$$

where $\lambda = \hbar/2\pi E_F \tau = \hbar/2\pi m D$ is a small dimensionless coupling constant, $g = U_0 \rho_d(\mu)/\hbar = m U_0 / \pi \hbar^3$ represents the strength of the interaction, and [36]

$$G_2(\omega_n, T) = \frac{2\pi}{\beta\hbar} \sum_{\omega_m < -\omega_n} \frac{1}{|\omega_m|} = \ln\left(\frac{\hbar\beta}{2\pi\tau}\right) - \psi\left(\frac{\hbar\beta\omega_n + 1}{2}\right), \tag{8.164}$$

where the latter function is the digamma function. In fact, each of the diagrams in Fig. 8.13 contributes, and $g = g_a + g_b - 2(g_c + g_d)$, where the subscript refers to the particular diagram of the figure. The density of states was given above in (8.87). The imaginary part of the Green's function involves the scattering time in the numerator. The presence of the additional scattering (indeed, of any scattering at all) reduces this lifetime and reduces the density of states at the Fermi energy. Normally, one is familiar with broadening of the density of states at a subband edge due to the scattering. However, at the Fermi energy at low temperature, the strong interaction among the dense number of electrons excited just above the Fermi energy creates an interaction that lowers the density of states at this point. This term has a logarithmic divergence with reduction of the temperature, as given by the first term of the last equality in (8.164). The sign difference in the overall quantity g, which measures the strength of the interaction, arises from the fact that the Hartree and exchange terms have different signs, and the factor of two accounts for the spin summation in the Hartree terms.

8.3.9 Electron–plasmon interactions and the phase-breaking time

Scattering from the (screened) potential of the other electrons, and from the collective plasmon modes, are both part of the total electron–electron interaction among the free carriers. Generally, it is not possible to consider the full Coulomb interaction beyond the lowest order of perturbation theory because of the long range of the potential associated with this interaction. As discussed above, the dielectric function has singularities at the plasmon frequency and at zero frequency, corresponding to the plasmon modes and the single-particle scattering, respectively. Formally, one can split the summation over $\mathbf{q}$ that appears in a Fourier transform of the potential into a short-range (in real space, which means large q) part, for which $q > q_c$, and a long-range (in real space, which means small q) part, for which $q < q_c$, where q_c is a cutoff wavevector defining this split. It was shown some time ago that the short-range part of the potential corresponds to the screened Coulomb interaction for single-particle

8.3 Electron–electron interaction effects

scattering [26], which was discussed in the previous sections. The long-range part, on the other hand, is responsible for scattering by the collective oscillations of the electron gas, which describe the motion of the electrons in the field produced by their own Coulomb potential, the *plasmons*. In fact, these are just the modes we discussed in the plasmon-pole approximation in Section 8.3.5. These collective oscillations are bosons, so their scattering rate can be calculated in much the same way as phonons – the distribution function that we include is that of the plasmons rather than the free electrons. The only difference from the formal approach of the phonons is that there is now a maximum q ($= q_c$) that can be involved in the scattering, and this cutoff is essentially the Debye wavevector. The general case, for three-dimensional electrons has been reviewed elsewhere [39]. Here, we want to consider the phase-breaking that occurs when coherent electrons interact with the plasmons in a low-dimensional system. We will find that, in general, there are two regimes: a low-temperature regime where the phase-breaking time is independent of temperature and a high-temperature regime where the phase-breaking time decreases as a power of the temperature. The transition region is where the plasmon energy is comparable to the thermal energy.

8.3.9.1 A quasi-two-dimensional system

Electrons that are in excited states (or states with energy above the average energy) will on average lose energy to the overall electron gas, as discussed above. This is true regardless of the dimensionality of the semiconductor. Here, we want to discuss this energy loss to the plasmon modes. In two dimensions, the plasma frequency is not constant, but is a function of the wavevector, hence approaches zero at $q = 0$. In the present section, we want to compute the scattering rate in this quasi-two-dimensional system for the electron–plasmon interaction. We will assume that the carriers are in the lowest subband, and that the wave function factors (8.137) are unity and can be ignored. An important factor that will become important is to limit the lower range of the integration over q (in a sense, this is using a long-range cutoff), just as is done for inelastic phonon scattering. Here, however, we will limit this value to the inverse of the mean free path, or $1/v\tau = 1/\sqrt{D\tau}$ in two dimensions, predicated on the fact that the scattering will break up any process that would emit a plasmon with lower momentum value (or that the coherence can not be maintained for more than a diffusion length).

We can now formulate the process of energy loss by an energetic electron in a quasi-two-dimensional electron gas, by treating a nearly free electron. Our treatment will be applicable to that of high-mobility carriers in e.g. a heterostructure with little impurity scattering, even though we will find essentially the same result as that of the disordered system. We can write the scattering rate from the self energy as

$$\frac{1}{\tau_\varphi} = \frac{2\pi}{\hbar} \sum_{\mathbf{q}} \int_{-\infty}^{\infty} \frac{d\omega}{2\pi} \coth\left(\frac{\hbar\omega}{2k_BT}\right) \left|\text{Im}\left\{\frac{V(q)}{\varepsilon(q,\omega)}\right\}\right| \delta\left(\omega - \frac{E_{k+q} - E_k}{\hbar}\right). \quad (8.165)$$

Here, we ignore the details of the Fermi factor that arises from the possibility that the final states are full, and we have combined the emission and absorption terms through

$$N_q + (N_q + 1) = \frac{1}{e^x - 1} + \frac{e^x}{e^x - 1} = \frac{e^{x/2} - e^{-x/2}}{e^{x/2} + e^{-x/2}} = \coth(x/2). \quad (8.166)$$

In two dimensions, the collective excitations (the plasmons) have a frequency that goes to zero as q goes to zero, and our basic interest lies in small frequency (small energy) exchange. Hence, first we will use the approximation that

$$\coth(x/2) \sim \frac{2}{x}. \quad (8.167)$$

For the dielectric function, we take the form (8.132) as

$$\varepsilon(\mathbf{q}, \omega) = \varepsilon\left[1 + \frac{\omega_p^2 \tau^2}{(1 + i\omega\tau)^2 + Dq^2\tau}\right] \quad (8.168)$$

with

$$\omega_p^2 = \frac{ne^2 q}{2m\varepsilon} = \chi_2 q. \quad (8.169)$$

In the low-frequency limit, we can write the imaginary part of the potential times the inverse dielectric functions as

$$-\frac{e^2}{2\varepsilon q} \frac{2\omega}{Dq^2 + \chi_2 q\tau}. \quad (8.170)$$

The integral over the frequency can now be written as

$$\int_{-\infty}^{\infty} \frac{d\omega}{2\pi} \frac{2k_BT}{\hbar\omega} \frac{e^2}{\varepsilon q} \frac{\omega}{Dq^2 + \chi_2 q\tau} \delta\left(\omega - \frac{E_{k+q} - E_k}{\hbar}\right) = \frac{e^2 k_BT}{\pi\hbar\varepsilon q(Dq^2 + \chi_2 q\tau)}. \quad (8.171)$$

Now, we can write the scattering rate as

$$\frac{1}{\tau_\varphi} = \frac{2\pi}{\hbar} \sum_{\mathbf{q}} \frac{e^2 k_BT}{\pi\hbar\varepsilon q(Dq^2 + \chi_2 q\tau)} = \frac{e^2 k_BT}{\pi\hbar^2 \varepsilon} \int_{q_{\min}}^{\infty} \frac{dq}{Dq^2 + \chi_2 q\tau}$$
$$\sim \frac{e^2 k_BT}{\pi\hbar^2 \varepsilon\chi_2\tau} \ln\left(1 + \chi_2\tau\sqrt{\frac{2\tau}{D}}\right). \quad (8.172)$$

Here, we have cut off the integration at a lower value for q, as discussed above. This form has a characteristic variation as $t_\varphi \sim 1/T$, which is typical for a two-dimensional system. We can estimate the value of t_φ for a GaAs/AlGaAs heterostructure at low temperature. We assume that the density is 4×10^{11} cm^{-2}, and that the mobility is 10^6 cm^2/Vs, typical of a high mobility structure. Then, $\tau \sim 3.8 \times 10^{-11}$ s, and $\chi_2 \sim 8.5 \times 10^{18}$ cm/s^2. This gives the value of $t_\varphi \sim 4.4 \times 10^{-10}$ s at 1 K. This value is comparable to actual measured values in such semiconductor structures at low temperatures.

At the low-temperature end of the range, we use the large argument limit for the hyperbolic cotangent, and replace (8.167) with

$$\coth(x/2) \sim 1. \tag{8.173}$$

Then, the phase-breaking time becomes

$$\frac{1}{\tau_\varphi} = \frac{2\pi}{\hbar} \sum_\mathbf{q} \int \frac{d\omega}{2\pi} \frac{e^2}{\varepsilon q} \frac{\omega}{Dq^2 + \chi_2 q\tau} \delta\left(\omega - \frac{E_{k+q} - E_k}{\hbar}\right). \tag{8.174}$$

Now, the obvious thing would be to integrate over the frequency, but this would give a cosine term which would then integrate to zero in the **q** integration. So, this has to be handled a little more carefully, and we integrate the angle within the **q** integration first, as

$$\int \frac{d\vartheta}{2\pi} \delta(\omega - qv\cos\vartheta) = \frac{1}{2\pi qv \sin\vartheta}\bigg|_{\cos\vartheta = \omega/qv} = \frac{1}{2\pi\sqrt{q^2v^2 - \omega^2}}. \tag{8.175}$$

Now, this leads to

$$\frac{1}{\tau_\varphi} = \frac{2\pi}{\hbar} \int \frac{qdq}{4\pi^2} \int \frac{d\omega}{2\pi} \frac{e^2}{\varepsilon q} \frac{\omega}{Dq^2 + \chi_2 q\tau} \frac{1}{\sqrt{q^2v^2 - \omega^2}}. \tag{8.176}$$

The frequency integration can now be carried out, but we should use an upper limit of $1/\tau$. Then, we have

$$\frac{1}{\tau_\varphi} = \frac{e^2}{2\pi^2 \varepsilon \hbar} \int \frac{dq}{q(Dq + \chi_2 \tau)} \left[qv - \sqrt{q^2v^2 - 1/\tau^2}\right], \tag{8.177}$$

and an approximate result is

$$\frac{1}{\tau_\varphi} = \frac{e^2}{4\pi^2 \varepsilon \hbar v \tau} \int \frac{dq}{q^2(Dq + \chi_2 \tau)}. \tag{8.178}$$

In this case, we are going to have to limit both the maximum value of q and the minimum value of q. For the former, we are guided by the equation itself, and take $q_{\max} = \chi_2/D\tau$, but this will only be used in the logarithmic term that results from the integral. Hence, (8.178) now gives

$$\frac{1}{\tau_\varphi} = \frac{e^2}{8\pi^2 \varepsilon \hbar v \tau \chi_2} \left[\sqrt{\frac{D}{\tau}} + \frac{2D}{\chi_2 \tau^2} \ln\left(\frac{\sqrt{D\tau}}{\chi_2}\right)\right]. \tag{8.179}$$

Here, it is clear that, in this limit, the phase-breaking time, due to plasmon interactions, is independent of the temperature. This is an effect that is seen in most mesoscopic systems where the phase-breaking time has been measured, such as in quantum dots attached to two-dimensional reservoirs [40,41].

In disordered, diffusive systems, where the mobility is relatively low, other forms need to be used, particularly to account for the interactions between the impurities and the electrons which can occur simultaneously. These are reviewed in Altshuler and Aronov [42] and Fukuyama [36].

8.3.9.2 A quasi-one-dimensional system

As discussed above, the principal rationale for this approach is that electrons that are in excited states will on average lose energy to the overall electron gas. This is true regardless of the dimensionality of the semiconductor. Here, we want to discuss this energy loss to the plasmon modes. In one dimension, the plasma frequency is not constant, but is a function of the wavevector, hence approaches zero at $q = 0$. In this present section, we want to compute the scattering rate in this quasi-one-dimensional system for the electron–plasmon interaction. We will assume that the carriers are in the lowest subband, and that the wave function factors (8.137) are unity and can be ignored. As above, we will have to limit the range of some of the integrations, just as is done for inelastic phonon scattering.

In this limit, we are really treating a quantum wire. Although we think of this as a one-dimensional system, in most cases they are really narrow two-dimensional systems, with the Fermi energy being determined by the two-dimensional reservoirs to which the wires are connected. In this case, the one-dimensional density is more properly given by $n_1 = n_2 W$, where W is the width of the wire. Nevertheless, we approach the electron–plasmon interaction as if it were a one-dimensional wire.

We can now formulate the process of energy loss by an energetic electron in a quasi-one-dimensional electron gas, by treating a nearly free electron. Our treatment will be applicable to that of high-mobility carriers in e.g. a heterostructure with little impurity scattering, even though we will find essentially the same result as that of the disordered system. We can write the scattering rate from the self-energy as

$$\frac{1}{\tau_\varphi} = \frac{2\pi}{\hbar} \sum_q \int_{-\infty}^{\infty} \frac{d\omega}{2\pi} \coth\left(\frac{\hbar\omega}{2k_BT}\right) \left|\mathrm{Im}\left\{\frac{V(q)}{\varepsilon(q,\omega)}\right\}\right| \delta\left(\omega - \frac{E_{k+q} - E_k}{\hbar}\right). \quad (8.180)$$

Here, we ignore the details of the Fermi factor that arises from the possibility that the final states are full, and we have combined the emission and absorption terms through

$$N_q + (N_q + 1) = \frac{1}{e^x - 1} + \frac{e^x}{e^x - 1} = \frac{e^{x/2} - e^{-x/2}}{e^{x/2} + e^{-x/2}} = \coth(x/2). \quad (8.181)$$

In two dimensions, the collective excitations (the plasmons) have a frequency that goes to zero as q goes to zero, and our basic interest lies in small frequency (small energy) exchange.

For a one-dimensional wire, we can write the Coulomb interaction in momentum space as

8.3 Electron–electron interaction effects

$$V(q) = \frac{e^2}{4\pi\varepsilon \ln(1+q_0^2/q^2)} \sim \frac{e^2}{4\pi\varepsilon}, \tag{8.182}$$

where q_0 is an appropriate cutoff (Debye or Fermi–Thomas screening). The logarithmic factor will cancel from an appropriate term in the dielectric function, so will be ignored. For the dielectric function, we take the form (8.132) as

$$\varepsilon(\mathbf{q},\omega) = \varepsilon\left[1 + \frac{\omega_p^2 \tau^2}{(1+i\omega\tau)^2 + Dq^2\tau}\right] \tag{8.183}$$

with

$$\omega_p^2 = \frac{n_1 e^2 q^2}{2m\varepsilon} = \chi_1 q^2. \tag{8.184}$$

In the low-frequency limit, we can write the imaginary part of the potential times the inverse dielectric functions as

$$-\frac{e^2}{4\pi\varepsilon}\frac{2\omega}{q^2(D+\chi_1\tau)}. \tag{8.185}$$

In evaluating (8.79), it will prove more convenient to carry out the integration over q prior to that over ω. The former will entail the delta function, and we will need to invoke some cutoffs on the ω integration. In fact, we are interested in frequencies that lie between $1/t_\varphi$ and $1/\tau$. We will take these as the upper and lower cutoffs, respectively, when we need to utilize such in the evaluation of the integrals. We can now write the q integration as

$$\sum_{\mathbf{q}}(\bullet) = \sum_{\pm}\int_0^\infty \frac{dq}{2\pi}\frac{e^2\omega}{2\pi\varepsilon q^2(D+\chi_1\tau)}\delta\left(\omega - \frac{\hbar q^2}{2m} \mp \frac{\hbar q k}{m}\right). \tag{8.186}$$

The summation that remains is over forward- and backward-scattering. In fact, the dominant contribution to the phase-breaking is by backward-scattering through plasmon emission, so that $q \gg k$, and the integration yields

$$\frac{e^2\omega}{4\pi^2\varepsilon(D+\chi_1\tau)}\frac{\hbar}{2m\omega}\sqrt{\frac{m}{2\hbar\omega}} = \frac{e^2 m}{4\pi^2\hbar\varepsilon(D+\chi_1\tau)}\left(\frac{\hbar}{2m}\right)^{3/2}\frac{1}{\sqrt{\omega}}. \tag{8.187}$$

We can now use this in the remainder of (8.180) as

$$\frac{1}{\tau_\varphi} = \frac{2\pi}{\hbar}\int_{-\infty}^\infty \frac{d\omega}{2\pi}\coth\left(\frac{\hbar\omega}{2k_B T}\right)\frac{e^2 m}{4\pi^2\hbar\varepsilon(D+\chi_1\tau)}\left(\frac{\hbar}{2m}\right)^{3/2}\frac{1}{\sqrt{\omega}}. \tag{8.188}$$

First, we will consider the high temperature limit, where the argument of the hyperbolic cotangent is not large, and use the approximation that

$$\coth(x/2) \sim \frac{2}{x}. \tag{8.189}$$

Then, we can write the scattering rate as

$$\frac{1}{\tau_\varphi} = \frac{e^2 m}{4\pi^2 \hbar^2 \varepsilon (D + \chi_1 \tau)} \left(\frac{\hbar}{2m}\right)^{3/2} \int \frac{d\omega}{\sqrt{\omega}} \frac{4k_B T}{\hbar \omega}$$

$$= \frac{e^2 m k_B T}{\pi^2 \hbar^3 \varepsilon (D + \chi_1 \tau)} \left(\frac{\hbar}{2m}\right)^{3/2} \int \frac{d\omega}{\omega^{3/2}} \quad (8.190)$$

$$\sim \frac{e^2 m k_B T}{\pi^2 \hbar^3 \varepsilon (D + \chi_1 \tau)} \left(\frac{\hbar}{2m}\right)^{3/2} \sqrt{\tau_\varphi} \, .$$

Here, we have cut off the integration at the upper value for ω, as discussed above. This equation can now be rearranged to give

$$\frac{1}{\tau_\varphi} = \left[\frac{e^2 m k_B T}{\pi^2 \hbar^3 \varepsilon (D + \chi_1 \tau)} \left(\frac{\hbar}{2m}\right)^{3/2}\right]^{2/3} . \quad (8.191)$$

This form has a characteristic variation as $\tau_\varphi \sim 1/T^{2/3}$, which is typical for a one-dimensional system. A very similar result was also found for disordered systems [43,44], although we have not accounted for disorder in the present approach. Using the same parameters as in the previous section, and a wire width of 6 nm, we find that the phase-breaking time is about 70 ps at 1 K.

At the low-temperature end of the range, we use the large argument limit for the hyperbolic cotangent, and replace (8.189) with

$$\coth(x/2) \sim 1. \quad (8.192)$$

Then, the phase-breaking time becomes

$$\frac{1}{\tau_\varphi} = \frac{e^2 m}{4\pi^2 \hbar^2 \varepsilon (D + \chi_1 \tau)} \left(\frac{\hbar}{2m}\right)^{3/2} \int \frac{d\omega}{\sqrt{\omega}}$$

$$= \frac{e^2 m k_B T}{\pi^2 \hbar^3 \varepsilon (D + \chi_1 \tau)} \left(\frac{\hbar}{2m}\right)^{3/2} \frac{1}{\sqrt{\tau}} . \quad (8.193)$$

Here, it is again clear that, in this limit, the phase-breaking time, due to plasmon interactions, is independent of the temperature. This result was found for quantum wires in semiconductors originally by Ikoma *et al.* [13], but has also been seen in metallic wires [45], and quantum dots coupled to quantum wires (where the phase breaking occurs through plasmon emission in the wires [46]). Using the values for the high-mobility heterostructure discussed above, the low-temperature value of the phase-breaking time is found to be about 0.93 ns.

8.4 Conductivity

At this point, it is useful to recompute the temperature dependence of the conductivity that arises merely from impurity scattering, without the complications of

8.4 Conductivity

the electron–electron interaction. There is not much change in the self-energy given earlier, so the diagonal conductivity is still given as (with the temperature Green's functions inserted in place of the zero-temperature functions)

$$\sigma = \frac{e^2 \hbar^2}{m^2} \int \frac{d^d\mathbf{k}}{(2\pi)^d} \frac{k^2}{\beta\hbar^2} \\ \times \sum_{\omega_n} \frac{1}{i\omega_n - (E_k - \mu)/\hbar - i/2\tau} \frac{1}{i\omega_n - (E_k - \mu)/\hbar + i/2\tau}. \quad (8.194)$$

There are two aspects to this equation that lead to differences from the earlier results. The first is that the frequency summation given here is actually the resolution of the delta function. The second is that the frequency sums are more difficult than those encountered in the previous section. This is because of the delta function that is incorporated into these sums. This delta function is broadened at finite temperature, and some effect must occur to lead to this behavior. The actual complication arises from the presence of the self-energy terms (the scattering terms) in the Green's functions of (8.142). The presence of these energy-dependent self-energy terms means that the complex integration used in (8.96) becomes more complicated in that there may be branch cuts that arise from the presence of the self-energy. Consequently, the best method of attacking the problem is to go back to the actual polarization bubble itself and use the limiting process

$$\sigma = -\lim_{\omega \to 0} \left[\frac{\text{Im}(\Pi_r)}{\omega} \right]. \quad (8.195)$$

In this sense, we can write the lowest-order bubble that contributes to the conductivity as

$$\Pi(i\omega'_m) = \frac{2e^2\hbar^2}{m^2} \int \frac{d^d\mathbf{k}}{(2\pi)^d} \frac{k^2}{\beta\hbar^2} \sum_{\omega_n} G(\mathbf{k}, i\omega_n) G(\mathbf{k}, i\omega_n + i\omega'_m), \quad (8.196)$$

where we use the general Green's functions to simplify the notation, and in which the spin summation has been carried out. The procedure to be followed is essentially the same as that of (8.96) in that we define a contour integral

$$\frac{\beta\hbar}{2\pi i} \oint_C \frac{dz}{e^{\beta\hbar z} + 1} g(z), \quad (8.197)$$

where $g(z)$ is "inspired" by the summation in (8.196). For simplicity, we will also use the reduced units $\xi = (E_k - \mu)/\hbar$. In this integration, however, the contributions from the simple pole of $g(z)$ that occurred in (8.96) must be replaced by the contributions from the two branch cuts at $z = \xi$ and $z = \xi - i\omega'_m$. The contour remains the basic circle of radius R in which the limit $R \to \infty$ is taken. This contour must be deformed, however, to create two line integrals, one above and one below each branch cut. Let us consider first the contribution from the branch cut at $z = \xi$. The contribution of this quantity to (8.197) is given by

$$\frac{\beta\hbar}{2\pi i} \int_{-\infty}^{\infty} d\xi G(\mathbf{k}, \xi + i\delta) G(\mathbf{k}, \xi + i\omega'_m) \int_{\infty}^{-\infty} d\xi G(\mathbf{k}, \xi - i\delta) G(\mathbf{k}, \xi + i\omega'_m), \quad (8.198)$$

where the order of the limits on the integrals is determined by the directions of the integrations. (The paths are both in the positive (counterclockwise) direction in closing the contour, which means that the path above the branch cut is in the direction of increasing energy, whereas the path below the branch cut is in the direction of decreasing energy.) Reversing the direction of integration in the second term allows us to rewrite this contribution as

$$\frac{\beta\hbar}{2\pi i} \int_{-\infty}^{\infty} d\xi [G(\mathbf{k}, \xi + i\delta) - G(\mathbf{k}, \xi - i\delta)] G(\mathbf{k}, \xi + i\omega'_m). \quad (8.199)$$

The first term in the square brackets is the retarded Green's function, and the second term is the advanced Green's function. The bracketed term contributes the spectral density. Contribution of the second branch cut, at $z = \xi - i\omega'_m$ can similarly be calculated (in this case, the second Green's function is expanded on either side of the branch cut). Thus, we can write the summation from (8.196) as

$$\begin{aligned} S &= \sum_{\omega_n} G(\mathbf{k}, i\omega_n) G(\mathbf{k}, i\omega_n + i\omega'_m) \\ &= \frac{\beta\hbar}{2\pi i} \int_{-\infty}^{\infty} d\xi A(\mathbf{k}, \xi) [G(\mathbf{k}, \xi + i\omega'_m) + G(\mathbf{k}, \xi - i\omega'_m)] f(\xi). \end{aligned} \quad (8.200)$$

To proceed further, it is time to take the analytic continuation of the last equation. This makes the change $i\omega'_m \to \omega + i\eta$, which again converts the two Green's functions in the square brackets into the retarded and advanced functions, respectively. Then, the imaginary parts of these two Green's functions (everything else is real) contribute just another factor of the spectral density. We make a change of variables in the second term, by shifting the axis of the ξ integration, and

$$\begin{aligned} \mathrm{Im}(S) &= -\mathrm{Im}\left\{ \frac{\beta\hbar}{2\pi i} \int_{-\infty}^{\infty} d\xi A(\mathbf{k}, \xi) [G(\mathbf{k}, \xi + \omega + i\eta) + G(\mathbf{k}, \xi - \omega - i\eta)] f(\xi) \right\} \\ &= \frac{\beta\hbar}{2\pi i} \int_{-\infty}^{\infty} d\xi A(\mathbf{k}, \xi) [A(\mathbf{k}, \xi + \omega) - A(\mathbf{k}, \xi - \omega)] f(\xi) \\ &= \frac{\beta\hbar}{2\pi i} \int_{-\infty}^{\infty} d\xi A(\mathbf{k}, \xi) A(\mathbf{k}, \xi + \omega) [f(\xi) - f(\xi + \omega)] \\ &\cong -\frac{\beta\hbar}{2\pi i} \int_{-\infty}^{\infty} d\xi A(\mathbf{k}, \xi) A(\mathbf{k}, \xi + \omega) \frac{\partial f(\xi)}{\partial \omega}. \end{aligned} \quad (8.201)$$

Using this in Eqs. (8.195) and (8.196) leads to the conductivity

$$\sigma = -\frac{e^2\hbar^2}{m^2}\int\frac{d^d\mathbf{k}}{(2\pi)^d}\frac{k^2}{2\pi\hbar}\int_{-\infty}^{\infty}d\xi A^2(\mathbf{k},\xi)\frac{\partial f}{\partial \xi}. \quad (8.202)$$

At low temperatures, the spectral density contributes to the delta function that occurred in the earlier forms. From (8.80), we know that the integral over the spectral density, with respect to energy, is unity (i.e., the density of states in momentum space is uniform). In a similar fashion,

$$\frac{1}{\pi}\int_{-\infty}^{\infty}dE A^2(\mathbf{k},E) = \frac{1}{\pi}\int_{-\infty}^{\infty}dE\left(\frac{\Sigma_i}{E^2 + \Sigma_i^2}\right)^2 = \frac{1}{\Sigma_i} = \frac{2\tau}{\hbar}. \quad (8.203)$$

To use this result, we assume that the scattering is weak so that the spectral density is sharply peaked, in which case we can bring the derivative of the distribution function out of the integral. Then, the conductivity can be written as

$$\sigma = -\frac{e^2\hbar^2}{m^2}\int\frac{d^d\mathbf{k}}{(2\pi)^d}k^2\tau\frac{\partial f}{\partial E_k} = -\frac{2e^2}{dm^2}\int dE_k\rho(E_k)E_k\tau\frac{\partial f}{\partial E_k}. \quad (8.204)$$

This last result is essentially the classical result obtained with the Boltzmann equation, which is reassuring, but it depends on a special interpretation of the result of using the frequency summations. We note that the angular integration factor normally found for impurity scattering, the factor $(1 - \cos\theta)$, which was shown in the last chapter to require the next higher-order terms, is missing here, since it must usually be put into the Boltzmann equation by hand. We can put the present result into the more usual form (see Chapter 2)

$$\sigma = \frac{ne^2\langle\tau\rangle}{m}, \quad \langle\tau\rangle = -\frac{2}{dn}\int dE_k\rho(E_k)E_k\tau\frac{\partial f}{\partial E_k}. \quad (8.205)$$

and we have used the fact that

$$n = -\frac{2}{d}\int dE_k\rho(E_k)E_k\frac{\partial f}{\partial E_k}. \quad (8.206)$$

As previously, the first set of ladder diagrams results in the addition of the $(1 - \cos\theta)$ term and the retention of the angular averages that have been already done in (8.196). From the previous subsection, it is also clear that the first correction that will arise from the interacting system is to modify the scattering time τ according to (8.163). Our main concerns are with the higher-order corrections represented by the diffuson and cooperon corrections.

It is important to note at this point that the main result of the integration contained in (8.205) is to take the effective zero-temperature conductivity and average it over an energy width of the order of k_BT at the Fermi energy. This latter is given by the width of the derivative of the Fermi function. In the low-temperature limit, this derivative approaches a delta function at the Fermi

Fig. 8.15 The terms arising from the exchange interaction contribution to the conductivity.

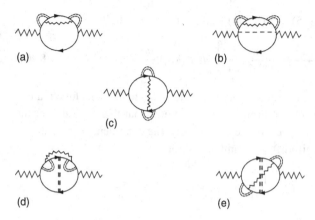

energy, and this leads to the low-temperature conductivity. However, this simple averaging result confirms the approach that was utilized in the first sections of this chapter to compute the temperature dependencies discussed there.

The lowest-order diagram for the higher-order corrections to the conductivity was shown in the last chapter as a bubble containing a major interaction to create the two-electron Green's function. The expansion of the terms that are to be considered here is shown in Fig. 8.15. Other diagrams that differ only in the directions of the electron lines are also important and should be considered, but it is the generic topology that is important [46]. When calculating the correction to the conductivity in a mixed interacting-impurity disordered system, it should be recalled that the dominant contributions will arise from the diagrams containing the maximum number of diffuson poles per segment. For this to happen, the imaginary parts of the Green's functions meeting at each vertex should have opposite signs, which implies that the Matsubara frequencies also should have opposite signs. This places a constraint on the ranges of the summations over the frequencies in each expression. By these arguments, the first three of these are quite small, since they are nested interaction and diffuson lines which act more or less independently of one another. The most important diagrams are Fig. 8.15(d,e), in which the two effects are nested and contain an additional pole besides the vertex corrections of the diffusons. Since the vertices have a vector nature, these latter two diagrams will vanish if the Green's functions to the right and left are taken in the small q limit. When the next-higher expansion terms are included for these Green's functions, an additional Dq^2 arises from the $\mathbf{q} \cdot \mathbf{v}$ terms, which actually cancels one of the diffuson poles, leaving only two such poles. Thus, only these two diagrams contribute to the corrections to the conductivity. These diagrams are those that arise from the lowest-order exchange interaction, incorporating a diffuson. There are comparable terms for the cooperon and for the Hartree terms. The method will be illustrated with the diagram of Fig. 8.15(d), for which the conductivity can be written as

$$\delta\sigma = \frac{e^2\hbar^2}{2\pi\rho(\mu)m^2} \frac{1}{\beta\hbar^2} \sum_{\omega_n} \int \frac{d^d\mathbf{k}}{(2\pi)^d} G^r(\mathbf{k}, i\omega_n) G^a(\mathbf{k}, i\omega_n)$$

$$\times \frac{1}{\beta\hbar^2} \sum_{\omega_m} \int \frac{d^d\mathbf{q}}{(2\pi)^d} G^r(\mathbf{k}+\mathbf{q}, i\omega_n+i\omega_m) \frac{V(\mathbf{q}, i\omega_n)}{[Dq^2+|\omega_n|]^3 \tau^3} \quad (8.207)$$

$$\times \int \frac{d^d\mathbf{k}'}{(2\pi)^d} \frac{kk' \cos\vartheta}{d} G^r(\mathbf{k}'+\mathbf{q}, i\omega_n+i\omega_m) G^a(\mathbf{k}', i\omega_n) G^r(\mathbf{k}', i\omega_n).$$

The first two Green's functions arise from the two terms connected to the input current tail, and the last two Green's functions are those on the right side of the diagram and are connected to the output current tail. The other two Green's functions are in the top leg on either side of the diffuson connection spanning the top and bottom legs of the diagram. In the potential term at the center of the expression, two of the diffuson lines arise from the vertex correction, and the third is that spanning the bubble diagram. At this point, we begin to work with only the two-dimensional situation. The integration can be simplified by considering the terms involved in the first momentum summation, as [47]

$$\mathbf{M}(\mathbf{q}) = \frac{\hbar}{m} \int \frac{d^2\mathbf{k}}{(2\pi)^2} \mathbf{k} G^r(\mathbf{k}, i\omega_n) G^a(\mathbf{k}, i\omega_n) G^r(\mathbf{k}+\mathbf{q}, i\omega_n+i\omega_m)$$

$$\cong \frac{\hbar}{m} \int \frac{d^2\mathbf{k}}{(2\pi)^2} \mathbf{k} G^r(\mathbf{k}, i\omega_n) G^a(\mathbf{k}, i\omega_n) G^{r2}(\mathbf{k}, i\omega_n+i\omega_m) \mathbf{v} \cdot \mathbf{q} \quad (8.208)$$

$$\cong \frac{2\tau^3 \mu}{\hbar} \mathbf{q},$$

where the leading term in the Taylor expansion for $G^r(\mathbf{k}+\mathbf{q}, i\omega_n+i\omega_m)$ has been ignored on the basis of the above arguments on size of the various terms (and the basic integral vanishes if the Green's functions are symmetric in the vector $\mathbf{k}$). The result (8.208) is obtained by converting the integration over momentum into an integration over energy. Then, it is noted that the electron–electron interaction basically requires an electron on one side of the Fermi energy to scatter from an electron on the other side of the Fermi energy. Thus, as discussed above, if $\omega_n > 0$, we require $\omega_n + \omega_m < 0$. This means that only two of the Green's functions lead to poles inside the contour; it is further assumed that the Fermi energy μ is the dominant energy in the resulting approximations. The integrands for Fig. 8.15(d,e) are the same, except that the integrand for (d) is proportional to $\mathbf{M}(\mathbf{q}) \cdot \mathbf{M}(\mathbf{q})$ and that for (e) is proportional to $\mathbf{M}(\mathbf{q}) \cdot \mathbf{M}(-\mathbf{q}) = -\mathbf{M}(\mathbf{q}) \cdot \mathbf{M}(\mathbf{q})$, so that these two diagrams are of opposite sign. Care must be taken in writing down the actual integration. Here, we follow the arguments of Altshuler et al. [47]. When $\omega_n < 0$, the contributions of these two diagrams cover the same region of integration, and the net result vanishes. On the other hand, when $\omega_n > 0$, the conditions for Fig. 8.15(d,e) are such that they add so long as the summation over ω_n is evaluated at ω_m. Then, with a factor of 2 for the two terms that contribute for positive frequency (they cancel for negative frequency), the conductivity can be written as

$$\sigma = \frac{2e^2\tau^2}{\hbar^2}\frac{1}{\beta\hbar^2}\sum_{\omega_n>0}\int\frac{d^2\mathbf{q}}{(2\pi)^2}\frac{V(\mathbf{q},i\omega_n)Dq^2}{[Dq^2+|\omega_n|]^3}. \qquad (8.209)$$

At this point, the fully screened Coulomb interaction is inserted from the general result of

$$V(\mathbf{q},i\omega_n) = \frac{e^2}{2\varepsilon q\left[1+\frac{\chi_2 Dq}{(\omega_n)+Dq^2}\right]} \sim \frac{e^2}{2\varepsilon q}\frac{(\omega_n)+Dq^2}{\chi_2 Dq}, \qquad (8.210)$$

so that we can now write

$$\sigma = \frac{2e^2\tau^2}{\hbar^2}\frac{e^2}{2\varepsilon\chi_2}\frac{1}{\beta\hbar^2}\sum_{\omega_n>0}\int\frac{d^2\mathbf{q}}{(2\pi)^2}\frac{1}{[Dq^2+\omega_n]^2}. \qquad (8.211)$$

There is another set of diagrams similar to Fig. 8.15(d,e), with the interaction line in the lower Green's function. This changes the limits of integration, but they are symmetric so we can add the results. We now note that since the expressions are valid only for $\omega\tau < 1$, we need to cut off the integration in frequency, and

$$\sigma = \frac{e^4\tau^2}{2\pi\varepsilon\chi_2\beta\hbar^4}\sum_{\omega_n>0}^{1/\tau}\int\frac{qdq}{[Dq^2+\omega_n]^2}. \qquad (8.212)$$

These can now be integrated to yield

$$\sigma \cong \frac{e^4\tau^2}{4\pi\varepsilon\chi_2\hbar^2 D}\ln\left(\frac{\tau}{\beta\hbar}\right). \qquad (8.213)$$

There are other contributions to the conductivity from other diagrams, such as the Hartree terms. Generally, these are of the same order but do not change the basic temperature dependence by very much. The point here is to demonstrate how this temperature dependence can be obtained. It is important to note that the result (8.213) is one form that has been obtained. Many different forms have been obtained, but the consistent equivalence lies in the logarithmic dependence of the temperature at low temperatures (for which this derivation is valid). At the lowest of temperatures, the additional terms here become important and can be quite important in evaluating the overall conductivity.

References

[1] S. Tarucha, T. Saku, Y. Hirayama, et al., in *Quantum Effect Physics, Electronics, and Applications*, eds. K. Ishmail, T. Ikoma, and H. I. Smith, IOP Conf. Ser. **127**, 127 (1992).
[2] R. A. Webb. S. Washburn, H. J. Haucke, et al., in *Physics and Technology of Submicron Structures*, eds. H. Heinrich, G. Bauer, and F. Kuchar (Berlin, Springer-Verlag, 1988), p. 98.

[3] W. J. Skocpol, P. M. Mankiewich, R. E. Howard, *et al.*, *Phys. Rev. Lett.* **56**, 2865 (1986); **58**, 2347 (1987).
[4] C. de Graaf, J. Caro, and S. Radelaar, *Phys. Rev. B* **46**, 12814 (1992).
[5] E. Abrahams, P. W. Anderson, P. A. Lee, and T. V. Ramakrishnan, *Phys. Rev. B* **24**, 6783 (1991).
[6] R. G. Wheeler, K. K. Choi, A. Goel, R. Wisnieff, and D. E. Prober, *Phys. Rev. Lett.* **49**, 1674 (1982).
[7] R. P. Taylor, P. C. Main, L. Eaves, *et al.*, *J. Phys.: Cond. Matter* **1**, 10413 (1989).
[8] P. A. Lee, A. D. Stone, and H. Fukuyama, *Phys. Rev. B* **35**, 1039 (1987).
[9] C. W. J. Beenakker and H. van Houten, *Phys. Rev. B* **37**, 6544 (1988).
[10] R. C. Dynes, *Physica B* **109+110**, 1857 (1982).
[11] K. K. Choi, D. C. Tsui, and K. Alavi, *Phys. Rev. B* **36**, 7751 (1987).
[12] J. P Bird, A. D. C. Grassie, M. Lakrami, *et al.*, *J. Phys.: Cond. Matter* **3**, 2897 (1991).
[13] T. Ikoma, T. Odagiri, and K. Hirakawa, in *Quantum Effect Phyics, Electronics, and Applications*, eds. K. Ishmail, T. Ikoma, and H. I. Smith, IOP Conf. Ser. **127**, 157 (1992).
[14] Y. K. Fukai, S. Yamada, and H. Nakano, *Appl. Phys. Lett.* **56**, 2133 (1990).
[15] J. H. Davies and J. A. Nixon, *Phys. Rev. B* **39**, 3423 (1989); J. A. Nixon, J. H. Davies, and H. U. Baranger, *Phys. Rev. B* **43**, 12638 (1991).
[16] Y. Takagaki and D. K. Ferry, *J. Phys.: Cond. Matter* **4**, 10421 (1992).
[17] D. J. Thouless, *J. Non-Cryst. Sol.* **35/36**, 3 (1980).
[18] R. E. Howard, L. D. Jackel, P. M. Mankiewich, and W. J. Skocpol, *Science* **231**, 346 (1986).
[19] J. Imry, *Europhys. Lett.* **1**, 249 (1986).
[20] P. Nozières and D. Pines, *Theory of Quantum Liquids*, (New York, Benjamin, 1966).
[21] S. Das Sarma, in *Quantum Transport in Ultrasmall Devices*, eds. D. K. Ferry, H. L. Grubin, C. Jacoboni, and A.-P. Jauho (New York, Plenum Press, 1995).
[22] See, for example, the discussion in P. A. Lee and T. V. Ramakrishnan, *Rev. Mod. Phys.* **57**, 287 (1985).
[23] B. L. Altshuler and A. G. Aronov, *Sol. State Commun.* **39**, 115 (1979); *Zh. Eksp.Teor. Fiz. Pis'ma Red.* [*JETP Lett.* **30**, 514 (1979)].
[24] D. K. Ferry, *Semiconductors* (New York, Macmillan, 1991).
[25] B. K. Ridley, *Quantum Processes in Semiconductors* (Oxford, Oxford University Press, 1982).
[26] O. Madelung, *Introduction to Solid State Theory* (Berlin, Springer-Verlag, 1978), pp. 104–9.
[27] P. Lugli and D. K. Ferry, *Appl. Phys. Lett.* **46**, 594 (1985); *IEEE Electron Dev. Lett.* **6**, 25 (1985).
[28] A. L. Fetter and J. D. Walecka, *Quantum Theory of Many-Particle Systems* (New York, McGraw-Hill, 1971).
[29] G. D. Mahan, *Many-Particle Physics* (New York, Plenum, 1981).
[30] C. P. Enz, *A Course on Many-Body Theory Applied to Solid State Physics* (Singapore, World Scientific Press, 1992).
[31] B. Vinter, *Phys. Rev. B* **13**, 4447 (1976).
[32] B. Vinter, *Phys. Rev. B* **15**, 3947 (1977).
[33] T. Ando, A. B. Fowler, and F. Stern, *Rev. Mod. Phys.* **54**, 437 (1982).
[34] H. Fukuyama, in *Electron-Electron Interactions in Disordered Systems*, eds. A. L. Efros and M. Pollak (Amsterdam, North-Holland, 1985).

[35] A. V. Chaplik, *Sov. Phys. JETP* **33**, 997 (1971).
[36] G. F. Giuliani and J. J. Quinn, *Phys. Rev. B* **26**, 4421 (1982).
[37] B. L. Altshuler, D. E. Khmelnitzkii, A. L. Larkin, and P. A. Lee, *Phys. Rev. Lett.* **44**, 1288 (1980).
[38] H. Fukuyama, *J. Phy. Soc. Jpn.* **49**, 644 (1980).
[39] D. K. Ferry, *Semiconductor Transport* (London, Taylor and Francis, 2000), Chapter 7.
[40] J. P. Bird, K. Ishibashi, D. K. Ferry, *et al.*, *Phys. Rev. B* **51**, 18037 (1995).
[41] C. Prasad, D. K. Ferry, A. Shailos, *et al.*, *Phys. Rev. B* **62**, 15356 (2000).
[42] B. L. Altshuler and A. G. Aronov, in *Electron-Electron Interactions in Disordered Systems*, eds. A. L. Efros and M. Pollak (Amsterdam, North-Holland, 1985).
[43] B. L. Altshuler, A. G. Aronov, and D. E. Khmelnitskii, *J. Phys. C* **15**, 7367 (1982).
[44] D. S. Golubev and A. D. Zaikin, *Phys. Rev. Lett.* **81**, 1074 (1998).
[45] P. Mohanty, E. M. Q. Jariwala, and R. A. Webb, *Phys. Rev. Lett.* **78**, 861 (1979).
[46] D. P. Pivin, A. Andresen, J. P. Bird, and D. K. Ferry, *Phys. Rev. Lett.* **82**, 4687 (1999).
[47] B. L. Altshuler, D. E. Khmelnitzkii, A. L. Larkin, and P. A. Lee, *Phys. Rev. B* **22**, 5142 (1980).

9
Nonequilibrium transport and nanodevices

The technological means now exists for approaching the fundamental limiting scales of solid-state electronics in which a single electron can, in principle, represent a single bit in an information flow through a device or circuit. The burgeoning field of single-electron tunneling (SET) effects, although currently operating at very low temperatures, has brought this consideration into the forefront. Indeed, the recent observations of SET effects in poly-Si structures *at room temperature* by a variety of authors has grabbed the attention of the semiconductor industry. While there remains considerable debate over whether the latter observations are really single-electron effects, the resulting behavior has important implications for future semiconductor electronics, regardless of the final interpretation of the physics involved. Indeed, the semiconductor industry is rapidly carrying out its own advance, with transistor gate lengths in the 20 nm range in production in 2009 (the so-called 35 nm node).

We pointed out in Chapter 1 that the semiconductor industry is following a linear scaling law that is expected to be fairly rigorous. With dimensions approaching 10 nm within another decade, there is a rapid search for possible new technologies that can supplement Si with the offer of improved performance. However, it is clear from a variety of considerations that the devices themselves may well not be the limitation on continued growth in device density within the integrated circuit chip. Factors such as resistance of metallization lines, time delays (and signal loss) in very long interconnects that must run across the entire chip or a significant part of the chip, clock skew across a chip, and leakage currents within the devices may impact the continued scaling to a much larger degree than the physics of the devices themselves. Does this mean that we should not be concerned about the device physics and studies of quantum devices? Thankfully, no. Architecture is proceeding at a pace comparable to that of circuit scaling, and there is considerable effort in seeking ways in which the insertion of quantum devices within each cell of the integrated circuit design will provide for enhanced functionality. This means that we must seek the manner in which quantum effects in open device structures will carry over to these applications.

How do we couple the physics of the last several chapters to that needed for understanding the future ultrasmall microdevices expected for approaching

Fig. 9.1 Random distribution of impurities in an ultrasmall device. (After Zhou and Ferry [1].)

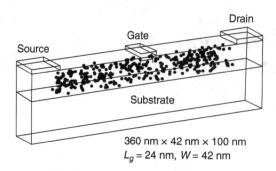

360 nm × 42 nm × 100 nm
L_g = 24 nm, W = 42 nm

future generations of integrated circuits? Consider the implications of a small device that is fabricated as a GaAs MESFET [1]. (The arguments are the same for MOSFETs or for HEMTs, so the approach with the MESFET will be adequate to illustrate several points.) A typical ultrasmall device can be considered with a gate length of 40 nm and a source-to-drain spacing of 100 nm. If the device is fabricated by growing a heavily doped, $n = 3 \times 10^{18}$ cm^{-3}, epitaxial layer 40 nm thick on a semi-insulating GaAs substrate, with an 80 nm width (twice the gate length), then there are (on average) only 384 impurity atoms under the gate, and fewer than 1000 in the entire device. These impurities are randomly located within the doped region, so there is a considerable variance in the actual number of dopant atoms under the gate (the rms variation under the gate is ±20 impurity atoms). This leads to a considerable inhomogeneity in the actual electron concentration at any point in the device. Consider for example Fig. 9.1, in which we plot the actual dopant atoms (one possible distribution of the atoms) in an ultrasmall device with a slightly smaller gate length. It is clear from this figure that there are regions where the density of dopant atoms is above the average value, and other regions where the actual density is much less than the average value. In Fig. 9.2, the resulting electron density (under bias with current flowing) obtained by self-consistently solving Poisson's equation is plotted. Again, there are many peaks in the density, which correlate with the regions where the number of dopant atoms is high, and other regions where the density is low. The highest peaks occur in the region between the source and the gate, a region known to be important for both metering the actual current flow through the device and for impacting the series "source" resistance in the device. The peaks are broader and lower in the region between the gate and the drain due to enhanced carrier heating in this region, but they are still significant. A simple calculation suggests that each of the peaks in the figure corresponds to a small region about 10 nm on a side (in three dimensions) in which there are roughly 10 electrons. That is, the doped region throughout the device appears to be formed of a large number of quantum boxes (quantum dots in three dimensions). Although the current has been calculated in a semiclassical manner it is clear that the real current in an actual device will be affected

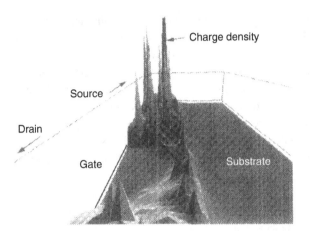

Fig. 9.2 The inhomogeneous carrier density found by self-consistent solutions to Poisson's equation for the impurity distribution of the previous figure. (After Zhou and Ferry [1].)

strongly by the potential barriers between the boxes, and that the actual potential shape (and value) will depend strongly on many-body effects within each quantum box. Will this transport be hopping transport from one box to another? Or will there be significant tunneling between the boxes? At this time, there are no answers to these questions. In quasi-two-dimensional systems such as HEMTs and MOSFETs, the carriers are localized at the interface (either between GaAs and AlGaAs or between Si and SiO_2). Nevertheless, the effects will continue to be there [2], except that the structures will now be essentially the quantum-dot structures discussed in an earlier chapter.

The theoretical structures discussed so far in this book have been described in terms of an equilibrium system, either at zero temperature or in thermal equilibrium with the surroundings. It is possible to describe the thermodynamics of an *open, nonequilibrium* system through the use of nonequilibrium Green's functions, the so-called real-time Green's functions developed by Schwinger and colleagues [3,4], by Kadanoff and Baym [5], and by Keldysh [6]. However, it has been pointed out that this interpretation is valid only when the entropy production (by dissipation) is relatively small and the system remains near to equilibrium [7], where, at least in lowest order, Liouville's equation remains valid as a equation of motion for the appropriate statistical ensemble representing the carrier transport under the applied fields. In *far-from-equilibrium* systems, however, there is usually strong dissipation, and the resulting statistical ensemble (even in steady state) is achieved as a balance between driving forces and dissipative forces; for example, it does not linearly evolve from the equilibrium state when the applied forces are "turned on." Enz [7] argues that there is no valid unitary operator (which describes the impact of perturbation theory) to describe the evolution through this symmetry-breaking transition to the dissipative steady state. As a consequence, there is no *general* formalism at this time that can describe the all-important far-from-equilibrium devices. On the other hand, the application of the real-time

Green's functions to what are surely very strongly far-from-equilibrium systems – the excitation of electron–hole pairs in intense femtosecond laser pulses – has yielded results that suggest that their use in these systems is quite reliable for studying both the transition to the semiclassical Boltzmann theory and to explain experimentally observed details [8,9,10]. As a consequence of these latter studies, as well as initial attempts to actually begin to model real devices with these Green's functions, the situation is believed to be much better than this pessimistic view would warrant. Nevertheless, it is essential that one move carefully with these approaches to assure that the problematic approach is really meaningful.

In this chapter, we will first review some of the experiments indicating nonequilibrium behavior in mesoscopic devices. We then focus on the industrially relevant Si nano-devices as well as some proposed alternative structures. We then turn to the theoretical aspects of trying to treat the transport in these devices. We first will illustrate the extension of the scattering matrix approach developed in Section 3.5.1 to three-dimensional devices and the dissipation through phonon scattering. This has been applied to small nanowire devices fabricated in silicon-on-insulator (SOI) and InAs structures. We then turn to the real-time Green's functions, and give their general derivation and properties. Then, we turn to quite new approaches which implement a full band structure along with the real-time Green's functions (this avoids using the effective mass approximation, which we have employed so far in this book) to treat heterojunctions more properly. Two such approaches are NEMO (for Nano-Electronic MOdeling), and TransSIESTA. In both of these efforts, local orbital wave functions have been adopted, and a final discussion will detail the use of delocalized plane wave orbitals.

9.1 Nonequilibrium transport in mesoscopic structures

Over the past two decades or so, our understanding of the transport of electrons (and holes) through mesoscopic systems has increased significantly. Most of this transport work, described in the previous chapters, is based on a system either at (assumed) zero temperature or at least in thermal equilibrium with its surroundings (the RTD is a notable exception). Yet in most devices of interest, this will not be the case. In the past few years there have been studies of mesoscopic systems that have probed the nonequilibrium behavior of these systems, at least to the lowest order. Here we examine some of these effects.

9.1.1 Nonequilibrium effects in tunnel barriers

First, let us consider a structure in which an AlGaAs barrier is inserted between a heavily doped (n^+) GaAs layer and a lightly doped (n^-) GaAs layer. A second heavily doped (n^+) layer is placed at the other end of the lightly doped GaAs layer. This would normally be an $n^+ n^- n^+$ structure, except the tunneling barrier is inserted at one end of the lightly doped region. It is this combination of structures that makes this example interesting. Hickmott *et al.* [11] were apparently the first to study such a structure, and they observed periodic oscillations in the applied voltage. These oscillations seemed to have a periodicity corresponding to the emission of a sequence of longitudinal (polar-) optical phonons. The basic structure is shown in Fig. 9.3 with an applied bias [12]. The second derivative of the applied voltage as a function of the reverse bias current ("reverse" is defined here in the sense that the electrons flow through the barrier into the lightly doped region) is shown in Fig. 9.4.

An understanding of the principle at work arises from the consideration that the voltage drop is across three distinct regions [11,13]: (i) the tunneling barrier, (ii) a fraction of the lightly doped (n^-) region that is depleted of carriers, and (iii) the lightly doped (n^-) region in which there are still a significant number of free carriers. Perhaps it is easiest to understand if we consider the constant current case instead of the constant potential case. Consider that a fixed number of electrons is being injected through the tunnel barrier into the depleted portion of the lightly doped region. The voltage dropped across this region is denoted as V_D. When this voltage is less than the optical phonon energy ($eV_D < \hbar\omega_{LO}$), a relatively large drop must arise across this region, and the region is terminated by the portion that is still populated with carriers. The current is determined by the velocity of the carriers and by the residual doping ($j = n(x)ev(x)$, which requires the density to be a decreasing function of x since the velocity is an increasing function of x). When $eV_D > \hbar\omega_{LO}$, however, the electron can emit an optical phonon and drop to a lower kinetic energy state (hence lower velocity

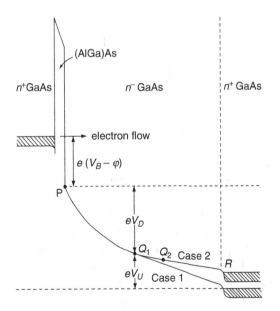

Fig. 9.3 The GaAs/AlGaAs/GaAs structure under a bias. (After Eaves *et al.*, in *Physics of Quantum Electron Devices*, ed. F. Capasso (New York, Springer-Verlag, 1990) [12], by permission.)

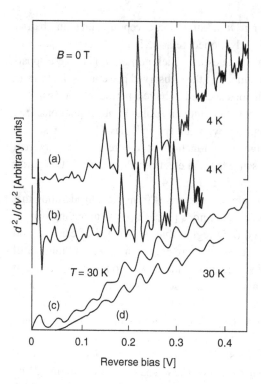

Fig. 9.4 The second derivative of the current with respect to voltage. [After Eaves *et al.*, in *Physics of Quantum Electron Devices*, ed. F. Capasso (New York, Springer-Verlag, 1990) [12], by permission.]

state). This requires that carriers move into the depleted region from the undepleted region to maintain the current constant. (More carriers at lower velocity are required to maintain the total current, which has been assumed constant.) This, in turn, tries to lower the total voltage drop that occurs across the device. In the end, the carrier movement actually tries to lower the voltage across the undepleted region as well, which leads to a feedback effect. In essence, these charge movements lead to the voltage swings required each time the voltage drop V_D passes the threshold for the emission of another optical phonon. The actual distribution of potential through the device must be found by solving Poisson's equation. The nonlinear feedback from the charge motion from the undepleted to the depleted region can lead to a kind of hysteresis in the current–voltage relationship, but it is this "switching" behavior in the actual charge distribution that leads to the observed oscillations.

The importance of the phonon emission is quite evident in resonant-tunneling diodes, where the carriers can be quite far from equilibrium in the quantum well. Normally, one expects that the carriers can pass ballistically through the double barriers (and wells) without losing any energy. However, it is also possible in many cases for the carriers to emit a phonon. This can occur when the bias is above that necessary for the normal peak current, so that the particles entering the well lie more than $\hbar\omega_{LO}$ above an acceptable output level (one in which tunneling through the second barrier leaves them in the conduction band of the

9.1 Nonequilibrium transport in mesoscopic structures

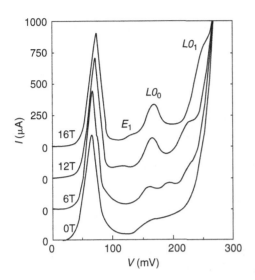

Fig. 9.5 The current–voltage characteristics for a resonant tunneling diode with a longitudinal magnetic field at 4 K. (After Eaves *et al.*, in *Physics of Quantum Electronic Devices*, ed. F. Capasso (New York, Springer-Verlag, 1990) [12], by permission.)

output side of the structure). This is enhanced significantly when the lower energy state is actually the bound state in the well, which leads to a potential distribution in which most of the potential drop is across the input barrier, while the bound state lies quite close to the Fermi level in the output barrier. Such a set of *I*–*V* curves is shown in Fig. 9.5 [12]. Here, the temperature is uniformly 4 K, and the resonant-tunneling diode is etched into a mesa shape of 100 μm diameter. The well is 11.7 nm wide, while the barriers are 5.6 nm thick. The principal resonant tunneling peak is evident at about 70 mV bias, and another peak (much smaller) appears at about 170 mV. This peak is much enhanced by the presence of a magnetic field oriented parallel to the current direction. Normally, an electron tunneling through a barrier conserves the transverse momentum. In the case of a phonon emission, however, this is no longer the case. The magnetic field helps to quantize the transverse momentum and highlight the phonon-assisted tunneling peak. The very weak peak marked E_1 in the figure is thought to be due to a non-resonant tunneling transition between the nth Landau level in the emitter and the $(n+1)$ Landau level in the well. Such a transition is thought to be due to the effect of ionized impurities or interface roughness. The peaks marked LO_p correspond to the transition from the nth Landau level in the emitter to the $(n+p)$th Landau level in the well through the emission of a single optical phonon. The relative amplitudes of the main resonant tunneling peak and these other peaks provide qualitative indications of the contribution of various charge transport processes to the measured current.

The magnetic field is quite important in the observation of these phonon-assisted tunneling currents. In Fig. 9.6, we show another series of measurements for a GaAs/AlGaAs resonant-tunneling diode, this time at 1.8 K [14]. The *I*–*V*

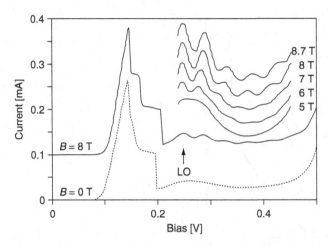

Fig. 9.6 The *I–V* characteristics at 1.8 K for a GaAs/AlGaAs resonant-tunneling diode for various values of the magnetic field. The peak marked LO is a phonon replica of the main peak. (After Yoo *et al.* [14].)

curves exhibit a main resonant peak at 144 meV and a first subsidiary peak, which has been marked as LO at 245 meV, in a magnetic field of 8 T. This subsidiary peak is attributed to tunneling with the assistance of a single-phonon-emission process. In the absence of the magnetic field, there remains an indication of the process, but the peak is much less pronounced and spread in energy. This peak is enhanced by the reduction in phase space that arises from the application of a magnetic field. The inset curves show how the peaks develop as the magnetic field is increased. Although there are several peaks in the spectrum, only the first peak does not shift with magnetic field; all the other peaks move in magnetic field. As in [14], these latter peaks are thought to arise from transitions that involve changes in the Landau level index during the tunneling process, as well as include the emission of the optical phonon.

9.1.2 Ballistic transport in vertical and planar structures

When carriers are accelerated, or injected, into a region in which they are subject to an electric field, it takes a certain period of time before the transport can be characterized by a mobility, and/or a drift velocity. In this short-time behavior (and also short-spatial dimension behavior), the carriers move quasi-ballistically. That is, they move either with an initial injection velocity or under direct acceleration of the electric field. The time over which this behavior occurs is essentially the momentum relaxation time. In a high-mobility quasi-two-dimensional electron (or hole) gas formed at the heterointerface between GaAs and AlGaAs, where the mobility can be several million at low temperatures, this time can be of the order of many tens of picoseconds, which corresponds to an elastic mean free path of tens of microns. Many people have tried to construct devices that would utilize (and measure) this quasi-ballistic transport.

Hayes et al. [15] utilized an $n^+p^-np^-n^+$ structure (known as a planar-doped-barrier transistor – PDBT), and Yokoyama et al. [16] and Heiblum et al. [17] replaced the lightly doped p-regions with hetero-barriers. Both Yokoyama et al. [16] and Levi et al. [18] observed indications of ballistic transport. The most convincing measurements, which allow for spectroscopy of the injected hot carriers, were provided by Heiblum et al.[19]. These latter structures utilized a tunneling injector to provide a very narrow (on the order of 8 meV wide) spectrum of injected electrons [20]. These devices are known as Tunneling Hot Electron Transfer Amplifiers (THETAs) and have the ability to use a collector heterobarrier to carry out energy spectral measurements of the injected carriers, as shown in Fig. 9.7. It may be noted from this picture that the base region forms a quantum well. The coherence of the injected beam of carriers can be estimated by the observation of oscillations in the transmission coefficient (from emitter to collector), which replicate the resonant levels in the well and the virtual resonances at energies above the barriers. This is shown in Fig. 9.8 for two different base widths and collector barrier heights (at the collection end of the collector in Fig. 9.7). Measuring the derivative of the emitter current with respect to the base-emitter voltage yields the transmission as a function of injection energy. With spectroscopic measurements of the injected beam, it is possible to identify the emission of LO phonons in the base region as an energy loss mechanism. However, this is not the major loss mechanism. Hollis et al. [21] have suggested that the dominant loss mechanism is due to the interaction of the injected electrons with the high density of cold electrons in the base region, and with plasmons (collective excitations) of this cold particle gas. Levi et al. [18] estimate that the mean free path in their PDBT is only about 30 nm due to this carrier–carrier interaction. There is significant evidence that this carrier–carrier scattering is also important in THETA devices.

More recent measurements have been carried out on structures in which a resonant-tunneling double barrier was inserted into the device at the collector edge of the gate [22]. The resonant level of this double-barrier structure can then be used to more effectively scan through the spectrum of injected carriers and hopefully could produce a negative differential conductance as in other double-barrier devices. However, no negative differential conductance was seen unless a high magnetic field was applied to the structures. The magnetic field constricts the phase space available for scattering and this greatly attenuates the carrier–carrier interaction. With a 30-nm-thick GaAs base, doped to $n = 10^{18}$ cm^{-3}, the carrier–carrier scattering limited mean free path is estimated to be about 30 nm [20].

In lateral devices, which are constructed using surface gates on a quasi-two-dimensional heterostructure layer, the above loss to the high density of base electrons (or holes in some structures) can be avoided [23]. Such a lateral

Fig. 9.7 Structure of a THETA device. (After Heiblum and Fischetti, in *Physics of Quantum Electron Devices,* ed. F. Capasso (New York, Springer-Verlag, 1990) [20] by permission.)

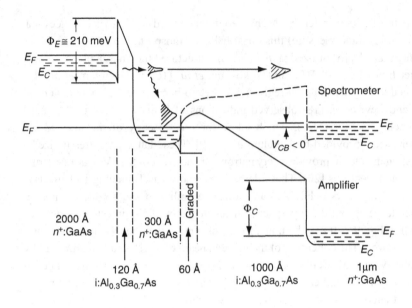

Fig. 9.8 Experimental and theoretical curves for the tunneling base current in a THETA device. (After Heiblum and Fischetti, in *Physics of Quantum Electron Devices,* ed. F. Capasso (New York, Springer Verlag, 1990) [20], by permission.)

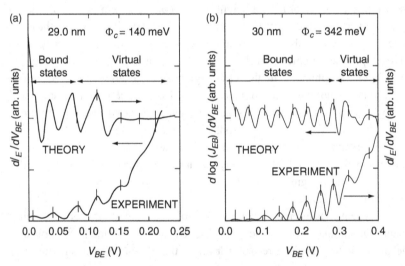

gate device is shown in Fig. 9.9. In these structures, further confinement of the electrons into a single, narrow lateral beam can be heightened by the use of side gates in the emitter region. Transfer efficiencies as high as $\alpha > 0.98$ can be achieved for injection energies below the optical phonon energy, and the elastic mean free path is estimated to be greater than 0.48 μm [24], easily in keeping with the earlier discussion. This value for the mean free path is comparable to that of the material itself (at low biases), so that it is conjectured that the value for the hot injected carriers is comparable to that of the cold background electrons, so long as the injection energy remains below the optical phonon threshold (all the measurements have been carried out at 4.2 K).

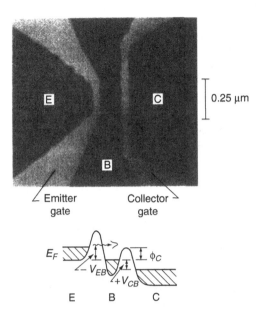

Fig. 9.9 SEM picture of a lateral THETA device and the schematic of the potential profile. (After Pavleski *et al.* [24], by permission.)

9.1.3 Thermopower in nanostructures

In general, when one passes a current through a semiconductor system, this electrical current is accompanied by a thermal current which constitutes the flow of energy through the conductor. In most cases, where the transport takes place in thermal equilibrium, this thermal current can safely be ignored. However, this is not always the case. The thermal current arises from the fact that the electrical current causes a temperature differential to exist in the semiconductor sample according to a relation in which the resulting voltage is given by

$$V = \frac{I}{G} + S\Delta T, \tag{9.1}$$

where S is known as the thermopower. Accompanying the temperature difference, one can define the thermal current as

$$Q = \Pi I - \kappa \Delta T, \tag{9.2}$$

where Π is known as the Peltier coefficient and κ is the total (electron plus lattice) thermal conductivity.

The earliest measurements of the non equilibrium-generated thermopower S seem to have been those of Galloway *et al.* [25]. These authors actually measured the fluctuations in the conductance and related them to heating effects in the quantum structure. In any case of diffusion-limited transport, such as that most mesoscopic systems exhibit, the thermopower can be expressed in terms of the energy dependence of the conductance through

Fig. 9.10 Schematic diagram of the sample used for thermopower measurements. (After Galloway et al. [25], by permission.)

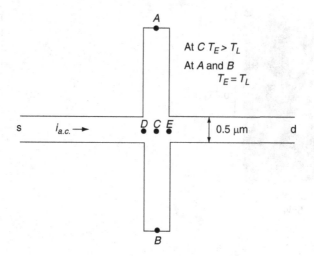

$$S = -\frac{\pi^2}{3}\frac{k_B^2 T}{eG}\frac{dG(E)}{dE}\bigg|_{E_F} \approx -\frac{\pi^2}{3}\frac{k_B^2 T}{e\Delta E}\frac{\delta G}{G}, \quad (9.3)$$

where the derivative in the last form has been replaced by the conductance fluctuation and the average energy scale, with $\Delta E = \max(k_B T, \hbar D/l_{in}^2, \hbar D/L^2)$, and the terms have their normal meanings used throughout this text. In the non-equilibrium case, the appropriate energy scale is the first of these quantities, and

$$S = -\frac{\pi^2}{3}\frac{k_B}{e}\frac{\delta G}{G}. \quad (9.4)$$

It should be noted that the coefficient is composed primarily of known constants, and this formula is readily checked against experiment. The coefficient $\pi^2 k_B/3e$ takes the value of 284 μV/K. Thus, a measurement of the thermopower as a function of the relative amplitude of the fluctuations in conductance gives a measure of this value.

Galloway *et al.* [25] made measurements on samples such as those shown in Fig. 9.10, in which quantum wires in a bulk GaAs structure were measured. The wires had a physical width of 0.5 μm and thickness of 50 nm. In these wires, it is thought that the thermal energy is larger than the other comparable energy-defining quantities listed under (9.3), at least for temperatures down to 0.1 K. The idea of the experiment is to heat the electrons by an applied current passed between the source s and drain d in the figure. While the electrons at the ends of the sidearm nominally will be at the lattice temperature, the carriers at point C will be at an elevated electron temperature T_E as a consequence of the heating by the joule current. The heating can be estimated by an energy balance equation, which semiclassically can be stated as [26]

$$\frac{\partial \langle E \rangle}{\partial t} = e\mathbf{J}\cdot\mathbf{F} - \frac{3k_B}{2\tau_E}(T_E - T) \quad (9.5)$$

9.1 Nonequilibrium transport in mesoscopic structures

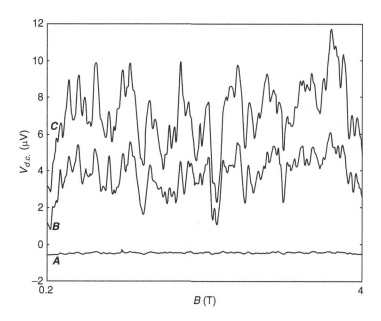

Fig. 9.11 The d.c. thermoelectric voltage fluctuations due to the electron temperature gradient. Curves *B* and *C* have been offset for clarity. (After Galloway *et al.* [25], by permission.)

in three dimensions, where **F** is the applied electric field. In steady state, the electron temperature can be found as

$$T_E = T + \frac{2e\tau_E \sigma}{3k_B} F^2. \tag{9.6}$$

We note that this heating is a heating of the electron system and not of the lattice as a whole. For a homogeneous conductor, the voltages V_{AC} and V_{CB} would cancel, but if the conductance fluctuations in these two sidearms are uncorrelated, then this will not be the case for the fluctuations. In these samples, it is felt that $l_\varphi \sim 0.25$ μm, while $l_{in} \sim 2.5$ μm, so that these voltages should be uncorrelated for the size of the sample used in the experiment. In Fig. 9.11, the voltage fluctuations measured at the three points *A*, *B*, and *C* (relative to point *A*) in the sample are shown as a function of the magnetic field. Here, the longitudinal field between points *D* and *E* is only 0.17 μV, and the scale on the fluctuations in the figure is 45 μV. The variation of the fluctuation voltage with the temperature differential $\Delta T = T_E - T$ is shown for a variety of lattice temperatures in Fig. 9.12. The linear dependence in this figure clearly confirms the expectations of (9.1) for a linear dependence of the fluctuating voltage on the temperature differential. The data give a slope of 220±20 μV/K, which is close to the theoretical value (within 20% in any case, considering that one needs to estimate the electron temperature from an indirect measurement, discussed below). These results seem to fit well to the theory of the thermopower in systems in which the electrons are heated by the conduction current.

In general, it is usually expected that carrier heating by the longitudinal current will produce voltages that are of odd order in the current, yet the above

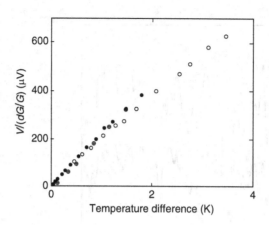

Fig. 9.12 The rms d.c. voltage fluctuation as a function of the temperature differential. The curves for open circles are at 1.08 K, those for the crosses at 2.0 K, and the solid circles at 4.0 K. (After Galloway *et al.* [25], by permission.)

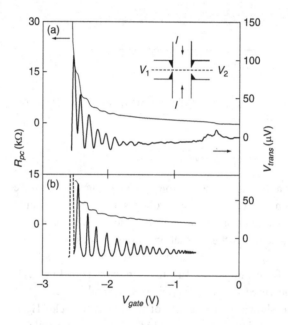

Fig. 9.13 Experimental traces of the conductance of the point contacts (thin curve) and the transverse voltage (thick curve). (After Molenkamp *et al.* [27], by permission.)

results are thought to be even in the current because they depend only on the temperature, which is quadratic in the current (or in the field as indicated above). Similar measurements have been shown by Molenkamp *et al.* [27] for the case in which the transverse voltage is coupled through point contacts. The sample is shown as an inset to Fig. 9.13, where the conductance through the transverse point contacts and the transverse voltage arising from the longitudinal heating current are shown. In this structure, the phase coherence length is small compared to the mean free path, so that the quantum-interference effects are not particularly important to the conductivity of the sample itself. The transport through the longitudinal channel is felt to be quasi-ballistic (e.g., not diffusive, but scattering is still considered to exist in order to create the enhanced electron temperature), and the two point contacts are not equivalent. If the two contacts were equivalent, then the total transverse voltage between the ends of the two sidearms would cancel, as discussed above. However, the inequality of the two point contacts breaks the inversion symmetry of the sample, which allows the observation of a transverse voltage (in the previous experiment above, only the fluctuations in the voltage could be observed), which is the difference in the thermopower-induced voltage across the two point contacts. The measurements are for a lattice temperature of 1.65 K, and the transverse voltage is plotted for a bias current of 5 mA and a fixed bias on one of the gates of -2 V. The curves in Fig. 9.13(b) are a calculation assuming an electron temperature of 4 K and a Fermi energy of 13 meV. This calculation assumes a quasi-equilibrium Fermi distribution for the carriers in the channel. The transverse voltage is even in the applied longitudinal current, and the large quantum oscillations occur as a result of the depopulation of the one-dimensional subbands in the point contacts. The actual voltage measured is proportional to the *difference* in the thermopower voltage of the two probes, as has been mentioned. Comparison of the peaks in the transverse voltage with the conductance through the point contacts confirms this behavior, in that the peaks are aligned (in voltage) with the steps in the conductance through the point contact. These authors have also measured the Peltier coefficient and the thermal conductivity of a point contact under nonequilibrium conditions [28] and some other thermoelectric effects [29].

The measurement of the thermopower can show other effects when it is used as a probe of adjacent quantum confined systems. In Fig. 9.14, the quantum wire is coupled to a quantum dot through one of the two transverse point contacts [30]. The active area of the dot is about 0.7×0.8 μm^2, which is adjacent to the 2-μm-wide, 20-μm-long channel. The dot area is tuned by the bias voltage applied to contact E. Heating is provided by a relatively large current flowing in the quantum wire, and the thermopower voltage is measured as the difference in voltages V_1 and V_2.

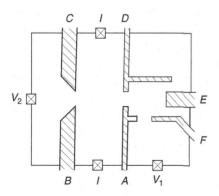

Fig. 9.14 Diagram of a 2-μm-wide channel coupled to a point contact (left) and a quantum dot (right). (After Molenkamp et al. [29], by permission.)

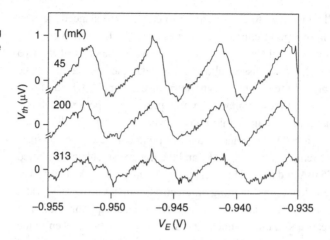

Fig. 9.15 Thermovoltage as a function of the tuning stub bias at various lattice temperatures. (After Molenkamp et al. [30], by permission.)

In the previous measurement, the oscillations in the transverse voltage (and in the thermopower) arose from the change in the number of modes propagated through the point contact. Here, this change in transmission will arise from the single-electron tunneling to the states of the quantum dot. This behavior is shown in Fig. 9.15 for a variety of lattice temperatures. Clearly, the heating current of 18 nA, used in this figure, is sufficient to raise the electron temperature in the channel significantly above the lattice temperature. The oscillations in this case have a pronounced triangular shape. The period of the oscillations is the same as that of the conductance to the dot and arises from the periodic depopulation of the dot by a single electron as the tuning bias V_E is varied. The triangular shape is decreased as the lattice temperature is raised, presumably due to thermal smearing of the capacitive charging induced shape.

9.1.4 Measuring the hot electron temperature and energy relaxation time

In the previous section, the measurements of the thermopower were generally dependent on knowing (or estimating) the electron temperature in the channel. One part of the experiment, which was not discussed, is the subsidiary measurements necessary to ascertain an estimate of the carrier temperature T_E in the channel for a given value of the bias current. Measurements of this type have been demonstrated by, for example, Ikoma et al. [31,32,33]. It may be noted from (9.6) that simple measurements of voltage, current, and conductance are insufficient to determine the electron temperature in the channel. One still must ascertain the energy-relaxation time τ_E that appears in this equation. Alternatively, one can probe the temperature itself and use (9.6) to determine the energy-relaxation time. This latter technique is based upon the measurements of the temperature dependence of the phase-breaking time τ_φ discussed in the

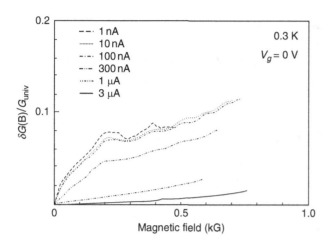

Fig. 9.16 Magneto-conductance spectra of a gated GaAs/AlGaAs quantum wire at 0.3 K for varying current levels. (After Ikoma *et al.* [33], by permission.)

last chapter. When carrier–carrier scattering dominates the phase-breaking time, then the dependence of τ_φ on carrier temperature is the same as its dependence on lattice temperature. By first measuring the lattice temperature dependence of the phase-breaking time, and then measuring the current (or field) dependence of the phase-breaking time, one can infer the electron temperature that corresponds to a given current. It should be pointed out, however, that such measurements tend to be *global* measurements, that is, measurements of the overall sample, whereas (9.6) is a *point* equation. Inhomogeneities in the field distribution, which are a natural consequence of the nonlinearity of the conductance with temperature, are a source of error in such measurements.

For the experimental measurements, a quantum wire of 350 nm width and 10 μm length was formed in a GaAs/AlGaAs heterostructure by using focused ion beams to define the wire [31–33]. In this approach, the ion beams produce damaged, low-mobility material adjacent to the defined wire. In Fig. 9.16, the magneto-conductance measured for a variety of bias currents is shown. These measurements are made at a lattice temperature of 0.3 K. As one can see, once the bias current exceeds about 0.3 μA, the conductance and the conductance fluctuations begin to decrease significantly. By fitting the conductance fluctuations (for magnetic fields above 0.2 T) with the theory of the previous chapters, one can determine the phase-breaking time. Here, one first determines the coherence length from the amplitude (and the correlation magnetic field) of the fluctuations, and the diffusion constant from the amplitude of the conductance itself. These two quantities then determine the phase-breaking time from $D\tau_\varphi = l_\varphi^2$. The values of the phase-breaking time determined in this manner are plotted in Fig. 9.17 for a variety of lattice temperatures. When the input power (I^2/G) exceeds 10^{-14} W, the phase-breaking time begins to decrease. In the previous chapter, it was determined how the lattice temperature affected the phase-breaking time. Using such lattice temperature measurements of the

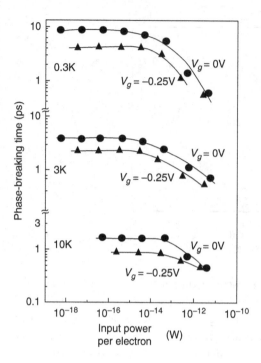

Fig. 9.17 The phase-breaking time as a function of the input power for various lattice temperatures. (After Ikoma *et al.* [33], by permission.)

phase-breaking time, one can now convert the values found in Fig. 9.17 to equivalent *electron* temperatures for each power level input to the device. In such a manner, it may be estimated that, for the case of $T = 3$ K and $P_e = 3.5 \times 10^{-13}$ W, the carrier temperature is 16 ± 5 K. This temperature corresponds to an energy relaxation time of 0.5 ns (determined by the temperature differential and input power through the energy balance equation), which is close to a predicted value for acoustic phonon relaxation in a GaAs/AlGaAs heterostructure [34]. This relaxation time for the energy is clearly three orders of magnitude larger than the phase-breaking time, which indicates that the phase-breaking mechanism is different than the energy relaxation mechanism. Presumably, carrier–carrier scattering is responsible for phase breaking, and phonon scattering is responsible for energy relaxation.

Another method of measuring the electron temperature arises from comparing variations in lattice temperature with variations in excitation current. Energy relaxation studies, conducted by comparing Shubnikov–de Haas (SdH) oscillations at various temperatures and currents in 2DEGs, can be used to extract the temperature dependence of the energy-loss rate. The difference between two terms corresponding to phonon emission at the electron temperature, T_e, and phonon absorption at the lattice temperature, T_{ph} [35,36], is used as a definition of this energy-loss rate. This is done for a quasi-two-dimensional electron gas in an InAlAs/InGaAs/InAlAs heterostructure [37]. The heterostructure material has a 30-nm *n*-doped (10^{18} cm^{-3}) In$_{0.52}$Al$_{0.48}$As donor region (and confinement barrier), a 10-nm-wide undoped $In_{0.52}Al_{0.48}As$ setback region and a 25-nm-wide $In_{0.53}Ga_{0.47}As$ quantum well region, and is lattice-matched to a GaAs substrate with a linearly graded match buffer. Two 2DEG samples were created and measured over two temperature ranges; one from 30 mK to 4.2 K and the other from 4.2 K to 30 K (in different cryostats).

Figure 9.18 shows the magneto-resistance data for this 2DEG, as a function of lattice temperature and sample current, and the correspondence between the two is clearly visible. The SdH oscillations have a dual-frequency nature that is attributed to the presence of two populated subbands in the system. Measured carrier densities of the two subbands (E_0/E_1) are $7.9 \times 10^{11}/cm^2$ and $5.0 \times 10^{11}/cm^2$, respectively, with a maximum measurement error of $\pm 4\%$. Simulation of the quantum well band structure carried out using a Green's function method with a 1D Schrödinger–Poisson solver verified the presence of two populated sub bands. The SdH oscillation amplitude is measured at multiple magnetic field points and by comparing these normalized amplitudes, the electron temperature as a function of the sample current is obtained, and from it, the temperature dependence of the energy-loss rate, which is defined as the difference between two terms corresponding to phonon emission at the electron temperature T_e, and phonon absorption at the lattice temperature T_{ph}. Both the phonon absorption and emission terms are assumed to have the same functional form $F(T)$, and the expression is given by [36]

$$P(T) = F(T_e) - F(T_{ph}) = A\left(T_e^n - T_{ph}^n\right). \qquad (9.7)$$

The energy-loss rate as a function of the electron temperature is shown in Fig. 9.19, and we see a single exponent of 3 with a coefficient of 9 eV/sK^3 over the entire regime. This exponent, n, has been thought of as an indicator of the phonon coupling and in the Bloch limit ($q_T < 2k_F$), values of 3, 5, and 7 are suggested [38] for unscreened piezoelectric, screened piezoelectric, and deformation potential coupling, respectively. Magnetic confinement of the electron–phonon interaction volume was suggested [39] as a possible cause for the exponent of 3 but no magnetic field dependence is seen in our data to confirm this. Others [40] tend to consider this as a crossover regime between low-temperature behavior and the equi-partition limit. At very low temperatures, the coupling to acoustic phonon modes is not sufficient to relax the energy of the electron gas and electron thermalization at the Ohmic contacts becomes important in cooling the hot carriers. However, more data is required at very low temperatures to verify this and the exponential suppression of power loss.

The power-loss rate given in (9.7) can be directly linked to the energy relaxation time through

$$\tau_E = \frac{k_B(T_e - T_{ph})}{P(T_e)}. \qquad (9.8)$$

Fig. 9.18 SdH magneto-resistance oscillations measured in the sample over a range of lattice temperatures and sample currents, where the arrows indicate increasing lattice temperature or sample current trends. This was measured for (a) 4.2, 6, 8, 10, 15, 20 and 30 K and (b) 0.5, 10, 36, 57, 85, 114, 150 and 247 μA.

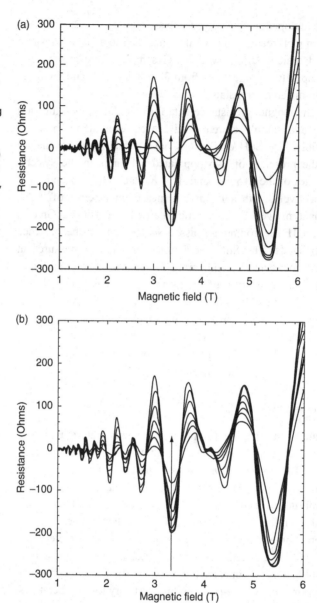

In Fig. 9.20, we plot the energy relaxation time as a function of the electron temperature for the above data, and at three different magnetic fields [41]. Generally, this follows a $1/T$ behavior up to a transition point, and then changes to a $1/T^3$ behavior. The latter behavior is often seen in clean systems where the relaxation process is dominated by the electron–phonon interaction. The lower-temperature behavior, with the slower decay, is not fully understood, nor is the apparent dependence upon the strength of the magnetic field.

9.1 Nonequilibrium transport in mesoscopic structures

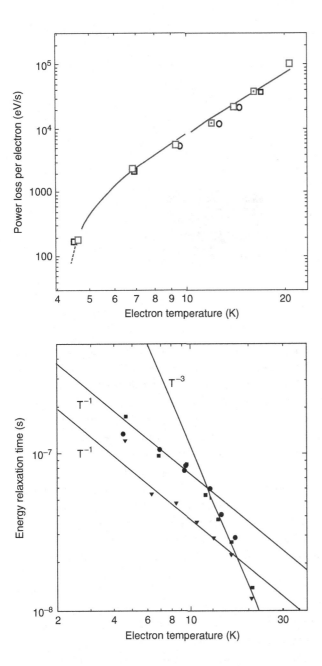

Fig. 9.19 The energy loss rate as a function of carrier temperature is shown, where the circles and squares represent data collected at 1.6 and 2.0 Tesla and the fitting equation is 8 $(T_e^3 - T_0^3)$.

Fig. 9.20 The energy relaxation time determined according to (9.8). The circles, squares, and triangles are for a magnetic field of 1.6, 2.0 and 3.2 T, respectively. (After Prasad *et al.* [41].)

Similar results can be obtained for quantum wires. Here, we discuss quantum wires fabricated in the same heterostructure quantum well layer as discussed above [42]. The wires are shown as the inset to Fig. 9.21. The wire electrical widths are obtained from the onset of the Shubnikov–de Haas (SdH) oscillations of the narrowest wire, which is then linearly extrapolated onto the wider wires to yield values of 215, 545, and 750 nm, indicating an edge depletion of 125 nm.

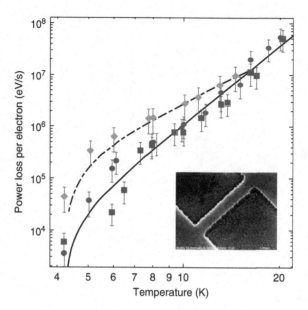

Fig. 9.21 Magnetic averaged power loss from wires of 750 nm (squares), 545 nm (circles) and 215 nm (diamonds) at $T_L = 4.2$ K. Fits to $A(T_e^n - T_L^n)$ are shown with $A = 11$ eV/sK^5 and $n = 5$ (solid line) and $A = 3000$ eV/sK^3 and $n = 3$ (dash-dot line). The inset shows an electron micrograph of a short quantum wire formed by electron beam lithography that was used in this study. (After Prasad et al. [42].)

The SdH oscillations have a dual-frequency nature that is attributed to the presence of two populated sub bands in the system, with nominal densities of 6.7×10^{11} cm^{-2} and 3.4×10^{11} cm^{-2}. The SdH oscillation amplitude is measured at various temperatures and at various currents and by comparing these normalized amplitudes, the electron temperature as a function of the sample current is obtained from (9.7). In the present measurements, wires of 545 nm and 750 nm effective widths show quasi-2D power-loss behavior that fits well to theoretical values for unscreened deformation potential coupling (Fig. 9.21). However, the 250 nm wire shows a larger power loss over the entire measurement range. This could not be fit with existing quasi-2D predictions and we believe that this wire is inherently quasi-1D in nature. Hence, a transition from quasi-2D to quasi-1D behavior is seen between 545 nm and 215 nm. The power-loss data is also used to estimate the energy relaxation time of the carriers through the equation

$$\tau_e = \frac{\pi^2 k_B^2}{6\varepsilon_F P_e}\left(T_e^2 - T_L^2\right). \tag{9.9}$$

This equation differs from (9.8) as it deals more with the heat capacity of the material rather than a simple energy relaxation equation. In Fig. 9.22, we plot the energy relaxation time that is found for these three wires.

It is clear in these last two figures that there is a great deal of similarity between the two larger width wires, while the narrower wire appears considerably different. The two wider wires can be fit with (9.7), with a value of $n = 5$, which is suggestive of screened piezoelectric scattering dominating the carrier heating. On the other hand, the narrow wire is best fit with a value $n = 3$, which is suggestive of unscreened piezoelectric scattering dominating the carrier

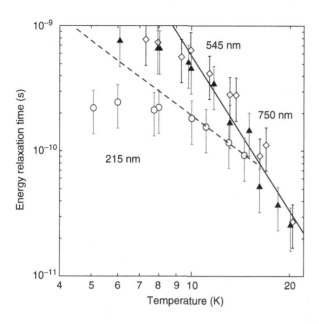

Fig. 9.22 The energy relaxation time that is found in each wire as a function of the temperature. The closed triangles are for the 750 nm wire, the open diamonds are for the 545 nm wire, and the open circles are for the 215 nm wire. Only the latter shows real 1D behavior.

heating [36,38]. The lack of screening is expected in a quasi-one-dimensional wire system, due to the very long range of the Coulomb interaction. It is our belief that the two wider wires are actually *narrow two-dimensional* systems in which the Fermi energy (relative to the bottom of the conduction band) is set by the contact reservoirs. In the one-dimensional, narrow wire, on the other hand, the depletion at the side walls has begun to interact so that the bottom of the conduction band is raised over that of the reservoirs. Hence, this becomes a true one-dimensional wire.

9.1.5 Hot carriers in quantum dots

Hot carriers are a common occurence in normal, large-scale semiconductor structures. Under the application of a large electric field, the electron temperature is raised, as has been discussed in the preceding sections [43]. On the other hand, most studies of quantum structures have concentrated on the near-equilibrium properties. The quantum dot, which is a form of a lateral resonant-tunneling structure, can also exhibit extreme nonequilibrium behavior. Consider the quantum dot, as shown in the inset to Fig. 9.23. The top, bottom, and right-central gate lines define a pair of "point-contact" barriers, which are marked *A* and *B*. The central region forms the quantum dot. In essence, the dot region corresponds to the quantum well defined between a pair of tunneling barriers. The actual barrier properties can be modified by varying the voltages applied to the three defining gates. On the other hand, the large gate left of center in the figure corresponds to a "plunger" that tunes the bound states and population of

Fig. 9.23 Schematic of the energy-band profile for a lateral double-constriction quantum dot. The inset shows the configuration of the bias gates. (After Goodnick et al. [43].)

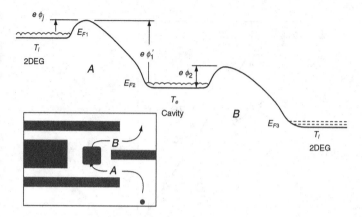

Fig. 9.24 Experimental and calculated current–voltage curves for two different bias configurations of the hot carrier quantum dot. (After Goodnick et al. [43].)

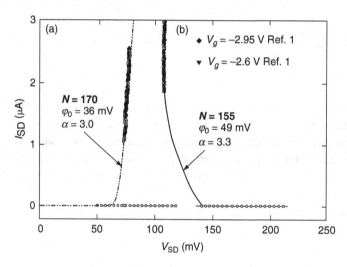

the quantum dot itself. Under bias, the potential structure can appear as shown in Fig. 9.23, with the potential drop partially occurring across the first barrier and partially across the second barrier. When the region under the control gates is depleted, the current path must pass through the two barrier regions and through the quantum dot.

When the bias voltage is sufficiently high across the coupled structure, the current–voltage characteristics show a marked *S*-type negative differential conductance (this is called *S*-type because the lines through the various branches seem to form an *S*, and the current is multi-valued for a given voltage) [44]. The transition from the low-current state to the high-current state is a strong function of the bias on the various gates in the structure. Typical current–voltage curves are shown in Fig. 9.24. In the figure are two different *I–V* curves. Each has a high-resistance line that is nearly flat at almost zero current (the extension to zero voltage is not shown for device (b)). Then, there is a high-current branch in which the voltage is almost constant, and for which negative differential resistance can be seen at lower current levels. As the voltage increases, the operating point follows the high-resistance branch until the switching voltage is reached, at which point operation jumps to the high-current branch. The controllable switching behavior in these structures can be explained by a balance approach in which energetic carriers injected through the emitter barrier will balance energy with the thermal bath of cool carriers residing in the quantum dot itself. Calculated current–voltage curves appear in Fig. 9.24 along with the data from the experiment [43].

Consider the potential barriers encountered by electrons as they traverse point A in Fig. 9.23. The barrier, as indicated in the figure, is created by depletion between the lower and right-central gates. This isolates the central region and allows the potential drop to occur between the dot and the outside regions. The separation between the Fermi level in the emitter region and the central dot region is given by

$$E_{F1} - E_{F2} = \frac{1}{2}eV_{SD} + \frac{eQ}{C}, \quad (9.10)$$

where C is the charging capacitance of the dot region and $Q = e(N_0 - N)$ is the excess charge in the dot (N is the number of electrons in the dot, and N_0 is the equilibrium number). Similarly, the difference in the Fermi energies of the dot and the collector region is given by

$$E_{F2} - E_{F3} = \frac{1}{2}eV_{SD} - \frac{eQ}{C}, \quad (9.11)$$

and it has been assumed that the input and output barriers are identical.

While the Fermi energies determine the equilibrium states, the barrier heights themselves determine the kinetic rates. Under bias, the input barrier is reduced in the amount

$$\varphi_1 = \varphi_0 - \left(\frac{1}{2}V_{SD} + \frac{Q}{C}\right)/\alpha, \quad (9.12)$$

and the factor α depends upon the abruptness of the barrier as discussed in Chapter 3. The output barrier is similarly lowered, but in the amount

$$\varphi_2 = \varphi_0 - \left(\frac{1}{2}V_{SD} - \frac{Q}{C}\right)/\alpha. \quad (9.13)$$

For a perfectly abrupt barrier, α would be infinite (neglecting any image force lowering of the barriers). In the present structures, the barrier abruptness is limited by the fact that they are essentially depletion barriers, so that the barrier potential varies over a distance roughly comparable to the channel width W. Since the channel lengths are of the same order as the widths, we may estimate α as 3 as discussed in Chapter 3. An equally important parameter is the excess energy of the injected electrons (into the well), which is shown in Fig. 9.23 as $e\varphi'_1$. The latter is related to the applied bias by

$$\varphi'_1 = \varphi_0 + \left(1 - \frac{1}{\alpha}\right)\left(\frac{1}{2}V_{SD} + \frac{Q}{C}\right). \tag{9.13}$$

One assumption to be made is that the dimensions of the quantum dot and the outlying regions are sufficiently large that we can treat them as quasi-two-dimensional regions. A second major assumption in determining the current–voltage relationship is that the injected electrons thermalize with the electrons already in the quantum dot region. This latter allows us to deal with the carriers in the dot through their Fermi energy and an equivalent electron temperature T_E. We can now write the current into the dot from the emitter as

$$I = eWv_{F1}\frac{mk_BT}{\pi\hbar^2}e^{-e\varphi_1/k_BT}\left[e^{E_{F1}/k_BT} - e^{E_{F2}/k_BT}\right], \tag{9.14}$$

where T is the lattice temperature. The bracketed term represents the difference in the Fermi factors for the currents into and out of the dot region. Once the current is known, and an equivalent amount of charge within the dot is determined along with the electron temperature, the input barrier can be determined from this equation. On the other hand, a reasonable approximation is to ignore the last term (large bias voltage) in the square brackets with respect to the first, and rewrite this as

$$I = eWv_{F1}\frac{mk_BT}{\pi\hbar^2}e^{-e\varphi_1/k_BT}\left[\exp\left(\frac{n_s\pi\hbar^2}{mk_BT}\right) - 1\right], \tag{9.15}$$

where

$$n_s = \frac{mk_BT}{\pi\hbar^2}\ln\left[1 + \exp\left(\frac{E_F}{k_BT}\right)\right], \tag{9.16}$$

is the two-dimensional density in a quasi-two-dimensional system, as given in Chapter 2. (The Fermi energy is measured from the conduction band edge in this latter expression.) In a similar manner, the current flowing through the second constriction can be written as

$$I = eWv_{F2}\frac{mk_BT_E}{\pi\hbar^2}e^{-e\varphi_1/k_BT}\left[e^{E_{F2}/k_BT} - e^{E_{F3}/k_BT}\right]$$

$$\sim eWv_{F2}\frac{mk_BT_E}{\pi\hbar^2}e^{-e\varphi_1/k_BT}\left[\exp\left(\frac{N\pi\hbar^2}{Amk_BT_E}\right) - 1\right], \tag{9.17}$$

and, in the last expression, the electron temperature corresponds approximately to that of a heated Maxwellian distribution. We have replaced the carrier density by the actual number of carriers N and the dot area A. The latter equation allows the determination of the barrier height of the second barrier under the same assumptions as mentioned above.

The energy flow into and out of the carriers in the quantum dot region can be defined by an energy balance equation. We can write this energy balance equation as

$$I\left(\varphi_1' - \varphi_2 - \frac{k_B(T_E - T)}{e}\right) = N\left\langle\frac{dE}{dt}\right\rangle\bigg|_{coll}, \quad (9.18)$$

where the last term represents the dissipation of energy due to collisions. The first two terms in the parentheses are just the net potential drop across the quantum dot, and the third term represents the energy carried out by the carriers leaving through the second dot, which removes energy from the dot region. The energy loss rate of hot carriers in a GaAs quantum well (between two AlGaAs barriers) has been measured by Shah et al. [45] using photoluminescence spectroscopy. Their data can be fit over a range of data by assuming both polar-optical phonon scattering and acoustic phonon scattering, described by

$$\left\langle\frac{dE}{dt}\right\rangle\bigg|_{coll} \sim \frac{\hbar\omega_{LO}}{\tau_{LO}} + \frac{k_B(T_E - T)}{\tau_{ac}}. \quad (9.19)$$

For polar-optical scattering, the nominal value of τ_{LO} is about 130 fs, while τ_{ac} is thought to be about 1.8 ns. However, it is likely that there are nonequilibrium polar-optical phonons as well [46], which would increase τ_{LO} to about 2 ps, which gives a better fit to the data of Shah et al.

The above equations are solved by first choosing a value of the current (since it is a multi-valued function of the voltage), and then iterating the above equations to find a consistent set of values for the various parameters. The results are plotted against the data in Fig. 9.24, and the fit is quite remarkable. Two different values of the gate voltage are shown in the figure, and it can be seen from this that a small change in the gate bias produces a significant change in the barriers and the resulting current–voltage characteristic. A width $W = 200$ nm was used for the constrictions. The capacitance was not determined from the experiment, but a value of 10^{-16} F gives good agreement with the experimental data. Switching from the high-impedance state to the low-impedance state is found to occur essentially when the input barrier is pulled down to the emitter Fermi energy ($E_{F1} = e\varphi_1$). If there is no excess charge in the dot prior to the switching, then the switching voltage is just found to be

$$V_s = 2\alpha\left(\varphi_0 - \frac{E_{F1}}{e}\right). \quad (9.20)$$

As the gate bias is made more negative, the barrier heights are increased, and this leads to an increase of the switching voltage, as observed in the figure.

The observation of hot electron effects for carriers passing through quantum dots is quite important to the structures discussed in the introduction to this chapter. The dots experimentally studied here are much larger than those envisaged in ultrasmall devices; the actual size enters into (9.20) only through the factor α. This represents the smoothness of the barrier potential and should be more or less the same regardless of the dot size. Consequently, one may expect to see significant switching and fluctuations in current transport within and between the dots that form, for example, in Fig. 9.2. In a device structure comparable to the latter figure, it is unlikely that the simple equations introduced here will be applicable, since the Fermi energies and temperatures will be varying throughout the structure. Approaches comparable to those of universal conductance fluctuations in Chapter 7 are thought to be necessary, since it will be necessary to determine long-range correlations throughout the structure.

9.1.6 Breakdown of the Landauer–Büttiker formula

Near-equilibrium transport occurs when the energy associated with the bias voltage across the quantum structure is smaller than either the subband separation or the Fermi energy. Transport in short waveguides, in which the transport is in near equilibrium, usually satisfies the Landauer equation, with a conductance that is quantized in units of $2e^2/h$, where the quantization level corresponds to the number of occupied subbands. When the source–drain bias across the quantum waveguide (or point contact as the case may be) becomes comparable to the energy separation of the one-dimensional subbands, the number of subbands that are occupied for forward and reverse transmission differs, and nonlinear behavior can result. In Chapter 3, we showed that for relatively small source–drain bias across a quantum point contact, this nonlinear transport is still governed by the multi-channel Landauer–Büttiker formula. The main experimental feature is an oscillatory behavior in the differential conductance versus source–drain voltage, which deviates from the near-equilibrium quantized conductance. This breakdown of the quantized conductance is due to the unequal population of the one-dimensional subbands as the Fermi energies on the left and right are moved up and down relative to one another. For larger source–drain bias, this oscillatory behavior damps out experimentally, whereas the Landauer–Büttiker formula (without explicit consideration of dissipative processes during transmission) still predicts structure in the differential conductance. As with resonant-tunneling diodes, the ideal picture of collisionless, coherent transport across a quantum point contact breaks down as the excess kinetic energy of the injected electrons is increased (relative to the Fermi energy in the absorbing contact) due to inter-carrier and phonon scattering processes. Moreover, it has been argued that the high bias regime in QPCs will result in a situation in which the higher occupied subbands will act somewhat like current *filaments*, and this will lead to the possibility of negative differential resistance (in this case, multiple values of the current for a given voltage, in contrast to the NDR effect of resonant tunneling diodes discussed in Chapter 3) [47]. Some forms of this negative differential resistance have been observed in pinched quantum-dot devices (discussed in the previous section), but this is attributed to

9.1 Nonequilibrium transport in mesoscopic structures

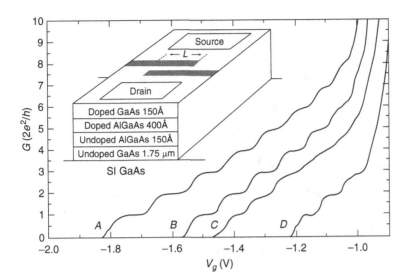

Fig. 9.25 The equilibrium conductance after light exposure of a zero overlap point contact, A, and the $L = 0.1$-µm (B), 0.2-µm (C), and 0.5-µm (D) waveguides. The inset is a schematic of the material and the devices. (After Berven *et al.* [50].)

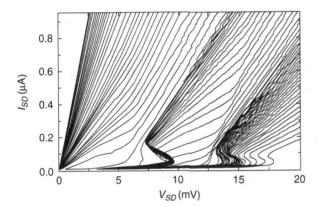

Fig. 9.26 The *I–V* characteristics of a 0.1-micron waveguide after illumination. The gate voltage increment is −20 mV. (After Berven *et al.* [50].)

the nature of nonlinearity in the injection process over the first set of barriers (forming the quantum dot).

Waveguides that exhibit negative differential conductance have been formed by creating two overlapping gates on a uniform quasi-two-dimensional GaAs/AlGaAs structure, as shown in the inset of Fig. 9.25. Here, the waveguide is formed in the region where the dots overlap, and this has been measured for overlap lengths of 0.1, 0.2, and 0.5 µm, which comprise relatively short wires. For low biases, the devices exhibit normal short-wire quantized conductances. In the shortest wire, 0.1 µm, conductance plateaus up to $n = 5$ are easily measured. In the longer wires, the quantization is not as well pronounced due to the role the inhomogeneous confining potential plays, as discussed already in Chapter 3 [48,49]. In Fig. 9.26, the *I–V* characteristics for large values of source–drain bias and various gate voltages are shown. At low bias, the *I–V*

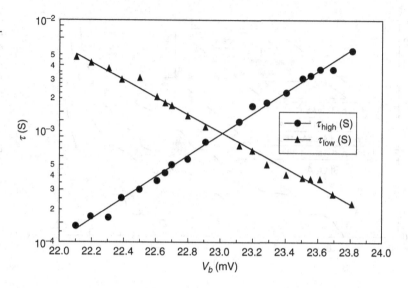

Fig. 9.27 The average times in the high- and low-current states as a function of source–drain bias for a fixed gate bias. (After Smith *et al.* [51].)

curves for various gate voltages are clustered around slopes corresponding to the quantized conductance. At higher bias, this particular sample exhibited two distinct regions of *S*-type negative differential resistance [50]. For the first, switching occurs for a bias of 6 meV, regardless of the applied potential on the two gates. The second switching occurs at a value of source–drain bias that is strongly dependent on the gate bias. The occurrence of this SNDC behavior is found to be strongly correlated with the thermal history of the samples, and could be induced by *in situ* exposure to infra-red light.

Time-dependent studies of the current–voltage characteristics reveal an even more surprising behavior. When biased in the region of instability, the current is observed to randomly switch between the off state and a current value along the d.c. *I–V* curve in the absence of SNDC [51]. This switching behavior has the characteristic behavior of random telegraph signals. However, the switching behavior of the present case is only observed in the region of instability. The average times in the low and high states follow one another such that their product is always constant over several orders of magnitude as shown in Fig. 9.27. Random telegraph signals originating from the independent fluctuations of an impurity do not generally exhibit this kind of lifetime behavior. Moreover, independent fluctuations of the quantum waveguide barrier should result in instabilities over the whole range of *I–V* characteristics rather than in localized bias regions.

Several transport mechanisms can be invoked to describe such an effect. One possible one is impact ionization of neutral donors. However, for this mechanism it would be expected that illumination should reduce the effect, but the opposite is observed – illumination enhances the negative differential resistance. Another possibility is the occurrence of Coulomb blockade in the wire. The current expected from Coulomb blockade, however, is orders of magnitude below that seen in the experiments. In the previous section, negative differential resistance in quantum-dot structures was shown to require carrier confinement in the quantum dots. While the wire structures discussed here do not formally have quantum dots, the inhomogeneous potential expected from the impurities leads to multiple barriers and regions of trapped charge [52]. A further possible explanation is that the formation of such barriers in these wires, which are operated near pinchoff, creates barriers and puddles of trapped charge between these barriers, and subsequent hot electron effects lead to the negative differential resistance [50]. However, the correspondence of random telegraph switching with such heating behavior would have to be provided by a dynamic rather than static model, as presented in the previous section [53].

9.2 Semiconductor nanodevices in the real world

Progress in the size of semiconductor devices that inhabit the real world, such as within modern integrated circuits like the Pentium chips, has brought these individual devices into the realm of real nanostructures. Minimum physical dimensions are usually associated with the thickness of the SiO_2 layer between the gate and the channel, but the more quoted value is the gate length. In today's circuits, this gate length is often only about half of the so-called "node size," which more often refers to width of interconnections. Today (2007), the physical gate length in production is only about 25 nm for the 45 nm node. Since many modern nano-technologies have appeared, and many of which claim to be the future *replacement* for the Si transistor, it is worthwhile to spend some time discussing the status and future of the Si transistor. Indeed, this is a multi-billion dollar industry which will not be easily replaced. Yet, some novel materials and structures may find use to supplement the normal Si transistor [54].

The scaling downward of individual semiconductor devices is coupled closely to the continual increase in the number of transistors on an individual chip. This behavior is known as Moore's law [55], and results in the general doubling of the number of transistors every 18 months. While some think that this results from the ability to continue to reduce the dimensions through scientific advances, the driving force is actually economics. This is because the cost (per unit area) of processed and packaged Si chips has not changed in real terms for almost four decades. Since the cost of any computational resource really lies in the number of functional units that are required, by putting more of these on a single integrated circuit chip, the cost per function is dramatically reduced. It is this exponentially lowering of the cost per function that drives Moore's law. Hence, the ability to reduce dimensions becomes an enabling technology, but the driving force is the economics of function cost. One can clearly see the results in the cost of a home computer over the past three decades.

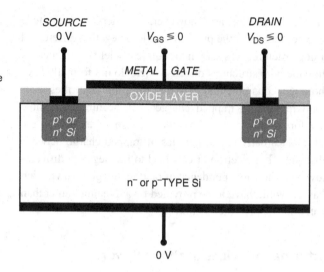

Fig. 9.28 Schematic illustration of a MOSFET. The notation for the relevant voltages is also defined in this figure. The "upper" notation is used for *p*-channel devices and the "lower" notation is used for *n*-channel devices.

In this section, we want to briefly review the operation of the Si metal-oxide-semiconductor field-effect transistor (MOSFET) which is the basis for nearly all current integrated circuits. This will allow us to study the scaling theory by which the size of such devices is reduced and to talk about limitations to future scaling. Then, we will briefly discuss some alternative devices and their limitations and a methodology for comparing such devices.

9.2.1 Scaling the MOSFET

The key factor of this semiconductor device is an insulator that is placed upon the surface, so that this insulator forms the central part of a metal-oxide-semiconductor (MOS) structure. The surface position of the Fermi energy is not strongly pinned and can be moved relatively easily. For example, in a MOS structure, such as sketched in Fig. 9.28, a positive voltage on the gate will pull electrons to the surface of the semiconductor. On the other hand, a negative voltage on the gate will push electrons away, and pull holes to the surface. If we use a *p*-type semiconductor, then a positive gate bias can even *invert* the surface to *n*-type material, by pulling a sufficient number of electrons to the surface. Hence, we can create a *p–n* junction at the surface of the semiconductor. The desirability of this MOS structure arises from two factors: (1) the ease of moving carriers at the surface to create different properties, and (2) the advantageous properties of the native oxide of Si – silicon dioxide, SiO_2. Silicon dioxide is one of the most common materials in nature, but it is one of the best insulators known in electronics. The fact that very high quality SiO_2 can be grown by simply putting a Si wafer in a high-temperature furnace is exceedingly important to the manufacturing of modern VLSI. The fact that we can make both *n*-channel and *p*-channel devices (in Si) makes the MOS field-effect transistor

(MOSFET) a far more desirable device structure for integrated circuits. Here, the channel is actually induced by the MOS structure, and this, in turn, provides the modulation of the conductance between a source and a drain that creates the transistor action.

Let us consider a *p*-type semiconductor material, as denoted in Fig. 9.28 by the lower set of parameters. While the bulk region is *p*-type, the source and drain contact regions are heavily doped *n*-type. When the semiconductor material is connected to the metal by a "back" contact, the Fermi level must line up so that it is invariant with position. In the MOS structure, we require semiconductors which have very good surfaces, as defined by very low numbers of surface (or interface when the oxide is grown) states. It is no problem, with proper processing, to reduce this density by more than two orders of magnitude at the Si/SiO$_2$ interface. This means that, for the MOS structure, one can actually move the Fermi energy at the interface relatively easily. When the gate is biased positively with respect to the semiconductor, electrons are drawn toward the interface with the oxide. At first, these electrons recombine with holes, leaving only the negatively charged ionized acceptors in a narrow region adjacent to the interface. For still larger values of the positive gate bias voltage, still more electrons are drawn to the interface, which at first broadens the depletion width throughout which the holes have been removed. In both cases, the conduction and valence bands are bent at the interface to account for this change in the number of free holes. Finally, at a critical bias voltage, known as the "turn-on" voltage, electrons begin to accumulate at the interface, creating an *n*-type region with mobile electrons. That is, we have *inverted* the type of the majority carrier at the surface, from holes to electrons. This *n*-type region "shorts" between the heavily *n*-type source and drain regions to create the conducting channel. Now, the conductance through this channel is controlled by the gate potential.

It is highly unlikely that the Fermi energy in the semiconductor will *exactly* match the position of the Fermi energy in the metal throughout the structure. A variation can occur due to charges in the oxide, as well as just due to variations of the Fermi level in the semiconductor with doping. As a consequence, one must actually apply a small voltage to the gate to bring the bands in the semiconductor into alignment. The amount of voltage that must be applied is known as the *flat-band voltage* V_{FB}. When this voltage is applied, then there is no net space charge at any point in the semiconductor *p*-type region. Any changes to this voltage causes a charge to accumulate at the interface between the semiconductor and the oxide. If the voltage is positive on the gate, then the charge is negative (electrons are pulled to the surface). If the voltage is negative, then the charge is positive (electrons are pushed away from the surface and holes are pulled to the surface). In general, we take the voltage at the source as the reference potential, and refer to V_{GS} as the voltage on the gate relative to the source. Hence, we can write the charge that accumulates at the semiconductor–oxide interface as

$$-Q_{int} = C_{ox}[V_{GS} - V_{FB}]. \tag{9.21}$$

It must be pointed out that the charge Q_{int} is actually the surface charge density in C/m^2. The oxide capacitance per unit area C_{ox} is ε_{ox}/t_{ox}, where t_{ox} is the thickness of the oxide layer and ε_{ox} is the permittivity of the oxide (approximately 3.4 ε_0 in SiO_2). We use the *specific* quantities (values per unit area) as we do not yet know the area of the oxide capacitor.

Once $V_{GS} > V_{FB}$, negative charge begins to accumulate at the interface between the oxide and the semiconductor. This charge is composed of two components. One is the ionized acceptors which are no longer compensated by the holes, as the latter have been pushed away. The second component is the electrons which have been drawn to the interface. In both cases, the actual charge at any position along the channel will depend upon the *local* voltage. The problem we have is that the charge at the source end of the channel is given by (9.21), while the charge at the drain end of the channel is given by

$$-Q_{int} = C_{ox}[V_{GS} - V_{FB} - V_{DS}]. \tag{9.22}$$

That is, the charge is determined by the potential between the gate and the drain, and the actual channel strength between the two varies with the local potential. In order to calculate a simple expression for the current, we have to know just how much band bending exists at each point along the channel. However, we are really interested in the amount of the mobile charge, due to the electrons.

The problem is approached by developing an equation for the small local voltage drop at an arbitrary point along the channel in terms of the current and the conductance at that point. This will then be integrated over the channel length to get a single expression for the current in terms of the terminal voltages indicated in Fig. 9.28. There is one difference here, and that is we deal with the sheet resistance R_s, sheet conductance G_s, and sheet density n_s (electrons per square meter) of electrons in keeping with the surface charge density Q_{int} defined in (9.21). Hence, we can write the incremental sheet resistance dR_s at an arbitrary point y as

$$dR_s = \frac{1}{Z\sigma_s}dy = \frac{1}{Zn_s e\mu_e}dy. \tag{9.23}$$

Here, Z is the width of the MOSFET in the direction normal to the plane of Fig. 9.28. Once the gate voltage is sufficiently large, electrons begin to accumulate at the interface, and their sheet density is given by

$$en_s = C_{ox}[V_{GS} - V_{FB} - V_T] \tag{9.24}$$

at the source end of the channel. The difference between this result and (9.21) is that we have introduced the *turn-on voltage* V_T and deal only with the electronic contribution to the surface charge. Away from the source end, the surface

9.2 Semiconductor nanodevices in the real world

potential is somewhere between the zero value at the source and V_{DS} at the drain. Thus, at our arbitrary point y, we must modify (9.24) to

$$en_s(y) = C_{ox}[V_{GS} - V_{FB} - V_T - V(y)]. \tag{9.25}$$

We can write the voltage drop at the point y as

$$dV(y) = IdR_s = \frac{Idy}{Zn_s(y)e\mu_e}. \tag{9.26}$$

Rearranging this equation, and inserting (9.25), we arrive at

$$Idy = ZC_{ox}\mu_e[V_{GS} - V_{FB} - V_T - V(y)]dV(y). \tag{9.27}$$

Before integrating (9.27), it is useful to introduce a reduced potential which combines the flat-band and turn-on voltages, as

$$V'_T = V_{FB} + V_T, \tag{9.28}$$

and this is usually referred to as the *threshold* voltage. Now, the limits on the integration of (9.27) are obviously $y = 0$ and $V(y) = 0$ at the source end of the channel, and $y = L_g$ and $V(y) = V_{DS}$ at the drain end of the channel. The integration can now be easily performed to give

$$I = \frac{ZC_{ox}\mu_e}{2L_g}\left\{[V_{GS} - V'_T]^2 - [V_{GS} - V'_T - V_{DS}]^2\right\}$$
$$= \frac{ZC_{ox}\mu_e}{L_g}\left[V_{GS} - V'_T - \frac{V_{DS}}{2}\right]V_{DS}. \tag{9.29}$$

This expression for the drain current is valid so long as the term in the square brackets is positive. Saturation in the current occurs for a sufficiently large V_{DS}. We can differentiate (9.29) with respect to V_{DS}, and then set this derivative to zero to find the saturation value of the drain voltage as

$$V_{DS,sat} \equiv V_{sat} = V_{GS} - V'_T. \tag{9.30}$$

For voltages above this value, the current is saturated at the value

$$I_{D,sat} = \frac{ZC_{ox}\mu_e}{2L_g}(V_{GS} - V'_T)^2. \tag{9.31}$$

The condition at which saturation sets in, as given by (9.30), is exactly the value discussed above for which the inversion density disappears at the drain end of the channel. Once the electrons get to this point they are injected into the depletion region, and subsequently move to the drain by a combination of drift and diffusion. The current remains governed by (9.31), as this is still the determining condition, which is set near the source end of the channel. That is, the current level is largely set by the properties of the local potential at the source-end of the channel, but saturation is initiated by drain-end properties. This is a general property of all field-effect transistors. We can estimate the size of the currents from the above equation, particularly (9.29). In the year 2000, the primary gate length being used for front line microprocessors was 0.25 μm. If the oxide thickness is 25 nm, then $C_{ox} = 1.2$ mF/m^2. If we operate with 1.5 V for V_{GS} and say 0.2 V for V_{DS} (the normal operation in logic circuits will have a low source–drain voltage for a high gate voltage, as we will see in a subsequent section). Usually then, the threshold voltage will be approximately 0.3 V, so that we can estimate the sheet density n_s as 9×10^{11} cm^{-12}. The field in the oxide is approximately 6×10^5 V/cm, for which the mobility will be about 500 cm^2/Vs. We can now compute the current density as about 53 mA/mm. If the drain voltage were higher, so that the device was in saturation, then the current would be 173 mA/mm, or about a factor of 3 higher. It is clear that the conditions for logic operation make the device operate more likely in the "linear" regime than in the saturated regime. On the other hand, this is the steady-state, and it is clear that the drain voltage is high when the gate potential is first applied.

Now, the general idea of progress in the integrated circuits world is that there is a scaling relationship for the individual transistor. The simplest approach, and the most useful, is to assume that we will scale the devices via a methodology that keeps the local electric fields within the device constant [56]. While other scaling laws are available [57], this so-called constant-field scaling remains the most important. Here, we assume that the area is reduced by a factor of 2 each generation. This means that each characteristic length L is reduced by a factor $\alpha = \sqrt{2}$ each generation (or about every 18 months according to the above discussion). In this way, the density on the chip doubles each generation. Thus, this means that the gate length, the oxide thickness, the device width, etc., are all reduced by α. At the same time, in order to maintain constant electric fields in the oxide, we must reduce the voltages by this same factor of α. This means that the area occupied by each transistor is reduced by a factor of 2 each generation. At the same time, we require that the depletion widths of the p–n junctions that form the source and drain (relative to the bulk material) junctions must also be reduced by this same factor. Since this depletion width is a result of Poisson's equation, we note that the $\nabla^2 V$ results in an increase of a factor of $1/\alpha$ (a factor of α in the numerator and a factor of a^2 in the denominator). This must be balanced on the right-hand side of Poisson's equation by an increase in the doping density in the source and drain by $1/\alpha$. From (9.29), we see that the current through the device decreases by a factor α. This means that the power density in the device decreases by a^2. Hence, the power density in the chip remains constant. However, since the chip area generally increases each generation, the total power dissipated on the chip actually increases. This is one of the major problems. Yet, the switching speed can be increased since the resistance and capacitance scale proportionately so that the speed can increase by $1/\alpha$. Yet, the problem is mainly to get the total power from the chip to an adequate heat sink. While this remains a crucial problem, it should be pointed out that this scaling theory has more or less worked well for more than three decades. In Fig. 9.29, we plot the crucial gate length according to the first year in which it

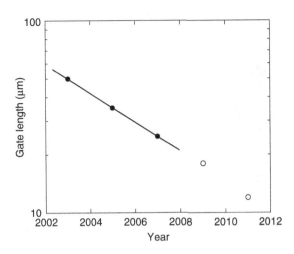

Fig. 9.29 Variation of the appropriate gate length as a function of the first year in which it went into production. For example, the 25-nm gate length went into production with the 45-nm node in 2007. The open circles are estimates for the future.

went into production (this gate length should be regarded as ±5 nm or so, as there is some question as to the exact gate lengths used at various "nodes" of technology).

9.2.2 Looking at other nanodevices

In the past few years, the continued advance in downscaling of semiconductor devices, discussed above, has led to interest in a vertical transistor in which current carrying channels can exist on either side of the vertical structure – the FinFET [58,59]. More recently, the tri-gate variant has appeared [60] and will apparently be the technology of choice for the 32 nm node [61]. As the fin width becomes small, the necessity for a high fin means that these two concepts will eventually merge in the future. In addition to the creation of more Si area for the transistor, a push toward adoption of FinFETs comes from the possibility of moving the inversion layer away from the Si surface, and more into the bulk of the fin, similar to that for thin SOI transistors [62]. By moving the carriers away from the surface, and into the bulk, of the fin, surface-roughness scattering is reduced with the result that the mobility is increased. Yet, without detailed simulation of any prospective device, there is no easy way to estimate key parameters (fin width, inversion density, etc.) for insuring that the carriers reside mainly in the bulk of the fin, with the resultant higher mobility. Thus, we have two reasons for moving to the FinFET, and in general transistors that are turned on their side. The first is the more effective use of silicon area and the second is an improvement in the mobility. In Fig. 9.30, we illustrate the ideas of the common FinFET.

We have developed an approximate technique which can yield the location of the charge in the fin in a simple and direct manner [63]. We considered only a double gate FinFET. The channels are considered as two-dimensional electron

Fig. 9.30 The conceptual idea of a vertical fin as the active part of the transistor. This FinFET will likely be the major technology for the 32-nm node (indicated as 2009 in Fig. 9.29).

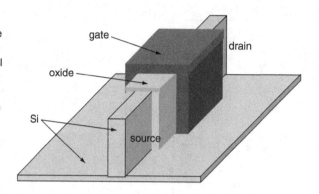

Fig. 9.31 The charge distribution along the fin width as a function of effective fin width. The electron density $n_s = 3 \times 10^{12}$ cm^{-2}.

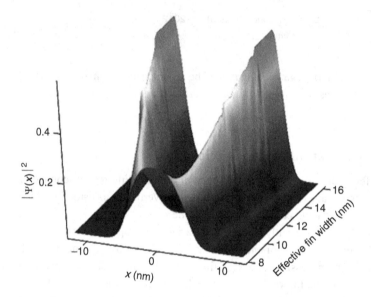

gas (2DEG) systems. For wider fins, channels form from surface inversion. But as the width decreases, the channels eventually overlap and lead to bulk inversion. This transition from surface to bulk inversion occurs gradually, as will be seen. Our approach is based upon the recognition that bulk inversion will lead to a confining potential that is quite close to that arising from a harmonic oscillator. On the other hand, surface inversion at the two sides of the fin will lead to a double-well potential, with the two wells close to the surface. In fact, these two potentials are partner potentials in supersymmetric quantum mechanics, which gives one the ability to smoothly scale between the two limits. In this discussion, we have ignored the variation in the density along the channel, as the interest is in the source–channel region which tends to set the current in the transistor [64]. In Fig. 9.31, we plot the position of the inversion charge as a function of the fin width. Clearly, for thin fins, the charge is mainly bulk inversion, whereas for thicker fins, we have surface inversion, with its consequent lower mobility.

In the sense of the devices we have talked about here, the FinFET is a *standard* device that is going to see production. In this sense, it does not qualify as a novel new device, even though it is quite novel and quite new. But, it suggests that nanowires will find a home as new and novel devices. In this case, the FinFET is composed of a relatively high nanowire, "nano" in the sense that the width of the fin is only a few nanometers. So, one can ask whether other nanowire transistors are possible. Indeed, the use of a small cross-section nanowire was (apparently) first suggested by Ono *et al.* [65]. Following this, nanowires were fabricated by Kedzierski *et al.* [66] and used as transistors. In this latter work, wires as small as 5 nm diameter were fabricated, although the performance of the transistors was not spectacular. Then, Lieber's group suggested the use of self-assembled nanowires as building elements for semiconductor logic circuits [67].

Self-assembled nanowires tend to be grown vertically, so using them in circuits often means putting them horizontally upon a prepared Si surface. Here, the Si is oxidized and metal contacts are deposited. Then, the self-assembled nanowire is placed across the contacts, with the back Si layer used as the gate. This is not very effective, and certainly is not a good use of Si surface area, as will be discussed in the next section. Nevertheless, transistors have been fabricated by a number of laboratories (we list only some representative groups) [68,69,70,71,72,73]. On the other hand, other materials, such as ZnO have been used to fabricate vertical transistors [74,75]. In a similar manner, vertical transistors have been fabricated with an InAs channel [76,77,78].

A natural nanowire is the carbon nanotube (CNT). Depending upon the chirality (the folding of the graphene sheet into a nanotube), the CNT can be either metallic or semiconducting. Of course, only the latter is of interest for a possible transistor, and it quite natural that studies have sought the transistor properties of CNTs laid on oxidized Si [79,80,81]. These devices are more complicated than the typical nanowire transistor in that the source and drain contacts are Schottky barriers and the transistor primarily operates by modulating these barriers [79,82,83]. Indeed, a form of scaling theory has been presented for CNT transistors [84]. Much like other nanowires, the CNT can exhibit multimode behavior in which more than a single quantum transverse mode can be propagating [85], so that the differences between the CNT and other nanowire transistors are not all that large.

Another novel device is the Y-branch switch, which quite naturally is a branched nanowire in which gates are used to switch the current from one branch to the other [86]. Such a switch has a number of advantages, not least of which is extremely high frequency of operation [87,88]. Generally, the Y-branch switch is a low-power device [89] and switching can even be below the thermal voltage via quantum effects [90].

Perhaps one of the most effective approaches to improve Si integrated circuits is simply to chose a semiconductor with a higher mobility and therefore a more likely improved current and speed. This approach has been followed for the mainline Si industry by Intel in which they have begun to study quantum well transistors based on e.g. InSb [54,91]. InSb offers a mobility a couple of orders of magnitude higher than Si devices, so should show considerably faster switching with the option of operation at considerably lower voltage. We will return to this point in the next section.

9.2.3 How do we compare different devices?

Each new device is predicted to become the so-called savior for the semiconductor industry and the ultimate replacement for the current complementary MOSFET (CMOS) logic circuits. In fact, many such saviors have come and gone, yet the reliable CMOS continues to be scaled and to reach ever higher performance levels. It is important, then, to try to ascertain what questions must be asked if a new transistor technology is to become useful. The answer to this question arises on several levels. First, and perhaps foremost, is that any new technology must ride upon the wave of the exceedingly large investment that has been made in CMOS integrated circuits. Modern day "fabs" cost upwards of $3B. This is a major investment, and such will not be set aside easily or lightly. The most likely scenario for a new transistor technology is one which will supplement the current CMOS circuits to provide enhanced performance for some part of the overall architecture of the computational chip.

If a new technology is to supplement the current approaches, then they must obey the first law of scaling as described by Moore's law. This is that the use of Si real estate must be optimized if the cost per function is going to continue to be reduced. We can illustrate this by considering the FinFET discussed above. In Fig. 9.32, we plot a single fin, in an array of FinFETs, and describe the various dimensions of this structure. Clearly, the peripheral distance, e.g. the net gate width of the transistor, is simply $2h + d$. Now, if we are to use the Si area to maximal effect, then we must have

$$2h + d > W, \tag{9.32}$$

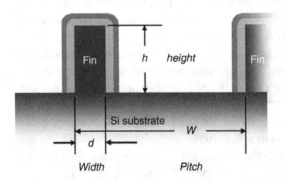

Fig. 9.32 Details of the height and thickness of a single fin within an array of FinFETs.

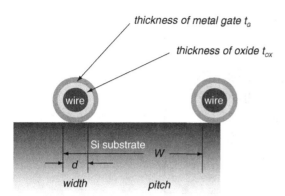

Fig. 9.33 The geometry of nanowires laid upon the Si surface to create planar transistors.

where W is the *pitch* of the transistors. Only if this equation is satisfied do we gain in Si area by use of the FinFET. However, for the case of the FinFET, this becomes a relatively easy condition to satisfy.

However, this same logic must be applied to concepts in which nanowires are used upon the Si surface. This will be case, whether the nanowires are self-assembled in vertical growth or are patterned in the Si surface. It will also be the case whether the nanowires are CNTs or are Si nanowires (or any other semiconductor). We illustrate the geometry in Fig. 9.33. The wire itself can be assumed to have a diameter d, but to this must be added the thickness of the oxide layer t_{ox} and the thickness of the gate layer t_G. Thus, to fit the wires within the pitch W, we must have

$$d + 2(t_{ox} + t_G) < W. \tag{9.33}$$

This requirement is just to fit the wires onto the planar surface. But, we cannot waste surface area – the law of cost of Si real estate. Hence, we must have more transistor periphery than W, or

$$\pi d > W. \tag{9.34}$$

These two equations then set a limit on how small the nanowire can be, which is given by

$$d > \frac{2(t_{ox} + t_G)}{\pi - 1}. \tag{9.35}$$

Generally, the gate must be about 3 nm or larger. If the oxide is then 2 nm thick, we are faced with the fact that the wire must have a diameter of 5 nm or more. Thus, we cannot make use of extremely small diameter nanowires. Note that this is a requirement just from Moore's law on the cost of Si real estate. It does not take into account any performance improvement in the nanowire. Indeed, for the performance to affect this argument, the nanowire would have to offer orders of magnitude performance enhancement over Si.

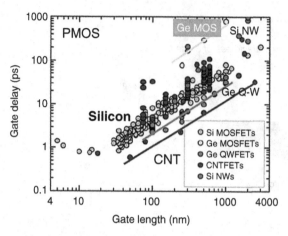

Fig. 9.34 Gate delay (the intrinsic speed CV/I) as a function of the physical gate length for *p*-channel MOSFETs (PMOS). (Source: R. Chau, Intel Corporation, http://download.intel.com/technology/silicon/Chau_DRC_062606.pdf, by permission.)

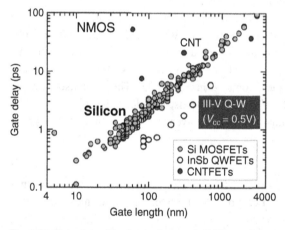

Fig. 9.35 Gate delay (the intrinsic speed CV/I) as a function of the physical gate length for *n*-channel MOSFETs (NMOS). (Source: R. Chau, Intel Corporation, http://download.intel.com/technology/silicon/Chau_DRC_062606.pdf, by permission.)

In addition to the above considerations, the new technology must operate as a proper transistor. Chau *et al.* [54] have outlined a set of benchmarks which describe quite well the operation of most transistors. These should be used in evaluating any new technology in order to assure that a level playing field is being used in all comparisons. There principally are four key metrics. These are: (i) intrinsic speed (CV/I, where C is the gate capacitance, V is the operation drain voltage, and I is the resulting drain current) versus gate length L_G, (ii) energy-delay product ($CV/I \cdot CV^2$) versus gate length L_G, (iii) transistor sub-threshold slope versus gate length L_G, and (iv) the intrinsic speed (CV/I) versus the I_{ON}/I_{OFF} ratio. In Figs. 9.34 and 9.35, plots of the intrinsic delay for Si MOSFETs

9.2 Semiconductor nanodevices in the real world

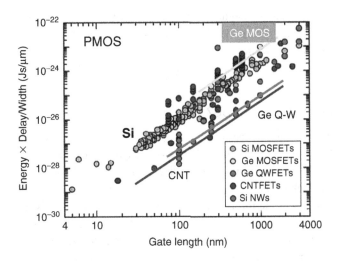

Fig. 9.36 Energy-delay product (the intrinsic speed CV/I times the energy CV^2) as a function of the physical gate length for *p*-channel MOSFETs (PMOS). (Source: R. Chau, Intel Corporation, http://download.intel.com/technology/silicon/Chau_DRC_062606.pdf, by permission.)

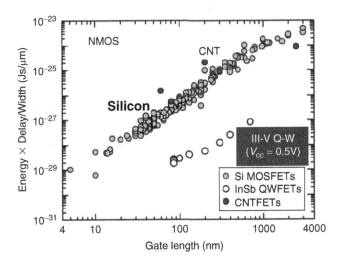

Fig. 9.37 Energy-delay product (the intrinsic speed CV/I times the energy CV^2) as a function of the physical gate length for *n*-channel MOSFETs (NMOS). (Source: R. Chau, Intel Corporation, http://download.intel.com/technology/silicon/Chau_DRC_062606.pdf, by permission.)

and for some alternative technologies are given [92]. In Figs. 9.36 and 9.37, plots of the energy-delay product for these same technologies are given [92].

For those devices which are plotted, there appears to be some prospects for CNTs, at least in *p*-channel devices. But, the most likely candidate is the III–V materials, primarily InSb quantum well channel transistors [91]. However, while the sub-threshold slopes are not plotted, they become equally as important when it is recognized that this relates to the ability to turn the transistor "off." If the transistor cannot be turned off, then it is not particularly useful, no matter how good the on-state properties are. Similarly, the importance of a smooth variation of the delay time with the on/off current ratio is also important to have a smoothly operating transistor. This is not the case in CNTs where the ambipolar behavior (both *n*-channel and *p*-channel behavior depending upon the gate voltage) can create a situation where there is no real off state. Hence, one must very carefully evaluate new concepts without resorting to hyperbole in performance claims.

9.2.4 Is ballistic transport good?

In this section, we will examine ballistic transport, and show that modern MOSFETs are probably not dominated by ballistic transport, and that this is likely a good thing! Ballistic transport in semiconductors is a relatively old idea, and here we mean the lack of *any* scattering. It was first discussed in regard to mesoscopic structures, where the mean free path was comparable to the device size, in connection with the Landauer formula [93]. In fact, the ideas of ballistic transport are even older, and derive from the earliest treatments of transport in vacuum diodes. The Langmuir–Child law describes the ballistic transport of electrons in a thermionic diode, with space charge built up near the cathode (corresponding to our source in a MOSFET), named after the two who developed it independently [94,95]. Both of these men derived the expression for the current, finding that

$$I \sim \frac{V^{3/2}}{L^2}, \qquad (9.36)$$

and it is this relationship that has become known as the Langmuir–Child law. More recently, Shur and Eastman proposed that device performance could be improved by utilizing ballistic transport in ultrasmall channel length semiconductor devices [96], but also showed that the current in an n^+-n-n^+ device would have the same space charge and current relationship as that of (9.36). It is important that the MOSFET has a space charge region, and potential maximum, between the source and the channel, and it is this that creates the connection to the Langmuir–Child law, as demonstrated by Shur and Eastman. In essence, the latter were suggesting use of high velocities due to transient velocity overshoot that can occur in many materials [97]. More recently, there have been many suggestions that ballistic transport can occur in short-channel devices, and might improve the performance [98]. This, in fact, is not the case, and a proper treatment of ballistic transport will show that it is detrimental to good device operation. In this section, we will outline the basic tenets that establish this point.

First, true ballistic transport occurs in the complete absence of scattering. This is, of course, found in vacuum tubes. There, electrons leave the cathode and form a space charge layer adjacent to this region. The solution of Poisson's equation for the region between the cathode and the plate yields the Langmuir–Child law (9.36). The importance of the Shur and Eastman result [95] is that *exactly the same behavior* is found in n^+-n-n^+ semiconductor structures, which is the structure that is found in the *n*-channel MOSFET. Electrons move out of the source into the channel, creating a space-charge region at the source–channel interface. It is modulation of this space-charge region by the gate potential that produces the normal device characteristics [99]. Variation of the space-charge region by the gate (or by the grid in the vacuum tube) leads to a family of triode-like curves obeying (9.36) with different (gate-voltage-dependent) coefficients. These triode curves are not good for either logic or high-frequency applications.

How then are the good characteristics, with current saturation, obtained? In the case of the vacuum tube, a "screen" grid (a metal grid) is inserted and held at a constant high potential so that the space-charge region is isolated from the anode potential. In the case of the MOSFET, similar screening occurs, but this time it is provided by the *scattering that occurs in the region between the space-charge layer and the drain*. In the MOSFET, one has scattering from the semiconductor phonons as well as from charges in the oxide and the semiconductor and from interface roughness. One may also have remote scattering from polar modes in the oxide. One normally does not connect scattering with screening, but this is a common occurrence in, for example, quantum transport. Moreover, it has been seen in detailed simulation that scattering has a large effect on the actual potential distribution and therefore on the device characteristics [100]. In fact, we can see this behavior in the simple device characteristics

$$\begin{aligned}I_{DS} &= \frac{\mu W C_{ox}}{L}\left(V_G - V_T - \frac{V_D}{2}\right)V_D \\ &= \frac{\mu W C_{ox}}{2L}\left[(V_G - V_T)^2 - (V_G - V_T - V_D)^2\right].\end{aligned} \quad (9.37)$$

Saturation occurs when the second term in the square brackets vanishes due to pinchoff at the drain end. Hence, in this situation, the drain potential does not affect the source operation.

However, when we begin to lose the scattering in the channel, e.g., when we begin to see quasi-ballistic transport, then we should begin to see a transition to triode-like characteristic curves, with a reduction in output drain resistance. This behavior is shown in Fig. 9.38. The saturation will disappear as this triode-like behavior becomes more and more prominent. The astute reader will notice that this behavior is exactly like drain-induced barrier lowering (DIBL). In fact, DIBL is the first precursor to ballistic transport. DIBL occurs when there is insufficient scattering to screen the space-charge region from the drain potential variations. Since it is generally accepted that DIBL is detrimental to good device operation, we may safely conclude that we really do not want to have ballistic transport occurring in our devices.

Given that ballistic behavior is detrimental to the devices, how can we ascertain that it is not really occurring? This is difficult to achieve experimentally, but not so difficult to investigate in meaningful device simulations. As we point out in the next section, it is quite likely that future devices may well be

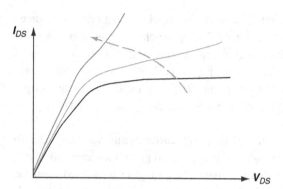

Fig. 9.38 Sketch of the expected change in the output characteristics as ballistic transport becomes important. As scattering is reduced, the curves will transition to triode-like behavior.

built around the concept of nanowires. To that end, it is logical to investigate whether there is any ballistic behavior in such devices. Kotylar *et al.* [101] examined classical scattering in a Si quantum wire and concluded that the mobility would not be improved in this structure, contrary to many expectations. We pursued a different approach and utilized a fully quantum-mechanical, and three-dimensional, simulation of small silicon quantum-wire MOSFETs [102], an approach which will be discussed in more detail in the next section. In this approach, the full Poisson equation solution is used to determine the local potential, and a recursive scattering matrix approach is used to determine the transport through the device. In this process, for each iteration from one transverse slice of the device to the next, a local Dyson's equation is solved with the slice Hamiltonian, a procedure equivalent to the scattering matrix solution of the Lippmann–Schwinger equation. This means that we can modify this Hamiltonian by the direct inclusion of a slice self-energy as well as a self-energy coupling between the slices where that is appropriate. This self-energy term describes the dissipation within the device [103]. We have computed this self-energy, to the lowest Born approximation, for all the normal phonon scattering processes expected to occur in a Si nanowire (impurity scattering is included directly through the random impurity potential). This self-energy is incorporated in the Hamiltonian to solve for the overall transport through the device. The general device characteristics that result from this simulation will be discussed in the next section. Here, we focus upon just the question of ballistic transport.

We can use the device structure, with varying gate length (and channel length) to study whether or not there is any ballistic behavior in the device. There is a good way to check this, and that is to vary the channel length at low drain bias. If the transport is ballistic, then Landauer's formula [92] tells us that the conductance should be constant, and therefore the resistance should be

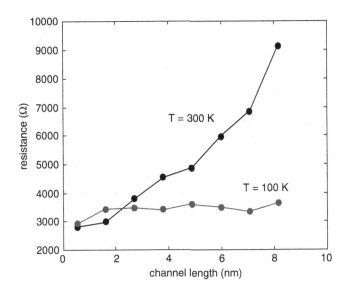

Fig. 9.39 Variation of the channel resistance, at a drain bias of 10 mV, as a function of the channel length. The constant behavior at 100 K is indicative of ballistic transport, while the linear rise at 300 K is indicative of diffusive transport.

independent of the length. On the other hand, if the transport is diffusive, then the resistance should depend linearly on the length of the channel. In Fig. 9.39, we show the results for a Si nanowire SOI MOSFET, in which we plot the resistance as a function of the channel length [104]. At low temperatures (100 K), the resistance is independent of the length of the channel. This result is expected for ballistic transport, which arises because the phonons are frozen out at this low temperature. At high temperatures (300 K), however, the resistance exhibits the expected Ohm's law linear dependence on device length. Below 2 nm, direct source–drain tunneling prohibits observation of the nanowire effects, and this is independent of temperature. Thus, it appears that there will be no real onset of important ballistic transport in future Si devices down to gate lengths below 5 nm, although there will continue to be problems with DIBL (while not shown here, the results of Fig. 9.39 are sensitive to the drain potential that is imposed).

But, the above is the case for silicon devices, which in general have relatively low mobilities and velocities. That is, the scattering in Si is quite strong, and the high-energy phonons give very good momentum randomization, all of which serves to strongly reduce the chances of ballistic behavior. Still, it is seen at low temperatures, as can be seen in Fig. 9.39. If we now move to a semiconductor with higher mobilities and velocities, and with scattering that is anisotropic, will the result change? In the III–V materials, the scattering is dominated by the polar LO phonon, which produces strongly anisotropic scattering, due to its Coulomb nature. We will see, in the next section, that for an InAs tri-gate MOSFET, the characteristics clearly show the trend toward the power-law behavior, which can be indicative of the onset of ballistic behavior. The mobility in InAs is almost two orders of magnitude larger than that of Si, so that a similar increase in mean free path is expected. Thus, a 2–4 nm limit in Si becomes several tens of nm in InAs, and the behavior seen in the figure is surely expected, even at room temperature. Indeed, recent experiments on InAs nanowires suggest that this ballistic length may be as long as 200 nm in such wires [105]. The behavior observed in this latter experiment is exactly that discussed above with regard to Fig. 9.39.

In this section, we have presented some thoughts on the future of semiconductor devices intended for use in VLSI chips, and the role that ballistic transport may play in the operation of these devices. As with other transport, it appears that the properties of the materials used in the device will play a role in whether or not ballistic behavior will appear. In Si devices, the mobility and the mean free path are sufficiently small that ballistic transport is not important down to gate lengths of 2–5 nm. On the other hand, in InAs devices, the mobility and mean free path are sufficiently large that ballistic behavior is seen in 30 nm devices, at least in the simulation of those devices. The role of parasitic source resistances on this behavior has not been considered, but may work to reduce the role of ballistic transport, as this tends to reduce transconductance which depends upon the mobility.

9.3 Quantum simulations via the scattering matrix

There have been many suggestions for different quantum methods to model ultrasmall semiconductor devices [106,107,108]. However, in each of these approaches, the length and the depth are modeled rigorously, while the third dimension (width) is usually included through the assumption that there is no interesting physics in this dimension (lateral homogeneity). Moreover, it is assumed that the mode does not change shape as it propagates from the source of the device to the drain of the device. Other simulation proposals have simply assumed that only one subband in the orthogonal direction is occupied, therefore making higher-dimensional transport considerations unnecessary. These may not be valid assumptions, especially as we approach devices whose width is comparable to the channel length, both of which may be less than 10 nm, such as the FinFET discussed above. It is important to consider all the modes that may be excited in the source (or drain) region, as this may be responsible for some of the interesting physics that we wish to capture. In the source, the modes that are excited are three dimensional in nature, even in a thin SOI device. These modes are then propagated from the source to the channel, and the coupling among the various modes will be dependent upon the details of the total confining potential at each point along the channel. Moreover, as the doping and the Fermi level in short-channel MOSFETs increases, we can no longer assume that there is only one occupied subband. In an effort to provide a more complete simulation method, we present here a full three-dimensional quantum simulation, based on the use of recursive scattering matrices, such as developed in Section 3.5.1.

The key first important factor is that we now use a two-dimensional plane at each slice of the simulation, instead of a one-dimensional line. The Schrödinger equation is now written as

$$\frac{-\hbar^2}{2}\left(\frac{1}{m_x}\frac{d^2}{dx^2}+\frac{1}{m_y}\frac{d^2}{dy^2}+\frac{1}{m_z}\frac{d^2}{dz^2}\right)\Psi + V(x,y,z)\Psi = E\Psi. \tag{9.38}$$

Since we now have L lines in the third dimension, then the Hamiltonian now becomes the $L \times L$ Bloch matrix

$$H = \begin{bmatrix} H_0(\mathbf{r}) & t_z & \ldots & 0 \\ t_z & H_0(\mathbf{r}) & \ldots & \ldots \\ \ldots & \ldots & \ldots & t_z \\ 0 & \ldots & t_z & H_0(\mathbf{r}) \end{bmatrix}, \tag{9.39}$$

where H is now an $ML \times ML$ matrix at each slice. Similarly, the hopping matrix is now written as

$$T_x = \begin{bmatrix} t_y & 0 & \ldots & 0 \\ 0 & t_y & \ldots & 0 \\ \ldots & \ldots & \ldots & \ldots \\ 0 & 0 & \ldots & t_y \end{bmatrix}. \tag{9.40}$$

With this setup of the matrices, the general procedure follows that laid out in the previous Chapter 3. One first solves the eigenvalue problem on slice 0 at the end of the source (away from the channel), which determines the propagating and evanescent modes for a given Fermi energy in this region. The wave function is thus written in a mode basis, but this is immediately transformed to the site basis, and one propagates from the drain end, using the scattering matrix iteration down to the drain end, and back again. Device characteristics are computed by integrating the assumed unit amplitude for each mode with the Fermi–Dirac distribution, and computing the flow from each end of the device.

Now, one addition we need to treat here is the inelastic phonon scattering that can occur between the modes of the device, and hence between the sites of the device. In Chapter 2, we computed the scattering between transverse modes i,j and i',j' in a quantum wire. For example, for the g-intervalley phonons in a Si nanowire, we found that

$$\left.\frac{1}{\tau}\right|_{i,j}^{i',j'} = \frac{2m*D_0^2}{\hbar^2 \rho \omega_q}I_{ij}^{i'j'}\left\{N_q \frac{1}{\sqrt{k^2+\Delta_+^2}}\right.$$

$$\left. +(N_q+1)\frac{1}{\sqrt{k^2+\Delta_-^2}}\right\}, \tag{9.41}$$

where

$$\Delta_\pm^2 = \frac{2m*}{\hbar^2}\left(E_k^{ij}-E_{k'}^{i'j'}\pm\hbar\omega_q\right) \tag{9.42}$$

is the offset between the two subbands. Here, this quantity can be positive or negative; it is highly unlikely that it will be zero at any point. Nevertheless, the Fourier transform can be carried out in an equivalent manner.

We now seek to compute the inverse Fourier transform, in order to get back to the real space basis (in the channel direction) used in the scattering matrix approach. Hence, we wish to do the Fourier transform

$$\frac{1}{\tau(x-x')}\bigg|_{ij}^{i'j'} \sim \int_{-\infty}^{\infty} \frac{e^{-ik(x-x')}dk}{\sqrt{k^2+\Delta^2}}. \tag{9.43}$$

Of course, the appropriate transform result will depend upon the sign of Δ^2. We will write $a = |\Delta|$ in the following, so that $a > 0$. We now have to consider the three possible cases:

$\Delta^2 > 0$:

Now, we can rewrite the Fourier transform in the new form, appropriate to this case,

$$\frac{1}{\tau(x-x')}\bigg|_{ij}^{i'j'} \sim \int_{-\infty}^{\infty} \frac{e^{-ik(x-x')}dk}{\sqrt{k^2+a^2}}. \tag{9.44}$$

This integral can be rewritten as

$$\frac{1}{\tau(\xi)}\bigg|_{ij}^{i'j'} \sim 2\int_{0}^{\infty} \frac{\cos(k\xi)dk}{\sqrt{k^2+a^2}} = 2\int_{0}^{\infty} \frac{\cos(u\xi a)du}{\sqrt{u^2+1}}. \tag{9.45}$$

$$= 2K_0(\xi a).$$

Hence, the leading terms in the series expansion for K_0 leads to

$$K_0(\xi a) \sim \sqrt{\frac{\pi}{2\xi a}} e^{-\xi a}\left(1 - \frac{1}{8\xi a} + \cdots\right). \tag{9.46}$$

$\Delta^2 < 0$:

Now, we can rewrite the Fourier transform in the new form, appropriate to this case,

$$\frac{1}{\tau(x-x')}\bigg|_{ij}^{i'j'} \sim 2\int_{a}^{\infty} \frac{e^{-ik(x-x')}dk}{\sqrt{k^2-a^2}}. \tag{9.47}$$

Again, this can be evaluated with standard known integrals, as

$$\frac{1}{\tau(x-x')}\bigg|_{ij}^{i'j'} \sim 2\int_{1}^{\infty} \frac{[\cos(u\xi a) + i\sin(u\xi a)]du}{\sqrt{u^2-1^2}} \tag{9.48}$$

$$= -\pi[Y_0(\xi a) - iJ_0(\xi a)] = i\pi H_0^{(1)}(\xi a).$$

9.3 Quantum simulations via the scattering matrix

Now, the leading term in the series is given as

$$iH_0^{(1)}(\xi a) \sim -\frac{2}{\pi}\ln(\xi a), \tag{9.49}$$

and, hence, is peaked around small ξa.
$\Delta^2 = 0$:

Now, we can rewrite the Fourier transform in the new form, appropriate to this case,

$$\left.\frac{1}{\tau(x-x')}\right|_{ij}^{i'j'} \sim \int_{-\infty}^{\infty} \frac{e^{-ik(x-x')}dk}{k}. \tag{9.50}$$

This can be easily rewritten as

$$\left.\frac{1}{\tau(\xi)}\right|_{ij}^{i'j'} \sim 2i\int_0^{\infty} \frac{\sin(k\xi)dk}{k} = i\pi. \tag{9.51}$$

Hence, only in the case of scattering within a single subband is there coupling between different slices of the solution space. However, this is an imaginary term, which means only a self-energy shift, which can be ignored.

Now, finally, we note that the integration over the momentum space (the Fourier transform) has removed a length from the final result for the scattering. That is, the dimensions are now 1/m-sec. Hence, we must carry out the integration over the length of the sample in order to recover this dimensional term. This is, in effect, integrating over the slice-conserving function that results from the Fourier transform. This leads to the result

$$\int \frac{1}{\tau(\xi)}d\xi \sim \begin{cases} \dfrac{2L}{\pi}, & \Delta^2 < 0, \\ 2\sqrt{\dfrac{2}{\pi}L}, & \Delta^2 > 0. \end{cases} \tag{9.52}$$

Both limits are taken for the case where $La = L|\Delta| < 1$.

The self-energy Σ has both real and imaginary parts, with the latter representing the dissipative interactions. In semiconductors, the scattering is weak, and is traditionally treated by first-order time-dependent perturbation theory, which yields the common Fermi golden rule for scattering rates. With such weak scattering, the real part of the self-energy can generally be ignored for the phonon interactions, and that part that arises from the carrier–carrier interactions is incorporated into the solutions of Poisson's equation by a local-density approximation, which approximately accounts for the Hartree–Fock corrections. In the many-body formulations of the self-energy, the latter is a two-site function in that it is written as

$$\Sigma(\mathbf{r}_1, \mathbf{r}_2). \tag{9.53}$$

In our case, where we are using transverse modes in the quantum wire, this may be rewritten as

$$\Sigma(i,j;i',j',x_1,x_2). \quad (9.54)$$

Here, the scattering accounts for transitions from transverse mode i,j at position x_1 to i', j' at position x_2. Generally, one then makes a center-of-mass transformation

$$X = \frac{x_1 + x_2}{2}, \quad \xi = x_1 - x_2, \quad (9.55)$$

and then Fourier transforms on the difference variable to give

$$\Sigma(i,j;i',j',X,k_x) = \frac{1}{2\pi}\int d\xi e^{i\xi k_x}\Sigma(i,j;i',j',X,\xi). \quad (9.56)$$

The center-of-mass position X remains in the problem as the mode structure may change as one moves along the channel. At this point, the left-hand side of (9.56) is the self-energy computed by the normal scattering rates, such as is done in quantum wells and quantum wires previously. However, these previous calculations usually used the Fermi golden rule, which is an evaluation of the bare self-energy in (9.56). In many-body approaches, one normally does not use the energy-conserving delta function that is the central part of the Fermi golden rule. Rather, this function is broadened into the *spectral density*, through the use of the self-consistent Born approximation. In this way, off-shell effects are taken into account through the broadening of the relationship between momentum and energy. In semiconductors, however, we have already noted that the scattering is weak. It has been pointed out that these off-shell corrections are only important in fast processes where we are interested in femtosecond response and their neglect introduces only slight errors for times large compared to the collision duration. Moreover, the broadening of the delta function will not be apparent when we reverse the Fourier transform of (4), as the area under the spectral density remains normalized to unity. Since our recursion is in the site representation, rather than in a mode representation, we have to reverse the Fourier transform in (9.56) to get the x-axis variation, and then do a mode-to-site unitary transformation to get the self-energy in the form necessary for the recursion. Hence, we begin by seeking the imaginary part of the self-energy, which is related to the scattering rate via

$$\text{Im}\{\Sigma(i,j;i',j',X,k_x)\} = \hbar\left(\frac{1}{\tau}\right)_{i,j}^{i',j'}, \quad (9.57)$$

and it is the latter scattering rate which we calculate. This result will be a function of the x-directed momentum (which is related, in turn, to the energy of the carrier) in the quantum wire. Finally, this scattering rate must be converted to the site representation with a unitary transformation

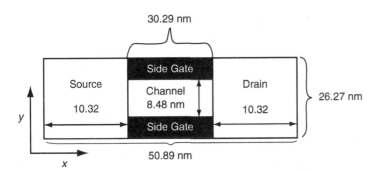

Fig. 9.40 Device structure in the x–y plane for the 30 nm InAs MOSFET device.

$$\Gamma_{ac} = \text{Im}\{\Sigma\} = U^+ \left(\frac{\hbar}{\tau}\right)^{i',j'}_{i,j} U, \qquad (9.58)$$

where U is a unitary mode-to-site transformation matrix. The unitary matrix U^+ results from the eigenvalue solutions in the transverse slice and are composed of the various eigenfunctions in the site basis. Hence, it represents a mode-to-slice transformation, as given by the extension of the approach in Chapter 3 to three dimensions.

9.3.1 InAs MOSFET transistors

In this section, we present the results of a three-dimensional, self-consistent, quantum-mechanical device simulation of a tri-gate InAs quantum wire MOSFET with discrete dopants. We examine some of the pertinent device characteristics and functional parameters surrounding the operation of devices with 30 nm and 10 nm channel lengths. In Fig. 9.40, we display a schematic of the device geometry for an InAs MOSFET with a 30 nm channel taken to lie in the x–y plane (the z-axis is normal to the plane shown). The exact device dimensions (multiples of the lattice constant) have been included in this simulation to aid in the inclusion of the discrete dopants. The thickness of the InAs layer is 9.09 nm. The source and drain of the device are n-type with a doping density of 6×10^{18} cm^{-3}, while the channel of the device is considered to be p-type, but undoped. The gate material is assumed to be platinum and the gate oxide on each side is composed of 1 nm of hafnium dioxide (HfO$_2$). Beneath the device, we have assumed a generic insulating substrate. The structure of the device with the 9.69 nm channel is exactly the same with the exception, of course, of the channel length. Once the device geometry is defined, the InAs lattice is scanned and the dopants are distributed according to the method presented in [109]. Following the distribution of the dopants, they are then mapped back onto the grid of the simulation mesh and the initial self-consistent Poisson solution is obtained. In this case, the full Coulomb potential of the dopants is incorporated.

Fig. 9.41 I_d–V_d curve for a 30 nm device with $V_g = 0.4$ V.

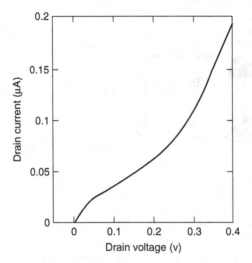

Then, the solution of Poisson's equation for the local potential is no longer smoothly varying in the source and the drain of the device. The inclusion of discrete dopants causes the formation of potential variation in the source and drain. The density throughout the device is calculated using the scattering matrix approach. The density is then updated using Broyden's method [110] and a new guess for the potential is obtained through the solution of Poisson's equation. The process is repeated until a desired level of convergence is obtained. It should be noted that all of the simulations presented here have been performed at 300 K and ballistic transport is assumed throughout the paper.

In Fig. 9.41, we plot the I_d–V_d curves for the 30 nm InAs quantum wire MOSFET. We find that the sub-threshold slope averages about 69 mV/dec and the threshold voltage is 0.46±0.02 V. The I_{on}/I_{off} ratio is 1.9×10^{10}. These numbers predict a transistor whose currents and charge control should far exceed those of its silicon counterpart. However, the threshold voltage for the devices is too large to be used in the present or the next generation technologies. This is due to the fact that in an InAs system, we have large quantization effects and gate work function dependence. This has both good and bad aspects. First, one normally expects InAs to have a Fermi level pinned roughly 0.4 eV into the conduction band. But, MOS operation has long been known to occur in this system [111]. Thus, proper choice of the metal gate is essential to selecting the proper threshold voltage. We have not optimized this process, but quantization in the channel helps to move the Fermi level out of the conduction band. However, in the present simulation, larger voltages must be applied to the system for conduction to occur. While the threshold voltage is too high in this system, the amount that the threshold voltage varies from device to device is minimal.

Now, putting in the scattering affects the output characteristics, but not significantly. In fact, the scattering rate for the polar longitudinal optical phonon is not sufficiently high to greatly affect the output characteristics. This is why the ballistic to diffusive crossover appears to be many tens of nanometers [104]. The curves in Fig. 9.41 curve upward dramatically, and seem to really have the ballistic characteristics discussed in the last section. This apparently is a result of the small electron effective mass and the relatively low scattering rate for the polar modes. Different behavior will appear for the Si devices in the next section.

9.3.2 Si MOSFET transistors

In modeling the Si MOSFET, it is important to consider all the modes that may be excited in the source (or drain) region, as this may be responsible for some of the interesting physics that we wish to capture. In the source, the modes that are excited are three dimensional (3D) in nature, even in a thin SOI device. These modes are then propagated from the source to the channel, and the coupling among the various modes will be dependent upon the details of the total confining potential at each point along the channel. Moreover, as the doping and the Fermi level in short-channel MOSFETs increases, we can no longer assume that there is only one occupied subband. In an effort to provide a more complete simulation method, we give here the scattering matrix version of a full 3D quantum simulation, which is being used in our group to simulate short-channel, fully depleted SOI MOSFET devices [112,113].

The device under consideration is a fully depleted SOI MOSFET structure. We orient the directions in order to correspond to the length and the height (thickness of the SOI layer) of the device respectively. The source and the drain contact regions are 10 nm in length and 18 nm in width (lateral direction, the y axis). In an actual device, the length of the source and the drain of a MOSFET would be much longer, but this length captures the important energy relaxation length. We implement open boundary conditions at the ends of the structure and on the sidewalls. The gate length of this device is 11 nm corresponding to a dimension that will allow the gate to fully control that channel of the device. The actual channel length of the device used in these simulations is 9 nm. The channel itself is 9 nm in width, so that the Si layer is a wide-narrow-wide structure as shown in Figure 9.42. The entire structure is on a silicon layer that is taken to be only 6 nm thick, with a 10 nm buried oxide (BOX) layer below this layer. The gate oxide is taken to be 2 nm thick.

An important point relates to the crystal orientation of the device, as indicated in Fig. 9.43. As is normal, we assume that the device is fabricated on a [100] surface of the Si crystal, and we then orient the channel so that the current will flow along the <110> direction. This direction is chosen as it is quite common in the industry to orient devices so that current flows in this direction. By this, we

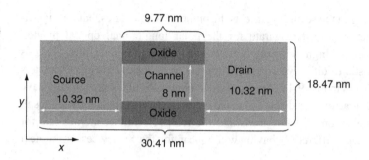

Fig. 9.42 Structure of the tri-gate Si MOSFET in the surface plane. The device is 8 nm thick.

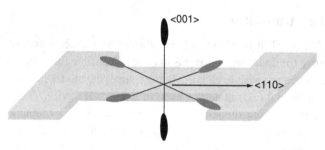

Fig. 9.43 Orientation of the channel of the simulation along the <110> crystalline direction and the manner in which the six equivalent ellipsoids of the Si conduction band are aligned with respect to this.

mean that the <110> direction lines up along the source–drain direction and the <001> direction lines up with the z direction normal to the plane of the device. Because this now separates the six equivalent ellipsoids into a two-fold set and a four-fold set, the basic Hamiltonian matrix and wave function matrix are now doubled in size. Hence, once more the matrix grows in size. Using this orientation complicates the wave function, but allows for simplicity in terms of the amount of memory needed to store the Hamiltonian and to construct the various scattering matrices (as well as the amount of computational time that is required).

The simulated device is shown in Fig. 9.42. It has identical source and drain dimensions, and the doping concentrations for the different regions are 6×10^{19} cm^{-3} in the source and drain, and 5×10^{18} cm^{-3} in the channel. Once a solution mesh is defined, the dopant atoms are distributed onto the grid points. Once the dopants are distributed, they are mapped back to the computational lattice and the resultant potential profile [114] for this device is shown in Fig. 9.44. It is clear that there are very few acceptors in the channel region. Full quantum-mechanical simulations are quite time consuming. This is compounded by the fact that in order to obtain an accurate representation of the transport in these ultra small devices, it is necessary to utilize a 3D simulation tool which mandates the use of additional grid points in the solution space. Using the 3D scattering matrix approach presented here, we obtain the quantum-mechanical solution for the

9.3 Quantum simulations via the scattering matrix

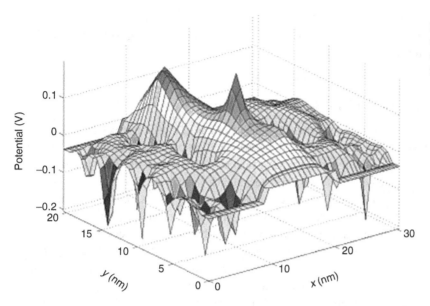

Fig. 9.44 Potential profile for the SOI MOSFET taken at a depth of 7 nm from the surface of the device.

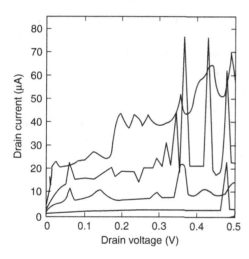

Fig. 9.45 I_d–V_d curves for the Si SOI MOSFET. From bottom to top the gate voltages are: 1.5 V, 2 V, 2.5 V, and 3 V. The peaks are evidence of the formation of resonant levels in the channel, and the consequent quantum interference.

electron density for positive and negatively directed k-states. It should be noted here that the solution scheme must be set up to find the solution for both the positive and negative k-states. The algorithm, assumes that there is only propagation in one direction. It is crucial to include both the positive and negative states as in an actual device there is not only injection from the source of the device to the drain, but there is injection from the drain end of the device as well.

We now examine results obtained from the simulation of the SOI MOSFET. In Fig. 9.45, we plot the I_d–V_d curves associated with one of the doping distributions. Peaks in the current give evidence of the formation of resonant levels in the channel based on the position of the dopants in the channel. We conclude that this resonant behavior persists even at greatly elevated drain voltages.

9.4 Real-time Green's functions

As we discussed in Chapter 2, considerations of transport in semiconductors has traditionally been described by the application of the Boltzmann transport equation or by reduced forms of this, such as the drift-diffusion approximation [115]. Powerful Monte Carlo techniques have been developed to solve this equation for quite complicated scattering processes and in very far-from-equilibrium circumstances [116,117]. The general feature of semiclassical far-from-equilibrium transport is that the distribution function deviates significantly from the equilibrium Maxwellian form (or Fermi–Dirac form, in degenerate systems) [118]. The form of this distribution must be found from a balance between the driving forces (electric and magnetic fields, for example) and the dissipative scattering processes. In many cases, such as in excitation of electron–hole pairs by an intense laser pulse, the resulting far-from-equilibrium distribution cannot be evolved from any equilibrium distribution, so a gradual evolutionary process to the final steady-state distribution cannot be found. In fact, the formation of the dissipative steady-state of the far-from-equilibrium distribution has been termed a type of phase transition [119].

In the case of quantum transport, we have utilized primarily the retarded and advanced Green's functions in the previous chapters. These two Green's functions describe the evolution of the single-particle energy–momentum relationship into a many-body description in which the delta function $\delta(E - \hbar^2 k^2 / 2m*)$, for a single parabolic band, evolves into the spectral density function $A(k, E)$. The latter is described by the imaginary parts of the retarded or advanced Green's functions. If we now move to the far-from-equilibrium dissipative steady-state, we need additional functions that will describe the quantum distribution function that evolves to describe the balance between driving forces and scattering processes. These additional Green's functions, which are actually proper correlation functions, are provided by the nonequilibrium, or real-time, Green's functions developed over the past few decades [3–6].

Previously, we introduced the retarded and advanced functions through the definitions

$$G_r(\mathbf{r}, t; \mathbf{r}', t') = -i\Theta(t - t')\langle\{\Psi(\mathbf{r}, t), \Psi^+(\mathbf{r}', t')\}\rangle, \tag{9.59}$$

$$G_a(\mathbf{r}, t; \mathbf{r}', t') = i\Theta(t' - t)\langle\{\Psi^+(\mathbf{r}', t'), \Psi(\mathbf{r}, t)\}\rangle. \tag{9.60}$$

The need for the new functions arises from the property of the unitary operator that introduces the perturbing potential, which is used to describe the interaction

9.4 Real-time Green's functions

of the carrier system with, for example, the dissipative mechanisms. We have defined this operator previously as

$$\exp\left[-\frac{i}{\hbar}\int_{t'}^{t} V(\eta)d\eta\right], \quad (9.61)$$

with the limits

$$t \to \infty, \quad t' \to -\infty. \quad (9.62)$$

These limits were acceptable because the system was assumed to be very near to equilibrium, so that the distribution function describing the carriers had the equilibrium form. Now, however, the first of these limits is unacceptable, since it assumes we know what happens in the infinite future. Since the distribution is not the equilibrium form, this limit can no longer be used. The second of the limits is problematical as well. To invoke such a limit means that the system evolution can be traced from the initial, equilibrium state. As remarked above, it is not clear that this procedure is valid either. At present, however, we will assume that such can be done, and we will return later to the treatment of failure of this process. We first address how to handle the upper limit of the integral in the unitary operator.

To avoid the need to proceed to $t \to \infty$ in the perturbation series, a new time path for the real-time functions was suggested by Blandin et al. [120]. (There may have been others who have suggested this approach, but this latter work seems to be the primary suggestion for the proper context.) This new contour is shown in Fig. 9.46. The evolution is assumed to begin with a thermal Green's function at $t' = t_0 - i\hbar\beta$ where β is the inverse temperature as introduced in a previous chapter. It then evolves into the nonequilibrium, noninteracting Green's function at t_0. The contour then extends forward in time to the maximum of (t, t') at which point it is returned *backward* in time to t_0. In many cases, the limit $t_0 \to -\infty$ is invoked when one is not interested in the initial condition (as in most far-from-equilibrium systems). The handling of the Green's functions when both time arguments are on either the upper branch or the lower branch is straightforward. On the other hand, when the two time arguments are on different branches of the time contour, the two new correlation functions must be defined [121]. These are the "less than" function

$$G^<(\mathbf{r},t;\mathbf{r}',t') = i\langle\Psi^+(\mathbf{r}',t')\Psi(\mathbf{r},t)\rangle \quad (9.63)$$

and the "greater than" function

$$G^>(\mathbf{r},t;\mathbf{r}',t') = -i\langle\Psi(\mathbf{r},t)\Psi^+(\mathbf{r}',t')\rangle. \quad (9.64)$$

In general, the four Green's functions defined here are all that are now needed, as the latter two will relate to the far-from-equilibrium distribution function, as

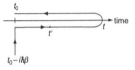

Fig. 9.46 The path of integration for the real-time Green's functions. The tail extending downward connects to the thermal equilibrium Green's functions, where appropriate.

we will see below. (This is true in lowest order, but higher-order multi-particle Green's functions may be needed also, as was the case in the previous chapters.) In practice, however, two further functions are useful. These two can be derived from the previous four, so they are merely combinations of the previous ones. The two new ones are the time-ordered Green's function

$$G_t(\mathbf{r},t;\mathbf{r}',t') = i\langle T[\Psi^+(\mathbf{r}',t')\Psi(\mathbf{r},t)]\rangle$$
$$= \Theta(t-t')G^>(\mathbf{r},t;\mathbf{r}',t') + \Theta(t'-t)G^<(\mathbf{r},t;\mathbf{r}',t'), \quad (9.65)$$

where T is the time-ordering operator, and the anti-time-ordered Green's function

$$G_{\bar{t}}(\mathbf{r},t;\mathbf{r}',t') = \Theta(t'-t)G^>(\mathbf{r},t;\mathbf{r}',t') + \Theta(t-t')G^<(\mathbf{r},t;\mathbf{r}',t'). \quad (9.66)$$

These are obviously related to the previous Green's functions, and the various quantities can be related through

$$G_r = G_t - G^< = G^> - G_{\bar{t}} = \Theta(t-t')(G^> - G^<), \quad (9.67)$$

$$G_a = G_t - G^> = G^< - G_{\bar{t}} = -\Theta(t'-t)(G^> - G^<). \quad (9.68)$$

For systems that have been driven out of equilibrium, the ensemble average, indicated by the angular brackets, no longer signifies thermodynamic averaging or an average over the zero-temperature ensemble distribution function. Instead, the brackets indicate that some average needs to be performed over the available states of the far-from-equilibrium system, but in which these states are weighted by the far-from-equilibrium distribution function. As in semiclassical transport in which the equation of motion is the Boltzmann equation, finding this far-from-equilibrium distribution is usually the most difficult part of the nonequilibrium problem. The quantum transport problem has been solved for the nonequilibrium situation in only a very few cases, and usually under stringent appoximations.

9.4.1 Equations of motion for the RTGF

It is obvious that solutions for the correlation functions will require that a set of equations of motion be developed for these quantities. With the added complications of four functions, however, these equations will be much more complicated than the simple Boltzmann equation, or the assumed forms used in previous chapters. Fortunately, only four of the six functions introduced above are independent in the general nonequilibrium situation.

Keldysh [6] introduced a general method of treating the set of Green's functions with a single matrix Green's function, and hence a single matrix equation of motion to be solved. This does not reduce the overall effort required, but it does simplify the equations in which these functions are described. We can illustrate this approach in the following manner, which will be done only for the functions defined on the two horizontal lines in Fig. 9.46. The approach can

9.4 Real-time Green's functions

certainly be extended to the third, vertical line if necessary [122], but this is usually not required in the far-from-equilibrium situation. The general Green's function is defined in terms of two field operators. Each of these field operators can be on the upper line or the lower line of the overall contour in Fig. 9.46. Our matrix is thus a 2 × 2 matrix. The rows of the matrix are defined by the operator $\Psi(\mathbf{r}, t)$. That is, row one of the matrix corresponds to when t is on the lower line, while row two is defined when t is on the upper part of the trajectory. Similarly, the operator $\Psi^+(\mathbf{r}', t')$ defines the columns of the matrix. Column one is defined when t' is on the lower part of the trajectory, while column two is defined when t' is on the upper part of the trajectory. Thus, for example, we can write the 11 element by assuming contour ordering (time increases in the positive sense around the contour) as

$$G_{11}(\mathbf{r}, t; \mathbf{r}', t') = \Theta(t - t')G^>(\mathbf{r}, t; \mathbf{r}', t') + \Theta(t' - t)G^<(\mathbf{r}, t; \mathbf{r}', t') = G_t. \quad (9.69)$$

In a similar manner the 22 term yields the $G_{\bar{t}}$ function. The 12 term is achieved by beginning with the "normal ordering" of the operators (in which products arising from the use of Wick's theorem are ordered with the creation operators to the left). Then

$$G_{12}(\mathbf{r}, t; \mathbf{r}', t') = i \langle T_C[\Psi^+(\mathbf{r}', t')\Psi(\mathbf{r}, t)] \rangle = G^<. \quad (9.70)$$

Similarly, the 21 term becomes $G^>$. It should be remarked that the definition of rows and columns is not unique and that the transpose is used equally as often. Thus, we may write the contour ordered Green's function matrix as

$$\mathbf{G}_C = \begin{bmatrix} G_t & G^< \\ G^> & G_{\bar{t}} \end{bmatrix}. \quad (9.71)$$

It is useful at this point to introduce a coordinate transformation, often called a *rotation in Keldysh space* [6]. This rotation removes some of the degeneracy of the various elements, separating out the retarded and advanced functions as well as simplifying the matrix itself. The rotation is a spin-like rotation, in which the non-unitary matrices are

$$\mathbf{L} = \frac{1}{\sqrt{2}} \begin{bmatrix} 1 & -1 \\ 1 & 1 \end{bmatrix}, \quad \mathbf{L}^+ = \frac{1}{\sqrt{2}} \begin{bmatrix} 1 & 1 \\ -1 & 1 \end{bmatrix}, \quad \tau = \begin{bmatrix} 1 & 0 \\ 0 & -1 \end{bmatrix}. \quad (9.72)$$

These are used to modify the matrix Green's function into the Keldysh form $\mathbf{G}_K = \mathbf{L}\tau\mathbf{\tau}_C\mathbf{L}^+$. The matrix Keldysh Green's function is then

$$\mathbf{G}_K = \begin{bmatrix} G_r & G_K \\ 0 & G_a \end{bmatrix}, \quad (9.73)$$

where $G_K = G^< + G^>$. (The reader is asked to pay particular attention to the fact that G_K is the matrix element representing the particular sum of the less-than and greater-than Green's functions, which is often called the Keldysh

Green's function, and $\mathbf{G}_K$, which is in bold, is the actual matrix of Green's functions.)

We can now develop the equations of motion for the noninteracting forms of these Green's functions, that is, the equations that the functions will satisfy in the absence of any applied potentials and/or perturbing interactions. For this, we assume that the individual field operators are based on the wave functions that satisfy the basic Schrödinger equation. However, in the case of single-point potentials, such as those following from the Poisson equation, this potential can be included in the Schrödinger equation. So, in the case of only such potentials, this leads to

$$\left(i\hbar\frac{\partial}{\partial t} - H_0(\mathbf{r}) - V(\mathbf{r})\right)\mathbf{G}_{K0} = \hbar\mathbf{I}, \tag{9.74}$$

$$\left(-i\hbar\frac{\partial}{\partial t'} - H_0(\mathbf{r}') - V(\mathbf{r}')\right)\mathbf{G}_{K0} = \hbar\mathbf{I}, \tag{9.75}$$

where $\mathbf{I}$ is the identity matrix (unity on the diagonal and zero off the diagonal). The zero subscript has been added to indicate the noninteracting form of the Green's function.

Transport, as has been stated earlier, arises as a balance between the driving forces and the dissipative forces. To achieve a description of transport with the Green's functions, it is necessary to add some interaction terms to the Hamiltonian. As before, these will lead to self-energy terms Σ. With the new contour, it is possible to construct a Feynman series expansion of the interaction terms with the new form of the contour-ordered unitary operator for the interaction. This procedure seems to continue to work in the case of the real-time Green's functions and has been pursued almost universally. The assumption is that the projection of the time axes back to the initial time allows the use of the pseudo-equilibrium to justify the use of an equivalent form of Wick's theorem. The various parts of the diagrams may then be regrouped into terms that represent the Green's function itself and terms that represent the interactions leading to the self-energy. The self-energy may also be expressed as a matrix, so it is possible to write the equation of motion for the full Green's function as

$$\left(i\hbar\frac{\partial}{\partial t} - H_0(\mathbf{r}) - V(\mathbf{r})\right)\mathbf{G}_K = \hbar\mathbf{I} + \mathbf{\Sigma}\mathbf{G}_K, \tag{9.76}$$

$$\left(-i\hbar\frac{\partial}{\partial t'} - H_0(\mathbf{r}') - V(\mathbf{r}')\right)\mathbf{G}_K = \hbar\mathbf{I} + \mathbf{G}_K\mathbf{\Sigma}, \tag{9.77}$$

where the self-energy matrix is now written in the Keldysh form as

$$\mathbf{\Sigma}_K = \begin{bmatrix} \Sigma_r & \Sigma_K \\ 0 & \Sigma_a \end{bmatrix}. \tag{9.78}$$

Generally, one would not expect the form of the self-energy to preserve the Keldysh form that is found here. In fact, if one applies the non-unitary transformations defined in and below (9.72) to the product functions (which are a convolution integral of the two Green's functions) that appear in the last terms of (9.77) and (9.78) before the Keldysh transform, the results differ greatly. However, it is important to note that care must be taken in correctly formulating the self-energy expressions in the contour-ordered Green's functions. The products that appear in (9.77) and (9.78) are integral products in the form

$$AB(1,3) \to \int d2 A(1,2) B(2,3), \tag{9.79}$$

where the shorthand notation $1 = (\mathbf{r},t)$ has been used, and the integration is over all internal variables. The proper way of writing these convolution integrals turns out to be that in which the Keldysh matrix Green's functions appear, as will be shown with the Langreth theorem in the next section.

9.4.2 The Langreth theorem

In truth, while we have written down the self-energy and the Green's function in Keldysh form in the last terms of (9.76) and (9.77), this result is not at all obvious. For example, when we begin with the easily understood Green's function matrix (9.71) and write the equations of motion, then the unit matrix on the right side of the equation is not appropriate. This is because the element in the 22 position should be $-\delta(t-t')\delta(\mathbf{r}-\mathbf{r}')$ [123]. The transform matrix $\boldsymbol{\tau}$ converts this to a proper identity matrix and then the remaining two transformation matrices leave it unchanged. Thus, the transformations suggested above can easily be applied to the equation of motion for the noninteracting forms of the Green's functions.

For the product of the self-energy and the Green's function, however, the application of the non-unitary transformations in Keldysh space fails. The problem lies in defining just what the terms in the two matrices $\boldsymbol{\Sigma G}$ and $\boldsymbol{G\Sigma}$ should be. The solution to this problem, however, is to compute directly the matrix elements in the Keldysh form. The problem arises in the convolution form of the integration over the internal time variable in these structures. The factor we are interested in has is problems in the time integration, and not in the spatial integration. Consider the product

$$C(t,t') = \int A(t,t'')B(t'',t') dt''. \tag{9.80}$$

The approach that has become used is due to Langreth [124]. Consider, for example, the case for $C^<$. For this case, t is on the outgoing (bottom) leg of the contour in Fig. 9.46, while t' is on the incoming (top) leg of the contour. There are two possibilities. Langreth first deformed the contour by taking the contour at time t and pulling it back to t_0, and then returning it out to *time t'*. This creates two time loop "spikes," and the result depends on which leg the time t'' resides. We can split the integral into two parts, one for t'' on each of the two new paths. We call these subcontours C_1 and C_2, respectively. We can then write (9.80) as

$$C^<(t,t') = \int_t^{-\infty} A(t,t'')B(t'',t')dt'' + \int_{-\infty}^t A(t,t'')B(t'',t')dt''. \quad (9.81)$$

Now, clearly in the first integral, which represents the C_1 contour, B is the less-than function, since its arguments always fold around the C_2 contour. The function A is the anti-time-ordered function, but since $t > t''$, the theta function inherent in this last function gives us just the less-than function. In the second integral, A is the time-ordered function, and under the same limits it yields just the greater-than function. Thus, we can write these two integrals as

$$C^<(t,t') = \int_{-\infty}^t \Theta(t-t'')(A^> - A^<)B^< dt''$$

$$= \int_{-\infty}^t A_r(t,t'')B^<(t'',t')dt''. \quad (9.82)$$

It is also possible to deform the contours not from t, but from t', which makes the A term always give the less-than function, and then the two paths contribute to yield the advanced function for B. These two possibilities must be combined to give the general result. Integrals such as this allow us to write the product formulas as

$$\begin{aligned} C^< &= A_r B^< + A^< B_a, \quad C_r = A_r B_r, \\ C^> &= A_r B^> + A^> B_a, \quad C_a = A_a B_a. \end{aligned} \quad (9.83)$$

Similar arguments can be used to form triple products. The key point of interest here is that the use of the Langreth theorems tells us that the matrices that form the products $\Sigma\mathbf{G}$ and $\mathbf{G}\Sigma$ should be written in the Keldysh form. Then, the proper matrix products are achieved in each case. One interesting point, however, is that the equation for the Keldysh Green's function G_K is actually two equations, which have been added for simplicity. In practice, however, the equations for the less-than and greater-than functions should be separated and treated as two disjoint equations of motion.

9.4.3 The Green–Kubo formula

The Kubo formula was developed in linear response to the applied fields, as the latter were represented by the vector potential. The use of the Kubo formula was a significant change from the normal treatment of the dominant streaming terms of the Boltzmann equation, or of the equivalent quantum transport equations, to the relaxation and/or scattering terms. With the Kubo formula, one concentrates on the relaxation processes through the correlation functions that describe the transport. Here, we talk about how the real-time Green's functions fit into the

9.4 Real-time Green's functions

Kubo formula, with the combination termed the Green–Kubo formula. Now, if we note that the quantum-mechanical current is described by

$$J = \frac{e\hbar}{2im*}\left[\Psi^+(\mathbf{r})\frac{\partial\Psi(\mathbf{r})}{\partial\mathbf{r}} - \frac{\partial\Psi^+(\mathbf{r})}{\partial\mathbf{r}}\Psi(\mathbf{r})\right], \quad (9.84)$$

it is not too difficult to then develop the Green's function form of this quantity to use in the Kubo formula. To begin with, we note that this latter equation can be written in terms of the Green's functions as

$$\langle j(\mathbf{r},t)\rangle = -\frac{e\hbar}{2m}\left(\frac{\partial}{\partial\mathbf{r}} - \frac{\partial}{\partial\mathbf{r}'}\right)G^<(\mathbf{r},t;\mathbf{r}',t')|_{\mathbf{r}',t'\to\mathbf{r},t^+} + i\frac{ne^2}{m\omega}F(\mathbf{r},t), \quad (9.85)$$

where the last term represents the displacement current and will be ignored (in this term, F is the applied field and will also appear in the current–current correlation function itself). The actual value of the current found from the Kubo formula is given by the current–current correlation function, as

$$\langle \mathbf{j}(\mathbf{r},t)\rangle = \frac{1}{\hbar}\int_0^t dt'\int d^3\mathbf{r}'\langle [j(\mathbf{r},t'),j(\mathbf{r}',t-t')]\rangle \cdot \mathbf{A}(\mathbf{r}',t'). \quad (9.86)$$

Here, $\mathbf{A}(\mathbf{r},t)$ within the integral is the vector potential describing the driving fields. Forming the two currents with the aid of (9.84) and (9.85) and using a different choice of variable for the internal integration, we then find the resultant current in terms of the real-time Green's functions as

$$\langle \mathbf{j}(\mathbf{r},t)\rangle = -\frac{ie^2\hbar}{4m^2}\lim_{\mathbf{r}',t'\to\mathbf{r},t^+}\left(\frac{\partial}{\partial\mathbf{r}} - \frac{\partial}{\partial\mathbf{r}'}\right)\int dt_s\int d^3\mathbf{s}\lim_{\mathbf{s}',t'_s\to\mathbf{s},t_s^+}\left(\frac{\partial}{\partial\mathbf{s}} - \frac{\partial}{\partial\mathbf{s}'}\right)$$
$$\times \left[G_r(\mathbf{r},t;\mathbf{s}',t'_s)G^<(\mathbf{s},t_s;\mathbf{r}',t') + G^<(\mathbf{r},t;\mathbf{s}',t'_s)G^<(\mathbf{s},t_s;\mathbf{r}',t')\right]\mathbf{A}(\mathbf{s},t_s). \quad (9.87)$$

In fact, the Green's function product should actually be a two-particle Green's function involving four field operators. However, this has been expanded into the lowest-order Green's functions. It should be remembered, though, that this is an approximation, and higher-order Green's function products may be needed to treat certain physical processes, as was also done to treat universal conductance fluctuations and weak localization. The result (9.87) can be Fourier-transformed to give the a.c. conductivity for a homogeneous system (required to take the spatial Fourier transforms), at least on the average scale of the response functions, giving (after a summation over spins)

$$\sigma(\mathbf{k},\omega) = -\frac{e^2\hbar}{m^2\omega}\int\frac{d^3\mathbf{k}'}{(2\pi)^3}\int\frac{d\omega'}{2\pi}\left(\mathbf{k}' + \frac{\mathbf{k}}{2}\right)\cdot\mathbf{k}$$
$$\times \left[G_r\left(\mathbf{k}' + \frac{\mathbf{k}}{2},\omega'\right)G^<\left(\mathbf{k}' - \frac{\mathbf{k}}{2},\omega' - \omega\right)\right.$$
$$\left. + G^<\left(\mathbf{k}' + \frac{\mathbf{k}}{2},\omega'\right)G_a\left(\mathbf{k}' - \frac{\mathbf{k}}{2},\omega' - \omega\right)\right]. \quad (9.88)$$

After a few further changes of variables, the conductivity can be found as

$$\sigma(\mathbf{k},\omega) = -\frac{e^2\hbar}{m^2\omega}\int\frac{d^3\mathbf{k}_1}{(2\pi)^3}\int\frac{d\omega'}{2\pi}\mathbf{k}_1\cdot\mathbf{k}$$
$$\times [G_r(\mathbf{k}_1,\omega')G^<(\mathbf{k}_1-\mathbf{k},\omega'-\omega)$$
$$+ G^<(\mathbf{k}_1,\omega')G_a(\mathbf{k}_1-\mathbf{k},\omega'-\omega)]. \qquad (9.89)$$

This particular form differs somewhat from that used earlier, in that the real-time functions appear here. It should also be noted that the higher-order two-particle Green's functions have not been included, so this formulation is that of the lowest-order Green's functions, assuming that Wick's theorem is perfectly valid. We will remark about this again later. The form of (9.89) has not been utilized very much in transport calculations based on the real-time Green's functions. Nevertheless, it is important to note that the conductance here is an integral (actually, a double integral) over the current–current correlation function.

It is now useful to rearrange the terms into those more normally found in the equilibrium, and zero-temperature, forms of the Green's functions. For this, we make the *ansatz*

$$G^<(\mathbf{k},\omega) = if(\omega)A(\mathbf{k},\omega) = f(\omega)[G_r(\mathbf{k},\omega) - G_a(\mathbf{k},\omega)]. \qquad (9.90)$$

This *ansatz* is known to be correct in the zero-temperature formulation, in the thermal-equilibrium formulation, and in the Airy-function formulation [125,126]. However, it is not known to be correct in all possible cases that may be found in the use of real-time Green's functions. There is no known proof of its correctness, and it is not in keeping with the various *ansätze* that have been proposed over the last decade or so for the decomposition of the less-than Green's function [127]. With this ansatz, however, we may rewrite (9.89) as

$$\sigma(\mathbf{k},\omega) = -\frac{e^2\hbar}{m^2\omega}\int\frac{d^3\mathbf{k}_1}{(2\pi)^3}\int\frac{d\omega'}{2\pi}\mathbf{k}_1\cdot\mathbf{k}[G_a(\mathbf{k}_1,\omega')G_a(\mathbf{k}_1-\mathbf{k},\omega'-\omega)f(\omega')$$
$$- G_r(\mathbf{k}_1,\omega')G_r(\mathbf{k}_1-\mathbf{k},\omega'-\omega)f(\omega'-\omega)$$
$$- G_r(\mathbf{k}_1,\omega')G_a(\mathbf{k}_1-\mathbf{k},\omega'-\omega)\{f(\omega')-f(\omega'-\omega)\}]. \qquad (9.91)$$

The first two products in the square brackets will cancel one another. This can be seen by changing the frequency variables as $\omega'' = \omega''' - \omega$, and then using the fact that $\sigma(\mathbf{k},\omega) = \sigma*(\mathbf{k},-\omega)$ and $G_r = G_a^*$. Thus, we are left with only the last term in the square brackets. For low frequencies (and we will go immediately to the long-time limit of $\omega = 0$), the distribution function can be expanded about ω', so that

$$\sigma(\mathbf{k},\omega) = -\frac{e^2\hbar}{m^2\omega}\int\frac{d^3\mathbf{k}_1}{(2\pi)^3}\int\frac{d\omega'}{2\pi}\mathbf{k}_1\cdot\mathbf{k}G_r(\mathbf{k}_1,\omega')G_a(\mathbf{k}_1-\mathbf{k},\omega'-\omega)\frac{\partial f(\omega')}{\partial\omega'}. \qquad (9.92)$$

Finally, at low frequencies and for homogeneous material, we arrive at the form

$$\sigma(\mathbf{k},\omega) = -\frac{e^2\hbar}{m^2}\int\frac{d^3\mathbf{k}_1}{(2\pi)^3}\int\frac{d\omega'}{2\pi}\mathbf{k}_1\cdot\mathbf{k}|G_r(\mathbf{k}_1,\omega')|^2\frac{\partial f(\omega')}{\partial\omega'}. \qquad (9.93)$$

In the case of very low temperatures, one arrives back at the form used extensively in previous chapters. That is, we replace the derivative of the distribution function with the negative of a delta function at the Fermi energy. The sum over the momentum counts the number of states that contribute to the conductivity and results in the density at the Fermi energy at low temperature. In mesoscopic systems, where only a single transverse state may contribute (in a quantum wire, for example), the Landauer formula can easily be recovered when one recognizes that $|G_r(\mathbf{k}_1,\omega')|^2$ represents the transmission of a particular mode. Even if there is no transverse variation, the integration over the longitudinal component of the wavevector will produce the difference in the Fermi energies at the two ends of the samples (as in, for example, the quantum Hall effect). The reduction of the less-than function has allowed a separation of the density–density correlation function from the current–current correlation function. This latter is represented by the polarization that appears as the magnitude squared of the retarded Green's function.

The approach (9.93) has been extensively utilized by the Purdue group to model mesoscopic systems with the equivalent Landauer formula for nonequilibrium Green's functions. For mesoscopic waveguides in the linear response regime, even with dissipation present, they have shown that this form can be extended to the use of a Wigner function, which can then be used to define a local thermodynamic potential, and that reasonable results are obtained so long as these potentials are defined over a volume comparable in size to the thermal de Broglie wavelength $\lambda_D = \sqrt{\pi\hbar^2/mk_BT}$ [128,129]. They have been particularly successful in probing inelastic tunneling in resonant-tunneling diodes through the emission of optical phonons. In general, however, the expression (9.93) is an approximation, in that the magnitude term in the retarded Green's function is really a lowest-order representation of the actual polarization that appears in the conductivity bubble. This was addressed earlier quite extensively, and the extended version becomes

$$\sigma(\mathbf{k},\omega) = -\frac{e^2\hbar}{m^2}\int\frac{d^3\mathbf{k}_1}{(2\pi)^3}\int\frac{d\omega'}{2\pi}\frac{\partial f(\omega')}{\partial\omega'}\Pi(\mathbf{k}_1,\omega'), \qquad (9.94)$$

where the polarization satisfies the (reduced) Bethe–Salpeter equation

$$\Pi(\mathbf{k},\omega) = G_r(\mathbf{k},\omega)G_a(\mathbf{k},\omega)\bigg\{\mathbf{k}_1\cdot\mathbf{k}\delta(\mathbf{k}_1-\mathbf{k})$$

$$+ \int\frac{d^3\mathbf{k}_1}{(2\pi)^3}\frac{\mathbf{k}_1\cdot\mathbf{k}}{k_1^2}T(\mathbf{k}_1-\mathbf{k})\Pi(\mathbf{k}_1,\omega)\bigg\}. \qquad (9.95)$$

In this last expression, we have summed the conductivity over the Fourier variables in position to get the spatially averaged conductivity (equivalent to $\mathbf{r} = 0$, but with the origin at any point in the sample), and made a change of variables on the momentum variables for convenience. The polarization is the general product of the retarded and advanced functions that was dealt with earlier, at least on a formal basis. For isotropic scattering processes, the second term actually vanishes, and the earlier result is the formally exact result.

While these latter results appear to be quite simple in form, it is important to point out that the form of the distribution function that appears in the equations still must be determined by the balance between the driving forces and the dissipative forces: in essence, the less-than Green's function must be determined in the nonequilibrium system, just as the distribution function must be determined for the use of the Boltzmann treatment introduced in Chapter 2.

9.4.4 Conductance in mesoscopic devices

Now, we want to follow a slightly different approach to the conductance through a mesoscopic system, still in the absence of many-body interactions and scattering. Here, we embed the active region, the mesoscopic device, between two contact regions [130]. The general idea is that there will be some coupling between the active region and the contacts, which can be described by a so-called self-energy Γ. In some sense, this approach follows the prototypical tunneling device formulated in the same manner [131,132]. Here, we take a set of fermion operators in the contacts, which are given by $c_{k\alpha}, c_{k\alpha}^+$, where α takes the values L and R for the left and right contact, respectively, and d_n, d_n^+ in the "active" region, which correspond to excitation (and de-excitation) of a particular energy level within the active region. For this approach to be valid, the basic "active region" must be closed in the quantum-mechanical sense. That is, the connections to the environment must be tunneling barriers which preserve the totality of the states within the active region. In fact, this approach can go beyond this limit, except that the so-called "open" quantum system has far fewer resonant levels, and the current carrying states do not have to correspond to eigenvalues of the active region Hamiltonian. Now, the overall Hamiltonian for this system can be written as

$$H = \sum_{\substack{k \\ \alpha \subset L,R}} \varepsilon_{k\alpha} c_{k\alpha}^+ c_{k\alpha} + H_{\text{int}}(\{d_n^+\};\{d_n\}) + \sum_{\substack{k \\ \alpha \subset L,R}} \left(V_{k\alpha,n} c_{k\alpha}^+ d_n + h.c.\right). \tag{9.96}$$

Here, we note that the first term on the right represents the kinetic energy in the two contact regions while the second term represents the kinetic energy in the active central region. The final term represents coupling between the contacts and the active region.

With the above Hamiltonian, it is now possible to compute the current flow through the device and the resulting conductance by direct application of our Green's functions techniques. Hence, the current density may be written as

9.4 Real-time Green's functions

$$J = \frac{ie}{\hbar} \sum_{\substack{k,n \\ \alpha=L}} \left(V_{k\alpha,n} \langle c_{k\alpha}^+ d_n \rangle - V_{k\alpha,n}^+ \langle d_n^+ c_{k\alpha} \rangle \right)$$

$$= \frac{e}{\hbar} \sum_{\substack{k,n \\ \alpha=L}} \int_{-\infty}^{\infty} \frac{d\omega}{2\pi} \left[V_{k\alpha,n} G_{n,k\alpha}^<(\omega) - V_{k\alpha,n}^+ G_{k\alpha,n}^<(\omega) \right]. \tag{9.97}$$

It is clear that this approach does not use the Kubo formula, but still it has become quite popular in the mesoscopic community. The two Green's functions that appear in the last line of (9.97) can be defined by

$$G_{k\alpha,n}^<(\omega) = \sum_m V_{k\alpha,m} \left[g_{k\alpha,k\alpha}^t(\omega) G_{m,n}^<(\omega) - g_{k\alpha,k\alpha}^<(\omega) G_{m,n}^{\bar{t}}(\omega) \right],$$

$$G_{n,k\alpha}^<(\omega) = \sum_m V_{k\alpha,m}^+ \left[g_{k\alpha,k\alpha}^<(\omega) G_{n,m}^t(\omega) - g_{k\alpha,k\alpha}^{\bar{t}}(\omega) G_{n,m}^<(\omega) \right]. \tag{9.98}$$

Now, the small g's are the equilibrium values of the various Green's functions. Each of these two Green's functions connects a transition from a contact to the active region, or vice versa. For example, the two correlation functions on the right-hand side can be written as

$$G^<(n,m;\omega) = i \langle d_m^+ d_n \rangle,$$

$$G^>(n,m;\omega) = -i \langle d_n d_m^+ \rangle. \tag{9.99}$$

Now, we need to write down a number of relationships between the various Green's functions, nearly all of which have been discussed previously. We rewrite these here, in order to make this section a little more coherent. These relationships become

$$G^>(\omega) + G^<(\omega) = G^t(\omega) + G^{\bar{t}}(\omega),$$

$$G^>(\omega) - G^<(\omega) = G^r(\omega) - G^a(\omega),$$

$$g_{kL,kL}^< = 2\pi i f_L(\omega) \delta(\omega - E_{kL}/\hbar),$$

$$g_{kL,kL}^> = -2\pi i [1 - f_L(\omega)] \delta(\omega - E_{kL}/\hbar). \tag{9.100}$$

With these relationships, the current (9.97) can be rewritten as

$$J = \frac{e}{\hbar} \sum_{\substack{k,n \\ \alpha=L}} \int_{-\infty}^{\infty} \frac{d\omega}{2\pi} \left[V_{k\alpha,n} G_{n,k\alpha}^<(\omega) - V_{k\alpha,n}^+ G_{k\alpha,n}^<(\omega) \right]$$

$$= \frac{e}{\hbar} \sum_{k,n} \int_{-\infty}^{\infty} \frac{d\omega}{2\pi} \left\{ V_{k\alpha,n} \sum_m V_{k\alpha,m}^+ \left[g_{kL,kL}^< G_{n,m}^t - g_{kL,kL}^{\bar{t}} G_{n,m}^< \right] \right.$$

$$\left. - V_{k\alpha,n}^+ \sum_m V_{k\alpha,m} \left[g_{kL,kL}^t G_{m,n}^< - g_{kL,kL}^< G_{m,n}^{\bar{t}} \right] \right\}. \tag{9.101}$$

This is our basic result, but with a little further manipulation, it can be put into a much more meaningful form. First, we replace the sum over wavevector (in the contacts) with a sum over the energy, which means introducing the density of states, as

$$J = \frac{ie}{\hbar} \sum_{k,n,m} V_{k\alpha,n} V^+_{k\alpha,m} \left[f(E_{kL}) A(\mathbf{k}, E_{kL}/\hbar) + G^<_{n,m} \right]$$

$$= \frac{ie}{\hbar} \sum_{n,m} \int dE \rho(E) V_{k\alpha,n} V^+_{k\alpha,m} \left[f(E_{kL}) \left(G^r_{n,m} - G^a_{n,m} \right) + G^<_{n,m} \right]. \quad (9.102)$$

Now, we introduce the couplings via the self-energies, which are defined as

$$\Gamma_L = \sum_{n,m} \Gamma^L_{n,m} = 2\pi \sum_{n,m} \rho(E) V_{kL,n} V^+_{kL,m} \quad (9.103)$$

for the left lead, and similarly for the right lead with obvious change of subscripts. With this definition, the left-derived current may be written as

$$J = \frac{ie}{2\pi\hbar} \int dE Tr \left\{ \Gamma^L_{nm} f(E_L) \left(G^r_{n,m} - G^a_{n,m} \right) - \Gamma^L_{nm} G^<_{n,m} \right\}. \quad (9.104)$$

A similar result arises naturally for the right-derived current. Then, the average current for both leads together may be written as

$$J = \frac{ie}{4\pi\hbar} \int dE Tr \left\{ \left[\Gamma^L_{nm} f(E_L) - \Gamma^R_{nm} f(E_R) \right] \left(G^r_{n,m} - G^a_{n,m} \right) - \left(\Gamma^L_{nm} - \Gamma^R_{nm} \right) G^<_{n,m} \right\}. \quad (9.105)$$

Now, in the interacting region, we may assume that the transport is largely ballistic, so that we can simplify the correlation functions in this region as

$$G^< = i f_L \text{Im}(A_L) + f_R \text{Im}(A_R) = i f_L G^r \Gamma^L G^a + i f_R G^r \Gamma^R G^a,$$

$$G^r - G^a = -2i \text{Im}(A_L + A_R) = 2 G^r (\Gamma^R + \Gamma^L) G^a. \quad (9.106)$$

Finally, we can achieve the final result for the current through the system as

$$J = \frac{ie}{4\pi\hbar} \int dE Tr \left\{ \left[\Gamma^L_{nm} f(E_L) - \Gamma^R_{nm} f(E_R) \right] \left(G^r_{n,m} - G^a_{n,m} \right) - \left(\Gamma^L_{nm} - \Gamma^R_{nm} \right) G^<_{n,m} \right\}$$

$$= \frac{e}{2\pi\hbar} \int dE Tr \Big\{ \Gamma^L_{nm} f(E_L) G^r_{n,m} (\Gamma^L_{nm} + \Gamma^R_{nm}) G^a_{n,m}$$

$$- \Gamma^R_{nm} f(E_R) G^r_{n,m} (\Gamma^L_{nm} + \Gamma^R_{nm}) G^a_{n,m}$$

$$- (\Gamma^L_{nm} - \Gamma^R_{nm}) \left[f(E_L) G^r_{n,m} \Gamma^L_{nm} G^a_{n,m} + f(E_R) G^r_{n,m} \Gamma^R_{nm} G^a_{n,m} \right] \Big\}$$

$$= \frac{e}{2\pi\hbar} \int dE [f(E_L) - f(E_R)] Tr \left\{ \Gamma^L_{nm} G^r_{n,m} \Gamma^R_{nm} G^a_{n,m} \right\}$$

$$\equiv \frac{e}{2\pi\hbar} \int dE [f(E_L) - f(E_R)] Tr \{tt^+\}. \quad (9.107)$$

Now, as mentioned, this last contribution results from ballistic transport as the only contribution to the self-energy comes from coupling between the contacts and the active region, which is assumed to be tunneling in nature. If there were proper scattering processes considered, then the approximations for the correlation functions would fail. Thus, this approach is limited to really ballistic devices, but cannot be applied to devices in which there is real dissipation occurring.

There have been many groups who have pursued this approach, or an earlier variant of it, for studying various mesoscopic systems. Among the earliest were the studies of single-electron tunneling into and out of a quantum dot [131]. But, this approach was also assumed in the computation of the earliest studies of transport through a molecule attached to a pair of metallic leads, in a Green's function formalism [133]. Here, the coupling between the molecule and the leads is treated as a rather mystical tunneling coefficient [134,135]. Although most of these approaches use Green's functions, they can as easily be formulated with the scattering matrix approach of Section 9.3 [136,137,138]. The key aspect of any of these approaches is that one must know the atomic structure in the molecule as well as in the metallic leads. In the treatment of Section 9.3, the atomic structure was found by assuming normal electron bands, as all the transport was in semiconductors. In the approach of these latter papers, however, the molecules are characterized by localized wave functions while the metallic leads have itinerant electron character. As a result, it is quite difficult to treat both the molecules and the leads with equal accuracy. The common approach is to use something like a tight-binding basis, but this really assumes localized orbitals, and it is difficult, for example, to accurately recover the work function of the metallic leads [137,139]. If the work function cannot be accurately computed by the atomic structure calculation, then there is no chance that the real potential barrier between the lead and the molecule will be determined in a realistic manner. Hence, the coupling between the two is no more than an adjustable parameter, and the fact that the current is still not computed correctly, even with this adjustable parameter, is disappointing. Nevertheless, it is imperative in these latter approaches to have a viable method of computing the atomic structure throughout the mesoscopic system. This becomes true even with the realm of nanowires. We turn to this in the next section, with one of the more successful approaches.

9.4.5 Nano-Electronic MOdeling (NEMO)

We now want to turn our attention to the problem of simulation of very small semiconductor (or other) structures. In the case that the transverse dimensions are small, it may be that the band structure appropriate to large volumes is no longer appropriate. Typical cases of this are very small resonant tunneling diodes, small diameter nanowires, carbon nanotubes, and molecules stretched between metallic (or semiconducting) leads. Here, the analytic band, effective mass approach used previously (such as with the scattering matrix, Lippmann–Schwinger equation approach to obtain the transmission) is no longer applicable. It is not the calculation of transport that is no longer applicable, but the energy structure of the system under consideration that is likely to be modified by the small transverse dimensions. There have been several methodologies developed to study this, most of which use localized atomic orbitals (or basis functions) to build up the energy structure in the medium. These include the atomic basis mentioned above [135], but more commonly accessible codes are available.

One of the earliest approaches which utilized the so-called tight-binding basis set of atomic functions was that of the Rome group [140]. This effort utilized the so-called exact exchange energy correction to the local density approximation to correct the band structure, and has been used to study semiconductor nanostructures [141], band mixing in these structures [142], molecules stretched between metallic conductors [143], and carbon nanotubes [144]. A somewhat more famous energy structure program is SIESTA, which also uses localized atomic orbitals (but numerically simulated ones) [145,146]. This program has been extended to computation of the conductance via Green's functions in the new package TranSIESTA [147,148]. These are all quite good approaches in this genre of simulation tools. However, to illustrate the approach, we will use another simulation tool, that of Nano-Electronic MOdeling (NEMO).

There is another important difference between NEMO and these other codes. The NEMO atomic structure is found from semi-empirical tight-binding structure [149]. The tight-binding orbitals are characterized by overlap integrals (matrix elements in the Hamiltonian) that are adjusted to reach a best fit to experimental data. On the other hand, SIESTA uses parameterized orbitals and these parameters are adjusted by a full density functional approach [150,151]. In the latter, the Hamiltonian uses nonlocal pseudopotentials and a local density approximation for exchange and correlation. Then, the overall computation is iterated self-consistently to produce the energies. A similar approach is used in the FIREBALL code [152,153], which we have used in our molecular conductor simulations mentioned above [137,138,154]. These latter approaches are more physically based, but have the well-known problem of producing a bandgap that is usually considerably smaller than seen experimentally. The empirical approach starts with at least getting the principal gaps correct, but the parameters are less physically based.

Some of the earliest work on the simulation of quantum structures was carried out at Purdue University under the direction of Supriyo Datta [155,156]. This was among the earliest Green's function work to incorporate inelastic processes in a realistic manner. This basis was quickly expanded when several of the former Purdue students wound up at Texas Instruments developing the NEMO simulation code. This approach depends upon detailed tight-binding atomic

9.4 Real-time Green's functions

structure and Green's functions for evaluation of the transport, and was developed and first applied for resonant tunneling diodes [157,158]. Here, it was very important to assure that the tight-binding band structures gave accepted values for the effective mass in large structures [159], that it gave acceptable values for the known warpage in the valence bands [160], and yielded a set of parameters that could be utilized between various semiconductors [161]. A byproduct of this careful development was the meeting of additional requirements in the presence of electromagnetic fields, which needed gauge invariance [162].

In any such code, it is important that it be applicable to realistic devices; e.g., those devices which are likely to be experimentally realizable. Beyond this, it must be extendable to higher dimensions and to other device structures. It must handle spatially varying potentials, the inclusion of detailed scattering processes, and must be able to handle some definition of the contacts or boundary conditions that fit to experimentally realizable devices. Of course, most of the codes mentioned above will satisfy these constraints when the self-energy is included within the Green's function formulation. Here, we describe primarily the NEMO code as an example (but a particularly useful example) of this approach.

We can write the general Hamiltonian for the device under consideration in the following form

$$H = H_0^D + H_0^L + H_0^R + H_0^{LD} + H_0^{RD}$$
$$+ H_{POP} + H_{ac} + H_{ir} + H_{al} + H_{id}. \qquad (9.108)$$

Here, the first five terms are for the central device (the quantum well and barriers in a resonant tunneling diode), the left contact region, the right contact region, and the coupling between the device and the left contact and right contact, respectively. The last five terms are for scattering via polar optical phonons, acoustic phonons, interface roughness, alloy disorder, and ionized dopants, respectively. These last five terms will be treated via perturbation theory as a self-energy correction to the total Hamiltonian. The two contact terms will be treated exactly, but will also lead to a self-energy correction to the reduced Hamiltonian for the central device itself. This latter self-energy term will represent the injection (and ejection) of particles to (from) the device region. The device itself is broken into slices, just as in Sections 9.3 and 3.5.1 for the scattering matrix approach. This means that the total Green's function will be built up just as the recursive Green's function was built in Section 3.5.2. Here, each individual layer in the device contains a cation layer and an anion layer. The position of the cation is designated by a vector which has a layer part and a transverse part indicating the cation location within the layer, as

$$\mathbf{R}^L = La\mathbf{a}_z + \mathbf{R}_t^L. \qquad (9.109)$$

The corresponding anion location is displaced by $(a/2)(111)$, where a here, and in (9.109), is the lattice constant. The localized anion and cation orbital wave functions are indicated by their locations as

$$|a, L, \mathbf{R}_t^L\rangle,$$
$$|c, L, \mathbf{R}_t^L\rangle. \tag{9.110}$$

The set $\{a\}$ and $\{c\}$ are the complete set of tight-binding orbitals at each anion and cation site, respectively. Thus for a set sp^3s^*, for example, each set in (9.110) contains five separate orbitals which differ by the index a or c. Most of the simulations, at least for the resonant tunneling diodes have used just this set, but some have incorporated five d levels as well [163,164].

From the tight-binding orbitals, one can now construct plane wave states by properly combining these orbitals. This leads to

$$|c, L, \mathbf{k}\rangle = \frac{1}{\sqrt{N}} \sum_{\mathbf{R}_t^L} e^{i\mathbf{k}\cdot\mathbf{R}_t^L} |c, L, \mathbf{R}_t^L\rangle,$$

$$|a, L, \mathbf{k}\rangle = \frac{1}{\sqrt{N}} \sum_{\mathbf{R}_t^L} e^{i\mathbf{k}\cdot(\mathbf{R}_t^L + \mathbf{v})} |a, L, \mathbf{R}_t^L\rangle, \mathbf{v} = \frac{a}{2}(111). \tag{9.111}$$

In this formulation, $\mathbf{k}$ is a purely transverse vector and lies witin the particular slice. We can now use these plane wave states to create the appropriate field operator

$$\hat{\psi}(\mathbf{r}) = \sum_{\mathbf{k},L} \left[\sum_c \langle \mathbf{r}|c, L, \mathbf{k}\rangle \xi_{c,L,\mathbf{k}} + \sum_a \langle \mathbf{r}|a, L, \mathbf{k}\rangle \xi_{a,L,\mathbf{k}} \right]. \tag{9.112}$$

In this equation, ξ is the annihilation operator for the particular plane wave state. A particularly important point here is that the Schrödinger discretization is *only* in the z-direction (along the length of the device). The transverse dimensions, whether only a second dimension or both x and y transverse dimensions, is a quasi-periodic array for which the transverse band structure is given by the tight-binding orbital overlap integrals. This is a traditional layer approach to systems in which the z-direction has broken symmetry (broken periodicity), such as surfaces or interfaces. Now, these overlap integrals, both in the slice and between slices, can be written as

$$\langle \alpha, L, \mathbf{k}|H_0|\alpha', L', \mathbf{k}\rangle = D_{\alpha,\alpha',L}(\mathbf{k})\delta_{LL'} - T_{\alpha,L;\alpha',L'}(\mathbf{k})\delta_{L',L\pm j\neq 0}. \tag{9.113}$$

The second term represents the layer-to-layer coupling between the orbitals, and each element of T is an $m \times m$ block, where it m is the number of atomic orbitals used. The span j tells us whether we are limiting ourselves to nearest neighbors ($j = 1$) or are going to more far-reaching interactions. The first term represents the interactions within a layer, and the diagonal $m \times m$ blocks contain, on the diagonal the orbital eigenenergies and the local potential, while the off-diagonal elements in each block represent the anion–cation interactions. The presence of off-diagonal blocks would be indicative of beyond nearest-neighbor interactions.

We can illustrate the nature of these various matrices by considering the simplest possible case first. In this case, we assume a single tight-binding orbital per site, with the cation and anion orbitals lumped into a single equivalent "band" described by an effective mass at each site (or on each slice). Then, the discretized Schrödinger equation becomes

$$H_0 = -\frac{\hbar^2}{2}\frac{d}{dz}\left(\frac{1}{m^*(z)}\frac{d}{dz}\right) + V_k(z) + \frac{\hbar^2 k^2}{2m_L}, \tag{9.114}$$

$$V_k(z) = V(z) + \frac{\hbar^2 k^2}{2m_L}\left(\frac{m_L}{m^*(z)} - 1\right), \tag{9.115}$$

$$T_{L,L'} = \frac{\hbar^2}{(m_L + m_{L'})a^2}, \tag{9.116}$$

$$D_L = \frac{\hbar^2}{2a^2}\left(\frac{1}{m^+} + \frac{1}{m^-}\right), \quad m^\pm = \frac{1}{2}(m_{L\pm 1} + m_L). \tag{9.117}$$

The only difference here from a normal discretization of the Schrödinger equation is the allowance for a non-uniform effective mass through the structure.

Now, when we do the sp^3s^* orbitals, each anion and cation has 5 orbitals, the four normal orbitals in the outer shell and the excited s-orbital. Hence, the Hamiltonian is block diagonal with 5×5 blocks. Each block is characterized by the cation (anion) eigenenergies on the diagonal and interactions among the cation (anion) orbitals, such as second-neighbor interactions or the spin-orbit interactions on the off-diagonals. The next blocks from the diagonal describe the interactions between the cation and the anion orbitals on nearest neighbors, and contain the Bloch sums over the lattice vectors in the transverse layer, which arise from the plane waves introduced above. The T-matrix is not block diagonal, but has 5×5 blocks describing interactions between cation (anion) orbitals in one layer with the anion (cation) orbitals in the adjacent layer.

We can understand the general approach by referring to Fig. 9.47, where we illustrate the potential profile for a resonant tunneling diode, with bias applied. The structure consists of two large reservoirs, left and right, and a short device. The left reservoir is described by layers $-\infty$ to 0. The central device is

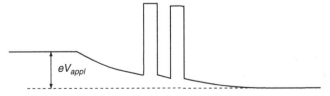

Fig. 9.47 A prototypical resonant tunneling diode. The structure is partitioned into two large reservoirs, left and right, and a short device. One should attempt to have the "device" include the entire quantum region, if possible.

characterized by layers 1 to N, and the right reservoir is characterized by layers $N + 1$ to ∞. Several self-energy terms are depicted in the figure. The self-energy Σ^{RB} accounts for loss from the central device to the contacts. The term $\Sigma^<$ accounts for injection from the contacts to the device. We can now write the less-than and greater-than correlation functions as

$$G^<_{\alpha,L;\alpha',L'}(\mathbf{k};t,t') = \frac{i}{\hbar}\left\langle \xi^+_{\alpha',L';\mathbf{k}}(t')\xi_{\alpha,L;\mathbf{k}}(t)\right\rangle,$$

$$G^>_{\alpha,L;\alpha',L'}(\mathbf{k};t,t') = -\frac{i}{\hbar}\left\langle \xi_{\alpha,L;\mathbf{k}}(t)\xi^+_{\alpha',L';\mathbf{k}}(t')\right\rangle. \quad (9.118)$$

We use the orbital notation, although α, α' could as easily denote transverse sites. Similarly, we can write the retarded and advanced Green's functions as

$$G^r_{\alpha,L;\alpha',L'}(\mathbf{k};t,t') = \vartheta(t-t')\left[G^>_{\alpha,L;\alpha',L'}(\mathbf{k};t,t') - G^<_{\alpha,L;\alpha',L'}(\mathbf{k};t,t')\right]$$

$$= -i\vartheta(t-t')A_{\alpha,L;\alpha',L'}(\mathbf{k};t,t'),$$

$$G^a_{\alpha,L;\alpha',L'}(\mathbf{k};t,t') = \vartheta(t'-t)\left[G^<_{\alpha,L;\alpha',L'}(\mathbf{k};t,t') - G^>_{\alpha,L;\alpha',L'}(\mathbf{k};t,t')\right]$$

$$= i\vartheta(t-t')A_{\alpha,L;\alpha',L'}(\mathbf{k};t,t'), \quad (9.119)$$

where the spectral density is given as

$$A_{\alpha,L;\alpha',L'}(\mathbf{k};t,t') = i\left[G^>_{\alpha,L;\alpha',L'}(\mathbf{k};t,t') - G^<_{\alpha,L;\alpha',L'}(\mathbf{k};t,t')\right]$$

$$= i\left[G^r_{\alpha,L;\alpha',L'}(\mathbf{k};t,t') - G^a_{\alpha,L;\alpha',L'}(\mathbf{k};t,t')\right]. \quad (9.120)$$

As we are concerned mainly with the long-time limit, or steady-state response, we can Fourier transform on the time difference and work with these transformed Green's functions

$$G^<_{\alpha,L;\alpha',L'}(\mathbf{k};\omega) = \int d(t-t')e^{i\omega(t-t')}G^<_{\alpha,L;\alpha',L'}(\mathbf{k};t-t'),$$

$$G^>_{\alpha,L;\alpha',L'}(\mathbf{k};\omega) = \int d(t-t')e^{i\omega(t-t')}G^>_{\alpha,L;\alpha',L'}(\mathbf{k};t-t'). \quad (9.121)$$

Some general relations that arise are the normal ones; that the advanced Green's function is the adjoint of the retarded Green's function, and that the spectral density is a real function. Some other general relationships are that

$$G^r(\mathbf{k};\omega) = \int \frac{d\omega'}{2\pi}\frac{A(\mathbf{k};\omega')}{\omega-\omega'} - \frac{i}{2}A(\mathbf{k};\omega), \quad (9.122)$$

where the first term is the principal part of the integral. In a similar manner, we can define the imaginary part of the retarded self energy as

$$\Gamma(\mathbf{k};\omega) = i\left[\Sigma^>(\mathbf{k};\omega) - \Sigma^<(\mathbf{k};\omega)\right]$$

$$= i\left[\Sigma^r(\mathbf{k};\omega) - \Sigma^a(\mathbf{k};\omega)\right], \quad (9.123)$$

and

$$\Sigma^r(\mathbf{k};\omega) = P\int \frac{d\omega'}{2\pi}\frac{\Gamma(\mathbf{k};\omega')}{\omega-\omega'} - \frac{i}{2}\Gamma(\mathbf{k};\omega). \tag{9.124}$$

Once the less-than correlation function is determined, then the density at slice L can be determined from

$$n_L = \frac{2i\hbar}{Sa}\sum_{\mathbf{k},\alpha}\int\frac{d\omega}{2\pi}G^<_{\alpha,L;\alpha L}(\mathbf{k},\omega), \tag{9.125}$$

where S is the cross-sectional area, and the current density through this slice is given by

$$J_L = \frac{2e}{S}\sum_{\mathbf{k},\alpha,\alpha'}\int\frac{d\omega}{2\pi}\sum_{L_1<L}\sum_{L_2>L}\left[T_{\alpha L;\alpha' L_2}G^<_{\alpha',L_2;\alpha L_1}(\mathbf{k},\omega) - T_{\alpha,L_2;\alpha' L_1}G^<_{\alpha',L_1;\alpha L_2}(\mathbf{k},\omega)\right]. \tag{9.126}$$

As mentioned above, the device is partitioned into left and right contact regions, which extend up to the active device region. Even with spatially varying potentials, these regions can be treated as reservoirs for the active device region. This approach allows for the injection of carriers from mixed, non-asymptotic states in the emitter (or collector), which is considered as the left (or right) contact region. This will also allow the inclusion of scattering in the contact regions. This is most easily done by inclusion of an imaginary potential in the Hamiltonian for the contact regions, a necessary step to account for bound states which may exist in these regions. The left and right contact regions are then collapsed into the appropriate self-energies, introduced above, which account for these regions on the active device regions.

To illustrate the approach with the minimum of extraneous notation, we will use the single-band model introduced in (9.114)–(9.117). Hence, we will keep only the layer indices, and change the notation to $L = i$, $L' = j$. For $i,j \subset \{1,\ldots,N\}$, we can write down Dyson's equation as

$$G^<_{i,j} = g^<_{i,j} + g^r_{i,1}(-T_{1,0})G^<_{0,j} + g^<_{i,1}(-T_{1,0})G^a_{0,j}$$
$$+g^r_{i,N}(-T_{N,N+1})G^<_{N+1,j} + g^<_{i,N}(-T_{N,N+1})G^a_{N+1,j}, \tag{9.127}$$

$$G^>_{i,j} = g^>_{i,j} + g^r_{i,1}(-T_{1,0})G^>_{0,j} + g^>_{i,1}(-T_{1,0})G^a_{0,j}$$
$$+g^r_{i,N}(-T_{N,N+1})G^>_{N+1,j} + g^>_{i,N}(-T_{N,N+1})G^a_{N+1,j}, \tag{9.128}$$

where the lower case Green's functions are the equilibrium values obtained with the T-matrices set to zero. The first line in each expression represents the coupling to the left contact region, while the second line represents the coupling to the right contact region. Each of these Green's functions still remain a function of both the transverse momentum $\mathbf{k}$ and the energy E. The Green's functions which cross the boundaries between the active device and the contacts also satisfy their own Dyson's equations through

$$G^<_{0,j} = g^r_{0,0}(-T_{0,1})G^<_{1,j} + g^<_{0,0}(-T_{0,1})G^a_{1,j}, \qquad (9.129)$$

$$G^a_{0,j} = g^a_{0,0}(-T_{0,1})G^a_{1,j}, \qquad (9.130)$$

$$G^>_{0,j} = g^r_{0,0}(-T_{0,1})G^>_{1,j} + g^>_{0,0}(-T_{0,1})G^a_{1,j}, \qquad (9.131)$$

$$G^<_{N+1,j} = g^r_{N+1,N+1}(-T_{N+1,N})G^<_{N,j} + g^<_{N+1,N+1}(-T_{N+1,N})G^a_{N,j}, \qquad (9.132)$$

$$G^a_{N+1,j} = g^a_{N+1,N+1}(-T_{N+1,N})G^a_{N,j}, \qquad (9.133)$$

$$G^>_{N+1,j} = g^r_{N+1,N+1}(-T_{N+1,N})G^>_{N,j} + g^>_{N+1,N+1}(-T_{N+1,N})G^a_{N,j}. \qquad (9.134)$$

Substituting these last equations into the set (9.127)–(9.128) leads to the results, after some algebra,

$$\begin{aligned}G^<_{i,j} &= g^r_{i,1}(T_{1,0})g^r_{0,0}(T_{0,1})G^<_{1,j} + g^r_{i,1}(T_{1,0})g^<_{0,0}(T_{0,1})G^a_{1,j}\\ &\quad + g^<_{i,1}(T_{1,0})g^a_{0,0}(T_{0,1})G^a_{1,j} + g^<_{i,N}(T_{N,N+1})g^a_{N+1,N+1}(T_{N+1,N})G^a_{N,j}\\ &\quad + g^r_{i,N}(T_{N,N+1})g^r_{N+1,N+1}(T_{N+1,N})G^<_{N,j}\\ &\quad + g^r_{i,N}(T_{N,N+1})g^<_{N+1,N+1}(T_{N+1,N})G^a_{N,j}. \end{aligned} \qquad (9.135)$$

We now introduce the self-energies as

$$\Sigma^{rB}_{1,1} = T_{1,0}g^r_{0,0}T_{0,1}, \quad \Sigma^{aB}_{1,1} = T_{1,0}g^a_{0,0}T_{0,1},$$

$$\Sigma^{<B}_{1,1} = T_{1,0}g^<_{0,0}T_{0,1}, \quad \Sigma^{<B}_{N,N} = T_{N,N+1}g^<_{N+1,N+1}T_{N+1,N},$$

$$\Sigma^{aB}_{N,N} = T_{N,N+1}g^a_{N+1,N+1}T_{N+1,N}, \quad \Sigma^{rB}_{N,N} = T_{N,N+1}g^r_{N+1,N+1}T_{N+1,N}. \qquad (9.136)$$

With these quantities, we can rewrite (9.135) as

$$\begin{aligned}G^<_{i,j} &= g^r_{i,1}\Sigma^{rB}_{1,1}G^<_{1,j} + g^r_{i,1}\Sigma^{<B}_{1,1}G^a_{1,j} + g^<_{i,1}\Sigma^{aB}_{1,1}G^a_{1,j}\\ &\quad + g^<_{i,N}\Sigma^{aB}_{N,N}G^a_{N,j} + g^r_{i,N}\Sigma^{rB}_{N,N}G^<_{N,j} + g^r_{i,N}\Sigma^{<B}_{N,N}G^a_{N,j}. \end{aligned} \qquad (9.137)$$

By definition, the actual contact regions are considered to be in equilibrium (we note that this is not the general case, and the Landauer formula has a correction to account for the nonequilibrium state adjacent to the actual device region, as discussed in Section 3.3). Hence, we can define the Green's functions for this area as

$$g^<_{0,0} = -f_e\left(g^r_{0,0} - g^a_{0,0}\right),$$

$$g^<_{N+1,N+1} = -f_c\left(g^r_{N+1,N+1} - g^a_{N+1,N+1}\right). \qquad (9.138)$$

Here, the subscripts "e" and "c" refer to the emitter (left contact) and the collector (right contact). We can also introduce the quantities

$$\Gamma^B_{1,1} = i\left(\Sigma^{rB}_{1,1} - \Sigma^{aB}_{1,1}\right)$$
$$\Gamma^B_{N,N} = i\left(\Sigma^{rB}_{N,N} - \Sigma^{aB}_{N,N}\right). \tag{9.139}$$

so that

$$\Sigma^{<B}_{1,1} = if_e \Gamma^B_{1,1}$$
$$\Sigma^{<B}_{N,N} = if_c \Gamma^B_{N,N}. \tag{9.140}$$

In the nearest-neighbor tight-binding model, all the boundary self-energy terms are zero except for the two cases $i = j = 1, i = j = N$. That is, the boundary self-energies only couple to the slices in the device that are adjacent to the contact regions. To get to the equation of motion for the Green's functions at the various slices, we operate with the bare (equilibrium) Green's function differential operator as

$$\left(\hbar\omega - H_0^D\right)G^<_{i,j} = \left(\hbar\omega - H_0^D\right)\left\{g^r_{i,1}\Sigma^{rB}_{1,1}G^<_{1,j} + g^r_{i,1}\Sigma^{<B}_{1,1}G^a_{1,j} + g^<_{i,1}\Sigma^{aB}_{1,1}G^a_{1,j}\right.$$
$$\left. + g^<_{i,N}\Sigma^{aB}_{N,N}G^a_{N,j} + g^r_{i,N}\Sigma^{rB}_{N,N}G^<_{N,j} + g^r_{i,N}\Sigma^{<B}_{N,N}G^a_{N,j}\right\}. \tag{9.141}$$

We can now use the general results that

$$\left(\hbar\omega - H_0^D\right)g^r_{i,i} = 1,$$
$$\left(\hbar\omega - H_0^D\right)g^<_{i,j} = 0, \tag{9.142}$$

for the case of no scattering within the active device region, we find that (9.141) becomes

$$\left(\hbar\omega - H_0^D\right)G^<_{i,j} = \delta_{i,1}\left[\Sigma^{rB}_{1,1}G^<_{1,j} + \Sigma^{<B}_{1,1}G^a_{1,j}\right]$$
$$+ \delta_{i,N}\left[\Sigma^{rB}_{N,N}G^<_{N,j} + \Sigma^{<B}_{N,N}G^a_{N,j}\right]. \tag{9.143}$$

Hence, for the diagonal blocks, we have the set of equations

$$\left(\hbar\omega - H_0^D - \Sigma^{rB}_{1,1}\right)G^<_{1,1} = \Sigma^{<B}_{1,1}G^a_{1,1},$$
$$\left(\hbar\omega - H_0^D - \Sigma^{rB}_{N,N}\right)G^<_{N,N} = \Sigma^{<B}_{N,N}G^a_{N,N}, \tag{9.144}$$
$$\left(\hbar\omega - H_0^D\right)G^<_{i,i} = 0, \quad i = 2,\ldots N-1.$$

As an example of the above approach, we consider the simple 3×3 block for a small device. Here, both i and j run from 1 to 3. Then, the matrices for the various Green's functions look like

$$[G^r] = \begin{bmatrix} \left(\hbar\omega - H_1 - \Sigma^{rB}_{1,1}\right) & T_{1,2} & 0 \\ T_{1,2} & \left(\hbar\omega - H_2\right) & T_{2,3} \\ 0 & T_{2,3} & \left(\hbar\omega - H_3 - \Sigma^{rB}_{3,3}\right) \end{bmatrix}^{-1}, \tag{9.145}$$

$$[G^<] = \begin{bmatrix} \left(\hbar\omega - H_1 - \Sigma^{rB}_{1,1}\right) & T_{1,2} & 0 \\ T_{1,2} & (\hbar\omega - H_2) & T_{2,3} \\ 0 & T_{2,3} & \left(\hbar\omega - H_3 - \Sigma^{rB}_{3,3}\right) \end{bmatrix}^{-1}$$ (9.146)

$$\begin{bmatrix} \Sigma^{<B}_{1,1} & 0 & 0 \\ 0 & 0 & 0 \\ 0 & 0 & \Sigma^{<B}_{3,3} \end{bmatrix} [G^r].$$ (9.146)

The boundary self-energies have been defined previously, and we need only evaluate these in terms of the boundary Green's functions. Here, the length of the boundaries will affect the computational time, and in order to illustrate this, we take only three slices for each boundary. Then, if the potential is constant in all three slices of the boundary, we may write the left contact retarded function as

$$g^r_{0,0} = \left(\begin{bmatrix} \left(\hbar\omega - H_{-2} - \Sigma^{rB}_{-2,-2}\right) & T_{-2,-1} & 0 \\ T_{-2,-1} & (\hbar\omega - H_{-1}) & T_{-1,0} \\ 0 & T_{-1,0} & \left(\hbar\omega - H_0 - \Sigma^{rB}_{0,0}\right) \end{bmatrix}^{-1}\right)_{0,0}.$$ (9.147)

Here, the self-energy at the left end is given as

$$\Sigma^{rB}_{-2,-2} = T_{-2,-3} g^r_{-3,-3} T_{-3,-2}$$
$$= T_{-2,-3}\left(\hbar\omega - H_{-3} + T_{-4,-3}\chi Z\chi^{-1}\right)^{-1} T_{-3,-2}.$$ (9.148)

In this last expression, χ is a matrix of Bloch states propagating toward the active device region, and Z is a diagonal matrix of the propagation factors. For the single-band model we are considering here, the T's are simple real scalars, which may be denoted as t_e for the emitter. Then,

$$g^r_{-3,-3} = \frac{1}{t_e} e^{i\gamma_e a}, \quad \Sigma^{rB}_{-2,-2} = -t_e e^{i\gamma_e a}.$$ (9.149)

In this case, the energy is related to a transverse dispersion, related to the D in this slice as

$$\hbar\omega = D_{-2} - 2t_e \cos(\gamma_e a).$$ (9.150)

Now, the retarded Green's function becomes

$$g^r_{0,0} = \left(\begin{bmatrix} -t_e e^{-i\gamma_e a} & T_{-2,-1} & 0 \\ T_{-2,-1} & (\hbar\omega - D_{-1}) & T_{-1,0} \\ 0 & T_{-1,0} & (\hbar\omega - D_0) \end{bmatrix}^{-1}\right)_{0,0}.$$ (9.151)

This is now easily solved to give the needed Green's functions for the solution for the device.

9.4 Real-time Green's functions

The current was given earlier in (9.126), which showed that we now need to compute the less-than correlation function in order to find the current through this device structure. The coupling from the emitter to the device leads to the less-than function across this boundary being given by

$$G^<_{e,d} = \sum_{e',d'} \left[g^r_{e,e'}(-t_{e',d'}) G^<_{d,d'} + g^<_{e,e'}(-t_{e',d'}) G^a_{d,d'} \right]$$
$$= -[G^<_{e,d}]^+. \tag{9.152}$$

Then, for only nearest-neighbor coupling, where the self-energy boundary term appears only in the first layer, we can write the current as

$$J_0 = \frac{2e}{S} \sum_{\mathbf{k}} \int \frac{d\omega}{2\pi} Tr \left\{ \Gamma^B_{1,1} \left[i\Gamma^B_{1,1} G^<_{1,1} + if_e A_{1,1} \right] \right\}. \tag{9.153}$$

If we now write

$$G^<_{1,1} = G^r_{1,1} \Sigma^{<B}_{1,1} G^a_{1,1} = if_e G^r_{1,1} \Gamma^B_{1,1} G^a_{1,1}, \tag{9.154}$$

we then have

$$J_0 = \frac{2e}{S} \sum_{\mathbf{k}} \int \frac{d\omega}{2\pi} f_e Tr \left\{ \Gamma^B_{1,1} \left[G^r_{1,1} \Gamma^B_{1,1} G^a_{1,1} + A_{1,1} \right] \right\}. \tag{9.155}$$

The term in the square brackets is just the tunneling coefficient through the active device, whereas the scattering operator within the trace operation plays the role of the velocity within the device. The Landauer formula requires that we take account of the back-propagating states, and this leads to the modification of (9.155) to

$$J_0 = \frac{2e}{S} \sum_{\mathbf{k}} \int \frac{d\omega}{2\pi} Tr \left\{ \Gamma^B_{1,1} \left[G^r_{1,1} \Gamma^B_{1,1} G^a_{1,1} + A_{1,1} \right] \right\} (f_e - f_c). \tag{9.156}$$

We can also use this approach to get the density in the slice from

$$iG^<_{L,L} = f_e G^r_{L,1} \Gamma^B_{1,1} G^a_{1,L} + f_c G^r_{L,N} \Gamma^B_{N,N} G^a_{N,L}$$
$$= f_e G^r_{L,1} \Gamma^B_{1,1} G^a_{1,L} + f_c \left[A_{L,L} - G^r_{L,1} \Gamma^B_{1,1} G^a_{1,L} \right]. \tag{9.157}$$

This last expression is now integrated over the momentum and energy to get the total density on the slice, and this is used in Poisson's equation for the local potential and field.

Let us now turn to how the recursion is carried out, as this set of equations described above is evaluated using the recursive Green's function approach first detailed in Section 3.5.2. The recursive Green's function approach is most useful in the nearest-neighbor tight-binding approach. In the absence of scattering, the term $g^{rL}_{L,L}$ takes into account *exactly* to the left while coupling to the next slice to the right. Similarly, $g^{rR}_{L,L}$ takes into account *exactly* to the right while coupling to the next slice to the left. Hence, one starts with the retarded function in the right lead, and propagates to the left as

$$g_{L,L}^{rR} = \left[\hbar\omega - D_L - t_{L,L+1}g_{L+1,L+1}^{rR}t_{L+1,L}\right]^{-1}. \tag{9.158}$$

Once we reach the left-hand side, we add the left boundary term to achieve

$$g_{1,1}^{r} = \left[\hbar\omega - D_1 - t_{1,0}g_{0,0}^{rL}t_{0,1} + t_{1,2}g_{2,2}^{rR}t_{2,1}\right]^{-1}. \tag{9.159}$$

If we needed only the current, we could stop here, as we have enough information in the ballistic case. However, to close the Poisson loop and compute the potential (and field) self-consistently, we also need to find the density. So we walk back to the right as

$$\begin{aligned}G_{L,L}^{r} &= g_{L,L}^{rR} + g_{L,L}^{rR}t_{L,L-1}G_{L-1,L-1}^{rR}t_{L-1,L}g_{L,L}^{rR},\\ G_{L,1}^{r} &= -g_{L,L}^{rR}t_{L,L-1}G_{L-1,1}^{r}.\end{aligned} \tag{9.160}$$

From this, we can compute the local density.

So, this is an advanced form of the recursive Green's function approach which builds in the nonequilibrium density approximations. However, there is no scattering in the system at present, so there has been no need to compute the complicated less-than correlation function accurately. Similarly, the only self-energy arises from the coupling to the contacts. However, we have established a powerful formalism that allows us to overcome these limitations. Yet, at this point the NEMO approach offers not much new except the inclusion of detailed band-structure calculations within the Green's function approach. Hence, we how have to turn to the case of scattering to move beyond this limitation.

When we introduce scattering into the problem, this appears as new self-energy terms that modify the equations of motion. Within the device, these now become

$$\begin{aligned}(\hbar\omega - H_0 - \Sigma_r)G^r &= 1,\\ (\hbar\omega - H_0 - \Sigma_r)G^< &= \left(\Sigma^< + \Sigma^{<B}\right)G^a.\end{aligned} \tag{9.161}$$

Thus, we have to account for both the new retarded self-energy and a less-than self-energy in these equations of motion. These terms represent the dissipative scattering processes. Each new self-energy is a sum over all the various processes which contribute to the scattering, but when introduced in this manner, these terms do not break current continuity. However, these terms do not appear in a simple manner. For example, let us consider elastic scattering, the most straightforward set of processes in a classical system. Here, the relaxation term can be written as

$$\Gamma = i(\Sigma^> - \Sigma^<) = \int \frac{d^2k}{(2\pi)^2} iM(G^> - G^<)$$
$$= \int \frac{d^2k}{(2\pi)^2} MA, \qquad (9.162)$$

where M is the kernel of the self-energy, containing the matrix element for the scattering, and A is the spectral density. Since A contains the function Γ, this expression must be solved self-consistently, and therein lies the difficulty. Thus, one usually adopts some approximation to evaluate this function. One such, discussed in Section 9.3, is to assume the Born approximation, in which the self-energy correction to A is ignored. This is valid only for weak scattering, which is the usual assumption in semiclassical scattering, as discussed in Chapter 2. NEMO goes beyond this, however, to account for off-shell processes that become allowed when the full form of A is used.

To illustrate the approach, we will use polar-optical phonon scattering. This process is inelastic, but the extensions to other scattering processes which are elastic can be found in a straightforward manner from this example. In general, the Hamiltonian for the electron–phonon interaction may be written as

$$H_{ep} = \frac{1}{\sqrt{V}} \sum_{L,\mathbf{k}} \sum_{\mathbf{q}} U_q e^{iq_z L} \left[e^{iq_z a/2} \sum_{i=\{a\}} \xi^+_{i,L,\mathbf{k}} \xi_{i,L,\mathbf{k}} \right.$$
$$\left. + \sum_{i=\{c\}} \xi^+_{i,L,\mathbf{k}} \xi_{i,L,\mathbf{k}} \right] \left(a^+_{\mathbf{q}} + a_{\mathbf{q}} \right). \qquad (9.163)$$

Here, the U_q is the matrix element, the ξ are electron operators and the $a_{\mathbf{q}}$ are the phonon operators. This Hamiltonian leads to the less-than self-energy term

$$\Sigma^<_{\alpha,L;\alpha',L'}(\mathbf{k},\omega) = \frac{1}{V} \sum_{\mathbf{q}} |U_{\mathbf{k}-\mathbf{q}}| e^{iq_z a(L-L'+\nu_{\alpha,\alpha'})}$$
$$\times \left[(N_q + 1) G^<_{\alpha,L;\alpha',L'}(\mathbf{q}_t, \omega - \omega_q) \right.$$
$$\left. + N_q G^<_{\alpha,L;\alpha',L'}(\mathbf{q}_t, \omega + \omega_q) \right]. \qquad (9.164)$$

The first term in the square brackets corresponds to phonon emission, while the second term corresponds to phonon absorption. The magnitude squared matrix element for polar-optical phonon scattering is given by

$$|U_q|^2 = \left[\frac{e^2 \hbar \omega_q}{2} \left(\frac{1}{\varepsilon_\infty} - \frac{1}{\varepsilon(0)} \right) \right] \frac{q^2}{(q^2 + q_D^2)^2}, \qquad (9.165)$$

where the ε's are the high-frequency and low-frequency dielectric constants and q_D is the Debye screening wave number. The latter assumes that the carrier density is sufficiently low that the material is nondegenerate so that Debye screening is the appropriate form of the screening to use. In the following, the terms within the square brackets will be denoted by the simpler β. With this form, (9.164) can be evaluated to yield the expression

$$\Sigma^<_{\alpha,L;\alpha',L'}(\mathbf{k},\omega) = \frac{\beta}{\pi}\int\frac{d^2q_t}{4\pi^2} Y(|L-L'+\nu_{\alpha,\alpha'}|,\mathbf{k},\mathbf{q}_t)$$

$$\times \Big[(N_q+1)G^<_{\alpha,L;\alpha',L'}(\mathbf{q}_t,\omega-\omega_q)$$

$$+ N_q G^<_{\alpha,L;\alpha',L'}(\mathbf{q}_t,\omega+\omega_q)\Big], \quad (9.166)$$

where

$$Y(x,\mathbf{k},\mathbf{q}_t) = \int_0^w dq_z \frac{\cos(q_z ax)}{\sqrt{(q^2+k^2+q_D^2)^2 - 4k^2 q_t^2}}$$

$$- \int_0^w dq_z \frac{\cos(q_z ax) q_D^2 (q^2+k^2+q_D^2)}{\big[(q^2+k^2+q_D^2)^2 - 4k^2 q_t^2\big]^{3/2}}. \quad (9.167)$$

The parameter w is a cutoff that is imposed upon the integral to assure that the integration does not pass through a singularity of the integrand. The retarded self-energy can be found in a similar manner, but we have to face the convolution integral, and the need for self-consistency, in each of these self-energies and their corresponding Green's function.

The full matrix solutions will require an enormous amount of memory to carry out to the full level. Thus, weak scattering is a viable assumption, except that here one goes beyond the simple Born approximation. The approach is illustrated with the retarded Green's function and self-energy, whereas the above used the less-than functions. The same approach can be applied to either, and we chose the retarded functions merely to expand the examples. The imaginary scattering function can be written as

$$\Gamma(\mathbf{k},\omega) = i[\Sigma^>(\mathbf{k},\omega) - \Sigma^<(\mathbf{k},\omega)] \sim i\Sigma^>(\mathbf{k},\omega)$$

$$\sim iM \otimes G^> \sim M \otimes A,$$

$$\Sigma^r \sim M \otimes G^r. \quad (9.168)$$

Here, the symbol $\otimes$ denotes the integration over transverse momentum, and sum over emission and absorption, with the Green's functions on the right offset by the phonon energy. Now, the equations for G^r and $G^<$ are uncoupled and can be solved independently. We solve for the retarded by use of a continued fraction expansion as

$$G^r = \left(\hbar\omega - H_0 - \Sigma^{rB} - \Sigma^r\right)^{-1}$$
$$= \left[\hbar\omega - H_0 - \Sigma^{rB} - M \otimes G^r\right]^{-1}$$
$$= \left\{\hbar\omega - H_0 - \Sigma^{rB} - M \otimes \right.$$
$$\left.\left[\hbar\omega - H_0 - \Sigma^{rB} - M \otimes \left(\hbar\omega - H_0 - \Sigma^{rB} - \Sigma^r\right)^{-1}\right]^{-1}\right\}^{-1} \quad (9.169)$$

to third order in the partial fraction expansion. It is usually sufficient to drop the retarded self-energy in this last term of the expansion, as convergence can be achieved with the subsequent summations. Having found the retarded function, the less-than function can now be found from (9.137), which itself is an iterated process. Finally, the current density is found from

$$J_0 = \frac{2e}{S} \sum_{\mathbf{k}} \int \frac{d\omega}{2\pi} Tr\left\{\Gamma^B_{1,1}\left(G^r_{1,1}\Gamma^B_{1,1}G^a_{1,1} - A_{1,1}\right)\right\}(f_e - f_c). \quad (9.170)$$

As previously, the Green's functions are built up using the scattering at each layer, and the recursion between layers to propagate from one end of the device to the other and back again. One interesting aspect of the simulation with the presence of phonons is the appearance of an inelastic tunneling peak (as opposed to the resonant tunneling peak which forms the current peak), which typically appears in the valley region and illustrates the contribution of the scattering processes to the valley current [165]. The phonon replica peak has been seen in experiments in InGaAs/InAlAs resonant tunneling diodes [166].

Hence, it is found that NEMO gives reasonably good agreement with experiment as all the correct physical processes are included. In fact, the agreement is sufficiently good that it can be an analytical tool to provide some feedback to the device fabricators about the actual dimensions of their structure. However, it is not perfect, but then no other simulation is perfect either. It is, however, a very good approach if used wisely in the hands of the knowledgeable.

The NEMO code has been used in recent years for simulations beyond the simple 1D model discussed above. It has been applied to alloy semiconductors [167], Si and GaAs nanowires [168,169,170], carbon nanotubes [171], and Ge/Si core-shell structures [172]. Various of these approaches have also been combined for alloy nanowires [173].

References

[1] J.-R. Zhou and D. K. Ferry, *IEEE Trans. Electron Dev.* **40**, 421 (1993).
[2] H. S. Wong and Y. Taur, *Proc. Int. Electron Dev. Mtg.* (New York, IEEE Press, 1193), p. 705.
[3] P. C. Martin, J. Schwinger, *Phys. Rev.* **115**, 1342 (1959).
[4] J. Schwinger, *J. Math. Phys.* **2**, 407 (1961).
[5] L. P. Kadanoff and G. Baym, *Quantum Statistical Mechanics* (New York, Benjamin, 1962).

[6] L. V. Keldysh, *Sov. Phys. JETP* **20**, 1018 (1965).

[7] C. P. Enz, *A Course on Many-Body Theory Applied to Solid-State Physics* (Singapore, World Scientific Press, 1992), p. 76.

[8] H. Haug, in *Quantum Transport in Ultrasmall Devices*, eds. D. K. Ferry, H. L. Grubin, C. Jacoboni, and A.-P. Jauho (New York, Plenum Press, 1995).

[9] T. Kuhn and F. Rossi, *Phys. Rev. B* **46**, 7496 (1992).

[10] H. Haug and A.-P. Jauho, *Quantum Kinetics in Transport and Optics of Semiconductors* (Berlin, Springer, 2004).

[11] T. W. Hickmott, P. M. Solomon, F. F. Fang, et al., *Phys. Rev. Lett.* **52**, 2053 (1984).

[12] L. Eaves, F. W. Sheard, and G. A. Toombs, in *Physics of Quantum Electron Devices*, ed. F. Capasso (New York, Springer-Verlag, 1990), pp. 107–46.

[13] L. Eaves, P. S. S. Guimaraes, B. R. Snell, D. C. Taylor, and K. E. Singer, *Phys. Rev. Lett.* **53**, 262 (1985).

[14] H. Yoo, S. M. Goodnick, J. R. Arthur, and M. A. Reed, *J. Vac. Sci. Technol. B* **8**, 370 (1990).

[15] J. R. Hayes, A. F. J. Levi, and W. Wiegmann, *Electron. Lett.* **20**, 851 (1984); *Phys. Rev. Lett.* **54**, 1570 (1985).

[16] N. Yokoyama, K. Imamura, T. Ohshima, et al., *Proc. Int. Electron Dev. Mtg.* (New York, IEEE Press, 1984), p. 532.

[17] M. Heiblum, D. C. Thomas, C. M. Knoedler, and M. I. Nathan, *Appl. Phys. Lett.* **47**, 1105 (1985).

[18] A. F. J. Levi, J. R. Hayes, P. M. Platzmann, and W. Wiegmann, *Phys. Rev. Lett.* **55**, 2071 (1985).

[19] M. Heiblum, M. I. Nathan, D. C. Thomas, and C. M. Knoedler, *Phys. Rev. Lett.* **55**, 2200 (1985).

[20] M. Heiblum and M. V. Fischetti, in *Physics of Quantum Electron Devices*, ed. F. Capasso (New York, Springer-Verlag, 1990), pp. 271–318.

[21] M. A. Hollis, S. C. Palmateer, L. F. Eastman, N. V. Dandekar, and P. M. Smith, *Electron. Dev. Lett.* **4**, 440 (1983).

[22] V. H. Y. Lam, M. W. Dellow, S. J. Bending, et al., *Semicond. Sci. Technol.* **10**, 110 (1995).

[23] A. Pavleski, M. Heiblum, C. P. Umbach, et al., *Phys. Rev. Lett.* **62**, 1776 (1989).

[24] A. Pavleski, C. P. Umbach, and M. Heiblum, *Appl. Phys. Lett.* **55**, 1421 (1989).

[25] T. Galloway, B. L. Gallagher, P. Beton, et al., *Surf. Sci.* **229**, 326 (1990).

[26] D. K. Ferry, *Semiconductors* (New York, Macmillan, 1991).

[27] L. W. Molenkamp, H. van Houten, C. W. J. Beenakker, R. Eppenga, and C. T. Foxon, *Phys. Rev. Lett.* **65**, 1052 (1990).

[28] L. W. Molenkamp, Th. Gravier, H. van Houten, et al., *Phys. Rev. Lett.* **68**, 3765 (1992).

[29] L. W. Molenkamp and M. J. M. de Jong, *Sol.-State Electron.* **37**, 551 (1994).

[30] L. W. Molenkamp, A. A. M. Staring, B. W. Alphanaar, H. van Houten, and C. W. J. Beenakker, *Semicond. Sci. Technol.* **9**, 903 (1994).

[31] T. Ikoma, K. Hirakawa, T. Hiramoto, and T. Odagiri, *Sol.-State Electron.* **32**, 1793 (1989).

[32] H. Hirakawa, T. Odagiri, T. Hiramoto, and T. Ikoma, in *Proc. Symp. New Phenomena in Mesoscopic Structures*, eds. S. Namba and C. Hamaguchi, December 1989.

[33] T. Ikoma and T. Hiramoto, in *Granular Nanoelectronics*, eds. D. K. Ferry, J. R. Barker, and C. Jacoboni (New York, Plenum Press, 1991), pp. 255–76.

[34] K. Hirakawa and H. Sasaki, *Appl. Phys. Lett.* **49**, 889 (1989).
[35] P. J. Price, *J. Appl. Phys.* **53**, 6863 (1982).
[36] R. Fletcher, Y. Feng, C. T. Foxon, J. J. Harris, *Phys. Rev. B* **61**, 2028 (2000).
[37] C. Prasad, D. K. Ferry, D. Vasileska, and H. H. Wieder, *Physica E* **19**, 215 (2003).
[38] Y. Ma, R. Fletcher, E. Zaremba, *et al.*, *Phys. Rev. B* **43**, 9033 (1991).
[39] N. J. Appleyard, J. T. Nicholls, M. Y. Simmons, W. R. Tribe, N. Pepper, *Phys. Rev. Lett.* **81**, 3491 (1998).
[40] S. Das Sarma, V. B. Campos, *Phys. Rev B* **47**, 3728 (1993).
[41] C. Prasad, D. K. Ferry, D. Vasileska, and H. H. Wieder, *J. Vac. Sci. Technol. B* **21**, 1936 (2003).
[42] C. Prasad, D. K. Ferry, and H. H. Wieder, *J. Vac. Sci. Technol.* **22**, 2059 (2004).
[43] S. M. Goodnick, J. C. Wu, M. N. Wybourne, and D. D. Smith, *Phys. Rev. B* **48**, 9150 (1993).
[44] J. C. Wu, M. N. Wybourne, C. Bervens, S. M. Goodnick, and D. D. Smith, *Appl. Phys. Lett.* **61**, 1 (1992).
[45] J. Shah, A. Pinczuk, A. C. Gossard, and W. Wiegmann, *Phys. Rev. Lett.* **54**, 2045 (1985).
[46] W. Pötz and P. Kocevar, in *Hot Carriers in Semiconductors: Physics and Applications* (Boston, Academic Press, 1992), pp. 87–120.
[47] R. J. Brown, M. J. Kelly, M. Pepper, H. Ahmed, D. G. Hasko, D. C. Peacock, J. E. F. Frost, D. A. Ritchie, and G. A. C. Jones, *J. Phys.: Cond. Matter* **1**, 6285 (1989).
[48] G. Timp, R. Behringer, S. Sampere, J. E. Cunningham, and R. E. Howard, in *Nanostructure Physics and Fabrication*, eds. M. A. Reed and W. P. Kirk (Boston, Academic Press, 1989), p. 31.
[49] Y. Takagaki and D. K. Ferry, *J. Phys.: Cond. Matter* **4**, 10421 (1992).
[50] C. Berven, M. N. Wybourne, A. Ecker, and S. M. Goodnick, *Phys. Rev. B* **50**, 14630 (1994).
[51] J. C. Smith, C. Berven, S. M. Goodnick, and M. N. Wybourne, *Physica B* **227**, 197 (1997).
[52] J. A. Nixon, J. H. Davies, and H. U. Baranger, *Phys. Rev. B* **43**, 12638 (1991).
[53] A. Wacker, *Phys. Rev. B* **49**, 16785 (1994).
[54] R. Chau, S. Datta, M. Doczy, *et al.*, *IEEE Trans. Nanotechn.* **4**, 153 (2005).
[55] G. Moore, *Electronics Mag.* **38**(8), 114 (April 1965).
[56] R. H. Dennard, F. H. Gaennslen, H. N. Yu, *et al.*, *IEEE Trans. Sol. State Circ.* **9**, 256 (1974).
[57] G. Baccarani, M. R. Wordeman, and R. H. Dennard, *IEEE Trans. Electron Dev.* **31**, 452 (1984).
[58] D. J. Frank, S. E. Laux, and M. Fischetti, *IEDM Tech. Dig.*, 1992, pp. 553–556.
[59] D. Hisamoto, W.-C. Lee, J. Kedzierski, *et al.*, *IEEE Trans. Electron Dev.* **47**, 2320 (2000).
[60] B. S. Doyle, S. Datta, M. Doczy, *et al.*, *IEEE Electron Dev. Lett.* **24**, 263 (2003).
[61] J. Kavalieros, B. Doyle, S. Datta, *et al.*, *VLSI Symp. Dig. Tech. Papers*, 2006, pp. 62–3.
[62] S. M. Ramey and D. K. Ferry, *IEEE Trans. Nanotech.* **2**, 110 (2003).
[63] R. Shishir and D. K. Ferry, *submitted for publication*.
[64] M. S. Lundstrom, *IEEE Electron Dev. Lett.* **18**, 361 (1997).
[65] Y. Ono, Y. Takahashi, K. Yamazaki, M. Nagase, H. Namatsu, K. Kurihara, and K. Murase, *Proc. Int. Electron Dev. Mtg.* (New York, IEEE Press, 1998), p. 123.
[66] J. Kedzierski, J. Bokor, and E. Anderson, *J. Vac. Sci. Technol. B* **17**, 3244 (1999).
[67] Y. Cui and C. M. Lieber, *Science* **291**, 851 (2001).

[68] Y. Cui, Z. Zhong, D. Wang, W. U. Wang, and C. M. Lieber, *Nano Lett.* **3**, 149 (2003).
[69] X. Duan, C. Niu, V. Sahi, *et al.*, *Nature* **425**, 274 (2003).
[70] S.-M. Koo, A. Fujiwara, J.-P. Han, *et al.*, *Nano Lett.* **4**, 2197 (2004).
[71] C. T. Black, *Appl. Phys. Lett.* **87**, 163116 (2005).
[72] Y. Ahn, J. Dunning, and J. Park, *Nano Lett.* **5**, 1367 (2005).
[73] J. Goldberger, A. I. Hochbaum, R. Fan, and P. Yang, *Nano Lett.* **6**, 973 (2006).
[74] P. Nguyen, H. T. Ng, T. Yamada, *et al.*, *Nano Lett.* **4**, 651 (2004).
[75] H. T. Ng, J. Han, T. Yamada, *et al.*, *Nano Lett.* **4**, 1247 (2004).
[76] T. Bryllert, L.-E. Wernersson, L. E. Fröberg, and L. Samuelson, *IEEE Electron Dev. Lett.* **27**, 323 (2006).
[77] T. Bryllert, L.-E. Wernersson, T. Löwgren, and L. Samuelson, *Nanotechnol.* **17**, S227 (2006).
[78] E. Lind, A. I. Persson, L. Samuelson, and L.-E. Wernersson, *Nano Lett.* **6**, 1842 (2006).
[79] S. J. Tans, A. Verschueren, and C. Dekker, *Nature* **393**, 49 (1998).
[80] R. Martel, T. Schmidt, H. R. Shea, T. Hertel, and Ph. Avouris, *Appl. Phys. Lett.* **73**, 2447 (1998).
[81] H. W. Ch. Postma, T. Leepen, Z. Yao, M. Grifoni, and C. Dekker, *Science* **293**, 76 (2001).
[82] R. Martel, V. Derycke, C. Lavoie, *et al.*, *Phys. Rev. Lett.* **87**, 256805 (2001).
[83] J. Appenzeller, J. Knoch, V. Derycke, *et al.*, *Phys. Rev. Lett.* **89**, 126801 (2002).
[84] M. Radosavljević, S. Heinze, J. Tersoff, and Ph. Avouris, *Appl. Phys. Lett.* **83**, 2435 (2003).
[85] J. Appenzeller, J. Knoch, M. Radosavljević, and Ph. Avouris, *Phys. Rev. Lett.* **92**, 226802 (2004).
[86] T. Palm and L. Thylén, *Appl. Phys. Lett.* **60**, 237 (1992).
[87] T. J. B. Janssen, J. C. Maan, J. Singleton, *et al.*, *J. Phys.: Cond. Matter* **6**, L163 (1994).
[88] Q. Hu, S. Verghese, R. A. Wyss, *et al.*, *Semicond. Sci. Technol.* **11**, 1888 (1996).
[89] L. Worschech, B. Weidner, S. Reitzenstein, and A. Forchel, *Appl. Phys. Lett.* **78**, 3325 (2001).
[90] G. M. Jones, C. H. Yang, M. J. Yang, and Y. B. Lyanda-Geller, *Appl. Phys. Lett.* **86**, 073117 (2005).
[91] S. Datta, T. Ashley, J. Brask, *et al.*, *Proc. ICIST*, Beijing, 2005.
[92] R. Chau, Intel Corporation, http://download.intel.com/technology/silicon/Chau_DRC_062606.pdf.
[93] R. Landauer, *IBM J. Res. Develop.* **1**, 223 (1957).
[94] C. D. Child, *Phys. Rev. (ser. 1)* **32**, 492 (1911).
[95] I. Langmuir, *Phys. Rev.* **2**, 450 (1913).
[96] M. S. Shur and L. F. Eastman, *IEEE Trans. Electron Dev.* **26**, 1677 (1979).
[97] J. R. Barker, D. K. Ferry, and H. L. Grubin, *IEEE Electron Dev. Lett.* **1**, 209 (1980).
[98] See, e.g., J. Saint Martin, A. Bournel, and P. Dollfus, *IEEE Trans. Electron Dev.* **51**, 1148 (2004).
[99] M. Lundstrom, Z. Ren, and S. Datta, *2000 SISPAD Tech. Dig.*, pp. 1–5, 2000.
[100] A. Svizhenko and M. P. Anantram, *IEEE Trans. Electron Dev.* **50**, 1459 (2003).
[101] R. Kotylar, B. Obradovic, P. Matagne, M. Stettler, and M. D. Giles, *Appl. Phys. Lett.* **84**, 5270 (2004).
[102] M. J. Gilbert and D. K. Ferry, *J. Appl. Phys.* **95**, 7954 (2004).

[103] M. J. Gilbert, R. Akis, and D. K. Ferry, *J. Appl. Phys.* **98**, 094303 (2005).
[104] R. Akis, M. J. Gilbert, and D. K. Ferry, *J. Phys. Conf. Series* **38**, 87 (2006).
[105] X. Zhou, S. A. Dayeh, D. Aplin, D. Wang, and E. T. Yu, *Appl. Phys. Lett.* **89**, 053113 (2006).
[106] F. G. Pikus, and K. K. Likharev, *Appl Phys Lett* **71**, 3661 (199).
[107] S. Datta, *Superlatt. Microstuct.* **28**, 253 (2000).
[108] J. Knoch, B. Lengeler, and J. Appenzeller, *IEEE Trans Elec Dev.* **49**, 1212 (2002).
[109] M. J. Gilbert and D. K. Ferry, *Superlatt. Microstruct.* **34**, 277 (2003).
[110] D. D. Johnson, *Phys. Rev. B* **38**, 12807 (1988).
[111] D. A. Baglee, D. K. Ferry, C. W. Wilmsen, and H. H. Wieder, *J. Vac. Sci. Technol. B* **17**, 1032 (1980).
[112] M. J. Gilbert and D. K. Ferry, *J. Appl. Phys.* **95**, 7954 (2004).
[113] M. J. Gilbert and D. K. Ferry, *IEEE Trans. Nanotechn.* **4**, 599 (2005).
[114] M. J. Gilbert and D. K. Ferry, *Superlatt. Microstruct.* **34**, 277 (2003).
[115] D. K. Ferry and H. L. Grubin, in *Solid State Physics* **49**, 283 (1995), eds. H. Ehrenreich and H. Turnbull (Boston, Academic Press, 1995).
[116] C. Jacoboni and L. Reggiani, *Rev. Mod. Phys.* **55**, 645 (1983).
[117] D. K. Ferry, M.-J. Kann, A. M. Kriman, and R. P. Joshi, *Comp. Phys. Commun.* **67**, 119 (1991).
[118] D. K. Ferry, in *Handbook of Semiconductors*, Vol. 1, 2nd edn., ed. P. T. Landsberg (Amsterdam, North-Holland, 1992), pp. 1039–78.
[119] I. Prigogine, *Acad. Roy. Belg., Bull. Classe Sci.* **31**, 600 (1945).
[120] A. Blandin, A. Nourtier, and D. W. Hone, *J. Physique* **37**, 369 (1976).
[121] J. Rammer and H. Smith, *Rev. Mod. Phys.* **58**, 323 (1986).
[122] M. Wagner, *Phys. Rev. B* **44**, 6104 (1991); **45**, 11595 (1992).
[123] G. D. Mahan, in *Quantum Transport in Semiconductors*, eds. D. K. Ferry and C. Jacoboni (New York, Plenum Press, 1992), pp. 101–40.
[124] D. C. Langreth, in *Linear and Nonlinear Electron Transport in Solids*, eds. J. T. Devreese and E. van Doren (New York, Plenum Press, 1976).
[125] R. Bertoncini, A. M. Kriman, and D. K. Ferry, *Phys. Rev. B* **41**, 1390 (1990).
[126] R. Bertoncini, A. M. Kriman, and D. K. Ferry, *Phys. Rev. B* **44**, 3655 (1991).
[127] P. Lipavský, V. pička, and B. Velický, *Phys. Rev. B* **34**, 3020 (1986).
[128] M. J. McLennan, Y. Lee, and S. Datta, *Phys. Rev. B* **43**, 13846 91991).
[129] Y. Lee, M. J. McLennan, G. Klimeck, R. K. Lake, and S. Datta, *Superlatt. Microstruct.* **11**, 137 (1992).
[130] Y. Meir and N. S. Wingreen, *Phys. Rev. Lett.* **68**, 2512 (1992).
[131] Y. Meir, N. S. Wingreen, and P. A. Lee, *Phys. Rev. Lett.* **66**, 3048 (1991).
[132] A. P. Jauho, N. S. Wingreen, and Y. Meir, *Phys. Rev. B* **50**, 5528 (1994).
[133] See, e.g., V. Mujica, M. Kemp, and M. A. Ratner, *J. Chem. Phys.* **101**, 6849 (1994), and references therein (we make no effort to be extensive in the list of references on this topic).
[134] W. Tian, S. Datta, S. Hong, R. Reifenberger, J. I. Henderson, and C. I. Kubiak, *J. Chem. Phys.* **109**, 2874 (1998).
[135] L. Hall, J. R. Reimers, N. S. Hush, and K. Silverbrook, *J. Chem. Phys.* **112**, 1510 (2000).
[136] M. Di Ventra, S. T. Pantelides, and N. D. Lang, *Phys. Rev. Lett.* **84**, 979 (2000).
[137] R. Dahlke and U. Schollwöck, *Phys. Rev. B* **69**, 085324 (2004).
[138] G. Speyer, R. Akis, and D. K. Ferry, *J. Vac. Sci. Technol. B* **24**, 1987 (2006).

[139] G. Speyer, R. Akis, and D. K. Ferry, *J. Phys. Conf. Series* **38**, 25 (2006).
[140] A. Di Carlo and P. Lugli, *Semicond. Sci. Technol.* **10**, 1673 (1995).
[141] A. Di Carlo, S. Pescetelli, M. Paciotti, P. Lugli, and M. Graf, *Sol. State Commun.* **98**, 803 (1996).
[142] A. Di Carlo, S. Pescetelli, A. Kavorkin, M. Vladimirova, and P. Lugli, *Phys. Stat. Sol. (b)* **204**, 275 (1997).
[143] G. C. Solomon, A. Gagliardi, A. Pecchia, *et al.*, *J. Chem. Phys.* **125**, 184702 (2006), and references contained therein.
[144] L. Latessa, A. Pecchia, and A. Di Carlo, *IEEE Trans. Nanotechnol.* **6**, 13 (2007).
[145] P. Ordejón, E. Artacho, and J. M. Soler, *Phys. Rev. B* **53**, 10441 (1996).
[146] J. M. Soler, E. Artacho, J. D. Gale, *et al.*, *J. Phys.: Cond. Matter* **14**, 2745 (2002).
[147] M. Brandbyge, J.-L. Mozos, P. Ordejón, J. Taylor, and K. Stokbro, *Phys. Rev. B* **65**, 165401 (2002).
[148] K. Stokbro, J. Taylor, M. Brandbyge, and P. Ordejón, *Ann. N. Y. Acad. Sci.* **1006**, 212 (2003).
[149] J. C. Slater and G. F. Koster, *Phys. Rev.* **94**, 1498 (1954).
[150] P. Hohenberg and W. Kohn, *Phys. Rev.* **136**, B864 (1964).
[151] W. Kohn and L. Sham, *Phys. Rev.* **140**, A1133 (1965).
[152] O. F. Sankey and D. J. Niklewski, *Phys. Rev. B* **40**, 3979 (1989).
[153] A. A. Demkov, J. Ortega, O. F. Sankey, and M. P. Grumbach, *Phys. Rev. B* **52**, 1618 (1995).
[154] G. Speyer, R. Akis, and D. K. Ferry, *IEEE Trans. Nanotechnol.* **4**, 403 (2005).
[155] R. Lake and S. Datta, *Phys. Rev. B* **45**, 6670 (1992).
[156] R. Lake, G. Klimeck, and S. Datta, *Phys. Rev. B* **47**, 6427 (1993).
[157] R. C. Brown, G. Klimeck, R. K. Lake, W. R. Frensley, and T. Moise, *J. Appl. Phys.* **81**, 3207 (1997).
[158] R. Lake, G. Klimeck, R. C. Bowen, and D. Javonivic, *J. Appl. Phys.* **81**, 7845 (1997).
[159] T. B. Boykin, G. Klimeck, R. C. Bowen, and R. Lake, *Phys. Rev. B* **56**, 4102 (1997).
[160] T. B. Boykin, L. J. Gamble, G. Klimeck, and R. C. Bowen, *Phys. Rev. B* **59**, 7301 (1999).
[161] G. Klimeck, R. C. Bowen, T. B. Boykin, and T. A. Cwik, *Superlatt. Microstruct.* **27**, 519 (2000).
[162] T. B. Boykin, R. C. Bowen, and G. Klimeck, *Phys. Rev. B* **63**, 245314 (2001).
[163] S. Lee, F. Oyafuso, P. von Allmen, and G. Klimeck, *Phys. Rev. B* **69**, 045316 (2004).
[164] Y. Zheng, C. Rivas, R. Lake, K. Alam, T. B. Boykin, and G. Klimeck, *IEEE Trans. Electron Dev.* **52**, 1097 (2005).
[165] T. Sandu, G. Klimeck, and W. P. Kirk, *Phys. Rev. B* **81**, 7845 (1997).
[166] A. Celeste, L. A. Cury, J. C. Portal, M. Allovon, D. K. Maude, L. Eaves, M. Davies, M. Heath, and M. Maldonado, *Sol. State Electron.* **32**, 1191 (1989).
[167] T. B. Boykin, N. Kharche, G. Klimeck, and M. Korkusinski, *J. Phys.: Cond. Matter* **19**, 036203 (2007).
[168] Y. Zheng, C. Rivas, R. Lake, *et al.*, *IEEE Trans. Electron Dev.* **52**, 1097 (2005).
[169] J. Wang, A. Rahman, A. Ghosh, G. Klimeck, and M. Lundstrom, *Appl. Phys. Lett.* **86**, 093113 (2005).
[170] M. Luisier, A. Schenk, W. Fichtner, and G. Klimeck, *Phys. Rev. B* **74**, 205323 (2006).
[171] G. Fiori, G. Iannaccone, and G. Klimeck, *IEEE Trans. Electron Dev.* **53**, 1782 (2006).
[172] G. Liang, J. Xiang, N. Kharche, *et al.*, *Nano Lett.* **7**, 642 (2007).
[173] T. B. Boykin, M. Luisier, A. Schenk, N. Kharche, and G. Klimeck, *IEEE Trans. Nanotechnol.* **6**, 43 (2007).

Index

0.7 feature 248, 272
1D–2D crossover 508
2D–3D crossover 509

acceleration 86
acoustic phonons 98
activated conductance 303
activation energy 415
addition energy 367
adiabatic 256
adiabatic approximation 256
advanced Green's function 171
Aharonov–Bohm 215
    effect 13, 167, 168, 392, 444, 491
    oscillations 31
    phase 240
    ring 389
Airy function 49
alloy disorder 128
Anderson 413, 420
    Hamiltonian 280, 287
    transition 427
Ando 197
annihilation operator 323
anti-bonding 373
antidot 216, 240
    superlattice 225
anyons 236
armchair nanotube 65
artifical atom 56
    atom 299
atomic-force microscope 210
autocovariance function 96
    length 96
avoided crossings 345

backscattering 261
    peak 467
ballistic 248

mean free path 4
    to diffusive crossover 617
    transport 249, 290, 443
bilayer system 194, 227–33
    total filling factor 229
binary alloys 423–4
Bloch frequency 480
Bohr magneton 69
Boltzmann transport equation 1, 15, 86, 329, 620
bonding 373
Bose–Einstein condensation 230
    statistics 98
bound state 39, 214
boundary conditions 117
    roughness 263
    scattering 75, 450
braid group 242
breakdown of the quantum Hall effect 204–6
Breit–Wigner formula 130
    resonance 391
bridge rectifier 293
broadening 71
Brownian motion 442
Büttiker 144, 149, 150, 193, 200

capacitance 301
carbon nanotube 22, 31, 61, 601
    chirality 22
    chirality 601
    single wall 22
    transistor 23
    transistor 601
carrier – carrier scattering, THETA device 571
carrier heating 14, 575
center coordinate 68
channel, length 20
channels 60, 249, 252

chaotic scattering 397
    sea 406
charge density wave 278
charge-coupled device 321
chemical potential 340
Chern class 223
Chern–Simons gauge transformation 224, 236, 237, 239
chiral spin states 237, 238
Chklovskii 208
CMOS circuit 602
coherence length 7, 432, 503
    decay of 494
coherence time, decay 499
coherent control 299
    superposition 37
    transport 131
    tunneling 124
collective modes 516
    molecular state 373
collision integral 86
composite fermion 194, 224–7
    fermion mass 225
compressible edge states 212, 219
    regions 207
conductance correlation function 464, 481
    fluctuations, lattice temperature 579
    quantization 11, 248
    mesoscopic device 630
conducting sphere 301
conductivity 5
    correction, impurity 558
    corrections for weak localization 444
    tensor 204
    classical 557
    Green–Kubo 628
    impurity scattering 554
    minimum metallic 6
conjugate variables 328

Index

constant-field scaling 598
   interaction model 341
contact resistance 255
contacts 9
continuum states 39
control of chaos 405
cooperon 469, 473, 480, 544, 545
   correction 469–73
correlated behavior 272
correlated tunneling 309, 331
correlation energy 464
   field 464, 466
   function in energy 481–6
   function in magnetic field 486–8
   functions 440
cotunneling 335, 356
Coulomb blockade 22, 61, 217, 276, 294, 299, 300, 308
   diamond 314, 348
   energy 280
   gap 303
   interaction 44, 528
      1D 552
      screened 560
   staircase 304, 334
coupled waveguide 167
creation operator 323
critical exponent, conductance 8
current bubble 473
   probe 202
   pump 294
   current correlation function 627
cyclotron frequency 67, 259, 342, 451
   orbit 74
   radius 19, 68
   resonance 227

Darwin–Fock spectrum 342
decoherence 397
   time 378
degeneracy 43
density matrix 15, 520
   of states 7, 10, 38, 60, 196, 254
   single particle 526
dephasing time 492
depletion width 47–8
depopulation 196, 203, 207
descreened frequency 533

detuning 376
developer 52
device simulation, quantum mechanical 615
   nanowire 601
   power density 598
   semiconductor 593
devices, ultrasmall 563
diagrammatic technique 322
dielectric function 76, 529, 532, 536, 539, 550
   1D 553
   lattice 533
   matrix 77, 82
   relaxation time 306
   response 76
differential conductance 269
diffusion 473, 480, 545
diffusion coefficient 87
   constant 7, 18, 491
   equation 443
   interaction 544
   particle 173
   poles 173, 472, 558
   propagator 473, 546
diffusive edge scattering 453
   scattering 450, 451
   transport 264
digamma function 455
dimensionality 5, 538
   crossover 508
dimensionless conductance 433
Dirac particle 62
   point 63, 64
dirty-metal regime 454
discrete dopants 615
   impurities 21
   spectrum 299
   semiconductors 413, 414–27
   systems 6, 543
dispersion relation 211
dissipative steady state 620
distribution function 15, 85, 124, 163
double-bubble diagram 481
drain-induced barrier lowering 607
dressing 179
Drude formula 180
dynamical phase space 398
Dyson's equation 178, 527, 639

edge states 75, 84, 193, 200–4, 249, 272, 451, 452
edge-state equilibration 209
effective field 102
   interaction 540
   interaction potential 527
   mass 40, 252
   mass approximation 28, 35, 85
   potential 257
eigenfunction 256
eigenstates 396
einselection 397
elastic scattering 90
electrochemical potential 201, 254, 272, 295, 364
electrometer 295, 386
electron drag 230
   energy loss 508
   2D 549
   lifetime 542
   temperature 575, 578
   beam lithography 52
   electron interaction 272, 506, 508, 543, 544
   electron propagator 480
   hole propagator 480,
   plasmon interaction 548, 571
   plasmon interaction, 1D 552
   spin resonance 388
electronics, molecular 23
electrostatic energy 301
energy balance equation, quantum dot 589
   loss rate, electron temperature 581
   loss rate, quantum well 589
   loss, 1D plasmons 552
   relaxation time 578, 580, 581
      quantum wire 584
   relaxation, length 4, 12
   relaxation, time 4
ensemble average 13
envelope function 28, 35–6, 40, 43, 58, 257
environment 396
equilibrium distribution function 133
equipartition 99, 254
erasable electrostatic lithography 54
ergodic hypothesis 480
evanescent states 28, 40
even-denominator plateau 223–7

exchange energy 45, 346, 353
   interaction 381
   scattering 354
   correlation energy 46
   correlation potential 35
excitation gap 221
excited state 368, 509
   spectroscopy 348
exponential localization 434, 435
extended states 6, 206, 413, 415, 416
extreme quantum limit 39

Fabry–Perot resonances 31
Fano effect 389
   formula 390
   Kondo effect 393
far-from-equilibrium system 565
Fermi edge singularity 506
   energy 39, 253
   golden rule 614
   velocity 5, 18
   wavelength 250
   wavevector 18
   Dirac distribution 124, 323
   Dirac distribution function 126, 522
Fermi's golden rule 89, 323, 324
ferromagnetic 284
ferromagnetism 278
field effect 19
   operator 520
filled state 340
filling factor 203, 223
fine structure constant 200
FinFET 109, 599, 602
   double gate 599
FIREBALL 634
flat-band voltage 595
fluctuations 4, 22, 494
   amplitude 506
   correlation function 504, 506, 516
   temperature decay 491, 503
   voltage 503
flux cancellation 456
focusing 283
form factor 78
four-probe configuration 459
four-terminal conductance 150–3
   measurements 10
FQHE-based quantum computing
   242–3

fractional charge 221, 223
   quantization 223
   quantum Hall effect 84, 194, 220–7
   states 237–9
   statistics 240–2
frequency doubling 216
Friedel oscillations 280
Frohlich interaction 102
fundamental conductance 8

gate length 3, 593
Gaussian decay 96
   pulse 332
geometric phase 168
giant backscattering resonance 218
Goldstone mode 230, 232
granular 283, 299
   charge 28
graphene 31, 34, 62
Green–Kubo formula 427, 626
Green's function 23, 541
   advanced 524, 620, 638
   anti-time ordered 622
   contour ordered 623
   greater than function 621, 638
   in disordered materials 467–88
   Keldysh matrix 623
   less than function 621, 638
   Matsubura 520, 523
   noninteracting 527
   nonequilibrium 565
   real time 565, 620
      equation of motion 622
   recursive 643
   retarded 524, 620, 638
   single particle 520
   temperature 520, 527
   time ordered 622
   two particle 544
ground state 368
group velocity 86
guiding center energy 211, 215
Gutzwiller formula 403

$h/2e$ oscillations 448
Haldane 221
half-integral QHE quantization 224
Hall coefficient 84
Hall conductivity 204
Hall resistance 201, 203, 213

Hall-bar geometry 201
Hamiltonian 67
harmonic oscillator 68, 72
   potential 252
Hartree correction 544
Hartree energy 45
   potential 35
heavy holes 37
Heisenberg temporal operator 522
heterostructure 1, 29
hop frequency 418
hopping 417–18
   energies 158–9
   matrix element 64
hot carrier spectroscopy 571
   quantum dot 585
hot electron effect, quantum dot 589
Hund's rule 342, 343, 346
hybrid frequency 342
   magneto-electric subbands 210
hybridization 63
   phonon–plasmon 533
hyperfine interaction 387

ideal covalent glass 415
III–V quantum well transistor 602, 605, 615
image potential 91
impurity averaging 176, 480
   distribution 460
   interaction 543
   potential 175
   scattering 175–81, 439
incompressibility 221, 237
incompressible regions 208, 210
inelastic 106
   mean free path 432
   scattering 220
inscattering 331
integer quantum Hall effect (IQHE) 193,
   194, 197–200, 249
integrated circuit 2, 593
inter-edge state scattering 212
interdigitated gates 293
interface mode 90, 98
interface roughness 128
interference 439
intersubband oscillations 81
   polarization 77
   scattering 88
intraband tunneling 122

intraband plasma oscillations 81
intrasubband polarization 77, 79
intrinsic bistability 131
  conduction 415
inversion layer 599
ionized impurities 90
ITRS roadmap 3

Jastrow wave function 238
Joule heating 574

Keldysh method 622
  space 623
kernel 171
kinetic equation 327
Knight shift 227
Kolmogorov master equation 329
  Arnold–Moser island 403
Kondo correlations 280
  effect 276, 299
  temperature 276, 354
Kronig–Penney model 43
Kubo formula 173, 197, 428, 440,
  481, 626

ladder diagram 182, 469, 545, 546
Landau gauge 68, 71, 163, 444
  index 197
  levels 69, 193, 196
Landauer approach 248
  formula 10, 28, 136–55, 193, 200, 203,
    255, 460, 606, 608, 629
  Büttiker formalism 249
  Büttiker formula 590
  Büttiker formula, multi-terminal 494
Landé g factor 69
Langmuir–Child law 606
Langreth theorem 625
laser ablation 31
lateral gate device 571
lattice Green's function 184
  Green's function method 157
Laughlin 221
Laughlin's wave function 238
LCAO model 43
length scale, temperature-dependent 504
level splitting 40–3
lever arm 271
lift-off process 53
light holes 37

Lindhard dielectric function 76, 78
  potential 538, 539
linear response 248
Liouville–von Neumann equation 15
Lippmann–Schwinger equation 156,
  157, 608
lithography 52
local adiabatic model 256
  chemical potential 464
  density approximation 35, 46
  magnetic moment 286
  spin density approximation 278
localization 9
  of electronic states 419–27
  Lorentzian distribution of energies 424–7
localized states 206, 415, 416, 417
  systems 413
logarithmic localization 435
long-range interaction 175
  scatterers 197
longitudinal conductivity 204
  resistance 203, 213
Lorentz force 67, 200
  number 285
Lorentzian function 97
Luttinger liquid 31, 85, 287

macroscopic quantum tunneling 335
magic numbers 342
magnetic correlation length 504
  decay time 453–7, 464
  depopulation 260
  field in weak localization 473–6
  field, correlation 504
  focusing 225
  length 19, 68, 200, 227–33
  lifetime 454
  moment 279
  time 446
magneto-conductance correction
  due to weak localization 455
  tunneling 352
Majorana fermions 239
many-body effect 274, 322
  interactions 248
Markovian behavior 173
master equation 304, 330
matrix element 91, 324
Matsubara frequency 523
maximally crossed diagrams 469, 545

Maxwellian distribution function 428
mean free path 250
  carrier–carrier scattering 571
  elastic 17–18, 570
  inelastic 1, 4, 18
mesoscopic 248
  conductor 5
  device 2, 4
  observables 9
  phenomena 273
metal organic chemical vapor
    deposition 29
  insulator transition 413, 427
  insulator transition in two dimensions
    435–9
method of images 302
minibands 43
minigaps 44
minimum metallic conductivity 427–31
mixed dynamics 398
mixed phase space 406
mobility 87, 89, 93
  edge 8, 420
mode-to-site transformation matrix 615
modes 11, 37, 60, 610
modulation doping 29, 46, 106
molecular beam epitaxy 29
  orbitals 362
momentum relaxation time 92, 195
Monte Carlo simulation 620
Moore's law 593, 602
MOS structure 594
MOSFET 594
  modeling 617
  scaling 594
  silicon on insulator (SOI) 617
  silicon on insulator modeling 619
Mott 413
multi-channel Landauer formula 140–7
multiple quantum wells 40

nanodevice 19, 563, 599
  semiconductor 593
nanomagnet 373
nanoribbon 62
nanoscale device 19, 22
nanostructure 1, 2, 19
nanowire 22, 33
  circuit 22
  self-assembled 601

Index

nearest-neighbor hopping 418
negative differential conductance 350
   differential resistance (NDR) 118, 119, 590
   differential resistance, quantum waveguide 591
   negative magneto-resistance 437
NEMO 633
non-Abelian phase 168
   statistics 243
non-dissipative transport 203
non-parabolicity 51
nonequilibrium system 565
   phonons 16
   transport 17
nonlocal properties 463
number operator 323
Numerov method 46
Nyquist noise 330

one-dimensional channel 200, 201
   weak localization 455
Onsager–Casimir symmetry 146
open boundary conditions 158
open quantum dot 166, 396
open systems 116
optical deformation potential 104
   modes 98
   phonon emission 567, 571
orbital angular momentum 344
orthodox model 304
outscattering 331
overlap energy 420
   factor 100
   integral 64
oxidation lithography 53

p bond 63
p-wave pairing 229
parabolic approximation 342
parasitic MESFET 48
particle–hole propagator 175
   particle correlator 441
   particle propagator 457
Pauli spin matrices 161
peak-to-valley ratio (PVR) 120, 122
Peierl's distortion 85
   substitution 14, 68
   phase factors 163
Peltier coefficient 573, 577

percolation 439
percolation transition 439
periodic orbits 403
Pfaffian 238
   Moore–Read wave function 239
   wave function 233, 238
phase-breaking length 2
   time 18, 443, 464, 499, 500, 548, 551
      carrier temperature 579
      electron temperature 579
      plasmons 551
      plasmons 554
      phase, breaking 1
   coherence 28, 194, 414, 416
   coherence length 18, 28, 37, 444, 497, 501
   coherence time 499
   coherent length 28
   interference 2, 12
   memory 442
   randomization 144
   space tunneling 404
phonon 98, 89
   assisted tunneling 569
   replica 647
photocurrent 379
photon-assisted tunneling 380
piezoelectric crystals 293
pinch-off 270
planar doped barrier transistor 571
plasma frequency 80, 81, 533
   one dimension 552
   two dimensions 549
plasmon 82, 83, 516, 519, 549
   frequency 548
   modes 548
   pole approximation 532, 533
Poincaré section 403
point contact 210
pointer state 167, 397
Poisson's equation 45
   discrete dopants 616
polar-optical phonon 102
polarization, Bethe–Salpeter equation 629
   charge 217, 301, 307
   field 76
   electronic 528, 530, 531
poly-methyl methacrylate (PMMAP) 53
potential barrier 251
   fluctuations 204

power-law localization 433
precession 51
probability of return 441
propagating states 28
propagator 171
pseudospin 227
   index 238
pumping configuration 379
pure-metal regime 454, 455
push–pull potential 291

quantization 250
quantized conductance, breakdown 590
   levels 340
quantum cellular automata 371
   computation 168, 381
   computing 242
   confined system, adjacent 577
   confinement 1, 29, 116
   control 381
   decoherence 397
   distribution function 135
   dot 30, 56, 214, 299
   field theory 23, 322
   fluctuations 310
   Hall effect 84
   interference 12, 31, 116, 299, 437, 459
   interference device 21
   kinetic theory 15
   mechanical current 627
   point contact 163, 210, 248
   point contact, current filament 590
   transport 620
   waveguide 37, 60
   well 36
   wire 55
quasi-ballistic regime 454
   bound state 119
   equilibrium 328
   particle mass 175
   two dimemsional electron gas 37
   two-dimensional system 539
   zero dimensional 299
qubit 242, 381

Rabi oscillation 375
radiative recombination in the FQHE 225
random charges 305
   fluctuations 94, 263

random charges (*Continued*)
  impurities 264, 564
   phase approximation 519, 539
   telegraph signal 592
Rashba effect 51
Rashba spin-orbit coupling 161
  spin-orbit interaction 161, 168
real-space transfer 48
reciprocity 292, 463
rectification 291
recursive Green's function 183–8
  scattering matrix 157–69
reentrant FQHE 227
regular scattering 397
regularization 175
relaxation time approximation 15, 87
renormalization 354
reservoirs 9
resist 52
resistance 5
resonant interband tunneling diode
   (RITD) 122–3
  tunneling 118–23
  tunneling diode (RTD) 22, 118,
   123–35, 568
  tunneling diode, GaAs/AlGaAs 569
retarded Green's function 171
  Langevin equations 440
ring diagrams 528
roughness fluctuations 56
  scattering 95

s bond 63
S-type negative differential
   conductance 586
  differential conductance 592
saddle-point energies 260
saddle potential 251
saturated current 597
scaling 431–5
  function 433
  parameter 436
  relationship 6
  theory 8, 413, 433
scanning gate microscopy (SGM) 163
  tunneling microscope 53
scarred wave functions 400
scarring 398
scattering, carrier–carrier 1D 498
  electron–electron 491, 496, 508
   impurity 544
   matrix 9, 322
   polar-optical phonon 645
   quantum device simulation 610
   rate, site representation 614
   real space basis 612
   single-particle 548
   spin-orbit 500
   surface roughness 599
   transverse modes 611
Schottky gate 57
Schrödinger equaiton 58
screening 175, 207, 516, 519, 544
  constant 80
  contact 630
  dynamic 536
  self-energy, lead 632
   matrix 624
   retarded 638
  low-dimensional 538
  momentum-dependent 534
  static 531
self gating 291
self-assembled structure 22
self-assembly 30
self-consistency 207
self-consistent Born approximation 178
  potential 207
  solutions 46
self-energy 191, 389, 178–80, 540, 541,
   542, 547, 548, 624
  impurity 543
semiconductor scaling 563, 593
sequential tunneling 129
short-range order 416
  scatterers 197
shot noise 283
Shubnikov–de Haas (SdH) effect
   194–7, 452
  oscillations 70, 194
   energy loss 580
   wires 583
shunt conductance 329
SIESTA 634
silicon dioxide 594
single-channel Landauer formula 139–40
single electron 21
  charging 28
  charging energy 21–2
  device 21
   pump 22, 320
   single-electron, transistor 22, 312
   tunneling 300, 563
   tunneling, quantum dot 578
   turnstile 22, 317
single-photon detection 194, 217
singlet–triplet 360
site energy 420
size effects in quantum wires 450–3
skipping orbits 74, 193, 200, 451, 452
  formation 455
slab modes 99
soliton 317
space-charge effects 131–2
  layer 251
spectral density 638, 614
  function 525, 526
specular scattering 451
spin 274
  blockade 350, 387
  degeneracy 38, 196, 203, 218, 274,
   344, 461
  degenerate 205
  dependent focusing 284
  filter 123, 281
  flip 353
  gap 275
  polarization 274, 278
  split subbands 261
  splitting 275
  flip scattering 386
  orbit Hamiltonian 161
   interaction 284
   parameter 168
spin-polarized current 218
spin-to-charge conversion 381
spinless fermion 224
spintronics 51
split gate 448
split-gate technique 57, 248, 249
spreading resistance 255
stability diagram 334
  plot 314
statistical phase 240
stochastic Coulomb blockade 370
Störmer 221
Stransky–Krastinov growth 30
stripe phase 233
strong localization 439
subband 37, 60, 248, 252

subthreshold slope 604
superfluid state 231
superlattice 43
surface acoustic wave 293
   inversion 595
   roughness 94
   scattering 9
   states 90, 595
symmetry breaking 292

technology node 599
temperature decay of coherence 493
terahertz regime 217
thermal conductance 285
   conductivity, quantum point
      contact 577
   current 573
   diffusion length 491, 492
   energy 300
   length 19, 504
   phase-breaking time 492
   smearing 219
thermopower 573
   conductance fluctuations 574
   nanostructures 573
Thouless length 18
   number 8, 433
three-terminal conductance 147–9
   measurement 154
threshold voltage 597
tight-binding basis 634
   model 43, 63
   orbitals 636
time-ordering operator 520
time-reversal symmetry 414, 439, 446
   path 414, 459, 467
Tomonanga–Luttinger liquid 287–8
total scattering time 195
transconductance 270
transducer 294
transfer Hamiltonian 322
   Hamiltonian method 118
   matrix technique 134
TranSIESTA 634

transistor, benchmark 604
   CMOS 20
   dual-gate 20
   energy-delay product 604
   intrinsic speed 604
   nanowire 601
   scaling 22, 598
   Si on Insulator 20
   SiGe 20
   tri-gate 20
transition density 437
transmission coefficient 10, 137, 144
transport, ballistic 3, 5, 9, 20, 570, 606
   in devices 610
   in MOSFET 606
transport, diffusive 6, 20, 609
   far-from-equilibrium 620
   localized 6
   low field 15
   nonequilibrium 17
   nonequilibrium in quantum
      structure 590
   nonlinear in quantum point contact 590
   quantum 23
traversing orbit 74
tri-gate transistor 109, 599
triangular-well approximation 49
triple point 365
Tsu–Esaki formula 127
Tsui 221
tunnel barriers, nonequilibrium effects 566
tunneling 117–35, 215, 258
Tunneling Hot Electron Transfer
   Amplifier 571
   probability 10
   resistance 325
   time 306
turn-on voltage 595, 596
turnstile 294
two-dimensional electron gas (2DEG)
   11, 193
   metallic state 436
two-particle Green's function 174–5
two-terminal conductance 147

unitarity limit 357
universal conductance fluctuations 17,
   398, 414, 448, 449, 459–67,
   480–8, 497, 503, 508
Usuki 159

vacuum tube 606
valley current 128
   degeneracy 38, 196, 461
vapor phase synthesis 32
   liquid-solid growth 32
variable-range hopping 427, 418–19
vector potential 14, 517
vertex correction 544
vertical transistor 599
virtual-crystal approximation 423
voltage fluctuation 461
   probe 202
von Klitzing constant 199
   Klaus 193, 198, 199

waveguide 168
   couplers 168
weak localization 397, 413, 414, 437,
   439–59, 467–80, 495
   in short wires 457–9
weakly disordered systems 413
Wick's theorem 176, 460, 623
Wiedemann–Franz law 284
Wiener–Kitchine theorem 96
Wigner crystal 229, 278
   function 629
wire, one-dimensional 21
wraparound gate 341

Y-branch 601, 20–1
   device 249, 290

Zeeman effect 275
   shift 287
   splitting 358
zero-bias anomaly 358
zigzag nanotube 65
zone folding 102

Printed in the United States
by Baker & Taylor Publisher Services